ELASTIC WAVES IN THE EARTH

FURTHER TITLES IN THIS SERIES

1 *F.A. VENING MEINESZ*
THE EARTH'S CRUST AND MANTLE

2 *T. RIKITAKE*
ELECTROMAGNETISM AND THE EARTH'S INTERIOR

3 *D.W. COLLINSON, K.M. CREER and S.K. RUNCORN*
METHODS IN PALAEOMAGNETISM

4 *M. BÅTH*
MATHEMATICAL ASPECTS OF SEISMOLOGY

5 *F.D. STACEY and S.K. BANERJEE*
THE PHYSICAL PRINCIPLES OF ROCK MAGNETISM

6 *L. CIVETTA, P. GASPARINI, G. LUONGO and A. RAPOLLA*
PHYSICAL VOLCANOLOGY

7 *M. BÅTH*
SPECTRAL ANALYSIS IN GEOPHYSICS

8 *O. KULHANEK*
INTRODUCTION TO DIGITAL FILTERING IN GEOPHYSICS

9 *T. RIKITAKE*
EARTHQUAKE PREDICTION

10 *N.H. RICKER*
TRANSIENT WAVES IN VISCO-ELASTIC MEDIA

Developments in Solid Earth Geophysics
11

ELASTIC WAVES IN THE EARTH

WALTER L. PILANT

Department of Earth and Planetary Sciences
University of Pittsburgh
Pittsburgh, Pa., U.S.A.

ELSEVIER SCIENTIFIC PUBLISHING COMPANY
Amsterdam — Oxford — New York 1979

ELSEVIER SCIENTIFIC PUBLISHING COMPANY
335 Jan van Galenstraat
P.O. Box 211, 1000 AE Amsterdam, The Netherlands

Distributors for the United States and Canada:

ELSEVIER/NORTH-HOLLAND INC.
52, Vanderbilt Avenue
New York, N.Y. 10017

Library of Congress Cataloging in Publication Data

Pilant, Walter L
Elastic waves in the earth.

(Developments in solid earth geophysics ; 11)
Bibliography: p.
Includes index.
1. Elastic waves. 2. Seismology. I. Title.
II. Series.
QE539.P54 551.2'2 79-11328
ISBN 0-444-41798-2

ISBN 0-444-41798-2 (Vol. 11)
ISBN 0-444-41799-0 (Series)

Printed in The Netherlands

PREFACE

This book is an outgrowth of a one-year course taught to graduate students at the University of Pittsburgh. As an advanced level course, the students were expected to have an appreciation of the general aspects of seismology and its relation to the larger field of geophysics. Consequently, the approach has been an axiomatic one, working from the elastodynamic equations of motion to derivative equations applicable to the propagation of elastic waves in the earth. As the course was presented, the students were expected to write two papers, one for each term. It was felt that most significant problems were of such a length that an in-depth treatment was better than a series of ad hoc exercises. Consequently no problem sets were included with the text. However, suitable problems may be found in several well known texts in geophysics and wave propagation.

The first sixteen chapters contain the basic material relating to the propagation of elastic waves and are covered in the first term. In the second half of the course, the style of presentation is altered. A chapter of theory is followed by a chapter of observational material. The theoretical chapters were limited by space in their depth of coverage; large numbers of references to the literature have been included to guide the reader to more extensive treatments. Observational citations have been limited to fairly recent material hoping that the reference list contained in the cited articles would be self-expanding. If extended papers or review articles were available, these were cited and only later work referenced.

A great deal of effort has been expended in trying to give a presentation with a unified notation. Probably the most difficult problem in the development of an understanding of material of the level presented here is that a reader of the literature has to sort through several different notations and conventions to obtain this understanding. I had hoped to ease this process by some notational changes, and the reaction of my students has been sufficient reward. As a result, some readers may find an unfamilar notation.

The major areas covered in detail in this book are:

- Fundamentals of elastodynamics
- Lamb's problem (including Cagniard-deHoop theory)
- Rays and modes in multilayered media (including Thomson-Haskell theory)
- Rays and modes in a radially inhomogeneous earth
- Elastic wave dissipation
- The seismic source
- Seismic noise
- Seismographs

Some subjects not treated are elastic wave propagation in non-linear, plastic, porous, or viscoelastic media. Most of these topics have little bearing on problems relating to the earth. The transmission of elastic waves in rods, plates, and shells have been more than adequately treated elsewhere and again these subjects find little application to the study of our planet earth.

I am grateful to a large number of students and colleagues who have offered constructive criticism of preliminary versions of the manuscript. Their suggestions have been greatly appreciated. Thanks too are due to the many seismologists who supplied material and permission for a large number of the figures. One unwritten "thank you" lies in the bibliography at the end of the book. Hundreds of researchers over the years have contributed to the sum of knowledge presented here. The choice of literature cited has been dictated by practicality and I would like to emphasize that the relative importance of any person's contributions cannot be judged by the number of citations listed. The early giants suffer particularly since their pioneering work has been distilled into what is almost common knowledge today. For a few, their life's work has found expression in one or more cited books.

Lastly, I would like to thank especially my wife Carol who did most of the artwork and much typing, and my daughter Michelle who spent many hours typing and correcting the intricate mathematical formulas in the text. Their contribution has been a labor of love and words are inadequate to express the part they played in the production of this book.

University of Pittsburgh
Pittsburgh, Pennsylvania

Walter L. Pilant
December 1978

TABLE OF CONTENTS

CHAPTER 1

SCALARS, VECTORS, AND TENSORS IN CARTESIAN COORDINATES

1.1. General.

The following material should be a review for most well prepared students. It represents only those aspects of formalism necessary to derive the results of later sections. We have adopted it in some places mainly as an economy measure, certainly not to make the derivations obscure. Most times, the economy could be effected by matrix notation, but the tensor calculus is necessary when derivative operators are involved.

For more complete, but still abbreviated, presentations, one can consult Jeffreys and Jeffreys (1956), Margenau and Murphy (1943), or Morse and Feshbach (1953). An alternative approach is taken by Officer (1974) where he develops this formalism in terms of dyadic notation. Such references should provide more than enough material for understanding what is to follow.

1.2. Transformation of Coordinates.

First of all, let us represent the cartesian system that we are talking about by the three coordinates x_1, x_2, x_3 as in Figure 1-1.

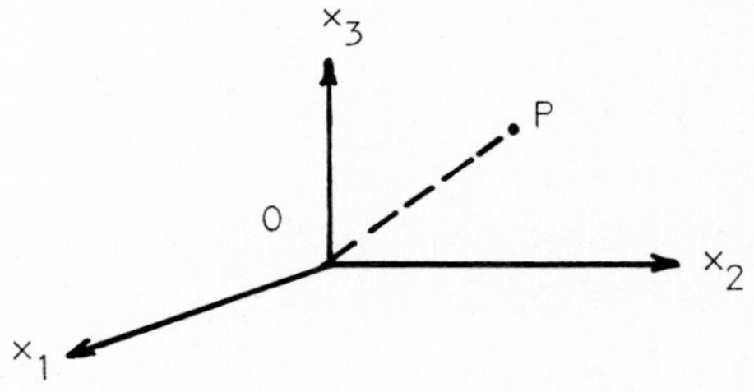

Figure 1-1. Cartesian coordinate system.

By definition, the three coordinate axes will be perpendicular to each other. Conventionally, they will be ordered in a right hand screw sense. The subscripted notation is used as a means of conserving letters; e.g., x, y, and z are usually used in this context, but now y and z are free to represent other quantities.

Now such a system of coordinates can be rather arbitrarily placed in space, that is, the origin 0 can be translated to a new position and the orientation of the coordinates can be chosen freely. In practice, the position of the origin may be chosen conveniently, such as the center of mass of a body or some fixed point in a body. However, the orientation of the coordinate system cannot be chosen so nicely in a great majority of problems, particularly in mechanics. Many times we will want to choose two or more coordinate systems, each simply oriented with regard to some natural direction. However, they will be at some arbitrary angle with respect to each other. Let us see how the coordinates of a given point P transform under a change in orientation (rotation) of the coordinate system x_1, x_2, x_3 to a new system $x_{1'}$, $x_{2'}$, $x_{3'}$.

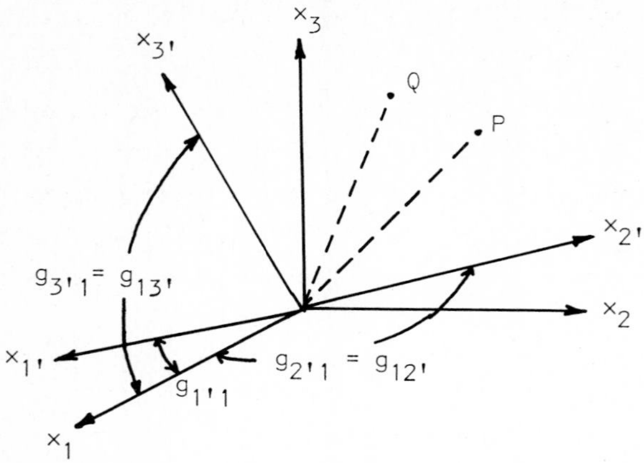

Figure 1-2. Two cartesian coordinate systems and the direction cosines relating them.

By projection we have

$$\begin{aligned} x_{1'} &= g_{1'1}x_1 + g_{1'2}x_2 + g_{1'3}x_3 = \sum_{i=1}^{3} g_{1'i}x_i, \\ x_{2'} &= g_{2'1}x_1 + g_{2'2}x_2 + g_{2'3}x_3 = \sum_{i=1}^{3} g_{2'i}x_i, \\ x_{3'} &= g_{3'1}x_1 + g_{3'2}x_2 + g_{3'3}x_3 = \sum_{i=1}^{3} g_{3'i}x_i, \end{aligned} \qquad (1\text{-}1)$$

where $g_{j'i}$ is the direction cosine of the angle between the axes $x_{j'}$ and x_i, the values of $\underline{j}'$ and $\underline{i}$ assuming any of the values 1, 2, 3.

We can achieve a certain economy of writing by using the summation convention. This involves dropping the explicit symbol $\sum_{i=1}^{3}$ from each of the right hand sides of (1-1) and using it implicitly whenever a term involving a repeated subscript appears. That is, whenever you see a term like $g_{1'i}x_i$, its value is to be computed by setting the repeated subscript $\underline{i}$ equal to all three values 1, 2, 3, and summing the three factors obtained. In special cases when repeated subscripts are not to be summed, one way out is to use capital letters. For example, g_{JJ} can take on the three values g_{11}, g_{22}, g_{33}, but summation is not to be performed.

Using the summation convention, we can then write

$$x_{j'} = g_{j'i}x_i \qquad (j' = 1, 2, 3), \qquad (1\text{-}1)'$$

and finally we can drop (while implicitly assuming) the notation (j' = 1, 2, 3) and obtain the desired form

$$x_{j'} = g_{j'i}x_i. \qquad (1\text{-}1)''$$

Similarly, by projecting the other way, we have

$$x_j = g_{ji'}x_{i'}. \tag{1-2}$$

It is obvious that $g_{ij'} = g_{j'i}$, however, $g_{ij'} \neq g_{i'j}$.

The direction cosine of an angle can also be thought of as the projection of a unit length along one axis upon another axis. Hence, from (1-1) we have

$$x_{1'} = g_{1'1}1, \qquad x_{2'} = g_{2'1}1, \qquad x_{3'} = g_{3'1}1,$$

where the "1" comes from a unit length along the x_1-axis. Now let $\overline{OP}$ be a unit length along the x_1, x_2, and x_3-axes in turn. Since the length of $\overline{OP}$ should remain <u>1</u> in the new system, we must have

$$(g_{1'1})^2 + (g_{2'1})^2 + (g_{3'1})^2 = 1,$$

and two more similar relations. In summation notation this can be written

$$g_{j'K}g_{j'K} = 1 \qquad (K = 1,2,3), \tag{1-3}$$

where the capitalized subscripts imply no summation. Likewise, going from unit lengths along each of the axes in the primed system, we have

$$g_{jK'}g_{jK'} = 1. \tag{1-4}$$

The condition that the primed axes be perpendicular to each other is given by considering the projection of the unit length $\overline{OP}$ upon the line $\overline{OQ}$ (by way of $\overline{OP}$ onto the coordinate axes x_1, x_2, x_3, and the sum of these projections onto $\overline{OQ}$), we have

$$g_{p'j}g_{jq'} = \text{Cos} \angle \text{POQ}.$$

In particular, if $\overline{OP}$ and $\overline{OQ}$ are any pair of the new axes $x_{1'}$, $x_{2'}$, $x_{3'}$, Cos ∠ POQ must equal zero, and we have

$$g_{p'j}g_{jq'} = 0 \qquad (p' \neq q'). \tag{1-5}$$

The six relations given by (1-4) and (1-5) can be written in a more compact form

$$g_{p'j}g_{jq'} = g_{jp'}g_{jq'} = \delta_{p'q'}, \tag{1-6}$$

where $\delta_{p'q'}$ is the <u>Kronecker delta</u> defined by

$$\begin{aligned} \delta_{p'q'} &= 0 \qquad (p' \neq q') \\ &= 1 \qquad (p' = q'). \end{aligned} \tag{1-7}$$

Similarly

$$g_{pj'}g_{j'q} = g_{j'p}g_{j'q} = \delta_{pq}. \tag{1-8}$$

Alternatively, we can think of the relationships (1-1) in matrix notation, i.e.,

$$\begin{pmatrix} x_{1'} \\ x_{2'} \\ x_{3'} \end{pmatrix} = \begin{pmatrix} g_{1'1} & g_{1'2} & g_{1'3} \\ g_{2'1} & g_{2'2} & g_{2'3} \\ g_{3'1} & g_{3'2} & g_{3'3} \end{pmatrix} \begin{pmatrix} x_1 \\ x_2 \\ x_3 \end{pmatrix} \tag{1-1)'''$$

or

$$\mathbf{x}' = \mathbf{g}\mathbf{x}.$$

Likewise

$$\begin{pmatrix} x_1 \\ x_2 \\ x_3 \end{pmatrix} = \begin{pmatrix} g_{11'} & g_{12'} & g_{13'} \\ g_{21'} & g_{22'} & g_{23'} \\ g_{31'} & g_{32'} & g_{33'} \end{pmatrix} \begin{pmatrix} x_{1'} \\ x_{2'} \\ x_{3'} \end{pmatrix} \qquad (1\text{-}2)'$$

or

$$\mathbf{x} = \mathbf{g}^{-1}\mathbf{x}'.$$

The appropriateness of the notation $\mathbf{g}^{-1}$ is apparent by multiplying both sides of (1-2)' by $\mathbf{g}$. We then have

$$\mathbf{g}\mathbf{x} = \mathbf{g}\mathbf{g}^{-1}\mathbf{x}' = \mathbf{x}'.$$

To see that this is true, consider the following:

$$\mathbf{g}\mathbf{g}^{-1} = \begin{pmatrix} g_{1'1} & g_{1'2} & g_{1'3} \\ g_{2'1} & g_{2'2} & g_{2'3} \\ g_{3'1} & g_{3'2} & g_{3'3} \end{pmatrix} \begin{pmatrix} g_{11'} & g_{12'} & g_{13'} \\ g_{21'} & g_{22'} & g_{23'} \\ g_{31'} & g_{32'} & g_{33'} \end{pmatrix} = g_{ik'}g_{k'j} = \delta_{ij},$$

where the matrix form of δ_{ij} is given by

$$\delta_{ij} = \begin{pmatrix} 1 & 0 & 0 \\ 0 & 1 & 0 \\ 0 & 0 & 1 \end{pmatrix} .$$

Here we have made use of (1-6).

1.3. Scalars.

A scalar is a single number associated with each point in space, e.g., temperature, density. In a scalar field, we have

$$\phi(x_1,x_2,x_3) = \phi(x_{1'},x_{2'},x_{3'}) \qquad (1\text{-}9)$$

as long as x_1, x_2, x_3, and $x_{1'}$, $x_{2'}$, $x_{3'}$ represent the same point. This will be true as long as x_i and $x_{i'}$ are related by (1-1) and (1-2).

1.4. Vectors.

A vector is a set of three components A_i that transforms as the coordinates of a point, and where the component in the direction n_i is given by $g_{ni}A_i$. That this is consistent with the idea of a vector being a quantity possessing both a magnitude and a direction can be seen in the following. Given a point $P(x_1,x_2,x_3)$, we can associate with it a magnitude $\overline{OP} = (x_1^2 + x_2^2 + x_3^2)^{\frac{1}{2}}$ and a direction given by the three direction cosines g_{p1}, g_{p2}, g_{p3}. Inasmuch as (1-3) and (1-4) preserve magnitude and directional relationships, they are ideal for preserving our physical notion of a vector.

A vector equation is written in the form $A_i = B_i$. On a rotation of coordinates, we would have

$$A_{i'} = g_{i'j}A_j,$$

$$B_{i'} = g_{i'j}B_j, \qquad (1\text{-}10)$$

and $A_{i'} = B_{i'}$.

In our new notation we can write the inner (or dot) product as

$$\mathbf{A}\cdot\mathbf{B} = |\mathbf{A}|\,|\mathbf{B}|\cos\theta = A_iB_i \qquad \text{(a scalar).} \qquad (1\text{-}11)$$

That this product is invariant to a rotation of axes may be seen from the following:

$$A_{i'}B_{i'} = g_{i'j}A_jg_{i'k}B_k = \delta_{jk}A_jB_k = A_jB_j.$$

The magnitude of a vector is defined as

$$(\mathbf{A}\cdot\mathbf{A})^{\frac{1}{2}} = |\mathbf{A}| = A = (A_iA_i)^{\frac{1}{2}}. \qquad (1\text{-}12)$$

In the primed system, we would have

$$(A_{i'}A_{i'})^{\frac{1}{2}} = (g_{i'j}A_jg_{i'k}A_k)^{\frac{1}{2}} = (\delta_{jk}A_jA_k)^{\frac{1}{2}} = (A_jA_j)^{\frac{1}{2}},$$

which is as it should be, invariant.

The vector (cross) product is defined by

$$\mathbf{A}\times\mathbf{B} = |\mathbf{A}|\,|\mathbf{B}|\sin\theta \qquad \text{(in a direction } \perp \text{ to } \mathbf{A} \text{ and } \mathbf{B}\text{).} \qquad (1\text{-}13)$$

The expression of this relationship is more complicated. First let us write $A_i = Al_i$ and $B_i = Bm_i$ where l_i and m_i are the direction cosines of the vectors A_i and B_i respectively. The direction cosines of a unit vector **n** perpendicular to A_i and B_i would be given by the solution of the following three equations:

$$l_1n_1 + l_2n_2 + l_3n_3 = 0,$$

$$m_1n_1 + m_2n_2 + m_3n_3 = 0, \qquad (1\text{-}14)$$

$$n_1^2 + n_2^2 + n_3^2 = 1.$$

Using the first two, we have

$$\frac{n_1}{l_2m_3 - l_3m_2} = \frac{n_2}{l_3m_1 - l_1m_3} = \frac{n_3}{l_1m_2 - l_2m_1} = K. \qquad (1\text{-}15)$$

Using the third, we have

$$\frac{(n_1^2 + n_2^2 + n_3^2)^{\frac{1}{2}}}{\{(l_2m_3 - l_3m_2)^2 + (l_3m_1 - l_1m_3)^2 + (l_1m_2 - l_2m_1)^2\}^{\frac{1}{2}}} = K = \frac{1}{\sin\theta}. \qquad (1\text{-}16)$$

Consequently,

$$n_1 = \frac{(l_2m_3 - l_3m_2)}{\sin\theta}, \quad n_2 = \frac{(l_3m_1 - l_1m_3)}{\sin\theta}, \quad n_3 = \frac{(l_1m_2 - l_2m_1)}{\sin\theta}. \qquad (1\text{-}17)$$

Hence

$$(\mathbf{A}\times\mathbf{B})_1 = |\mathbf{A}|\,|\mathbf{B}|\sin\theta\, n_1 = AB(l_2m_3 - l_3m_2),$$

with two more equations like it for the other components of $\mathbf{A}\times\mathbf{B}$. A shorthand way of writing this is

$$(\mathbf{A}\times\mathbf{B})_i = AB\varepsilon_{ijk}l_j m_k = \varepsilon_{ijk}A_j B_k, \tag{1-18}$$

where

$$\varepsilon_{ijk} = 1 \quad \text{if } \underline{ijk} \text{ are a cyclic permutation of } 123,$$
$$= -1 \quad 132, \tag{1-19}$$
$$= 0 \quad \text{otherwise.}$$

Note that the vector product is a proper vector only in orthogonal coordinates.

The triple scalar product defined by

$$V = \mathbf{A}\cdot\mathbf{B}\times\mathbf{C} \tag{1-20}$$

is a scalar which represents the volume enclosed by a parallelopiped formed by the vectors **A**, **B**, and **C**. This can be written as

$$V = \varepsilon_{ijk}A_i B_j C_k. \tag{1-21}$$

1.5. Tensors.

Quite often, we will find that a vector A_i is a simple function of a scalar variable such as the time t. This means that $A_1 = A_1(t)$, $A_2 = A_2(t)$, $A_3 = A_3(t)$. The functional form of A_1, A_2, A_3, however, need not be the same. If the vector A_i were a linear function of the time, we could have

$$A_1 = at, \quad A_2 = bt, \quad A_3 = ct.$$

On the other hand, we can have the case where one vector is a linear (or more complicated) function of another vector, e.g.,

$$B_1 = \mu_{11}H_1 + \mu_{12}H_2 + \mu_{13}H_3 \qquad \text{(Electromagnetic Theory)}$$

etc, and

$$L_1 = I_{11}\omega_1 + I_{12}\omega_2 + I_{13}\omega_3, \qquad \text{(Rotational Mechanics)}$$

where **B** is the magnetic induction, **H** the magnetic field, **μ** the magnetic permeability in the first case, and **L** the angular momentum, **I** the inertia, and **ω** the angular velocity in the second. A vector is a linear function of another vector if each component of the first is a linear function of the three components of the second, i.e., we can write

$$A_1 = C_{11}B_1 + C_{12}B_2 + C_{13}B_3,$$
$$A_2 = C_{21}B_1 + C_{22}B_2 + C_{23}B_3, \tag{1-22}$$
$$A_3 = C_{31}B_1 + C_{32}B_2 + C_{33}B_3.$$

Using the summation convention, we can write this as

$$A_i = C_{ij}B_j, \tag{1-23}$$

or in matrix notation

$$\mathbf{A} = \mathbf{C}\mathbf{B}. \tag{1-24}$$

If these relationships are to be useful, they should be independent of the cartesian coordinate system in which they are expressed. For a simple rotation we would have

$$A_i = g_{ij'}A_{j'} = C_{ij}g_{jk'}B_{k'}. \tag{1-25}$$

Multiplying each member by $g_{l'i}$, we have

$$g_{l'i}g_{ij'}A_{j'} = A_{l'} = (g_{l'i}C_{ij}g_{jk'})B_{k'}, \qquad (1\text{-}26)$$

since $g_{l'i}g_{ij'} = \delta_{l'j'}$ and $\delta_{l'j'}A_{j'} = A_{l'}$. If the quantity $g_{l'i}C_{ij}g_{jk'}$ is called $C_{l'k'}$, we will then have

$$A_{l'} = C_{l'k'}\ B_{k'} \qquad (1\text{-}27)$$

in the rotated system. That is, the relationship expressed by (1-23) is independent of the coordinate system when the nine constants C_{ij} relating A_i and B_j transform in the following manner:

$$C_{l'k'} = g_{l'i}C_{ij}g_{jk'}. \qquad (1\text{-}28)$$

A <u>tensor</u> is a general name given to quantities that transform as in (1-9), (1-10), or (1-28). A scalar is a tensor of rank 0, inasmuch as it is independent of the particular coordinate system. A vector is a tensor of rank 1 and a quantity such as C_{ij} is a tensor of rank 2. The rank of a tensor is associated with the number of indices in the indicial notation. For tensors of rank higher than 2, we cannot use the pre- and post-factor matrix oriented notation of (1-28) and must use a more fundamental notation making use of the symmetry of the $g_{ij'}$ expressions. For example, in the stress-strain relationship, we have the stress tensor t_{ij} related to the strain tensor e_{kl} by

$$t_{ij} = c_{ijkl}e_{kl}$$

where t_{ij} and e_{kl} transform as in (1-28). However, the c_{ijkl} transform as

$$c_{m'n'p'q'} = g_{m'i}g_{n'j}g_{p'k}g_{q'l}c_{ijkl}\ . \qquad (1\text{-}29)$$

This is the transformation law for a tensor of 4th rank in cartesian coordinates.

Tensors of all ranks may be added or subtracted by adding or subtracting similar elements. Multiplication is performed in accordance with (1-23) and in other ways not described here. Division of tensors is not defined. Instead we use multiplication by reciprocal tensors as on page 4.

In this new shorthand notation, an additional part is the use of the "," to denote differentiation. The summation convention remains in effect. Some important relations are given by the following. The <u>gradient of a scalar</u> is given by

$$(\text{grad}\ \phi)_i \equiv (\nabla\phi)_i = \frac{\partial\phi}{\partial x_i} = \phi,_i. \qquad (1\text{-}30)$$

That this is a proper tensor may be seen from

$$\frac{\partial\phi}{\partial x_{i'}} = \frac{\partial\phi}{\partial x_j}\frac{\partial x_j}{\partial x_{i'}} = g_{i'j}\phi,_j. \qquad (1\text{-}31)$$

Likewise the <u>divergence of a vector</u> is written as

$$\text{div}\,\mathbf{A} \equiv \nabla\cdot\mathbf{A} = \frac{\partial A_1}{\partial x_1} + \frac{\partial A_2}{\partial x_2} + \frac{\partial A_3}{\partial x_3} = \frac{\partial A_i}{\partial x_i} = A_i,_i. \qquad (1\text{-}32)$$

Again

$$A_{i',i'} = \frac{\partial A_{i'}}{\partial x_j}\frac{\partial x_j}{\partial x_{i'}} = \frac{\partial}{\partial x_j}(g_{i'k}A_k)\, g_{i'j}$$

$$= g_{i'k}g_{i'j}\frac{\partial A_k}{\partial x_j} = \delta_{kj}\frac{\partial A_k}{\partial x_j} = A_{k,k}\,. \tag{1-33}$$

From (1-30) and (1-32) we have

$$\nabla^2\phi = \nabla\cdot(\nabla\phi) = \phi_{,ii}\,. \tag{1-34}$$

The <u>Green-Gauss Theorem</u>, namely

$$\int_{vol} \nabla\cdot\mathbf{A}\, dV = \int_{surf} \mathbf{A}\cdot\mathbf{n}\, dS$$

takes the form

$$\int_{vol} A_{i,i}\, dV = \int_{surf} A_i n_i\, dS, \tag{1-35}$$

where dS is an element of area, dV is an element of volume, and n_i the exterior normal to the surface S. If we set $A_i = \phi_{,i}$, then we obtain the identity

$$\int_{vol} \nabla^2\phi\, dV = \int_{surf} \frac{\partial\phi}{\partial n}\, dS$$

in the form

$$\int_{vol} \phi_{,ii}\, dV = \int_{surf} \phi_{,i} n_i\, dS\,. \tag{1-36}$$

<u>Green's Second Identity</u> (<u>Green's Theorem</u>)

$$\int_{vol} (\psi\nabla^2\phi - \phi\nabla^2\psi)\, dV = \int_{surf} (\psi\frac{\partial\phi}{\partial n} - \phi\frac{\partial\psi}{\partial n})\, dS$$

becomes

$$\int_{vol} (\psi\phi_{,ii} - \phi\psi_{,ii})\, dV = \int_{surf} (\psi\phi_{,i} - \phi\psi_{,i}) n_i\, dS\,. \tag{1-37}$$

The <u>curl of a vector</u> can be defined as follows:

$$(\nabla\times\mathbf{A})_i = \varepsilon_{ijk}A_{j,k}\,. \tag{1-38}$$

Then <u>Stokes' Theorem</u>

$$\int_{surf} (\nabla\times\mathbf{A})\cdot d\mathbf{S} = \int_{curve} \mathbf{A}\cdot d\mathbf{l}$$

becomes

$$\int_{surf} \varepsilon_{ijk}A_{j,k}n_i\, dS = \int_{curve} A_i\,(d\mathbf{l})_i\,. \tag{1-39}$$

1.6. Symmetry Operations on a Tensor of Rank 2.

If we have a tensor of rank 2 such that C_{ij}, $= C_{ji}$ it is said to be symmetric. If $C_{ij} = -C_{ji}$, it is antisymmetric. An antisymmetric tensor has zero diagonal components. Any 2nd rank tensor may be represented by the sum of a symmetric (S_{ij}) part and an antisymmetric (A_{ij}) part, both of which are tensors. It is left to the student to show that this may be done in one and only one way, namely

$$C_{ij} = \tfrac{1}{2}(C_{ij} + C_{ji}) + \tfrac{1}{2}(C_{ij} - C_{ji}) = S_{ij} + A_{ij}. \qquad (1\text{-}40)$$

We can relate an anti-symmetric tensor of rank 2 to a vector by the following. Consider A_{ij} where

$$A_{ij} = \begin{pmatrix} 0 & A_{12} & -A_{13} \\ -A_{12} & 0 & A_{23} \\ A_{13} & -A_{23} & 0 \end{pmatrix} .$$

It can be seen that there are only three independent components. Hence we see that a vector B_i defined by

$$B_i = \tfrac{1}{2}\varepsilon_{ijk}A_{jk} \qquad (1\text{-}41)$$

has the requisite transformation properties, and consequently an anti-symmetric tensor of rank 2 can be represented by a vector (tensor of rank 1).

CHAPTER 2

THE ANALYSIS OF STRAIN

2.1. The Deformation of an Elastic Body and the Strain Components.

In our analysis of strain, we will assume that the material with which we are working is continuous, i.e., that the smallest element of volume has the same mechanical properties as the whole. By this definition we will disregard the crystalline structure of matter.

A body is said to be strained when the relative position of points in a continuous body is altered. This excludes rigid body displacements and rotations from the definition of strain.

Deformation is the change in the relative position of points in the body. A precise measure of deformation is given by the strain components defined in the following.

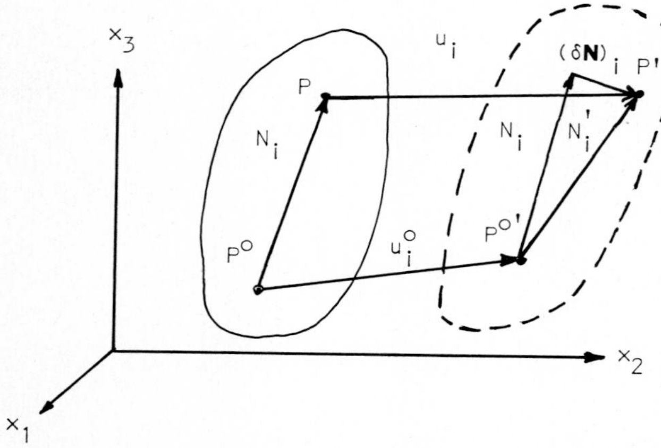

Figure 2-1. A description of local deformation.

In the deformation of an elastic body we will be using one material point P^o as a local origin for a cartesian coordinate system. We then wish to look at the displacements of another point P which is to be regarded as an arbitrary point in the neighborhood of P^o. In the deformed state we will call the same two points $P^{o'}$ and P'. The coordinates of point P are given by

$$P = (x_1, x_2, x_3),$$

and of P' by

$$P' = (x_1', x_2', x_3').$$

The equations describing the deformation can be written as

$$x_i' = x_i'(x_1, x_2, x_3) \equiv x_i'(\mathbf{x}). \tag{2-1}$$

Since we are working with a continuous medium, we will be considering only continuous deformations, i.e., no gaps or superpositions. Such an assumption will guarantee that there will be a unique, single valued inverse:

$$x_i = x_i(x_1', x_2', x_3') \equiv x_i(\mathbf{x}'). \qquad (2\text{-}2)$$

and the transformations of all points will be one-to-one.

A nearby point $P(\mathbf{x})$ can be related to $P^o(\mathbf{x}^o)$ by a neighborhood vector N_i where

$$N_i = x_i - x_i^o. \qquad (2\text{-}3)$$

On deformation the point P^o will move to $P^{o'}$ by a displacement

$$u_i^o(x_1^o, x_2^o, x_3^o) = x_i^{o'} - x_i^o. \qquad (2\text{-}4)$$

Similarly the displacement of P is given by

$$u_i(x_1, x_2, x_3) = u_i(x_1^o+N_1, x_2^o+N_2, x_3^o+N_3) = x_i' - x_i. \qquad (2\text{-}5)$$

Upon deformation, we obtain an altered neighborhood vector

$$N_i' = x_i' - x_i^{o'}, \qquad (2\text{-}6)$$

and a measure of the local deformation is given by

$$\begin{aligned}(\delta\mathbf{N})_i &= N_i' - N_i \\ &= (x_i' - x_i^{o'}) - (x_i - x_i^o) \\ &= (x_i' - x_i) - (x_i^{o'} - x_i^o) \\ &= (u_i - u_i^o) = (\delta\mathbf{u})_i. \qquad (2\text{-}7)\end{aligned}$$

If we expand (2-5) in a Taylor's series about the point P^o, keeping only terms to first order, and then subtract (2-4), we obtain

$$\begin{aligned}(\delta\mathbf{N})_i &= u_i(x_1, x_2, x_3) - u_i^o(x_1^o, x_2^o, x_3^o) = \\ &= u_i^o(x_1^o, x_2^o, x_3^o) + \frac{\partial u_i^o}{\partial x_1^o} N_1 + \frac{\partial u_i^o}{\partial x_2^o} N_2 + \frac{\partial u_i^o}{\partial x_3^o} N_3 + \cdots - u_i^o(x_1^o, x_2^o, x_3^o).\end{aligned}$$

The ⋯ stand for second and higher order terms we are going to neglect. Substituting subscript notation for $\partial u_i^o/\partial x_j^o$, we have

$$(\delta\mathbf{N})_i = u_{i,j} N_j. \qquad \text{FIRST ORDER LINEAR THEORY} \qquad (2\text{-}8)$$

The neglected terms are usually quite small in the propagation of elastic waves. For an adequate theory of second order effects, one should refer to texts on nonlinear elasticity, plasticity, rheology, etc.

The quantities $u_{i,j}$ may be shown to be tensor quantities in the following manner. Now in a rotated coordinate system we would have

$$u_i = g_{ij'} u_{j'} \text{ and } x_{l'} = g_{l'k} x_k,$$

using the law of transformation of vectors. Then

$$u_{i',k} = g_{ij'}\frac{\partial u_{j'}}{\partial x_k} = g_{ij'}\frac{\partial u_{j'}}{\partial x_{l'}}\frac{\partial x_{l'}}{\partial x_k} = g_{ij'}u_{j',l'}g_{l'k},$$

which is indeed the law of transformation for tensors. Hence the vector $(\delta\mathbf{N})_i$ will be a proper vector function of the vector N_j. Also, the linear combinations of the $u_{i,j}$'s which we shall be computing shortly will be proper tensors.

We can now write the general displacement of a point P in the neighborhood of P^o as

$$u_i = u_i^o + (\delta\mathbf{N})_i,$$

Where the components of $(\delta\mathbf{N})_i$ have been given by (2-8). The term u_i^o represents a translation (rigid body motion) of the local origin. Let us see if there are any further rigid body motions remaining. A rigid body motion is characterized by the fact that distances between all pairs of points remain constant, i.e.,

$$N^2 = N_iN_i = \text{constant}. \tag{2-9}$$

That is, we must have that

$$\tfrac{1}{2}\delta(N^2) = N_i(\delta\mathbf{N})_i = 0.$$

for all N_i. But

$$N_i(\delta\mathbf{N})_i = u_{i,j}N_iN_j. \tag{2-10}$$

Since N_i and N_j are completely arbitrary, we must have that

$$u_{i,j} + u_{j,i} = 0. \tag{2-11}$$

This implies that the symmetric parts of the tensor $u_{i,j}$ are zero in a rigid body motion, hence it is the antisymmetric parts that are to be identified with a rigid body motion.

It thus serves our purpose to separate (as in (1-40)) the tensor $u_{i,j}$ into symmetric and antisymmetric components by writing

$$u_{i,j} = e_{ij} + w_{ij},$$

where

$$e_{ij} = \tfrac{1}{2}(u_{i,j} + u_{j,i}),\ w_{ij} = \tfrac{1}{2}(u_{i,j} - u_{j,i}). \tag{2-12}$$

By construction we have that $e_{ij} = e_{ji}$ and $w_{ij} = -w_{ji}$.

Then

$$(\delta\mathbf{N})_i = e_{ij}N_j + w_{ij}N_j$$
$$= \delta N_i^s + \delta N_i^r. \tag{2-13}$$

Where δN_i^s is the relative displacement due to strain, and δN_i^r that due to rotation. The quantities e_{ij} are involved in the deformation of the medium and are known as <u>strain components</u> and will be described in detail. As we shall see in the next section, the quantities w_{ij} are involved in a local rotation of the medium and are known as the <u>rotation components</u>.

2.2. The Geometrical Interpretation of the Components w_{ij}.

Because of the antisymmetry of the components w_{ij}, we can write δN_i^r as (1-41)

$$\delta N_i^r = \varepsilon_{ijk}\Omega_j N_k \qquad (\delta \mathbf{N}^r = \mathbf{\Omega}\times\mathbf{N}), \tag{2-14}$$

where $\Omega_1 = w_{32}$, $\Omega_2 = w_{13}$, $\Omega_3 = w_{21}$; i.e.,

$$\Omega_i = \tfrac{1}{2}\varepsilon_{ijk}u_{j,k} \qquad (\mathbf{\Omega} = \tfrac{1}{2}\nabla\times\mathbf{u}). \tag{2-15}$$

Now in Figure 2-2 we see that for a small rotation about the x_3 axis we have that

$$u_2 = r d\theta \cos\theta = x_1 d\theta,$$
$$u_1 = -r d\theta \sin\theta = -x_2 d\theta.$$

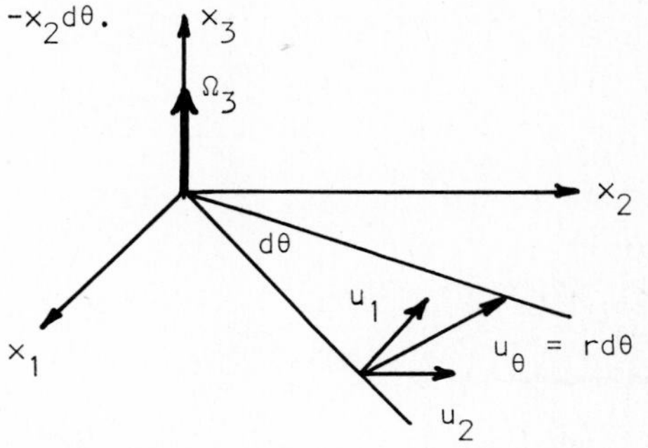

Figure 2-2. Displacements in a simple rotation.

On direct calculation we obtain

$$\Omega_3 = w_{21} = \tfrac{1}{2}(u_{2,1} - u_{1,2}) = d\theta. \tag{2-16}$$

In a similar manner the other components of Ω_i can be identified with small rotations about the other axes.

2.3. The Geometrical Interpretation of the Components e_{ij}.

If we divide the deformational part of (2-13) by the factor N, we obtain

$$(\delta\mathbf{N})_i/N = e_{ij}N_j/N. \tag{2-17}$$

If N_i were initially parallel to the x_1-axis, then $N_2 \equiv 0$, $N_3 \equiv 0$, and (2-17) becomes

$$\delta N_1/N_1 = e_{11}. \tag{2-18}$$

That is, e_{11} represents the extension or change in length per unit length of a vector originally parallel to the x_1-axis. Similar considerations hold for e_{22} and e_{33}. Thus the diagonal strain components represent extensions.

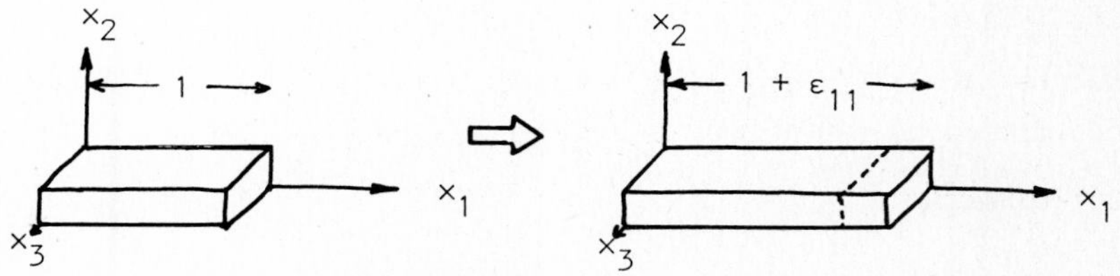

Figure 2-3. A simple extension along the x_1-axis.

Let us consider another experiment. On deformation of the parallelopiped formed by the vectors A_1, B_2, and C_3, we would find that

$$A_i' = (\delta\mathbf{A})_i + \delta_{i1}A_1,$$
$$B_j' = (\delta\mathbf{B})_i + \delta_{j2}B_2, \tag{2-19}$$
$$C_k' = (\delta\mathbf{C})_k + \delta_{k3}C_3,$$

where δ_{i1} etc., represent Kronecker delta symbols.

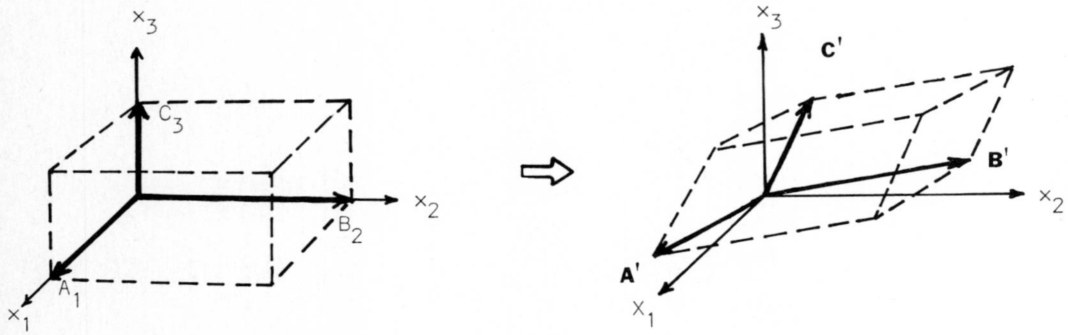

Figure 2-4. Deformation of a parallelopiped.

Now the volume enclosed by the parallelopiped is given by (1-21)

$$V' = \varepsilon_{ijk}A_i'B_j'C_k' \tag{2-20}$$
$$= \{A_1+(\delta\mathbf{A})_1\}\ \{B_2+(\delta\mathbf{B})_2\}\ \{C_3+(\delta\mathbf{C})_3\} + \text{terms of higher order}$$
$$\simeq A_1B_2C_3\ (1+e_{11})\ (1+e_{22})\ (1+e_{33}).$$

We see that to terms of first order the fractional change in volume of a parallelopiped is

$$\delta V/V = (V' - V)/V \simeq e_{11} + e_{22} + e_{33} = e_{ii} = \Theta. \tag{2-21}$$

This quantity (the sum of the three diagonal strain components) is called the dilatation and represents the fractional volume change. An expansion is arbitrarily signed positively while a compression is negative. It will be shown later that the dilatation is invariant to a rotation of coordinates.

To interpret the off-diagonal (or mixed) terms, we need to consider two neighborhood vectors, e.g., A_2 and B_3. After deformation we have

$$A_i' = (\delta\mathbf{A})_i + \delta_{i2}A_2,$$

$$B_i' = (\delta\mathbf{B})_i + \delta_{i3}B_3.$$

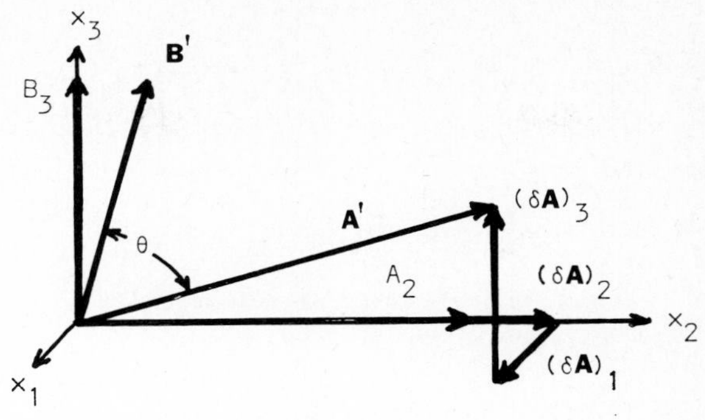

Figure 2-5. Deformation of the vectors defining the parallelopiped.

From (1-11) we have

$$\begin{aligned} \cos\theta &= A_i B_i / AB \\ &\simeq \{A_2(\delta\mathbf{B})_2 + B_3(\delta\mathbf{A})_3\}/A_2B_3 \\ &= (\delta\mathbf{B})_2/B_3 + (\delta\mathbf{A})_3/A_2, \end{aligned} \qquad (2\text{-}22)$$

where θ is the angle between A_i' and B_i'. Since the only terms appearing in this order of approximation are the 2- and 3-terms, we can redraw a plane figure as in Figure 2-6. Remembering that $A_1 = 0$, $A_3 = 0$, $B_1 = 0$, and $B_2 = 0$, and for the time being considering only displacement fields in which $w_{23} = 0$, we have from (2-13) that

$$\begin{aligned} (\delta\mathbf{B})_2 &= e_{23}B_3, \\ (\delta\mathbf{A})_3 &= e_{32}A_2 = e_{23}A_2. \end{aligned} \qquad (2\text{-}23)$$

Such a displacement is known as <u>pure</u> <u>shear</u>.

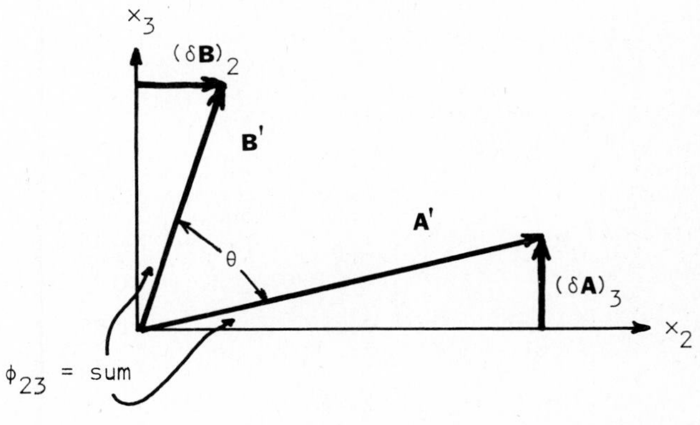

Figure 2-6. Planar relationships in the deformation of a parallelopiped (pure shear).

Now

$$\cos\theta = \cos(\pi/2 - \phi_{23}) = \sin\phi_{23} \approx \phi_{23}$$

$$\approx (\delta\mathbf{B})_2/B_3 + (\delta\mathbf{A})_3/A_2 = 2e_{23} = 2e_{32}. \tag{2-24}$$

That is, a positive value of e_{23} represents a decrease in the right angle between A_2 and B_3. This will be true even if $w_{23} \neq 0$, since

$$(\delta\mathbf{B})_2 = e_{23}B_3 + w_{23}B_3 \text{ and } (\delta\mathbf{A})_3 = e_{32}A_2 + w_{32}A_2$$

and

$$\phi_{23} \approx e_{23} + e_{32} + \cancel{w_{23}} + \cancel{w_{32}} = 2e_{23}. \tag{2-25}$$

By a rotation of the previous plane figure, we can obtain the deformation shown in Figure 2-7. The displacements corresponding to this deformation are

$u_1 = 0,\ u_2 = \phi x_3,\ u_3 = 0.$

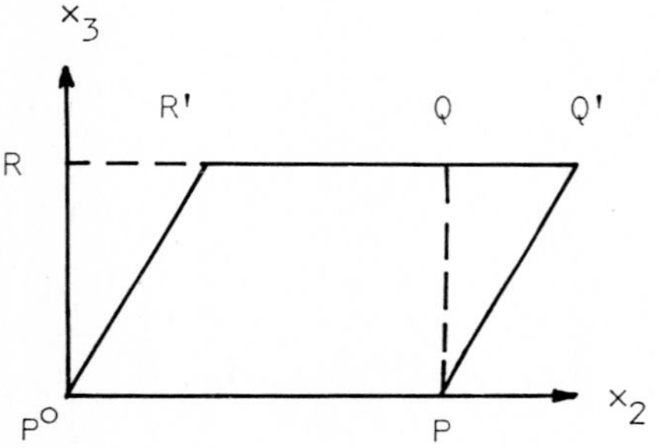

Figure 2-7. Deformation plus rotation — simple shear.

By direct computation

$$e_{23} = \tfrac{1}{2}\left(\frac{\partial u_2}{\partial x_3} + \frac{\partial u_3}{\partial x_2}\right) = \tfrac{1}{2}\phi;$$

likewise

$$w_{23} = \tfrac{1}{2}\left(\frac{\partial u_2}{\partial x_3} - \frac{\partial u_3}{\partial x_2}\right) = \tfrac{1}{2}\phi = -w_{32} = -\Omega_1.$$

The term w_{23} represents a counterclockwise rotation about the x_1-axis of $\tfrac{1}{2}\phi$ inasmuch as for small angles $\tan\phi \approx \phi$. Likewise the strain e_{23} is also equal to the quantity $\tfrac{1}{2}\phi$. Thus in the deformation diagrammed above, half of the distortional angle ϕ is represented by rotation, the other half by e_{23}. The deformation shown in Figure 2-7 is called simple shear. The mixed strain components are likewise referred to as components of shearing strain, even though simple shear includes some rotation.

It should be clear from Figure 2-7 that the area of the rectangle P^oPQR and of the parallelogram $P^oPQ'R'$ are equal. (This is true only to first order, i.e., as long as the angle ϕ is small.) Likewise a cube in three dimensions would be deformed into a parallelopiped of equal volume.

2.4. The Strain Quadric of Cauchy.

Let us look more closely at the change of the neighborhood vector itself. To first order, the change in length of the vector N_i is given by $N_i(\delta\mathbf{N})_i/N$ (inasmuch as components at right angles to N_i only cause a rotation). After dividing by another factor of N, we obtain the fractional change in length or a generalized extension <u>e</u>. This can be written for any direction

$$e = N_i(\delta\mathbf{N})_i/N^2 = e_{ij}N_iN_j/N^2.$$

Rewriting, we obtain

$$eN^2 = e_{ij}N_iN_j = 2Q(N_1,N_2,N_3), \tag{2-26}$$

where Q is a quadratic form in three variables. Constant values of this function, such as

$$e_{ij}N_iN_j = \pm k^2, \tag{2-27}$$

will determine a quadric surface in three-dimensional N-space. The shape of this surface will depend upon the strain field, with the sign of k^2 so chosen that the surface is real. Such a representation has several useful properties. First of all we see that

$$e = \pm k^2/N^2, \tag{2-28}$$

i.e., the fractional change in length of any line (from a local origin P^o to a point P) is inversely proportional to the square of the radius vector running along that line to the quadric surface. From the geometry in Figure 2-8, we can see that the maximum and minimum fractional changes in length will be in the directions of the axes of the quadric surface. From the definition of Q (2-26), we also see that

$$\partial Q/\partial N_i = e_{ij}N_j = (\delta\mathbf{N})_i. \tag{2-29}$$

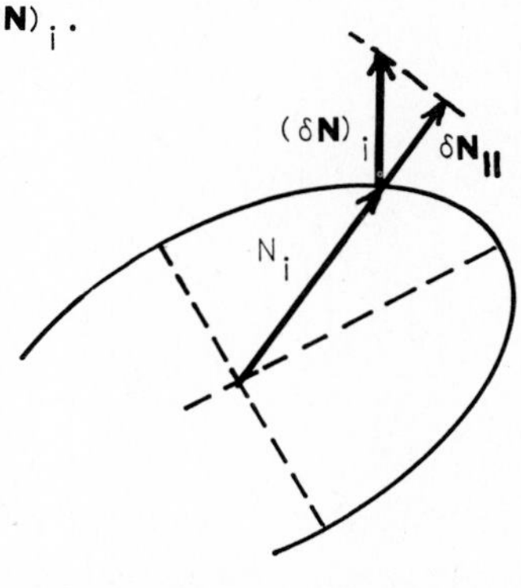

Figure 2-8. The strain quadric of Cauchy.

But the gradient of Q defined by (2-29) lies along the normal to the quadric surface; therefore the vector $(\delta\mathbf{N})_i$ is in the direction of the normal to the quadric surface at the point where N_i touches it.

These geometrical properties are useful, but by far the most useful property is that this quadric surface is invariant to a rotation of coordinates. To show this, we compute (using the tensor properties of e_{kl})

$$e_{i'j'}N_{i'}N_{j'} = g_{i'k}e_{kl}g_{lj'}g_{i'm}N_m g_{j'n}N_n$$
$$= \delta_{km}\delta_{ln}e_{kl}N_m N_n$$
$$= e_{kl}N_k N_l = \pm k^2. \tag{2-30}$$

Thus the value of the quadratic form is independent of the coordinates in which it is expressed. Inasmuch as Figure 2-8 relates to purely geometric quantities, this should be so. Once again we see the value of casting our equations in tensor form.

Note: This surface is not the strain ellipsoid of rock mechanics and structural geology. In some sense, it is inverse to it. The situation is compounded even further because in those fields a compression (rather than an expansion) is signed positively. The strain ellipsoid can be briefly described as the shape a sphere assumes under small deformations.

2.5. Principal Strains.

We have seen from the geometry of Figure 2-8 that there should be directions in which the deformation is only in that direction, i.e., it is a pure extension as in (2-18). Mathematically, this means that we should look for vectors N_i such that

$$(\overset{n}{\delta\mathbf{N}})_i = \overset{n}{E}\delta_{ij}\overset{n}{N}_j. \tag{2-31}$$

If this were true, we would then be able to write, from (2-13),

$$(\overset{n}{\delta\mathbf{N}})_i = e_{ij}\overset{n}{N}_j = \overset{n}{E}\delta_{ij}\overset{n}{N}_j, \tag{2-32}$$

and consequently we must have

$$(e_{ij} - \overset{n}{E}\delta_{ij})\overset{n}{N}_j = 0 \tag{2-33}$$

for some vectors (n of them) $\overset{n}{N}_j$. This will be so only if

$\mathrm{Det}|e_{ij} - \overset{n}{E}\delta_{ij}| = 0$. That is,

$$\begin{vmatrix} e_{11} - \overset{n}{E} & e_{12} & e_{13} \\ e_{21} & e_{22} - \overset{n}{E} & e_{23} \\ e_{31} & e_{32} & e_{33} - \overset{n}{E} \end{vmatrix} = 0. \tag{2-34}$$

In general, this determinant will have three distinct roots. The case of degenerate roots will be discussed later. Let us assume that we have found the roots — and with each root $\overset{n}{E}$ we have an associated direction given by $\overset{n}{N}_i$. Then for $\overset{1}{E}$ we have

$$\overset{1}{E}\overset{1}{N}_i = e_{ik}\overset{1}{N}_k.$$

Multiplying both sides by $\overset{2}{N}_i$ and summing, we have

$$\overset{1}{E}\,\overset{1}{N}_i\overset{2}{N}_i = e_{ik}\overset{1}{N}_k\overset{2}{N}_i.$$

Similarly

$$\overset{2}{E}\,\overset{1}{N}_i\overset{2}{N}_i = e_{ik}\overset{1}{N}_i\overset{2}{N}_k = e_{ki}\overset{1}{N}_k\overset{2}{N}_i = e_{ik}\overset{1}{N}_k\overset{2}{N}_i.$$

Then, on subtraction

$$(\overset{1}{E} - \overset{2}{E})\,\overset{1}{N}_i\overset{2}{N}_i = 0. \qquad (2\text{-}35)$$

Hence, if the roots are distinct, $\overset{1}{N}_i$ and $\overset{2}{N}_i$ are orthogonal. Although not proved here, the roots $\overset{n}{E}$ can be shown to be real. In general we may write

$$\overset{m}{N}_i\overset{n}{N}_i = 0 \qquad (m \neq n). \qquad (2\text{-}36)$$

The directions $\overset{m}{N}_i$ are called the <u>principal directions of strain</u> and the strains $\overset{m}{E}$ are the extensions of the vector N_i in the principal directions and are termed the <u>principal strains</u>. Since $(\delta\mathbf{N})_i$ is always normal to the quadric surface, the principal directions must coincide with its principal axes. If two or more of the principal strains $\overset{n}{E}$ are equal, then the principal directions are undefined, but one can always choose a set that is orthogonal.

Let us form a matrix of these vectors in the principal directions where we represent $\overset{n}{N}_j$ by P_{jn}. We then have

$$e_{ij}P_{jn} = \overset{n}{E}\,\delta_{ij}\,P_{jn} = \overset{n}{E}P_{in}. \qquad (2\text{-}37)$$

We may then multiply both sides by the <u>inverse</u> <u>(reciprocal)</u> <u>matrix</u>

$$P^{-1}_{ij} \equiv \frac{P^{ji}}{\mathrm{Det}|P_{ij}|}, \qquad (2\text{-}38)$$

where

P^{ji} = transposed matrix of cofactors,

$\mathrm{Det}|P_{ij}|$ = determinant of matrix with elements P_{ij}.

It is left to the student to show that such a definition of P^{-1}_{ij} is consistent, i.e., $P_{ij}P^{-1}_{jk} = \delta_{ik}$. We then obtain

$$P^{-1}_{mi}e_{ij}P_{jn} = P^{-1}_{mi}\,\overset{n}{E}\,P_{in} = \delta_{mn}\overset{n}{E}. \qquad (2\text{-}39)$$

Equation (2-39) is the rule for obtaining the diagonalized form of the strain tensor with the principal strains appearing on the diagonal. We shall give an example shortly. If we take the principal axes as our coordinate axes, the off-diagonal terms of (2-26) disappear and 2Q takes the form

$$2Q(P_1,P_2,P_3) = \overset{1}{E}\, P_1^2 + \overset{2}{E}\, P_2^2 + \overset{3}{E}\, P_3^2 = \pm k^2. \qquad (2\text{-}40)$$

A final note on the strain components. The six components of strain are not completely independent. Six relationships, the equations of compatibility, can be derived between their second derivatives. These equations are necessary for the uniqueness of the displacements u_i obtained upon integrations of the equations

$$e_{ij} = \tfrac{1}{2}(u_{i,j} + u_{j,i}).$$

A full discussion may be found in Sokolnikoff (1956, pp. 25-29).

As an example of the rotation of the principal axes of strain, let us consider the following two dimensional problem in pure shear. The strain matrix can be written

$$e_{ij} = \begin{pmatrix} 0 & e_{23} \\ e_{23} & 0 \end{pmatrix}, \qquad (2\text{-}41)$$

and (2-33) becomes

$$\begin{pmatrix} -E & e_{23} \\ e_{23} & -E \end{pmatrix}\begin{pmatrix} N_2 \\ N_3 \end{pmatrix} = 0$$

for some N_2 and N_3. Then we must solve

$$\text{Det}\begin{vmatrix} -E & e_{23} \\ e_{23} & -E \end{vmatrix} = E^2 - e_{23}^2 = 0,$$

finding that

$$E = \pm e_{23}. \qquad (2\text{-}42)$$

For the positive root we have, from (2-33),

$$-N_2 + N_3 = 0,$$
$$N_2 - N_3 = 0,$$

or

$$N_3 = N_2. \qquad (2\text{-}43)$$

This says that the first principal direction is in the upper right hand quadrant making equal angles with the x_2 and x_3 axes. For the negative root we have

$$N_2 + N_3 = 0,$$
$$N_2 + N_3 = 0,$$

or

$$N_3 = -N_2. \qquad (2\text{-}44)$$

The second principal direction is then in the upper left hand quadrant making equal angles with the negative x_2- and positive x_3-axes. In the new coordinates we have the following picture.

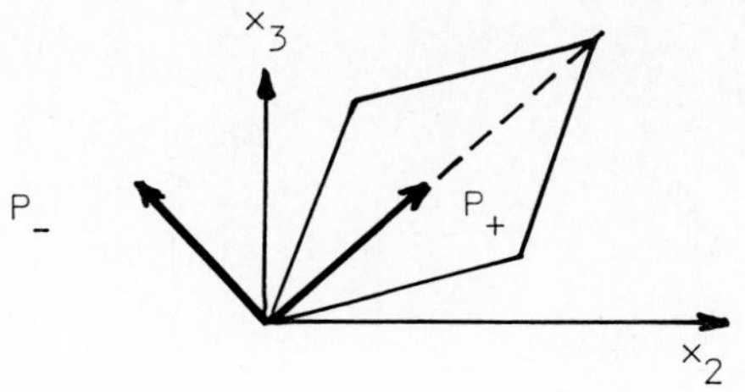

Figure 2-9. Principal directions of strain — pure shear.

Following the prescription given by (2-39), we form the following matrices (normalized to give $\det|P_{ij}| = 1$):

$$P_{ij} = \overset{\quad P_+ \qquad\quad P_-}{\begin{pmatrix} 2^{-\frac{1}{2}} & -2^{-\frac{1}{2}} \\ 2^{-\frac{1}{2}} & 2^{-\frac{1}{2}} \end{pmatrix}} \; ; \qquad P^{-1}_{ij} = \begin{pmatrix} 2^{-\frac{1}{2}} & 2^{-\frac{1}{2}} \\ -2^{-\frac{1}{2}} & 2^{-\frac{1}{2}} \end{pmatrix} . \tag{2-45}$$

We see that $P_{ij}P^{-1}_{jk} = \delta_{ik}$, as it should. In this simplified case (2-39) becomes

$$\begin{pmatrix} 2^{-\frac{1}{2}} & 2^{-\frac{1}{2}} \\ -2^{-\frac{1}{2}} & 2^{-\frac{1}{2}} \end{pmatrix} \begin{pmatrix} 0 & e_{23} \\ e_{23} & 0 \end{pmatrix} \begin{pmatrix} 2^{-\frac{1}{2}} & -2^{-\frac{1}{2}} \\ 2^{-\frac{1}{2}} & 2^{-\frac{1}{2}} \end{pmatrix} =$$

$$\begin{pmatrix} e_{23} & 0 \\ 0 & -e_{23} \end{pmatrix} = \begin{pmatrix} e_{++} & 0 \\ 0 & -e_{--} \end{pmatrix}, \tag{2-46}$$

and the strain quadric (2-26) becomes

$$e_{23}\, P_+^{\,2} - e_{23}\, P_-^{\,2} = \pm k^2. \tag{2-47}$$

Likewise in the new system we have

$$\begin{aligned} \delta P_+ &= e_{23}\, P_+ \\ \delta P_- &= -e_{23}\, P_- \end{aligned} \tag{2-48}$$

and the deformation of pure shear is equivalent to a lengthening of the diagonal along the P_+ axis and a shortening of the diagonal along the P_- axis. This may be seen in Figure 2-9.

Finally, a word about strain invariants. The cubic represented (2-34) can be written as

$$\begin{aligned} C(\overset{1}{E},\overset{2}{E},\overset{3}{E}) &= (\overset{1}{E} - E)\,(\overset{2}{E} - E)\,(\overset{3}{E} - E) \\ &= -E^3 + (\overset{1}{E} + \overset{2}{E} + \overset{3}{E})E^2 - (\overset{23}{EE} + \overset{31}{EE} + \overset{12}{EE})\,E + \overset{123}{EEE} \\ &= -E^3 + I_1 E^2 - I_2 E + I_3 = 0. \end{aligned} \tag{2-49}$$

But (2-34) can be expanded directly, and we have as the coefficients I_1, I_2, and I_3 in (2-49):

$$I_1 = e_{11} + e_{22} + e_{33},$$

$$I_2 = \begin{vmatrix} e_{22} & e_{23} \\ e_{23} & e_{33} \end{vmatrix} + \begin{vmatrix} e_{11} & e_{13} \\ e_{13} & e_{33} \end{vmatrix} + \begin{vmatrix} e_{11} & e_{12} \\ e_{12} & e_{22} \end{vmatrix}, \qquad (2\text{-}50)$$

$$I_3 = \begin{vmatrix} e_{11} & e_{12} & e_{13} \\ e_{12} & e_{22} & e_{23} \\ e_{13} & e_{23} & e_{33} \end{vmatrix}.$$

Now a given set of roots for a cubic equation uniquely gives the coefficients of that cubic. Since the roots $\overset{n}{E}$ represent geometric quantities, they should be invariant, and consequently so should the coefficients I_1, I_2, and I_3. Let us show that these roots are invariant by direct computation. In the original system we had

$$(e_{ij} - \delta_{ij}\overset{n}{E})\,\overset{n}{N}_j = 0. \qquad (2\text{-}33)$$

Pre-multiplying all terms by $g_{k'i}$ and inserting the expanded form of the Kronecker delta symbol between factors we have

$$g_{k'i}(e_{ij} - \delta_{ij}\overset{n}{E})g_{jl'}g_{l'm}\overset{n}{N}_m = 0. \qquad (2\text{-}51)$$

But

$$g_{k'i}e_{ij}g_{jl'} = e_{k'l'}, \quad g_{l'm}\overset{n}{N}_m = \overset{n}{N}_{l'}, \quad g_{k'i}\delta_{ij}g_{jl'} = g_{k'j}g_{jl'} = \delta_{k'l'}.$$

Hence (2-33) becomes

$$(e_{k'l'} - \delta_{k'l'}\overset{n}{E})\overset{n}{N}_{l'} = 0. \qquad (2\text{-}52)$$

We see that in the new system the roots of the cubic $\overset{n}{E}$ remain unchanged. It follows then that the factors I_1, I_2, I_3 are invariant combinations of these roots. Only I_1 has a specific interpretation. In (2-21) this was identified as Θ, the dilatation, proving the statement on page 14.

2.6. Strain Components in Curvilinear Coordinates.

In the interest of economy, we will not be deriving the vector and tensor relations of Chapter 1 in curvilinear coordinates. The reader is referred to standard texts on applied mathematics. However, for convenience, we include the forms of the strain components in two commonly encountered coordinate systems.

In cylindrical coordinates we have:

$$
\begin{aligned}
e_{rr} &= \frac{\partial u_r}{\partial r} \\
e_{\theta\theta} &= \frac{1}{r}\frac{\partial u_\theta}{\partial \theta} + \frac{u_r}{r} \\
e_{zz} &= \frac{\partial u_z}{\partial z} \\
e_{r\theta} = e_{\theta r} &= \tfrac{1}{2}\left[r\frac{\partial}{\partial r}\left(\frac{u_\theta}{r}\right) + \frac{1}{r}\frac{\partial u_r}{\partial \theta}\right] \\
e_{rz} = e_{zr} &= \tfrac{1}{2}\left[\frac{\partial u_z}{\partial r} + \frac{\partial u_r}{\partial z}\right] \\
e_{\theta z} = e_{z\theta} &= \tfrac{1}{2}\left[\frac{\partial u_\theta}{\partial z} + \frac{1}{r}\frac{\partial u_z}{\partial \theta}\right].
\end{aligned}
\tag{2-53}
$$

In spherical coordinates we have:

$$
\begin{aligned}
e_{rr} &= \frac{\partial u_r}{\partial r} \\
e_{\theta\theta} &= \frac{1}{r}\frac{\partial u_\theta}{\partial \theta} + \frac{u_r}{r} \\
e_{\phi\phi} &= \frac{u_r}{r} + \frac{u_\theta}{r}\cot\theta + \frac{1}{r\sin\theta}\frac{\partial u_\phi}{\partial \phi} \\
e_{r\theta} = e_{\theta r} &= \tfrac{1}{2}\left[\frac{1}{r}\frac{\partial u_r}{\partial \theta} + r\frac{\partial}{\partial r}\left(\frac{u_\theta}{r}\right)\right] \\
e_{r\phi} = e_{\phi r} &= \tfrac{1}{2}\left[\frac{1}{r\sin\theta}\frac{\partial u_r}{\partial \phi} + r\frac{\partial}{\partial r}\left(\frac{u_\phi}{r}\right)\right] \\
e_{\theta\phi} = e_{\phi\theta} &= \tfrac{1}{2}\left[\frac{1}{r\sin\theta}\frac{\partial u_\theta}{\partial \phi} + \frac{\sin\theta}{r}\frac{\partial}{\partial \theta}\left(\frac{u_\phi}{\sin\theta}\right)\right].
\end{aligned}
\tag{2-54}
$$

CHAPTER 3

THE ANALYSIS OF STRESS

3.1. The Stress Vector and Stress Components.

Imagine our elastic medium to be divided by a plane through any point P. The direction of this plane is described by its unit normal vector n_i. The stress vector ${}^nT_i$ across the plane is the force exerted by the portion of the material on the positive side of the plane upon the material on the negative side of the plane,

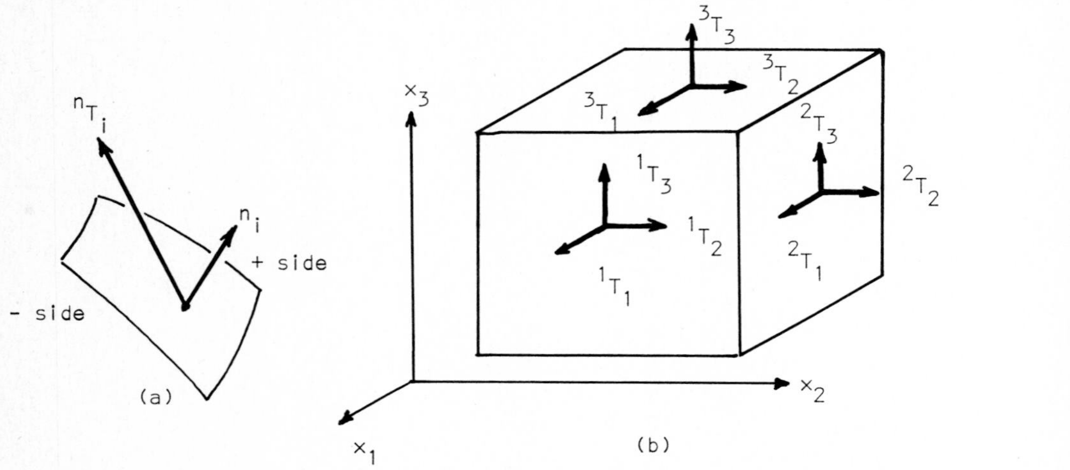

Figure 3-1(a). The stress vector ${}^nT_i$ acting on a surface element.
(b). Decomposition of the stress vectors acting at the surfaces of a cube.

divided by the surface area of the plane of separation. Usually the stress vector ${}^nT_i$ is resolved into components directed normally and tangentially to the plane separating the two portions of the medium. The former will be designated as a tension if it is positive and a pressure if it is negative. (This convention is by no means universal, particularly in rock mechanics and structural geology where the opposite is true.) The normal and tangential components of the stress vector are also known as tensile stress and shearing stress, respectively. Reversal of the direction of the unit normal vector n_i changes the old positive side into the negative side and vice versa. However, equilibrium requires that the force on the opposite side of the surface will have the opposite sign. Consequently, a tension remains a tension, or a pressure a pressure, regardless of the arbitrary sense of n_i.

As an example of this nomenclature, consider the stress vectors operating on the surfaces of a cube oriented along the axes of the x_1, x_2, x_3 system. They will be described as in figure 3-1(b). A shorthand notation can be developed for these components by setting

$$^{1}T_1 = t_{11}\ ,\ ^{1}T_2 = t_{12}\ ,\ ^{2}T_3 = t_{23}\ ,$$

or in general terms

$$^{i}T_j = t_{ij}\ . \tag{3-1}$$

Here the first subscript of t_{ij} will refer to the orientation of the surface and the second subscript to the component of force.

Let us now determine the relationship of the <u>stress vector</u> $^{n}T_j$, for a surface with unit normal n_i, to the <u>stress components</u> t_{ij} along the axes set x_1, x_2, x_3. To do this, we consider a tetrahedron formed by the surface of interest and the planes perpendicular to the three coordinate axes as in Figure 3-2. If the tetrahedron is sufficiently small, the body forces (proportional to the volume) can be

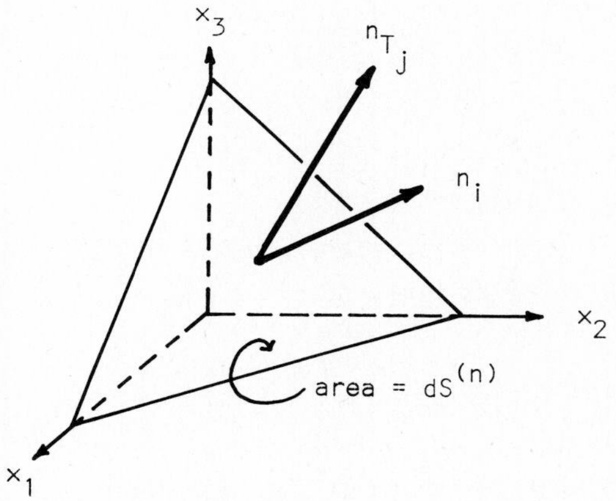

Figure 3-2. An arbitrarily oriented stress vector.

neglected in comparison to the surface forces. Consequently, for equilibrium we must have

$$^{n}T_j dS^{(n)} = g_{ni} t_{ij} dS^{(n)},$$

or

$$^{n}T_j = g_{ni}\ t_{ij}. \tag{3-2}$$

Now the orientation of the surface dS is totally arbitrary and its normal can be chosen to lie along each of a new set of axes ($x_{1'}$, $x_{2'}$, $x_{3'}$) in turn. We would then have

$$^{k'}T_j = g_{k'i} t_{ij}.$$

Resolving each of these force components in the unprimed system along the primed axes, and adding, we have

$$^{k'}T_{l'} = g_{k'i} t_{ij} g_{jl'}. \tag{3-3}$$

But we could just as well have written ${}^{k'}T_{l'} = t_{k'l'}$, giving

$$t_{k'l'} = g_{k'i}t_{ij}g_{jl'} \;. \qquad (3\text{-}4)$$

This is just the law of tensor transformation (1-28) so the stress components defined as in (3-1) are indeed the components of a tensor of rank 2.

3.2. Equations of Equilibrium.

Consider a small but non-vanishing portion of material body which is in equilibrium. This would mean that the resultant force and torque upon this element must vanish. Mathematically, the condition of equiibrium of forces requires that

$$\int_{vol} F_j \, dV + \int_{surf} t_{ij}n_i \, dS = 0. \qquad (3\text{-}5)$$

Here F_j represents the totality of all forces acting upon the volume of material, including inertial ones. The terms $t_{ij}n_i$ represent the forces acting upon the surfaces of the element. Both F_j and t_{ij} are assumed to be "nice" functions, i.e., they are continuous and have continuous derivatives of sufficient order to insure the validity of all mathematical operations that we will perform. Using the Green-Gauss theorem (1-35), we obtain

$$\int_{vol} (F_j + t_{ij},_i)dV = 0. \qquad (3\text{-}6)$$

Since our region of integration is arbitrary and the integrand is "nice," the latter must vanish identically. Hence at every point we may write

$$F_j = -t_{ij},_i . \qquad (3\text{-}7)$$

This equation says that the force vector on the interior of an elastic medium is to be computed as the negative divergence of the stress tensor.

Next we must see what the vanishing of the resultant torque on such an element of volume implies. We have

$$L_i = \int_{vol} \varepsilon_{ijk}x_jF_k \, dV + \int_{surf} \varepsilon_{ijk}x_jt_{lk}n_l \, dS = 0, \qquad (3\text{-}8)$$

where L_i is the resultant torque vector. Applying (1-35) to the second term of (3-8), we obtain

$$\int_{surf} \varepsilon_{ijk}x_jt_{lk}n_l \, dS = \int_{vol} (\varepsilon_{ijk}x_jt_{lk}),_l dV$$

$$= \int_{vol} \varepsilon_{ijk}(x_jt_{lk},_l + \delta_{jl}t_{lk})dV$$

$$= \int_{vol} \varepsilon_{ijk}(-x_jF_k + t_{jk})dV.$$

Hence, we find that

$$L_i = \int_{vol} \varepsilon_{ijk}t_{jk}dV = 0. \qquad (3\text{-}9)$$

Again the "niceness" of the integrand and the arbitrary selection of the element of volume require that

$$\varepsilon_{ijk}t_{jk} = 0 \qquad (3\text{-}10)$$

everywhere. The first component of (3-10) can be written as

$$\varepsilon_{123}t_{23} + \varepsilon_{132}t_{32} = (1)t_{23} + (-1)t_{32} = 0,$$

or

$$t_{23} = t_{32}.$$

Similarly for the other components. Hence we can write

$$t_{ij} = t_{ji}. \quad (3\text{-}11)$$

That is, the stress tensor is required to be symmetric by the physical conditions of equilibrium and does not have to be made symmetric as in the case of strain.

3.3 The Stress Quadric, Principal Stresses, and Stress Invariants.

The normal component of the stress vector may be written as

$$N = {}^nT_jn_j = (n_it_{ij})n_j. \quad (3\text{-}12)$$

If we let $x_i = An_i$ be a coordinate vector of (as yet) unspecified length in the direction of n_i, we can write

$$NA^2 = t_{ij}x_ix_j = 2Q'(x_1,x_2,x_3). \quad (3\text{-}13)$$

If we then constrain x_i to lie on the surface $2Q' = \pm k^2$, we have

$$t_{ij}x_ix_j = \pm k^2, \quad (3\text{-}14)$$

where again the sign of k^2 is chosen such that the surface is real.

In a manner similar to the strain quadric, we have

$$N = \pm k^2/A^2 \quad (3\text{-}15)$$

i.e., the normal component of stress is inversely proportional to the square of the radius vector in the direction n_i.

Likewise (using the relation $t_{ij} = t_{ji}$) we have

$$\partial Q'/\partial x_i = t_{ij}x_j = An_jt_{ji} = A\ {}^nT_i, \quad (3\text{-}16)$$

i.e., the stress vector is perpendicular to the stress quadric. Consequently the stress vector is given by a knowledge of its normal component. On construction of the stress quadric, N and ${}^nT_i$ occur as in Figure 3-3. Similarly to the strain quadric, the stress quadric Q' is invariant to a rotation of the coordinate axes.

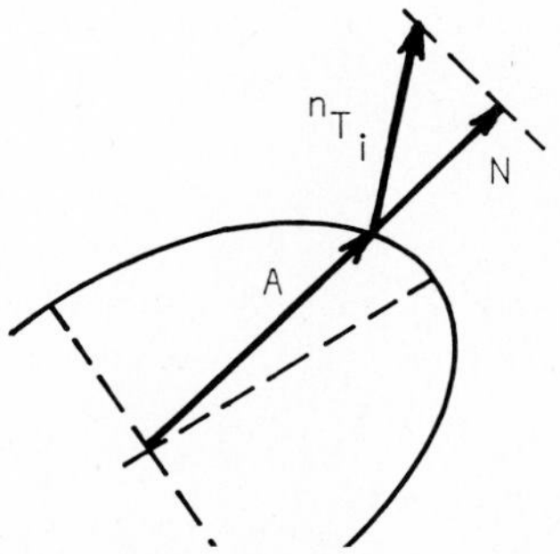

Figure 3-3. The stress quadric of Cauchy.

We also find that there are directions in which the stress vector lies only in that direction, i.e.,

$$ {}^{n}T_j = \overset{n}{T}\delta_{ij} g_{ni}. \tag{3-17} $$

But,

$$ {}^{n}T_j = g_{ni} t_{ij}. \tag{3-2} $$

So in these particular directions we must have

$$ (t_{ij} - \overset{n}{T}\delta_{ij})\, g_{ni} = 0. \tag{3-18} $$

This will be true if

$$ \det \begin{vmatrix} t_{11} - \overset{n}{T} & t_{12} & t_{13} \\ t_{21} & t_{22} - \overset{n}{T} & t_{23} \\ t_{31} & t_{32} & t_{33} - \overset{n}{T} \end{vmatrix} = 0. \tag{3-19} $$

The three roots $\overset{1}{T}$, $\overset{2}{T}$, $\overset{3}{T}$ will be the <u>principal stresses</u>. The directions g_{1i}, g_{2i}, g_{3i} will be the three <u>principal directions of stress</u>, and the corresponding planes will be called the <u>principal planes of stress</u>. In a coordinate system based on these three directions (arbitrary orthogonalization may be necessary in the case of degenerate roots) the stress quadric will take the form

$$ 2Q'(x_1, x_2, x_3) = \overset{1}{T}x_1^2 + \overset{2}{T}x_2^{\,2} + \overset{3}{T}x_3^{\,2} = \pm k^2. \tag{3-20} $$

By reasoning similar to that leading to (2-52) we can show that

$$ \overset{1}{T} + \overset{2}{T} + \overset{3}{T} = \Theta_t = \text{invariant}. \tag{3-21} $$

THE EQUATIONS OF ELASTICITY

4.1. Work and the Strain Energy Function.

The first law of thermodynamics states that

$$\int_{vol} \frac{dU}{dt}\, dV + \int_{vol} \frac{dK}{dt}\, dV = \int_{vol} \frac{\delta W}{dt}\, dV + \int_{vol} \frac{\delta Q}{dt}\, dV, \tag{4-1}$$

where dU is the increase in potential (internal) energy per unit volume of material, dK is the increase in kinetic energy, δW is the work done by all external forces f_j and δQ is the heat supplied. Equation (4-1) has been written to show that dU and dK are perfect differentials and that δW and δQ are not. Although δQ is not a perfect differential, the differential dS (S = entropy) is, where dS = δQ/T, and T is the absolute temperature. Now

$$\int_{vol} \frac{\delta W}{dt}\, dV = \int_{vol} f_j \dot{u}_j\, dV + \int_{surface} t_{ij} n_i \dot{u}_j\, dS. \tag{4-2}$$

Using the Green-Gauss theorem, we can write the 2nd term on the right hand side of (4-2) as

$$\int_{vol} (t_{ij}\dot{u}_j)_{,i}\, dV = \int_{vol} t_{ij,i}\dot{u}_j\, dV + \int_{vol} t_{ij}\dot{u}_{j,i}\, dV$$

$$= \int_{vol} t_{ij,i}\dot{u}_j\, dV + \int_{vol} t_{ij}(\dot{e}_{ji} + \dot{w}_{ji})\, dV$$

$$= -\int_{vol} F_j \dot{u}_j\, dV + \int_{vol} t_{ij}\dot{e}_{ji}\, dV,$$

where the rotational terms $\dot{w}_{ji}$ have summed out. Now the external forces f_j are defined by

$$f_j = F_j + \rho \ddot{u}_j .$$

Hence

$$\int_{vol} \frac{\delta W}{dt}\, dV = \int_{vol} (\ddot{u}_j \dot{u}_j)\, \rho dV + \int_{vol} t_{ij}\dot{e}_{ji}\, dV. \tag{4-3}$$

Now

$$\int_{vol} \frac{dK}{dt}\, dV = \frac{d}{dt}\left\{ \tfrac{1}{2} \int_{vol} (\dot{u}_j \dot{u}_j) \rho dV \right\} = \int_{vol} \ddot{u}_j \dot{u}_j \rho dV.$$

Thus we see that (4-1) can be written as

$$\int_{vol} \frac{dU}{dt}\, dV = \int_{vol} t_{ij}\dot{e}_{ji}\, dV + \int_{vol} \frac{\delta Q}{dt}\, dV.$$

If the change is accomplished adiabatically, i.e., δQ = 0 , then $t_{ij}de_{ji}$ must be a perfect differential equal to dU. On the other hand, if T = constant, then δQ = TdS is a perfect differential and $t_{ij}de_{ji}$ must be also. If $t_{ij}de_{ji}$ is a perfect differential, then there must exist a function X such that

$$\partial X/\partial e_{ij} = t_{ij}. \tag{4-4}$$

This function X is called the <u>strain energy function</u>.

To simplify the writing of such a function, let us write

$$t_1 = t_{11},\ t_2 = t_{22},\ t_3 = t_{33},\ t_4 = t_{23},\ t_5 = t_{31},\ t_6 = t_{12}, \tag{4-5}$$

$$e_1 = e_{11},\ e_2 = e_{22},\ e_3 = e_{33},\ e_4 = 2e_{23},\ e_5 = 2e_{31},\ e_6 = 2e_{12}.$$

The factor 2 appears in the terms e_4, e_5, e_6 for convenience in later calculations. If we now expand X in powers of the strain components up to second order, we have

$$X = c_o + c_i e_i + \tfrac{1}{2} c_{ij} e_i e_j.$$

Taking partial derivatives of the strain energy function, we find that

$$t_i \equiv \frac{\partial X}{\partial e_i} = c_i + \tfrac{1}{2}(c_{ij} + c_{ji}) e_j. \tag{4-6}$$

From the form of (4-6) we see that c_i must be zero, since when there is no strain $t_i(0) = 0$. We can also see that the antisymmetric parts of c_{ij} will not count, consequently we need only symmetric coefficients c_{ij}, i.e., we can write

$$c_{ij} = c_{ji}. \tag{4-7}$$

Taking account of these observations, (4-6) becomes

$$t_i \equiv \frac{\partial X}{\partial e_i} = c_{ij} e_j. \tag{4-8}$$

and the strain energy function becomes

$$X = \tfrac{1}{2} c_{ij} e_i e_j = \tfrac{1}{2} t_i e_i = \tfrac{1}{2} t_{ij} e_{ij}. \tag{4-9}$$

This formula can be used in computing the strain energy in an elastic deformation. Because of (4-7), the relationship between t_i and e_j involves at most 21 of the 36 c_{ij}'s as independent constants.

The relationship (4-8) is a special case of a more general linear relationship between t_{ij} and e_{kl}, which may be written as

$$t_{ij} = c_{ijkl} e_{kl}. \tag{4-10}$$

This is known as the <u>generalized Hooke's Law</u> and states that each component of stress is linearly proportional to each component of strain. Here we have 81 c_{ijkl}'s, but the symmetry of t_{ij} and e_{kl} reduce these to 36. The existence of the strain energy function reduces these to 21. Although the existence of this function has been shown only in the adiabatic and isothermal cases, experimental evidence has shown no need for more than these 21, hence such a function probably exists in the general case.

A word now about the experimental determination of these constants. They are generally determined by static tensile and torsional experiments or by the determination of elastic wave velocities (the relationship between the constants and these velocities will be derived later). In static experiments, the constants determined are the isothermal ones, inasmuch as the changes are so slow that all temperature differences have a chance to equalize. The thermal regime during the passage of an elastic wave needs more consideration. In the thermal cooling of a solid, we have as a characteristic time

$$t_{cool} \simeq L^2/\kappa,$$

where L is a characteristic dimension and κ is the thermal diffusivity. For an elastic wave, we have as a characteristic dimension

$$L = v \cdot \text{Period},$$

where $\underline{v}$ is the velocity of propagation of elastic waves. The ratio

$$\frac{t_{cool}}{\text{Period}} \simeq v^2 \cdot \text{Period}/\kappa$$

gives a rough idea of the thermal regime in a particular case. Some reasonable numbers for rocks are:

$$\kappa = 10^{-2},$$

$$v = 5 \times 10^5 \text{ cm/sec},$$

$$\text{Period} = 1 \text{ sec}.$$

Hence

$$\frac{t_{cool}}{\text{Period}} \simeq 2.5 \times 10^{13}.$$

This number is unrealistically high; other processes set in well before reaching it. However, for frequencies up to several megacycles per second, the thermal regime for elastic waves is adiabatic, and one determines the adiabatic values of the c_{ijkl}'s from wave propagation experiments whether in the earth or in seismic models.

4.2. Hooke's Law for a Homogeneous Isotropic Medium.

In such a medium (4-8) should not depend upon the labeling of axes, and hence should have the form

$$\begin{aligned}
t_{11} &= Ae_{11} + B(e_{22} + e_{33}) + 2Ce_{23} + 2D(e_{12} + e_{13}),\\
t_{22} &= Ae_{22} + B(e_{11} + e_{33}) + 2Ce_{13} + 2D(e_{12} + e_{23}),\\
t_{33} &= Ae_{33} + B(e_{11} + e_{22}) + 2Ce_{12} + 2D(e_{13} + e_{23}),\\
t_{12} &= Ee_{33} + F(e_{11} + e_{22}) + 2Ge_{12} + 2H(e_{13} + e_{23}),\\
t_{13} &= Ee_{22} + F(e_{11} + e_{33}) + 2Ge_{13} + 2H(e_{12} + e_{23}),\\
t_{23} &= Ee_{11} + F(e_{22} + e_{33}) + 2Ge_{23} + 2H(e_{12} + e_{13}).
\end{aligned} \qquad (4\text{-}11)$$

However, these equations should remain invariant upon a reflection of one of the axes, i.e., changing to a system of left-handed symmetry. In such a system the mixed terms involving the reflected axes will change sign, but the others will remain the same. Consequently C, D, E, F and H must be zero and we have only three coefficients left. If we now consider a rotation of θ about the x_3-axis, then for all angles θ we must have

$$t_{1'2'} = 2Ge_{1'2'}. \qquad (4\text{-}12)$$

But

$$t_{1'2'} = g_{1'i} t_{ij} g_{j2'}$$
$$= (t_{22} - t_{11})\cos\theta \sin\theta + t_{12}\cos 2\theta$$
$$= \{A(e_{22} - e_{11}) + B(e_{11} - e_{22})\}\cos\theta \sin\theta + 2Ge_{12}\cos 2\theta,$$

and

$$e_{1'2'} = (e_{22} - e_{11})\cos\theta \sin\theta + e_{12}\cos 2\theta.$$

If (4-12) is to hold for all angles, then we must have

$$B = A - 2G. \tag{4-13}$$

Now A is usually written as $\lambda + 2\mu$, G as μ and hence $B = \lambda$. The constants λ and μ are known as the Lamé constants. With these substitutions (4-11) becomes

$$t_{ij} = \lambda\Theta\delta_{ij} + 2\mu e_{ij}. \tag{4-14}$$

This is Hooke's Law for a homogeneous isotropic medium. The strain energy function for such a medium can be written from (4-9) as

$$X = \tfrac{1}{2}\{\lambda\Theta^2 + 2\mu(e_{11}^2 + e_{22}^2 + e_{33}^2 + 2e_{12}^2 + 2e_{13}^2 + 2e_{23}^2)\}. \tag{4-15}$$

4.3. Three Simple Experiments.

To get a better notion of the relationships between stress and strain, let us recall three experiments from elementary physics.

Simple Shear:

We have shown in Section 2.3 that

$$t_{21} = 2\mu e_{21} = \mu\phi. \tag{4-16}$$

The quantity μ is called the shear modulus. The geometric relationships are given in Figure 4-1.

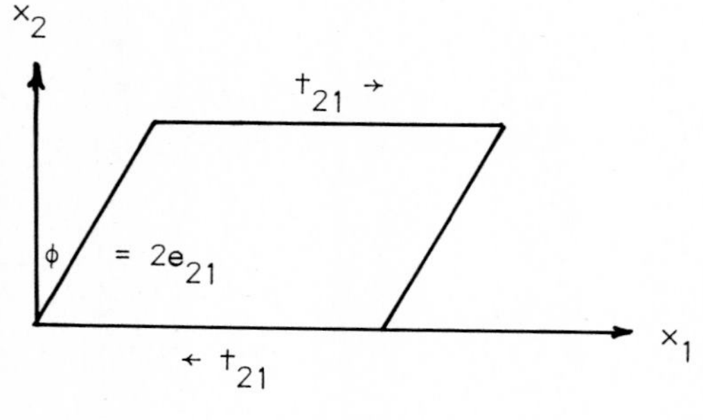

Figure 4-1. Simple Shear.

Uniform Pressure:

From symmetry considerations we must have $t_{11} = t_{22} = t_{33} = -p = -\text{pressure}$. Adding these three stress components we find that

$$-p = (\lambda + 2\mu/3)\Theta = k\, dV/V, \tag{4-17}$$

where $k = \lambda + 2\mu/3$ is called the bulk modulus.

Simple Extension:

In this case we have only one tensile stress component acting. All other stress components are zero. If we call the one acting t_{11}, we can solve for zero values of t_{22} and t_{33} giving

$$(\lambda + 2\mu)e_{22} + \lambda e_{33} = -\lambda e_{11},$$
$$\lambda e_{22} + (\lambda + 2\mu)e_{33} = -\lambda e_{11}.$$

The solution to this pair of equations is

$$e_{22} = e_{33} = \frac{-\lambda(\lambda + 2\mu) + \lambda^2}{(\lambda + 2\mu)^2 - \lambda^2} e_{11} = \frac{-\lambda}{2(\lambda + \mu)} e_{11} = -\sigma e_{11}.$$

Substituting these values for the strains into (4-14), we find that

$$t_{11} = \{(\lambda + 2\mu) - \frac{2\lambda^2}{2(\lambda + \mu)}\}e_{11} = \frac{\mu(3\lambda + 2\mu)}{\lambda + \mu} e_{11} = Ee_{11}. \quad (4\text{-}18)$$

The quantity $E = \mu(3\lambda + 2\mu)/(\lambda + \mu)$ is Young's Modulus and relates extensional strain to tensile stress. The quantity $\sigma = \lambda/2(\lambda + \mu)$ is known as Poisson's ratio. It is the negative ratio of the lateral contraction to the longitudinal extension in a simple extensional experiment. Its value can range between 0.0 (very large rigidity) and 0.5 (fluid), lying generally between 0.25 and 0.33 for most solids including the rocks of the earth.

4.4. Strain as a Function of Stress.

For homogeneous isotropic media equations (4-4) can be inverted to solve for the strain as a function of the stress. If we do so, we find that

$$e_{ij} = \frac{1}{2\mu} (t_{ij} - \frac{\lambda}{3\lambda + 2\mu} \Theta_t \delta_{ij}), \quad (4\text{-}19)$$

where $\Theta_t = t_{ii}$, is the first stress invariant. The strain energy function can also be written as

$$X = \frac{1}{4\mu}\{(t_{11}^2 + t_{22}^2 + t_{33}^2 + 2t_{12}^2 + 2t_{13}^2 + 2t_{23}^2) - \frac{\lambda}{3\lambda + 2\mu} \Theta_t^{\,2}\} . \quad (4\text{-}20)$$

The strain energy function is a quadratic in either the stress or the strain when one is working with the linear theory of elasticity.

CHAPTER 5

EQUATIONS OF MOTION

5.1. General.

We saw that the total force on a portion of an elastic medium was given by

$$F_j = -t_{ij,i}, \qquad (3\text{-}7)$$

where F_j represented the totality of forces acting on the medium. We can write

$$F_j = f_j - \rho \ddot{u}_j, \qquad (5\text{-}1)$$

where f_j are the body forces acting such as gravity, and $\rho \ddot{u}_j$ is the inertial force acting. From (3-7) we can then write

$$f_j + t_{ij,i} - \rho \ddot{u}_j = 0. \qquad (5\text{-}2)$$

This is the elastic equation of motion in its most general form. We will now make a few comments upon the nature of this equation in homogeneous, anisotropic media and in isotropic, inhomogeneous media before specializing to isotropic, homogeneous media.

5.2. Homogeneous, anisotropic media.

Probably the most frequently encountered research papers involving anisotropic media are related to acoustic resonators, waveguides, and transducers. Some recent monographs in this field are Auld (1973) and Musgrave (1970). Many review articles will be found in the continuing series on current topics in physical acoustics edited by Mason (1976).

In the earth, however, anisotropy has generally proven to be a second order effect. Although most of the crystalline materials that make up rocks are intrinsically anisotropic, their ordinarily random orientation gives rise to rocks which behave nearly isotropically when waves which are long compared to the crystal dimensions propagate through them. On a larger scale, nevertheless, the real earth is generally stratified (vertically or radially) with elastic properties and density which vary continuously or step-wise with increasing depth. If the propagating waves are long enough compared to the characteristic dimensions of these varying strata, then another averaging effect takes place which can lead to transverse anisotropy if conditions are right. Backus (1962) derives much of the relevant theory and gives several references to previous work. In addition to this transverse anisotropy, crustal properties have been seen to differ with direction of propagation (Forsyth, 1975; and Raitt et al, 1969). The effects are small, generally less than 5%, but their magnitude is obscured by other effects including noise. There seems to be a fair amount of disagreement as to the amount of azimuthal anisotropy present. A special issue of the Geophysical Journal of the Royal Astronomical Society edited by

D. Bamford (Vol. 49, pp. 1-243, 1977) has been devoted to a discussion of this subject. The leadoff article (Bamford and Crampin, 1977) contains a useful review.

Anisotropy can also be found in the vicinity of large deformations. Green, Rivlin, and Shield (1952) showed that such anisotropy could be due to non-linear effects. Nur (1971) investigated the effects of stress-induced cracks. Both conditions can be found in the vicinity of earthquake rupture zones and near the blast region of nuclear devices. Such anisotropy should die off rather rapidly away from the strongly deformed regions and should not significantly influence wave propagation thereafter. However, source radiation patterns could be altered depending on the amount of anisotropy present.

Lastly, Anderson and Toksöz (1963) found that earth-flattening approximations used to convert spherical-earth models into flat-earth models introduce anisotropy (as well as inhomogeneity) into the flat-earth model when it was not originally present in the spherical-earth model.

Let us now look at the equations governing the propagation of elastic waves in homogeneous anisotropic media. Such media can be described by the twenty-one quantities c_{ijkl}. If we replace the stress components of (5-2) with their equivalents in terms of displacement derivatives, using (4-10) and (2-12), we obtain:

$$\begin{aligned} 0 &= f_j + \tfrac{1}{2}c_{ijkl}(u_{k,l} + u_{l,k})_{,i} - \rho\ddot{u}_j \\ &= f_j + c_{ijkl}\,u_{k,li} - \rho\ddot{u}_j. \end{aligned} \tag{5-3}$$

In a two dimensional case, the first component of (5-3) is

$$\begin{aligned} f_1 &+ \left\{c_{1111}\frac{\partial^2}{\partial x_1^2} + (c_{1112} + c_{2111})\frac{\partial^2}{\partial x_1\,\partial x_2} + c_{2112}\frac{\partial^2}{\partial x_2^2}\right\} u_1 \\ &+ \left\{c_{1121}\frac{\partial^2}{\partial x_1^2} + (c_{1122} + c_{2121})\frac{\partial^2}{\partial x_1\,\partial x_2} + c_{2122}\frac{\partial^2}{\partial x_2^2}\right\} u_2 - \rho\ddot{u}_1 = 0. \end{aligned}$$

A similar equation holds for the second component. We can write this in a more compact form as

$$\begin{aligned} f_1 + L_{11}u_1 + L_{12}u_2 &= 0, \\ f_2 + L_{21}u_1 + L_{22}u_2 &= 0, \end{aligned} \tag{5-4}$$

where the factors

$$L_{11} = \left\{c_{1111}\frac{\partial^2}{\partial x_1^2} + (c_{1112} + c_{2111})\frac{\partial^2}{\partial x_1\,\partial x_2} + c_{2112}\frac{\partial^2}{\partial x_2^2} - \rho\frac{\partial^2}{\partial t^2}\right\},$$

$$L_{12} = \left\{c_{1121}\frac{\partial^2}{\partial x_1^2} + (c_{1122} + c_{2121})\frac{\partial^2}{\partial x_1\,\partial x_2} + c_{2122}\frac{\partial^2}{\partial x_2^2}\right\},$$

etc., are second order differential operators. The pair (5-4) represents a coupled set of second order partial differential equations. Equations (5-4) can be solved in many different ways; see Kraut (1963) for a review. However, we wish to illustrate one technique he used that is conceptually simple. Multiplying the first line of (5-4) by L_{22} and the second by L_{12} and then subtracting, we obtain

$$(L_{21}L_{12} - L_{22}L_{11})u_1 = (L_{22}f_1 - L_{12}f_2). \qquad (5\text{-}5)$$

Similarly, multiplying the first by L_{21} and the second by L_{11}, we get

$$(L_{21}L_{12} - L_{22}L_{11})u_2 = (L_{21}f_1 - L_{11}f_2). \qquad (5\text{-}6)$$

These are now fourth order partial differential equations, but they are uncoupled. Usual transform methods may be used to solve them. The algebra gets extremely messy, but the techniques are essentially the same as we will use in the solution of Lamb's problem. Note that the homogeneous portions are the same in both (5-5) and (5-6). One has to be careful inasmuch as this method of uncoupling using higher order derivatives quite often introduces extraneous solutions. The solutions obtained for (5-5) and (5-6) must be checked by substitution back into (5-4).

This is all we shall say for the present about homogeneous anisotropic media. However, some comments will appear from time to time throughout the text. A number of selected references relating to propagation in anistropic media are given in Appendix A. References appearing there have not appeared elsewhere in the text.

5.3. Isotropic, Inhomogeneous Media.

As noted above, the properties of the earth vary to first order in a vertical or a radial direction (vertical or radial inhomogeneity). They also vary to lesser degree with changes in latitude and longitude (lateral inhomogeneity). Up to the present, asymptotic ray theory — to be developed in Chapter 21 — has been used quite successfully in treating the propagation of elastic body waves through the real earth. The work of Richards (1974) suggests why this is so. He derives exact equations in which "difficult" terms can be neglected, given reasonable models of elastic properties and density within a radially inhomogeneous earth. All that remains are three Helmholtz equations with radially varying wave-numbers. A large body of literature exists for the solution of such equations and we shall consider some of the less complicated solutions in Chapter 21.

The smaller variations of lateral inhomogeneity can be separated into two groups. On the one hand there is a regional variation between continents and oceans and between stable and active regions within each. We also find smaller local variations especially in active tectonic regions. The former can often be treated by a method of averaging and results have been obtained which are characteristic of each type of region. The latter inhomogeneities tend to act as scattering centers and the techniques of scattering theory must be brought to bear with the local inhomogeneities being treated as perturbations of an otherwise homogeneous media. We shall see in Chapter 24 that such perturbations cause the frequencies of oscillation of a vibrating earth to lose the degeneracy which they would have in the case of a radially inhomogeneous earth.

If we now consider the equations governing wave propagation in this type of media, we have that λ and μ will be functions of the spatial coordinates. We then have to write

$$\begin{aligned} t_{ij,i} &= (\lambda e_{kk}\delta_{ij})_{,i} + (2\mu e_{ij})_{,i} \qquad (5\text{-}7) \\ &= (\lambda u_{k,k}\delta_{ij})_{,i} + \{\mu(u_{i,j} + u_{j,i})\}_{,i} \\ &= (\lambda u_{k,k}\delta_{ij})_{,i} + \{\mu(u_{i,j} - u_{j,i})\}_{,i} + 2(\mu u_{j,i})_{,i} \\ &= \nabla(\lambda\nabla\cdot\mathbf{u}) + \nabla\times(\mu\nabla\times\mathbf{u}) + 2(\nabla\cdot\mu\nabla)\mathbf{u} \ . \end{aligned}$$

But

$$\begin{aligned} (\nabla\cdot\mu\nabla)\mathbf{u} = \nabla(\mu\nabla\cdot\mathbf{u}) - \nabla\times(\mu\nabla\times\mathbf{u}) + (\nabla\mu\cdot\nabla)\mathbf{u} \\ -(\nabla\mu)(\nabla\cdot\mathbf{u}) + (\nabla\mu)\times(\nabla\times\mathbf{u}). \end{aligned}$$

Combining terms, we have on rewriting (5-2) in vector form:

$$\begin{aligned} \mathbf{f} + \nabla\{(\lambda+2\mu)\nabla\cdot\mathbf{u}\} - \nabla\times(\mu\nabla\times\mathbf{u}) \qquad (5\text{-}8) \\ + 2\{(\nabla\mu\cdot\nabla)\mathbf{u} - (\nabla\mu)\nabla\cdot\mathbf{u} + (\nabla\mu)\times(\nabla\times\mathbf{u})\} - \rho\ddot{\mathbf{u}} = 0. \end{aligned}$$

This is the equation of motion for isotropic inhomogeneous media.

References to some solutions of this equation in particular cases will be found in Appendix B. This is in addition to the approximate methods of Chapters 21 and 22. Additional references relating to scattering from local inhomogeneities will be listed in Appendix C. The body of literature relating to the theory of propagation in inhomogeneous media is much smaller than that relating to the theory of propagation in anisotropic media. The reason is obvious; equation (5-8) is much more difficult to solve than (5-4).

5.4. Isotropic, Homogeneous Media.

If λ and μ in (5-8) are constants, these equations reduce to

$$\mathbf{f} + (\lambda + 2\mu)\nabla(\nabla\cdot\mathbf{u}) - \mu\nabla\times(\nabla\times\mathbf{u}) - \rho\ddot{\mathbf{u}} = 0. \qquad (5\text{-}9)$$

This is the elastic equation of motion for an isotropic homogeneous medium. Most of the remainder of this book will be concerned with the solutions of this equation under a variety of boundary and initial conditions.

5.5. Other Types of Media.

Propagation in anisotropic media is a linear process and the equations governing propagation in inhomogeneous media can be made linear if both the material deformations and the variations in elastic parameters are small. But we also have other types of media in which the propagation of elastic waves is either linear or may be made linear. For example, propagation in porous media has significance in the case of sediments lying at the bottom of the oceanic basins. Exploration geophysicists are vitally interested in earth strata which have their pore space filled with fluid hydrocarbons. The possibility of partial melting in the upper mantle is also a problem relating to porous media. A fourth case arises with the dilatancy-diffusion model wherein fluids filling up cracks in stressed rock are intimately linked to the mechanics of faulting. If the fluid filling the pore space is viscous, the attenuation of propagating waves can be affected.

Some other linear processes include thermoelasticity and viscoelasticity. In most cases the deformations are sufficiently small (away from the source regions) that the former is not important. The latter is more applicable to solids such as pitch, tar, wax, and some plastics rather than earth materials.

Yet another group of linear processes include elastic interactions with the electromagnetic field. These processes are anisotropic and have already been alluded to in Section 5.2. Here we have the electro-acoustic and magneto-acoustic effects on wave propagation as well as design considerations in the fabrication and use of electrostrictive (piezo-electric) and magnetostrictive transducers. A small number of papers have been written concerning the interaction of seismic waves and the earth's magnetic field. Visible light can also be refracted as in the use of plexiglas sheets to show dynamic stress patterns (photo-elasticity) and visible light can be reflected at varying angles from surfaces of elastic materials excited by traveling or standing waves.

Non-linear processes are practically limited in scope since most take place near the source region where the actual mechanics are poorly understood at present. Many theoretical solutions exist to special cases. One could additionally introduce any or all of anisotropy, inhomogeneity, viscosity or some other loss-mechanism, as well as electromagnetic interactions. This would result in rather formidable problems.

The literature on the above types of wave propagation is scattered and most lies outside journals of interest to geophysicists. No bibliography has been included with this text, although a start may be made in Eringen and Suhubi (1974), Achenbach (1973), Auld (1973), White (1965), Ewing, Jardetzky, and Press (1957) and the continuing series edited by Mason (1976).

GENERAL SOLUTIONS OF THE ISOTROPIC, HOMOGENEOUS MEDIUM EQUATIONS OF MOTION

6.1. Reduction by Wave Potentials.

The equation of motion

$$(\lambda + 2\mu)\nabla(\nabla\cdot\mathbf{u}) - \mu\nabla\times(\nabla\times\mathbf{u}) - \rho\ddot{\mathbf{u}} = -\mathbf{f} \tag{5-9}$$

is more complicated than any other partial differential equation frequently met in classical physics, save those of magneto-fluid dynamics which are even more formidable. In the homogeneous case (where $\mathbf{f} = 0$), we can use the Helmholtz vector decomposition theorem (Morse and Feshbach, 1953, pp. 52-54) to suggest a solution of the form:

$$\mathbf{u}(x,y,z,t) = \nabla\Phi(x,y,z,t) + \nabla\times\boldsymbol{\Psi}(x,y,z,t); \tag{6-1}$$

i.e., we represent the vector displacement $\mathbf{u}$ as the sum of the gradient of a scalar potential Φ and the curl of a vector potential $\boldsymbol{\Psi}$. The quantities Φ and $\boldsymbol{\Psi}$ are called the Lamé potentials. If we substitute this representation of $\mathbf{u}$ into (5-9), we are left with something more easily handled. Now (prove it, if you don't believe it):

$$\nabla\times(\nabla\Phi) \equiv 0 \ ; \ \nabla\cdot(\nabla\times\boldsymbol{\Psi}) \equiv 0. \tag{6-2}$$

On substitution of (6-1) and using (6-2), equation (5-9) becomes (for $\mathbf{f} = 0$)

$$(\lambda+2\mu)\nabla\{\nabla^2\Phi - \frac{\rho}{\lambda+2\mu}\frac{\partial^2\Phi}{\partial t^2}\} - \mu\nabla\times\{\nabla\times\nabla\times\boldsymbol{\Psi} + \frac{\rho}{\mu}\frac{\partial^2\boldsymbol{\Psi}}{\partial t^2}\} = 0. \tag{6-3}$$

Now it is sufficient (but not necessary) for a solution of (6-3) that the following two equations hold:

$$\nabla^2\Phi - \frac{\rho}{\lambda+2\mu}\frac{\partial^2\Phi}{\partial t^2} = 0, \qquad \text{SCALAR WAVE EQUATION} \tag{6-4}$$

$$\nabla\times\nabla\times\boldsymbol{\Psi} + \frac{\rho}{\mu}\frac{\partial^2\boldsymbol{\Psi}}{\partial t^2} = 0. \qquad \text{VECTOR WAVE EQUATION I} \tag{6-5}$$

Inasmuch as the addition of $-\nabla(\nabla\cdot\boldsymbol{\Psi})$ inside the second brace pair of (6-3) does not change its value, we can have an alternative form of the vector wave equation:

$$-\nabla\times\nabla\times\boldsymbol{\Psi} + \nabla(\nabla\cdot\boldsymbol{\Psi}) - \frac{\rho}{\mu}\frac{\partial^2\boldsymbol{\Psi}}{\partial t^2} = 0.$$

That is,

$$\nabla^2\boldsymbol{\Psi} - \frac{\rho}{\mu}\frac{\partial^2\boldsymbol{\Psi}}{\partial t^2} = 0. \qquad \text{VECTOR WAVE EQUATION II} \tag{6-6}$$

This second form of the equation is often useful when we seek one-component solutions for the $\boldsymbol{\Psi}$ potential.

A more sophisticated treatment of these matters may be found in Eringen and Suhubi (1975, Sec. 5.4). They consider the case where $\mathbf{f}$ is not zero and in which $\mathbf{u}$ can be derived from the Somigliana potential. This latter solution has not proved especially useful because it is of a more complicated structure than the Lamé solutions given above. They also discuss questions of completeness and the relation of both solutions to the familiar solutions of static elasticity.

6.2. Solutions of the Scalar Wave Equation — P-Waves.

Consider the one-dimensional case of (6-4), i.e.,

$$\frac{\partial^2 \Phi}{\partial x^2} - \frac{\rho}{\lambda+2\mu}\frac{\partial^2 \Phi}{\partial t^2} = 0.$$

One solution is $\Phi = \Phi(x \pm v_p t)$, where $v_p = \{(\lambda+2\mu)/\rho\}^{\frac{1}{2}}$. That this is a solution can be verified upon direct substitution into (6-4). Now with such a solution Φ, we find that

$$u_x = \frac{\partial \Phi}{\partial x}(x,t);\ u_y \equiv 0;\ u_z \equiv 0.$$

That is, there is only a component of motion in the x-direction, the direction in which the scalar field Φ varies. The particle motion is perpendicular to the wave front. Hence another name given to P-waves (so-called because they occur as the first arrival on a seismogram, i.e., the primary waves) is longitudinal waves because of their polarization. In general, the particle motion of P-waves is perpendicular to the wave fronts, or nearly so; but not always, as we shall see later (Secs. 7.2, 7.3). This longitudinal motion is illustrated in Figure 6-1.

We note also that, since $\nabla\times\nabla\Phi \equiv 0$,

$$\nabla\times\mathbf{u}^P \equiv 0 \qquad (\text{where } \mathbf{u}^P = \nabla\Phi); \qquad (6\text{-}7)$$

i.e., the curl of the displacement associated with the scalar potential vanishes identically. We have seen that the rotation $\mathbf{\Omega} = \frac{1}{2}\nabla\times\mathbf{u}$, hence the waves associated with v_p are irrotational. While P-waves are not always longitudinally polarized, they are always irrotational by virtue of (6-7).

Investigating the divergence of P-motion, we find that

$$\nabla\cdot\mathbf{u}^P \equiv \Theta^P \neq 0 \qquad (\text{in general}),$$

and these waves are also called dilatational or compressional waves. If one takes the divergence of the equation of motion (5-9) in the case $\mathbf{f} = 0$, noting that $\nabla\cdot\{\nabla\times(\nabla\times\mathbf{u})\} \equiv 0$, it can be seen from the resulting equation that the dilatational part of the motion always travels with the velocity v_p.

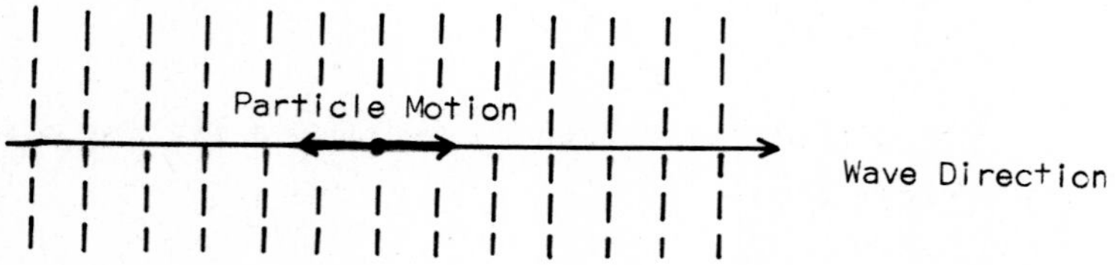

Figure 6-1. Longitudinal Polarization of P-waves.

6.3. Solutions of the Vector Wave Equation — S-Waves.

We now try as a one-dimensional solution of (6-5), the following vector potential:

$$\boldsymbol{\Psi} = \mathbf{e}_x\Psi_x(x-v_st) + \mathbf{e}_y\Psi_y(x-v_st) + \mathbf{e}_z\Psi_z(x-v_st),$$

where $v_s = (\mu/\rho)^{\frac{1}{2}}$. It is to be noted that v_s is always less than v_p, hence the name secondary (S) waves for these waves which always arrive after the primary (P) waves. Now the motion associated with the vector wave potential is given by

$$\mathbf{u}^s = \nabla\times\boldsymbol{\Psi} = -\mathbf{e}_y \frac{\partial\Psi_z}{\partial x} + \mathbf{e}_z \frac{\partial\Psi_y}{\partial x}.$$

We see that given a set of three scalar fields which vary in the x-direction, the vector wave potential $\boldsymbol{\Psi}$ generates components of motion only in the y- and z-directions. These are transverse to the variation of the fields. Thus S-waves are sometimes called transverse waves.

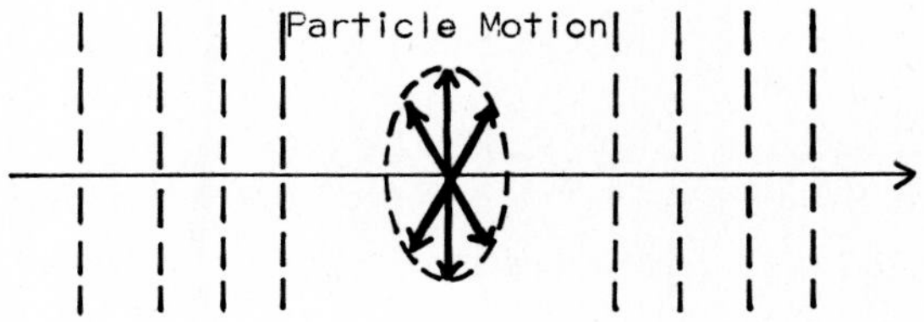

Figure 6-2. Transverse Polarization of S-waves.

In general, the particle motion of S-waves is parallel or nearly parallel to the wave front, but not always (for some exceptions, see Sec. 7.2, 7.3). This transverse particle motion is illustrated in Figure 6-2.

Since the scalar component Ψ_x doesn't contribute any motion, we will drop it and continue the computation. Then

$$\nabla\times\nabla\times\boldsymbol{\Psi} = -\mathbf{e}_y \frac{\partial^2\Psi_y}{\partial x^2} - \mathbf{e}_z \frac{\partial^2\Psi_z}{\partial x^2},$$

and the vector wave equation becomes

$$\mathbf{e}_y\left(-\frac{\partial^2\Psi_y}{\partial x^2} + \frac{\rho}{\mu}\frac{\partial^2\Psi_y}{\partial t^2}\right) + \mathbf{e}_z\left(-\frac{\partial^2\Psi_z}{\partial x^2} + \frac{\rho}{\mu}\frac{\partial^2\Psi_z}{\partial t^2}\right) = 0.$$

Thus a three component (one of which turns out to be zero) plane-wave solution satisfies (6-5), Vector Wave Equation I. In such a form, each component will be a solution of the scalar wave equation

$$\frac{\partial^2\Psi_i}{\partial x^2} - \frac{\rho}{\mu}\frac{\partial^2\Psi_i}{\partial t^2} = 0.$$

We note again that since $\nabla\cdot(\nabla\times\mathbf{\Psi}) \equiv 0$, then

$$\nabla\cdot\mathbf{u}^S \equiv \Theta^S = 0. \tag{6-8}$$

Since the divergence of the vector $\mathbf{u}^S$ vanishes, the waves associated with the velocity v_S are dilatationless, or, more commonly, equivoluminal. This property of S-waves is always true, even though they are not always polarized transverse to the wave direction. Although the quantity $\mathbf{\Omega}^S(x,y,z,t) = \frac{1}{2}\nabla\times\mathbf{u}^S \neq 0$, in general, these waves are not called rotational waves. We can show, however, that the rotational part of their motion always travels with the velocity v_S. This can be done by taking the curl of the equation of motion (5-9) (with $\mathbf{f} = 0$), using the fact that $\nabla\times\nabla(\nabla\cdot\mathbf{u}) \equiv 0$.

One also finds that the shearing strains, i.e.,

$$e^S_{xy} = e^S_{yx} = \tfrac{1}{2}\left(\frac{\partial u^S_x}{\partial y} + \frac{\partial u^S_y}{\partial x}\right); \quad e^S_{yz}; \quad e^S_{xz};$$

etc., also are not, in general, zero. Hence these waves have traditionally been called shear waves, or distortional waves.

6.4. Independent Shear Wave Components — SH and SV Waves.

We have seen that the fundamental reason for the introduction of wave potentials, i.e., equation (6-1), is that there are two basic types of propagation, and such a decomposition allows us to separate these two types of motion. We have seen above how one component of the shear potential did not contribute anything. This left us with two independent (but degenerate) components. (Note: This degeneracy disappears in an anisotropic medium.) Inasmuch as there are three components of motion, one would expect only three independent components of potential. One of these three components is the irrotational potential Φ. We now need a method for obtaining the two additional independent shear components.

Since the separation of the elastic displacement vector into an irrotational part and a equivoluminal part proved so useful in the preceding analysis, let us try and extend this type of separation further. The irrotational waves need no further analysis. However, since the shear waves are rotational, we can help ourselves by separating this rotation into two parts. The first rotational part will be such that all motion is locally parallel to the bounding surface, i.e., the instantaneous axis of rotation is normal to the surface. When the medium is a half-space, such waves are called SH-waves since all the motion is in the plane of the surface (horizontal in the case of the earth). The second rotational part will be such that the instantaneous axis of rotation is parallel to the bounding surface. Again, in the half-space problem, such waves are called SV-waves since all their motion is in a plane normal to the bounding surface (a vertical plane in the case of the earth) containing both vertical and horizontal components of motion.

Eringen and Suhubi (1975, Sec. 8.1) show us that we can write a general solution of the vector wave equation (6-6) as

$$\mathbf{L} = \nabla\Phi, \tag{6-9}$$
$$\mathbf{M} = \nabla\times\mathbf{e}_1 w(x_1)\Lambda, \tag{6-10}$$
$$\mathbf{N} = -\ell\nabla\times\nabla\times\mathbf{e}_1 w(x_1)\Psi. \tag{6-11}$$

Here $\mathbf{e}_1$ is a unit vector perpendicular to the bounding surface, $w(x_1)$ is a function supplied to make Λ and Ψ solutions of the scalar wave equation, and ℓ is a dimensional factor to give the same dimensions to Φ, Λ, and Ψ. The first solution, (6-9), is an extraneous solution for (6-6) which arises from the addition of the term $-\nabla\nabla\cdot\Psi$ to (6-5). As we have already mentioned, all dilatational motion such as (6-9) travels with the velocity v_p. The second solution (6-10) represents SH-waves and the third solution (6-11) represents SV-waves.

Such a beautiful separation as (6-9) through (6-11) comes with some strings. Eringen and Suhubi (ibid) go on to show that this separation can be used in only six coordinate systems. The three most common are cartesian, where $\mathbf{e}_1$ can be any of $\mathbf{e}_x$, $\mathbf{e}_y$, or $\mathbf{e}_z$ and $w = 1$; circular cylindrical, where $\mathbf{e}_1$ is $\mathbf{e}_z$ and $w = 1$; and spherical where $\mathbf{e}_1$ is $\mathbf{e}_r$ and $w = r$. The other three are elliptic cylindrical, parabolic cylindrical and conical. They go on to note that if azimuthal symmetry is present, this separation also works in the parabolic, prolate spheroidal, and oblate spheroidal systems.

If the bounding surfaces do not coincide with one of these geometries, the above separation is not useful. All is not lost, but one must resort to rather brute force methods to obtain solutions. However, the three common coordinate systems listed above are sufficient to solve problems of half-space, plates, rods, and spheres, which are good models for a vast number of problems associated with the real earth.

It will be left to the student to show that the displacement components given in the following paragraphs are indeed the appropriate solutions of the vector wave equation (6-6). For reference in later chapters, both displacement and stress components associated with Φ,Λ, and Ψ will be written out at this point, even though the discussion of SH- and SV-waves does not require it.

<u>Cartesian Coordinates</u>. For convenience we shall take $\mathbf{e}_1 = \mathbf{e}_z$. (The other two coordinate directions for $\mathbf{e}_1$ offer nothing new; only a re-labeling.) We then have

$$\mathbf{u} = \nabla\Phi - \ell\nabla\times\nabla\times\mathbf{e}_z\Psi + \nabla\times\mathbf{e}_z\Lambda, \tag{6-12}$$

with

$$u_x = u_x^P + u_x^{SV} + u_x^{SH}$$
$$= \frac{\partial\Phi}{\partial x} - \ell\frac{\partial^2\Psi}{\partial x\partial z} + \frac{\partial\Lambda}{\partial y}, \tag{6-13}$$

$$u_y = u_y^P + u_y^{SV} + u_y^{SH}$$

$$= \frac{\partial\Phi}{\partial y} - \ell\frac{\partial^2\Psi}{\partial y\partial z} - \frac{\partial\Lambda}{\partial x}, \tag{6-14}$$

$$u_z = u_z^P + u_z^{SV}$$

$$= \frac{\partial\Phi}{\partial z} + \ell\left(\frac{\partial^2\Psi}{\partial x^2} + \frac{\partial^2\Psi}{\partial y^2}\right) \tag{6-15}$$

$$= \frac{\partial\Phi}{\partial z} + \ell\left(\nabla^2\Psi - \frac{\partial^2\Psi}{\partial z^2}\right).$$

The functions Φ, Ψ, and Λ satisfy the following wave equations:

$$\nabla^2\Phi - \frac{1}{v_P^2}\frac{\partial^2\Phi}{\partial t^2} = 0, \tag{6-16}$$

$$\nabla^2\Psi - \frac{1}{v_S^2}\frac{\partial^2\Psi}{\partial t^2} = 0, \tag{6-17}$$

$$\nabla^2\Lambda - \frac{1}{v_S^2}\frac{\partial^2\Lambda}{\partial t^2} = 0. \tag{6-18}$$

Using (4-14) and (2-12), we find the following stress components:

$$t_{xx} = \lambda\nabla^2\Phi + 2\mu\left(\frac{\partial^2\Phi}{\partial x^2} - \ell\frac{\partial^3\Psi}{\partial x^2\partial z} + \frac{\partial^2\Lambda}{\partial x\partial y}\right), \tag{6-19}$$

$$t_{yy} = \lambda\nabla^2\Phi + 2\mu\left(\frac{\partial^2\Phi}{\partial y^2} - \ell\frac{\partial^3\Psi}{\partial y^2\partial z} - \frac{\partial^2\Lambda}{\partial x\partial y}\right), \tag{6-20}$$

$$t_{zz} = \lambda\nabla^2\Phi + 2\mu\left[\frac{\partial^2\Phi}{\partial z^2} + \ell\frac{\partial}{\partial z}\left(\nabla^2\Psi - \frac{\partial^2\Psi}{\partial z^2}\right)\right], \tag{6-21}$$

$$t_{xy} = t_{yx} = \mu\left[2\frac{\partial^2\Phi}{\partial x\partial y} - 2\ell\frac{\partial^3\Psi}{\partial x\partial y\partial z} + \left(\frac{\partial^2\Lambda}{\partial y^2} - \frac{\partial^2\Lambda}{\partial x^2}\right)\right], \tag{6-22}$$

$$t_{xz} = t_{zx} = \mu\left[2\frac{\partial^2\Phi}{\partial x\partial z} + \ell\frac{\partial}{\partial x}\left(\nabla^2\Psi - 2\frac{\partial^2\Psi}{\partial z^2}\right) + \frac{\partial^2\Lambda}{\partial y\partial z}\right], \tag{6-23}$$

$$t_{yz} = t_{zy} = \mu\left[2\frac{\partial^2\Phi}{\partial y\partial z} + \ell\frac{\partial}{\partial y}\left(\nabla^2\Psi - 2\frac{\partial^2\Psi}{\partial z^2}\right) - \frac{\partial^2\Lambda}{\partial x\partial z}\right]. \tag{6-24}$$

Now it will be shown in Chapter 8 that the P- and SV-components couple together through the boundary conditions at a free surface. However, we can show here that it is possible to cause the stresses at the surface z = constant to vanish in the case of SH-motion without the introduction of any other potentials. Looking at those parts of the stresses which derive from the Λ-potential, we have

$$t_{zx}^{SH} = \mu\frac{\partial^2\Lambda}{\partial y\partial z}; \qquad t_{zy}^{SH} = -\mu\frac{\partial^2\Lambda}{\partial x\partial z}; \qquad t_{zz}^{SH} \equiv 0.$$

Hence we see that the normal component of stress vanishes identically and that the tangential components can be made to vanish by simply requiring the normal derivative of Λ to vanish at the surface z = constant (Neumann conditions).

The set of potentials (6-12) allow us to solve plane-wave problems in layered media where the z-axis is perpendicular to the layering.

In the case of two-dimensional propagation (let y be the ignorable coordinate), some simplification arises. We can set $\Psi^{\dagger} = \ell\partial\Psi/\partial x$ in **N** and obtain the following simplified forms:

$$u_x = \frac{\partial\Phi}{\partial x} - \frac{\partial\Psi^{\dagger}}{\partial z} , \tag{6-25}$$

$$u_y = -\frac{\partial\Lambda}{\partial x} , \tag{6-26}$$

$$u_z = \frac{\partial\Phi}{\partial z} + \frac{\partial\Psi^{\dagger}}{\partial x} , \tag{6-27}$$

that is

$$\mathbf{u} = \nabla\Phi + \nabla\times\mathbf{e}_y\Psi^{\dagger} + \nabla\times\mathbf{e}_z\Lambda . \tag{6-28}$$

Only x- and z-partial derivatives are to be taken in (6-28). The associated two dimensional stresses are given by:

$$t_{xx} = \lambda\nabla^2\Phi + 2\mu\left(\frac{\partial^2\Phi}{\partial x^2} - \frac{\partial^2\Psi^{\dagger}}{\partial x\partial z}\right) , \tag{6-29}$$

$$t_{yy} = \lambda\nabla^2\Phi , \tag{6-30}$$

$$t_{zz} = \lambda\nabla^2\Phi + 2\mu\left(\frac{\partial^2\Phi}{\partial z^2} + \frac{\partial^2\Psi^{\dagger}}{\partial x\partial z}\right) , \tag{6-31}$$

$$t_{xy} = t_{yx} = -\mu\frac{\partial^2\Lambda}{\partial x^2} , \tag{6-32}$$

$$t_{xz} = t_{zx} = \mu\left[2\frac{\partial^2\Phi}{\partial x\partial z} + \left(\frac{\partial^2\Psi^{\dagger}}{\partial x^2} - \frac{\partial^2\Psi^{\dagger}}{\partial z^2}\right)\right] , \tag{6-33}$$

$$t_{yz} = t_{zy} = -\mu\frac{\partial^2\Lambda}{\partial x\partial z} . \tag{6-34}$$

The set (6-25) to (6-34) provide solutions for plane-wave and line-source problems in layered media with the line-source parallel to the layering.

Cylindrical Coordinates. In cylindrical coordinates we have only one choice for our base vector. We must take $\mathbf{e}_1$ as $\mathbf{e}_z$. We then have

$$\mathbf{u} = \nabla\Phi - \ell\nabla\times\nabla\times\mathbf{e}_z\Psi + \nabla\times\mathbf{e}_z\Lambda , \tag{6-35}$$

with

$$u_r = u_r^P + u_r^{SV} + u_r^{SH}$$
$$= \frac{\partial\Phi}{\partial r} - \ell\frac{\partial^2\Psi}{\partial r\partial z} + \frac{1}{r}\frac{\partial\Lambda}{\partial\theta} , \tag{6-36}$$

$$u_\theta = u_\theta^P + u_\theta^{SV} + u_\theta^{SH}$$

$$= \frac{1}{r}\frac{\partial\Phi}{\partial\theta} - \frac{\ell}{r}\frac{\partial^2\Psi}{\partial\theta\partial z} - \frac{\partial\Lambda}{\partial r}, \tag{6-37}$$

$$u_z = u_z^P + u_z^{SV}$$

$$= \frac{\partial\Phi}{\partial z} + \frac{\ell}{r}\frac{\partial}{\partial r}\left(r\frac{\partial\Psi}{\partial r}\right) + \frac{\ell}{r^2}\frac{\partial^2\Psi}{\partial\theta^2}$$

$$= \frac{\partial\Phi}{\partial z} + \ell\left(\nabla^2\Psi - \frac{\partial^2\Psi}{\partial z^2}\right). \tag{6-38}$$

Again Φ, Λ, and Ψ satisfy the wave equations (6-16) - (6-18). Using (4-14), (2-53), and (2-12), we find the following stress components:

$$t_{rr} = \lambda\nabla^2\Phi + 2\mu\left[\frac{\partial^2\Phi}{\partial r^2} - \ell\frac{\partial^3\Psi}{\partial r^2\partial z} + \frac{\partial}{\partial r}\left(\frac{1}{r}\frac{\partial\Lambda}{\partial\theta}\right)\right], \tag{6-39}$$

$$t_{\theta\theta} = \lambda\nabla^2\Phi + 2\mu\left[\begin{array}{l}\frac{1}{r}\left(\frac{\partial\Phi}{\partial r} + \frac{1}{r}\frac{\partial^2\Phi}{\partial\theta^2}\right)\\ -\frac{\ell}{r}\frac{\partial}{\partial z}\left(\frac{\partial\Psi}{\partial r} + \frac{1}{r}\frac{\partial^2\Psi}{\partial\theta^2}\right)\\ +\frac{1}{r}\left(\frac{1}{r}\frac{\partial\Lambda}{\partial\theta} - \frac{\partial^2\Lambda}{\partial r\partial\theta}\right)\end{array}\right], \tag{6-40}$$

$$t_{zz} = \lambda\nabla^2\Phi + 2\mu\left[\frac{\partial^2\Phi}{\partial z^2} + \ell\frac{\partial}{\partial z}\left(\nabla^2\Psi - \frac{\partial^2\Psi}{\partial z^2}\right)\right], \tag{6-41}$$

$$t_{r\theta} = t_{\theta r} = \mu\left[\begin{array}{l}2\left(\frac{1}{r}\frac{\partial^2\Phi}{\partial r\partial\theta} - \frac{1}{r^2}\frac{\partial\Phi}{\partial\theta}\right)\\ -\frac{2\ell}{r}\frac{\partial}{\partial z}\left(\frac{\partial^2\Psi}{\partial r\partial\theta} - \frac{1}{r}\frac{\partial\Psi}{\partial\theta}\right)\\ +\frac{1}{r^2}\frac{\partial^2\Lambda}{\partial\theta^2} - r\frac{\partial}{\partial r}\left(\frac{1}{r}\frac{\partial\Lambda}{\partial r}\right)\end{array}\right], \tag{6-42}$$

$$t_{rz} = t_{zr} = \mu\left[2\frac{\partial^2\Phi}{\partial r\partial z} + \ell\frac{\partial}{\partial r}\left(\nabla^2\Psi - 2\frac{\partial^2\Psi}{\partial z^2}\right) + \frac{1}{r}\frac{\partial^2\Lambda}{\partial\theta\partial z}\right], \tag{6-43}$$

$$t_{\theta z} = t_{z\theta} = \mu\left[\frac{2}{r}\frac{\partial^2\Phi}{\partial\theta\partial z} + \frac{\ell}{r}\frac{\partial}{\partial\theta}\left(\nabla^2\Psi - 2\frac{\partial^2\Psi}{\partial z^2}\right) - \frac{\partial^2\Lambda}{\partial r\partial z}\right]. \tag{6-44}$$

Looking at the SH-associated stresses on surfaces of z = constant, we have

$$t_{zr}^{SH} = \frac{\mu}{r}\frac{\partial^2\Lambda}{\partial\theta\partial z}; \qquad t_{z\theta}^{SH} = -\mu\frac{\partial^2\Lambda}{\partial r\partial z}; \qquad t_{zz}^{SH} \equiv 0 .$$

Hence in this system, the tangential components of stress can be made to vanish by again requiring the normal derivative of Λ to vanish at the surface, and SH-motion will not couple into the P-SV motion.

The set of potentials (6-35) will allow us to work with point and line-source problems in layered media with the line-source perpendicular to the plane of layering. We will discuss the case of propagation in cylindrical structures after we look at the special case where azimuthal symmetry is present.

If we have such symmetry, there is some simplification. We can set $\Psi^{\dagger} = \ell\partial\Psi/\partial r$ in **N** and work out the following expressions:

$$u_r = \frac{\partial\Phi}{\partial r} - \frac{\partial\Psi^{\dagger}}{\partial z} \quad , \tag{6-45}$$

$$u_\theta = -\frac{\partial\Lambda}{\partial r} \, , \tag{6-46}$$

$$u_z = \frac{\partial\Phi}{\partial z} + \frac{1}{r}\frac{\partial}{\partial r}(r\Psi^{\dagger}), \tag{6-47}$$

that is,

$$\mathbf{u} = \nabla\Phi + \nabla\times\mathbf{e}_\theta\Psi^{\dagger} + \nabla\times\mathbf{e}_z\Lambda \quad . \tag{6-48}$$

Here only $\underline{r}$- and $\underline{z}$-partial derivatives are to be taken. The associated two dimensional stresses are:

$$t_{rr} = \lambda\nabla^2\Phi + 2\mu\left(\frac{\partial^2\Phi}{\partial r^2} - \frac{\partial^2\Psi^{\dagger}}{\partial r\partial z}\right) \quad , \tag{6-49}$$

$$t_{\theta\theta} = \lambda\nabla^2\Phi + 2\mu\left(\frac{1}{r}\frac{\partial\Phi}{\partial r} - \frac{1}{r}\frac{\partial\Psi^{\dagger}}{\partial z}\right) \quad , \tag{6-50}$$

$$t_{zz} = \lambda\nabla^2\Phi + 2\mu\left[\frac{\partial^2\Phi}{\partial z^2} + \frac{1}{r}\frac{\partial}{\partial r}\left(r\frac{\partial\Psi^{\dagger}}{\partial z}\right)\right], \tag{6-51}$$

$$t_{r\theta} = t_{\theta r} = -\mu r\frac{\partial}{\partial r}\left(\frac{1}{r}\frac{\partial\Lambda}{\partial r}\right) \quad , \tag{6-52}$$

$$t_{rz} = t_{zr} = \mu\left[2\frac{\partial^2\Phi}{\partial r\partial z} + \frac{\partial}{\partial r}\left\{\frac{1}{r}\frac{\partial}{\partial r}(r\Psi^{\dagger})\right\} - \frac{\partial^2\Psi^{\dagger}}{\partial z^2}\right] \quad , \tag{6-53}$$

$$t_{\theta z} = t_{z\theta} = -\mu\frac{\partial^2\Lambda}{\partial r\partial z} \quad . \tag{6-54}$$

This system of displacements and stresses allows the solution of axially directed point and line-source problems in layered media with the line-source perpendicular to the plane of layering.

When the bounding surfaces are concentric cylinders, things become much more complicated. From (6-36) you can see that neither Λ nor Ψ gives rise to motion that is purely parallel to cylindrical surfaces. Only in the particular case of azimuthal symmetry where (6-48) is valid can we make such a separation. In this case, too, SH is derived from Λ and SV is derived from Ψ. There is P-SV coupling at the cylindrical surfaces. The SH motion is identified with torsional waves in a rod, while the P-SV motion is identified with axially symmetric longitudinal waves propagating in the rod (Young's modulus waves in the long-wavelength limit).

If azimuthal symmetry is not present, then no P-SV-SH separation is possible, as can be seen from (6-36) - (6-38). It can also be shown that the potentials Φ, Λ, and Ψ are triply coupled at the cylindrical boundaries. One simple case of this type motion is that of flexural vibrations of a rod or tube. Another simple, but not common, case is that of thickness-shear vibrations where all motion is parallel to the z-axis. More complicated motions have no simple identification.

An extensive treatment of axial wave propagation in cylinders will be found in Eringen and Suhubi (1975, Sec. 8.10 - 8.11). The principal application to seismology in the earth for this type geometry is the modeling of elastic waves propagating down cylindrical structures such as void or fluid-filled boreholes typically found in oilfields. White (1965, Chap. 4) gives details.

Spherical Coordinates. In Spherical coordinates we must take as our base vector, $\mathbf{e}_1 = \mathbf{e}_r$. We must also use $w(r) = r$. The displacement vector can be written

$$\mathbf{u} = \nabla\Phi - \ell\nabla\times\nabla\times\mathbf{e}_r(r\Psi) + \nabla\times\mathbf{e}_r(r\Lambda) \quad . \tag{6-55}$$

For the radial component of motion we have

$$\begin{aligned} u_r &= u_r^P + u_r^{SV} \\ &= \frac{\partial\Phi}{\partial r} + \frac{\ell}{r^2 \mathrm{Sin}\theta}\left[\frac{\partial}{\partial\theta}\{\mathrm{Sin}\theta\frac{\partial}{\partial\theta}(r\Psi)\} + \frac{1}{\mathrm{Sin}\theta}\frac{\partial^2}{\partial\phi^2}(r\Psi)\right] \\ &= \frac{\partial\Phi}{\partial r} + \ell\left[r\nabla^2\Psi - \frac{\partial^2}{\partial r^2}(r\Psi)\right] \\ &\left(= \frac{\partial\Phi}{\partial r} - \frac{\ell}{r}n(n+1)\Psi_n^m\right) \quad . \end{aligned} \tag{6-56}$$

The expression in parenthesis is valid if $\Psi(r, \theta,\phi)$ is given in terms of spherical harmonics, $\Psi_n^m(r)Y_n^m(\theta,\phi)$. The remaining two components of motion are:

$$\begin{aligned} u_\theta &= u_\theta^P + u_\theta^{SV} + u_\theta^{SH} \\ &= \frac{1}{r}\frac{\partial\Phi}{\partial\theta} - \frac{\ell}{r}\frac{\partial}{\partial r}(r\frac{\partial\Psi}{\partial\theta}) + \frac{1}{\mathrm{Sin}\theta}\frac{\partial\Lambda}{\partial\phi} \ , \end{aligned} \tag{6-57}$$

$$\begin{aligned} u_\phi &= u_\phi^P + u_\phi^{SV} + u_\phi^{SH} \\ &= \frac{1}{r\mathrm{Sin}\theta}\frac{\partial\Phi}{\partial\phi} - \frac{\ell}{r\mathrm{Sin}\theta}\frac{\partial}{\partial r}(r\frac{\partial\Psi}{\partial\phi}) - \frac{\partial\Lambda}{\partial\theta} \quad . \end{aligned} \tag{6-58}$$

The functions Φ,Λ, and Ψ must satisfy the wave equations given in (6-16) - (6-18). Using (4-14), (2-54) and (2-12), we generate the associated stress components:

$$t_{rr} = \lambda\nabla^2\Phi + 2\mu\left[\frac{\partial^2\Phi}{\partial r^2} + \ell\frac{\partial}{\partial r}\{ r\nabla^2\Psi - \frac{\partial^2}{\partial r^2}(r\Psi) \}\right] \ , \tag{6-59}$$

$$t_{\theta\theta} = \lambda\nabla^2\Phi + 2\mu\left[\begin{array}{l} \frac{1}{r}(\frac{\partial\Phi}{\partial r} + \frac{1}{r}\frac{\partial^2\Phi}{\partial\theta^2}) + \frac{1}{r}\frac{\partial}{\partial\theta}(\frac{1}{\mathrm{Sin}\theta}\frac{\partial\Lambda}{\partial\phi}) \\ + \frac{\ell}{r}\{ r\nabla^2\Psi - \frac{\partial^2}{\partial r^2}(r\Psi)\} - \frac{\ell}{r^2}\frac{\partial}{\partial r}(r\frac{\partial^2\Psi}{\partial\theta^2}) \end{array}\right] \ , \tag{6-60}$$

$$t_{\phi\phi} = \lambda\nabla^2\Phi + 2\mu \left[\begin{array}{l} \frac{1}{r}\frac{\partial\Phi}{\partial r} + \frac{\text{Cot}\theta}{r^2}\frac{\partial\Phi}{\partial\theta} + \frac{1}{r^2\text{Sin}^2\theta}\frac{\partial^2\Phi}{\partial\phi^2} \\ + \frac{\ell}{r}\{ r\nabla^2\Psi - \frac{\partial^2}{\partial r^2}(r\Psi) \} - \frac{\ell\text{Cot}\theta}{r^2}\frac{\partial}{\partial r}(r\frac{\partial\Psi}{\partial\theta}) \\ - \frac{\ell}{r^2\text{Sin}^2\theta}\frac{\partial}{\partial r}(r\frac{\partial^2\Psi}{\partial\phi^2}) + \frac{\text{Cot}\theta}{r\text{Sin}\theta}\frac{\partial\Lambda}{\partial\phi} - \frac{1}{r\text{Sin}\theta}\frac{\partial^2\Lambda}{\partial\theta\partial\phi} \end{array} \right], \tag{6-61}$$

$$t_{r\theta} = t_{\theta r} = \mu \left[\begin{array}{l} \frac{2}{r}(\frac{\partial^2\Phi}{\partial r\partial\theta} - \frac{1}{r}\frac{\partial\Phi}{\partial\theta}) + \frac{r}{\text{Sin}\theta}\frac{\partial}{\partial r}(\frac{1}{r}\frac{\partial\Lambda}{\partial\phi}) \\ + \frac{\ell}{r}\frac{\partial}{\partial\theta}\{ r\nabla^2\Psi - 2\frac{\partial^2}{\partial r^2}(r\Psi) \} + \frac{2\ell}{r^2}\frac{\partial}{\partial r}(r\frac{\partial\Psi}{\partial\theta}) \end{array} \right], \tag{6-62}$$

$$t_{r\phi} = t_{\phi r} = \mu \left[\begin{array}{l} \frac{2}{r\text{Sin}\theta}(\frac{\partial^2\Phi}{\partial r\partial\phi} - \frac{1}{r}\frac{\partial\Phi}{\partial\phi}) - r\frac{\partial}{\partial r}(\frac{1}{r}\frac{\partial\Lambda}{\partial\theta}) \\ + \frac{\ell}{r\text{Sin}\theta}\frac{\partial}{\partial\phi}\{ r\nabla^2\Psi - 2\frac{\partial^2}{\partial r^2}(r\Psi) \} + \frac{2\ell}{r^2\text{Sin}\theta}\frac{\partial}{\partial r}(r\frac{\partial\Psi}{\partial\phi}) \end{array} \right], \tag{6-63}$$

$$t_{\theta\phi} = t_{\phi\theta} = \mu \left[\begin{array}{l} \frac{2}{r^2\text{Sin}\theta}(\frac{\partial^2\Phi}{\partial\theta\partial\phi} - \text{Cot}\theta\frac{\partial\Phi}{\partial\phi}) \\ - \frac{2\ell}{r^2\text{Sin}\theta}\{ \frac{\partial}{\partial r}(r\frac{\partial^2\Psi}{\partial\theta\partial\phi}) - \text{Cot}\theta\frac{\partial}{\partial r}(r\frac{\partial\Psi}{\partial\phi})\} \\ + \frac{1}{r}(\frac{1}{\text{Sin}^2\theta}\frac{\partial^2\Lambda}{\partial\phi^2} - \frac{\partial^2\Lambda}{\partial\theta^2} + \text{Cot}\theta\frac{\partial\Lambda}{\partial\theta}) \end{array} \right]. \tag{6-64}$$

Again we would like to show that SH is not coupled to P or SV at a spherical surface. To do this we need to look at the following three stress components associated with SH:

$$t^{SH}_{r\theta} = \mu r\frac{\partial}{\partial r}(\frac{1}{r\text{Sin}\theta}\frac{\partial\Lambda}{\partial\phi}); \qquad t^{SH}_{r\phi} = -\mu r\frac{\partial}{\partial r}(\frac{1}{r}\frac{\partial\Lambda}{\partial\theta}); \qquad t^{SH}_{rr} \equiv 0.$$

The first two expressions can be made zero by setting $\partial(\Lambda/r)/\partial r = 0$ at the surface. This is not the Neumann condition, but similar to it. Consequently, there is no coupling between Λ and the other potentials. It might be noted here that Eringen and Suhubi (1975, p. 807) have a slightly different (but equivalent) form for $t_{r\theta}$ and $t_{r\phi}$. The form used here simplifies the decoupling analysis.

These potentials are useful in describing the motion of a spherically layered earth where flat earth approximations will not suffice. This will be discussed in more detail in Chapter 24.

As in the previous two coordinate systems, there is simplification when azimuthal symmetry is present. We have, on setting $\Psi^{\dagger} = \ell r^{-1}\partial\Psi/\partial\theta$ in **N**, the following simplified forms of (6-56) - (6-64):

$$u_r = \frac{\partial\Phi}{\partial r} + \frac{1}{\mathrm{Sin}\theta}\frac{\partial}{\partial\theta}(\Psi^{\dagger}\mathrm{Sin}\theta), \tag{6-65}$$

$$u_\theta = \frac{1}{r}\frac{\partial\Phi}{\partial\theta} - \frac{1}{r}\frac{\partial}{\partial r}(r^2\Psi^{\dagger}), \tag{6-66}$$

$$u_\phi = -\frac{\partial\Lambda}{\partial\theta}, \tag{6-67}$$

that is,

$$\mathbf{u} = \nabla\Phi + \nabla\times\mathbf{e}_\phi(r\Psi^{\dagger}) + \nabla\times\mathbf{e}_r(r\Lambda). \tag{6-68}$$

Here only r- and θ-partial derivatives are to be taken. The associated azimuthally symmetric stresses are:

$$t_{rr} = \lambda\nabla^2\Phi + 2\mu\left[\frac{\partial^2\Phi}{\partial r^2} + \frac{1}{\mathrm{Sin}\theta}\frac{\partial}{\partial\theta}\left(\mathrm{Sin}\theta\frac{\partial\Psi^{\dagger}}{\partial r}\right)\right], \tag{6-69}$$

$$t_{\theta\theta} = \lambda\nabla^2\Phi + 2\mu\left[\begin{array}{l}\frac{1}{r}\frac{\partial\Phi}{\partial r} + \frac{1}{r^2}\frac{\partial^2\Phi}{\partial\theta^2} + \frac{1}{r\mathrm{Sin}\theta}\frac{\partial}{\partial\theta}(\Psi^{\dagger}\mathrm{Sin}\theta) \\ -\frac{1}{r^2}\frac{\partial}{\partial r}\left(r^2\frac{\partial\Psi^{\dagger}}{\partial\theta}\right)\end{array}\right], \tag{6-70}$$

$$t_{\phi\phi} = \lambda\nabla^2\Phi + 2\mu\left[\begin{array}{l}\frac{1}{r}\frac{\partial\Phi}{\partial r} + \frac{\mathrm{Cot}\theta}{r^2}\frac{\partial\Phi}{\partial\theta} + \frac{1}{r\mathrm{Sin}\theta}\frac{\partial}{\partial\theta}(\Psi^{\dagger}\mathrm{Sin}\theta) \\ -\frac{\mathrm{Cot}\theta}{r^2}\frac{\partial}{\partial r}(r^2\Psi^{\dagger})\end{array}\right], \tag{6-71}$$

$$t_{r\theta} = t_{\theta r} = \mu\left[\begin{array}{l}\frac{2}{r}\left(\frac{\partial^2\Phi}{\partial r\partial\theta} - \frac{1}{r}\frac{\partial\Phi}{\partial\theta}\right) - r\frac{\partial}{\partial r}\left\{\frac{1}{r^2}\frac{\partial}{\partial r}(r^2\Psi^{\dagger})\right\} \\ \frac{1}{r}\frac{\partial}{\partial\theta}\left\{\frac{1}{\mathrm{Sin}\theta}\frac{\partial}{\partial\theta}(\Psi^{\dagger}\mathrm{Sin}\theta)\right\}\end{array}\right], \tag{6-72}$$

$$t_{r\phi} = t_{\phi r} = -\mu r\frac{\partial}{\partial r}\left(\frac{1}{r}\frac{\partial\Lambda}{\partial\theta}\right), \tag{6-73}$$

$$t_{\theta\phi} = t_{\phi\theta} = \frac{\mu}{r}\left(\mathrm{Cot}\theta\frac{\partial\Lambda}{\partial\theta} - \frac{\partial^2\Lambda}{\partial\theta^2}\right). \tag{6-74}$$

These displacements and stresses can describe the azimuthally symmetric motions of a spherically layered earth such as would be caused by radially directed forces and torques.

6.5. The Fundamental Elastic Velocities and their Measurement.

The velocities of P- and S-waves may be determined directly from hand samples by the simple expedient of placing an acoustic transmitter and receiver on opposite sides of suitably prepared specimens and measuring the transit time over a known distance. However, there are many factors that prevent more than a few percent accuracy from such a determination. Two significant ones are finite pulse rise-time and finite transit-time through the transducer coupling medium. To avoid these two problems, one can look at later reflections between planar surfaces. These reflections have traversed a longer distance and the effects noted above are reduced. Even more accuracy can be obtained by superimposing the echo pulses or by comparing the phases between two pulsed sine-wave bursts. Carrying these ideas to the limit, one can get indirect data through the measurement of resonant phenomena in rods, bars, and spheres. The relationship of the resonant frequencies to the fundamental elastic velocities has to be worked out separately. Two recent works outlining the methods involved are Schreiber, Anderson, and Soga (1973) and Papadakas (1976). Collections of data relating to v_p and v_s can be found in Birch (1966) and Press (1966). The latter two references contain data from rocks under various conditions of pressure, temperature and fluid saturation.

CHAPTER 7

SOURCE FUNCTIONS IN INFINITE MEDIA

7.1. General.

The material in this section is intimately linked with two applications. The first is primarily mathematical and is the development of representation theorems in which the elastic wave motion is expressed in terms of volume and surface integrals involving the source function and its derivatives. This will be discussed in detail in Chapter 8.

The second application is more physical in nature, in that we will try to find a mathematical model for the source mechanism of earthquakes. Postulated mechanisms are force systems, stress differences, and the collapse or expansion of cavities among others. A discussion of source mechanisms for earthquakes will be put off until Chapter 28, however.

The material in this Chapter is among the most complicated in the book, rather more than one would like. However, even more detail is necessary to give a rigorous development. Both Achenbach (1973, Chapter 3) and Eringen and Suhubi (1975, Chapter 5) give a more detailed description than we do here. They also develop source functions from several other points of view. These other formulations may prove useful at times and the reader should be aware of them.

7.2. One Dimension.

One cannot usefully speak of sources in only one dimension. However, the wavefields appropriate here are plane waves, and they may be thought of as the limiting form of spherical or cylindrical waves as the source region is removed to very great distance. In this case, the near-field terms have long since decayed. Consequently, plane waves are very special, and the resulting phenomena from their use do not fully represent the physical situation. The relationships derived using plane waves will most likely remain true, but quite often are interfered with or overridden by additional effects when the source function remains at a distance close to the region of interest.

The simplification afforded by plane wave, in spite of the above obstacles, makes them useful in obtaining partial answers. This is especially true in problems where elastic waves are scattered from surface or volume irregularities. In some scattering problems, we will find two plane wave expansions of special importance. The first expands a plane wave into cylindrical coordinates. In this case the plane wave is going in the direction of $\mathbf{k}$ with velocity $\underline{v}$ (where $k = \omega/v$), and the x-axis is chosen to lie in the direction of the projection of $\mathbf{k}$ onto the horizontal plane.

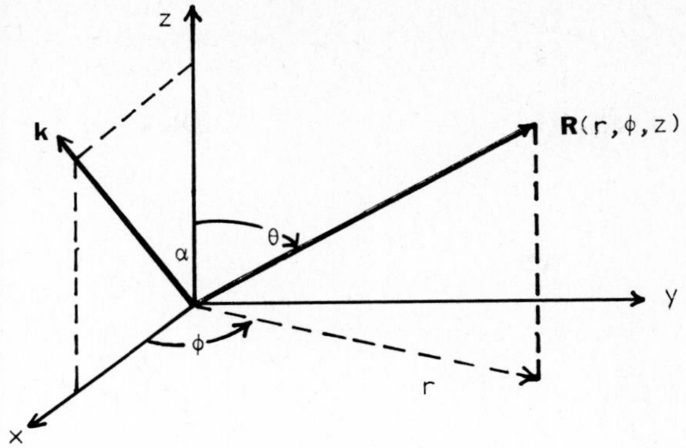

Figure 7-1. Plane Wave geometry.

For example, a plane P-wave can be represented in cartesian coordinates (where $x = r \cos\phi$, $y = r \sin\phi$), as

$$\Phi = e^{ik_o\left[r\sin\alpha\cos\phi + z\cos\alpha\right] - i\omega t} \tag{7-1}$$

$$= \left[\sum_{n=-\infty}^{\infty} i^n J_n(k_o r\sin\alpha)e^{in\phi}\right] e^{ik_o z\cos\alpha - i\omega t} ,$$

(Stratton, 1941, page 372). Likewise, for a plane P-wave going in the z-direction, we have the following expansion in spherical coordinates (ibid. p. 409):

$$\Phi = e^{i(k_o R\cos\theta - \omega t)} = e^{-i\omega t}\sum_{n=0}^{\infty} i^n (2n+1)\, j_n(k_o R)\, P_n(\cos\theta). \tag{7-2}$$

7.3. Two-Dimensional Point Sources.

Two-dimensional problems arise in two ways. The first is when you have a line-source in three dimensions and the axial direction may be ignored. In elastodynamics, SH motion would be parallel to the axis of the line-source and the greater part of the P-SV motion would lie in a plane perpendicular to that axis. This P-SV motion would then be in a state of plane strain (Sokolnikoff, 1956, p. 250). The second arises when elastic waves are propagating in thin sheets of material (not flexural waves). Here the stresses normal to the faces of the sheets vanish on the faces and are very small on the interior. The stress vector lies largely in the plane of the sheet. Such motion is in a state of plane stress (ibid, p. 253). For motion in a thin sheet, SH and SV are defined with respect to the edges, not the faces.

Scalar Media — Time Harmonic Source. We wish to find a solution of the inhomogeneous scalar wave equation;

$$\frac{\partial^2\Phi}{\partial x^2} + \frac{\partial^2\Phi}{\partial z^2} - \frac{1}{v^2}\frac{\partial^2\Phi}{\partial t^2} = -2\pi\delta(x)\delta(z)e^{-i\omega t} = -\frac{\delta(r)}{r}e^{-i\omega t}, \tag{7-3}$$

where the factor -2π is inserted for normalization, $r = (x^2 + z^2)^{\frac{1}{2}}$, $\delta(x)$, $\delta(z)$ are Dirac δ-functions, and the time variation of the source is given by the factor $e^{-i\omega t}$. If we set

$$\Phi(x,z,t) = \overline{\Phi}(x,z)e^{-i\omega t}, \tag{7-4}$$

then (7-3) becomes

$$\frac{\partial^2\overline{\Phi}}{\partial x^2} + \frac{\partial^2\overline{\Phi}}{\partial z^2} + k_o^2\overline{\Phi} = -2\pi\delta(x)\delta(z), \tag{7-5}$$

where $k_o = \omega/v$. We then take a Fourier transform with respect to $\underline{x}$, i.e.,

$$\overline{\overline{\Phi}}(k,z) = \int_{-\infty}^{\infty} \overline{\Phi}(x,z)e^{-ikx}\,dx \quad ; \qquad \overline{\Phi} = \frac{1}{2\pi}\int_{-\infty}^{\infty} \overline{\overline{\Phi}}(k,z)e^{ikx}\,dk; \tag{7-6}$$

giving

$$-k^2\overline{\overline{\Phi}} + \frac{\partial^2\overline{\overline{\Phi}}}{\partial z^2} + k_o^2\overline{\overline{\Phi}} = -2\pi\delta(z) \quad . \tag{7-7}$$

Lastly, we take a two-sided Laplace transform with respect to $\underline{z}$, i.e.,

$$\overline{\overline{\overline{\Phi}}}(k,p) = \int_{-\infty}^{\infty} \overline{\overline{\Phi}}(k,z)e^{-pz}\,dz; \qquad \overline{\overline{\Phi}} = \frac{1}{2\pi i}\int_{\alpha-i\infty}^{\alpha+i\infty} \overline{\overline{\overline{\phi}}}(k,p)e^{pz}\,dp, \tag{7-8}$$

giving

$$\{p^2 - (k^2 - k_o^2)\}\,\overline{\overline{\overline{\Phi}}} = -2\pi. \tag{7-9}$$

Inverting with respect to $\underline{p}$ gives

$$\overline{\overline{\Phi}} = \frac{\pi e^{-(k^2 - k_o^2)^{\frac{1}{2}}|z|}}{(k^2 - k_o^2)^{\frac{1}{2}}} \quad . \tag{7-10}$$

Inverting with respect to $\underline{k}$, we obtain

$$\overline{\Phi}(x,z) = \frac{1}{2}\int_{-\infty}^{\infty} \frac{e^{-(k^2 - k_o^2)^{\frac{1}{2}}|z|}\,e^{ikx}\,dk}{(k^2 - k_o^2)^{\frac{1}{2}}} = \int_{0}^{\infty} \frac{e^{-(k^2 - k_o^2)^{\frac{1}{2}}|z|}\,\text{Cos}\,kx\,dk}{(k^2 - k_o^2)^{\frac{1}{2}}}$$

$$= \frac{i\pi}{2}H_o^{(1)}(k_o r), \tag{7-11}$$

where $H_o^{(1)}(k_o r)$ is the Hankel function of the first kind (Morse and Feshbach, 1953, page 823) with properties

$$H_o^{(1)}(k_o r) = J_o(k_o r) + iN_o(k_o r) \xrightarrow[r\to\infty]{} \frac{2}{(\pi k_o r)^{\frac{1}{2}}}e^{i(k_o r - \pi/4)} \quad . \tag{7-12}$$

On the other hand, we could have reversed the Fourier and Laplace transforms obtaining

$$\frac{\partial^2\overline{\overline{\Phi}}}{\partial x^2} - k^2\overline{\overline{\Phi}} + k_o^2\overline{\overline{\Phi}} = -2\pi\delta(x), \tag{7-13}$$

and

$$\{p^2 - (k^2 - k_o^2)\}\,\overline{\overline{\overline{\Phi}}} = -2\pi. \tag{7-14}$$

Then inverting with respect to p we have

$$\overline{\overline{\Phi}} = \frac{\pi e^{-(k^2 - k_o^2)^{\frac{1}{2}}|x|}}{(k^2 - k_o^2)^{\frac{1}{2}}} . \tag{7-15}$$

Finally inverting with respect to k, we get

$$\overline{\Phi}(x,z) = \frac{1}{2}\int_{-\infty}^{\infty} \frac{e^{-(k^2 - k_o^2)^{\frac{1}{2}}|x|}\, e^{ikz}\, dk}{(k^2 - k_o^2)^{\frac{1}{2}}} = \int_{o}^{\infty} \frac{\text{Cos } kz\; e^{-(k^2 - k_o^2)^{\frac{1}{2}}|x|}\, dk}{(k^2 - k_o^2)^{\frac{1}{2}}} . \tag{7-16}$$

It should be pointed out that (7-11) and (7-16) are just two plane-wave expansions for the same function, even though they look quite different.

Lastly, adding in the time term, we have

$$\Phi(x,z,t) = \frac{i\pi}{2} H_o^{(1)}(k_o r) e^{-i\omega t} . \tag{7-17}$$

Scalar Media — Impulsive Source. In this case, we wish to find a solution of the inhomogeneous scalar wave equation

$$\frac{\partial^2\Phi}{\partial x^2} + \frac{\partial^2\Phi}{\partial z^2} - \frac{1}{v^2}\frac{\partial^2\Phi}{\partial t^2} = -2\pi\delta(x)\delta(z)\delta(t) = -\frac{\delta(r)\delta(t)}{r} . \tag{7-18}$$

Taking a Laplace transform with respect to time, we obtain

$$\frac{\partial^2\overline{\Phi}}{\partial x^2} + \frac{\partial^2\overline{\Phi}}{\partial z^2} - \frac{s^2}{v^2}\overline{\Phi} = -2\pi\delta(x)\delta(z), \tag{7-19}$$

where

$$\overline{\Phi} = \int_{o}^{\infty} \Phi(x,z,t) e^{-st} dt. \tag{7-20}$$

In order to simplify what is to come, we shall take a slightly modified Fourier transform with respect to x, i.e.,

$$\overline{\overline{\Phi}}(q,z,s) = \int_{-\infty}^{\infty} \overline{\Phi}(x,z,s) e^{-isqx/v} dx , \tag{7-21}$$

with inverse

$$\overline{\Phi} = \frac{1}{2\pi}\int_{-\infty}^{\infty} \overline{\overline{\Phi}}(q,z,s) e^{isqx/v} d(sq/v) . \tag{7-22}$$

This gives

$$-(sq/v)^2\overline{\overline{\Phi}} + \partial^2\overline{\overline{\Phi}}/\partial z^2 - (s/v)^2\overline{\overline{\Phi}} = -2\pi\delta(z). \tag{7-23}$$

Finally, taking a two-sided Laplace transform with respect to z, we have

$$\{p^2 - (s/v)^2(q^2 + 1)\}\,\overline{\overline{\overline{\Phi}}} = -2\pi , \tag{7-24}$$

where

$$\overline{\overline{\overline{\Phi}}} = \int_{-\infty}^{\infty} \overline{\overline{\Phi}}(q,z,s) e^{-pz} dz .$$

Inverting with respect to $\underline{p}$, we have

$$\bar{\bar{\Phi}} = (\pi v/s)e^{-(s/v)(q^2+1)^{\frac{1}{2}}|z|}(q^2+1)^{-\frac{1}{2}} . \qquad (7\text{-}25)$$

Inverting with respect to $\underline{q}$, we obtain

$$\bar{\Phi}(x,z,s) = \frac{1}{2}\int_{-\infty}^{\infty} e^{-(s/v)(q^2+1)^{\frac{1}{2}}|z|}(q^2+1)^{-\frac{1}{2}}e^{isqx/v}\,dq = K_o(sr/v). \qquad (7\text{-}26)$$

The expression (7-26) is just the integral representation of the Macdonald function $K_o(sr/v)$. Now the Macdonald and Hankel functions of zero order are related by

$$K_o(x) \equiv (i\pi/2)H_o^{(1)}(ix) \xrightarrow[x\to\infty]{} (\pi/2x)^{\frac{1}{2}}e^{-x}, \qquad (7\text{-}27)$$

(Stratton, 1941, pp. 390-391). Then using the well-known correspondence, $s \longleftrightarrow -i\omega$, relating Laplace and Fourier transforms, one can show that (7-26) corresponds to (7-11) and (7-16).

To return to the time domain, we would then have to write

$$\Phi = \frac{1}{2\pi i}\int_{\alpha-i\infty}^{\alpha+i\infty} \bar{\Phi}(x,z,s)e^{st}\,ds, \qquad (7\text{-}28)$$

i.e., Φ would be represented by a double integral over the two variables $\underline{q}$ and $\underline{s}$, or one integral over $\underline{s}$ with the kernel $K_o(sr/v)$. How to evaluate such an integral is not immediately obvious; however, a special technique allows us to obtain the solution to (7-18) as well as more complex problems. This method is known as the

Cagniard - deHoop Transformation. This technique (deHoop, 1960) involves the following change of variable:

$$\text{Cos}\theta\,(q^2+1)^{\frac{1}{2}} - iq\text{Sin}\theta = \tau = vt/r, \qquad (7\text{-}29)$$

where $r\text{Cos}\theta = z$, $r\text{Sin}\theta = x$, and τ is the reduced time variable.

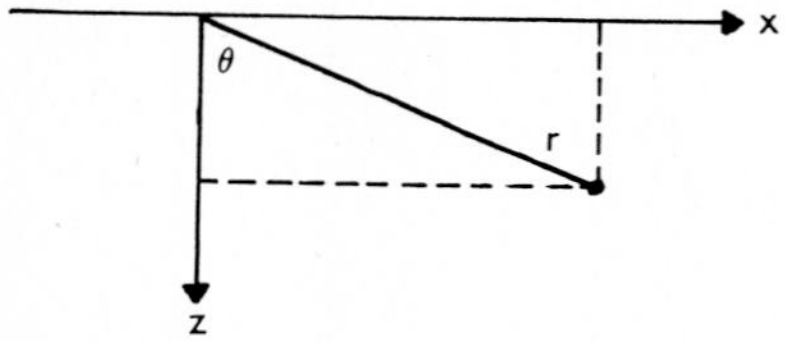

Figure 7-2. Two-dimensional coordinate systems. Note that the r - θ system is not standard cylindrical coordinates.

The inverse of this transformation is

$$q(\tau) = i\tau\text{Sin}\theta + \text{Cos}\theta\,(\tau^2-1)^{\frac{1}{2}}; \qquad (7\text{-}30)$$

and for later use we also compute

$$\frac{dq}{d\tau} = i\text{Sin}\theta + \frac{\tau\text{Cos}\theta}{(\tau^2-1)^{\frac{1}{2}}} = \frac{(q^2+1)^{\frac{1}{2}}}{(\tau^2-1)^{\frac{1}{2}}}, \qquad (7\text{-}31)$$

the last expression coming from solving for $(q^2+1)^{\frac{1}{2}}$ from (7-29) while substituting (7-30) for $\underline{q}$. Taking account of the symmetry of the real and imaginary parts of $\exp\{isqx/v\}$, we can write (7-26) as

$$\overline{\Phi} = \mathrm{Re}\left[\int_0^\infty \frac{e^{-(s/v)(q^2+1)^{\frac{1}{2}}|z| + isqx/v}}{(q^2+1)^{\frac{1}{2}}}\,dq\right] , \qquad (7\text{-}32)$$

where Re $\left[\quad\right]$ stands for the real part of the bracketed expression. We can now write this in terms of the new variable "τ" and obtain

$$\overline{\Phi}(x,z,s) = \mathrm{Re}\left[\int_?^? \frac{e^{-st}}{(q^2+1)^{\frac{1}{2}}} \frac{dq}{d\tau}\frac{v}{r}\,dt\right] \qquad (7\text{-}33)$$

$$= \mathrm{Re}\left[\int_?^? \frac{e^{-st}}{(\tau^2-1)^{\frac{1}{2}}} \frac{v}{r}\,dt\right] ,$$

where we have used (7-31). Equation (7-33) can now be recognized as the Laplace Transform of the function

$$\mathrm{Re}\left[\frac{1}{(\tau^2-1)^{\frac{1}{2}}}\frac{v}{r}\right]$$

looked at as a function of the time variable "τ". However, we have to look at a few details before we can say that this identification is valid and place proper limits on the integral. First of all, we want to look at the path $\underline{q}$ takes as we let the variable τ run from 0 to ∞. For $\tau = 0$, we have that $q = -i\mathrm{Cos}\theta$ where the sign has been chosen in (7-30) to satisfy (7-29). The variable $\underline{q}$ then moves up the imaginary axis to $q = i\mathrm{Sin}\theta$, and then branches out into the first quadrant along a hyperbola as defined by (7-30) and along an asymptote at an angle θ as in Figure 7-3(a). Inasmuch as the singularities of (7-32) are branch points at $q = \pm i$, we see that the original path can be deformed into the dashed line path as in Figure 7-3(b). However, on the vertical segment from 0 to $i\mathrm{Sin}\,\theta$ we see that the integrand of (7-32) has no real part. Consequently the limits on (7-33) may be written

$$\overline{\Phi}(x,z,s) = \mathrm{Re}\left[\int_{r/v}^{\infty} \frac{e^{-st}}{(\tau^2-1)^{\frac{1}{2}}}\frac{v}{r}\,dt\right] . \qquad (7\text{-}34)$$

By inspection we have that

$$\Phi = \frac{1}{(t^2-r^2/v^2)^{\frac{1}{2}}} H(t-r/v) , \qquad (7\text{-}35)$$

where H is the Heaviside Unit Step Function defined by

$$\begin{aligned} H(x) &= 1, \quad x > 0 \\ &= \tfrac{1}{2}, \quad x = 0 \\ &= 0, \quad x < 0 . \end{aligned} \qquad (7\text{-}36)$$

There is a sharp wavefront associated with the response to a delta-function source, but in two dimensions we also have a tail associated with the waveform in contrast to the delta-function which has zero width.

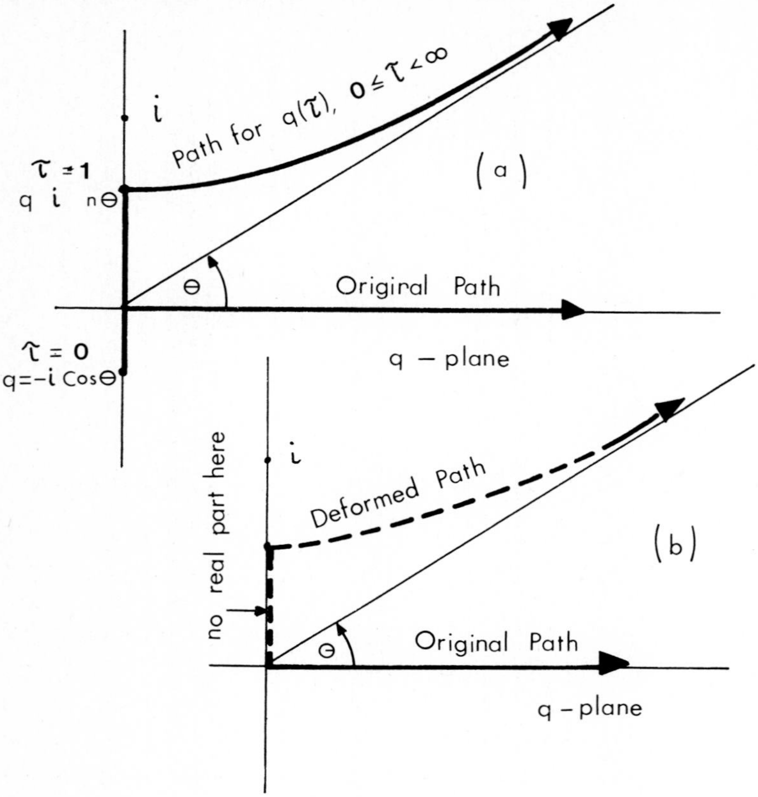

Figure 7-3(a). The relationship between the original path of integration in (7-32) and the path which q takes as τ varies between zero and infinity.
(b). The relationship between the original path and the deformed path in (7-33) which leads to (7-34). The deformed path is sometimes known as the Cagniard path.

Some final remarks on the Cagniard-deHoop technique. It is necessary to be able to invert the transformation (7-29) in analytical form to obtain a closed form of the solution. Secondly, it is necessary that the Laplace transform variable $\underline{s}$ not appear in the integrand of expressions such as (7-26) except in the exponential terms or as a multiplicative factor of the whole integrand. That is, the integrand (save the exponential factor) must be homogeneous in $\underline{s}$. This occurs rather naturally in most elastodynamic problems without any loss mechanism. However, if a dissipative term appears in (7-18), for example, one would have a different exponential expression after following through the analysis to an equivalent of (7-26). One would also have the variable $\underline{s}$ appearing in the integrand. The path on which τ was real could still be found, but the time function could not be separated out from the integrand as we did in going from (7-34) to (7-35). This means that the power of the Cagniard-deHoop technique is lost for wave-propagation problems in lossy media. However, the effect of dissipation can be determined from plane-wave problems where solutions are possible.

Elastic Media — Line Source of Dilatation. Here Φ can be any of the scalar source functions discussed above. Then we have

$$\mathbf{u}^p = \nabla\Phi = \mathbf{e}_r \frac{\partial\Phi}{\partial r} . \tag{7-37}$$

Elastic Media — Line Source of Shear. In this case, Ψ can be any one of the scalar sources discussed above. Then we have

$$\mathbf{u}^s = \nabla\times\mathbf{e}_r\Psi_r \equiv 0 . \tag{7-38}$$

$$\mathbf{u}^s = \nabla\times\mathbf{e}_\theta\Psi_\theta = \mathbf{e}_z \frac{1}{r}\frac{\partial}{\partial r}(r\Psi_\theta) . \qquad \text{AXIAL SHEAR} \tag{7-39}$$

$$\mathbf{u}^s = \nabla\times\mathbf{e}_z\Psi_z = \mathbf{e}_\theta \frac{\partial\Psi_z}{\partial r} . \qquad \text{TORSIONAL SHEAR} \tag{7-40}$$

Here the form of (7-38) tells us that a radial component of the shear vector potential is the non-contributory one. Thus (7-37), (7-39), and (7-40) are our three independent potentials. For axial shear and torsional shear as defined above, coordinates are standard cylindrical.

Elastic Media — Line Force. The elastic equation of motion (5-9) for a line force acting in an infinite medium can be written as

$$\frac{\lambda + 2\mu}{\rho}\nabla\nabla\cdot\mathbf{w} - \frac{\mu}{\rho}\nabla\times\nabla\times\mathbf{w} - \frac{\partial^2\mathbf{w}}{\partial t^2} = -\mathbf{e}_z\frac{f(t)\delta(r)}{\rho 2\pi r} , \tag{7-41}$$

where $r = (x^2 + z^2)^{\frac{1}{2}}$, and $\mathbf{e}_z$ is a unit vector in the direction of the force f(t). We will return again to the special two-dimensional coordinate system of Figure 7-2. If we take a Fourier transform of both sides of this equation, we obtain

$$v_p^2\nabla\nabla\cdot\mathbf{W} - v_s^2\nabla\times\nabla\times\mathbf{W} + \omega^2\mathbf{W} = -\mathbf{e}_z\frac{F(\omega)\delta(r)}{\rho 2\pi r} , \tag{7-42}$$

where

$$\mathbf{W}(x,z,\omega) = \int_{-\infty}^{\infty} \mathbf{w}(x,z,t)e^{i\omega t}\,dt \quad ; \quad F(\omega) = \int_{-\infty}^{\infty} f(t)e^{i\omega t}\,dt .$$

The method of solving (7-42) is definitely not intuitive. A number of ad hoc steps are introduced to achieve the solution. Although the computation to follow is not completely straightforward, it is relatively simple. I also think it lends to better understanding of the physics involved with the equivoluminal and irrotational parts of the solution. First of all we write

$$\mathbf{W} = \nabla(\nabla\cdot\mathbf{A}_p) - \nabla\times(\nabla\times\mathbf{A}_s) , \tag{7-43}$$

i.e., we define a scalar potential $\nabla\cdot\mathbf{A}_p$ and a vector potential $\nabla\times\mathbf{A}_s$. In two dimensions we have that

$$-\mathbf{e}_z \frac{F\delta(r)}{\rho 2\pi r} = -\mathbf{e}_z \frac{F}{2\pi\rho} \nabla^2(\ln r) \tag{7-44}$$

$$= \nabla\nabla\cdot \left[-\mathbf{e}_z \frac{F \ln r}{2\pi\rho}\right] - \nabla\times\nabla\times \left[-\mathbf{e}_z \frac{F \ln r}{2\pi\rho}\right] ,$$

where we have used the vector identity

$$\nabla^2 \mathbf{A} \equiv \nabla\nabla\cdot\mathbf{A} - \nabla\times\nabla\times\mathbf{A} \; .$$

That this is true may be seen by evaluating the following:

$$\frac{1}{2\pi}\int \nabla^2(\ln r)dS = \frac{1}{2\pi}\int \frac{\partial}{\partial r}(\ln r)dl = \frac{1}{2\pi}\int \frac{rd\theta}{r} = 1 = \int \frac{\delta(r)dS}{2\pi r}$$

per unit thickness of slab as in Figure 7-4. Substituting (7-43) and (7-44)

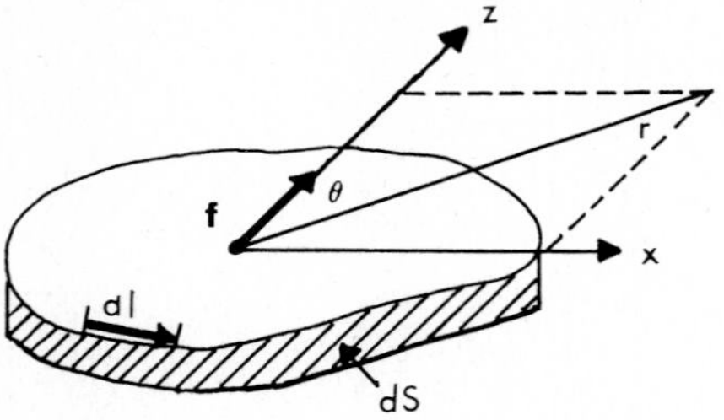

Figure 7-4. Two-dimensional geometry. The force **f** lies in the plane of the slab along the z-axis.

into (7-42), we obtain

$$v_p^2 \nabla\nabla\cdot \left[\nabla\nabla\cdot\mathbf{A}_p + k_p^2 \mathbf{A}_p + \mathbf{e}_z \frac{F \ln r}{2\pi\rho v_p^2}\right] + v_s^2 \nabla\times\nabla\times \left[\nabla\times\nabla\times\mathbf{A}_s - k_s^2 \mathbf{A}_s - \mathbf{e}_z \frac{F \ln r}{2\pi\rho v_s^2}\right] = 0.$$

We can add the term $-\nabla\times\nabla\times\mathbf{A}_p$ to the first bracketed portion and the term $-\nabla\nabla\cdot\mathbf{A}_s$ to the second in that they do not change the value of the whole expression. Adding these two terms, we then have

$$v_p^2 \nabla\nabla\cdot \left[\nabla\nabla\cdot\mathbf{A}_p - \nabla\times\nabla\times\mathbf{A}_p + k_p^2 \mathbf{A}_p + \mathbf{e}_z \frac{F \ln r}{2\pi\rho v_p^2}\right] \tag{7-45}$$

$$+ v_s^2 \nabla\times\nabla\times \left[\nabla\times\nabla\times\mathbf{A}_s - \nabla\nabla\cdot\mathbf{A}_s - k_s^2 \mathbf{A}_s - \mathbf{e}_z \frac{F \ln r}{2\pi\rho v_s^2}\right] = 0 \; .$$

We then have a solution if

$$\nabla^2 \mathbf{A}_p + k_p^2 \mathbf{A}_p = -\mathbf{e}_z \frac{F \ln r}{2\pi\rho v_p^2} , \tag{7-46}$$

and

$$\nabla^2 \mathbf{A}_s + k_s^2 \mathbf{A}_s = -\mathbf{e}_z \frac{F \ln r}{2\pi\rho v_s^2} . \tag{7-47}$$

We can then write $\mathbf{A}_p = \mathbf{e}_z A_p$ and $\mathbf{A}_s = \mathbf{e}_z A_s$, and we will still have a solution if

$$\frac{\partial^2 A_p}{\partial x^2} + \frac{\partial^2 A_p}{\partial z^2} + k_p^2 A_p = -\frac{F \ln r}{2\pi\rho v_p^2} , \tag{7-48}$$

and

$$\frac{\partial^2 A_s}{\partial x^2} + \frac{\partial^2 A_s}{\partial z^2} + k_s^2 A_s = - \frac{F \ln r}{2\pi\rho v_s^2} , \tag{7-49}$$

where we have written out the components of the Laplacian operator ∇^2. We have reduced the problem of (7-42) to two identical scalar equations, one for A_p and one for A_s. There are still some difficulties however.

We would like to use transform methods to solve (7-48) and (7-49), but the singular nature of ln(r) prevents us from doing this directly. However, we can go at it indirectly for we have seen that

$$\nabla^2(\ln r) = \delta(r)/r = 2\pi\delta(x)\delta(z).$$

Taking a Fourier transform with respect to x, and a two-sided Laplace transform with respect to z, we have

$$(p^2 - k^2)\, \overline{\overline{\ln r}} = 2\pi \quad ,$$

or

$$\overline{\overline{\ln r}} = \frac{2\pi}{p^2 - k^2} \quad .$$

We now take the same transform pair of the left-hand side of (7-48) and obtain $(k_p^2 + p^2 - k^2)A_p$. Combining, we have

$$\overline{\overline{A_p}} = - \frac{F}{\rho v_p^2} \frac{1}{p^2 - k^2} \frac{1}{k_p^2 + p^2 - k^2} \quad . \tag{7-50}$$

Expanding, we have

$$\overline{\overline{A_p}} = - \frac{F}{\rho\omega^2} \left[\frac{1}{p^2 - k^2} - \frac{1}{k_p^2 + p^2 - k^2} \right] \quad .$$

But we recognize the first term in brackets as the double transform of ln (r/2) and the second term as the double transform of $-iH_o^{(1)}(k_p r)/4$. Hence we can write

$$A_p = \frac{-F}{2\pi\rho\omega^2} \left[\ln r + \frac{i\pi}{2} H_o^{(1)}(k_p r) \right] \quad . \tag{7-51}$$

Similarly

$$A_s = \frac{-F}{2\pi\rho\omega^2} \left[\ln r + \frac{i\pi}{2} H_o^{(1)}(k_s r) \right] \quad . \tag{7-52}$$

Now

$$\begin{aligned} \nabla\times\nabla\times\mathbf{A}_s &= \nabla\nabla\cdot\mathbf{A}_s - \nabla^2\mathbf{A}_s \\ &= \nabla\nabla\cdot\mathbf{A}_s + k_s^2\mathbf{A}_s + \mathbf{e}_z \frac{F \ln r}{2\pi\rho v_s^2} \\ &= \nabla\nabla\cdot\mathbf{A}_s - \mathbf{e}_z \frac{iF}{4\rho v_s^2} H_o^{(1)}(k_s r) , \end{aligned}$$

and from (7-43) we finally obtain

$$\mathbf{W} = \nabla\nabla\cdot(\mathbf{A}_p - \mathbf{A}_s) + \mathbf{e}_z \frac{iF}{4\rho v_s^2} H_o^{(1)}(k_s r) \tag{7-53}$$

$$= -\frac{iF}{4\rho\omega^2}\left\{ \nabla\nabla\cdot\left[\mathbf{e}_z H_o^{(1)}(k_p r) - \mathbf{e}_z H_o^{(1)}(k_s r)\right] - \mathbf{e}_z k_s^2 H_o^{(1)}(k_s r)\right\} .$$

Making the correspondence between Laplace and Fourier transforms, we see that the transformed solution for an impulsive source would be

$$\mathbf{W} = \frac{F}{2\pi\rho}\left\{\frac{1}{s^2}\nabla\nabla\cdot\left[\mathbf{e}_z K_o(\frac{sr}{v_p}) - \mathbf{e}_z K_o(\frac{sr}{v_s})\right] + \frac{1}{v_s^2}\mathbf{e}_z K_o(\frac{sr}{v_s})\right\} , \tag{7-54}$$

where F is the strength of the impulse in the z-direction.

This expression can be analyzed further. Taking the divergence, we have

$$\mathbf{W} = \frac{F}{2\pi\rho}\left\{\frac{1}{s^2}\nabla\frac{\partial}{\partial z}\left[K_o(\frac{sr}{v_p}) - K_o(\frac{sr}{v_s})\right] + \frac{1}{v_s^2}\mathbf{e}_z K_o(\frac{sr}{v_s})\right\}$$

$$= \frac{F}{2\pi\rho}\left\{\frac{1}{s^2}\nabla\left[\frac{s\,\mathrm{Cos}\theta}{v_p}K_o'(\frac{sr}{v_p}) - \frac{s\,\mathrm{Cos}\theta}{v_s}K_o'(\frac{sr}{v_s})\right] + \frac{1}{v_s^2}\mathbf{e}_z K_o(\frac{sr}{v_s})\right\}$$

$$= \frac{F}{2\pi\rho}\left\{\begin{array}{l} \mathbf{e}_r\mathrm{Cos}\theta\left[v_p^{-2}K_o''(\frac{sr}{v_p}) - v_s^{-2}K_o''(\frac{sr}{v_s}) + v_s^{-2}K_o(\frac{sr}{v_s})\right] \\ + \mathbf{e}_\theta\mathrm{Sin}\theta\left[-v_p^{-2}(\frac{v_p}{sr})K_o'(\frac{sr}{v_p}) + v_s^{-2}(\frac{v_s}{sr})K_o'(\frac{sr}{v_s}) - v_s^{-2}K_o(\frac{sr}{v_s})\right]\end{array}\right\} .$$

Here we have used

$$\partial K_o/\partial z = s v_p^{-1}\mathrm{Cos}\theta K_o'(sr/v_p),$$

where

$$\mathbf{e}_z = \mathbf{e}_r\mathrm{Cos}\theta - \mathbf{e}_\theta\mathrm{Sin}\theta .$$

Also we note that

$$K_o'(\xi) = -K_1(\xi) \qquad ; \qquad K_o''(\xi) = K_o(\xi) - \xi^{-1}K_1(\xi) .$$

Continuing to work with the expression for **W**, we obtain

$$\mathbf{W} = \frac{F}{2\pi\rho}\left\{\begin{array}{l} \mathbf{e}_r\mathrm{Cos}\theta\left[v_p^{-2}K_o(\frac{sr}{v_p}) + (\frac{v_p}{sr})K_1(\frac{sr}{v_p}) - (\frac{v_s}{sr})K_1(\frac{sr}{v_s})\right] \\ + \mathbf{e}_\theta\mathrm{Sin}\theta\left[v_p^{-2}(\frac{v_p}{sr})K_1(\frac{sr}{v_p}) - v_s^{-2}K_o(\frac{sr}{v_s}) - v_s^{-2}(\frac{v_s}{sr})K_1(\frac{sr}{v_s})\right]\end{array}\right\} .$$

From Erdelyi et al (1954, Inverse transform #5.15(11)), we have that

$$(\frac{v}{sr})K_1(\frac{sr}{v}) \Longleftrightarrow \frac{v^2}{r^2}(t^2 - r^2/v^2)^{\frac{1}{2}}H(t - r/v) , \tag{7-55}$$

and from (7-35) and (7-26) above, we have that

$$K_o(\frac{sr}{v}) \Longleftrightarrow (t^2 - r^2/v^2)^{-\frac{1}{2}}H(t - r/v) . \tag{7-56}$$

Returning to the time dimension, we calculate that

$$\mathbf{W} = \frac{F}{2\pi\rho}\left\{\begin{aligned} &\mathbf{e}_r \cos\theta \left[\frac{H(t - r/v_p)}{v_p^2(t^2 - r^2/v_p^2)^{\frac{1}{2}}} + \frac{(t^2 - r^2/v_p^2)^{\frac{1}{2}}}{r^2} H(t-r/v_p) - \frac{(t^2 - r^2/v_s^2)^{\frac{1}{2}}}{r^2} H(t-r/v_s)\right] \\ &+ \mathbf{e}_\theta \sin\theta \left[\frac{(t^2 - r^2/v_p^2)^{\frac{1}{2}}}{r^2} H(t-r/v_p) - \frac{H(t-r/v_s)}{v_s^2(t^2 - r^2/v_s^2)^{\frac{1}{2}}} - \frac{(t^2 - r^2/v_s^2)^{\frac{1}{2}}}{r^2} H(t-r/v_s)\right] \end{aligned}\right\} \quad (7\text{-}57)$$

This agrees with a solution worked out by Eason, Fulton and Sneddon (1956) and expressed in cartesian coordinates.

This analysis shows us two things. To begin with, the first order term of w_r travels with velocity v_p, that is, at large distances the greatest part of the displacement motion associated with v_p (P-motion) is perpendicular to the wave front, i.e., it is a longitudinal wave. However, it is neither true that all P-motion is perpendicular to the wave front, nor true that all motion perpendicular to the wave front is P-motion. Similarly with S-motion. The first order term of w_θ travels with velocity v_s and consequently the greatest part of the S-motion is parallel to the wave front. However, some P-motion is parallel to the wave front and some S-motion is perpendicular to the wave front.

Secondly, we note that the response $(t^2-r^2/v^2)^{-\frac{1}{2}}$ is the characteristic impulse response in two-dimensions. We see that terms like $r^{-2}(t^2-r^2/v^2)^{\frac{1}{2}}$ not only fall off faster with distance, but are time-integrals of the first order P- and S-wave terms. This is a rather frequent occurrence in elastic wave theory. One finds that diffracted or near field terms fall off one or more powers of _r_ faster than the direct waves and at the same time they are time-integrals of the direct wave time function.

Elastic Media — Line Source of Stress. In infinite media, the line source of stress and the line force are equivalent in that one chooses the stress to go to infinity in such a way that the product of stress times area (equals force) remains finite. Consequently one does not have to distinguish between them.

However, when a bounding surface is introduced, and forces act on this surface, it is quite often more convenient to build the source function into the boundary conditions rather than introduce it as a singularity into the differential equation. We will see an example of this in two dimensions in the discussion of Lamb's Problem (Chapter 11).

7.4. Three-Dimensional Point Sources.

Scalar Media — Time Harmonic Source. We wish to find a solution of the inhomogeneous scalar wave equation

$$\frac{1}{r}\frac{\partial}{\partial r}\left(r\frac{\partial \Phi}{\partial r}\right) + \frac{\partial^2 \Phi}{\partial z^2} - \frac{1}{v^2}\frac{\partial^2 \Phi}{\partial t^2} = -4\pi\delta(x)\delta(y)\delta(z)e^{-i\omega t}$$

$$= -2\,\frac{\delta(r)}{r}\,\delta(z)e^{-i\omega t} \qquad (7\text{-}58)$$

$$= -\frac{\delta(R)}{R^2}\,e^{-i\omega t} \quad ,$$

where the factor -4π is inserted for normalization and $R = (x^2 + y^2 + z^2)^{\frac{1}{2}}$. We then set

$$\Phi(r,z,t) = \overline{\Phi}(r,z)e^{-i\omega t} \quad , \qquad (7\text{-}59)$$

and (7-58) becomes on substitution of this time harmonic form:

$$\frac{1}{r}\frac{\partial}{\partial r}\left(r\frac{\partial \overline{\Phi}}{\partial r}\right) + \frac{\partial^2 \overline{\Phi}}{\partial z^2} + k_o^2\overline{\Phi} = -2\,\frac{\delta(r)}{r}\,\delta(z) \quad , \qquad (7\text{-}60)$$

where again $k_o = \omega/v$. To eliminate the r-dependence, we take the Fourier-Bessel (Hankel) transform with respect to r defined by:

$$\overline{\overline{\Phi}}(k,z) = \int_o^\infty \overline{\Phi}(r,z)J_o(kr)rdr \quad ; \quad \overline{\Phi} = \int_o^\infty \overline{\overline{\Phi}}(k,z)J_o(kr)kdk. \qquad (7\text{-}61)$$

Now

$$\int_o^\infty \frac{1}{r}\frac{\partial}{\partial r}\left(r\frac{\partial \overline{\Phi}}{\partial r}\right)J_o(kr)rdr = r\,\frac{\partial \overline{\Phi}}{\partial r}\,J_o(kr)\Big|_o^\infty - \int_o^\infty r\,\frac{\partial}{\partial r}\,J_o(kr)\,\frac{\partial \overline{\Phi}}{\partial r}\,dr$$

$$= -\left[r\,\frac{\partial}{\partial r}\,J_o(kr)\right]\overline{\Phi}\,\Big|_o^\infty + \int_o^\infty \frac{\partial}{\partial r}\left[r\,\frac{\partial}{\partial r}\,J_o(kr)\right]\overline{\Phi}dr$$

$$= -k^2\int_o^\infty \overline{\Phi}J_o(kr)rdr = -k^2\overline{\overline{\Phi}} \, .$$

Thus (7-60) reduces to

$$-k^2\overline{\overline{\Phi}} + \frac{\partial^2\overline{\overline{\Phi}}}{\partial z^2} + k_o^2\overline{\overline{\Phi}} = -2\delta(z) \; . \qquad (7\text{-}62)$$

We now take a Laplace transform with respect to z, giving

$$(k_o^2 - k^2 + p^2)\overline{\overline{\overline{\Phi}}} = -2 \; . \qquad (7\text{-}63)$$

Inverting with respect to p gives

$$\overline{\overline{\Phi}} = \frac{e^{-(k^2 - k_o^2)^{\frac{1}{2}}|z|}}{(k^2 - k_o^2)^{\frac{1}{2}}} \; . \qquad (7\text{-}64)$$

Inverting with respect to k gives

$$\overline{\Phi} = \int_o^\infty \frac{e^{-(k^2 - k_o^2)^{\frac{1}{2}}|z|}}{(k^2 - k_o^2)^{\frac{1}{2}}}\,J_o(kr)kdk \qquad (7\text{-}65)$$

$$= e^{ik_oR}/R$$

$$= ik_o h_o^{(1)}(k_oR),$$

where $h_o^{(1)}(k_oR)$ is a spherical Hankel function of the first kind. The identification (7-65) is given by Watson (1952, p. 416, equation (4)).

We cannot continue as in the two-dimensional case by reversing the Fourier-Bessel and Laplace transforms inasmuch as the equivalent of a Laplace type Bessel Transform has a singularity at the origin. However, we can come up with formulas equivalent to (7-17) by deforming the branch cuts in (7-65) and using the following two identities:

$$J_o(x) = \tfrac{1}{2}\ \{H_o^{(1)}(x) + H_o^{(2)}(x)\} \quad , \tag{7-66}$$

and

$$H_o^{(2)}(x) = -H_o^{(1)}(-x) \quad . \tag{7-67}$$

Now the branch cuts and the path of integration of (7-65) are as in Figure 7-5(a). Using the above identities we can write

$$\frac{e^{ik_oR}}{R} = \tfrac{1}{2}\int_{-\infty}^{\infty} \frac{e^{-(k^2-k_o^2)^{\frac{1}{2}}|z|}}{(k^2-k_o^2)^{\frac{1}{2}}}\, H_o^{(1)}(kr)kdk \ , \tag{7-68}$$

where the path of integration is shown in Figure 7-5(b). Since there are no

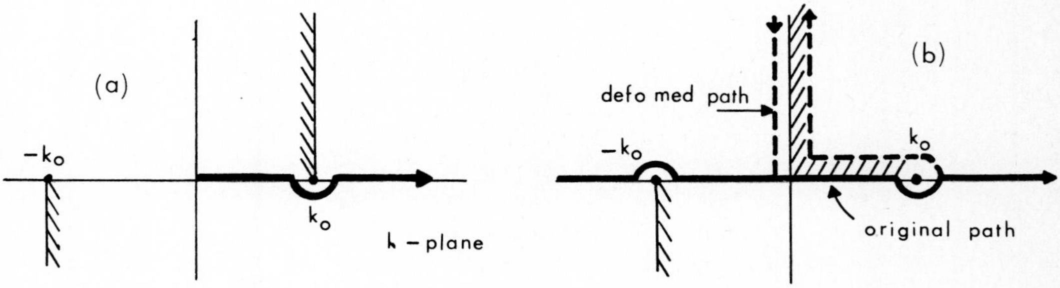

Figure 7-5(a). Path of integration in equation (7-65).
(b). Path of integration in equation (7-68), with deformed path where the new variable, $q = \pm(k_o^2-k^2)^{\frac{1}{2}}$ is real.

singularities in the upper half-plane, the contour along the real axis can be deformed to surround the upper branch cut which is itself deformed to lie along the real axis from 0 to $+k_o$ and along the positive imaginary axis from 0 to ∞. We find that along the new contour we can define

$$q = \pm(k_o^2-k^2)^{\frac{1}{2}} \quad ;$$

the sign dependent on which side of the branch cut the path is on. Then

$$dq = \frac{ikdk}{(k^2-k_o^2)^{\frac{1}{2}}} \ ,$$

and we finally obtain

$$\frac{e^{ik_oR}}{R} = \frac{i}{2}\int_{-\infty}^{\infty} e^{iqz}\, H_o^{(1)}\ \{(k_o^2-q^2)^{\frac{1}{2}}r\}\ dq \quad , \tag{7-69}$$

where the path of integration is shown in Figure 7-6.

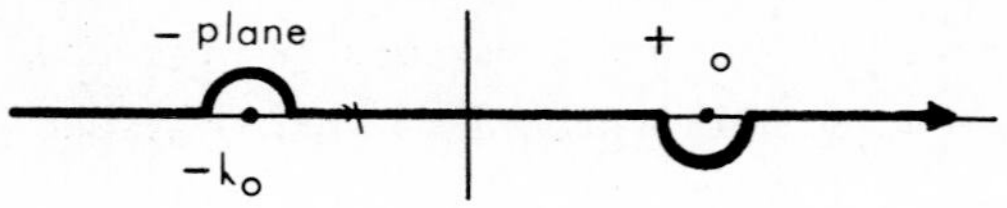

Figure 7-6. Path of integration for equation (7-69).

Scalar Media — Impulsive Source. The Cagniard-deHoop technique does not work in this case because of the complexity of the kernels in (7-65), that is, the non-exponential character of J_o and $H_o^{(1)}$. However, the change $s \Longleftrightarrow -i\omega$ gives

$$\bar{\Phi} = \frac{e^{-sR/v}}{R} = \frac{s}{v} k_o(sR/v) \quad , \tag{7-70}$$

as the solution of

$$\frac{1}{r}\frac{\partial}{\partial r}\left(r\frac{\partial\bar{\Phi}}{\partial r}\right) + \frac{\partial^2\bar{\Phi}}{\partial z^2} - \frac{s^2}{v^2}\bar{\Phi} = -4\pi\delta(x)\delta(y)\delta(z) = -2\delta(r)\delta(z)/r = -\delta(R)/R^2 \quad . \tag{7-71}$$

Equation (7-71) is in turn the Laplace transform of

$$\frac{1}{r}\frac{\partial}{\partial r}\left(r\frac{\partial\Phi}{\partial r}\right) + \frac{\partial^2\Phi}{\partial z^2} - \frac{1}{v^2}\frac{\partial^2\Phi}{\partial t^2} = -4\pi\delta(x)\delta(y)\delta(z)\delta(t) \quad , \tag{7-72}$$

in which the time function is the impulse function $\delta(t)$. However, the inverse of (7-70) is well known, being

$$\Phi(x,y,z,t) = \frac{\delta(t - R/v)}{R} \quad , \tag{7-73}$$

i.e., the time part of the source is repeated at the "retarded time" $t = R/v$, and diminished by the spherical divergence factor, $1/R$. NOTE: deHoop (1960) makes a transformation of variables after starting with multiple transforms of cartesian coordinates. This is not properly the Cagniard-deHoop technique although the change of variables is based upon it.

Elastic Media — Point Source of Dilatation. Again Φ can be any of the three-dimensional scalar source functions described in the preceding paragraphs. Thus we have

$$\mathbf{u}^P = \nabla\Phi = \mathbf{e}_R \frac{\partial\Phi}{\partial R} \quad . \tag{7-74}$$

Elastic Media — Point Source of Shear. Here the potentials $\Psi_R, \Psi_\theta, \Psi_\phi$ can be any of the three-dimensional scalar source functions above. Then we would have

$$\mathbf{u}^S = \nabla\times\mathbf{e}_R\Psi_R = \frac{1}{R^2\text{Sin}\theta}\begin{vmatrix} \mathbf{e}_R & R\mathbf{e}_\theta & R\text{Sin}\,\theta\mathbf{e}_\phi \\ \frac{\partial}{\partial R} & 0 & 0 \\ \Psi_R & 0 & 0 \end{vmatrix} \equiv 0 \quad . \tag{7-75}$$

Obviously, Ψ_R contributes nothing, so it is the scalar potential we can ignore. However, the other two source components do give reasonable sources. That is,

$$\mathbf{u}^s = \nabla\times\mathbf{e}_\theta\Psi_\theta = \mathbf{e}_\phi \frac{1}{R}\frac{\partial}{\partial R}(R\Psi_\theta) \qquad \text{TORSIONAL SHEAR (Axial Twist)} \qquad (7\text{-}76)$$

and

$$\mathbf{u}^s = \nabla\times\mathbf{e}_\phi\Psi_\phi = -\mathbf{e}_\theta \frac{1}{R}\frac{\partial}{\partial R}(R\Psi_\phi). \qquad \text{'DOUGHNUT' SHEAR} \qquad (7\text{-}77)$$

Elastic Media — Point Force. The elastodynamic equation of motion (5-9) for a point force acting in an infinite medium can be written as

$$\left(\frac{\lambda+2\mu}{\rho}\right)\nabla\nabla\cdot\mathbf{w} - \frac{\mu}{\rho}\nabla\times\nabla\times\mathbf{w} - \frac{\partial^2\mathbf{w}}{\partial t^2} = -\mathbf{e}_z \frac{f(t)}{\rho}\frac{\delta(R)}{R^2}\ , \qquad (7\text{-}78)$$

where $R = (x^2 + y^2 + z^2)^{\frac{1}{2}}$, and the direction of the force is taken conveniently as the z-axis. Taking the Fourier transform of (7-78), noting that

$$\frac{1}{4\pi}\int \nabla^2\left(\frac{1}{R}\right)dV = \frac{1}{4\pi}\int \frac{\partial}{\partial R}\left(\frac{1}{R}\right)dS = -\frac{1}{4\pi}\iint \frac{R^2 \sin\theta\, d\theta\, d\phi}{R^2} = -1 = -\frac{1}{4\pi}\int \frac{\delta(R)}{R^2}\,dV,$$

and following the same procedure as in (7-42) through (7-45), we can write:

$$\begin{aligned} v_p^2\nabla\nabla\cdot &\left[\nabla\nabla\cdot\mathbf{A}_p - \nabla\times\nabla\times\mathbf{A}_p + k_p^2\mathbf{A}_p - \frac{\mathbf{e}_z F}{4\pi\rho v_p^2}\left(\frac{1}{R}\right)\right] \\ &+ v_s^2\nabla\times\nabla\times\left[\nabla\times\nabla\times\mathbf{A}_s - \nabla\nabla\cdot\mathbf{A}_s - k_s^2\mathbf{A}_s + \frac{\mathbf{e}_z F}{4\pi\rho v_s^2}\left(\frac{1}{R}\right)\right] = 0\ . \end{aligned} \qquad (7\text{-}79)$$

This equation will be satisfied if $\mathbf{A}_p$ and $\mathbf{A}_s$ satisfy

$$\nabla^2\mathbf{A}_p + k_p^2\mathbf{A}_p = \frac{\mathbf{e}_z F}{4\pi\rho v_p^2}\left(\frac{1}{R}\right)\ , \qquad (7\text{-}80)$$

and

$$\nabla^2\mathbf{A}_s + k_s^2\mathbf{A}_s = \frac{\mathbf{e}_z F}{4\pi\rho v_s^2}\left(\frac{1}{R}\right)\ . \qquad (7\text{-}81)$$

Again, this will be true if we write $\mathbf{A}_p = \mathbf{e}_z A_p$ and $\mathbf{A}_s = \mathbf{e}_z A_s$, where A_p and A_s satisfy

$$\nabla^2 A_p + k_p^2 A_p = \frac{F}{4\pi\rho v_p^2}\left(\frac{1}{R}\right), \qquad (7\text{-}82)$$

and

$$\nabla^2 A_s + k_s^2 A_s = \frac{F}{4\pi\rho v_s^2}\left(\frac{1}{R}\right). \qquad (7\text{-}83)$$

As in the two-dimensional case, we solve this pair indirectly. First of all we note that

$$\nabla^2\left(\frac{1}{R}\right) = -\frac{\delta(R)}{R^2} = -4\pi\delta(x)\delta(y)\delta(z)\ . \qquad (7\text{-}84)$$

Taking the Fourier transform with respect to <u>x</u> and <u>y</u>, and a two-sided Laplace transform with respect to <u>z</u>, we have

$$(p^2 - k^2 - l^2)(\overline{\overline{\overline{\tfrac{1}{R}}}}) = -4\pi \quad ,$$

or

$$(\overline{\overline{\overline{\tfrac{1}{R}}}}) = \frac{-4\pi}{p^2 - k^2 - l^2} \quad .$$

This is essentially the transform of the right-hand side of (7-82). Taking the same transform triplet of the left-hand side of (7-82) gives

$$\overline{\overline{\overline{A}}}_p = \frac{-F}{\rho v_p^2} \frac{1}{(p^2 - k^2 - l^2)} \frac{1}{(k_p^2 + p^2 - k^2 - l^2)} \quad . \tag{7-85}$$

Expanding into partial fractions, we have

$$\overline{\overline{\overline{A}}}_p = \frac{-F}{\rho\omega^2} \left[\frac{1}{(p^2 - k^2 - l^2)} - \frac{1}{(k_p^2 + p^2 - k^2 - l^2)} \right] \quad .$$

But we recognize the first bracketed term as the transform of $-1/(4\pi R)$ and the second as the transform of $e^{ik_pR}/(4\pi R)$. Hence we can write:

$$A_p = \frac{F}{4\pi\rho\omega^2} \left[\frac{1}{R} - \frac{e^{ik_pR}}{R} \right] \quad , \tag{7-86}$$

and

$$A_s = \frac{F}{4\pi\rho\omega^2} \left[\frac{1}{R} - \frac{e^{ik_sR}}{R} \right] \quad . \tag{7-87}$$

Now

$$\begin{aligned} \nabla\times\nabla\times\mathbf{A}_s &= \nabla\nabla\cdot\mathbf{A}_s - \nabla^2\mathbf{A}_s \\ &= \nabla\nabla\cdot\mathbf{A}_s + k_s^2\mathbf{A}_s - \frac{\mathbf{e}_z F}{4\pi\rho v_s^2} (\frac{1}{R}) \\ &= \nabla\nabla\cdot\mathbf{A}_s - \frac{\mathbf{e}_z F}{4\pi\rho v_s^2} (\frac{e^{ik_sR}}{R}), \end{aligned}$$

and from (7-43) we have

$$\begin{aligned} \mathbf{W} &= \nabla\nabla\cdot(\mathbf{A}_p - \mathbf{A}_s) + \frac{\mathbf{e}_z F}{4\pi\rho v_s^2} (\frac{e^{ik_sR}}{R}) \\ &= \frac{-F(\omega)}{4\pi\rho\omega^2} \left[\nabla\nabla\cdot(\mathbf{e}_z \frac{e^{ik_pR}}{R} - \mathbf{e}_z \frac{e^{ik_sR}}{R}) - \mathbf{e}_z k_s^2 \frac{e^{ik_sR}}{R} \right] \quad . \end{aligned} \tag{7-88}$$

As with the two-dimensional solution, we can perform the indicated vector operations. Taking the divergence we have

$$\mathbf{W} = -\frac{F}{4\pi\rho\omega^2} \left[\nabla \frac{\partial}{\partial z} (\frac{e^{ik_pR}}{R} - \frac{e^{ik_sR}}{R}) - \mathbf{e}_z k_s^2 \frac{e^{ik_sR}}{R} \right]$$

$$= -\frac{F}{4\pi\rho\omega^2}\left[\nabla\begin{pmatrix} ik_p\text{Cos}\theta \dfrac{e^{ik_pR}}{R} - \text{Cos}\theta \dfrac{e^{ik_pR}}{R^2} \\ -ik_s\text{Cos}\theta \dfrac{e^{ik_sR}}{R} + \text{Cos}\theta \dfrac{e^{ik_sR}}{R^2}\end{pmatrix} - \mathbf{e}_z k_s^2 \frac{e^{ik_sR}}{R}\right]$$

$$= -\frac{F}{4\pi\rho\omega^2}\left\{\begin{array}{l} \mathbf{e}_R\left[\begin{array}{l} -k_p^2\text{Cos}\theta \dfrac{e^{ik_pR}}{R} - 2ik_p\text{Cos}\theta \dfrac{e^{ik_pR}}{R^2} + 2\text{Cos}\theta \dfrac{e^{ik_pR}}{R^3} \\ \qquad + 2ik_s\text{Cos}\theta \dfrac{e^{ik_sR}}{R^2} - 2\text{Cos}\theta \dfrac{e^{ik_sR}}{R^3}\end{array}\right] \\ +\mathbf{e}_\theta\left[\begin{array}{l} \qquad - ik_p\text{Sin}\theta \dfrac{e^{ik_pR}}{R^2} + \text{Sin}\theta \dfrac{e^{ik_pR}}{R^3} \\ +k_s^2\text{Sin}\theta \dfrac{e^{ik_sR}}{R} + ik_s\text{Sin}\theta \dfrac{e^{ik_sR}}{R^2} - \text{Sin}\theta \dfrac{e^{ik_sR}}{R^3}\end{array}\right]\end{array}\right\}.$$

This then gives

$$W_R = \frac{F(\omega)\text{Cos}\theta}{4\pi\rho}\left[\begin{array}{l} \dfrac{1}{v_p^2}\dfrac{e^{ik_pR}}{R} + \dfrac{2i}{v_p\omega}\dfrac{e^{ik_pR}}{R^2} - \dfrac{2}{\omega^2}\dfrac{e^{ik_pR}}{R^3} \\ \text{NO FIRST ORDER TERM} - \dfrac{2i}{v_s\omega}\dfrac{e^{ik_sR}}{R^2} + \dfrac{2}{\omega^2}\dfrac{e^{ik_sR}}{R^3}\end{array}\right] \tag{7-89}$$

and

$$W_\theta = \frac{F(\omega)\text{Sin}\theta}{4\pi\rho}\left[\begin{array}{l} \text{NO FIRST ORDER TERM} + \dfrac{i}{v_p\omega}\dfrac{e^{ik_pR}}{R^2} - \dfrac{1}{\omega^2}\dfrac{e^{ik_pR}}{R^3} \\ -\dfrac{1}{v_s^2}\dfrac{e^{ik_sR}}{R} - \dfrac{i}{v_s\omega}\dfrac{e^{ik_sR}}{R^2} + \dfrac{1}{\omega^2}\dfrac{e^{ik_sR}}{R^3}\end{array}\right]. \tag{7-90}$$

Again we see that not all P-motion is in the R-direction, nor all S-motion in the θ-direction. It should also be noted that the θ-variation in (7-89) and (7-80) is the same as for (7-57). This happens in every case that I have seen and provides one additional reason for solving problems in two dimensions. The first reason being that the power of the Cagniard-deHoop technique usually allows solution in closed form. Most problems in three dimensions have a final solution containing an integral that cannot be evaluated in closed form.

We see that the terms which fall off more rapidly with distance also fall off more rapidly with increasing frequency. This happens because the dimensionality of each term must be the same. The dimensions of the first order term are

$$\frac{1}{v_p^2 R} = \frac{[T]^2}{[L]^3} ,$$

so for each order of increase in the inverse power of R, we have to lose a power of v_p (or v_s) and substitute a power of ω. Now the theory of Fourier transforms tells us that

$$- \frac{i}{\omega} F(\omega) \Longleftrightarrow \int f(t)dt,$$

so we recognize the second and third order terms as first and second time-integrals of the first order term. This is to be contrasted with the two-dimensional case where only the first time-integral appears.

We have already identified the fundamental impulse response as $\delta(t - R/v)/R$ in (7-73). The first and second time-integrals would then be step- and ramp-functions with appropriate time delays and geometrical divergences. Several alternative time solutions are given by White (1965, Chapter 5), Achenbach (1973, Chapter 3), and Erigen and Suhubi (1975, Chapter 5).

Had we taken the Laplace transform of (7-78) with respect to time, we would have found that

$$\mathbf{W} = \frac{F(s)}{4\pi\rho} \left[\frac{1}{s^2} \nabla\nabla\cdot\left(\mathbf{e}_z \frac{e^{-sR/v_p}}{R} - \mathbf{e}_z \frac{e^{-sR/v_s}}{R}\right) + \mathbf{e}_z \frac{1}{v_s^2} \frac{e^{-sR/v_s}}{R} \right] . \qquad (7\text{-}91)$$

Infinite Media — Point Source of Stress. The same comments pertain here as in the two-dimensional case. Pekeris (1955) gives an example of the three-dimensional stress source in the solution of Lamb's Problem with a surface source.

CHAPTER 8

BOUNDARY CONDITIONS, UNIQUENESS, RECIPROCITY, AND A REPRESENTATION THEOREM

8.1. Boundary Conditions.

Solutions of the elastic wave equation are subject to three types of boundary conditions, occuring when the solid body is in contact with another solid body or a viscous fluid, a perfect fluid, or is bounded by a vacuum.

Solid-Solid Interface (welded) or Solid-Viscous Fluid Interface (no cavitation).

$$\text{SOLID \#1} \qquad \begin{aligned} u_i^{(1)} &= u_i^{(2)} \\ t_{ij}^{(1)} &= t_{ij}^{(2)} \end{aligned} \qquad \text{SOLID \#2} \tag{8-1}$$

Here we have continuity of all stress components inasmuch as we must not have infinite forces (the result of differentiating a discontinuity) at non-singular locations. We also have continuity of all displacement components inasmuch as we cannot have gaps or overlaps. Gaps are prohibited by "welding" the two solids together or by the lack of cavitation in the solid-viscous fluid case. Overlaps cannot occur because two bodies cannot occupy the same space.

Solid-Perfect Fluid Interface (no cavitation).

$$\text{SOLID} \qquad \begin{aligned} u_{\text{normal}}^{(1)} &= u_{\text{normal}}^{(2)} \\ u_{\text{tangential}}^{(1)} & \text{ is unspecified} \\ t_{\text{normal}}^{(1)} &= t_{\text{normal}}^{(2)} \\ t_{\text{tangential}}^{(1)} &= 0 \end{aligned} \qquad \text{PERFECT FLUID} \tag{8-2}$$

Here we have continuity of the normal stress component as before; however, the physics of a perfect fluid require that it cannot sustain shearing stress. Consequently, continuity of stress at the interface requires that the tangential component of stress be zero. The tangential component of fluid motion is unrelated to the tangential component of solid motion. However, we still have enough information to solve for all unknowns.

A variation on the solid-perfect fluid interface is the lubricated crack in an otherwise homogeneous solid. Here we have for each portion of the solid, the solid-fluid boundary conditions. A number of results have been obtained relating to the lubricated crack, but they are somewhat unrealistic from a physical point of view. This is because all fluids have some viscosity, and a very thin lubricating layer would then transmit tangential stress just as though it were not there. Only in a thick fluid layer are the "boundary layer" phenomena in the fluid such that we can consider the fluid as a perfect fluid.

Solid-Vacuum Interface.

$$\text{SOLID} \quad \begin{matrix} u_i^{(1)} \text{ is unspecified} \\ t_{ij}^{(1)} = 0 \end{matrix} \quad \text{VACUUM} \qquad (8\text{-}3)$$

Inasmuch as a vacuum cannot sustain any stress, we have that all components of stress vanish at the free surface of a solid. The displacemnt components are unspecified, but there is again sufficient information to solve for all unknowns.

A more realistic situation is an elastic body bounded by air. However, there is little effect of the air (due mostly to its extremely low density) unless the velocity of propagation for elastic waves is approximately the same as for air. At standard conditions, c_{air} = .34 km/sec, while elastic waves in rock generally travel at velocities greater than 3 km/sec. One case where coupling does occur, however, is in the propagation of dispersive Rayleigh-waves in the weathered zone of the earth's surface. Here the phase velocity of such waves is near the sound velocity in air (Ewing, Jardetzky, and Press, 1957, p. 230).

8.2. Type Boundary Value Problems.

We find four Type boundary value problems in elastic wave theory. The first is a good approximation to physical conditions, while the second most often occurs with the first in a mixed problem. The last two are included inasmuch as a number of elastic wave propagation problems are solvable with these boundary conditions, but not with either of the first two. We shall demonstrate why this happens shortly. In all four cases we wish to find the distribution of stress and displacements on the interior of a region upon whose surface certain boundary conditions are prescribed.

TYPE BOUNDARY VALUE PROBLEMS

a. Type I.

Prescribed: body forces within the interior of the region and surface stresses on the boundary.

b. Type II.

Prescribed: body forces, surface displacements.

c. Type III.

Prescribed: body forces, normal stresses, tangential displacements.

d. Type IV.

Prescribed: body forces, tangential stresses, normal displacements.

There are a limitless number of mixed cases in which the surface of the region is divided into a finite number of subsurfaces, and on each subsurface one of the Type boundary conditions are prescribed. For instance, we may have the stresses equal to zero over most of a body's surface while over the remaining small area the displacement is prescribed as in the case of a punch embossing a surface.

In Section 8.3 we shall show that solutions to these four Type problems are unique. There cannot be a unique solution if only normal stresses and displacements or only tangential stresses and displacements are prescribed.

Let us now take a look at the meaning of these four Type boundary value problems in terms of the elastic potentials introduced in Chapter 6. For simplicity, we will consider only P-SV motion in the two-dimensional case. Here we have

$$u_x = \overset{\text{ODD}}{\frac{\partial\Phi}{\partial x}} - \overset{\text{EVEN}}{\frac{\partial\Psi^\dagger}{\partial z}} \quad \text{in } x, \tag{6-25}$$

$$u_z = \overset{\text{EVEN}}{\frac{\partial\Phi}{\partial z}} + \overset{\text{ODD}}{\frac{\partial\Psi^\dagger}{\partial x}} \quad \text{in } x, \tag{6-27}$$

and

$$t_{xx} = \overset{\text{EVEN}}{\lambda\nabla^2\Phi} + 2\mu\left(\overset{\text{EVEN}}{\frac{\partial^2\Phi}{\partial x^2}} - \overset{\text{ODD}}{\frac{\partial^2\Psi^\dagger}{\partial x\partial z}}\right) \quad \text{in } x, \tag{6-29}$$

$$t_{xz} = \mu\left[2\overset{\text{ODD}}{\frac{\partial^2\Phi}{\partial x\partial z}} + \left(\overset{\text{EVEN}}{\frac{\partial^2\Psi^\dagger}{\partial x^2}} - \overset{\text{EVEN}}{\frac{\partial^2\Psi^\dagger}{\partial z^2}}\right)\right] \quad \text{in } x. \tag{6-33}$$

Now let us look at the Type boundary value problems when we have let the x-axis be normal to the bounding surface and the z-axis lie in the surface as in Figure 8.1.

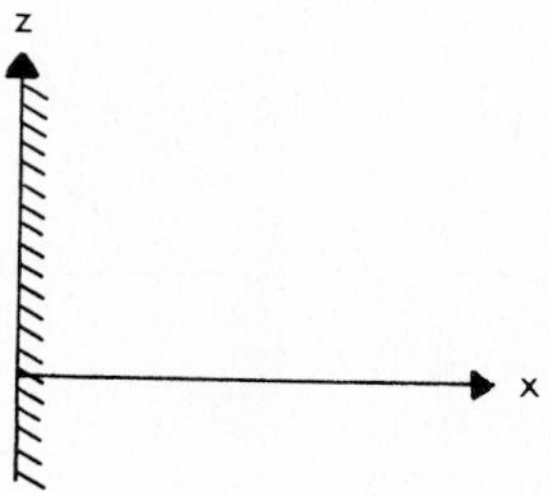

Figure 8-1. Coordinate geometry for discussion of Type boundary value problems.

Let us now consider each Type problem in turn. Type I requires the specification of the two stresses t_{xx} and t_{xz} on the bounding surface. These stresses have mixed parity (evenness or oddness) for both Φ and $\Psi^\dagger$. For Type II, we consider the two displacement components u_x and u_z. Again Φ and $\Psi^\dagger$ appear with mixed parity. For Type III we consider u_z and t_{xx}. In this case we have that Φ appears with even parity and $\Psi^\dagger$ appears with odd parity. Lastly in Type IV, the components u_x and t_{xz} have Φ appearing with odd parity and $\Psi^\dagger$ appearing with odd parity.

Let us now sketch out two physical problems involving a free surface. If we have a force f_x acting on the interior of an unbounded medium, as in Figure 8-2, the resulting displacements are shown schematically with small arrows. We will now add

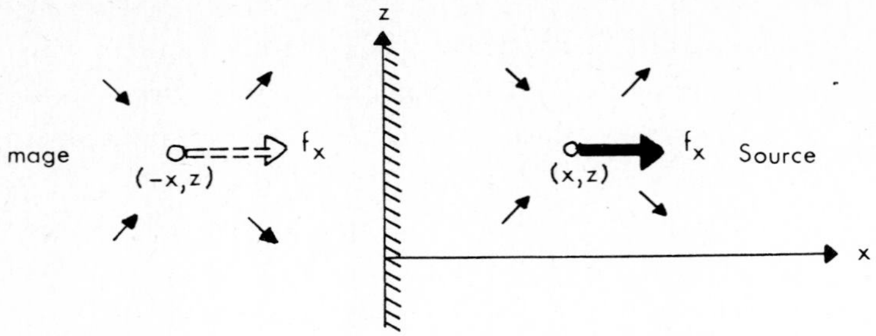

Figure 8-2. X-force and image resulting in Type III or IV boundary conditions.

an image force at (-x,z) with schematic displacements as shown. If we consider as a possible displacement field the sum of the fields of the force and its image, we see that u_z and t_{xx} are odd in x while u_x and t_{xz} are even in x. For a positive image (force in the same direction) we would have u_z and t_{xx} vanishing at the surface $x = 0$, with u_x and t_{xz} taking on non-zero values. For a negative image we would have the reverse. Hence one of these two image pairs will solve a Type III or IV boundary value problem if the prescribed boundary values are zero everywhere.

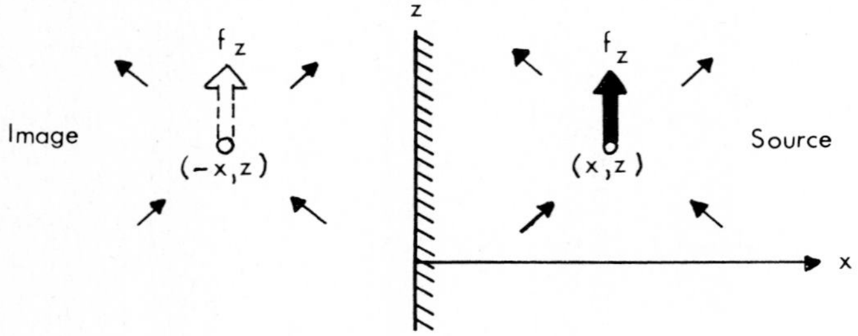

Figure 8-3. Z-force and image resulting in Type III or IV boundary conditions.

Likewise, let us consider the physical problem represented in Figure 8-3. Here we have u_x and t_{xz} odd in x while u_z and t_{xx} are even. Hence u_x and t_{xz} would vanish on the surface $x = 0$ for a positive image pair and u_z and t_{xx} would vanish for a negative image pair.

Inasmuch as the parity is mixed up in Types I and II, simple images will not solve these boundary value problems. The solutions are more complicated.

8.3. Uniqueness.

In Section 4.1, we showed that when the strain energy function exists, then

$$\int_{vol} \frac{dW}{dt} = \int_{vol} (\ddot{u},\dot{u})\rho dV + \int_{vol} t_{ij}\dot{e}_{ji} dV = \int_{vol} \frac{dK}{dt} dV + 2 \int_{vol} \frac{dX}{dt} dV. \quad (4-3)$$

If $\int_{vol} \frac{dW}{dt} dV = 0$, then $\int_{vol} (K + 2X)dV$ must be a constant. If we start from equilibrium, we can set that constant equal to zero. Inasmuch as K is a positive definite function of the velocities u_j, and X is a positive definite function of the strains (4-15), we then have that K = 0 and X = 0 separately. Consequently the only displacement system that can give rise to a zero value of $\int_{vol} \frac{dW}{dt} dV$ must be identically zero.

Let us now consider the existence of two possible solutions of (5-9) for a given set of boundary conditions of Types I to IV. If such is the case, we would have

$$t_{ij,i}\{\mathbf{u}^{(1)}\} - \rho\ddot{u}_j^{(1)} = -f_j, \qquad (8\text{-}4)$$

$$t_{ij,i}\{\mathbf{u}^{(2)}\} - \rho\ddot{u}_j^{(2)} = -f_j.$$

Subtracting, we have

$$t_{ij,i}\{\delta\mathbf{u}\} - \rho\delta\ddot{u}_j = 0.$$

That is, the difference displacement field satisfies the elastic wave equation with zero body forces. If we compute the rate of work being done in achieving the difference displacement configuration, we have (from (4-2))

$$\int_{vol} \frac{dW}{dt} dV = \int_{surf} t_{ij} n_i \delta\dot{u}_j dS, \qquad (8\text{-}5)$$

where $t_{ij} = t_{ij}\{\delta\mathbf{u}\} = t_{ij}^{(1)} - t_{ij}^{(2)}$. However, in Types I and II boundary value problems, we have that either the stress difference or the displacement difference vanishes on the bounding surface. Hence $\int_{vol} \frac{dW}{dt} dV$ must vanish, and we see that $e_{ij}^{(1)}$ must equal $e_{ij}^{(2)}$ and $u_i^{(1)}$ must equal $u_i^{(2)}$ (up to a rigid body displacement in a Type I problem).

For problems of Types III and IV we have that

$$t_{ij}\delta\dot{u}_j = t_{i\,normal}\delta\dot{u}_{normal} + t_{i\,tangential}\delta\dot{u}_{tangential}.$$

But the boundary conditions are such that one member of each term on the right-hand side vanishes. Consequently $\int_{vol} \frac{dW}{dt} dV$ again vanishes, and we must have $u_i^{(1)} = u_i^{(2)}$.

We can see immediately from (8-6) why prescribing $t_{i\,normal}$ and u_{normal} would not give us a unique solution. In this case

$$\int_{vol} \frac{dW}{dt} dV = \int_{surf} t_{i\,tangential}\, n_i \delta\dot{u}_{tangential}\, dS \neq 0,$$

and consequently $X\{e_{ij}^{(1)} - e_{ij}^{(2)}\}$ need not vanish. Then $e_{ij}^{(1)}$ need not equal $e_{ij}^{(2)}$ and we have no unique solution. Similarly for $t_{i\,tangential}$ and $u_{tangential}$.

Eringen and Suhubi (1975, Sec. 5.7) point out that this proof of uniqueness rests on the positive definite nature of X which in turn says something about the nature of the elastic constants. Although they present no physical counter-examples, their proof is somewhat less restrictive. They also consider the case of an unbounded medium and give the mathematical niceties necessary for a more rigorous proof. In the above, and in the analysis to follow, we shall take the more simplistic approach that the domain of interest shall be bounded but large enough so that all phenomena occur without interference with the distant boundary.

8.4. Reciprocity.

Consider the same force acting at two different points in an elastic medium. We would then have for the Fourier transforms of the two equations of motion:

$$T_{ij},_i(\mathbf{U}) + \rho\omega^2 U_j(\mathbf{x}|\mathbf{x}^o;\omega) = -F_j(\omega)\delta(\mathbf{x} - \mathbf{x}^o), \tag{8-6}$$

$$S_{ij},_i(\mathbf{W}) + \rho\omega^2 W_j(\mathbf{x}|\mathbf{x}';\omega) = -F_j(\omega)\delta(\mathbf{x} - \mathbf{x}'). \tag{8-7}$$

In (8-6) the observation point is given by $\mathbf{x}$ and the source point by $\mathbf{x}^o$. In (8-7) the observation point is $\mathbf{x}$ and the source point is $\mathbf{x}'$. If we multiply (8-6) by W_j and (8-7) by U_j, integrate all terms over an arbitrary volume and subtract, we obtain

$$\int_{vol} (W_j T_{ij},_i - U_j S_{ij},_i) dV = - \int_{vol} \left[F_j W_j \delta(\mathbf{x} - \mathbf{x}^o) - F_j U_j \delta(\mathbf{x} - \mathbf{x}') \right] dV$$

$$= -F_j W_j(\mathbf{x}^o|\mathbf{x}';\omega) + F_j U_j(\mathbf{x}'|\mathbf{x}^o;\omega). \tag{8-8}$$

But

$$(W_j T_{ij},_i - U_j S_{ij},_i) = (W_j T_{ij} - U_j S_{ij}),_i - (W_j,_i T_{ij} - U_j,_i S_{ij}). \tag{8-9}$$

Now the second term on the right-hand side of (8-9) becomes

$$\tfrac{1}{2} W_j,_i c_{ijkl}(U_k,_l + U_l,_k) - \tfrac{1}{2} U_j,_i c_{ijkl}(W_k,_l + W_l,_k), \tag{8-10}$$

But $c_{ijkl} = c_{ijlk}$ and (8-10) becomes

$$W_j,_i c_{ijkl} U_l,_k - U_j,_i c_{ijkl} W_l,_k = W_j,_i c_{ijkl} U_l,_k - U_l,_k c_{klij} W_j,_i, \tag{8-11}$$

since the altered subscripts are only dummy ones. We also have

$$c_{klij} = c_{ijkl}, \tag{4-7}$$

and (8-11) becomes

$$W_j,_i c_{ijkl} U_l,_k - U_l,_k c_{ijkl} W_j,_i = 0.$$

Consequently we see that (8-9) reduces to

$$\int_{vol} (W_j T_{ij},_i - U_j S_{ij},_i) dV = \int_{vol} (W_j T_{ij} - U_j S_{ij}),_i dV. \tag{8-12}$$

Using the Green-Gauss Theorem, we can write (8-8) as

$$\int_{surf} (W_j T_{ij} - U_j S_{ij}) n_i dS = -F_j W_j(\mathbf{x}^o|\mathbf{x}';\omega) + F_j U_j(\mathbf{x}'|\mathbf{x}^o;\omega). \tag{8-13}$$

Now the left-hand side of (8-13) can be equal to zero for a number of reasons:

a) fields die off sufficiently rapidly as the surface recedes to infinity,
b) Type I - Type IV boundary conditions are prescribed equal to zero. If such is the case we then have

$$F_j(\mathbf{x}',\omega) U_j(\mathbf{x}'|\mathbf{x}^o;\omega) = F_j(\mathbf{x}^o,\omega) W_j(\mathbf{x}^o|\mathbf{x}';\omega). \tag{8-14}$$

However, since the variable $\underline{\omega}$ does not appear explicitly anywhere, we can write

$$f_j(\mathbf{x}',t)u_j(\mathbf{x}'|\mathbf{x}^o;t) = f_j(\mathbf{x}^o,t)w_j(\mathbf{x}^o|\mathbf{x}';t). \qquad (8\text{-}15)$$

Let us now introduce a Green's tensor for elastic media by

$$u_j(\mathbf{x}'|\mathbf{x}^o;t) = f_i(\mathbf{x}^o,t)g_{ij}(\mathbf{x}'|\mathbf{x}^o;t), \qquad (8\text{-}16)$$

where $g_{ij}(\mathbf{x}'|\mathbf{x}^o;t)$ is the $\underline{j}$-component of displacement at the position $\mathbf{x}'$ due to the $\underline{i}$-component of force acting at the position $\mathbf{x}^o$. We then have

$$f_j(\mathbf{x}',t)f_i(\mathbf{x}^o,t)g_{ij}(\mathbf{x}'|\mathbf{x}^o;t) = f_j(\mathbf{x}^o,t)f_i(\mathbf{x}',t)g_{ij}(\mathbf{x}^o|\mathbf{x}';t).$$

if the forces f_j are momentarily restricted to lie along each of the cartesian axes in turn, we see that

$$g_{ij}(\mathbf{x}'|\mathbf{x}^o;t) = g_{ji}(\mathbf{x}^o|\mathbf{x}';t). \qquad (8\text{-}17)$$

It should be noted that in some cases at least,

$$g_{ij}(\mathbf{x}'|\mathbf{x}^o;t) \neq g_{ij}(\mathbf{x}^o|\mathbf{x}';t). \qquad (8\text{-}18)$$

What this says is the following: If an arbitrary force $\mathbf{f}(t)$ is acting in the i-direction at a point $\mathbf{x}'$ and gives a displacement $\mathbf{u}(t)$ in the j-direction at a point $\mathbf{x}^o$, then the same force $\mathbf{f}(t)$ acting in the $\underline{j}$-direction at point $\mathbf{x}^o$ will give the same displacement $\mathbf{u}(t)$ in the i-direction at the point $\mathbf{x}'$.

This relationship holds for an elastic medium with arbitrary boundaries, inhomogeneity, and anisotropy. In special cases, similar relationships hold for the interchange of force and dilatation, torque and rotation etc., but apparently no general relationship other than (8-17) can be proved. Some examples are shown in Figure 8-4 below. Knopoff and Gangi (1959) show some nice seismic model experiments demonstrating reciprocity.

As with the case of uniqueness, additional details are given by Eringen and Suhubi (1975, Sec. 5.8). An even more extended treatment may be found in Wheeler and Sternberg (1968). The latter reference gives an extensive bibliography of early work and rather more mathematical treatments of questions of uniqueness, reciprocity, and the derivation of representation theorems.

If we consider the motion at one point due to a force acting at another point in an infinite medium, we find that further conditions apply. In cartesian coordinates we had

$$\mathbf{W} = -\frac{F(\omega)}{4\pi\rho\omega^2}\left[\nabla\nabla\cdot(\mathbf{e}_z\frac{e^{ik_pR}}{R} - \mathbf{e}_z\frac{e^{ik_sR}}{R}) - \mathbf{e}_z k_s^2\frac{e^{ik_sR}}{R}\right]. \qquad (7\text{-}88)$$

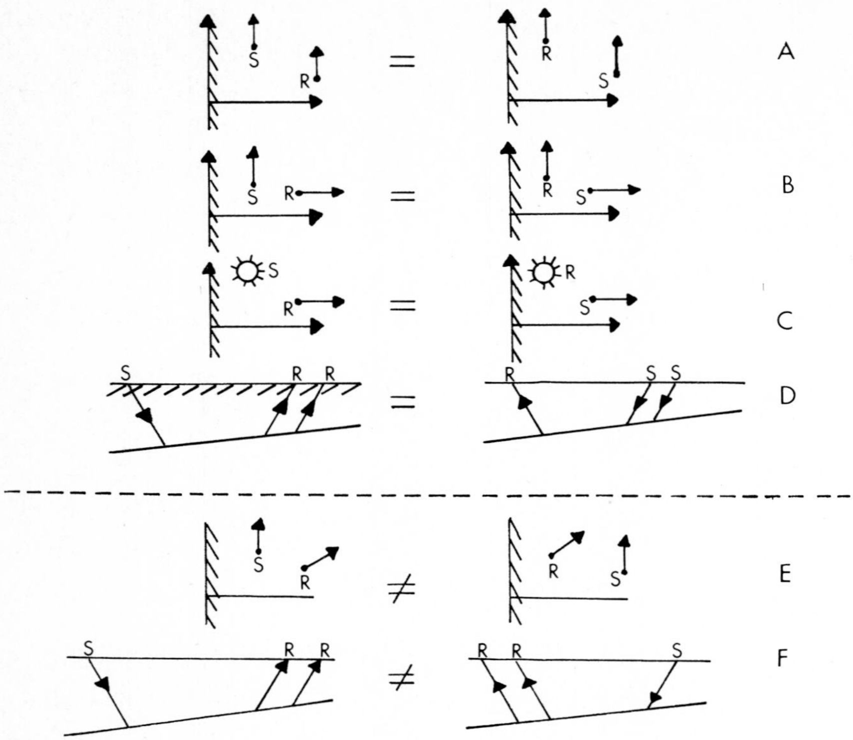

Figure 8-4. Examples of reciprocal and non-reciprocal situations. S = source, R = receiver, particle motion given by arrows, ray-paths by solid lines. In case C, the sun-burst figure represents an explosive source or a pressure receiver. General Rule: KEEP THE SAME DIAGRAM, RE-LABEL S AND R (reverse ray-path directions if appropriate).

If we let the force be directed along each of the coordinate axes in turn, and remembering (8-16), we can write

$$W_j = F_i(\omega)\Gamma_{ij}, \tag{8-19}$$

where

$$\Gamma_{ij} = -\frac{1}{4\pi\rho\omega^2}\left[\frac{\partial}{\partial x_j}\frac{\partial}{\partial x_i}\left(\frac{e^{ik_pR}}{R} - \frac{e^{ik_sR}}{R}\right) - k_s^2\frac{e^{ik_sR}}{R}\delta_{ij}\right], \tag{8-20}$$

is the <u>infinite-space Green's tensor for elastic waves</u>. But (8-20) is obviously symmetric in the subscripts <u>i</u> and <u>j</u>. Consequently we have

$$\Gamma_{ij}(\mathbf{x}'|\mathbf{x}^o;\omega) = \Gamma_{ji}(\mathbf{x}'|\mathbf{x}^o;\omega). \tag{8-21}$$

Remember, this is a <u>special case</u>. If we now apply (8-17) we can write

$$\Gamma_{ij}(\mathbf{x}'|\mathbf{x}^{o};\omega) = \Gamma_{ji}(\mathbf{x}^{o}|\mathbf{x}';\omega) = \Gamma_{ij}(\mathbf{x}^{o}|\mathbf{x}';\omega), \tag{8-22}$$

and we see that the infinite-space Green's tensor is symmetric with respect to interchanges of both its subscripts and its arguments as well as obeying the more general law of reciprocity (8-17). One must be careful in using this symmetry in orthogonal curvilinear coordinate systems. The component interchange has to be investigated in more detail. However, the argument interchange is independent of the coordinate system.

8.5. A Representation Theorem.

We want to find the solution of

$$T_{ij,i}(\mathbf{U}) + \rho\omega^{2}U_{j}(\mathbf{x},\omega) = -F_{j}(\mathbf{x},\omega), \tag{8-23}$$

in a general region. To do this, we need to invoke the solution of

$$S_{ij,i}(\mathbf{W}) + \rho\omega^{2}W_{j}(\mathbf{x}|\mathbf{x}^{o};\omega) = -e_{j}\delta(\mathbf{x} - \mathbf{x}^{o}), \tag{8-24}$$

where S_{ij} is the stress tensor associated with the displacement field W_j and e_j is a unit vector in the j-direction. If we multiply (8-23) by W_j and (8-24) by U_j, subtract and integrate over the volume of the region, we have

$$\int_{vol}(W_{j}T_{ij,i} - U_{j}S_{ij,i})dV = e_{j}U_{j}(\mathbf{x}^{o},\omega) - \int_{vol}F_{j}(\mathbf{x},\omega)W_{j}(\mathbf{x}|\mathbf{x}^{o};\omega)dV. \tag{8-25}$$

Using the reasoning leading up to (8-13), this becomes

$$e_{j}U_{j}(\mathbf{x}^{o},\omega) = \int_{vol}F_{j}(\mathbf{x},\omega)W_{j}(\mathbf{x}|\mathbf{x}^{o};\omega)dV + \int_{surf}(W_{j}T_{ij} - U_{j}S_{ij})n_{i}dS. \tag{8-26}$$

Let us now consider the representation of W_j in terms of the Green's tensor for the region, i.e.,

$$W_{k} = e_{j}G_{jk}(\mathbf{x}|\mathbf{x}^{o};\omega). \tag{8-27}$$

The stresses associated with W_k are (using (4-10))

$$S_{kl} = e_{j}R_{jkl} = e_{j}c_{klpq}g_{jp,q} \tag{8-28}$$

$$\left[= e_{j}\{\lambda G_{jp,p}\delta_{kl} + \mu(G_{jk,l} + G_{jl,k})\},\right]$$

where the bracketed form holds for isotropic, homogeneous media. If we put these two representations into (8-26), we obtain

$$e_{j}U_{j} = e_{j}\int_{vol}F_{k}G_{jk}dV + e_{j}\int_{surf}(G_{jl}T_{kl} - U_{l}R_{jkl})n_{k}dS. \tag{8-29}$$

Now the unit vector e_j is completely arbitrary, so we may drop it leaving

$$U_{j}(\mathbf{x}^{o},\omega) = \int_{vol}F_{k}(\mathbf{x},\omega)G_{jk}(\mathbf{x}|\mathbf{x}^{o};\omega)dV + \int_{surf}(G_{jl}T_{kl} - U_{l}R_{jkl})n_{k}dS. \tag{8-30}$$

In the form (8-30), the arguments of the Green's tensor $G_{jk}(\mathbf{x}|\mathbf{x}^{o};\omega)$ and its associated stress function R_{jkl} are reversed. That is, the integration is over the first (observation point) argument. We can use (8-17) to reverse the arguments, but the form for $R_{jkl} = c_{klpq}G_{jp',q}$ needs further interpretation. We can write

$$\begin{aligned} U_l n_k c_{klpq} G_{jp',q}(\mathbf{x}|\mathbf{x}^o;\omega) &= G_{jp',q}(\mathbf{x}|\mathbf{x}^o;\omega)c_{pqkl}U_l n_k \\ &= G_{pj',q}(\mathbf{x}^o|\mathbf{x};\omega)c_{pqkl}U_l n_k \\ &= G_{pj',q}(\mathbf{x}^o|\mathbf{x};\omega)C_{pq}(\mathbf{x},\omega), \end{aligned} \tag{8-31}$$

where

$$C_{pq} = C_{qp} \equiv c_{pqkl}U_l n_k \tag{8-32}$$
$$\left[= \lambda U_m n_m \delta_{pq} + \mu(U_p n_q + U_q n_p). \right]$$

The bracketed form holds for an isotropic, homogeneous medium. In the first step of (8-31) we have used the fact that $c_{klpq} = c_{pqkl}$ (from (4-7)); in the second we have reversed arguments and subscripts of the Green's function; and in the third we have defined the new quantity C_{pq} which is of a form similar to the stress form, but composed of displacement elements and the normal direction cosines. We can then write:

$$\begin{aligned} U_j(\mathbf{x}^o,\omega) = & \int_{vol} F_k(\mathbf{x},\omega)G_{kj}(\mathbf{x}^o|\mathbf{x};\omega)dV \\ & + \int_{surf} T_{kl}(\mathbf{x},\omega)G_{lj}(\mathbf{x}^o|\mathbf{x};\omega)n_k dS \\ & - \int_{surf} C_{kl}(\mathbf{x},\omega)G_{lj',k}(\mathbf{x}^o|\mathbf{x};\omega)\,dS. \end{aligned} \tag{8-33}$$

In this formulation, the arguments and subscripts are in the right order, i.e., the integration is now over the second argument (source point). The first integral on the right hand side is the volume integral of the forces over the region weighted by the tensor Green's function which is the j-component of motion due to a point force in the k-direction. The second is the surface integral of the stress vector ($T_{kl}n_k dS$) again weighted by this tensor Green's function. The last integral involves the factor $C_{kl}dS$ weighted by a spatial derivative of the Green's tensor. It will be shown in Chapter 28 that $C_{kl}dS$ can be interpreted as a local couple. Achenbach (1973, Chapter 3) and Eringen and Suhubi (1975, Chapter 5) give additional details. Burridge and Knopoff (1964) present a formulation of the representation theorem which is particularly suitable for physical interpretation of seismic source mechanisms. Gangi (1970) derives the representation theorem starting with a general reciprocity relationship.

Equation (8-30) is useful in several cases. If we are solving boundary value problems of Type I, and have found the tensor Green's function satisfying zero stresses on the surface of the region, then (8-30) becomes

$$U_j = \int_{vol} F_k G_{kj} dV + \int_{surf} T_{kl} G_{lj} n_k dS, \tag{8-34}$$

and the problem is reduced to the evaluation of the integrals appearing in (8-34).

Likewise, if we have a boundary value problem of Type III, we should find the tensor Green's function satisfying boundary conditions of vanishing tangential displacement and normal stress. In this case (8-30) becomes

$$U_j = \int_{vol} F_k g_{kj} dV + \int_{surf} T_{k\ normal} G_{normal\ j} n_k dS - \int_{surf} U_{tan} R_{jk\ tan} n_k dS. \quad (8\text{-}35)$$

In (8-34) and (8-35) the tensor Green's function and its associated stresses assume the role of an "influence function" or "weighting function" in a superposition of all forces, stresses and diplacements acting. However, in most instances, it is quite hard to find the proper tensor Green's function, so carrying the problem as far as (8-30) is usually the less difficult step.

A second use for (8-30) appears in the approximation of solutions for which exact answers cannot be obtained. For instance in a boundary value problem of Type I with zero stresses on the surface of the region, the second integral would vanish identically leaving

$$U_j = \int_{vol} F_k G_{kj} dV - \int_{surf} U_l R_{jkl} n_k dS. \quad (8\text{-}36)$$

This give U_j everwhere in terms of the values of U_l on the surface of the region and of forces acting. If one can come up with a reasonable value for U_l on the surface, this can be substituted into the surface integral giving a correction to the assumed value of U_l. In theory, this can be repeated, and if such a sequence converges (it had better do so usefully) it will probably lead to the right answer. The hesitation here comes from the non-rigorous development of (8-30). Sequences leading to wrong answers are usually so wrong, they are obviously incorrect. A number of scattering problems using this type of solution will be found in Appendix C.

CHAPTER 9

PLANE WAVES INCIDENT UPON A PLANE FREE SURFACE

9.1. Plane Wave Solutions of the Scalar Wave Equation.

First of all, we shall consider only two dimensions, inasmuch as the y-axis can be oriented along the intersection of the wave front with the plane free surface. Then x gives the lateral distance along the surface and z gives the depth. The z-direction is usually taken positive downwards for convenience.

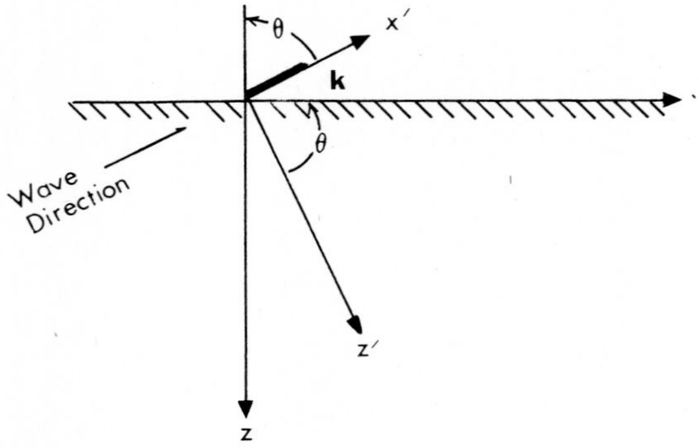

Figure 9-1. Plane wave coordinate geometry.

When a plane P-wave is traveling in the x'-direction, it can be written as

$$\Phi = e^{i(k_p x' - \omega t)}, \tag{9-1}$$

where $k_p = \omega/v_p$. But

$$x' = x\mathrm{Sin}\theta - z\mathrm{Cos}\theta. \tag{9-2}$$

So we can write (9-1) in rotated coordinates as

$$\Phi = e^{i\{k_p(x\mathrm{Sin}\theta - z\mathrm{Cos}\theta) - \omega t\}}. \tag{9-3}$$

For the remainder of this chapter, we will always have the same time factor $e^{-i\omega t}$, so we will suppress it and represent our plane wave solution as

$$\Phi = e^{ik_p(x\mathrm{Sin}\theta - z\mathrm{Cos}\theta)}. \tag{9-4}$$

Note that along the planes of constant phase (of the exponential) the amplitude is constant. We shall soon meet another solution of the scalar wave equation where this is not true.

9.2. P-waves Incident Upon a Free Surface.

We shall take as our representation of the incident wave

$$\Phi_i = (1.0)e^{ik_p(x\mathrm{Sin}\theta_o - z\mathrm{Cos}\theta_o)}, \tag{9-5}$$

of the reflected P-wave

$$\Phi_r = R_{pp} e^{ik_p(x\mathrm{Sin}\theta_p + z\mathrm{Cos}\theta_p)}, \tag{9-6}$$

and of the reflected S-wave

$$\Psi_r^{\dagger} = R_{ps} e^{ik_s(x\mathrm{Sin}\theta_s + z\mathrm{Cos}\theta_s)}. \tag{9-7}$$

The coordinates are as in Figure 9-2.

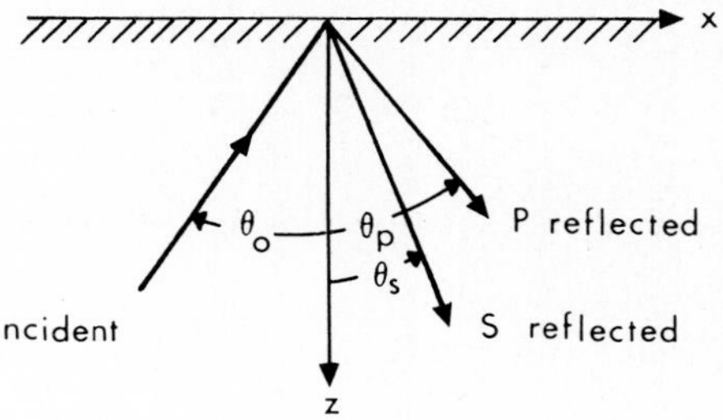

Figure 9-2. Coordinate geometry — incident P-waves.

When a plane P-wave strikes a free surface, it is necessary that there be a reflected S-wave as well as the expected P-wave. This could be demonstrated directly by showing that incident and reflected P-waves will not satisfy the free-surface boundary conditions. However, we will proceed by showing that the reflection coefficient for S-waves is non-zero. The uniqueness theorem then says that S-waves must be present.

The appropriate two-dimensional displacement vector is given by

$$\mathbf{u} = \nabla\Phi + \nabla\times\mathbf{e}_y\Psi_r^{\dagger} = \nabla\Phi + \frac{1}{ik_s}\nabla\times\mathbf{e}_y\frac{\partial\Psi}{\partial x}; \tag{6-28}$$

or

$$u_x = \frac{\partial\Phi}{\partial x} - \frac{\partial\Psi_r^{\dagger}}{\partial z}, \quad u_z = \frac{\partial\Phi}{\partial z} + \frac{\partial\Psi_r^{\dagger}}{\partial x};$$

where $\Phi = \Phi_i + \Phi_r$ and we have set "ℓ" equal to $1/ik_s$ for convenience. We also need the stresses on planes of z = constant. From (6-31) we have

$$\frac{t_{zz}}{\mu} = -k_s^2\Phi + 2\left(\frac{\partial^2\Psi^{\dagger}}{\partial x\partial z} - \frac{\partial^2\Phi}{\partial x^2}\right), \tag{9-8}$$

where we have used the relationship

$$\frac{1}{v_s^2}\frac{\partial^2\Phi}{\partial t^2} = -\frac{\omega^2}{v_s^2}\Phi = -k_s^2\Phi.$$

From (6-33) we have

$$\frac{t_{zx}}{\mu} = 2\frac{\partial^2\Phi}{\partial z\partial x} + \frac{\partial^2\Psi^{\dagger}}{\partial x^2} - \frac{\partial^2\Psi^{\dagger}}{\partial z^2}. \tag{9-9}$$

If we are to have any solution at all, the functional forms of (9-5) - (9-7) must match along the free surface z = 0. This means that

$$k_p \operatorname{Sin}\theta_o \text{ (must) } = k_p \operatorname{Sin}\theta_p \text{ (must) } = k_s \operatorname{Sin}\theta_s \ , \qquad (9\text{-}10)$$

or in terms of velocities

$$\frac{\operatorname{Sin}\theta_o}{v_p} = \frac{\operatorname{Sin}\theta_p}{v_p} = \frac{\operatorname{Sin}\theta_s}{v_s} \ . \qquad (9\text{-}11)$$

This is simply Snell's Law, derived in this case from a consideration of the particular form of the wave solution given by (9-1). Henceforth, we will let $\theta_o = \theta_p$.

We now apply the boundary conditions of zero stress at the free surfaces to the wave field given by (9-5) - (9-7). On evaluating at z = 0 and suppressing the x factors (which are all the same because of (9-10)), the following equations must hold:

$$-k_s^2(1 + R_{pp}) + 2k_p^2\operatorname{Sin}^2\theta_p(1 + R_{pp}) - 2k_s^2\operatorname{Sin}\theta_s\operatorname{Cos}\theta_s R_{ps} = 0,$$

$$2k_p^2\operatorname{Sin}\theta_p\operatorname{Cos}\theta_p(1 - R_{pp}) - k_s^2\operatorname{Sin}^2\theta_s R_{ps} + k_s^2\operatorname{Cos}^2\theta_s R_{ps} = 0.$$

Regrouping, we have

$$R_{pp}\left[2k_p^2\operatorname{Sin}^2\theta_p - k_s^2\right] + R_{ps}\left[-2k_s^2\operatorname{Sin}\theta_s\operatorname{Cos}\theta_s\right] = -(2k_p^2\operatorname{Sin}^2\theta_p - k_s^2),$$

$$R_{pp}\left[-2k_p^2\operatorname{Sin}\theta_p\operatorname{Cos}\theta_p\right] + R_{ps}\left[k_s^2(\operatorname{Cos}^2\theta_s - \operatorname{Sin}^2\theta_s)\right] = -2k_p^2\operatorname{Sin}\theta_p\operatorname{Cos}\theta_p. \qquad (9\text{-}12)$$

Let us now divide these through by k_s^2 and set

$$k_p^2/k_s^2 = v_s^2/v_p^2 = a^2. \qquad (9\text{-}13)$$

Note that $a^2<1$. With a liberal use of the relation (9-10) in the form

$$\operatorname{Sin}\theta_s = a\operatorname{Sin}\theta_p, \qquad \operatorname{Cos}\theta_s = (1 - a^2\operatorname{Sin}^2\theta_p)^{\frac{1}{2}}, \qquad (9\text{-}14)$$

(9-12) can be rewritten as

$$R_{pp}\left[-1 + 2a^2\operatorname{Sin}^2\theta_p\right] + R_{ps}\left[-2a\operatorname{Sin}\theta_p(1 - a^2\operatorname{Sin}^2\theta_p)^{\frac{1}{2}}\right] = 1 - 2a^2\operatorname{Sin}^2\theta_p,$$

and

$$R_{pp}\left[-2a^2\operatorname{Sin}\theta_p\operatorname{Cos}\theta_p\right] + R_{ps}\left[1 - 2a^2\operatorname{Sin}^2\theta_p\right] = -2a^2\operatorname{Sin}\theta_p\operatorname{Cos}\theta_p \ . \qquad (9\text{-}15)$$

Solving we find

$$R_{pp} = -\frac{(1 - 2a^2\operatorname{Sin}^2\theta_p)^2 - 4a^3\operatorname{Sin}^2\theta_p\operatorname{Cos}\theta_p(1 - a^2\operatorname{Sin}^2\theta_p)^{\frac{1}{2}}}{(1 - 2a^2\operatorname{Sin}^2\theta_p)^2 + 4a^3\operatorname{Sin}^2\theta_p\operatorname{Cos}\theta_p(1 - a^2\operatorname{Sin}^2\theta_p)^{\frac{1}{2}}} \qquad (9\text{-}16)$$

and

$$R_{ps} = -\frac{4a^2\operatorname{Sin}\theta_p\operatorname{Cos}\theta_p(1 - 2a^2\operatorname{Sin}^2\theta_p)}{(1 - 2a^2\operatorname{Sin}^2\theta_p)^2 + 4a^3\operatorname{Sin}^2\theta_p\operatorname{Cos}\theta_p(1 - a^2\operatorname{Sin}^2\theta_p)^{\frac{1}{2}}} \ . \qquad (9\text{-}17)$$

Note that the reflection coefficients are independent of frequency since there is no characteristic length involved. This proves to be extremely useful in that the shape of an impulsive wave-form is not altered for reflection at less than the critical angle. For reflections beyond the critical angle, the wave form can change shape due to phase shifts. Section 9.5 will discuss this in more detail.

If we look at the rate of energy transport we can check these results as well as demonstrate the meaning of these coefficients. The rate of energy transport is given as the energy per unit volume (twice the Kinetic Energy) times the velocity of propagation times the cross-sectional area of a bundle of rays. This can be seen in Figure 9-3, where the cross sectional area is proportional to Cosθ.

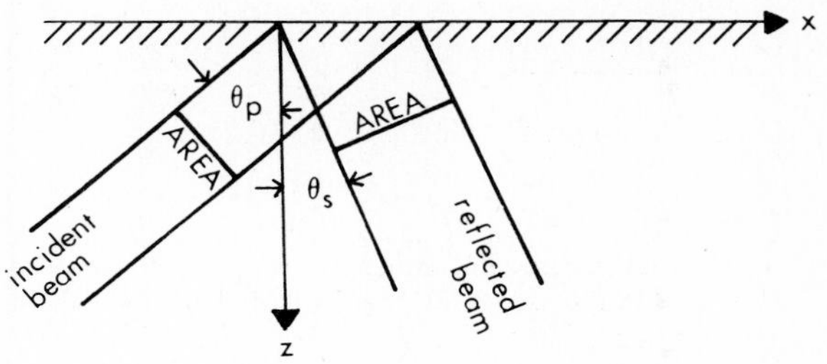

Figure 9-3. Geometric relationships for energy transport calculation.

In the case of incident P-waves we have ("*" stands for the complex conjugate):

$$2K.E. = \rho(\dot{u}_x^p \dot{u}_x^{p*} + \dot{u}_z^p \dot{u}_z^{p*})$$

$$= \rho\omega^2 k_p^2 (\mathrm{Sin}^2\theta_p + \mathrm{Cos}^2\theta_p)\, 1^2 = \rho\omega^2 k_p^2 1^2 \quad . \tag{9-18}$$

For conservation of energy flux, we then have

$$1 k_p^2 v_p \mathrm{Cos}\theta_p = |R_{pp}|^2 k_p^2 v_p \mathrm{Cos}\theta_p + |R_{ps}|^2 k_s^2 v_s \mathrm{Cos}\theta_s,$$

or

$$1 = |R_{pp}|^2 + \frac{1}{a}\frac{\mathrm{Cos}\theta_s}{\mathrm{Cos}\theta_p}|R_{ps}|^2 = |R_{pp}|^2 + \frac{\mathrm{Tan}\theta_p}{\mathrm{Tan}\theta_s}|R_{ps}|^2. \tag{9-19}$$

The student can demonstrate to himself that (9-16) and (9-17) satisfy (9-19) by direct calculation. A sketch of the reflection relationships for P-waves incident upon a free surface is given in Figure 9-4.

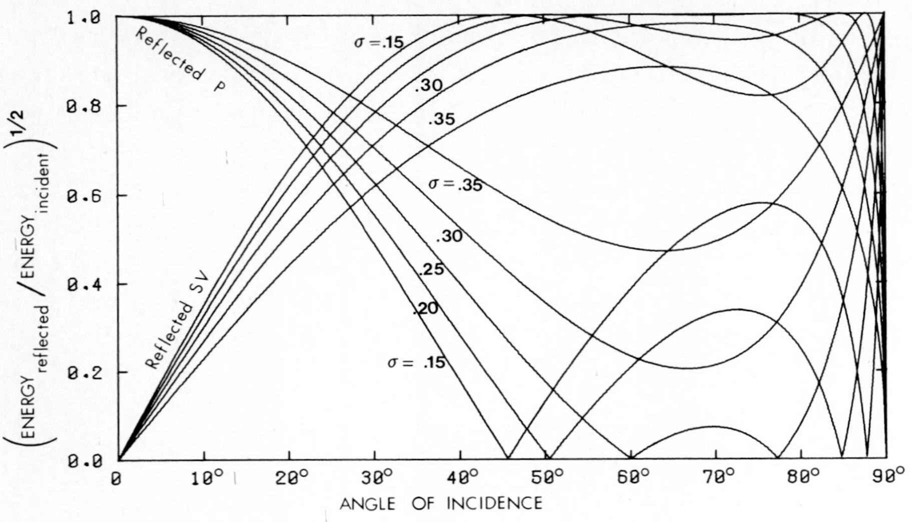

Figure 9-4. Square root of energy ratios for reflected P- and S-waves with an incident P-wave. Poisson's ratio σ varies from 0.15 to 0.35.

9.3. SV-waves Incident Upon a Free Surface

For this case we represent the wave fields by

$$\Phi = R_{sp} e^{ik_p (x\mathrm{Sin}\theta_p + z\mathrm{Cos}\theta_p)}, \tag{9-20}$$

$$\Psi^{\dagger} = (1.0) e^{ik_s (x\mathrm{Sin}\theta_s - z\mathrm{Cos}\theta_s)} + R_{ss} e^{ik_s (x\mathrm{Sin}\theta_s + z\mathrm{Cos}\theta_s)}. \tag{9-21}$$

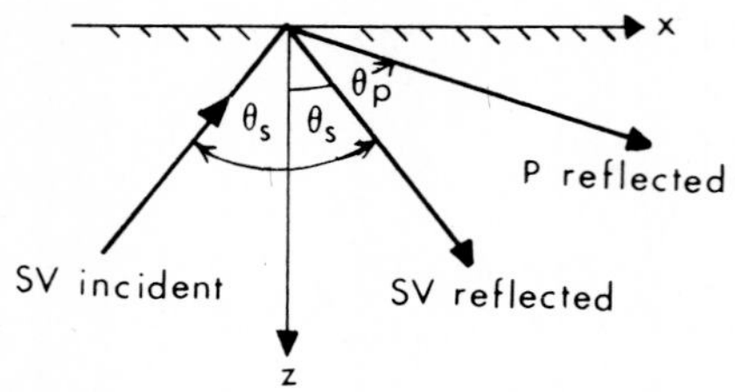

Figure 9-5. Coordinate geometry — incident SV-waves.

The application of the boundary conditions of zero stress leads to the two determining equations:

$$-k_s^2 R_{sp} + 2k_p^2 \mathrm{Sin}^2\theta_p R_{sp} + 2k_s^2 \mathrm{Sin}\theta_s \mathrm{Cos}\theta_s (1 - R_{ss}) = 0,$$

$$-2k_p^2 \mathrm{Sin}\theta_p \mathrm{Cos}\theta_p R_{sp} - k_s^2 \mathrm{Sin}^2\theta_s (1 + R_{ss}) + k_s^2 \mathrm{Cos}^2\theta_s (1 + R_{ss}) = 0.$$

Regrouping and writing in terms of θ_s and noting that $\mathrm{Cos}\theta_p = (1 - \mathrm{Sin}^2\theta_s/a^2)^{\frac{1}{2}}$, we are led to the pair

$$R_{sp}\left[-1 + 2\mathrm{Sin}^2\theta_s\right] + R_{ss}\left[-2\mathrm{Sin}\theta_s \mathrm{Cos}\theta_s\right] = -2\mathrm{Sin}\theta_s \mathrm{Cos}\theta_s,$$

$$R_{sp}\left[-2\mathrm{Sin}\theta_s (a^2 - \mathrm{Sin}^2\theta_s)^{\frac{1}{2}}\right] + R_{ss}\left[1 - 2\mathrm{Sin}^2\theta_s\right] = -1 + 2\mathrm{Sin}^2\theta_s. \tag{9-22}$$

Solving we find that

$$R_{sp} = \frac{4\mathrm{Sin}\theta_s \mathrm{Cos}\theta_s (1 - 2\mathrm{Sin}^2\theta_s)}{(1 - 2\mathrm{Sin}^2\theta_s)^2 + 4\mathrm{Sin}^2\theta_s \mathrm{Cos}\theta_s (a^2 - \mathrm{Sin}^2\theta_s)^{\frac{1}{2}}}, \tag{9-23}$$

$$R_{ss} = \frac{(1 - 2\mathrm{Sin}^2\theta_s)^2 - 4\mathrm{Sin}^2\theta_s \mathrm{Cos}\theta_s (a^2 - \mathrm{Sin}^2\theta_s)^{\frac{1}{2}}}{(1 - 2\mathrm{Sin}^2\theta_s)^2 + 4\mathrm{Sin}^2\theta_s \mathrm{Cos}\theta_s (a^2 - \mathrm{Sin}^2\theta_s)^{\frac{1}{2}}}. \tag{9-24}$$

The energy relation in this case turns out to be

$$1 = |R_{ss}|^2 + \frac{\mathrm{Tan}\theta_s}{\mathrm{Tan}\theta_p}|R_{sp}|^2. \tag{9-25}$$

However, this case of incident SV requires futher discussion. At an angle of

$$\theta_s = \mathrm{Sin}^{-1} a = \theta_c , \tag{9-26}$$

we find that $\theta_p = 90^\circ$. At this angle the factor $(a^2 - \mathrm{Sin}^2\theta_s)^{\frac{1}{2}} = a\mathrm{Cos}\theta_p$ becomes zero, and for $\theta_s > \theta_c$ this factor must become imaginary, but the sign must be chosen. Looking back at (9-21)

$$\Phi = e^{ik_p(x\mathrm{Sin}\theta_p + z\mathrm{Cos}\theta_p)},$$

we see that since the solution must remain bounded for all k_p and positive z, then the sign must be chosen so that

$$(a^2 - \mathrm{Sin}^2\theta_s)^{\frac{1}{2}} \longrightarrow i(\mathrm{Sin}^2\theta_s - a^2)^{\frac{1}{2}} . \tag{9-27}$$

This angle is called the critical angle and it is the angle of incidence of the SV-wave at which the P-wave first becomes refracted along the free surface of the elastic medium. Equation (9-27) means that there is a phase-shift of magnitude

$$\theta_{ss} = -2\mathrm{Tan}^{-1}\left[\frac{4\mathrm{Sin}^2\theta_s \mathrm{Cos}\theta_s (\mathrm{Sin}^2\theta_s - a^2)^{\frac{1}{2}}}{(1 - 2\mathrm{Sin}^2\theta_s)^2}\right] \tag{9-28}$$

associated with the totally reflected S-wave. The relationships given by (9-23), (9-24), and (9-28) are shown graphically by Figure 9-6. The significance of these phase shifts will be discussed in Section 9.5.

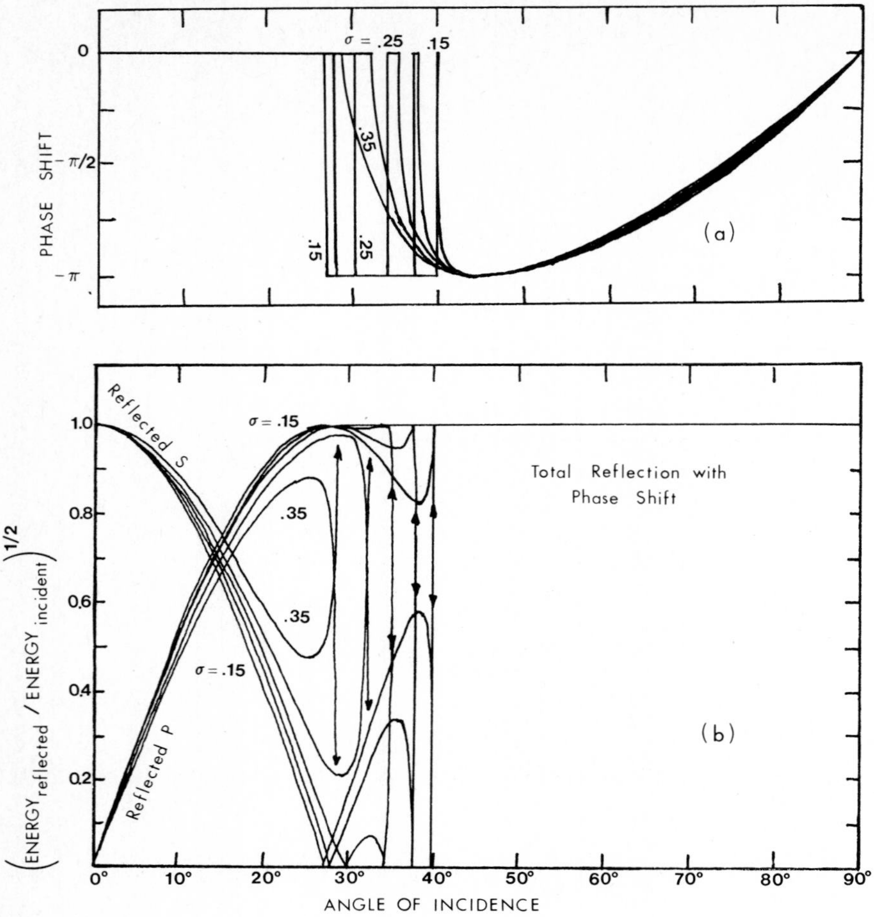

Figure 9-6(a). Phase shifts of reflected S-waves. Note that all phase curves go through $-\pi$ at $\theta_s = 45^o$.

(b). Square root of energy ratio for reflected P- and S-waves with an incident S-wave. Poisson's ratio varies from 0.15 to 0.35.

9.4. SH-Waves Incident Upon a Free Surface.

Here it will be found that we need only one reflected SH-wave, and so we describe the wave field by

$$\Lambda = (1.0)e^{ik_s(x\text{Sin}\theta_s - z\text{Cos}\theta_s)} + R_{ss}e^{ik_s(x\text{Sin}\theta_s + z\text{Cos}\theta_s)} \quad . \tag{9-29}$$

However, in this case we have

$$\mathbf{u} = \nabla\times\mathbf{e}_z\Lambda = -\mathbf{e}_y\frac{\partial\Lambda}{\partial x} \quad . \tag{6-28}$$

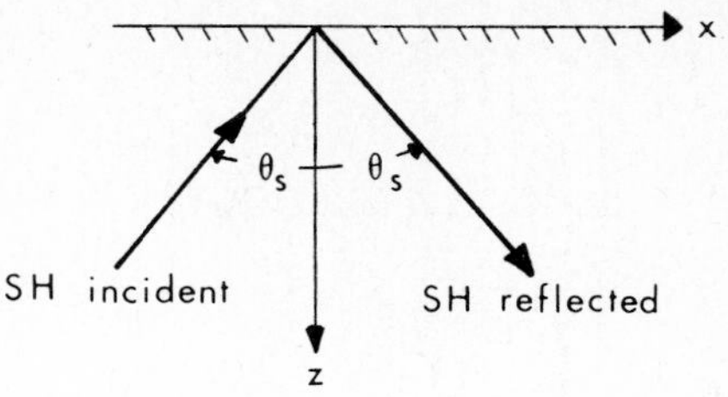

Figure 9-7. Coordinate geometry — incident SH-waves.

Hence we see that

$$t_{zz}/\mu = (\lambda/\mu)\nabla\cdot\mathbf{u} + \frac{\partial u_z}{\partial z} \equiv 0 \quad ; \quad t_{xz}/\mu = \frac{\partial u_x}{\partial z} + \frac{\partial u_z}{\partial x} \equiv 0 \quad ; \tag{9-30}$$

and the only stress component not identically zero is

$$t_{yz}/\mu = \frac{\partial u_y}{\partial z} = -\frac{\partial^2\Lambda}{\partial x\partial z} . \tag{9-31}$$

Setting t_{yz} equal to zero at the surface $z = 0$ requires that

$$-k_s^2 \mathrm{Sin}\theta_s \mathrm{Cos}\theta_s (1 - R_{ss}) = 0 ,$$

i.e.,

$$R_{ss} = 1 \tag{9-32}$$

for all angles of incidence.

9.5. Phase Shifts, Allied Functions, and Hilbert Transforms.

To determine the effect of the phase shift upon the propagation of a plane wave as in (9-28), we need to look more closely at some of the properties of the Fourier Transform. Our starting point will be the Fourier Integral Theorem:

$$f(t) = \frac{1}{2\pi}\int_{-\infty}^{\infty} d\omega \int_{-\infty}^{\infty} f(t')e^{i\omega(t'-t)} dt' . \tag{9-33}$$

For "nice" $f(t')$ we can reverse the order of integration giving

$$f(t) = \frac{1}{2\pi}\int_{-\infty}^{\infty} f(t')dt' \int_{-\infty}^{\infty} e^{i\omega(t'-t)} d\omega = \frac{1}{\pi}\int_{-\infty}^{\infty} f(t')dt' \int_{0}^{\infty} \mathrm{Cos}\left[\omega(t'-t)\right] d\omega,$$

inasmuch as the odd part of the inner integral vanishes. Again reversing the order of integration we have

$$f(t) = \frac{1}{\pi}\int_{0}^{\infty} d\omega \int_{-\infty}^{\infty} f(t')\mathrm{Cos}\omega(t' - t)dt' . \tag{9-34}$$

Let us now consider what the insertion of a 90° negative phase shift in the outer integral of (9-33) will do. We define

$$\hat{f}(t) = \frac{1}{2\pi}\int_{-\infty}^{\infty} \frac{\mathrm{Sgn}\,\{\omega\}}{i} d\omega \int_{-\infty}^{\infty} f(t')e^{i\omega(t' - t)} dt' . \tag{9-35}$$

By the same steps as led from (9-33) to (9-34) we have

$$\hat{f}(t) = \frac{1}{\pi} \int_0^{\infty} d\omega \int_{-\infty}^{\infty} f(t')\mathrm{Sin}\left[\omega(t' - t)\right] dt' . \qquad (9\text{-}36)$$

The function $\hat{f}(t)$ is called the <u>allied function</u> to f(t). Reversing the order of integration once again we have

$$\hat{f}(t) = \frac{1}{\pi} \int_{-\infty}^{\infty} f(t')dt' \int_0^{\infty} \mathrm{Sin}\left[\omega(t' - t)\right] d\omega . \qquad (9\text{-}37)$$

If we now put a convergence factor $e^{-\alpha\omega}$ into the inner integral (intending to take the limit as $\alpha \to 0$ later) we find that (9-37) becomes

$$\hat{f}(t) = \frac{1}{\pi} \int_{-\infty}^{\infty} f(t')dt' \int_0^{\infty} e^{-\alpha\omega} \mathrm{Sin}\left[\omega(t' - t)\right] d\omega .$$

But

$$\int_0^{\infty} e^{-\alpha\omega} \mathrm{Sin}\,\omega(t' - t)d\omega = \frac{t' - t}{\alpha^2 + (t' - t)^2} \xrightarrow[\alpha\to 0]{} \frac{1}{t' - t} .$$

Substituting this result back into (9-37) we have

$$\hat{f}(t) = \frac{1}{\pi} P \int_{-\infty}^{\infty} \frac{f(t')dt'}{t' - t} . \qquad (9\text{-}38)$$

Written in this form $\hat{f}(t)$ is just the <u>Hilbert transform</u> of f(t). "P" stands for the principal value of the following integral. From this analysis we see that the effect of a negative phase shift of 90° (corresponding to a time advance) is to convert f(t) into $\hat{f}(t)$ where we have two ways, (9-36) and (9-38), to compute $\hat{f}(t)$. A few illustrative examples are given in Figure 9-8. Additional considerations regarding the nature of allied functions can be found in Titchmarsh (1937) and Hudson (1962). The latter reference and Arons and Yennie (1950) treat the problem of phase shifts in the case of plane waves incident upon a plane interface.

In the first three examples of Figure 9-8, we see that even though the input wave forms f(t) have no value for $t < 0$, the three output phase-shifted wave forms have non-zero values for $t < 0$. This occurs because a constant phase shift is a non-causal operator. Such an operator violates well known physical principles. However we will see that when phase shift operators are present, there is always a precursor wave in the real case, and that the phase shift is obtained as one takes the high-frequency limit. In the plane wave reflection coefficients computed above, the source is assumed to be at infinity so that all results obtained are the high-frequency limit. In Lamb's problem, we shall see in more detail what happens when one passes from the total solution to the high frequency limit.

The four examples of Figure 9-8 show that rather large changes in shape are possible for an impulse when all frequencies are shifted by minus 90°. Let us now see what will be the effect of a phase shift of an arbitrary amount upon a less

$f(t) = \delta(t)$

$$\hat{f}(t) = \frac{1}{\pi} P \int_{-\infty}^{\infty} \frac{\delta(t')dt'}{t' - t} = - \frac{1}{\pi t}$$

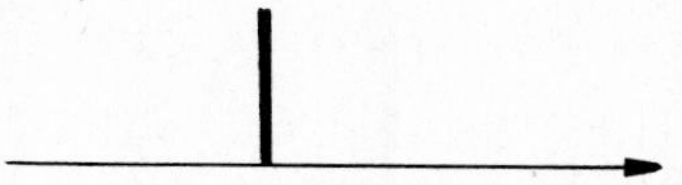

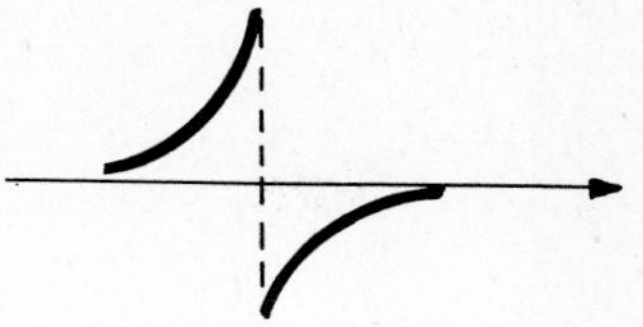

$f(t) = H(t)$

$$\hat{f}(t) = \frac{1}{\pi} P \int_{-\infty}^{\infty} \frac{H(t')dt'}{t' - t} = - \frac{1}{\pi} \ln|t|$$

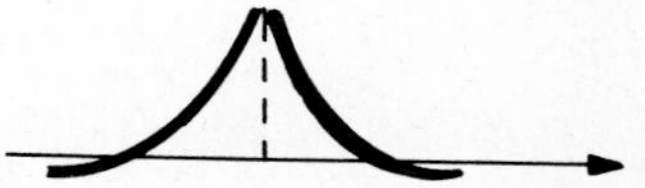

$f(t) = e^{-\alpha t} H(t)$

$$\hat{f}(t) = \frac{1}{\pi} P \int_{-\infty}^{\infty} \frac{e^{-\alpha t'} H(t')dt'}{t' - t}$$

$$= e^{-\alpha t} P \int_{-t}^{\infty} e^{-\alpha \xi} \frac{d\xi}{\xi} = -e^{-\alpha t} Ei(\alpha t)$$

(Abramovitz and Stegun, 1970, p.228.)

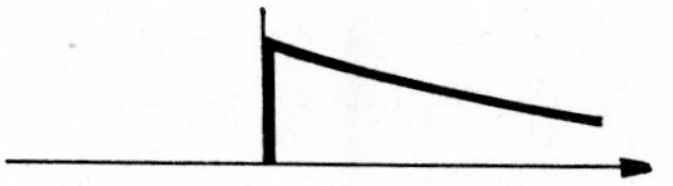

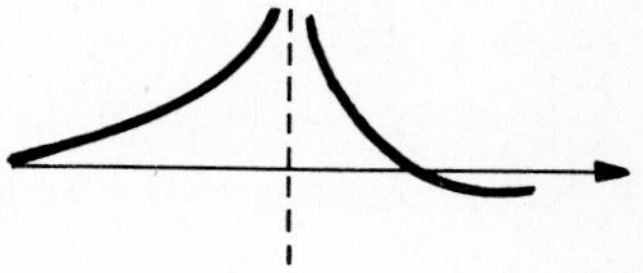

$$f(t) = \frac{t_o}{t^2 + t_o^2}$$

$$\hat{f}(t) = \frac{1}{\pi} P \int_{-\infty}^{\infty} \frac{t_o}{t'^2 + t_o^2} \frac{dt'}{t' - t} = \frac{-t}{t^2 + t_o^2}$$

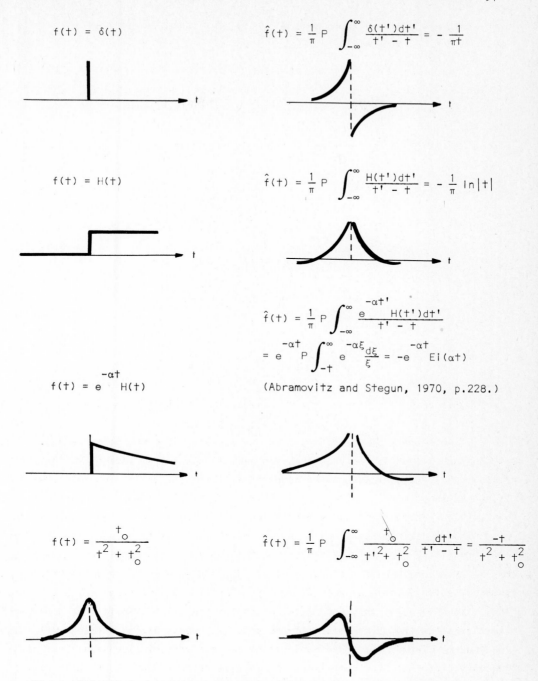

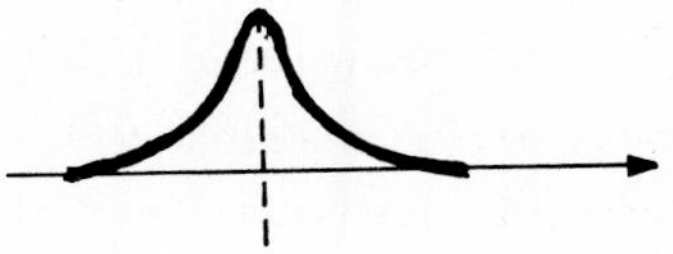

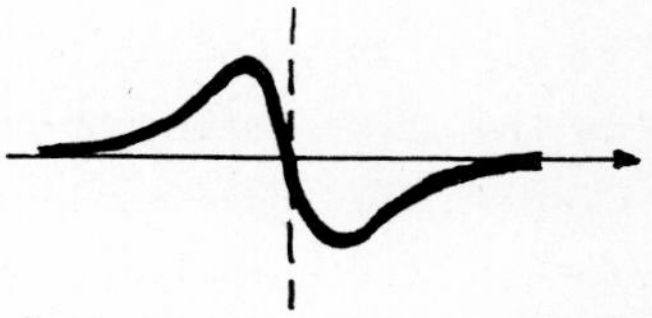

Figure 9-8. A number of common time functions and their associated allied functions (or Hilbert transforms).

pathalogic waveform. If all frequencies making up an input function $f(t)$ are given an arbitrary constant phase shift Φ sgn $\{\omega\}$ (corresponding to a time delay of Φ/ω units of time at each frequency), we then see that the resulting output function $F(t)$

will be

$$F(t) = f(t)\cos\Phi - \hat{f}(t)\sin\Phi. \qquad (9\text{-}39)$$

For a seismic pulse that often appears, we demonstrate the effects of phase shifting in Figure 9-9. The frequency content of all the waveforms in Figure 9.9 is the same. The sharper rise-time in the center of the 90°-shifted pulse might make one think otherwise.

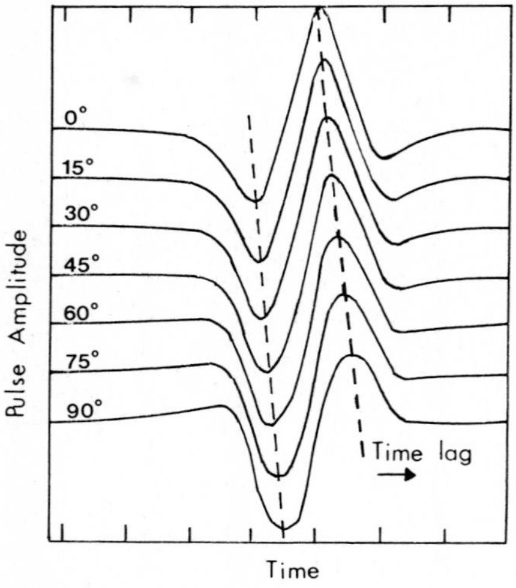

Figure 9-9. Changes in waveform due to positive phase shift.

The origin-time of a phase-shifted pulse is masked by the change in waveform. The dashed lines in the figure show this apparent time shift. Clayton, McClary, and Wiggins (1976) suggest ways to overcome this difficulty using advanced data-processing techniques.

Besides phase shifts in pulses reflected at angles beyond the critical, phase shifts are found in a number of other situations. In waves propagating in stratified media there is a constant phase shift as they go through a caustic (focusing region). This will be considered in Chapters 21 and 22. Surface waves traveling over a spherical earth go through a focusing at the antipodal points halfway around the globe. At this point they undergo a constant phase shift. Additional phase shifts are incurred as the waves return to the epicenter and continue through it. This is discussed in Chapter 24. Waves undergoing dispersion (velocity is a function of frequency), whether of a geometrical or physical nature, suffer phase shifts. In this case the shift is frequency-dependent. This will be examined in Chapters 20 and 26.

9.6. Particle Motion at the Free Surface.

Now that we have evaluated the coefficients R_{pp}, R_{ps}, R_{ss} and R_{sp}, we can determine the particle motion at the free surface. This motion is (for P-waves incident upon the free surface)

$$\mathbf{u}^p = \nabla(\Phi_o + \Phi_r) + \nabla\times\mathbf{e}_y\Psi_r^\dagger, \tag{9-40}$$

where the superscript p in $\mathbf{u}^p$ stands for an incident P-wave and Φ_o, Φ_r, and $\Psi_r^\dagger$ are given by (9-5) to (9-7). Evaluating (9-40) gives

$$\begin{aligned} u_z^p &= -ik_p\mathrm{Cos}\theta_p(1 - R_{pp}) + ik_s\mathrm{Sin}\theta_s R_{ps} \\ &= -ik_p\mathrm{Cos}\theta_p\left(1 - R_{pp} - \frac{\mathrm{Sin}\theta_p}{\mathrm{Cos}\theta_p}R_{ps}\right) \\ &= -2ik_p\left[\frac{\mathrm{Cos}\theta_p(1 - 2a^2\mathrm{Sin}^2\theta_p)}{(1 - 2a^2\mathrm{Sin}^2\theta_p)^2 + 4a^3\mathrm{Sin}^2\theta_p\mathrm{Cos}\theta_p(1 - a^2\mathrm{Sin}^2\theta_p)^{\frac{1}{2}}}\right] \\ &= -2ik_p D_z^p(\theta) , \end{aligned} \tag{9-41}$$

and

$$\begin{aligned} u_x^p &= ik_p\mathrm{Sin}\theta_p(1 + R_{pp}) - ik_s\mathrm{Cos}\theta_s R_{ps} \\ &= ik_p\mathrm{Sin}\theta_p\left\{1 + R_{pp} - \frac{(1 - a^2\mathrm{Sin}^2\theta_p)^{\frac{1}{2}}}{a\mathrm{Sin}\theta_p}R_{ps}\right\} \\ &= 2ik_p\left[\frac{2a\mathrm{Sin}\theta_p\mathrm{Cos}\theta_p(1 - a^2\mathrm{Sin}^2\theta_p)^{\frac{1}{2}}}{(1 - 2a^2\mathrm{Sin}^2\theta_p)^2 + 4a^3\mathrm{Sin}^2\theta_p\mathrm{Cos}\theta_p(1 - a^2\mathrm{Sin}^2\theta_p)^{\frac{1}{2}}}\right] \\ &= 2ik_p D_x^p(\theta) . \end{aligned} \tag{9-42}$$

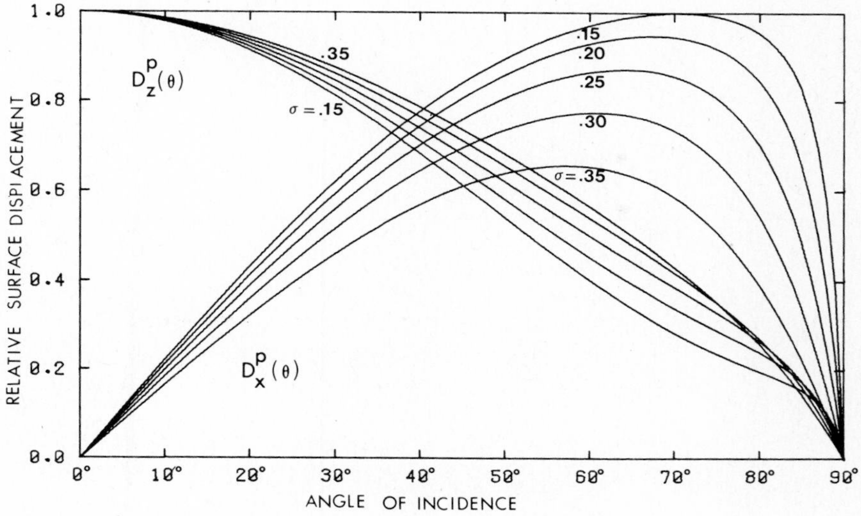

Figure 9-10. Directivity functions $D_z^p(\theta)$ and $D_x^p(\theta)$ for plane P-waves incident upon a free surface. Poisson's ratio σ varies from 0.15 to 0.35.

The notation $D_z^P(\theta)$ is introduced for the <u>directivity function</u> of a point surface receiver (z-component) for P-waves incident upon a free surface. Similarly for $D_x^P(\theta)$. Figure 9-10 gives the directivity patterns for incident P-waves as Poisson's ratio varies from 0.15 to 0.35.

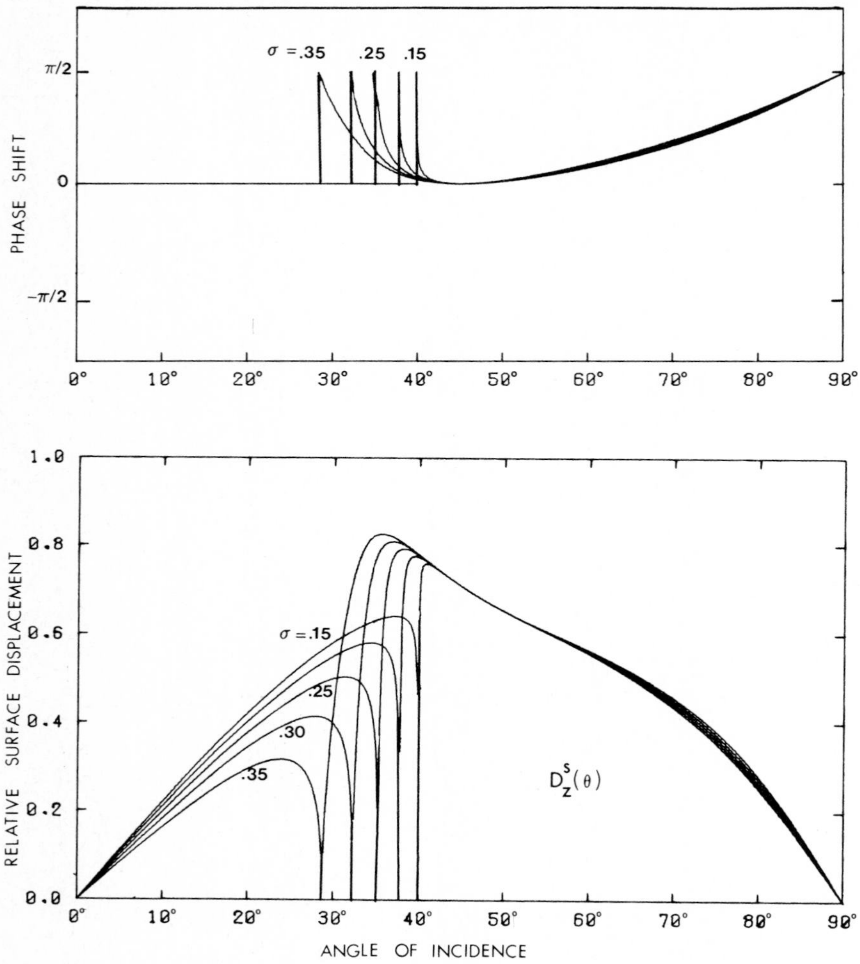

Figure 9-11. Directivity function $D_z^S(\theta)$ for plane S-waves incident upon a free surface. Poisson's ratio σ varies from 0.15 to 0.35.

OPPOSITE PAGE

Figure 9-12. Directivity function $D_x^S(\theta)$ for plane S-waves incident upon a free surface. Poisson's ratio σ varies from 0.15 to 0.35.

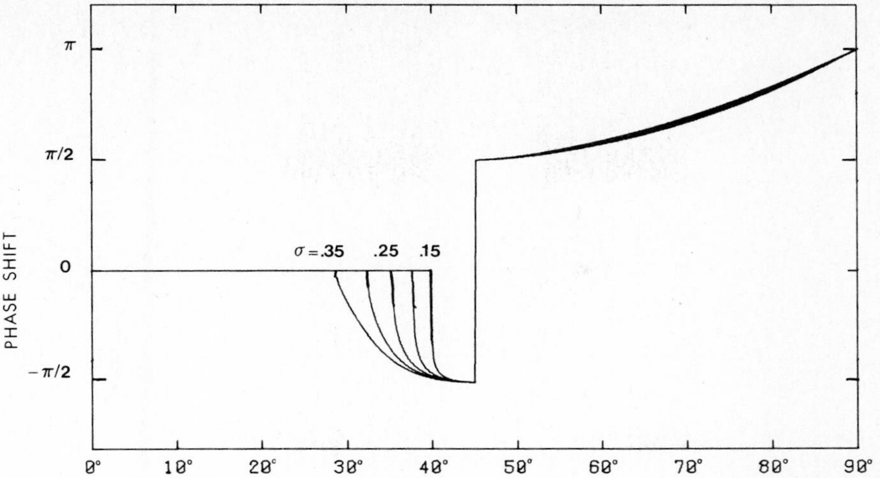
π
π/2
0
−π/2
PHASE SHIFT
σ = .35
.25
.15
0°
10°
20°
30°
40°
50°
60°
70°
80°
90°

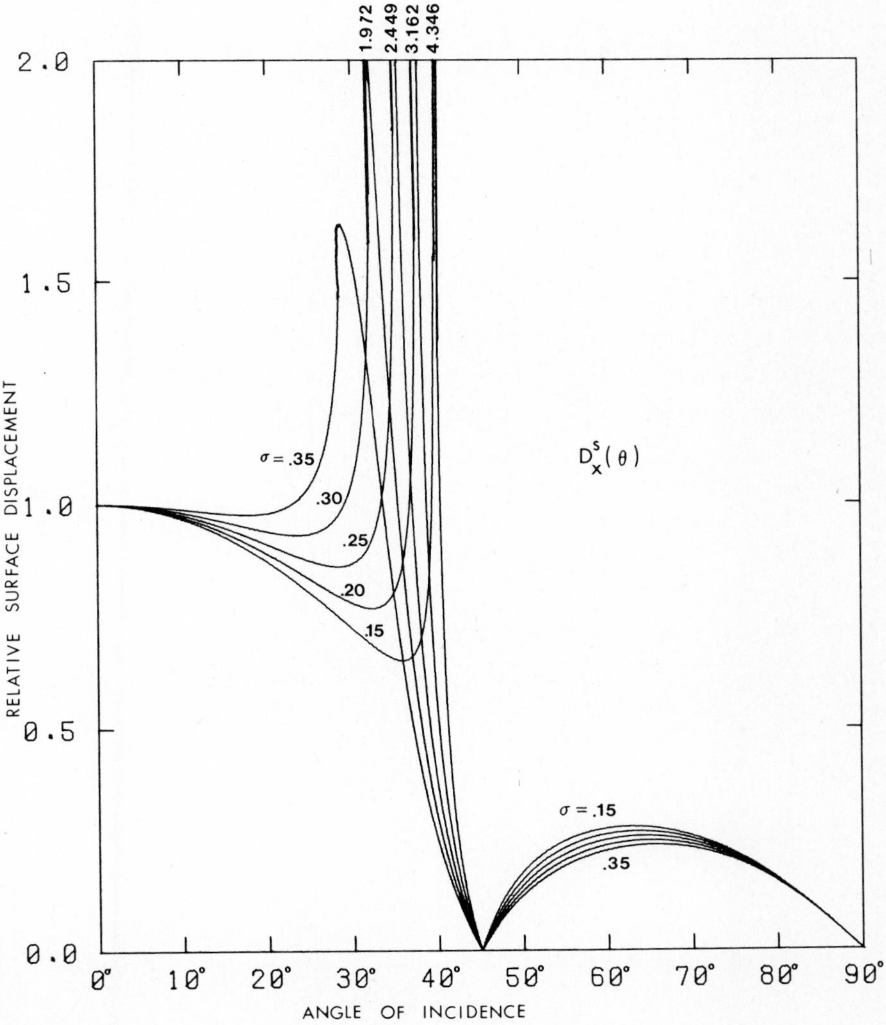
1.972
2.449
3.162
4.346
2.0
1.5
1.0
0.5
0.0
RELATIVE SURFACE DISPLACEMENT
σ = .35
.30
.25
.20
.15
$D_x^s(\theta)$
σ = .15
.35
0°
10°
20°
30°
40°
50°
60°
70°
80°
90°
ANGLE OF INCIDENCE

For incident SV-waves we have

$$\mathbf{u}^s = \nabla\Phi_r + \nabla\times\mathbf{e}_y(\Psi_o^\dagger + \Psi_r^\dagger), \tag{9-43}$$

where Φ_r, Ψ_o and $\Psi_r^\dagger$ are given by (9-20) and (9-21). Evaluating we have

$$\begin{aligned} u_z^s &= ik_p\mathrm{Cos}\theta_p R_{sp} + ik_s\mathrm{Sin}\theta_s(1 + R_{ss}) \\ &= ik_s\mathrm{Sin}\theta_s\left[\frac{a(a^2 - \mathrm{Sin}^2\theta_s)^{\frac{1}{2}}}{\mathrm{Sin}\theta_s}R_{sp} + 1 + R_{ss}\right] \\ &= 2ik_s\left[\frac{2\mathrm{Sin}\theta_s\mathrm{Cos}\theta_s(a^2 - \mathrm{Sin}^2\theta_s)^{\frac{1}{2}}}{(1 - 2\mathrm{Sin}^2\theta_s)^2 + 4\mathrm{Sin}^2\theta_s\mathrm{Cos}\theta_s(a^2 - \mathrm{Sin}^2\theta_s)^{\frac{1}{2}}}\right] \\ &= 2ik_s D_z^s(\theta), \end{aligned} \tag{9-44}$$

and

$$\begin{aligned} u_x^s &= ik_p\mathrm{Sin}\theta_p R_{sp} + ik_s\mathrm{Cos}\theta_s(1 - R_{ss}) \\ &= ik_s\mathrm{Cos}\theta_s\left(\frac{\mathrm{Sin}\theta_s}{\mathrm{Cos}\theta_s}R_{sp} + 1 - R_{ss}\right) \\ &= 2ik_s\left[\frac{\mathrm{Cos}\theta_s(1 - 2\mathrm{Sin}^2\theta_s)}{(1 - 2\mathrm{Sin}^2\theta_s)^2 + 4\mathrm{Sin}^2\theta_s\mathrm{Cos}\theta_s(a^2 - \mathrm{Sin}^2\theta_s)^{\frac{1}{2}}}\right] \\ &= 2ik_s D_x^s(\theta). \end{aligned} \tag{9-45}$$

Figures 9-11 and 9-12 give the directivity patterns for incident S-waves.

Directivity patterns for additional angles are given by Knopoff et al (1957) although their phase angle for θ_x^s does not have the 180° discontinuity indicated by Figure 9-12. White (1965, p. 227) gives the directivity patterns in polar coordinates.

CHAPTER 10

RAYLEIGH WAVES — FREE SURFACE PHENOMENA

10.1. Straight-Crested or Inhomogeneous Waves.

We have already seen that

$$\Phi = e^{i\left[k_p(x\mathrm{Sin}\theta + z\mathrm{Cos}\theta) - \omega t\right]} \qquad (10\text{-}1)$$

is a solution of the scalar wave equation. This represents a plane wave traveling in the direction indicated in Figure 10-1. However, we ran into a problem when the

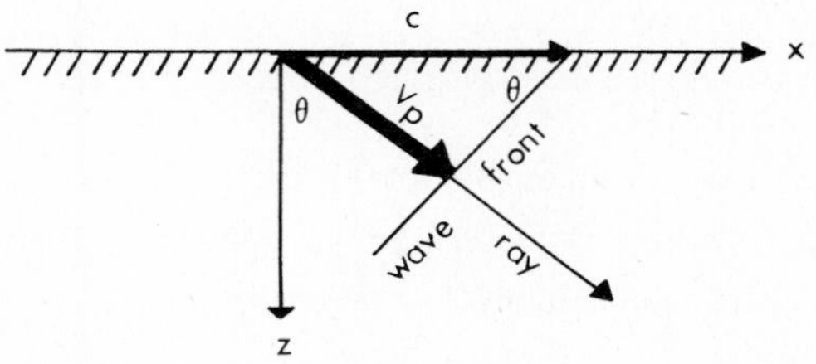

Figure 10-1. Geometrical relationships for the propagation of plane-waves.

factor Cosθ became imaginary. How should we interpret such a solution of the wave equation? First of all, we write

$$k = \frac{\omega}{c} = k_p\mathrm{Sin}\theta \ , \qquad (10\text{-}2)$$

which implies

$$(k_p^2 - k^2)^{\frac{1}{2}} = k_p\mathrm{Cos}\theta \ . \qquad (10\text{-}3)$$

We can see from the geometry of Figure 10-1 that $c = v_p/\mathrm{Sin}\theta$ is the apparent wave velocity of the plane wave along the surface. Completing the rewriting of (10-1) in terms of the new variable k, we have

$$\Phi = e^{i\left[kx + (k_p^2 - k^2)^{\frac{1}{2}}z - \omega t\right]} \ , \qquad (10\text{-}4a)$$

or

$$\Phi = e^{ikx - (k^2 - k_p^2)^{\frac{1}{2}}z - i\omega t} \ , \qquad (10\text{-}4b)$$

where in the second expression we have reversed the order of the arguments of the square root. By direct substitution, either form of (10-4) will be found to be a solution of the scalar wave equation. When $k < k_p$, the first form will be applicable and this can be seen to be the plane wave described by (10-1) when one makes the substitution of (10-2) and (10-3).

When $k > k_p$, i.e., $c < v_p$, we must use the form indicated by (10-4b). In this case we can rewrite slightly and obtain

$$\Phi = e^{-(k^2 - k_p^2)^{\frac{1}{2}} z} e^{i(kx - \omega t)} \quad . \tag{10-5}$$

This is the representation of a wave traveling in the x-direction (all the way from z = 0 to z = ∞), but one which is amplitude modulated in the z-direction. For such a wave, any plane of x = constant is also a plane of constant phase (as long as $\underline{k}$ is real). Consequently two names have been given to waves of the type (10-5): straight-crested waves or equi-phase waves. The feature distinguishing them from plane waves is the modulation in the z-direction. This gives rise to the name inhomogeneous plane waves. Generalizing, we can write an expression for this type wave in the form

$$\Phi = e^{\pm(k^2 - k_p^2)^{\frac{1}{2}} z} e^{i(\pm kx - \omega t)} , \tag{10-6}$$

where the choice of the four combinations of signs will be dictated by the conditions of a particular problem. For convenience, we shall also retain the names straight-crested or inhomogeneous plane waves for solutions of the type (10-6) even if $\underline{k}$ is complex and planes of $\underline{x}$ are no longer planes of constant phase.

10.2. Rayleigh Waves.

For the moment, let us suppress the factor $e^{-i\omega t}$ and try to find elastic potentials of the form

$$\begin{aligned} \Phi &= A(k)e^{ikx - (k^2 - k_p^2)^{\frac{1}{2}} z} , \\ \Psi^{\dagger} &= \frac{1}{ik}\frac{\partial \Psi}{\partial x} = B(k)e^{ikx - (k^2 - k_s^2)^{\frac{1}{2}} z} , \end{aligned} \tag{10-7}$$

such that the boundary conditions on the free surface of an elastic half-space are satisfied. From (6-31) and (6-33) we have that

$$\begin{aligned} \frac{t_{zz}}{\mu} &= -k_s^2 \Phi + 2\left(\frac{\partial^2 \Psi^{\dagger}}{\partial x \partial z} - \frac{\partial^2 \Phi}{\partial x^2} \right) = 0, \\ \frac{t_{zx}}{\mu} &= 2\frac{\partial^2 \Phi}{\partial z \partial x} + \left(\frac{\partial^2 \Psi^{\dagger}}{\partial x^2} - \frac{\partial^2 \Psi^{\dagger}}{\partial z^2} \right) = 0, \end{aligned} \tag{10-8}$$

where we have used (6-16), and the coordinates are as in Figure 10-2.

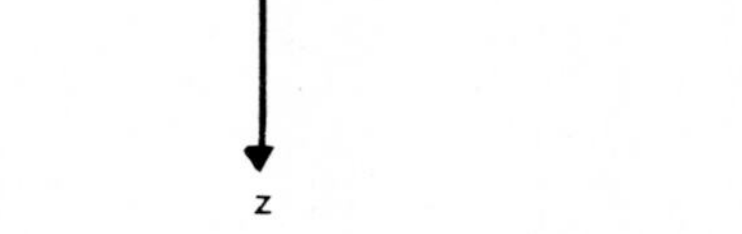

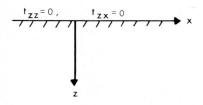

Figure 10-2. Geometrical relationships for straight-crested Rayleigh-waves propagating in a half-space.

Substituting equations (10-7) into these two conditions requires that

$$(2k^2 - k_s^2)A(k) - 2ik(k^2 - k_s^2)^{\frac{1}{2}}B(k) = 0 ,$$

and

$$-2ik(k^2 - k_p^2)^{\frac{1}{2}}A(k) - (2k^2 - k_s^2)B(k) = 0. \tag{10-9}$$

If this pair of equations is to hold, the determinant of the coefficients of A and B must vanish, i.e.,

$$R(k) = (2k^2 - k_s^2)^2 - 4k^2(k^2 - k_p^2)^{\frac{1}{2}}(k^2 - k_s^2)^{\frac{1}{2}} = 0. \tag{10-10}$$

We shall come upon this form again in the denominators of equations (11-15) and (11-16) in the solution of Lamb's problem in the next chapter. We have already seen it in the denominators of (9-16), (9-17), (9-23), and (9-24). In the first two we have to make the identification $k = k_p \text{Sin}\theta$ and in the second two we make the identification $k = k_s \text{Sin}\theta$. Equation (10-10) is known as the Rayleigh Denominator. Rewritten in terms of velocities, (10-10) becomes

$$(2 - c^2/v_s^2)^2 - 4(1 - c^2/v_p^2)^{\frac{1}{2}}(1 - c^2/v_s^2)^{\frac{1}{2}} = 0. \tag{10-11}$$

For real _k_ (implying real _c_) there is only one root of (10-10) with the signs of the square roots as given. For complex _k_, the situation is more complicated and will be discussed later in the analysis of Lamb's problem. This velocity ratio varies as a function of Poisson's ratio as in Figure 10-3.

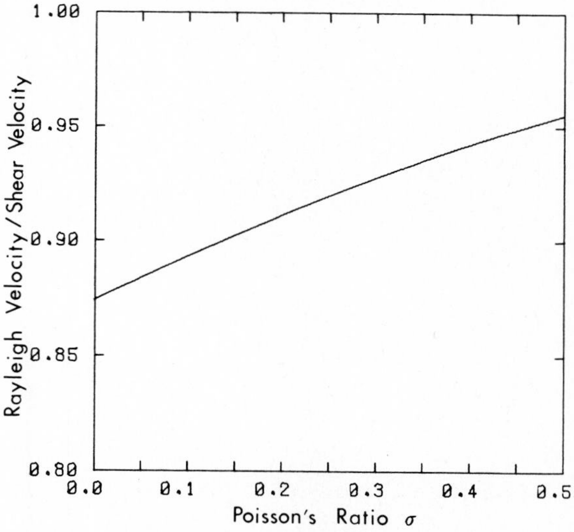

Figure 10-3. The ratio of Rayleigh-wave velocity to shear-wave velocity as a function of Poisson's ratio.

If _k_ is a root of R(k), we can write

$$\frac{A}{B} = \frac{2ik(k^2 - k_s^2)^{\frac{1}{2}}}{2k^2 - k_s^2} = - \frac{2k^2 - k_s^2}{2ik(k^2 - k_p^2)^{\frac{1}{2}}} . \tag{10-12}$$

Then we have from (6-25) and (6-27)

$$u_x = \left[ike^{-(k^2-k_p^2)^{\frac{1}{2}}z} + \frac{2k^2-k_s^2}{2ik} e^{-(k^2-k_s^2)^{\frac{1}{2}}z} \right] Ae^{i(kx-\omega t)} ,$$

$$u_z = \left[-(k^2-k_p^2)^{\frac{1}{2}} e^{-(k^2-k_p^2)^{\frac{1}{2}}z} + \frac{2k^2-k_s^2}{2(k^2-k_s^2)^{\frac{1}{2}}} e^{-(k^2-k_s^2)^{\frac{1}{2}}z} \right] Ae^{i(kx-\omega t)} . \tag{10-13}$$

For $\lambda = \mu$, $\sigma = 0.25$, $c = 0.9194\ V_s$ and (10-13) becomes

$$u_x = (e^{-.8475kz} - 0.5774e^{-.3933kz})ikAe^{i(kx-\omega t)} ,$$

$$u_z = (-.8475e^{-.8475kz} + 1.4679e^{-.3933kz})kAe^{i(kx-\omega t)} . \tag{10-14}$$

At $z = 0$ this represents a retrograde ellipse with the x-motion lagging 90° behind the z-motion. Both u_x and u_z decay exponentially with depth, u_x having a change in sign when $e^{-.4542kz} = .5774$, or at $z/L = .192$, where L is the wavelength. Below this

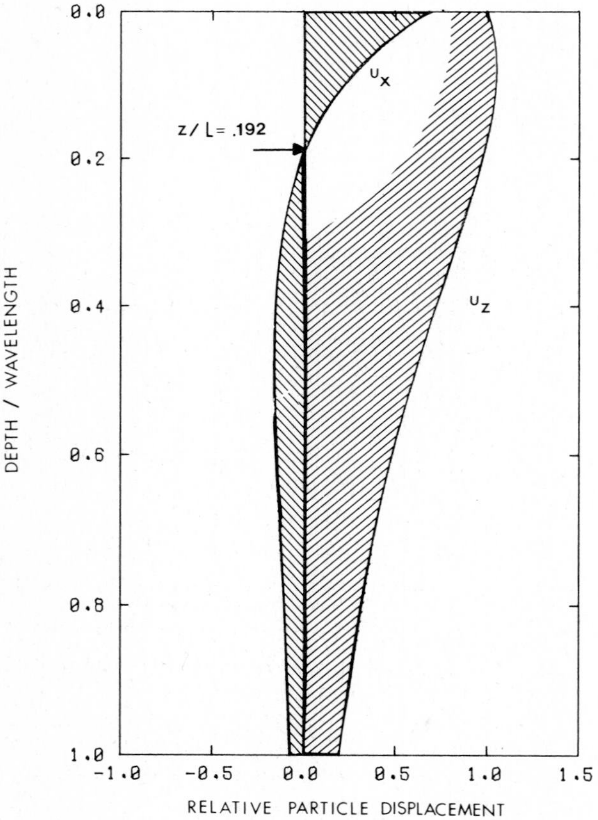

Figure 10-4. Normalized Rayleigh-wave particle displacement vs. depth in an isotropic, homogeneous medium.

depth, the ellipse changes to prograde. At the surface $u_z/u_x = 1.47$. Note that at all points on the surface where $t_{zz} = 0$ and $t_{zx} = 0$, the ratios (10-12) hold no matter how complicated A(k) and B(k) might become as a result of rather general stress distributions on the surface. This fact sometimes will prove useful in analyzing more complicated motions, for example, as a check on the algebra.

10.3. Rayleigh Waves in Cylindrical Coordinates.

The above analysis has shown the existence of Rayleigh waves in the form of straight-crested waves. However, this does not imply their existence when we have propagation with axial symmetry (Figure 10-5). Such waves are called circular-crested waves.

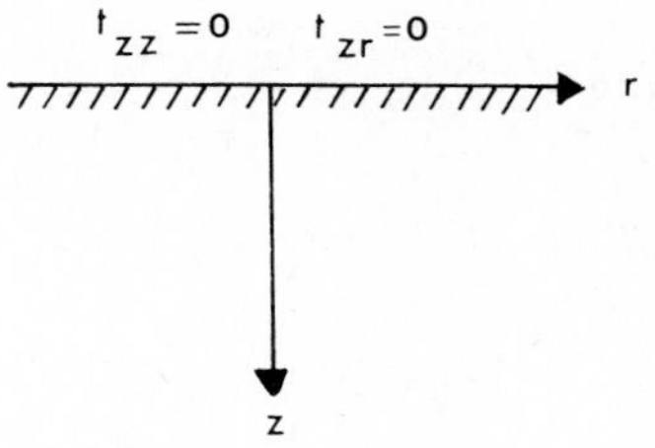

Figure 10-5. Geometrical relationships for circular-crested Rayleigh-waves propagating in a half-space.

For P- and SV-waves we have from (6-48)

$$\mathbf{u} = \nabla\Phi + \ell\nabla\times\mathbf{e}_\theta \frac{\partial\Psi}{\partial r}$$
$$= \nabla\Phi + k^{-1}\nabla\times\mathbf{e}_\theta \frac{\partial\Psi}{\partial r}, \tag{10-15}$$

where, for convenience, we have set "ℓ" equal to k^{-1} (= $L/2\pi$). The potentials Φ and Ψ are solutions of the scalar-wave equations (6-16) and (6-17). Appropriate general solutions are

$$\Phi = A(k)e^{-(k^2-k_p^2)^{\frac{1}{2}}z} J_o(kr),$$
$$\Psi = -B(k)e^{-(k^2-k_p^2)^{\frac{1}{2}}z} J_o(kr), \tag{10-16}$$

for waves that do not have a singularity at the origin, and where we have again assumed a time factor of $e^{-i\omega t}$. We then have (noting $J_o' = -J_1$)

$$u_r = \frac{\partial\Phi}{\partial r} - \frac{1}{k}\frac{\partial^2\Psi}{\partial r\partial z} = -\left[kAe^{-(k^2-k_p^2)^{\frac{1}{2}}z} - (k^2-k_s^2)^{\frac{1}{2}}Be^{-(k^2-k_s^2)^{\frac{1}{2}}z}\right]J_1(kr),$$
$$u_z = \frac{\partial\Phi}{\partial z} + \frac{1}{kr}\frac{\partial}{\partial r}\left(r\frac{\partial\Psi}{\partial r}\right) = \left[-(k^2-k_p^2)^{\frac{1}{2}}Ae^{-(k^2-k_p^2)^{\frac{1}{2}}z} + kBe^{-(k^2-k_s^2)^{\frac{1}{2}}z}\right]J_o(kr); \tag{10-17}$$

also

$$\frac{t_{zz}}{\mu} = \frac{\lambda}{\mu}\,\nabla\cdot\mathbf{u} + 2\,\frac{\partial u_z}{\partial z} = \left[(2k^2 - k_s^2)Ae^{-(k^2-k_p^2)^{\frac{1}{2}}z} - 2k(k^2-k_s^2)^{\frac{1}{2}}Be^{-(k^2-k_s^2)^{\frac{1}{2}}z}\right]J_o(kr),$$

$$\frac{t_{zr}}{\mu} = \frac{\partial u_z}{\partial r} + \frac{\partial u_r}{\partial z} = \left[+\,2k(k^2-k_p^2)^{\frac{1}{2}}Ae^{-(k^2-k_p^2)^{\frac{1}{2}}z} - (2k^2-k_s^2)Be^{-(k^2-k_s^2)^{\frac{1}{2}}z}\right]J_1(kr). \tag{10-18}$$

If we now require that t_{zz} and t_{zr} vanish on the surface $z = 0$, we have

$$(2k^2 - k_s^2)A(k) - 2k(k^2-k_s^2)^{\frac{1}{2}}B(k) = 0,$$

$$2k(k^2-k_p^2)^{\frac{1}{2}}A(k) - (2k^2-k_s^2)B(k) = 0. \tag{10-19}$$

For a solution we must have

$$R(k) = (2k^2-k_s^2)^2 - 4k^2(k^2-k_p^2)^{\frac{1}{2}}(k^2-k_s^2)^{\frac{1}{2}} = 0. \tag{10-20}$$

In this case we find that

$$\frac{A}{B} = \frac{2k(k^2-k_s^2)^{\frac{1}{2}}}{2k^2-k_s^2} = \frac{2k^2-k_s^2}{2k(k^2-k_p^2)^{\frac{1}{2}}}\,. \tag{10-21}$$

The ratios (10-21) are somewhat different than in the case of straight-crested Rayleigh-waves. The same comments hold however, regardless of the complexity of azimuthally symmetric stress distributions over the free surface.

10.4. Other Considerations.

One of the intriguing problems of elastic wave theory is the nature of Rayleigh-waves. Knowles (1966) has formulated the Rayleigh-wave problem in terms of a single potential; however, he gives no physical model for constraining the elastic energy to a surficial region. This single potential formulation has been extended somewhat by Eringen and Suhubi (1975, Sec. 7.6). Alsop, Goodman, and Gregersen (1974) have identified Rayleigh-waves as the constructive interference of P and SV inhomogeneous waves. Again, no physical model is given. Now, the effective elastic constants on the surface of a half-space are less than those on the interior because the missing half-space provides no restoring forces. It might be possible to conceive of Rayleigh-waves as energy trapped in this surficial region because their velocities might be lower due to the reduced elastic constants. However, no results in this direction have been forthcoming.

The propagation of Rayleigh-waves on a flat surface is non-dispersive, i.e., the propagation is independent of frequency. We shall see in Chapter 24, however, that Rayleigh-waves propagating on a curved surface are dispersive.

Rayleigh-waves can be looked at as a special case of the propagation of interface waves along the contact between a material body and a vacuum. We shall consider the more general case of interface waves in Chapter 14.

CHAPTER 11

LAMB'S PROBLEM

11.1. General.

The solution of the problem known as Lamb's (1904) was one of the earliest unravelings of a non-trivial question in the propagation of elastic waves. Without the power of the Cagniard-deHoop technique, Lamb found the impulsive response in two dimensions and sketched it in three dimensions. In Section 9 of his paper, he anticipated this method but did not recognize its generality. In his monograph, Cagniard (1939) treated the problem of Lamb's as well as a more general one. He also considered the propagation of waves in a medium consisting of two elastic half-spaces joined together. This work was based on methods which he developed in the mid-1930's. An independent development was given by Pekeris (1941). These early developments lay dormant until the 1950's when there was a general interest in impulsive solutions. Mencher (1953), Dix (1954), Pekeris (1955a and 1955b), and Gilbert (1956) used the Cagniard technique for three-dimensional problems. Garvin (1956) found a complete solution for the two-dimensional buried line-source problem using the Cagniard technique with a Laplace transformation in time. Gilbert (1956) independently solved the same problem using a Fourier transformation in time. This latter work demonstrates that the Laplace transformation in time is not required.

In fact, Lamb's problem can be solved for the impulsive response without using Cagniard's technique at all. For example, Sauter (1950) and Sherwood (1958) solved the two-dimensional problem, and Pinney (1954) worked with the three-dimensional problem. These works used general integral transform methods of solution. A rather different approach is that of using self-similar solutions (Eringen and Suhubi, 1975, Sec.'s 7.14 - 7.16). These methods were developed in elastodynamics by Sobolev (1932) and Smirnov and Sobolev (1932). Eringen and Suhubi (ibid.) present in Section 7.16 a treatment of Lamb's problem with an arbitrarily oriented surface line-source. They also give additional references to the work of Smirnov and Sobolev as well as more modern treatments. This particular method of solution has a special beauty of interpretation but it is not commonly presented as part of an applied mathematics curriculum. So, we will not consider it here, referring the reader to Eringen and Suhubi. A fairly complete summary of early papers (through 1961) using many methods to obtain impulsive solutions has been put together by Flinn and Dix (Cagniard, 1962, Appendix III).

Lastly deHoop (1960 and 1961) refined Cagniard's method while applying it to seismic source problems and to the three-dimensional Lamb's problem. Because of these simplifying modifications, we now refer to the direct inversion procedure as the Cagniard-deHoop technique. We shall investigate the two-dimensional problem next.

11.2. Motion Due to a Surface Vertical Force.

If we take the Laplace transform with respect to time of the elastic wave equation for isotropic homogeneous media (5-9), we obtain

$$v_p^2\nabla\nabla\cdot\mathbf{U} - v_s^2\nabla\times\nabla\times\mathbf{U} = s^2\mathbf{U}, \tag{11-1}$$

where

$$\mathbf{U}(x,z,s) = \int_0^\infty e^{-st}\,\mathbf{u}(x,z,t)dt. \tag{11-2}$$

The transformed displacement vector $\mathbf{U}$ can be related to the potentials Φ and $\Psi^\dagger$. Its components are:

$$U_x = \frac{\partial\Phi}{\partial x} - \frac{\partial\Psi^\dagger}{\partial z}\ , \tag{11-3}$$

and

$$U_z = \frac{\partial\Phi}{\partial z} + \frac{\partial\Psi^\dagger}{\partial x}\ , \tag{11-4}$$

where from page 45, we have used

$$\Psi^\dagger = \ell\,\frac{\partial\Psi}{\partial x}\ . \tag{11-5}$$

The potential Φ satisfies the transformed wave equation

$$\nabla^2\Phi - \frac{s^2}{v_p^2}\,\Phi = 0. \tag{11-6}$$

Elemental solutions of (11-6) can be written as

$$\Phi = e^{(s/v_s)\{\pm iqx \pm (q^2+a^2)^{\frac{1}{2}}z\}}\ ,$$

where q is a dimensionless separation parameter introduced for later convenience. If we now integrate over the parameter q, we can write as a general solution in the positive z, positive x quarter-space:

$$\Phi(x,z,s) = \int_{-\infty}^{\infty} A(q)e^{(s/v_s)\{iqx - (q^2+a^2)^{\frac{1}{2}}z\}}dq. \tag{11-7}$$

Likewise the potential Ψ satisfies the transformed wave equation

$$\nabla^2\Psi - \frac{s^2}{v_s^2}\,\Psi = 0, \tag{11-8}$$

with elemental solutions

$$\Psi = e^{(s/v_s)\{\pm iqx \pm (q^2+1)^{\frac{1}{2}}z\}}\ .$$

Because of (11-5), we shall write our general solution as

$$\Psi(x,z,s) = \int_{-\infty}^{\infty} \frac{B(q)}{iq}\,e^{(s/v_s)\{iqx - (q^2+1)^{\frac{1}{2}}z\}}dq. \tag{11-9}$$

Then on setting $\ell = v_s/s$, we find that

$$\Psi^\dagger(x,z,s) = \int_{-\infty}^{\infty} B(q)e^{(s/v_s)\{iqx - (q^2+1)^{\frac{1}{2}}z\}}dq. \tag{11-10}$$

Now the conditions of the problem are sketched out in Figure 11-1. The source

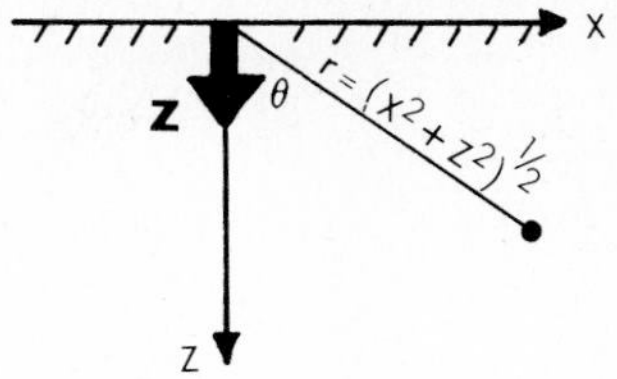

Figure 11-1. Geometrical relationships for a vertical force on the surface of a half-space.

function in Figure 11-1 is a line source of stress, but since it acts on the boundary, we will put it into the mathematical formulation through the boundary conditions. A delta-function in normal stress can be represented as

$$-Z\delta(x) = -\frac{Z}{2\pi}\int_{-\infty}^{\infty} e^{i(sq/v_s)x}\, d(sq/v_s). \qquad (11\text{-}11)$$

This represents the action of a point force of magnitude Z acting downwards on the surface. Since the Laplace transform variable $\underline{s}$ disappears on performing the indicated integration of (11-11), the appropriate time function for our source is that of a delta-function in time. The transformed surface stresses in terms of Φ and $\Psi^{\dagger}$ are obtained from (6-31) and (6-33) as

$$\frac{t_{zz}}{\mu} = \frac{\lambda}{\mu}\nabla\cdot\mathbf{U} + 2\frac{\partial U_z}{\partial z} = \frac{s^2}{v_s^2}\Phi + 2\left(\frac{\partial^2\Psi^{\dagger}}{\partial x\partial z} - \frac{\partial^2\Phi}{\partial x^2}\right) = -Z\delta(x), \qquad (11\text{-}12)$$

$$\frac{t_{zx}}{\mu} = \frac{\partial U_z}{\partial x} + \frac{\partial U_x}{\partial z} = \frac{\partial^2\Psi^{\dagger}}{\partial x^2} + 2\frac{\partial^2\Phi}{\partial x\partial z} - \frac{\partial^2\Psi^{\dagger}}{\partial z^2} = 0, \qquad (11\text{-}13)$$

where we have made use of (11-6). Putting (11-7), (11-10) and (11-11) into the boundary conditions, we find

$$A(q) = -\frac{Z}{2\pi\mu}\frac{v_s}{s}\frac{(2q^2+1)}{R(q)},$$

$$B(q) = \frac{Z}{2\pi\mu}\frac{v_s}{s}\frac{2iq(q^2+a^2)^{\frac{1}{2}}}{R(q)},$$

where

$$R(q) = (2q^2+1)^2 - 4q^2(q^2+a^2)^{\frac{1}{2}}(q^2+1)^{\frac{1}{2}}. \qquad (11\text{-}14)$$

Hence

$$\Phi = -\frac{Z}{2\pi\mu}\frac{v_s}{s}\int_{-\infty}^{\infty}\frac{2q^2+1}{R(q)}\, e^{(s/v_s)\{iqx - (q^2+a^2)^{\frac{1}{2}}z\}}\, dq, \qquad (11\text{-}15)$$

and

$$\Psi^{\dagger} = \frac{Z}{2\pi\mu}\frac{v_s}{s}\int_{-\infty}^{\infty}\frac{2iq(q^2+a^2)^{\frac{1}{2}}}{R(q)}\, e^{(s/v_s)\{iqx - (q^2+1)^{\frac{1}{2}}z\}}\, dq. \qquad (11\text{-}16)$$

In general, factors such as

$$(2q^2+1)/R(q) \quad \text{and} \quad 2iq(q^2+a^2)^{\frac{1}{2}}/R(q),$$

are characteristic of the boundary conditions, while the factors

$$e^{(s/v_s)\{iqx - (q^2+a^2)^{\frac{1}{2}}z\}} \quad \text{and} \quad e^{(s/v_s)\{iqx - (q^2+1)^{\frac{1}{2}}z\}}$$

are characteristic of the transformed wave equations (11-6) and (11-8). From (11-3) and (11-4) we then have

$$U_x(x,z,s) = -\frac{iZ}{2\pi\mu}\int_{-\infty}^{\infty}\left[\begin{array}{l}(2q^2+1)e^{-(s/v_s)(q^2+a^2)^{\frac{1}{2}}z} \\ \\ -2(q^2+a^2)^{\frac{1}{2}}(q^2+1)^{\frac{1}{2}}e^{-(s/v_s)(q^2+1)^{\frac{1}{2}}z}\end{array}\right]\frac{e^{(s/v_s)iqx}}{R(q)}\,qdq, \tag{11-17}$$

$$U_z(x,z,s) = \frac{Z}{2\pi\mu}\int_{-\infty}^{\infty}\left[\begin{array}{l}(2q^2+1)e^{-(s/v_s)(q^2+a^2)^{\frac{1}{2}}z} \\ \\ -2q^2e^{-(s/v_s)(q^2+1)^{\frac{1}{2}}z}\end{array}\right]\frac{e^{(s/v_s)iqx}}{R(q)}(q^2+a^2)^{\frac{1}{2}}dq. \tag{11-18}$$

Let us rewrite the integrands of (11-17) and (11-18) in terms of r and θ, where

$$x = r\sin\theta; \qquad z = r\cos\theta. \tag{11-19}$$

In obtaining the following expressions, we have also used the evenness and oddness of the integrands.

$$U_x^P(r,\theta,s) = \frac{Z}{\pi\mu}\,\mathrm{Im}\int_0^{\infty}\frac{(2q^2+1)e^{(sr/v_s)\{iq\sin\theta - (q^2+a^2)^{\frac{1}{2}}\cos\theta\}}}{R(q)}\,qdq, \tag{11-20}$$

$$U_x^S(r,\theta,s) = -\frac{Z}{\pi\mu}\,\mathrm{Im}\int_0^{\infty}\frac{2(q^2+a^2)^{\frac{1}{2}}(q^2+1)^{\frac{1}{2}}e^{(sr/v_s)\{iq\sin\theta - (q^2+1)^{\frac{1}{2}}\cos\theta\}}}{R(q)}\,qdq, \tag{11-21}$$

$$U_z^P(r,\theta,s) = \frac{Z}{\pi\mu}\,\mathrm{Re}\int_0^{\infty}\frac{(2q^2+1)e^{(sr/v_s)\{iq\sin\theta - (q^2+a^2)^{\frac{1}{2}}\cos\theta\}}}{R(q)}(q^2+a^2)^{\frac{1}{2}}dq, \tag{11-22}$$

$$U_z^S(r,\theta,s) = -\frac{Z}{\pi\mu}\,\mathrm{Re}\int_0^{\infty}\frac{2q^2e^{(sr/v_s)\{iq\sin\theta - (q^2+1)^{\frac{1}{2}}\cos\theta\}}}{R(q)}(q^2+a^2)^{\frac{1}{2}}dq. \tag{11-23}$$

We note that for the convergence of these and other integrals to follow (assuming $\theta \geq 0$), we must have $\mathrm{Im}\,\{q\} \geq 0$, $\mathrm{Re}\,\{(q^2+a^2)^{\frac{1}{2}}\} \geq 0$, and $\mathrm{Re}\,\{(q^2+1)^{\frac{1}{2}}\} \geq 0$.

We now make use of the Cagniard-deHoop technique developed in Section 7.3, and make the following change of variable,

$$\cos\theta(q^2+a^2)^{\frac{1}{2}} - iq\sin\theta = \tau = v_s t/r. \tag{11-24}$$

The inverse of this transformation is

$$q(\tau) = i\tau\sin\theta + \cos\theta(\tau^2 - a^2)^{\frac{1}{2}}, \tag{11-25}$$

and for later use we also compute

$$dq/d\tau = i\sin\theta + \tau\cos\theta(\tau^2 - a^2)^{-\frac{1}{2}}. \tag{11-26}$$

Rewriting (11-20) in terms of the new variable t, we have

$$U_x^P(r,\theta,s) = \frac{Z}{\pi\mu} \; Im \int_?^? \frac{q(t)\;\{2q^2(t) + 1\}\; e^{-st}}{R\;\{q(t)\}} \; \frac{dq(t)}{dt} \; dt. \tag{11-27}$$

Equation (11-27) can now be recognized as the Laplace Transform of the factor

$$\frac{Z}{\pi\mu} \; Im \left[\frac{q(t)\;\{2q^2(t) + 1\}}{R\;\{q(t)\}} \; \frac{dq(t)}{dt} \right]$$

looked at as a function of the time variable $\underline{t}$.

However, we have to look at a few details before we can call this recognition valid and place proper limits on the integral. First of all, we want to look at the path $\underline{q}$ takes as we let the variable $\underline{t}$ run from 0 to ∞. From (11-25) we see that for

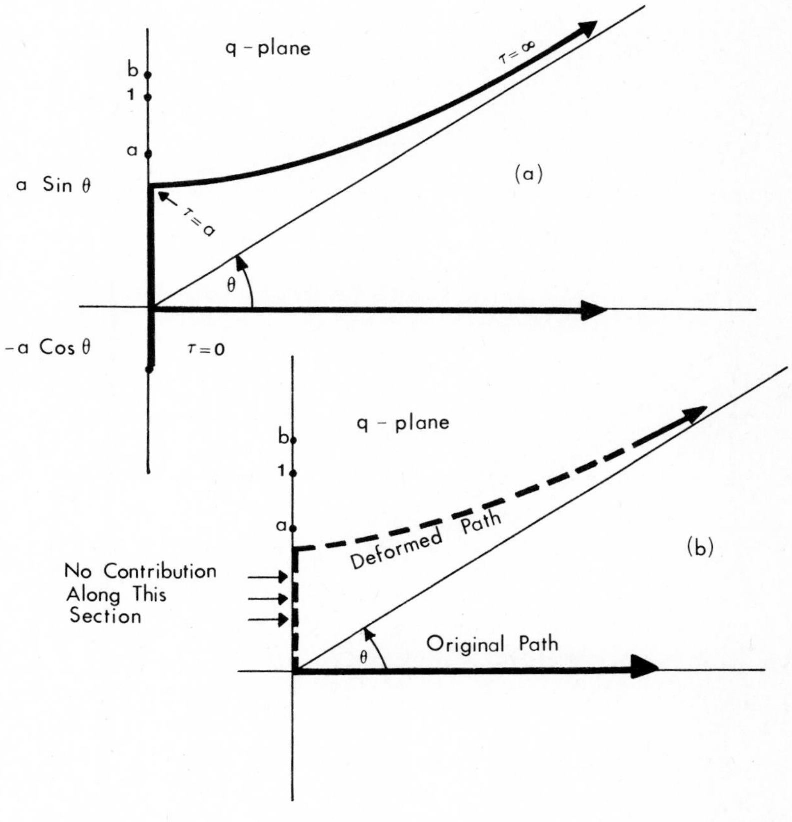

Figure 11-2(a). The relationship the original path of integration in (11-20) and the path which $\underline{q}$ takes as τ varies between zero and infinity. Singularities include branch points at $\underline{ia}$ and $\underline{i}$ plus a pole at $\underline{ib}$.

(b). The relationship between the original path and the deformed path in (11-27) which leads to (11-28).

τ = 0, q = -iaCosθ. The derivative (11-26) changes at τ = a, i.e., the variable $\underline{q}$ runs up the imaginary axis to q = iaSinθ, and then branches out into the first quadrant on a hyperbola as defined by equation (11-25) and along an asymptote as in Figure 11-2(a). Inasmuch as the singularities of (11-20) are branch points at $q = ia$, $q = i$, and a pole of the integrand at $q = ib$, we see that the original path can be deformed into a new path as shown in Figure 11-2b. However, on the vertical segment from 0 to ia, we see that (11-20) has no imaginary part. Consequently the lower limit on (11-27) will be r/v_p and the upper limit will be ∞. We can then write

$$U_x^P(r,\theta,s) = \frac{Z}{\pi\mu}\,\mathrm{Im}\int_{r/v_p}^{\infty} \frac{q(t)\{2q^2(t)+1\}}{R\{q(t)\}}\,\frac{dq(t)}{dt}\,e^{-st}\,dt. \qquad (11\text{-}28)$$

With the limits now defined, we recognize the following as the inverse transform of (11-28):

$$u_x^P(r,\theta,t) = \frac{Zv_s}{\pi\mu r}\,\mathrm{Im}\left[\frac{q(2q^2+1)}{R(q)}\,\frac{dq(\tau)}{d\tau}\right]H(\tau - a). \qquad (11\text{-}29)$$

Likewise

$$u_z^P(r,\theta,t) = \frac{Zv_s}{\pi\mu r}\,\mathrm{Re}\left[\frac{(q^2+a^2)^{\frac{1}{2}}(2q^2+1)}{R(q)}\,\frac{dq(\tau)}{d\tau}\right]H(\tau - a). \qquad (11\text{-}30)$$

We can change the above expressions to polar coordinates by using the following relationships:

$$u_r = u_x\mathrm{Sin}\theta + u_z\mathrm{Cos}\theta, \qquad (11\text{-}31)$$

$$u_\theta = u_x\mathrm{Cos}\theta - u_z\mathrm{Sin}\theta,$$

and

$$\mathrm{Re}\,\{-i(\xi + i\eta)\} = \mathrm{Im}\,\{\xi + i\eta\} = \eta, \qquad (11\text{-}32)$$

$$\mathrm{Im}\,\{i(\xi + i\eta)\} = \mathrm{Re}\,\{\xi + i\eta\} = \xi.$$

From (11-29) and (11-30) we can compute the radial component of P-motion:

$$u_r^P(r,\theta,t) = \frac{Zv_s}{\pi\mu r}\,\mathrm{Re}\left[\{(q^2+a^2)^{\frac{1}{2}}\mathrm{Cos}\theta - iq\mathrm{Sin}\theta\}\,\frac{2q^2+1}{R(q)}\,\frac{dq}{d\tau}\right]H(\tau - a)$$

$$= \frac{Zv_s}{\pi\mu r}\,\mathrm{Re}\left[\frac{\tau(2q^2+1)}{R(q)}\,\frac{dq}{d\tau}\right]H(\tau - a), \qquad (11\text{-}33)$$

where we have used (11-24). Equation (11-33) is exact; no approximations have been made to this point. We can also compute the transverse component of P-motion:

$$u_\theta^P(r,\theta,t) = \frac{Zv_s}{\pi\mu r}\,\mathrm{Im}\left[\{-i(q^2+a^2)^{\frac{1}{2}}\mathrm{Sin}\theta + q\mathrm{Cos}\theta\}\,\frac{2q^2+1}{R(q)}\,\frac{dq}{d\tau}\right]H(\tau - a).$$

But

$$-i(q^2+a^2)^{\frac{1}{2}}\mathrm{Sin}\theta + q\mathrm{Cos}\theta = -i\left(\frac{\tau + iq\mathrm{Sin}\theta}{\mathrm{Cos}\theta}\right)\mathrm{Sin}\theta + q\mathrm{Cos}\theta = \frac{-i\tau\mathrm{Sin}\theta + q}{\mathrm{Cos}\theta} = (\tau^2-a^2)^{\frac{1}{2}},$$

where we have used (11-25). Hence,

$$u_\theta^P(r,\theta,t) = \frac{Zv_s}{\pi\mu r}\,\mathrm{Im}\left[\frac{(\tau^2-a^2)^{\frac{1}{2}}(2q^2+1)}{R(q)}\,\frac{dq}{d\tau}\right]H(\tau - a). \qquad (11\text{-}34)$$

We now have exact expressions for the motion anywhere on the interior of the half-space, in both cartesian and polar coordinates. Expressions (11-29), (11-30), (11-33), and (11-34) can be evaluated on a modest computer. This was done in preparing the examples shown in Figures 11-6.

Approximations that are both useful and instructive may be made after a method developed by Knopoff (1958) called the First Motion Approximation. This amounts to considering only those motions caused by the singularities of the four expressions listed above and neglecting the slowly varying portions. In this sense, it is a high-frequency approximation. However, we always have the exact formulation when this approximation is not useful. Investigation of (11-33) and (11-34) shows that the functions vary quite smoothly except for the factor $dq/d\tau$ near the scaled time $\tau = a$, and the factor $R(q)$ near $q = ib$ when θ is near 90°. We shall look at the former singularity, but put off the latter until we have developed the S-motion. The value of q corresponding to $\tau = a$ is

$$q_o = ia\mathrm{Sin}\theta \; . \tag{11-35}$$

Expanding about this singular point we find that

$$\frac{dq}{d\tau} \simeq \frac{\tau\mathrm{Cos}\theta}{(\tau^2 - a^2)^{\frac{1}{2}}} \simeq \frac{a^{\frac{1}{2}}}{2^{\frac{1}{2}}}\,\frac{\mathrm{Cos}\theta}{(\tau - a)^{\frac{1}{2}}} \; . \tag{11-36}$$

We can then write

$$u_r^P(r,\theta,t) \simeq \frac{Zv_s^{\frac{1}{2}}}{\pi\mu r^{\frac{1}{2}}}\,\frac{a^{3/2}}{2^{\frac{1}{2}}}\left[\frac{\mathrm{Cos}\theta(1 - 2a^2\mathrm{Sin}^2\theta)}{(1-2a^2\mathrm{Sin}^2\theta)^2+4a^3\mathrm{Sin}^2\theta\mathrm{Cos}\theta(1-a^2\mathrm{Sin}^2\theta)^{\frac{1}{2}}}\right]\frac{H(t - r/v_p)}{(t-r/v_p)^{\frac{1}{2}}}$$

$$\simeq \frac{Zv_s^{\frac{1}{2}}}{\pi\mu r^{\frac{1}{2}}}\,\frac{a^{3/2}}{2^{\frac{1}{2}}}\,D_z^P(\theta)\,\frac{H(t - r/v_p)}{(t - r/v_p)^{\frac{1}{2}}} \; , \tag{11-37}$$

where we have used (9-41). As θ tends towards 90° the bracketed expression goes to zero. We will say more about this later. In (11-34) everything is real except for $dq/d\tau$, so we must use

$$\frac{dq}{d\tau} \simeq i\mathrm{Sin}\theta \; ,$$

and we find that

$$u_\theta^P(r,\theta,t) \simeq \frac{Zv_s^{3/2}2^{\frac{1}{2}}a^{\frac{1}{2}}}{\pi\mu r^{3/2}}\left[\frac{\mathrm{Sin}\theta(1 - 2a^2\mathrm{Sin}^2\theta)}{(1 - 2a^2\mathrm{Sin}^2\theta)^2 + 4a^3\mathrm{Sin}^2\theta\mathrm{Cos}\theta(1 - a^2\mathrm{Sin}^2\theta)^{\frac{1}{2}}}\right]$$

$$\cdot\,(t - r/v_p)^{\frac{1}{2}}\,H(t - r/v_p). \tag{11-38}$$

The θ-motion is thus the time-integral of the r-motion and drops off a factor of $1/r$ faster. Again we see that all P-motion is not perpendicular to the wave-front. However, the bracketed expression in (11-38) goes to zero at $\theta = 0^{\circ}$ as it should from symmetry considerations. At 90° it does not go to zero, so we see that at the surface (11-38) will contribute to the P-motion. But more about this later. At great distances and for angles less than 90°, the only singular behavior of (11-33) and (11-34) occurs at $\tau = a$. Consequently if the input signal were a pulse of short duration, then its directivity pattern would be governed by the bracketed expressions in (11-37) and (11-38). By short, we mean that the pulse length in time is much less than the time until the next arrival, either S or P-S (to be defined later).

For the S-motion, we have to use a slightly different change of variable. Here we have

$$\mathrm{Cos}\theta(q^2+1)^{\frac{1}{2}} - iq\mathrm{Sin}\theta = \tau = v_s t/r. \tag{11-39}$$

Inversely

$$q(\tau) = i\tau\mathrm{Sin}\theta + \mathrm{Cos}\theta(\tau^2-1)^{\frac{1}{2}}, \tag{11-40}$$

and

$$dq/d\tau = i\mathrm{Sin}\theta + \tau\mathrm{Cos}\theta(\tau^2-1)^{-\frac{1}{2}}. \tag{11-41}$$

Applying the same reasoning that led to (11-29) and (11-30) we obtain

$$u_x^s(r,\theta,t) = -\frac{Zv_s}{\pi\mu r}\,\mathrm{Im}\left[\frac{2q(q^2+a^2)^{\frac{1}{2}}(q^2+1)^{\frac{1}{2}}}{R(q)}\frac{dq(\tau)}{d\tau}\right]H(\tau-1) \quad (\theta<\theta_c) \tag{11-42}$$

$$= -\frac{Zv_s}{\pi\mu r}\,\mathrm{Im}\left[\frac{2q(q^2+a^2)^{\frac{1}{2}}(q^2+1)^{\frac{1}{2}}}{R(q)}\frac{dq(\tau)}{d\tau}\right]H(\tau-\tau_c) \quad (\theta>\theta_c), \tag{11-43}$$

and

$$u_z^s(r,\theta,t) = -\frac{Zv_s}{\pi\mu r}\,\mathrm{Re}\left[\frac{2q^2(q^2+a^2)^{\frac{1}{2}}}{R(q)}\frac{dq}{d\tau}\right]H(\tau-1) \quad (\theta<\theta_c) \tag{11-44}$$

$$= -\frac{Zv_s}{\pi\mu r}\,\mathrm{Re}\left[\frac{2q^2(q^2+a^2)^{\frac{1}{2}}}{R(q)}\frac{dq}{d\tau}\right]H(\tau-\tau_c) \quad (\theta>\theta_c), \tag{11-45}$$

where

$$\tau_c \equiv \tau(ia\mathrm{Sin}\theta) = (1-a^2)^{\frac{1}{2}}\mathrm{Cos}\theta + a\mathrm{Sin}\theta = \mathrm{Cos}(\theta-\theta_c). \tag{11-46}$$

The forms (11-43) and (11-45) occur because there is a contribution along the vertical segment from $q = ia$ to $q = i\mathrm{Sin}\theta$ as in Figure 11-3. For $\theta < \theta_c$ we have

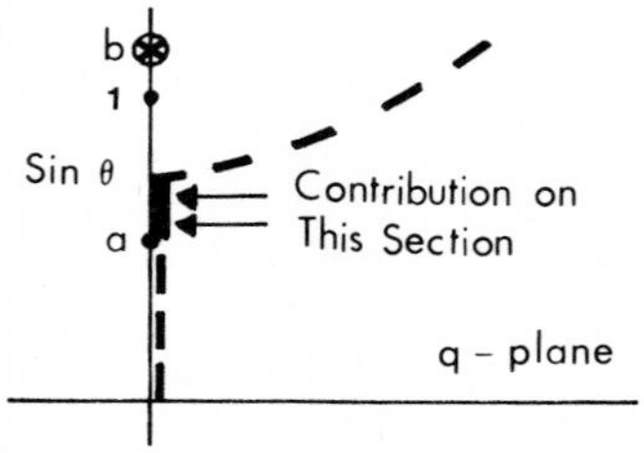

Figure 11-3. The deformed path of integration for S-motion. The motion immediately preceding the S-arrival (head-wave) comes from the imaginary axis between $q = ia$ and $q = i\mathrm{Sin}\theta$.

$$u_\theta^s(r,\theta,t) = -\frac{Zv_s}{\pi\mu r}\,\mathrm{Im}\left[\{(q^2+1)^{\frac{1}{2}}\mathrm{Cos}\theta - iq\mathrm{Sin}\theta\}\frac{2q(q^2+a^2)^{\frac{1}{2}}}{R(q)}\frac{dq}{d\tau}\right]H(\tau-1) \tag{11-47}$$

$$= -\frac{Zv_s}{\pi\mu r}\,\mathrm{Im}\left[\frac{2\tau q(q^2+a^2)^{\frac{1}{2}}}{R(q)}\frac{dq}{d\tau}\right]H(\tau-1) \quad (\theta<\theta_c).$$

For $\theta>\theta_c$, this becomes

$$u_\theta^s(r,\theta,t) = -\frac{Zv_s}{\pi\mu r}\,\mathrm{Im}\left[\frac{2\tau q(q^2+a^2)^{\frac{1}{2}}}{R(q)}\frac{dq}{d\tau}\right]H(\tau-\tau_c) \quad (\theta>\theta_c). \tag{11-48}$$

Also

$$u_r^s(r,\theta,t) = -\frac{Zv_s}{\pi\mu r}\,\mathrm{Re}\left[\{-i(q^2+1)^{\frac{1}{2}}\mathrm{Sin}\theta + q\mathrm{Cos}\theta\}\frac{2q(q^2+a^2)^{\frac{1}{2}}}{R(q)}\frac{dq}{d\tau}\right]H(\tau-1) \quad (11\text{-}49)$$

$$= -\frac{Zv_s}{\pi\mu r}\,\mathrm{Re}\left[\frac{(\tau^2-1)^{\frac{1}{2}}2q(q^2+a^2)^{\frac{1}{2}}}{R(q)}\frac{dq}{d\tau}\right]H(\tau-1) \qquad (\theta<\theta_c),$$

and

$$u_r^s(r,\theta,t) = -\frac{Zv_s}{\pi\mu r}\,\mathrm{Re}\left[\frac{(\tau^2-1)^{\frac{1}{2}}2q(q^2+a^2)^{\frac{1}{2}}}{R(q)}\frac{dq}{d\tau}\right]H(\tau-\tau_c) \qquad (\theta>\theta_c). \quad (11\text{-}50)$$

Again, expressions (11-42) through (11-45) and (11-47) through (11-50) give exact expressions for the components of motion in both the rectangular and polar coordinate directions. We will analyze the second set of expressions for the significant portions of the motion.

As before, the functions (11-47) and (11-49) are smoothly varying except for the factor $dq/d\tau$ at $\tau = 1$. However, (11-48) and (11-50) have a discontinuity at $\tau = 1$, a change in slope at $\tau = \tau_c$, and a large bump due to the smallness of $R(q)$ near $q = ib$ when θ is close to 90°. As the S-motion due to the singularity $\tau = 1$ occurs whether or not $\theta<\theta_c$, we can analyze for it and take account of the value of θ in the formulation. At this singular point, we must evaluate the product

$$q\frac{dq}{d\tau} = \left[i\tau\mathrm{Sin}\theta + (\tau^2-1)^{\frac{1}{2}}\mathrm{Cos}\theta\right]\left[i\mathrm{Sin}\theta + \frac{\tau\mathrm{Cos}\theta}{(\tau^2-1)^{\frac{1}{2}}}\right]$$

$$= \tau(\mathrm{Cos}^2\theta - \mathrm{Sin}^2\theta) + i\left[(\tau^2-1)^{\frac{1}{2}} + \frac{\tau^2}{(\tau^2-1)^{\frac{1}{2}}}\right]\mathrm{Sin}\theta\mathrm{Cos}\theta. \quad (11\text{-}51)$$

If the singular factor $(\tau^2-1)^{\frac{1}{2}}$ in the denominator is not cancelled out, we have

$$q\frac{dq}{d\tau} \simeq \frac{i\mathrm{Sin}\theta\mathrm{Cos}\theta}{2^{\frac{1}{2}}(\tau-1)^{\frac{1}{2}}} = -\frac{\mathrm{Sin}\theta\mathrm{Cos}\theta}{2^{\frac{1}{2}}(1-\tau)^{\frac{1}{2}}}. \quad (11\text{-}52)$$

Then for $\tau > 1$, $0 < \theta < \pi/2$, we have

$$u_\theta^{s+}(r,\theta,t) \simeq -\frac{Zv_s^{\frac{1}{2}}}{\pi\mu r^{\frac{1}{2}}}\frac{1}{2^{\frac{1}{2}}}\,\mathrm{Re}\left[\frac{2\mathrm{Sin}\theta\mathrm{Cos}\theta(a^2-\mathrm{Sin}^2\theta)^{\frac{1}{2}}}{(1-2\mathrm{Sin}^2\theta)^2 + 4\mathrm{Sin}^2\theta\mathrm{Cos}\theta(a^2-\mathrm{Sin}^2\theta)^{\frac{1}{2}}}\right]\frac{H(t-r/v_s)}{(t-r/v_s)^{\frac{1}{2}}}$$

$$\simeq -\frac{Zv_s^{\frac{1}{2}}}{\pi\mu r^{\frac{1}{2}}}\frac{1}{2^{\frac{1}{2}}}\,\mathrm{Re}\left[D_z^s(\theta)\right]\frac{H(t-r/v_s)}{(t-r/v_s)^{\frac{1}{2}}}, \quad (11\text{-}53)$$

where the s+ superscript stands for the approximate motion after the S-arrival. When $\theta > \theta_c$ the square root factor of (11-48) and (11-50) takes the form

$$(a^2-\mathrm{Sin}^2\theta)^{\frac{1}{2}} = i(\mathrm{Sin}^2\theta - a^2)^{\frac{1}{2}}, \quad (11\text{-}54)$$

so with this change (11-53) holds for all angles between 0° and 90°. The angular factor in (11-53) is the same as in (9-44) which measured the vertical component of motion due to incident S-waves. We also have

$$u_r^{s+} \simeq -\frac{Zv_s^{3/2}2^{\frac{1}{2}}}{\pi\mu r^{3/2}}\,\mathrm{Re}\left[\frac{2(a^2-\mathrm{Sin}^2\theta)^{\frac{1}{2}}(\mathrm{Cos}^2\theta - \mathrm{Sin}^2\theta)}{(1-2\mathrm{Sin}^2\theta)^2 + 4\mathrm{Sin}^2\theta\mathrm{Cos}\theta(a^2-\mathrm{Sin}^2\theta)^{\frac{1}{2}}}\right]$$

$$\cdot\,(t-r/v_s)^{\frac{1}{2}}H(t-r/v_s). \quad (11\text{-}55)$$

Again the factor $(a^2 - \mathrm{Sin}^2\theta)^{\frac{1}{2}}$ can go imaginary for $\theta > \theta_c$. Again we have the time-integral of u_θ^{s+} and a $1/r$ greater dropoff with distance.

For $\tau < 1$, $\theta < \theta_c$ there is no imaginary part to (11-47) or real part to (11-49). However, for $\tau < 1$, $\theta_c < \theta < \pi/2$ we have

$$u_\theta^{s-} \simeq \frac{Zv_s^{\frac{1}{2}}}{\pi\mu r^{\frac{1}{2}}} \frac{1}{2^{\frac{1}{2}}} \mathrm{Re}\left[\frac{2\mathrm{Sin}\theta\mathrm{Cos}\theta(\mathrm{Sin}^2\theta - a^2)^{\frac{1}{2}}}{(1-2\mathrm{Sin}^2\theta)^2 + 4i\mathrm{Sin}^2\theta\mathrm{Cos}\theta(\mathrm{Sin}^2\theta - a^2)^{\frac{1}{2}}} \right] \frac{H(r/v_s - t)}{(r/v_s - t)^{\frac{1}{2}}}$$

$$\simeq - \frac{Zv_s^{\frac{1}{2}}}{\pi\mu r^{\frac{1}{2}}} \frac{1}{2^{\frac{1}{2}}} \mathrm{Im}\left[D_z^s(\theta) \right] \frac{H(r/v_s - t)}{(r/v_s - t)^{\frac{1}{2}}} . \tag{11-56}$$

Also

$$u_r^{s-} \simeq - \frac{Zv_s^{3/2} 2^{\frac{1}{2}}}{\pi\mu r^{3/2}} \mathrm{Re}\left[\frac{2(\mathrm{Sin}^2\theta - a^2)^{\frac{1}{2}}(\mathrm{Cos}^2\theta - \mathrm{Sin}^2\theta)}{(1-2\mathrm{Sin}^2\theta)^2 + 4i\mathrm{Sin}^2\theta\mathrm{Cos}\theta(\mathrm{Sin}^2\theta - a^2)^{\frac{1}{2}}} \right] \tag{11-57}$$

$$\cdot\ (r/v_s - t)^{\frac{1}{2}} H(r/v_s - t).$$

<u>Head Waves</u> (<u>Refraction Arrival</u>). To analyze the arrival at $\tau = \tau_c$ corresponding to $q = ia$, we will go to the components u_x^s and u_z^s given by (11-43) and (11-45). In the vicinity of $q = ia$ we will let

$$q = i(a + \varepsilon). \tag{11-58}$$

Then from (11-40) we have

$$\left.\frac{d\tau}{dq}\right|_{ia} = \left.\frac{q\mathrm{Cos}\theta}{(q^2+1)^{\frac{1}{2}}} - i\mathrm{Sin}\theta\right|_{ia} = \frac{i\ \{a\mathrm{Cos}\theta - \mathrm{Sin}\theta(1-a^2)^{\frac{1}{2}}\}}{(1-a^2)^{\frac{1}{2}}} = \frac{-i\mathrm{Sin}(\theta-\theta_c)}{(1-a^2)^{\frac{1}{2}}} .$$

Consequently

$$\left.\frac{dq}{d\tau}\right|_{ia} = \frac{i(1 - a^2)^{\frac{1}{2}}}{\mathrm{Sin}(\theta-\theta_c)} . \tag{11-59}$$

Also

$$\tau - \tau_c \simeq i\varepsilon \frac{d\tau}{dq} = \frac{\varepsilon\mathrm{Sin}(\theta-\theta_c)}{(1 - a^2)^{\frac{1}{2}}} ,$$

or

$$\varepsilon = \frac{(\tau - \tau_c)(1 - a^2)^{\frac{1}{2}}}{\mathrm{Sin}(\theta-\theta_c)} . \tag{11-60}$$

Lastly

$$(q^2 + a^2)^{\frac{1}{2}} \simeq (-2a\varepsilon)^{\frac{1}{2}} = i(2a\varepsilon)^{\frac{1}{2}} = \frac{i(2a)^{\frac{1}{2}}(\tau - \tau_c)^{\frac{1}{2}}(1 - a^2)^{\frac{1}{4}}}{\mathrm{Sin}^{\frac{1}{2}}(\theta-\theta_c)} . \tag{11-61}$$

Then we have (liberally using $a = \mathrm{Sin}\theta_c$, $(1 - a^2)^{\frac{1}{2}} = \mathrm{Cos}\theta_c$)

$$u_x^{P-S} \simeq \frac{Zv_s^{3/2}(2a)^{3/2}}{\pi\mu r^{3/2}} \left[\frac{(1-a^2)^{3/4}\mathrm{Cos}\theta_c}{(1-2a^2)^2\mathrm{Sin}^{3/2}(\theta-\theta_c)} \right] (t - t_c)^{\frac{1}{2}} H(t - t_c), \tag{11-62}$$

and

$$u_z^{P-S} \simeq - \frac{Zv_s^{3/2}(2a)^{3/2}}{\pi\mu r^{3/2}} \left[\frac{(1-a^2)^{3/4}\mathrm{Sin}\theta_c}{(1-2a^2)^2\mathrm{Sin}^{3/2}(\theta-\theta_c)} \right] (t - t_c)^{\frac{1}{2}} H(t - t_c), \tag{11-63}$$

or

$$\mathbf{u}^{P-S} \approx \mathbf{e}_{\theta_c} \frac{Zv_s^{3/2}(2a)^{3/2}}{\pi\mu r^{3/2}} \left[\frac{(1-a^2)^{3/4}}{(1-2a^2)^2 \mathrm{Sin}^{3/2}(\theta-\theta_c)} \right] (t-t_c)^{\frac{1}{2}} H(t-t_c). \qquad (11\text{-}64)$$

Now from (11-46)

$$t_c = r/v_s \,\mathrm{Cos}(\theta-\theta_c).$$

This is the time necessary to travel the length $\overline{ad}$ in Figure 11-4 with the velocity v_s. However, from the definition of the critical angle, the time for a wave to go from a to b along the surface with velocity v_p equals the time to go from a to c with velocity v_s. Hence t_c is the time necessary to travel the path $\overline{abe}$ along $\overline{ab}$ with velocity v_p and along $\overline{be}$ with velocity v_s.

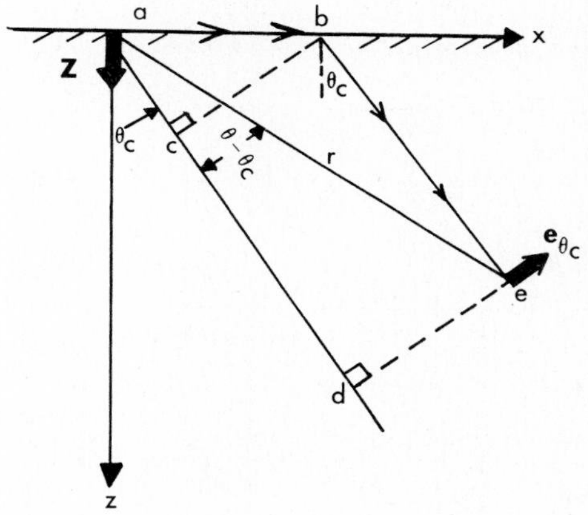

Figure 11-4. Geometric relationships for head waves in Lamb's Problem.

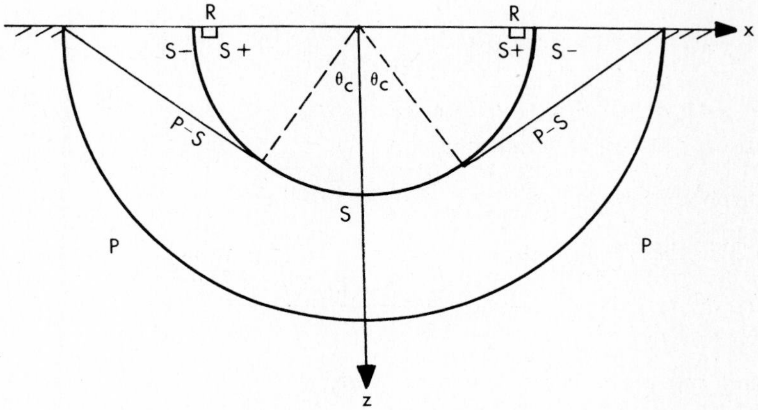

Figure 11-5. Wavefronts for Lamb's problem. These are the same regardless of whether a normal or tangential impulse is applied.

<u>Wave Fronts</u>. Depending upon the angle from the source, we can have three distinct arrivals (plus the surface motion due to the root of the Rayleigh Denominator). These occur at $t = r/v_p$, $t = r/v_s$, and $t = (r/v_s)\mathrm{Cos}(\theta-\theta_c)$. The corresponding wavefronts are sketched in Figure 11-5. The origin of the P-S motion can be explained as a continuous generation of S by the P-wave as it travels along and interacts with the free surface. The P-S wave then travels in the direction defined by θ_c so as to maintain the proper phase relationship with P along the surface. This conversion of energy is compatible with the fact that the angular part of (11-37) goes to zero at $\theta = 90^\circ$ corresponding to the free surface. The same wavefronts are developed in the case of a horizontal force which will be considered shortly.

<u>Motion on the Epicentral Line</u>. For $\theta = 0$, there is a simplification and one can analyze this case more fully. We can work most conveniently starting with (11-29) and (11-30). For P-motion we have at $\theta = 0$ (from (11-25) and (11-26))

$$q = (\tau^2 - a^2)^{\frac{1}{2}}; \quad \frac{dq}{d\tau} = \frac{\tau}{(\tau^2 - a^2)^{\frac{1}{2}}} . \tag{11-65}$$

This says that the path of q lies entirely along the real axis for $\tau > a$. Then (11-29) becomes

$$u_x^P(r,0^\circ,t) = -\frac{Zv_s}{\pi\mu r}\,\mathrm{Im}\left[\frac{\{2(\tau^2-a^2)+1\}\tau}{\{2(\tau^2-a^2)+1\}^2 - 4(\tau^2-a^2)(\tau^2-a^2+1)^{\frac{1}{2}}\tau}\right]H(\tau-a) \equiv 0, \tag{11-66}$$

as it should from consideration of symmetry. We also see that (11-30) becomes

$$u_z^P(r,0^\circ,t) = \frac{Zv_s}{\pi\mu r}\left[\frac{\tau^2\{2(\tau^2-a^2)+1\}}{\{2(\tau^2-a^2)+1\}^2 - 4(\tau^2-a^2)(\tau^2-a^2+1)^{\frac{1}{2}}\tau}\right]\frac{H(\tau-a)}{(\tau^2-a^2)^{\frac{1}{2}}} . \tag{11-67}$$

Expanding about the point $\tau = a$, we find that

$$u_z^P(r,0^\circ,t) \simeq \frac{Zv_s^{\frac{1}{2}}a^{3/2}}{\pi\mu r^{\frac{1}{2}}2^{\frac{1}{2}}}\,\frac{H(t - r/v_p)}{(t - r/v_p)^{\frac{1}{2}}} , \tag{11-68}$$

which is in agreement with (11-37) for $\theta = 0^\circ$. For S-motion we also have that

$$q = (\tau^2 - 1)^{\frac{1}{2}}, \qquad \frac{dq}{d\tau} = \frac{\tau}{(\tau^2 - 1)^{\frac{1}{2}}} ,$$

and again the path of integration lies along the real axis. Making the proper substitutions into (11-43) we see that

$$u_x^S(r,0^\circ,t) \equiv 0, \tag{11-69}$$

and that (11-45) becomes

$$u_z^S(r,0^\circ,t) = -\frac{Zv_s}{\pi\mu r}\,\frac{2\tau(\tau^2-1)^{\frac{1}{2}}(\tau^2-1+a^2)^{\frac{1}{2}}}{(2\tau^2-1)^2 - 4\tau(\tau^2-1)(\tau^2-1+a^2)^{\frac{1}{2}}}\,H(\tau-1). \tag{11-70}$$

Expanding about the point $\tau = 1$, we find that

$$u_z^S(r,0^\circ,t) \simeq -\frac{Zv_s^{3/2}2^{3/2}}{\pi\mu r^{3/2}}\,a(t - r/v_s)^{\frac{1}{2}}\,H(t - r/v_s), \tag{11-71}$$

which is a change in slope and is in agreement with (11-55) for $\theta = 0$. Here again we have S-motion which is purely normal to the wavefront. The total motion on the epicentral line is shown at the top of Figure 11-6. (Note that $u_z = u_r$.)

Motion on the Free Surface. We note also that (11-37) and (11-53) also vanish as θ tends to 90°, i.e., as $\underline{z}$ tends to zero. Rewriting (11-17) for $z = 0$, we have

$$U_x(r,90^\circ,s) = \frac{Z}{\pi\mu}\,\mathrm{Im}\int_0^\infty \frac{(2q^2+1) - 2(q^2+a^2)^{\frac{1}{2}}(q^2+1)^{\frac{1}{2}}}{R(q)}\, e^{(s/v_s)iqr}\, q\,dq. \quad (11\text{-}72)$$

The Cagniard-deHoop transformation in this case becomes

$$q = iv_s t/r = i\tau \quad \text{and} \quad dq/d\tau = i. \quad (11\text{-}73)$$

Then

$$U_x(r,90^\circ,s) = -\frac{Zv_s}{\pi\mu r}\,\mathrm{Im}\left[\begin{aligned} &\int_{t_p}^{t_r-\varepsilon} \frac{(1-2\tau^2) - 2i(\tau^2-a^2)^{\frac{1}{2}}(1-\tau^2)^{\frac{1}{2}}}{(1-2\tau^2)^2 + 4i\tau^2(\tau^2-a^2)^{\frac{1}{2}}(1-\tau^2)^{\frac{1}{2}}}\,\tau e^{-st}\,dt \\ &+ \underbrace{\int_{t_r+\varepsilon}^{\infty} \frac{(1-2\tau^2) + 2(\tau^2-a^2)^{\frac{1}{2}}(\tau^2-1)^{\frac{1}{2}}}{(1-2\tau^2)^2 - 4\tau^2(\tau^2-a^2)^{\frac{1}{2}}(\tau^2-1)^{\frac{1}{2}}}\,\tau e^{-st}\,dt}_{\to\,0} \\ &+ \frac{r\pi}{v_s}\,\frac{(1-2\tau_r^2) + 2(\tau_r^2-a^2)^{\frac{1}{2}}(\tau_r^2-1)^{\frac{1}{2}}}{R'(i\tau_r)}\,\tau_r e^{-st_r} \end{aligned}\right].$$

The second term vanishes as it has no imaginary part. Now $R'(i\tau_r)$ is positive imaginary inasmuch as R is positive at $i(1+\varepsilon)$, negative at $i\infty$, hence δR will be negative and δq will be positive imaginary. Then

$$u_x(r,90^\circ,t) = -\frac{Zv_s}{\pi\mu r}\,\mathrm{Im}\left[\frac{(1-2\tau^2) - 2i(\tau^2-a^2)^{\frac{1}{2}}(1-\tau^2)^{\frac{1}{2}}}{(1-2\tau^2)^2 + 4i\tau^2(\tau^2-a^2)^{\frac{1}{2}}(1-\tau^2)^{\frac{1}{2}}}\,\tau\right]\left[H(t-r/v_p) - H(t-r/v_s)\right]$$

$$+\frac{Z}{\mu}\,\frac{(1-2\tau_r^2) + 2(\tau_r^2-a^2)^{\frac{1}{2}}(\tau_r^2-1)^{\frac{1}{2}}}{|R'(i\tau_r)|}\,\tau_r\,\delta(t-t_r), \quad (11\text{-}74)$$

since the bracketed portion of the first line of (11-74) becomes real for $\tau > 1$. Expanding about $\tau = a$, we have

$$u_x(r,90^\circ,t) \simeq \frac{Zv_s^{3/2}}{\pi\mu r^{3/2}}\,\frac{(2a)^{3/2}(1-a^2)^{\frac{1}{2}}}{(1-2a^2)^3}\,(t-r/v_p)^{\frac{1}{2}}\,H(t-r/v_p). \quad (11\text{-}75)$$

Expanding about $\tau = 1$, we have

$$u_x(r,90^\circ,t) \simeq -\frac{Zv_s^{3/2}}{\pi\mu r^{3/2}}\,2^{3/2}(1-a^2)^{\frac{1}{2}}(r/v_s - t)^{\frac{1}{2}}\,H(r/v_s - t). \quad (11\text{-}76)$$

It can be seen from (11-74) that when $\tau = 2^{-\frac{1}{2}}$, the bracketed portion of the first line is purely real and

$$u_x(r,90^\circ,2^{-\frac{1}{2}}r/v_s) = 0. \quad (11\text{-}77)$$

The x-component of motion is shown at the bottom of Figure 11-6. (Note that $u_x = u_r$.)

Going back to (11-18), we can rewrite it for $z = 0$ in the following form:

$$U_z(r,90^\circ,s) = \frac{Z}{\pi\mu}\,\mathrm{Re}\int_0^\infty \frac{(q^2+a^2)^{\frac{1}{2}} e^{(s/v_s)iqr}}{R(q)}\,dq. \tag{11-78}$$

Again in this case we have

$$q = i\,\frac{v_s t}{r} = i\tau, \qquad \frac{dq}{d\tau} = i. \tag{11-79}$$

Then

$$U_z(r,90^\circ,s) = -\frac{Zv_s}{\pi\mu r}\,\mathrm{Re}\left[\int_{t_p}^{t_r-\varepsilon} \frac{(\tau^2-a^2)^{\frac{1}{2}} e^{-st}\,dt}{(1-2\tau^2)^2 + 4i\tau^2(\tau^2-a^2)^{\frac{1}{2}}(1-\tau^2)^{\frac{1}{2}}} + \int_{t_r+\varepsilon}^{\infty} \frac{(\tau^2-a^2)^{\frac{1}{2}} e^{-st}\,dt}{(1-2\tau^2)^2 - 4\tau^2(\tau^2-a^2)^{\frac{1}{2}}(\tau^2-1)^{\frac{1}{2}}} + \frac{r\pi}{v_s}\cdot\frac{(\tau_r^2-a^2)^{\frac{1}{2}} e^{-st_r}}{R'(i\tau_r)} \to 0\right].$$

We then have

$$u_z(r,90^\circ,t) = -\frac{Zv_s}{\pi\mu r}\,\mathrm{Re}\left[\frac{(\tau^2-a^2)^{\frac{1}{2}}}{(1-2\tau^2)^2 + 4i\tau^2(\tau^2-a^2)^{\frac{1}{2}}(1-\tau^2)^{\frac{1}{2}}}\right] H(\tau-a). \tag{11-80}$$

Expanding about $\tau = a$, we find that

$$u_z(r,90^\circ,t) \simeq -\frac{Zv_s^{3/2}}{\pi\mu r^{3/2}}\,\frac{(2a)^{\frac{1}{2}}}{(1-2a^2)^2}\,(t-r/v_p)^{\frac{1}{2}}\,H(t-r/v_p). \tag{11-81}$$

For $\tau = 2^{-\frac{1}{2}}$, the bracketed portion of (11-80) is pure imaginary and we find that

$$u_z(r,90^\circ,2^{-\frac{1}{2}}r/v_s) = 0. \tag{11-82}$$

Expanding about $\tau = 1$, we find that

$$u_z(r,90^\circ,t) \simeq -\frac{Zv_s^{3/2}}{\pi\mu r^{3/2}}(1-a^2)\,2^{5/2}(t-r/v_s)^{\frac{1}{2}}\,H(t-r/v_s). \tag{11-83}$$

For τ in the neighborhood of <u>b</u>, we have that

$$u_z(r,90^\circ,t) = -\frac{Zv_s}{\pi\mu r}\,\frac{(\tau^2-a^2)^{\frac{1}{2}}}{(1-2\tau^2)^2 - 4\tau^2(\tau^2-a^2)^{\frac{1}{2}}(\tau^2-1)^{\frac{1}{2}}}. \tag{11-84}$$

For $\tau = 1^+$ the denominator is positive. The z-component of motion is shown at the bottom of Figure 11-6. Note that $u_z = -u_\theta$.

Additional considerations, and additional figures are given by Sherwood (1958) who used a different method to obtain his results. He also diagrams the P- and S-components of motion which we do not.

<u>Reciprocity and the Buried Source</u>.

The reciprocity theorem of Section 8.4 tells us that the vertical component of motion on the surface for a buried vertical source is the same as (11-30) plus either (11-44) or (11-45) depending on the angle θ. The vertical component of motion on

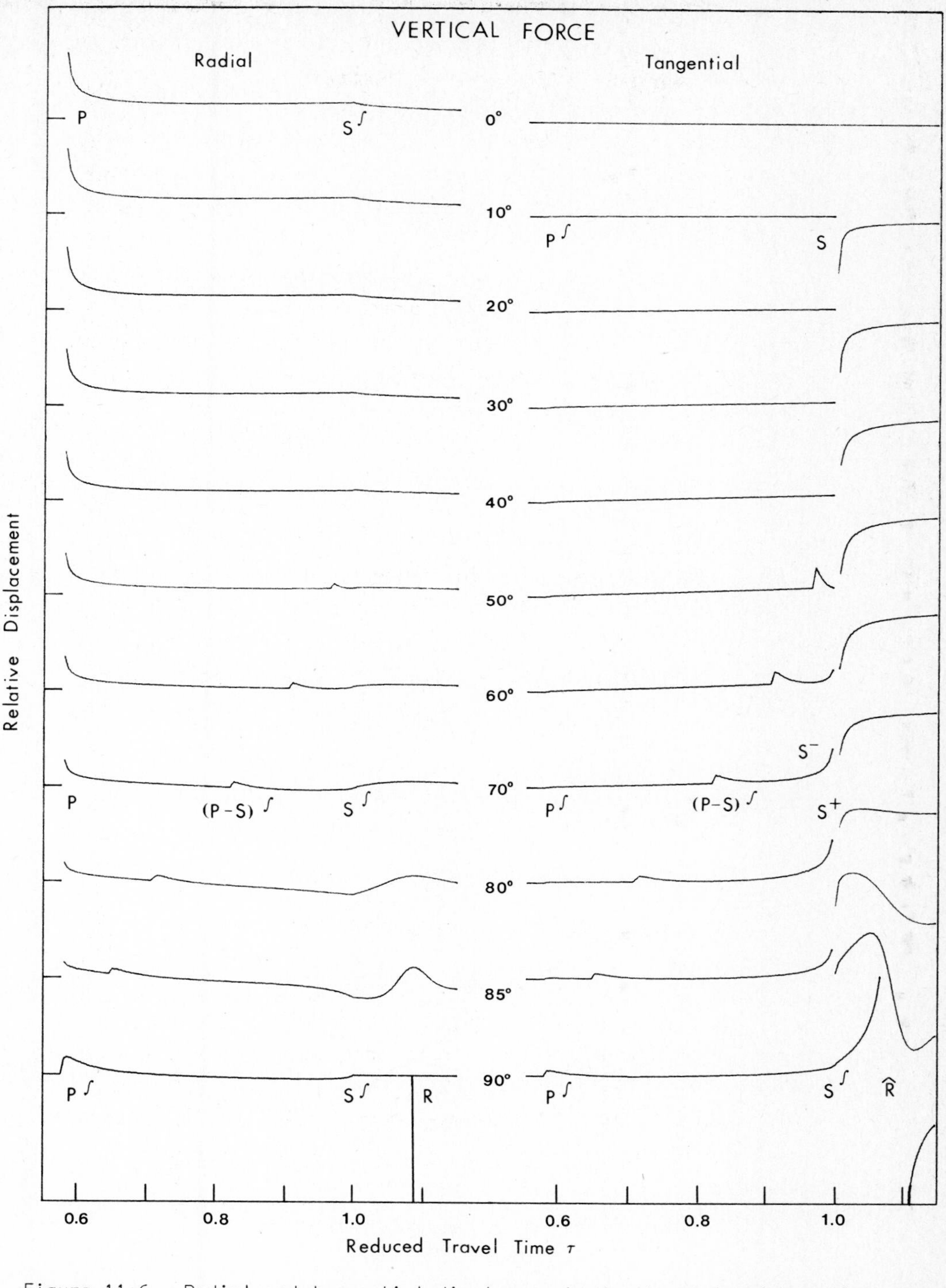

Figure 11-6. Radial and tangential displacements due to an impulse acting normally to the surface of a half-space. "∫" implies time-integral. "^" implies allied function.

the surface for a buried horizontal source would be the same as (11-29) plus the appropriate one of the pair (11-42) or (11-43). The approximate formulation worked out for the P-, S-, and Rayleigh arrivals would also hold.

Reciprocity again tells us of the equivalence between the horizontal surface motion due to a buried vertical source with (11-93) and one of (11-100) or (11-101). Also equivalent are the horizontal surface motion due to a buried horizontal source and (11-92) plus one of (11-98) or (11-99). The relations mentioned here will be derived in the next section.

Unfortunately, reciprocity won't help us when we wish to evaluate the motion on the interior of the half-space due to a buried source. Although one can set up the problem in exactly the same manner, there is one exasperating detail. For the PS reflection and the SP reflection the formal expression for $\tau = \tau(q)$ can be obtained, but it can be inverted only numerically by some iteration scheme or other in the complex plane. Consequently the new features one would hope to see are not immediately apparent.

11.3. A Surface Horizontal Force.

In this case we start with the same potentials (11-9) and (11-10) that we used in the vertical force case. The boundary conditions, however, are changed. The pertinent geometrical relations are shown in Figure 11-7. In this case, we have a horizontal force acting in the plus x-direction. This is equivalent to a negative line-stress distribution of t_{zx} at the origin. We write as the boundary conditions:

$$t_{zz} \equiv 0,$$

$$t_{zx} = -\frac{X}{2\pi\mu}\int_{-\infty}^{\infty} e^{i(sq/v_s)x}\, d(sq/v_s) = -X\delta(x). \qquad (11\text{-}85)$$

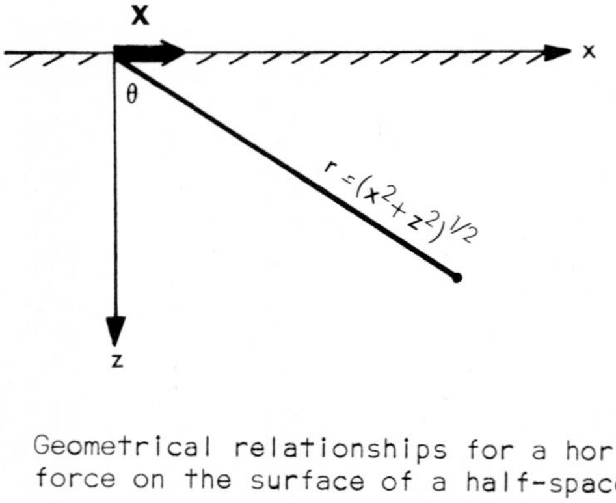

Figure 11-7. Geometrical relationships for a horizontal force on the surface of a half-space.

The time dependence of the second expression is a delta function. Evaluating the coefficients A(q) and B(q) upon substitution into (11-85), we find that

$$A(q) = \frac{X}{2\pi\mu}\frac{v_s}{s}\frac{2i(q^2+1)^{\frac{1}{2}}}{R(q)},$$

$$B(q) = \frac{X}{2\pi\mu}\frac{v_s}{s}\frac{2q^2+1}{R(q)},$$

where again,

$$R(q) = (2q^2+1)^2 - 4q^2(q^2+a^2)^{\frac{1}{2}}(q^2+1)^{\frac{1}{2}}.$$

The potentials may then be written

$$\Phi(x,z,s) = \frac{X}{2\pi\mu}\frac{v_s}{s}\int_{-\infty}^{\infty}\frac{2iq(q^2+1)^{\frac{1}{2}}}{R(q)}e^{(s/v_s)\{iqx - (q^2+a^2)^{\frac{1}{2}}z\}}dq. \tag{11-86}$$

$$\Psi^{\dagger}(x,z,s) = \frac{X}{2\pi\mu}\frac{v_s}{s}\int_{-\infty}^{\infty}\frac{2q^2+1}{R(q)}e^{(s/v_s)\{iqx - (q^2+1)^{\frac{1}{2}}z\}}dq. \tag{11-87}$$

Calculating the displacements from (11-4) and introducting polar coordinates we obtain

$$U_x^P(r,\theta,s) = -\frac{X}{\pi\mu}\,\mathrm{Re}\int_0^{\infty}\frac{2q^2(q^2+1)^{\frac{1}{2}}}{R(q)}e^{(s/v_s)\{iq\mathrm{Sin}\theta - (q^2+a^2)^{\frac{1}{2}}\mathrm{Cos}\theta\}}dq. \tag{11-88}$$

$$U_x^S(r,\theta,s) = \frac{X}{\pi\mu}\,\mathrm{Re}\int_0^{\infty}\frac{(2q^2+1)(q^2+1)^{\frac{1}{2}}}{R(q)}e^{(s/v_s)\{iq\mathrm{Sin}\theta - (q^2+1)^{\frac{1}{2}}\mathrm{Cos}\theta\}}dq. \tag{11-89}$$

$$U_z^P(r,\theta,s) = \frac{X}{\pi\mu}\,\mathrm{Im}\int_0^{\infty}\frac{2q(q^2+a^2)^{\frac{1}{2}}(q^2+1)^{\frac{1}{2}}}{R(q)}e^{(s/v_s)\{iq\mathrm{Sin}\theta - (q^2+a^2)^{\frac{1}{2}}\mathrm{Cos}\theta\}}dq. \tag{11-90}$$

$$U_z^S(r,\theta,s) = -\frac{X}{\pi\mu}\,\mathrm{Im}\int_0^{\infty}\frac{q(2q^2+1)}{R(q)}e^{(s/v_s)\{iq\mathrm{Sin}\theta - (q^2+1)^{\frac{1}{2}}\mathrm{Cos}\theta\}}dq. \tag{11-91}$$

Using the Cagniard-deHoop transformation (11-24), we obtain

$$u_x^P(r,\theta,t) = -\frac{Xv_s}{\pi\mu r}\,\mathrm{Re}\left[\frac{2q^2(q^2+1)^{\frac{1}{2}}}{R(q)}\frac{dq}{d\tau}\right]H(\tau - a), \tag{11-92}$$

$$u_z^P(r,\theta,t) = \frac{Xv_s}{\pi\mu r}\,\mathrm{Im}\left[\frac{2q(q^2+1)^{\frac{1}{2}}(q^2+a^2)^{\frac{1}{2}}}{R(q)}\frac{dq}{d\tau}\right]H(\tau - a). \tag{11-93}$$

Using (11-31) to transform to polar components, we obtain

$$u_r^P(r,\theta,t) = \frac{Xv_s}{\pi\mu r}\,\mathrm{Im}\left[\frac{2\tau q(q^2+1)^{\frac{1}{2}}}{R(q)}\frac{dq}{d\tau}\right]H(\tau - a), \tag{11-94}$$

$$u_\theta^P(r,\theta,t) = -\frac{Xv_s}{\pi\mu r}\,\mathrm{Re}\left[\frac{2(\tau^2-a^2)^{\frac{1}{2}}q(q^2+1)^{\frac{1}{2}}}{R(q)}\frac{dq}{d\tau}\right]H(\tau - a). \tag{11-95}$$

Expanding about the critical point $\tau = a$, we find that

$$u_r^P(r,\theta,t) \simeq \frac{Xv_s^{\frac{1}{2}}a^{3/2}}{\pi\mu r^{\frac{1}{2}}2^{\frac{1}{2}}}\left[\frac{2a\mathrm{Sin}\theta\mathrm{Cos}\theta(1 - a^2\mathrm{Sin}^2\theta)^{\frac{1}{2}}}{(1-2a^2\mathrm{Sin}^2\theta)^2 + 4a^3\mathrm{Sin}^2\theta\mathrm{Cos}\theta(1 - a^2\mathrm{Sin}^2\theta)^{\frac{1}{2}}}\right]\frac{H(t-r/v_p)}{(t-r/v_p)^{\frac{1}{2}}}$$

$$\simeq \frac{Xv_s^{\frac{1}{2}}}{\pi\mu r^{\frac{1}{2}}}\frac{a^{3/2}}{2^{\frac{1}{2}}}D_x^P(\theta)\frac{H(t - r/v_p)}{(t - r/v_p)^{\frac{1}{2}}}, \tag{11-96}$$

where we have used (9-42). Also we have

$$u_\theta^p(r,\theta,t) \simeq -\frac{Xv_s^{3/2}2^{\frac{1}{2}}a^{\frac{1}{2}}}{\pi\mu r^{3/2}}\left[\frac{2a(\mathrm{Cos}^2\theta - \mathrm{Sin}^2\theta)(1-a^2\mathrm{Sin}^2\theta)^{\frac{1}{2}}}{(1-2a^2\mathrm{Sin}^2\theta)^2 + 4a^3\mathrm{Sin}^2\theta\mathrm{Cos}\theta(1-a^2\mathrm{Sin}^2\theta)^{\frac{1}{2}}}\right]$$

$$\cdot\ (t - r/v_p)^{\frac{1}{2}}H(t - r/v_p). \qquad (11\text{-}97)$$

Again we see the time-integral and the $r^{-3/2}$ dropoff with distance. In obtaining (11-97) use was made of the relation

$$qdq/d\tau = \tau(\mathrm{Cos}^2\theta - \mathrm{Sin}^2\theta) + i\{(\tau^2 - a^2)^{\frac{1}{2}} + \tau^2(\tau^2 - a^2)^{-\frac{1}{2}}\}\mathrm{Cos}\theta\mathrm{Sin}\theta.$$

Using the Cagniard-deHoop transformation (11-39) we obtain from (11-89) and (11-91)

$$u_x^s(r,\theta,t) = \frac{Xv_s}{\pi\mu r}\mathrm{Re}\left[\frac{(2q^2 + 1)(q^2+ 1)^{\frac{1}{2}}}{R(q)}\frac{dq}{d\tau}\right]H(\tau - 1) \qquad (\theta<\theta_c), \qquad (11\text{-}98)$$

$$= \frac{Xv_s}{\pi\mu r}\mathrm{Re}\left[\frac{(2q^2+ 1)(q^2+ 1)^{\frac{1}{2}}}{R(q)}\frac{dq}{d\tau}\right]H(\tau - \tau_c) \qquad (\theta>\theta_c), \qquad (11\text{-}99)$$

and

$$u_z^s(r,\theta,t) = -\frac{Xv_s}{\pi\mu r}\mathrm{Im}\left[\frac{q(2q^2+ 1)}{R(q)}\frac{dq}{d\tau}\right]H(\tau - 1) \qquad (\theta<\theta_c), \qquad (11\text{-}100)$$

$$= -\frac{Xv_s}{\pi\mu r}\mathrm{Im}\left[\frac{q(2q^2+ 1)}{R(q)}\frac{dq}{d\tau}\right]H(\tau - \tau_c) \qquad (\theta>\theta_c). \qquad (11\text{-}101)$$

Transforming to polar components gives

$$u_\theta^s(r,\theta,t) = \frac{Xv_s}{\pi\mu r}\mathrm{Re}\left[\frac{\tau(2q^2+ 1)}{R(q)}\frac{dq}{d\tau}\right]H(\tau - 1) \qquad (\theta<\theta_c), \qquad (11\text{-}102)$$

$$= \frac{Xv_s}{\pi\mu r}\mathrm{Re}\left[\frac{\tau(2q^2+ 1)}{R(q)}\frac{dq}{d\tau}\right]H(\tau - \tau_c) \qquad (\theta>\theta_c), \qquad (11\text{-}103)$$

and

$$u_r^s(r,\theta,t) = -\frac{Xv_s}{\pi\mu r}\mathrm{Im}\left[\frac{(\tau^2- 1)^{\frac{1}{2}}(2q^2+ 1)}{R(q)}\frac{dq}{d\tau}\right]H(\tau - 1) \qquad (\theta<\theta_c), \qquad (11\text{-}104)$$

$$= -\frac{Xv_s}{\pi\mu r}\mathrm{Im}\left[\frac{(\tau^2- 1)^{\frac{1}{2}}(2q^2+ 1)}{R(q)}\frac{dq}{d\tau}\right]H(\tau - \tau_c) \qquad (\theta>\theta_c), \qquad (11\text{-}105)$$

where as before

$$\tau_c = (1 - a^2)^{\frac{1}{2}}\mathrm{Cos}\theta + a\mathrm{Sin}\theta = a\mathrm{Cos}(\theta - \theta_c). \qquad (11\text{-}46)$$

Expanding about the critical point $\tau = 1$, we have for $0 \le \theta \le \pi/2$

$$u_\theta^{s+}(r,\theta,t) \simeq \frac{Xv_s^{\frac{1}{2}}}{\pi\mu r^{\frac{1}{2}}2^{\frac{1}{2}}}\mathrm{Re}\left[\frac{\mathrm{Cos}\theta(1-2\mathrm{Sin}^2\theta)}{(1-2\mathrm{Sin}^2\theta)^2 + 4\mathrm{Sin}^2\theta\mathrm{Cos}\theta(a^2- \mathrm{Sin}^2\theta)^{\frac{1}{2}}}\right]\frac{H(t-r/v_s)}{(t-r/v_s)^{\frac{1}{2}}}$$

$$\simeq \frac{Xv_s^{\frac{1}{2}}}{\pi\mu r^{\frac{1}{2}}}\frac{1}{2^{\frac{1}{2}}}\mathrm{Re}\left[D_x^s(\theta)\right]\frac{H(t - r/v_s)}{(t - r/v_s)^{\frac{1}{2}}}, \qquad (11\text{-}106)$$

where we have used (9-45). We also have

$$u_r^{s+}(r,\theta,t) \simeq -\frac{Xv_s^{3/2}2^{\frac{1}{2}}}{\pi\mu r^{3/2}}\mathrm{Re}\left[\frac{(1 - 2\mathrm{Sin}^2\theta)\mathrm{Sin}\theta}{(1 - 2\mathrm{Sin}^2\theta)^2 + 4\mathrm{Sin}^2\theta\mathrm{Cos}\theta(a^2- \mathrm{Sin}^2\theta)^{\frac{1}{2}}}\right]$$

$$\cdot\ (t - r/v_s)^{\frac{1}{2}}H(t - r/v_s). \qquad (11\text{-}107)$$

For $\tau < 1$, $\theta > \theta_c$ we have that

$$u_\theta^{s-}(r,\theta,t) \simeq -\frac{Xv_s^{\frac{1}{2}}}{\pi\mu r^{\frac{1}{2}}2^{\frac{1}{2}}}\,\mathrm{Im}\left[\frac{\mathrm{Cos}\theta(1-2\mathrm{Sin}^2\theta)}{(1-2\mathrm{Sin}^2\theta)^2+4i\mathrm{Sin}^2\theta\mathrm{Cos}\theta(\mathrm{Sin}^2\theta-a^2)^{\frac{1}{2}}}\right]\frac{H(r/v_s-t)}{(r/v_s-t)^{\frac{1}{2}}},$$

$$\simeq -\frac{Xv_s^{\frac{1}{2}}}{\pi\mu r^{\frac{1}{2}}}\frac{1}{2^{\frac{1}{2}}}\,\mathrm{Im}\left[D_x^s(\theta)\right]\frac{H(r/v_s-t)}{(r/v_s-t)^{\frac{1}{2}}}\,. \qquad (11\text{-}108)$$

Finally

$$u_r^{s-}(r,\theta,t) \simeq -\frac{Xv_s^{3/2}2^{\frac{1}{2}}}{\pi\mu r^{3/2}}\,\mathrm{Im}\left[\frac{(1-2\mathrm{Sin}^2\theta)\mathrm{Sin}\theta}{(1-2\mathrm{Sin}^2\theta)^2+4i\mathrm{Sin}^2\theta\mathrm{Cos}\theta(\mathrm{Sin}^2\theta-a^2)^{\frac{1}{2}}}\right] \qquad (11\text{-}109)$$

$$\cdot\,(r/v_s-t)^{\frac{1}{2}}\,H(r/v_s-t).$$

Head-waves. Substituting relations (11-59), (11-60), and (11-61) into (11-99) and (11-101) we obtain

$$u_x^{P-S} = \frac{Xv_s^{3/2}(2a)^{5/2}}{\pi\mu r^{3/2}}\,\frac{(1-a^2)^{5/4}(t-t_c)^{\frac{1}{2}}\mathrm{Cos}\theta_c}{(1-2a^2)^3\mathrm{Sin}^{3/2}(\theta-\theta_c)}\,H(t-t_c),$$

and

$$u_z^{P-S} = -\frac{Xv_s^{3/2}(2a)^{5/2}}{\pi\mu r^{3/2}}\,\frac{(1-a^2)^{5/4}(t-t_c)^{\frac{1}{2}}\mathrm{Sin}\theta_c}{(1-2a^2)^3\mathrm{Sin}^{3/2}(\theta-\theta_c)}\,H(t-t_c).$$

Hence

$$\mathbf{u}^{P-S} = \mathbf{e}_{\theta_c}\frac{Xv_s^{3/2}}{\pi\mu r^{3/2}}\,\frac{(2a)^{5/2}(1-a^2)^{5/2}(t-t_c)^{\frac{1}{2}}H(t-t_c)}{(1-2a^2)^3\mathrm{Sin}^{3/2}(\theta-\theta_c)}\,. \qquad (11\text{-}110)$$

Wave-Fronts. The wavefronts in the case of a tangential force are the same as in the case of a normal force (Figure 11-5); however, the amplitude distribution over the wave fronts differs.

Motion on the Epicentral Line. Starting with (11-92), (11-93) and using (11-65) we have that

$$u_x^P(r,0^\circ,t) = -\frac{Xv_s}{\pi\mu r}\left[\frac{2(\tau^2-a^2)^{\frac{1}{2}}\,(\tau^2-a^2+1)^{\frac{1}{2}}\,\tau}{\{2(\tau^2-a^2)+1\}^2-4(\tau^2-a^2)(\tau^2-a^2+1)^{\frac{1}{2}}\tau}\right]H(\tau-a), \qquad (11\text{-}111)$$

and

$$u_z^P(r,0^\circ,t) = \frac{Xv_s}{\pi\mu r}\,\mathrm{Im}\left[\frac{2(\tau^2-a^2+1)^{\frac{1}{2}}\tau^2}{\{2(\tau^2-a^2)+1\}^2-4(\tau^2-a^2)(\tau^2-a^2+1)^{\frac{1}{2}}\tau}\right]H(\tau-a) \equiv 0. \qquad (11\text{-}112)$$

Expanding (11-111) about the critical point $\tau = a$, we have

$$u_x^P(r,0^\circ,t) \simeq -\frac{Xv_s^{3/2}}{\pi\mu r^{3/2}}(2a)^{3/2}(t-r/v_p)^{\frac{1}{2}}\,H(t-r/v_p), \qquad (11\text{-}113)$$

which is in agreement with (11-97). Starting with (11-98) and (11-100) and the relations following (11-68) we have

$$u_x^s(r,0^\circ,t) = \frac{Xv_s}{\pi\mu r}\left[\frac{(2\tau^2-1)\tau^2}{\{(2\tau^2-1)^2 - 4\tau(\tau^2-1)(\tau^2-1+a^2)^{\frac{1}{2}}\}(\tau^2-1)^{\frac{1}{2}}}\right] H(\tau-1). \qquad (11\text{-}114)$$

As with (11-112) we also see that

$$u_z^s(r,0^\circ,t) \equiv 0. \qquad (11\text{-}115)$$

Expanding (11-114) about the critical point $\tau = 1$, we obtain

$$u_x^s(r,0^\circ,t) \simeq \frac{Xv_s^{\frac{1}{2}}}{\pi\mu r^{\frac{1}{2}}2^{\frac{1}{2}}}\frac{H(t - r/v_s)}{(t - r/v_s)^{\frac{1}{2}}}, \qquad (11\text{-}116)$$

which is in agreement with (11-106).

Motion at the Free Surface. Adding (11-88) and (11-89) we obtain

$$U_x(r,90^\circ,s) = \frac{X}{\pi\mu}\,\mathrm{Re}\int_0^\infty \frac{(q^2+1)^{\frac{1}{2}}}{R(q)}\,e^{(s/v_s)iqr}\,dq. \qquad (11\text{-}117)$$

The Cagniard-deHoop transformation in this case is

$$q = iv_s t/r = i\tau, \quad dq/d\tau = i; \qquad (11\text{-}79)$$

and we find that

$$U_x(r,90^\circ,s) = \frac{Xv_s}{\pi\mu r}\,\mathrm{Re}\left[\int_{t_p}^{t_r-\varepsilon}\frac{i(1-\tau^2)^{\frac{1}{2}}e^{-st}\,dt}{(1-2\tau^2)^2+4i\tau^2(\tau^2-a^2)^{\frac{1}{2}}(1-\tau^2)^{\frac{1}{2}}} + \int_{t_r+\varepsilon}^{\infty}\frac{i(1-\tau^2)^{\frac{1}{2}}e^{-st}\,dt}{(1-2\tau^2)^2+4i\tau^2(\tau^2-a^2)^{\frac{1}{2}}(1-\tau^2)^{\frac{1}{2}}} - \frac{r\pi}{v_s}\underbrace{\frac{(\tau_r^2-1)^{\frac{1}{2}}e^{-st_r}}{R'(i\tau_r)}}_{\to\, 0}\right].$$

Hence

$$u_x(r,90^\circ,t) = \frac{Xv_s}{\pi\mu r}\,\mathrm{Re}\left[\frac{i(1-\tau^2)^{\frac{1}{2}}}{(1-2\tau^2)^2+4i\tau^2(\tau^2-a^2)^{\frac{1}{2}}(1-\tau^2)^{\frac{1}{2}}}\right]H(\tau-a). \qquad (11\text{-}118)$$

Expanding about the point $\tau = a$, we find that

$$u_x(r,90^\circ,t) \simeq \frac{Xv_s^{3/2}}{\pi\mu r^{3/2}}\frac{2^{5/2}a^{3/2}(1-a^2)}{(1-2a^2)^4}(t - r/v_p)^{\frac{1}{2}}\,H(t - r/v_p). \qquad (11\text{-}119)$$

At $\tau = 2^{\frac{1}{2}}$, the bracketed portion of (11-118) is pure real and non-zero. Expanding about the point $\tau = 1$, we have that

$$u_x(r,90^\circ,t) \simeq \frac{Xv_s}{\pi\mu r}(1-\tau^2)4(1-a^2)^{\frac{1}{2}}, \qquad (11\text{-}120)$$

which is a simple zero crossing. Lastly, in the vicinity of $q = ib$, we have

$$u_x(r,90^\circ,t) = -\frac{Xv_s}{\pi\mu r}\frac{(\tau^2-1)^{\frac{1}{2}}}{(1-2\tau^2)^2-4\tau^2(\tau^2-a^2)^{\frac{1}{2}}(\tau^2-1)^{\frac{1}{2}}}. \qquad (11\text{-}121)$$

Examining the z-component of motion, we have

$$U_z(r,90^\circ,s) = -\frac{X}{\pi\mu}\,\mathrm{Im}\int_0^\infty \frac{(2q^2+1)-2(q^2+a^2)^{\frac{1}{2}}(q^2+1)^{\frac{1}{2}}}{R(q)}\,e^{-(s/v_s)iqr}\,q\,dq, \qquad (11\text{-}122)$$

which is just the negative of (11-72). We see here that the z-motion for a force in the x-direction is the negative of the x-motion for a force in the z-direction. That this pair is not reciprocal can be seen from the diagrams in Figure 11-8. You see that the picture going from (a) to (b) is not merely relabeled, but that the vectors

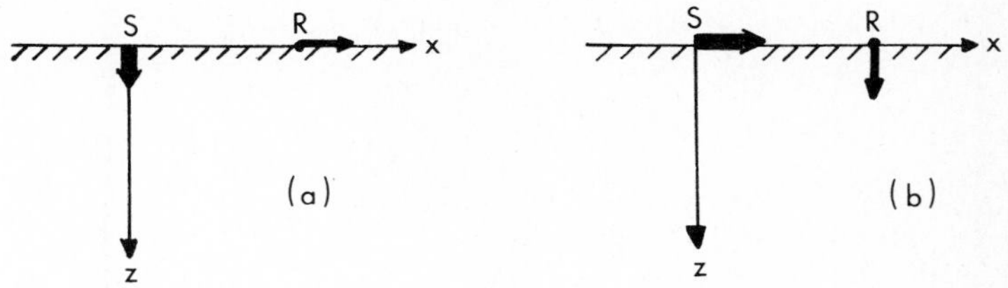

Figure 11-8. Relative geometries of the vertical- and horizontal-sources in Lamb's problem. Note that the configuration is not reciprocal.

have exchanged left for right also. The end result is that the response waveform is the same, but the sense is reversed.

The total motion for a horizontal-source is shown in Figure 11-9. Motion on the epicentral line corresponds to that for 0°, while the surface motion corresponds to that for 90°. Note that $u_{\theta} = -u_z$ on the free surface.

11.4. Pseudo-Waves.

It was discovered by Gilbert and Laster (1962) during the evaluation of exact expressions similar to those obtained above, that when Poisson's ratio was in the neighborhood of $\sigma = 0.45$ an unexplained bump in the wave motion occured for times shortly after P when θ was near 90°. Further analysis revealed that the cause of this disturbance was a root of the denominator near the branch point $q = ia$. We can investigate this by looking at the nature of the Rayleigh denominator in the vicinity of the imaginary axis.

Because of the multivalued nature of the factor $(q^2 + a^2)^{\frac{1}{2}}$, we must introduce a cut somewhere in the q-plane. In the analysis previous to this, we introduced it along the imaginary q-axis as in Figure 11-10(a). This ensured that the factor $\exp\{-(q^2 + a^2)^{\frac{1}{2}}z\}$ would not blow up as long as $\underline{q}$ stayed in the first quadrant. Fortunately values of $\underline{q}$ along the Cagniard-deHoop path did just this. However, the branch cut could have been placed most anywhere as long as it did not intersect the Cagniard-deHoop path. For example, it can be deformed to the ad hoc curve shown in Figure 11-10(b). Now if in (b) one goes about the point $q = ia$ in a counter-clockwise manner, the value of the factor $(a^2 - \xi^2)^{\frac{1}{2}}$ reverses phase by 180°, i.e., it

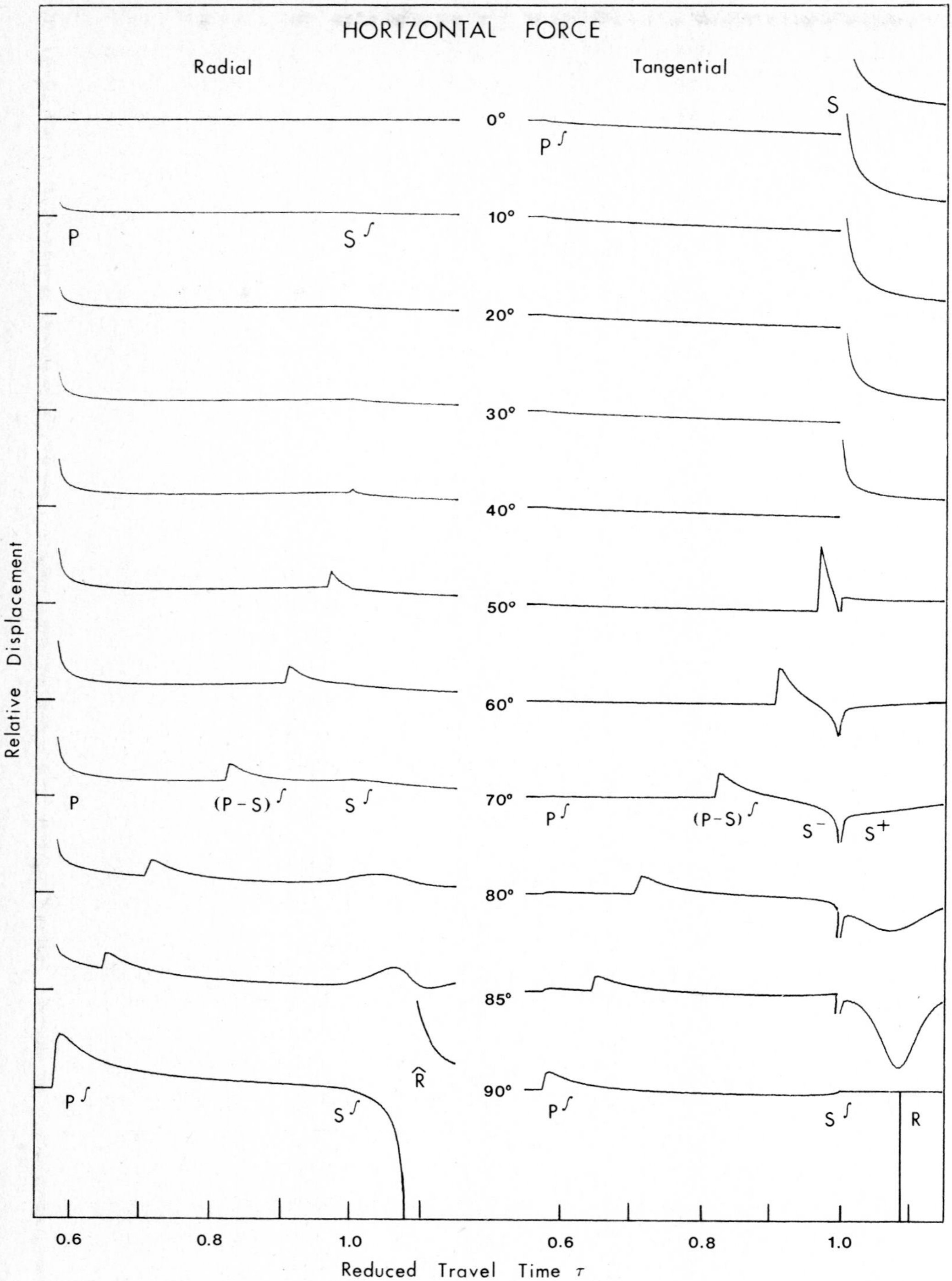

Figure 11-9. Radial and tangential displacements due to an impulse acting tangentially on the surface of a half-space. "∫" implies time-integral. "^" implies allied function.

assumes the value $-(a^2 - \xi^2)^{\frac{1}{2}}$. Here we have written $\xi = iq$. In this case the Rayleigh denominator may be written as

$$(1 - 2\xi^2)^2 - 4\xi^2(1 - \xi^2)^{\frac{1}{2}}(a^2 - \xi^2)^{\frac{1}{2}}, \qquad (11\text{-}123)$$

for values of $\xi < a$. This is in contrast to the case where $\xi > 1$ and we pass both points <u>ia</u> and <u>i</u> on the right as we go up the imaginary axis. In this case one has

$$(1 - 2\xi^2)^2 - 4\xi^2(\xi^2 - 1)^{\frac{1}{2}}(\xi^2 - a^2)^{\frac{1}{2}}. \qquad (11\text{-}124)$$

Now we have already found the root of (11-124) which lies at $q = ib$. However, (11-123) has roots as well. For $a^2 > \sim 0.32$, there are two roots on the imaginary

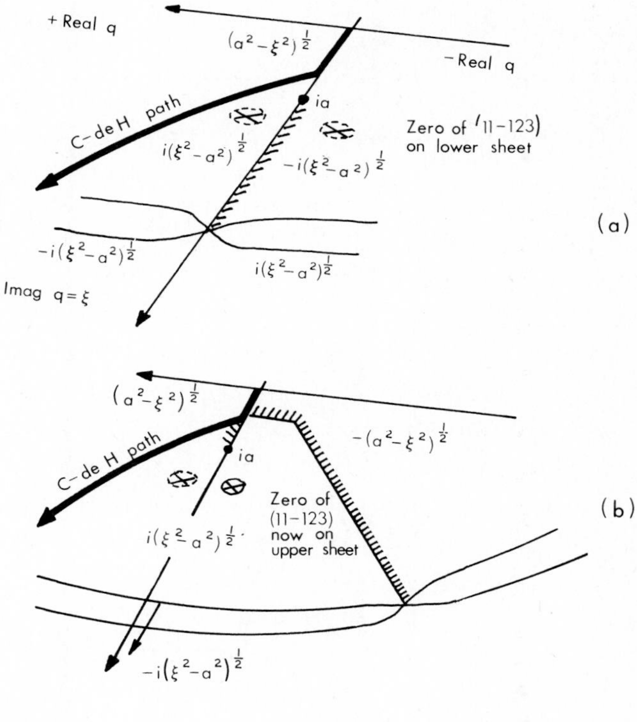

Figure 11-10. Lower sheet zeros of the Rayleigh Denominator. For simplicity, the upper sheet zero at $q = ib$ and the branch point at $q = i$ have not been shown.

q-axis lying below the point <u>ia</u>. For $a^2 < \sim 0.32$, the roots become complex, one now lying in the second quadrant of the upper sheet and the other lying in the first quadrant of the lower sheet. When the root lies on the imaginary q-axis below the point <u>ia</u>, then the effect of the pole is absorbed by the branch point. However, when the root moves out into the second quadrant, it will contribute to the motion. This can be seen in Figure 11-11. An expression such as (11-33) can be looked at in the following manner. The factor $\tau(2q^2+1)/R(q)$ can be looked at as the topography of the q-plane and the factor $dq/d\tau$ can be looked at as the velocity of travel in the q-plane. The seismic motion can be interpreted as the product of the topography times the velocity of travel. Consequently, we get noticeable motion whenever we

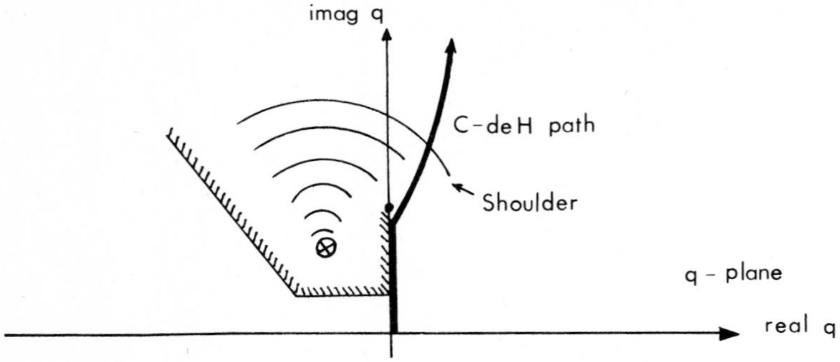

Figure 11-11. Topography in the q-plane.

pass over a large bump at a slow rate, or over a small bump at a fast rate. When the pole associated with (11-123) becomes complex, it sends a shoulder out into the first quadrant, and if the Cagniard-deHoop path is sufficiently close, i.e., θ is approximately 90°, then we get a bump.

This impulse is called a pseudo-wave because it does not partake of a wave-like nature, i.e., no singularity is propagated. The Rayleigh disturbance is also a pseudo-wave in this sense for all angles except $\theta = 90^\circ$. Both of these disturbances lose amplitude as they travel due to the factor $Zv_s/\pi\mu r$ in front of (11-33). Only where $R(q)$ or $dq/d\tau$ is sufficiently singular, do we get less of a dropoff with distance. There is no dropoff in the case of the Rayleigh wave along the surface, and $r^{-\frac{1}{2}}$ dropoff in the case of body waves in the half-space.

We shall see the occurrence of hidden roots of the denominator affecting our results again and again, each time in a slightly different way. Here the effect has been the introduction of a slight bump into the motion following the P-wave for $\sigma \sim 0.45$ and $\theta \sim 90^\circ$.

11.5. Other Considerations.

We have just treated, in some detail, the two-dimensional Lamb's problem. This choice was made because we could obtain an exact solution and then use the first motion approximation as needed. In three dimensions, Lamb's problem can be set up in a similar manner, but the solution comes out in terms of an integral. The answer is again exact, but numerical or first-motion methods must be used to evaluate the resulting expressions. Only on the epicentral line and on the free surface do some simplifications occur. Pekeris and Lifson (1957) give the impulsive response curves for a buried vertical force. Mooney (1974) gives vertical and horizontal motion curves for a surface vertical source with a more realistic time variation. Johnson (1974), in a very substantial paper, derives a Green's function for Lamb's problem and gives a number of response curves. In addition, he works out the spatial derivatives of the Green's Function which are necessary for inclusion in a Representation Theorem such as (8-33).

In all three of the papers, however, the one remaining integral causes problems in getting a feeling for what is taking place. If one works at the three-dimensional solution from a time-harmonic formulation, one is able to use asymptotic expansions for the Bessel functions involved and obtain directivity patterns for P- and S-waves as well as the strengths of Rayleigh waves. Two examples of the time-harmonic approach are Miller and Pursey (1954) and Cherry (1962). The directivity patterns they obtained are the same as we obtained above and in Chapter 9. In the time-harmonic case, one must then use a convolution integral to get the impulse response. Again a portion of the solution is hidden from explicit consideration. The most illuminating method to circumvent these difficulties, to my mind at least, is the first-motion approach adapted to three dimensions by Knopoff and Gilbert (1959). They carefully considered the mathematical complexities in making appropriate asymptotic expansions so that first motions could be obtained.

Miller and Pursey (1954) noted that the time-harmonic directivity patterns were the same for line-sources and axially symmetric point sources. Thau and Pao (1970) and Dampney (1971) develop relationships between the impulsive responses of two- and three-dimensional sources. Again, the three-dimensional solution is given as an integral over two-dimensional solutions.

The problem of moving sources has been extensively treated by Eringen and Suhubi (1975, Chaps. 7 and 8) in both two and three dimensions. References to the pertinent literature will be found there.

Lastly, the half-space problem may be complicated by considering more intricate distributions of sources, both on and below the surface. Adding surficial layers changes the whole nature of the problem (see Chapters 15-19), although some authors refer to a half-space with a thin layer as Lamb's problem. We shall not list the many available papers treating such complications, but merely note that these additional factors must be considered when modeling a realistic earthquake source.

CHAPTER 12

REFLECTION AND TRANSMITTION OF BODY-WAVES AT A PLANE INTERFACE

12.1. P-Waves Incident Upon an Interface Between Two Elastic Media.

The simplest case of this has been considered in Chapter 9 where the second medium was a void. In this more complicated situation we have an incident P-wave, reflected P- and S-waves, and transmitted P- and S-waves as in Figure 12-1. The

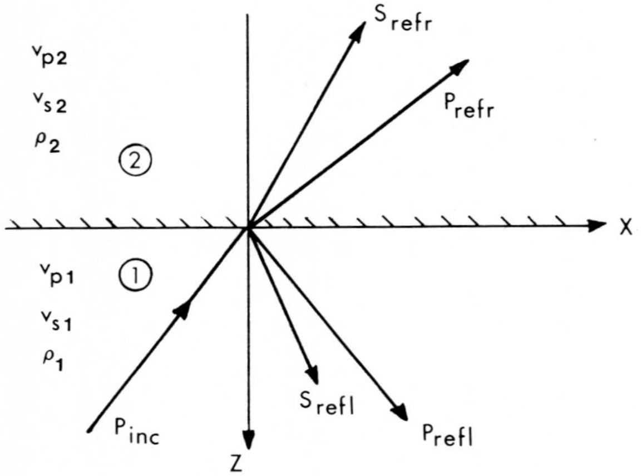

Figure 12-1. Incident P-waves on an interface between two elastic media.

boundary conditions require continuity of both components of displacement and both components of stress across the interface. The potential representations of the pertinent wave fields may be written

$$
\begin{aligned}
\Phi_o &= (1.0)e^{ik_{p1}(x\mathrm{Sin}\theta_{p1} - z\mathrm{Cos}\theta_{p1})}, \\
\Phi_1 &= R_{pp}e^{ik_{p1}(x\mathrm{Sin}\theta_{p1} + z\mathrm{Cos}\theta_{p1})}, \\
\Psi_1^{\dagger} &= R_{ps}e^{ik_{s1}(x\mathrm{Sin}\theta_{s1} + z\mathrm{Cos}\theta_{s1})}, \\
\Phi_2 &= T_{pp}e^{ik_{p2}(x\mathrm{Sin}\theta_{p2} - z\mathrm{Cos}\theta_{p2})}, \\
\Psi_2^{\dagger} &= T_{ps}e^{ik_{s2}(x\mathrm{Sin}\theta_{s2} - z\mathrm{Cos}\theta_{s2})},
\end{aligned}
\qquad (12\text{-}1)
$$

where a factor of $e^{-i\omega t}$ has been suppressed, $k_{p1} = \omega/v_{p1}$, $k_{p2} = \omega/v_{p2}$, etc. As in

Chapter 9, if we are to have any solution at all, Snell's law must hold, i.e.,

$$k_{p1}\mathrm{Sin}\theta_{p1} = k_{s1}\mathrm{Sin}\theta_{s1} = k_{p2}\mathrm{Sin}\theta_{p2} = k_{s2}\mathrm{Sin}\theta_{s2}. \qquad (12\text{-}2)$$

Because of these relationships, we see that there can be two critical angles for incident P if the velocities are in the order $v_{s1}<v_{p1}<v_{s2}<v_{p2}$; one critical angle if the velocities are in the order $v_{s1}<v_{s2}<v_{p1}<v_{p2}$; and no critical angles if v_{p1} is the greatest velocity. Let us define the following quantities:

$$\frac{v_{s1}}{v_{p1}} = a; \quad \frac{v_{s1}}{v_{s2}} = b; \quad \frac{v_{s1}}{v_{p2}} = c; \quad \frac{\rho_1}{\rho_2} = d. \qquad (12\text{-}3)$$

From (12-2) we then have

$$\mathrm{Sin}\theta_{s1} = a\mathrm{Sin}\theta_{p1},$$

$$a\mathrm{Cos}\theta_{p1} = (a^2 - \mathrm{Sin}^2\theta_{s1})^{\frac{1}{2}} = i(\mathrm{Sin}^2\theta_{s1} - a^2)^{\frac{1}{2}}, \qquad (12\text{-}4)$$

because the reflected S-wave must remain finite for finite $\underline{z}$ with k_{p1} tending to infinity. Likewise we have

$$\mathrm{Sin}\theta_{s1} = b\mathrm{Sin}\theta_{s2},$$

$$b\mathrm{Cos}\theta_{s2} = (b^2 - \mathrm{Sin}^2\theta_{s1})^{\frac{1}{2}} = i(\mathrm{Sin}^2\theta_{s1} - b^2)^{\frac{1}{2}}. \qquad (12\text{-}5)$$

Here $\underline{-z}$ is a positive quantity and the transmitted S-wave must also remain finite as k_{s2} tends to infinity. Similarly,

$$\mathrm{Sin}\theta_{s1} = c\mathrm{Sin}\theta_{p2},$$

$$c\mathrm{Cos}\theta_{p2} = (c^2 - \mathrm{Sin}^2\theta_{s1})^{\frac{1}{2}} = i(\mathrm{Sin}^2\theta_{s1} - c^2)^{\frac{1}{2}}. \qquad (12\text{-}6)$$

Using (6-31) and (6-33), the boundary conditions lead to the following four equations:

$$ik_{p1}\mathrm{Sin}\theta_{p1}(1 + R_{pp}) - ik_{s1}\mathrm{Cos}\theta_{s1}R_{ps} = ik_{p2}\mathrm{Sin}\theta_{p2}T_{pp} + ik_{s2}\mathrm{Cos}\theta_{s2}T_{ps},$$

$$-ik_{p1}\mathrm{Cos}\theta_{p1}(1 - R_{pp}) + ik_{s1}\mathrm{Sin}\theta_{s1}R_{ps} = -ik_{p2}\mathrm{Cos}\theta_{p2}T_{pp} + ik_{s2}\mathrm{Sin}\theta_{s2}T_{ps},$$

$$\mu_1(2k_{p1}^2\mathrm{Sin}^2\theta_{p1} - k_{s1}^2)(1 + R_{pp}) - 2\mu_1k_{s1}^2\mathrm{Sin}\theta_{s1}\mathrm{Cos}\theta_{s1}R_{ps} = \qquad (12\text{-}7)$$

$$\mu_2(2k_{p2}^2\mathrm{Sin}^2\theta_{p2} - k_{s2}^2)T_{pp} + 2\mu_2k_{s2}^2\mathrm{Sin}\theta_{s2}\mathrm{Cos}\theta_{s2}T_{ps},$$

$$2\mu_1k_{p1}^2\mathrm{Sin}\theta_{p1}\mathrm{Cos}\theta_{p1}(1 - R_{pp}) - \mu_1k_{s1}^2(\mathrm{Sin}^2\theta_{p1} - \mathrm{Cos}^2\theta_{p1})R_{ps} =$$

$$2\mu_2k_{p2}^2\mathrm{Sin}\theta_{p2}\mathrm{Cos}\theta_{p2}T_{pp} - \mu_2k_{s2}^2(\mathrm{Sin}^2\theta_{p2} - \mathrm{Cos}^2\theta_{p2})T_{ps}.$$

Regrouping and dividing the first two by k_{s1} and the second two by $\mu_1k_{s1}^2$, we have (since $\mathrm{Sin}^2\theta - \mathrm{Cos}^2\theta = 2\mathrm{Sin}^2\theta - 1$, and $\mu_1/\mu_2 = b^2d$),

$$R_{pp}\left[a\sin\theta_{p1}\right] + R_{ps}\left[-\cos\theta_{s1}\right] + T_{pp}\left[-c\sin\theta_{p2}\right] + T_{ps}\left[-b\cos\theta_{s2}\right] = -a\sin\theta_{p1},$$

$$R_{pp}\left[a\cos\theta_{p1}\right] + R_{ps}\left[\sin\theta_{s1}\right] + T_{pp}\left[c\cos\theta_{p2}\right] + T_{ps}\left[-b\sin\theta_{s2}\right] = a\cos\theta_{p1},$$

$$R_{pp}\left[2a^2\sin^2\theta_{p1}-1\right] + R_{ps}\left[-2\sin\theta_{s1}\cos\theta_{s1}\right] + T_{pp}\left[\frac{-2c^2\sin^2\theta_{p2}+b^2}{b^2d}\right] \qquad (12\text{-}8)$$

$$+ T_{ps}\left[\frac{-2b^2\sin\theta_{s2}\cos\theta_{s2}}{b^2d}\right] = -2a^2\sin^2\theta_{p1}+1.$$

$$R_{pp}\left[-2a^2\sin\theta_{p1}\cos\theta_{p1}\right] + R_{ps}\left[-2\sin^2\theta_{s1}+1\right] + T_{pp}\left[\frac{-2c^2\sin\theta_{p2}\cos\theta_{p2}}{b^2d}\right]$$

$$+ T_{ps}\left[\frac{2b^2\sin^2\theta_{s2} - b^2}{b^2d}\right] = -2a^2\sin\theta_{p1}\cos\theta_{p1}.$$

Then

$$R_{pp} = \frac{\Delta_{p1}}{\Delta_p}, \quad R_{ps} = \frac{\Delta_{p2}}{\Delta_p}, \quad T_{pp} = \frac{\Delta_{p3}}{\Delta_p}, \quad T_{ps} = \frac{\Delta_{p4}}{\Delta_p}; \qquad (12\text{-}9)$$

where Δ_{p1} is the determinant $\Delta_p(\theta_{p1})$ with the first column replaced by the column

$$\begin{bmatrix} -a\sin\theta_{p1} \\ a\cos\theta_{p1} \\ -2a^2\sin^2\theta_{p1} + 1 \\ -2a^2\sin\theta_{p1}\cos\theta_{p1} \end{bmatrix}, \qquad (12\text{-}10)$$

Δ_{p2} is $\Delta_p(\theta_{p1})$ with the second column replaced by (12-10), etc., and $\Delta_p(\theta_{p1})$ is given by:

$$\Delta_p(\theta_{p1}) = \qquad (12\text{-}11)$$

$$\begin{vmatrix} a\sin\theta_{p1} & -(1-a^2\sin^2\theta_{p1})^{\frac{1}{2}} & -a\sin\theta_{p1} & -(b^2-a^2\sin^2\theta_{p1})^{\frac{1}{2}} \\ a\cos\theta_{p1} & a\sin\theta_{p1} & (c^2-a^2\sin^2\theta_{p1})^{\frac{1}{2}} & -a\sin\theta_{p1} \\ 2a^2\sin^2\theta_{p1}-1 & -2a\sin\theta_{p1}(1-a^2\sin^2\theta_{p1})^{\frac{1}{2}} & \dfrac{-2a^2\sin^2\theta_{p1}+b^2}{b^2d} & \dfrac{-2a\sin\theta_{p1}(b^2-a^2\sin^2\theta_{p1})^{\frac{1}{2}}}{b^2d} \\ -2a^2\sin\theta_{p1}\cos\theta_{p1} & -2a^2\sin^2\theta_p+1 & \dfrac{-2a\sin\theta_{p1}(c^2-a^2\sin^2\theta_{p1})^{\frac{1}{2}}}{b^2d} & \dfrac{2a^2\sin^2\theta_{p1}-b^2}{b^2d} \end{vmatrix}$$

If v_{p1} is greater than only v_{s1}, then we see that the factors $(c^2 - a^2\sin^2\theta_{p1})^{\frac{1}{2}}$ and $(b^2 - a^2\sin^2\theta_{p1})^{\frac{1}{2}}$ both become imaginary for certain (separate) values of θ_{p1}. If v_{p1} is less than only v_{p2}, then only the first factor becomes imaginary. If v_{p1} is the greatest velocity, then all factors remain real and there are no critical angles.

Using the same reasoning leading to (9-19), we find that the following energy relationship holds:

$$1 = |R_{pp}|^2 + \frac{\mathrm{Tan}\theta_{p1}}{\mathrm{Tan}\theta_{s1}} |R_{ps}|^2 + \frac{\rho_2}{\rho_1} \frac{\mathrm{Tan}\theta_{p1}}{\mathrm{Tan}\theta_{p2}} |T_{pp}|^2 + \frac{\rho_2}{\rho_1} \frac{\mathrm{Tan}\theta_{p1}}{\mathrm{Tan}\theta_{s2}} |T_{ps}|^2. \qquad (12\text{-}12)$$

The relationships (12-9) are extremely complicated and best used symbolically. Equations (12-8) can be solved directly by a computer program set up to solve complex-variable simultaneous equations. I have done this to obtain the curves in Figures 12-2, -3, and -4. Reflection and transmission coefficients may also be calulated from formulations given elsewhere. Young and Braile (1976) present a computer program to calculate the potential reflection and transmission coefficients, Zoeppritz's displacement amplitude coefficients, and Knott's energy coefficients. This program is based on the equations given by Cerveny and Ravindra (1971, p. 63). Figure 2 of the Young and Braile paper is of the Knott energy coefficients. They do not present phase information, however. I have chosen to show the reflection and transmission data after Gutenberg (1944) in the form of the square root of the energy ratio. This has the advantage of being more nearly related to the seismic response (in terms of relative particle displacements), yet being easily converted to energy ratios. The model parameters chosen were those of Young and Braile to facilitate comparison of results.

Very little phase-shift information has been published, and that has been the subject of much controversy. Young and Braile (1976) reference the several papers involved. The results of Figures 12-2, -3, and -4 have been checked against those of Young and Braile and against those of Costain et al (1963). Although Young and Braile do not have phase information, it is apparent from later work of Costain et al (1965) where they attempted to correct an assumed mistake, that the choice of signs as in (12-4), (12-5), and (12-6) are apparent in the energy ratios. My results agree with those of Young and Braile. Additional calculations agree with the first paper of Costain et al (1963) with one minor exception. My phase-shift for transmitted P-waves is 180° out of phase with theirs over the whole range.

It should noted that the sense of the phase-shift is dependent on the form of the incident plane-wave. In this book, we are using plane-waves of the form $e^{i(kx-\omega t)}$. If the incident plane-wave were represented by $e^{i(\omega t-kx)}$, then all phase-shifts would be reversed in sense from those shown above. Also note that all phase-shifts are applicable to potential reflection and transmission coefficients only. Spatial derivatives involved in getting particle displacements can alter these by 180°.

When θ is zero, certain simplifications occur, and one is able to obtain useful relationships. If we look at R_{pp} when $\theta = 0$, we find that

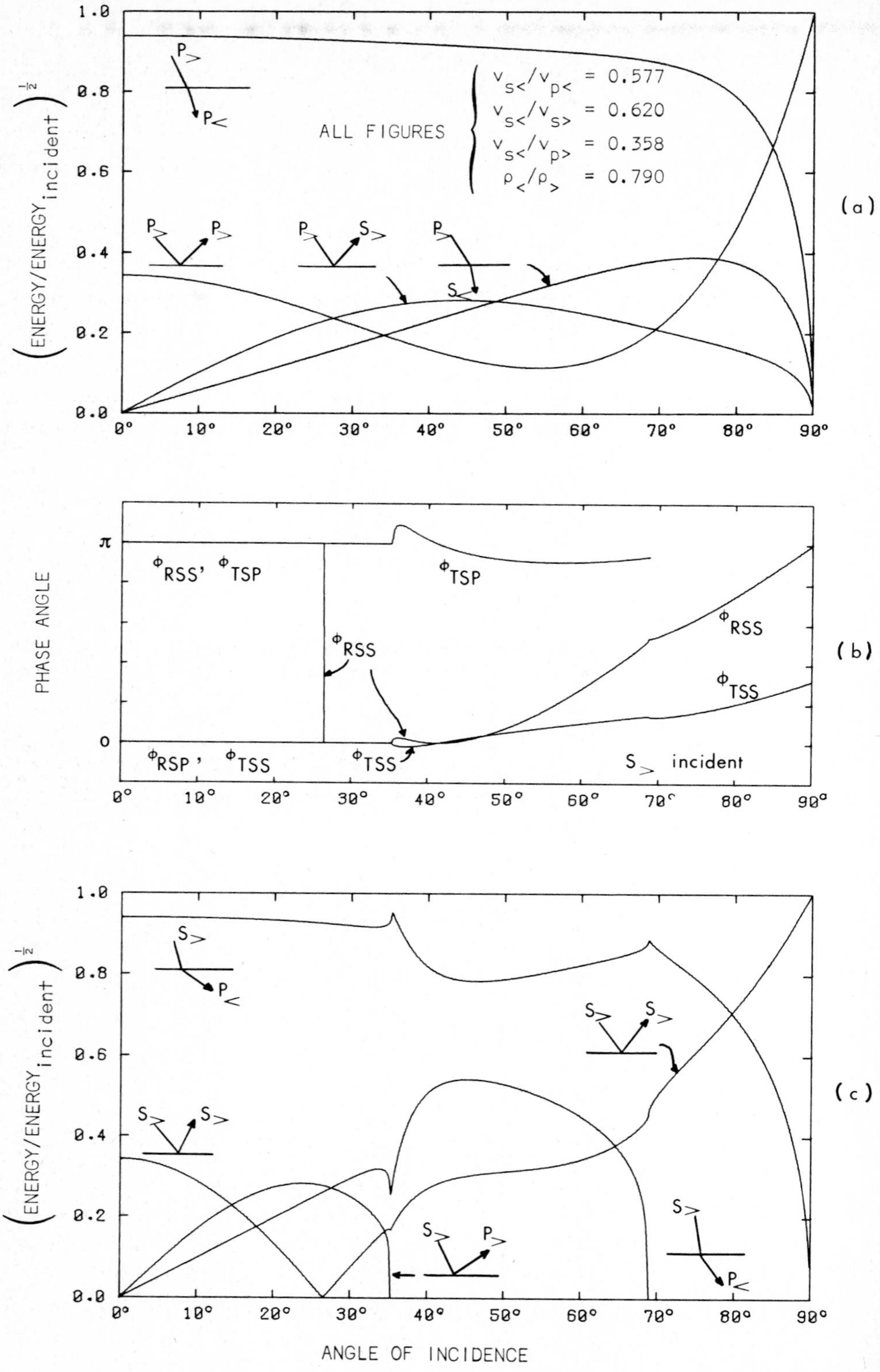
(a)
(b)
(c)
ALL FIGURES
$v_{s<}/v_{p<}$ = 0.577
$v_{s<}/v_{s>}$ = 0.620
$v_{s<}/v_{p>}$ = 0.358
$\rho_{<}/\rho_{>}$ = 0.790
$(\text{ENERGY/ENERGY}_{\text{incident}})^{\frac{1}{2}}$
PHASE ANGLE
ϕ_{RSS}, ϕ_{TSP}
ϕ_{TSP}
ϕ_{RSS}
ϕ_{TSS}
ϕ_{RSP}, ϕ_{TSS}
$S_{>}$ incident
ANGLE OF INCIDENCE

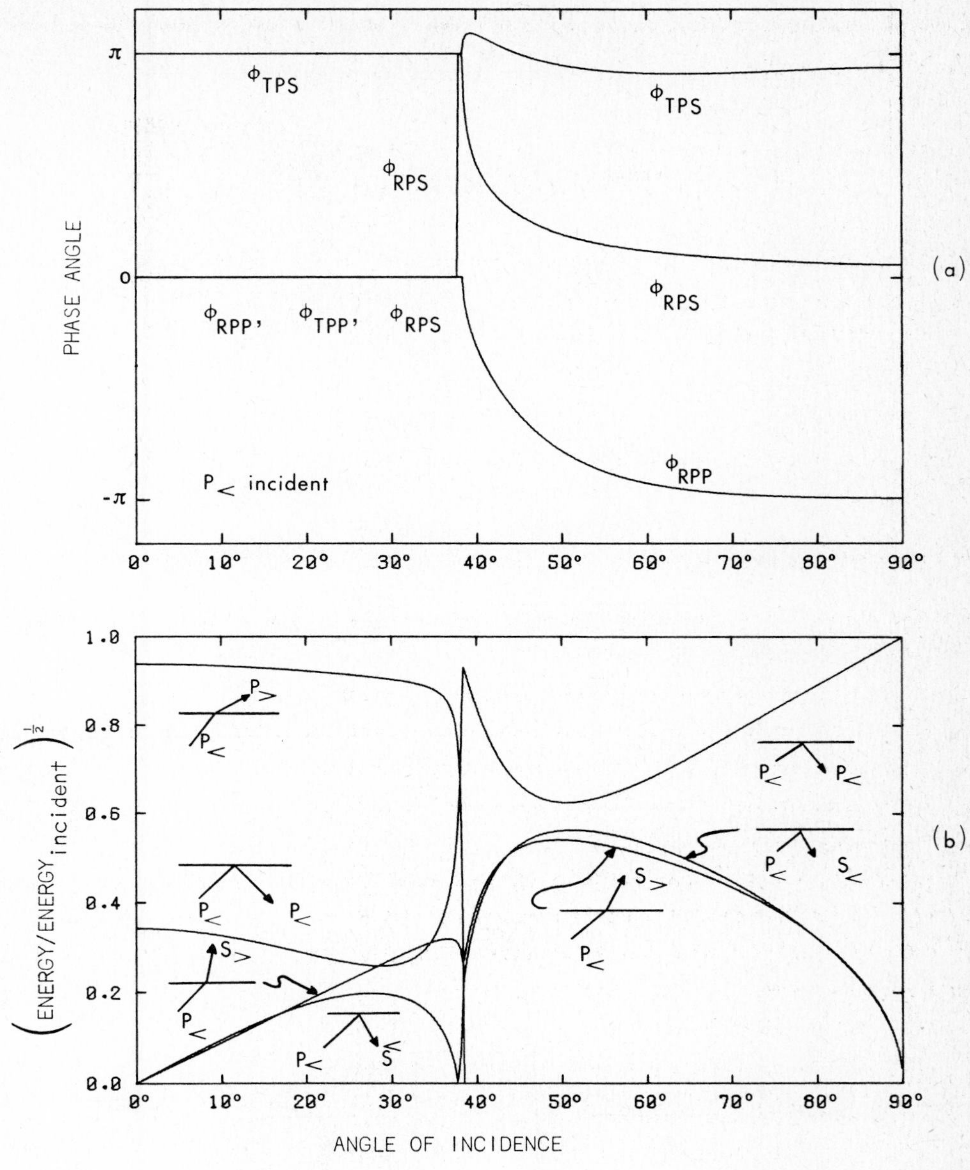

Figure 12-3(a). Phase-shifts for slower P incident on a plane interface.

(b). Square root of energy ratios for slower P incident on a plane interface.

OPPOSITE PAGE

Figure 12-2(a). Square root of energy ratios for faster P incident on a plane interface.

(b). Phase-shifts for faster S incident on a plane interface.

(c). Square root of energy ratios for faster S incident on a plane interface.

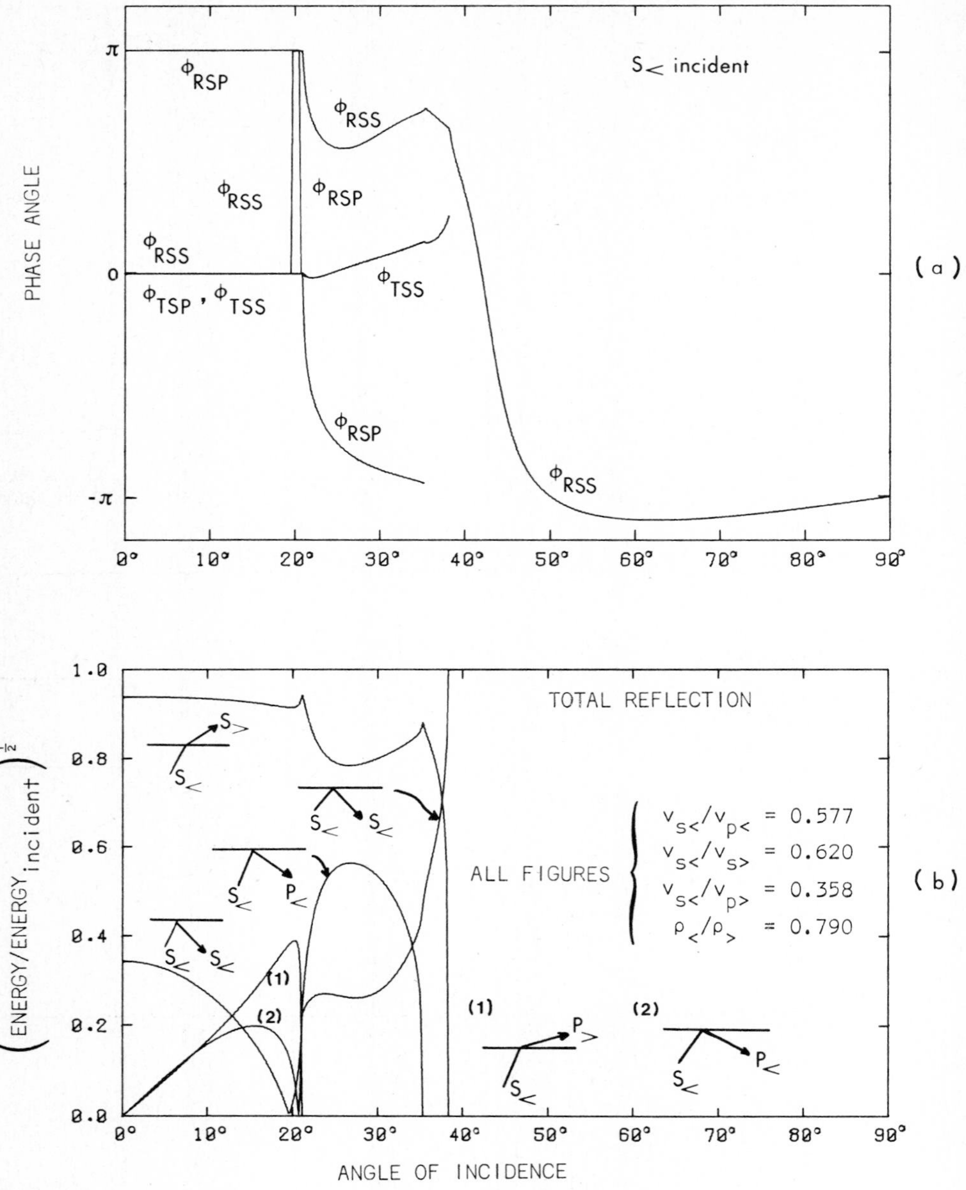

Figure 12-4(a). Phase-shifts for slower S incident on a plane interface.

(b). Square root of energy ratios for slower S incident on a plane interface.

$$R_{pp} = \frac{\begin{vmatrix} 0 & -1 & 0 & -b \\ a & 0 & c & 0 \\ 1 & 0 & 1/d & 0 \\ 0 & 1 & 0 & -1/d \end{vmatrix}}{\begin{vmatrix} 0 & -1 & 0 & -b \\ a & 0 & c & 0 \\ -1 & 0 & 1/d & 0 \\ 0 & 1 & 0 & -1/d \end{vmatrix}} = \frac{(a/d - c)(1/d + b)}{(a/d + c)(1/d + b)} = \frac{\rho_2 v_{p2} - \rho_1 v_{p1}}{\rho_2 v_{p2} + \rho_1 v_{p1}} . \qquad (12\text{-}13)$$

Similarly

$$T_{pp} = \frac{\begin{vmatrix} 0 & -1 & 0 & -b \\ a & 0 & a & 0 \\ -1 & 0 & 1 & 0 \\ 0 & 1 & 0 & -1/d \end{vmatrix}}{\Delta} = \frac{2a}{a/d + c} = \frac{2\rho_1 v_{p2}}{\rho_2 v_{p2} + \rho_1 v_{p1}} . \qquad (12\text{-}14)$$

Plugging into (12-12) we find that

$$\left[\frac{\rho_2 v_{p2} - \rho_1 v_{p1}}{\rho_2 v_{p2} + \rho_1 v_{p1}}\right]^2 + \frac{\rho_2 v_{p1}}{\rho_1 v_{p2}} \left[\frac{2\rho_1 v_{p2}}{\rho_2 v_{p2} + \rho_1 v_{p1}}\right]^2 = 1,$$

as it should. This at least partially checks (12-9). For $\theta = 0$, we also find that R_{ps} and T_{ps} are identically zero. That is, the case of normally incident P-waves is satisfied by a reflected and a transmitted P-wave with coefficients given by (12-13) and (12-14). Remember these are not displacement, but potential amplitude coefficients when comparing results. These relationships are most useful in determining the behavior of P-waves traveling through a "layer-cake" structure at nearly normal incidence. Looking at the Figures 12-2 and 12-3, we see that this is not too bad an approximation for θ less than 20° or so.

When the velocity contrasts are small, an additional approximation is often used in reflection seismology. That is

$$R_{pp} \simeq \frac{d(\rho v)}{2\rho v} = \frac{1}{2}\, d\left[\ln(\rho v)\right],$$

and

$$T_{pp} \simeq 1. \qquad (12\text{-}15)$$

12.2. <u>SV-Waves Incident Upon An Interface Between Two Elastic Media.</u>

In this case we may write (see Figure 12-5)

$$\begin{aligned}
\Phi_1 &= R_{sp} e^{ik_{p1}(x\sin\theta_{p1} + z\cos\theta_{p1})}, \\
\Psi_o^{\dagger} &= (1.0) e^{ik_{s1}(x\sin\theta_{s1} - z\cos\theta_{s1})}, \\
\Psi_1^{\dagger} &= R_{ss} e^{ik_{s1}(x\sin\theta_{s1} + z\cos\theta_{s1})}, \qquad (12\text{-}16) \\
\Phi_2 &= T_{sp} e^{ik_{p2}(x\sin\theta_{p2} - z\cos\theta_{p2})}, \\
\Psi_2^{\dagger} &= T_{ss} e^{ik_{s2}(x\sin\theta_{s2} - z\cos\theta_{s2})}.
\end{aligned}$$

Substitution into the boundary conditions gives:

$$ik_{p1}\text{Sin}\theta_{p1}R_{sp} + ik_{s1}\text{Cos}\theta_{s1}(1 - R_{ss}) = ik_{p2}\text{Sin}\theta_{p2}T_{sp} + ik_{s2}\text{Cos}\theta_{s2}T_{ss},$$

$$ik_{p1}\text{Cos}\theta_{p1}R_{sp} + ik_{s1}\text{Sin}\theta_{s1}(1 + R_{ss}) = -ik_{p2}\text{Cos}\theta_{p2}T_{sp} + ik_{s2}\text{Sin}\theta_{s2}T_{ss},$$

$$\mu_1(2k_{p1}^2\text{Sin}^2\theta_{p1} - k_{s1}^2)R_{sp} + 2\mu_1 k_{s1}^2\text{Sin}\theta_{s1}\text{Cos}\theta_{s1}(1 - R_{ss}) = \mu_2(2k_{p2}^2\text{Sin}^2\theta_{p2} - k_{s2}^2)T_{sp} + 2\mu_2 k_{s2}^2\text{Sin}\theta_{s2}\text{Cos}\theta_{s2}T_{ss}, \qquad (12\text{-}17)$$

$$-2\mu_1 k_{p1}^2\text{Sin}\theta_{p1}\text{Cos}\theta_{p1}R_{sp} - \mu_1 k_{p1}^2(\text{Sin}^2\theta_{p1} - \text{Cos}^2\theta_{p1})R_{ss} = 2\mu_2 k_{p2}^2\text{Sin}\theta_{p2}\text{Cos}\theta_{p2}T_{sp} - \mu_2 k_{s2}^2(\text{Sin}^2\theta_{s2} - \text{Cos}^2\theta_{s2})T_{ss}.$$

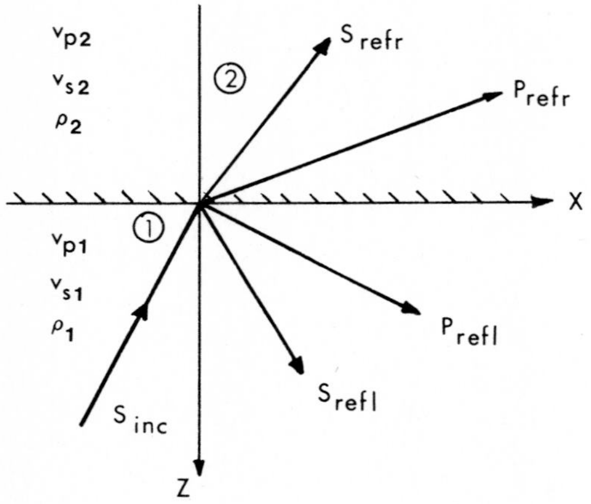

Figure 12-5. Incident SV-Waves on an interface between two elastic media.

Regrouping, dividing the first two by k_{s1} and the second two by $\mu_1 k_{s1}^2$, we have

$$R_{sp}\left[a\text{Sin}\theta_{p1}\right] + R_{ss}\left[-\text{Cos}\theta_{s1}\right] + T_{sp}\left[-c\text{Sin}\theta_{p2}\right] + T_{ss}\left[-b\text{Cos}\theta_{s2}\right] = -\text{Cos}\theta_{s1},$$

$$R_{sp}\left[a\text{Cos}\theta_{p1}\right] + R_{ss}\left[\text{Sin}\theta_{s1}\right] + T_{sp}\left[c\text{Cos}\theta_{p2}\right] + T_{ss}\left[-b\text{Sin}\theta_{s2}\right] = -\text{Sin}\theta_{s1},$$

$$R_{sp}\left[2a^2\text{Sin}^2\theta_{p1} - 1\right] + R_{ss}\left[-2\text{Sin}\theta_{s1}\text{Cos}\theta_{s1}\right] + T_{sp}\left[\frac{-2c^2\text{Sin}^2\theta_{p2} + b^2}{b^2d}\right] + T_{ss}\left[\frac{-2b^2\text{Sin}\theta_{s2}\text{Cos}\theta_{s2}}{b^2d}\right] = -2\text{Sin}\theta_{s1}\text{Cos}\theta_{s1}, \qquad (12\text{-}18)$$

$$R_{sp}\left[-2a^2\text{Sin}\theta_{p1}\text{Cos}\theta_{p1}\right] + R_{ss}\left[-2\text{Sin}^2\theta_{s1} + 1\right] + T_{sp}\left[\frac{-2c^2\text{Sin}\theta_{p2}\text{Cos}\theta_{p2}}{b^2d}\right] + T_{ss}\left[\frac{2b^2\text{Sin}^2\theta_{s1} - b^2}{b^2d}\right] = 2\text{Sin}^2\theta_{s1} - 1.$$

Solving for the coefficients we have

$$R_{sp} = \frac{\Delta_{s1}}{\Delta_s}, \quad R_{ss} = \frac{\Delta_{s2}}{\Delta_s}, \quad T_{sp} = \frac{\Delta_{s3}}{\Delta_s}, \quad T_{ss} = \frac{\Delta_{s4}}{\Delta_s}, \tag{12-19}$$

where the replacement column is

$$\begin{bmatrix} -\mathrm{Cos}\theta_{s1} \\ -\mathrm{Sin}\theta_{s1} \\ -2\mathrm{Sin}\theta_{s1}\mathrm{Cos}\theta_{s1} \\ 2\mathrm{Sin}^2\theta_{s1} - 1 \end{bmatrix}; \tag{12-20}$$

and the denominator determinant is given by:

$$\Delta_s(\theta_{s1}) = \tag{12-21}$$

$$\begin{vmatrix} \mathrm{Sin}\theta_{s1} & -\mathrm{Cos}\theta_{s1} & -\mathrm{Sin}\theta_{s1} & -(b^2-\mathrm{Sin}^2\theta_{s1})^{\frac{1}{2}} \\ (a^2-\mathrm{Sin}^2\theta_{s1})^{\frac{1}{2}} & \mathrm{Sin}\theta_{s1} & (c^2-\mathrm{Sin}^2\theta_{s1})^{\frac{1}{2}} & -\mathrm{Sin}\theta_{s1} \\ 2\mathrm{Sin}^2\theta_{s1}-1 & -2\mathrm{Sin}\theta_{s1}\mathrm{Cos}\theta_{s1} & \dfrac{-2\mathrm{Sin}^2\theta_{s1}+b^2}{b^2d} & \dfrac{-2\mathrm{Sin}\theta_{s1}(b^2-\mathrm{Sin}^2\theta_{s1})^{\frac{1}{2}}}{b^2d} \\ -2\mathrm{Sin}\theta_{s1}(a^2-\mathrm{Sin}^2\theta_{s1})^{\frac{1}{2}} & -2\mathrm{Sin}^2\theta_{s1}+1 & \dfrac{-2\mathrm{Sin}\theta_{s1}(a^2-\mathrm{Sin}^2\theta_{s1})^{\frac{1}{2}}}{b^2d} & \dfrac{2\mathrm{Sin}^2\theta_{s1}-b^2}{b^2d} \end{vmatrix}$$

In this case the energy relationship is

$$1 = |R_{ss}|^2 + \frac{\mathrm{Tan}\theta_{s1}}{\mathrm{Tan}\theta_{p1}}|R_{sp}|^2 + \frac{\rho_2}{\rho_1}\frac{\mathrm{Tan}\theta_{s1}}{\mathrm{Tan}\theta_{s2}}|T_{ss}|^2 + \frac{\rho_2}{\rho_1}\frac{\mathrm{Tan}\theta_{s1}}{\mathrm{Tan}\theta_{p2}}|T_{sp}|^2 . \tag{12-22}$$

Results for incident S-waves are given in Figures 12-3 and 12-4.

We note that simplifying relationships such as (12-15) have never found much use in reflection seismology using S, inasmuch as the S-arrival is usually cluttered up with multiple reflections of P.

12.3. SH-Waves Incident Upon An Interface Between Two Elastic Media.

In this case we may write

$$\begin{aligned} \Lambda_o &= (1.0)e^{ik_{s1}(x\mathrm{Sin}\theta_{s1} - z\mathrm{Cos}\theta_{s1})} \\ \Lambda_1 &= R_{ss}e^{ik_{s1}(x\mathrm{Sin}\theta_{s1} + z\mathrm{Cos}\theta_{s1})} \\ \Lambda_2 &= T_{ss}e^{ik_{s2}(x\mathrm{Sin}\theta_{s2} - z\mathrm{Cos}\theta_{s2})} \end{aligned} \tag{12-23}$$

The conditions upon R_{ss} and T_{ss} so that (12-23) may satisfy the boundary conditions are:

$$\begin{aligned} &k_{s1}\mathrm{Sin}\theta_{s1}(1 + R_{ss}) = k_{s2}\mathrm{Sin}\theta_{s2}T_{ss}, \\ &-\mu_1k_{s1}^2\mathrm{Sin}\theta_{s1}\mathrm{Cos}\theta_{s1}(1 - R_{ss}) = -\mu_2k_{s2}^2\mathrm{Sin}\theta_{s2}\mathrm{Cos}\theta_{s2}T_{ss}. \end{aligned} \tag{12-24}$$

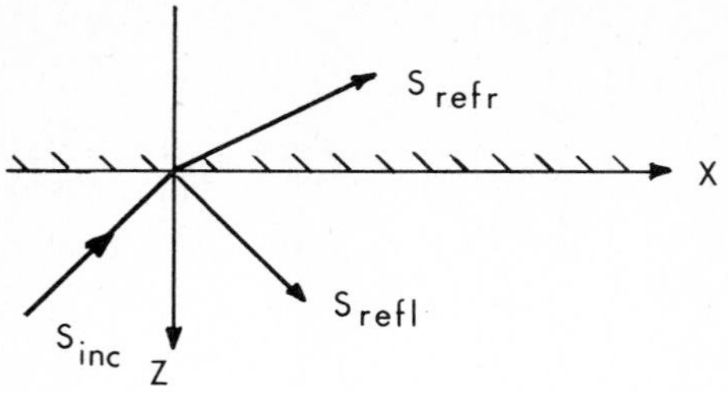

Figure 12-6. Incident SH-waves on an interface between two elastic media.

Using Snell's law we have

$$R_{ss} - T_{ss} = -1$$

$$\text{Cos}\theta_{s1} R_{ss} + \frac{(b^2 - \text{Sin}^2\theta_{s1})^{\frac{1}{2}}}{b^2 d} T_{ss} = \text{Cos}\theta_{s1} \, .$$

Then

$$R_{ss} = \frac{\begin{vmatrix} -1 & -1 \\ \text{Cos}\theta_{s1} & \dfrac{(b^2-\text{Sin}^2\theta_{s1})^{\frac{1}{2}}}{b^2 d} \end{vmatrix}}{\begin{vmatrix} 1 & -1 \\ \text{Cos}\theta_{s1} & \dfrac{(b^2-\text{Sin}^2\theta_{s1})^{\frac{1}{2}}}{b^2 d} \end{vmatrix}} = \frac{-(b^2-\text{Sin}^2\theta_{s1})^{\frac{1}{2}} + b^2 d\text{Cos}\theta_{s1}}{(b^2-\text{Sin}^2\theta_{s1})^{\frac{1}{2}} + b^2 d\text{Cos}\theta_{s1}} \, , \tag{12-25}$$

$$T_{ss} = \frac{\begin{vmatrix} 1 & -1 \\ \text{Cos}\theta_{s1} & \text{Cos}\theta_{s1} \end{vmatrix}}{\begin{vmatrix} 1 & -1 \\ \text{Cos}\theta_{s1} & \dfrac{(b^2-\text{Sin}^2\theta_{s1})^{\frac{1}{2}}}{b^2 d} \end{vmatrix}} = \frac{2b^2 d\text{Cos}\theta_{s1}}{(b^2-\text{Sin}^2\theta_{s1})^{\frac{1}{2}} + b^2 d\text{Cos}\theta_{s1}} \, . \tag{12-26}$$

For $\theta_s = 0$, these reduce to

$$R_{ss} = \frac{\rho_1 v_{s1} - \rho_2 v_{s2}}{\rho_1 v_{s1} + \rho_2 v_{s2}} \, , \tag{12-27}$$

$$T_{ss} = \frac{2\rho_1 v_{s1}}{\rho_1 v_{s1} + \rho_2 v_{s2}} \, . \tag{12-28}$$

Going back to (9-18), we must have for the conservation of energy flux

$$\rho_1 k_{s1}^2 \text{Sin}^2\theta_{s1} v_{s1} \text{Cos}\theta_{s1} 1^2 = \rho_1 k_{s1}^2 \text{Sin}^2\theta_{s1} v_{s1} \text{Cos}\theta_{s1} |R_{ss}|^2 + \rho_2 k_{s2}^2 \text{Sin}^2\theta_{s2} v_{s2} \text{Cos}\theta_{s2} |T_{ss}|^2 ,$$

or

$$1 = |R_{ss}|^2 + \frac{\rho_2 \mathrm{Tan}\theta_{s1}}{\rho_1 \mathrm{Tan}\theta_{s2}} |T_{ss}|^2 . \qquad (12\text{-}29)$$

It is easily shown that (12-25) and (12-26) satisfy this relation.

12.4. Reflection And Transmission From A Generalized Interface.

It has been shown that the reflection and transmission coefficients at a plane interface are frequency independent. However, if there is some structure to the interfacial zone, the analysis we have been using breaks down and one has to proceed more carefully. Two cases are of particular interest. They are illustrated in Figure 12-7.

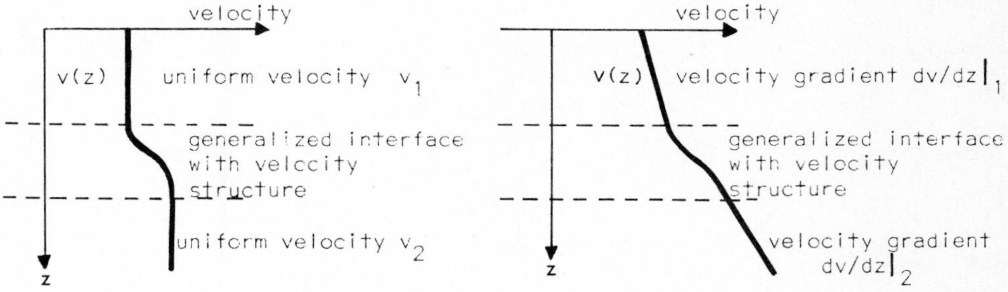

Figure 12-7. Velocities in the vicinity of a general interface.

The first case represents a smooth transition between two velocities at the interface, rather than the jump-discontinuity considered in the previous three sections. The second case adds the complication of inhomogeneous media. It will be covered more completely in Chapter 21. Although we have sketched only a velocity profile, the density structure could vary as well leading to similar problems.

The first case is a generalization to elastic media of a well-studied problem in scalar waves where many methods of approximate solution have been set forth. In a series of three papers (Lapwood, Hudson, and Kembhavi, 1973 and 1975; Lapwood and Hudson, 1975), a variational approach has been taken to determine approximate reflection and transmission coefficients. Gupta (1966a and 1966b) treats the linear transition case. In the first paper an exact solution is found; in the second an approximate solution is found by letting the transition be represented by many homogeneous layers. He then compares the two results.

For the second case, we note the following: in Chapter 10, the wave-number $\underline{k}$ has been identified with $k_p\mathrm{Sin}\theta$, and $(k_p^2 - k^2)^{\frac{1}{2}}$ with $k_p\mathrm{Cos}\theta$. Likewise in Chapter 11, we have identified (in the First Motion Approximation) q_o with $ia\mathrm{Sin}\theta$, $(q_o^2 + a^2)^{\frac{1}{2}}$ with $a\mathrm{Cos}\theta$. This suggests that reflection and transmission coefficients might be written in terms of these wave-number (separation parameter) variables. We shall indeed do this in Chapter 13. However, in the second case, where we have inhomogeneous media outside the generalized interface, these separation parameters can become functions of the spatial coordinate z, if indeed we can even separate the elastic wave equation for inhomogeneous media. Now, the reflection and transmission coefficients arise from requiring continuity of stress and displacement at the interface: Sinθ from the x-spatial derivative and Cosθ from the z-spatial derivative. Recognizing this, Richards (1976) has worked out generalized transmission and reflection coefficients and their approximations when θ is near 90°: either at classical turning points (see Chapter 21) or when a ray is almost critically refracted.

CHAPTER 13

GENERALIZED PLANE-WAVE THEORY AND HEAD-WAVES

13.1. A Point Source Near A Fluid-Fluid Interface.

The geometry is given in Figure 13-1. The potential representations for the

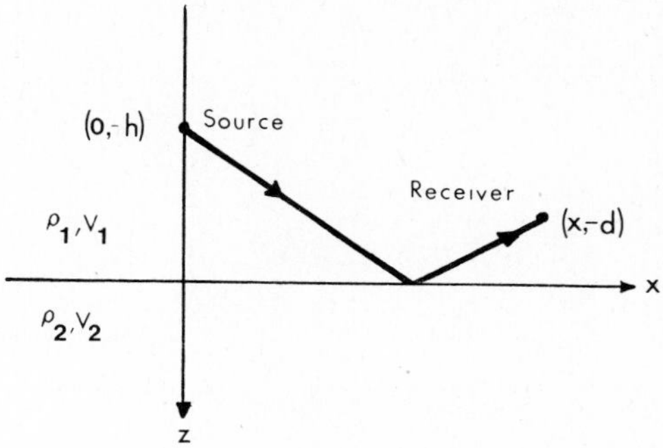

Figure 13-1. Geometric relationships in the description of head-waves. Source and receiver are both on the same side of the interface.

source (see Sec. 7.3), the reflected solution, and the transmitted solution are:

$$\Phi_o = \frac{1}{s}\int_{-\infty}^{\infty} \frac{e^{-(s/v_1)(q^2+1)^{\frac{1}{2}}(z+h) + (s/v_1)iqx}}{2(q^2+1)^{\frac{1}{2}}}\,dq = \frac{1}{s}K_o(sr/v_1),$$

$$\Phi_1 = \frac{1}{s}\int_{-\infty}^{\infty} \frac{A(q)e^{(s/v_1)(q^2+1)^{\frac{1}{2}}z + (s/v_1)iqx}}{2(q^2+1)^{\frac{1}{2}}}\,dq, \qquad \text{(Reflected)} \qquad (13\text{-}1)$$

$$\Phi_2 = \frac{1}{s}\int_{-\infty}^{\infty} \frac{B(q)e^{-(s/v_1)(q^2+a^2)^{\frac{1}{2}}z + (s/v_1)iqx}}{2(q^2+a^2)^{\frac{1}{2}}}\,dq. \qquad \text{(Transmitted)}$$

The boundary conditions at the interface z = 0 are:

$$U_z = \frac{\partial\Phi}{\partial z} \text{ is continuous,}$$

and $\qquad\qquad$ (13-2)

$$T_{zz} = \lambda\nabla\cdot\mathbf{U} = \frac{\lambda s^2\Phi}{v^2} = \rho s^2\Phi \text{ is continuous.}$$

Substituting (13-1) into (13-2) and solving the resulting equations for A and B gives:

$$A = \left[\frac{\rho_2(q^2+1)^{\frac{1}{2}} - \rho_1(q^2+a^2)^{\frac{1}{2}}}{\rho_2(q^2+1)^{\frac{1}{2}} + \rho_1(q^2+a^2)^{\frac{1}{2}}}\right] e^{-(s/v_1)(q^2+1)^{\frac{1}{2}}h},$$

and

$$B = \left[\frac{2\rho_1 (q^2 + a^2)^{\frac{1}{2}}}{\rho_2 (q^2 + 1)^{\frac{1}{2}} + \rho_1 (q^2 + a^2)^{\frac{1}{2}}} \right] e^{-(s/v_1)(q^2+1)^{\frac{1}{2}} h} .$$

Then

$$\Phi_1 = \frac{1}{s} \int_{-\infty}^{\infty} \left[\frac{\rho_2 (q^2 + 1)^{\frac{1}{2}} - \rho_1 (q^2 + a^2)^{\frac{1}{2}}}{\rho_2 (q^2 + 1)^{\frac{1}{2}} + \rho_1 (q^2 + a^2)^{\frac{1}{2}}} \right] \cdot \frac{e^{(s/v_1)(q^2+1)^{\frac{1}{2}}(z - h) + (s/v_1)iqx}}{2(q^2 + 1)^{\frac{1}{2}}} \, dq, \qquad (13\text{-}3)$$

and

$$\Phi_2 = \frac{1}{s} \int_{-\infty}^{\infty} \left[\frac{2\rho_1 (q^2 + a^2)^{\frac{1}{2}}}{\rho_2 (q^2 + 1)^{\frac{1}{2}} + \rho_1 (q^2 + a^2)^{\frac{1}{2}}} \right] \cdot \frac{e^{-(s/v_1)\{(q^2+a^2)^{\frac{1}{2}} z + (q^2+1)^{\frac{1}{2}} h\} + (s/v_1)iqx}}{2(q^2 + a^2)^{\frac{1}{2}}} \, dq. \qquad (13\text{-}4)$$

13.2. Plane-Wave Theory.

Now a plane wave in Laplace transform notation looks like (for region 1):

$$e^{s\left(t - \frac{x\mathrm{Sin}\theta}{v_1} + \frac{z\mathrm{Cos}\theta}{v_1}\right)} .$$

Hence we may identify $(q^2 + 1)^{\frac{1}{2}}$ with $\mathrm{Cos}\theta$ and $-iq$ with $\mathrm{Sin}\theta$. With this identification, we see that the bracketed portion of (13-3) becomes (for $q < 1$)

$$\left[\frac{\rho_2 \mathrm{Cos}\theta_1 - a\rho_1 \mathrm{Cos}\theta_2}{\rho_2 \mathrm{Cos}\theta_1 + a\rho_1 \mathrm{Cos}\theta_2} \right] .$$

But this is just the reflection coefficient for P-waves in the physical system described by Figure 13-1. Likewise the bracketed portion of (13-4) is the transmission coefficient for such a system.

Hence, for a given wave system, one can construct the answer for the problem (at least in terms of an integral representation) by taking the source spectrum, multiply under the integral sign by the reflection coefficient (or transmission coefficient) in a suitable representation, multiply by a phase factor proportional to the vertical distance traveled, and finally multiply by a phase factor to account for the horizontal distance traveled. Source factors are:

$$\frac{1}{s} \int_{-\infty}^{\infty} \frac{dq}{2(q^2 + 1)^{\frac{1}{2}}} \text{ (region 1)}; \quad \frac{1}{s} \int_{-\infty}^{\infty} \frac{dq}{2(q^2 + a^2)^{\frac{1}{2}}} \text{ (region 2)}.$$

Phase factors for vertical distances are:

$$e^{\pm(s/v_1)(q^2+1)^{\frac{1}{2}} z} \text{ (region 1)}; \quad e^{\pm(s/v_1)(q^2+a^2)^{\frac{1}{2}} z} \text{ (region 2)}.$$

The phase factor for horizontal distance is $e^{(s/v_1)iqx}$.

Gilbert (1956, Chapter 6) and Spencer (1960) give a much more detailed discussion of this ad hoc formulation. As we shall see below, the simple reflection coefficient, when integrated over all angles (both real and complex) gives information about the head-wave and also about any interface phenomenon that may be present. This is quite amazing and any physical interpretation must be difficult indeed!

13.3. Head-Waves.

Since $\mathbf{U} = \nabla\Phi$, we have

$$U_x^{(1)} = -\frac{1}{v_1}\,\mathrm{Im}\left\{\int_0^\infty\left[\frac{\rho_2(q^2+1)^{\frac{1}{2}} - \rho_1(q^2+a^2)^{\frac{1}{2}}}{\rho_2(q^2+1)^{\frac{1}{2}} + \rho_1(q^2+a^2)^{\frac{1}{2}}}\right] \cdot \frac{e^{(s/v_1)(q^2+1)^{\frac{1}{2}}(z-h) + (s/v_1)iqx}}{(q^2+1)^{\frac{1}{2}}}\,q\,dq\right\}, \tag{13-5}$$

and

$$U_z^{(1)} = \frac{1}{v_1}\,\mathrm{Re}\left\{\int_0^\infty\left[\frac{\rho_2(q^2+1)^{\frac{1}{2}} - \rho_1(q^2+a^2)^{\frac{1}{2}}}{\rho_2(q^2+1)^{\frac{1}{2}} + \rho_1(q^2+a^2)^{\frac{1}{2}}}\right] \cdot e^{(s/v_1)(q^2+1)^{\frac{1}{2}}(z-h) + (s/v_1)iqx}\,dq\right\}. \tag{13-6}$$

If we now introduce the Cagniard-deHoop transformation (putting the point of observation at $z = -d$), we have

$$\begin{aligned}
&(q^2+1)\mathrm{Cos}\theta - iq\mathrm{Sin}\theta = v_1 t/r = \tau,\\
&q(\tau) = i\tau\mathrm{Sin}\theta + \mathrm{Cos}\theta(\tau^2-1)^{\frac{1}{2}}, \qquad\qquad (13\text{-}7)\\
&\frac{dq}{d\tau} = i\mathrm{Sin}\theta + \frac{\tau\mathrm{Cos}\theta}{(\tau^2-1)^{\frac{1}{2}}},
\end{aligned}$$

where $r = \{(d+h)^2 + x^2\}^{\frac{1}{2}}$. Then we find that (for angles $\theta > \theta_c$)

$$u_x^{(1)} = -\frac{1}{r}\,\mathrm{Im}\left\{\left[\frac{\rho_2(q^2+1)^{\frac{1}{2}} - \rho_1(q^2+a^2)^{\frac{1}{2}}}{\rho_2(q^2+1)^{\frac{1}{2}} + \rho_1(q^2+a^2)^{\frac{1}{2}}}\right]\frac{q}{(q^2+1)^{\frac{1}{2}}}\frac{dq}{d\tau}\right\} H(\tau-\tau_c), \tag{13-8}$$

and

$$u_z^{(1)} = +\frac{1}{r}\,\mathrm{Re}\left\{\left[\frac{\rho_2(q^2+1)^{\frac{1}{2}} - \rho_1(q^2+a^2)^{\frac{1}{2}}}{\rho_2(q^2+1)^{\frac{1}{2}} + \rho_1(q^2+a^2)^{\frac{1}{2}}}\right]\frac{dq}{d\tau}\right\} H(\tau-\tau_c); \tag{13-9}$$

where

$$\tau_c = (1-a^2)^{\frac{1}{2}}\mathrm{Cos}\theta + a\mathrm{Sin}\theta = \mathrm{Cos}(\theta-\theta_c), \tag{13-10}$$

$$t_c = r/v_1\mathrm{Cos}(\theta-\theta_c).$$

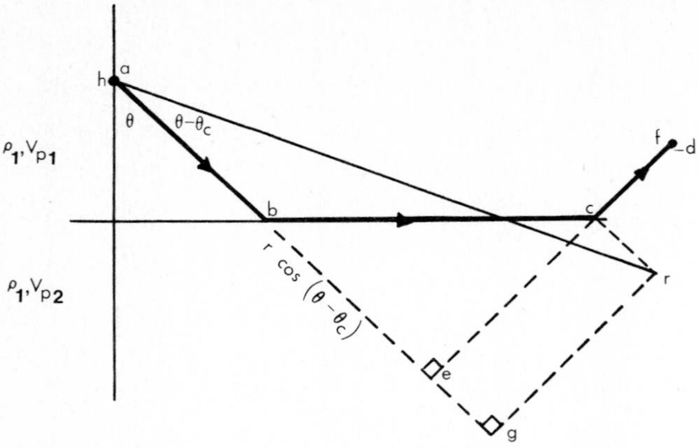

Figure 13-2. Geometric relationships for the Head-Wave.

The time t_c can be interpreted by means of a diagram. The time to travel $\overline{ag}$ with velocity v_{p1} is equal to the time necessary to travel $\overline{ab}$ plus $\overline{be}$ plus $\overline{eg}$, all with velocity v_{p1}. But to travel $\overline{be}$ with velocity v_{p1} takes the same time as to travel $\overline{bc}$ with velocity v_{p2}. Hence the time t_c is just the time necessary to travel the refracted path $\overline{abcf}$ in Figure 13-2.

To determine the amplitude and character of the arrival, we expand about the point $q = ia$. This gives

$$u_x^{(1)} \simeq \frac{1}{r}\,\mathrm{Im}\left[\frac{\rho_2(1 - a^2)^{\frac{1}{2}} - \rho_1(q^2 + a^2)^{\frac{1}{2}}}{\rho_2(1 - a^2)^{\frac{1}{2}} + \rho_1(q^2 + a^2)^{\frac{1}{2}}}\right]\frac{\mathrm{Sin}\theta_c}{\mathrm{Sin}(\theta - \theta_c)}\,H(t - t_c), \qquad (13\text{-}11)$$

and

$$u_z^{(1)} \simeq -\frac{1}{r}\,\mathrm{Im}\left[\frac{\rho_2(1 - a^2)^{\frac{1}{2}} - \rho_1(q^2 + a^2)^{\frac{1}{2}}}{\rho_2(1 - a^2)^{\frac{1}{2}} + \rho_1(q^2 + a^2)^{\frac{1}{2}}}\right]\frac{\mathrm{Cos}\theta_c}{\mathrm{Sin}(\theta - \theta_c)}\,H(t - t_c), \qquad (13\text{-}12)$$

where we have used relations similar to those of (11-58) through (11-60). Defining

$$\mathbf{e}_{r_c} = \mathbf{e}_x\mathrm{Sin}\theta_c - \mathbf{e}_z\mathrm{Cos}\theta_c,$$

and noting that

$$(q^2 + a^2)^{\frac{1}{2}} \simeq \frac{i(2a)^{\frac{1}{2}}(\tau - \tau_c)^{\frac{1}{2}}(1 - a^2)^{\frac{1}{4}}}{\mathrm{Sin}^{\frac{1}{2}}(\theta - \theta_c)}, \qquad (11\text{-}61)$$

we find that

$$\mathbf{u}^{\mathrm{Head}} \simeq \frac{-\mathbf{e}_{r_c}}{r^{3/2}\mathrm{Sin}^{3/2}(\theta - \theta_c)}\,\frac{2\rho_1 v_{p1}^{\frac{1}{2}}(2a)^{\frac{1}{2}}}{\rho_2(1 - a^2)^{\frac{1}{4}}}\,(t - t_c)^{\frac{1}{2}}\,H(t - t_c); \qquad (13\text{-}13)$$

that is, the motion is along the ray path $\overline{cf}$ in Figure 13-2. It also has the time integral of the reflected ray motion. Figures 13-3(a) to 13-3(c) show what is happening in terms of wave fronts.

The analysis derived above has been for a sharp transition layer. The real

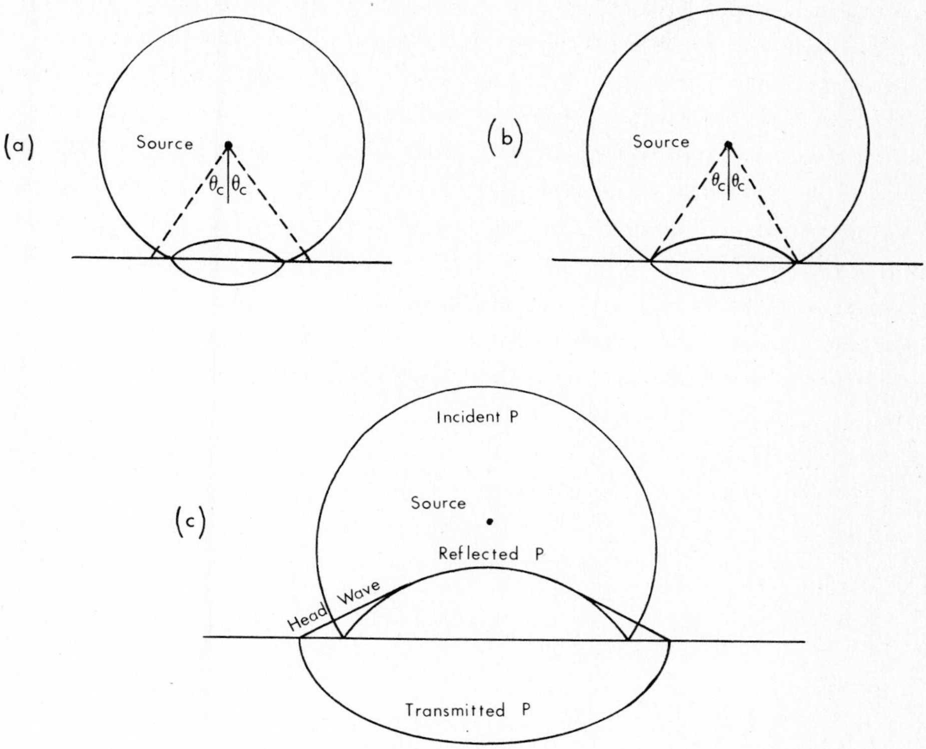

Figure 13-3. Wave-fronts at an interface (schematic).
(a). Incident wave has not reached the critical distance.
(b). Incident wave at the critical distance. (Head-waves start to form here.)
(c). Incident wave beyond the critical distance. (Head-waves fully developed.)

earth differs from this model in two important ways. It consists of solid rather than fluid materials, and the velocity transitions are probably not very sharp. As noted earlier, the elastic wave equation in media with varying velocity is rather difficult to solve, so studies on gradual transitions have been carried out mostly on fluid layers. Two recent papers are Merzer (1971) and (1974) dealing with the properties of head-waves formed at a gradual transition region.

13.4. Applications To Seismology.

In treating the real earth, one can include solid materials but is generally restricted to sharp transition layers. Cagniard (1939) was one of the first to deal with head-waves (he called them conical waves) in a quantitative manner. More recently, a whole book devoted to seismic head-waves has been published. Cerveny and Ravindra (1971) have investigated a number of reasonable models for the earth. They have considered a single interface, multilayered media, curved and dipping interfaces, as well as interference head-waves which occur when a velocity gradient is present in the critically refracting layer. This extensive work also includes a large bibliography pertaining to head-waves. In their final chapter, the approximate methods developed earlier in their book are compared with the few exact solutions available.

Head-waves must be included in the analysis of multilayered media, whether of flat or spherical layers, and computations of synthetic seismograms in such media must include appropriate head-wave arrivals. Such computations will be considered in greater detail in Chapters 18 and 23.

WAVES ALONG A SOLID-SOLID INTERFACE

14.1. Real Interface Waves (Stoneley Waves).

In the same way that we related the denominator determinant of plane-wave reflection coefficients in a half-space to the Rayleigh denominator, we can relate the denominator determinant of plane-wave reflection and transmission coefficients at a solid-solid interface to a determinant which gives rise to the propagation of waves along such an interface. We again use the following substitutions:

$$k_p \mathrm{Sin}\theta_p = k, \qquad k_p \mathrm{Cos}\theta_p = (k_p^2 - k^2)^{\frac{1}{2}} = i(k^2 - k_p^2)^{\frac{1}{2}}, \qquad k_s \mathrm{Cos}\theta_s = (k_s^2 - k^2)^{\frac{1}{2}} = i(k^2 - k_s^2)^{\frac{1}{2}}. \tag{14-1}$$
$$k_s \mathrm{Sin}\theta_s = k,$$

Eliminating a common factor of frequency and multiplying by the wave-velocity $\underline{c}$, we get the following correspondences:

$$k_p \mathrm{Sin}\theta_p \rightarrow 1 \qquad k_p \mathrm{Cos}\theta_p \rightarrow i(1 - c^2/v_p^2)^{\frac{1}{2}} = iA,$$
$$k_s \mathrm{Sin}\theta_s \rightarrow 1 \qquad k_s \mathrm{Cos}\theta_s \rightarrow i(1 - c^2/v_s^2)^{\frac{1}{2}} = iB, \tag{14-2}$$
$$2k_p^2 \mathrm{Sin}^2\theta_p - k_s^2 \rightarrow 2 - c^2/v_s^2 = D.$$

Relations (12-7) then give rise to the following determinant:

$$\begin{vmatrix} 1 & -iB_1 & -1 & -iB_2 \\ iA_1 & 1 & iA_2 & -1 \\ \mu_1 D_1 & -2i\mu_1 B_1 & -\mu_2 D_2 & -2i\mu_2 B_2 \\ -2i\mu_1 A_1 & -\mu_1 D_1 & -2i\mu_2 A_2 & \mu_2 D_2 \end{vmatrix} = 0. \tag{14-3}$$

Expanding this determinant, and using the relation $\mu = \rho v_s^2$, one obtains the equation for the existence of real interface waves first obtained by Stoneley (1924). This is

$$c^4 \left[(\rho_2 - \rho_1)^2 - (\rho_1 A_2 + \rho_2 A_1)(\rho_1 B_2 + \rho_2 B_1) \right]$$
$$+ 4c^2(\rho_1 v_{s1}^2 - \rho_2 v_{s2}^2)(\rho_1 A_2 B_2 - \rho_2 A_1 B_1 - \rho_1 + \rho_2) \tag{14-4}$$
$$+ 4(\rho_1 v_{s1}^2 - \rho_2 v_{s2}^2)^2 (A_1 B_1 - 1)(A_2 B_2 - 1) = 0.$$

For Poisson's ratio $\sigma = 0.25$, Scholte(1947) determined the range of existence for a real root of (14-4), and Yamaguchi and Sato (1955) evaluated the value of this real root as well as determined the particle motion. Some additional considerations were given by Ginzbarg and Strick (1958). The real root exists in the region labeled "STONELEY" in Figure 14-1. Outside this region, the real root becomes complex. We shall consider this more general case in the next section. Region $\underline{2}$ is arbitrarily taken as the medium of greater density. The interface wave velocity is seen to lie between the slower shear wave velocity and the Rayleigh wave velocity in the denser medium. These papers also showed that the particle motion was retrograde with respect to the denser medium.

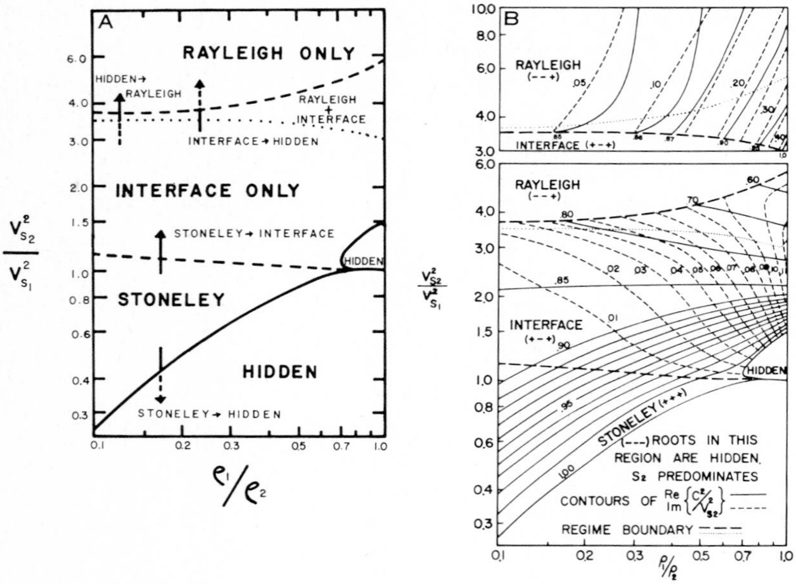

Figure 14-1(A). Stoneley, Interface, Rayleigh regimes and associated transitions.

(B). Contour values of the complex squares of the normalized phase velocities for the two types of waves associated with interfaces. The upper portion of the figure is repeated so that the contours for Interface and Rayleigh waves may be separated. (After Pilant, 1972).

14.2. Generalized Interface Waves.

There are at least two empirical behavior patterns governing roots of equations like (14-4) in the complex plane. The first is a law of "Conservation of Roots". This says that roots of such an equation are neither created nor destroyed, i.e., you can follow them around the complex plane (albeit many-sheeted) as a function of the system parameters. The second says that movement along a real or an imaginary axis continues along that axis until the root collides with either another root or with a branch point of a factor such as A_1. (This second rule is generally true, but some degenerate cases exist in which double roots moving through each other have been found.) Using this sort of information, Pilant (1972) investigated in detail the behavior of the interface root for physical parameter sets outside the existence range delimited by Scholte. He found that there are eight independent Riemann sheets associated with (14-4) on which lie sixteen independent roots. Only two of these roots have any great physical significance.

As an example, consider the case where both solid media have Poisson's ratios equal to 0.25 (i.e., $v_p = 3^{\frac{1}{2}} v_s$) and the denser Medium 2 has a density twice that of Medium 1. The behavior of the branch points, associated branch cuts, and roots are sketched (not to scale) in Figure 14-2. Contours of the root values for a range of parameters are given in Figure 14-1. Returning to Figure 14-2, we can follow the behavior of the interface root as the velocity of Medium 1 is varied relative to that in Medium 2. In Figure 14-2(a), the Stoneley wave root lies hidden on a lower sheet, and no interface pulse is propagated. However, this case would be difficult to achieve experimentally inasmuch as the medium with the higher density has the lower velocity. As the two velocities approach each other, the Stoneley root migrates onto

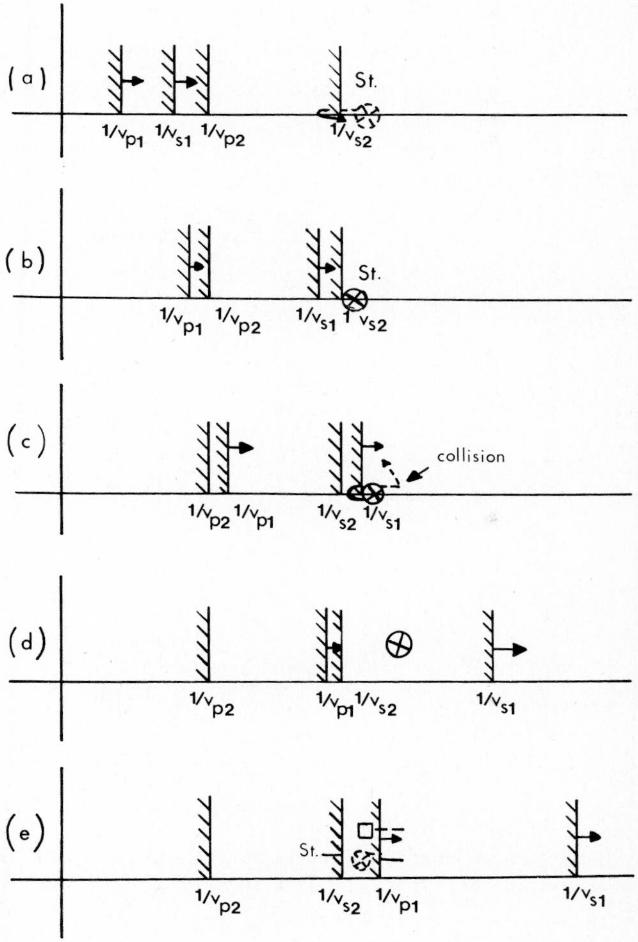

Figure 14-2. Behavior of the complex roots and branch cuts in the complex reciprocal velocity (c^{-1}) — plane. Density $\rho_2 = 2\rho_1$; $\sigma_2 = \sigma_1 = 0.25$. Figure not to scale.

the upper sheet as in Figure 14-2(b). In this position, it contributes to a pulse propagating along the interface. As v_{s1} continues to decrease relative to v_{s2} the two branch points interchange places. The real Stoneley root moves slowly away from the branch point at $(v_{s2})^{-1}$. See Figure 14-2(c). As v_{s1} further decreases, the branch point at $(v_{s1})^{-1}$ overtakes the Stoneley root, and it migrates onto a hidden sheet. On this hidden sheet, it soon collides with another root (of the 16) on the real axis and becomes complex. Further decrease of v_{s1} uncovers this root as in Figure 14-2(d). While the interface root is complex, a pulse-like phenomenon can still propagate down the interface; however, it dies off more rapidly than a purely real interface pulse would. This extra loss of energy is due to the radiation of energy away from the interface. Thus this complex root is another "pseudo-wave" as discussed in Section 11.4.

Finally, as v_{s1} becomes vanishingly small, the branch cut associated with $(v_{p1})^{-1}$ overtakes the complex interface root and it moves onto a lower sheet. See

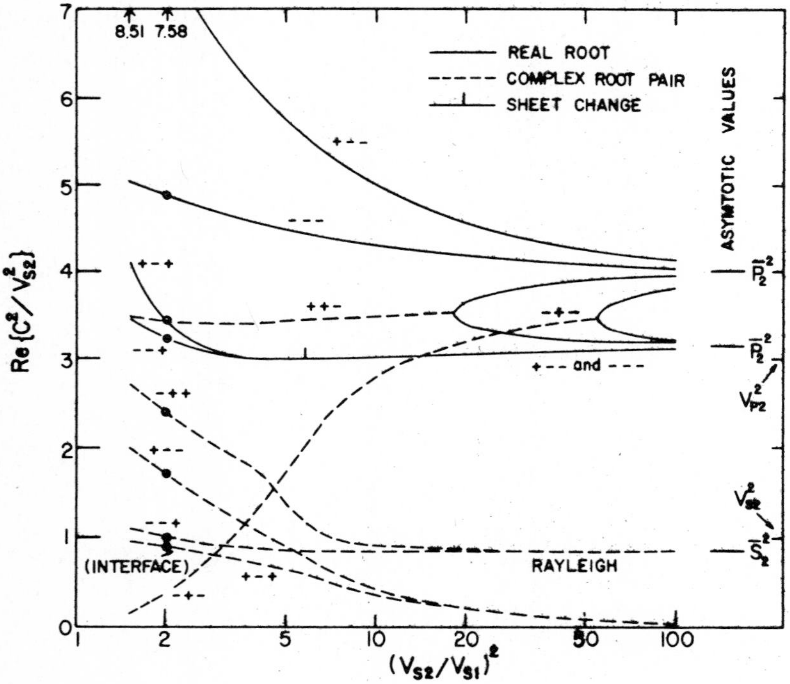

Figure 14-3. Behavior of the Stoneley equation roots in the vacuum limit (elastic half-space). The two asymptotic velocities $\overline{P_2}$ are those given by Gilbert and Laster (1962). $\overline{S_2}$ is the velocity associated with Rayleigh-waves in a half-space. (After Pilant 1972).

Figure 14-2(e). Shortly before this takes place, another root (the Rayleigh root), which had lain hidden on a lower sheet, is exposed by the rapidly moving branch cut associated with $(v_{p1})^{-1}$. For a relatively small range of parameters, both complex roots contribute pulses propagating along the interface. The complex Rayleigh pulse has the greater attenuation. If v_{s1} is ultimately decreased to zero along with ρ_1, the imaginary part of the Rayleigh root vanishes giving the normal Rayleigh value for $\sigma = 0.25$. Hence, in a rather complex manner, we see that in the limit of one solid becoming a vacuum, the Stoneley pulse does not become the Rayleigh pulse. This detailed behavior was not found by many authors. They wrongly identified the Rayleigh pulse as the limit of the complex Stoneley pulse. More accurate values of the sixteen roots in this limiting case as the velocities in Medium 1 go to zero are shown in Figure 14-3. The region of existence for the simultaneous presence of both complex pulses is so narrow that it would be difficult to design an experiment to exhibit the details described above.

14.3. The Impulsive Solution.

Although the early work of Stoneley (1924) paved the way for an understanding of the interface wave propagating between two elastic media, early interest in impulsive solutions centered on the fluid-solid interface. This was of some importance to exploration geophysics in that it modeled the sediment-rock interface in shallow seas. A detailed study was made by Strick (1959a and 1959b) in order to explain events seen by seismic parties. Accompanying these papers was a model study by Roever and Vining

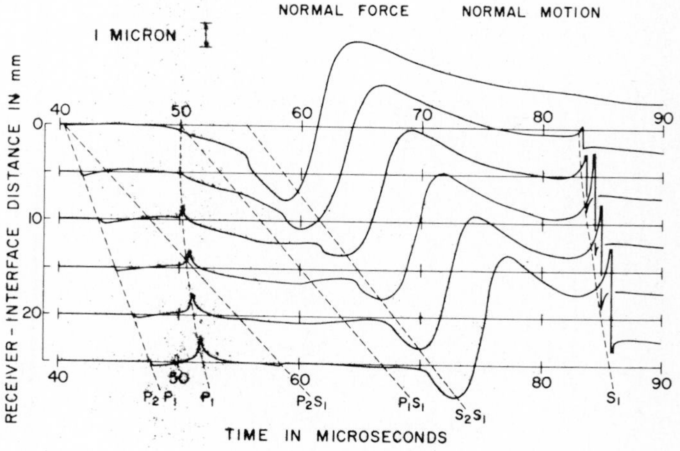

Figure 14-4. Details of the "Pseudo-Stoneley" wave (complex interface). Material parameters are: $v_{p1} = 2.0$ km/sec, $v_{s1} = 1.2$ km/sec, $\rho_1 = 2.5$ gm/cm^3; $v_{p2} = 2.5$ km/sec, $v_{s2} = 1.8$ km/sec, $\rho_2 = 3.0$ gm/cm^3. (After Dampney, 1972)

(1959) which showed well the nature of the impulse propagating along the fluid-solid interface. A similar work for the solid-solid interface was part of the paper by Gilbert and Laster (1962) which we have already considered in the case of pseudo-waves in a half-space. They present several theoretical seismograms for various conditions of propagation both in the real interface (Stoneley) regime as well as the complex interface regime. Gilbert et al (1962) gave model observations of both types of pulses. Dampney (1972) constructed theoretical seismograms for the complex interface regime which were somewhat more detailed than those of Gilbert and Laster (ibid.). His results are shown in Figure 14-4. Tanyi (1967a) considered the case of generalized Stoneley waves at a fluid-solid interface with the added complication that the interface was made spherical to model a real earth. In another paper Tanyi (1967b) gives a somewhat different formulation for the generalized Rayleigh and Stoneley-wave problems.

CHAPTER 15

ONE LAYER OVER A HALF-SPACE — RAY THEORY

15.1. General

We have seen how an exact formulation of boundary value problems leads to extremely complicated integrals as the number of interfaces goes up. We have looked at an infinite medium, a half-space, and at an interface between two half-spaces. The next order of complication is introduced by the addition of a layer of elastic material of thickness H over a half-space, giving two media and two interfaces. We shall look at this problem first of all from a ray-theory point of view and then from a surface-wave point of view. Ray-analysis of body-waves introduces little in the way of new behavior, whereas surface-waves must be examined in an entirely new way.

For body-waves in a layer over a half-space, then, we have only to formulate methods of analyses to interpret such a simple structural model. Because of the complicated nature of the seismogram after the arrival of the initial P-phases, one usually restricts the analysis to P-waves. The seismic reflection method detailed in Section 15.2 and in Chapters 17 and 18 is largely based on geometrical construction of minimum-time ray paths. We use Snell's law to develop travel-time curves (time of arrival of the various body-wave events versus source-receiver distance for each event). The use of amplitudes is somewhat limited by practical difficulties discussed below. The seismic refraction method is outlined in Section 15.3 and also in Chapters 17 and 18. Again, it is largely based on minimum-time ray paths with a large part of the source-receiver distance being traversed by the head-wave (critically refracted ray) of Chapter 13. Here, too, amplitude analysis is limited somewhat by practical difficulties but a greater attempt has been made to use amplitude data in determining earth structure than in the case of reflected rays.

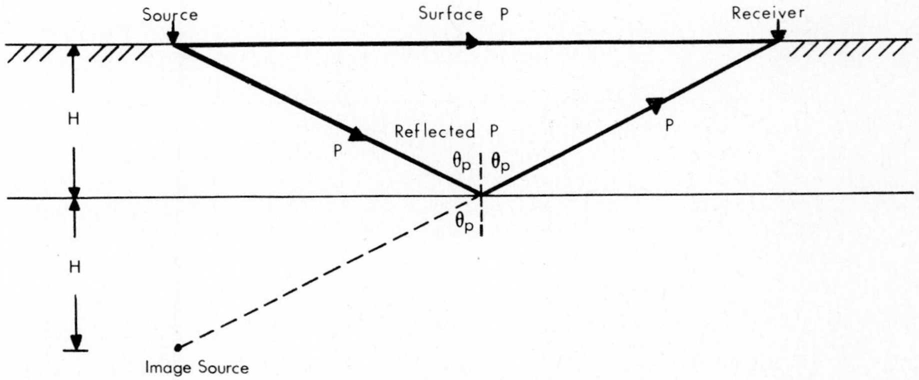

Figure 15-1. Reflection of a P-wave in a layer overlying a half-space.

15.2. Travel-Time Curves For Reflected Rays.

In working with ray theory in a layered medium, one generally only considers P-waves because, as first arrivals, they are not interfered with and their onsets can be more clearly defined. Figure 15-1 shows a ray emanating from a surface source, being reflected from the interface at depth H and received at a distance x. The total path length traversed by the ray is then

$$\{x^2 + (2H)^2\}^{\frac{1}{2}}.$$

Hence the time necessary for a P-wave to travel the path indicated in Figure 15-1 is

$$T_p = \{x^2 + (2H)^2\}^{\frac{1}{2}}/v_{p1}. \qquad (15\text{-}1)$$

This is the equation of a hyperbola with assymptotes $1/v_{p1}$ and $-1/v_{p1}$, and intercept $2H/v_{p1}$. The travel-time curve is plotted in Figure 15-2. If one squares and re-arranges (15-1) one obtains

$$T_p^{\,2} = (x^2 + 4H^2)/v_{p1}^2, \qquad (15\text{-}2)$$

which is a straight line in the variables T_p^2 and x^2 with slope $1/v_{p1}^2$. Such straight line plots are routinely used to determine the layer velocity. The intercept $4H^2/v_{p1}^2$ may be used to determine the layer thickness H.

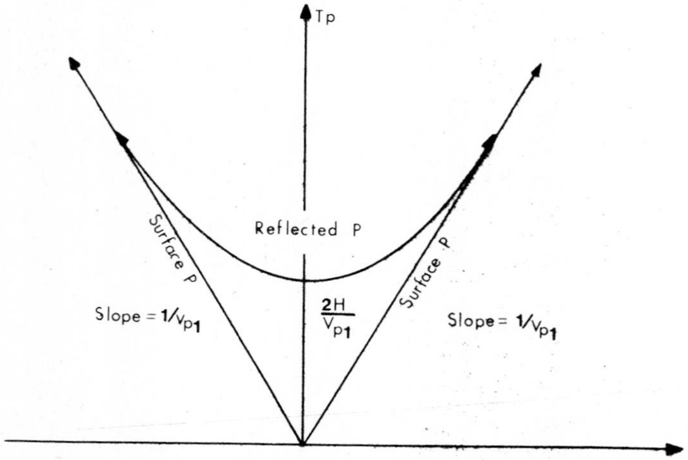

Figure 15-2. Travel-time curve for reflected P-waves in a layer overlying a half-space. The intercept time $2H/v_{p1}$ is a measure of the thickness of the layer.

One may use the reflection and directivity coefficients obtained in Chapters 12 and 9 to say something about how the amplitude of a seismic pulse will vary with distance. However, in the practical case, this amplitude does not follow these nice theoretical values except in a general way. This is due to imperfect coupling of the source and receiver, inhomogeneities in the earth, and a variable attenuation in these anomalous regions. It should be noted that for inhomogeneous media plane-wave theory is not appropriate. More general considerations will be given in Chapters 21 and 22. Unless absolute amplitude information is obtainable, one can only determine the P-wave velocity in the layer and the thickness of the layer. No information about the lower layer is available from (15-1) or (15-2).

One can introduce an additional complication by having the interface dipping with respect to the free surface as in Figure 15.3. In this case the distance

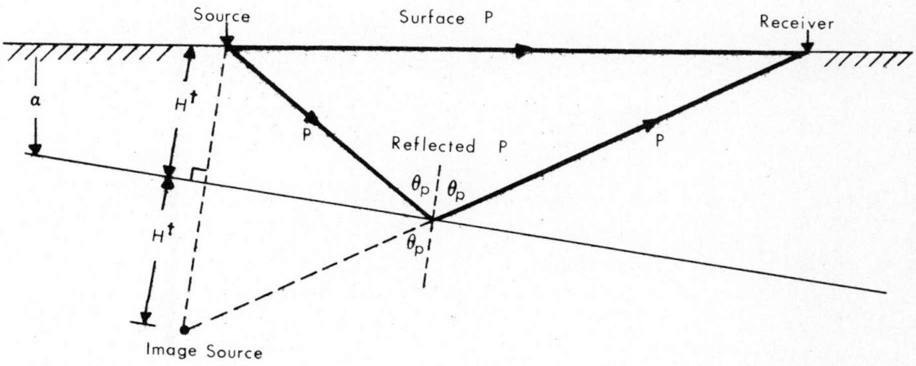

Figure 15-3. Reflection of a P-wave from a dipping interface.

traveled by the reflected ray is given by

$$\begin{aligned} D^2 &= x^2 + (2H^\dagger)^2 - 4xH^\dagger \mathrm{Cos}(\pi/2 + \alpha) \\ &= x^2 + (2H^\dagger)^2 + 4xH^\dagger \mathrm{Sin}\alpha \\ &= (x + 2H^\dagger \mathrm{Sin}\alpha)^2 + (2H^\dagger \mathrm{Cos}\alpha)^2, \end{aligned}$$

where $H^\dagger$ is the shortest distance to the interface (measured from the source perpendicularly to the interface, not along the vertical) and α is the angle of dip. The travel-time is given by

$$T_p = \{(x + 2H^\dagger \mathrm{Sin}\alpha)^2 + (2H^\dagger \mathrm{Cos}\alpha)^2\}^{\frac{1}{2}}/v_{p1}. \tag{15-3}$$

This is the equation of an offset hyperbola with center located at $-2H^\dagger \mathrm{Sin}\alpha$, intercept at $2H^\dagger/v_{p1}$, and assymptotes from the center of slopes $\pm 1/v_{p1}$.

Rewriting (15-3), we can obtain

$$y = v_{p1}^2 T_p^2 - x^2 = 4xH^\dagger \mathrm{Sin}\alpha + 4H^{\dagger 2}. \tag{15-4}$$

This is a linear equation of the form $y = ax + b$, where

$$a = 4H^\dagger \mathrm{Sin}\alpha, \qquad b = 4H^{\dagger 2}. \tag{15-5}$$

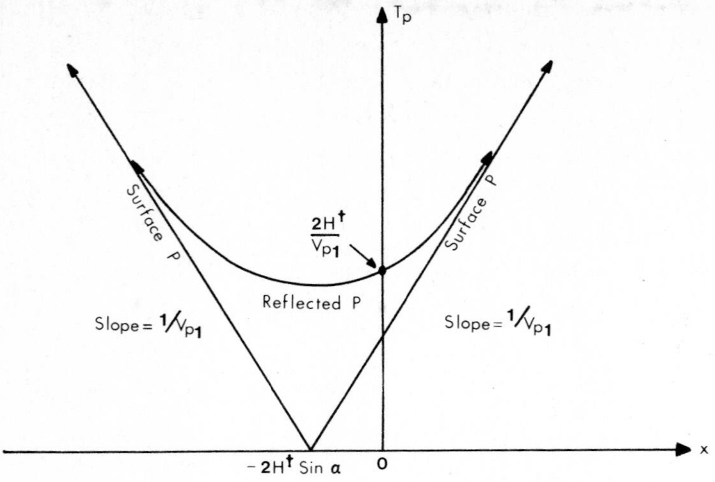

Figure 15-4. Travel-time curves for reflected P-waves in a layer with a dipping interface. The intercept time $2H^{\dagger}/v_{pi}$ is a measure of the slant depth, while the assymetry of the hyperbola about $x = 0$ implies a dipping interface.

We then see that the $H^{\dagger}$ is given by

$$H^{\dagger} = b^{\frac{1}{2}}/2, \tag{15-6}$$

and the dip angle α by

$$\text{Sin}\alpha = \tfrac{1}{2}a/b^{\frac{1}{2}}. \tag{15-7}$$

The vertical depth is given by

$$H = H^{\dagger}/\text{Cos}\alpha. \tag{15-8}$$

The evaluation of $H^{\dagger}$, α, and finally H depends in this analysis upon a knowledge of v_{p1}. It can be obtained from observation of surface P, or from trial and error substitution into (15-4) until a straight line is obtained. This is in contrast to (15-2) where a straight line may be plotted and the velocity v_{p1} obtained directly.

15.3. Travel-Time Curves For Refracted Rays.

The use of the critically refracted ray along the interface between Medium 1 and Medium 2 will give the velocities of both media as well as the depth to the interface. The travel-time of the ray indicated in Figure 15-5 can be computed as follows:

$$T = T_{ab} + T_{bc} + T_{cd} = \frac{2H}{v_{p1}\text{Cos}\theta_c} + \frac{x - 2H\text{Tan}\theta_c}{v_{p2}} \tag{15-9}$$

$$= \frac{x}{v_{p2}} + \frac{2H\text{Cos}\theta_c}{v_{p1}} = \frac{x}{v_{p2}} + \frac{2H(v_{p2}^2 - v_{p1}^2)^{\frac{1}{2}}}{v_{p1}v_{p2}} \qquad (x > 2H\text{Tan}\theta_c).$$

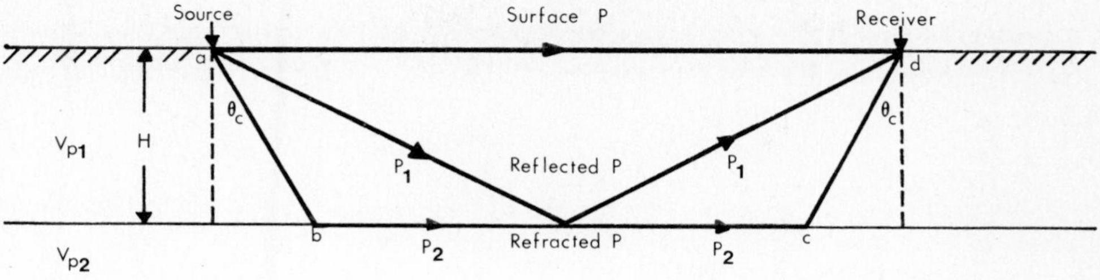

Figure 15-5. Critical refraction of a P-wave in a layer overlying a half-space.

The travel-time curve for this ray is the straight line $\overline{efg}$ in Figure 15-6 which joins on to the travel-time curve for the reflected ray at point e. It has a slope of $1/v_{p2}$. At distances less than that of point e, the critical distance, there is no critically refracted ray; the angle of the incident P-wave being less than θ_c. This distance is given by

$$x_{crit} = 2H\text{Tan}\,\theta_c.$$

Inasmuch as the exact point of separation of the critically refracted ray from the reflected ray is difficult to determine, little use can be made of this particular equation.

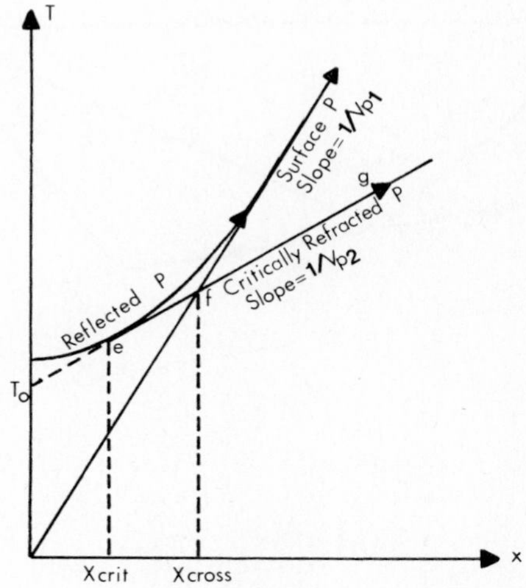

Figure 15-6. Travel-time curves for P-waves in a layer overlying a half-space.

The intercept time T_o of the refracted ray's travel-time curve may be obtained by setting $x = 0$ in (15-9). This value is given by

$$T_o = \frac{2H(v_{p2}^2 - v_{p1}^2)^{\frac{1}{2}}}{v_{p1}v_{p2}} . \quad (15\text{-}10)$$

At the crossover distance, $x = x_{cross}$, (corresponding to the point $\underline{f}$ on the travel-time curve) the times of arrival of the direct P along the surface and the refracted P are the same, i.e.,

$$\frac{X_{cross}}{v_{p2}} + \frac{2H(v_{p2}^2 - v_{p1}^2)^{\frac{1}{2}}}{v_{p1}v_{p2}} = \frac{X_{cross}}{v_{p1}} ,$$

or

$$X_{cross} = 2H\left[\frac{v_{p2} + v_{p1}}{v_{p2} - v_{p1}}\right]^{\frac{1}{2}} . \quad (15\text{-}11)$$

Having found the two P-wave velocities, v_{p1} and v_{p2}, from the inverse slopes of the two straight-line portions of the travel-time curves in Figure 15-6, both (15-10) or (15-11) can be inverted to find the layer thickness H.

It is apparent from the above that the refraction method gives somewhat more information than does the reflection method. From the data which Figure (15-6) represents, one can obtain v_{p1}, v_{p2}, and H. That is, we have added to our knowledge the velocity of the half-space. There is a price to be paid however. The critically refracted ray loses energy more rapidly than does the reflected ray for the same source-receiver distance. Consequently much more energetic sources must be used in refraction work.

Again the complication of a dipping interface can be introduced. Ray paths are as in Figure 15-7. The travel-time can be separated into three component parts. We

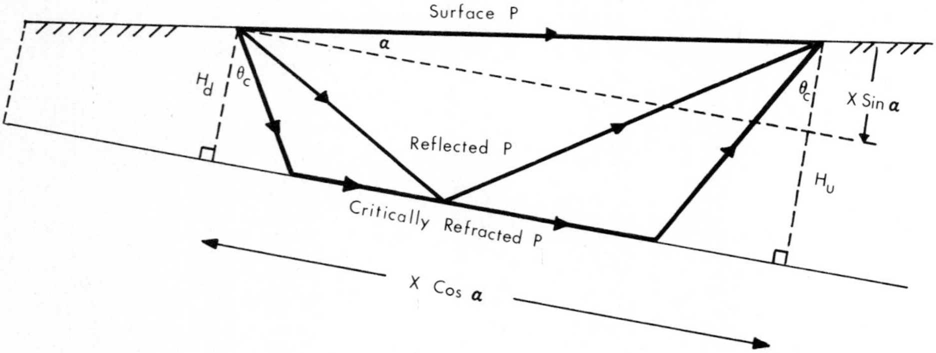

Figure 15-7. Critical refraction of a P-wave in a dipping layer overlying a half-space.

then have for the down-dip travel-time:

$$T_d = \frac{H_d}{v_{p1}\mathrm{Cos}\theta_c} + \frac{H_d + x\mathrm{Sin}\alpha}{v_{p1}\mathrm{Cos}\theta_c} + \frac{x\mathrm{Cos}\alpha - H_d\mathrm{Tan}\theta_c - (H_d + x\mathrm{Sin}\alpha)\mathrm{Tan}\theta_c}{v_{p2}}$$

$$= \frac{2H_d\mathrm{Cos}\theta_c}{v_{p1}} + \frac{x\mathrm{Sin}(\theta_c + \alpha)}{v_{p1}} . \quad (15\text{-}12)$$

Similarly, the up-dip travel-time can be computed as

$$T_u = \frac{2H_u \cos\theta_c}{v_{p1}} + \frac{x \sin(\theta_c - \alpha)}{v_{p1}} . \tag{15-13}$$

Sketches of these two relationships are shown in Figure 15-8.

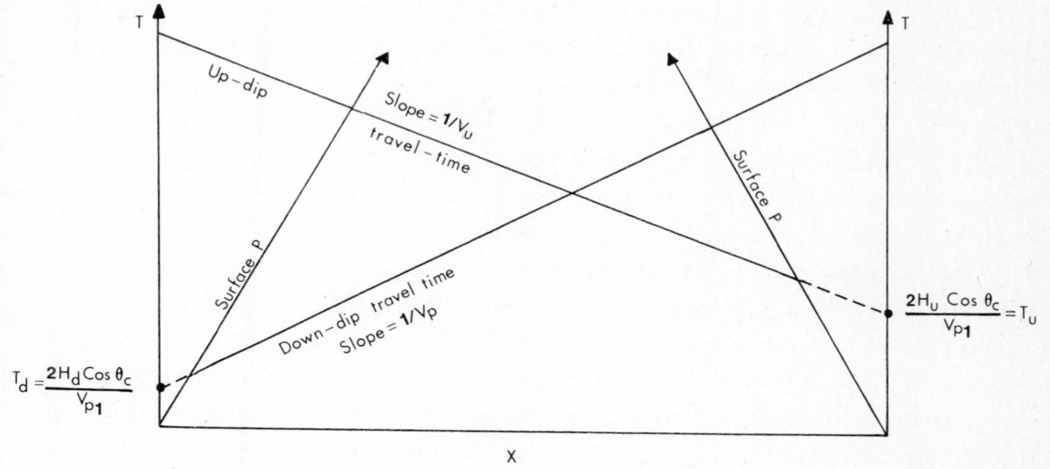

Figure 15-8. Travel-time curves for P-waves in a dipping layer overlying a half-space. Note that time intercepts and slopes of the two refraction travel-time curves are not the same.

The student can show that (15-12) and (15-13) are equal (as they should be considering reciprocity). It can also be seen that the dip angle α is linked to the calculated velocity in the half-space. We can separate them as follows. Let

$$\sin(\theta_c + \alpha)/v_{p1} = 1/v_d, \tag{15-14}$$

be the apparent down-dip velocity, and

$$\sin(\theta_c - \alpha)/v_{p1} = 1/v_u, \tag{15-15}$$

be the apparent up-dip velocity. Combining we have

$$\frac{1}{2}\left[\frac{1}{v_d} + \frac{1}{v_u}\right] = \frac{1}{2v_{p1}}\left[\sin(\theta_c + \alpha) + \sin(\theta_c - \alpha)\right]$$

$$= \frac{1}{v_{p1}} \sin\theta_c \cos\alpha = \frac{\cos\alpha}{v_{p2}} . \tag{15-16}$$

If α is small enough, the cosine can be approximated by unity, and the reciprocal of the true second medium velocity is equal to the average of up- and down-dip reciprocal apparent velocities. This shows why refraction work should be double-ended to be meaningful, unless one knows from other information that the refracting layer is horizontal. Otherwise, one obtains only the apparent velocities given by (15-14) or (15-15).

ONE LAYER OVER A HALF-SPACE — MODE THEORY

16.1. General.

The introduction of a layer over a half-space presents additional phenomena that were not present previously. Whereas a uniform half-space supports only a Rayleigh surface-wave (coupled P- and SV-motion), a medium having a generally increasing velocity with depth will also support a wave of SH-type. This is known as a Love-wave. Local low-velocity layers or regions do not preclude its existance; however, in the case of a single layer overlying a half-space, the shear-velocity in the half-space must be the greater.

A second phenomenon is that both Rayleigh- and Love-waves are dispersive, i.e., the wave-velocity is a function of the frequency (or alternatively, the wavelength). The impulse response for a simple source in a dispersive medium is not localized in time. Rather, the energy is spread out over an interval in the form of a wave-train which is both amplitude and frequency modulated. This requires a specialized analysis that is considerably more complicated than that involved with the interpretation of travel-time curves. We will have to introduce the concept of phase- and group-velocities in Section 16.4. The interpretation of a single-layered structure from a dispersed wave-train is conceptually simple, but practically difficult in that one is generally limited to a rather narrow band of frequencies composing the surface-wave signal.

We shall find that hidden roots on lower sheets also play a significant role here in the behavior of surface waves.

16.2. Love Modes (SH).

The geometrical relationships are as in Figure 16-1. For the case of SH-motion in a layer over a half-space, we find that there are surface waves possible as long

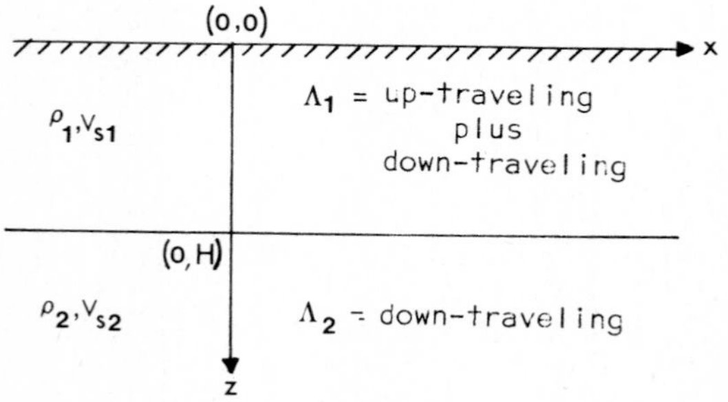

Figure 16-1. Geometrical relationships for Love-Waves in a layer overlying a half-space.

as the shear-wave velocity of the half-space exceeds the shear-wave velocity in the layer. A mathematical model for this situation is given by:

$$\mathbf{u} = -\mathbf{e}_y \frac{\partial \Lambda}{\partial x}, \tag{6-26}$$

$$t_{yz} = -\mu \frac{\partial^2 \Lambda}{\partial x \partial z}, \tag{6-34}$$

where

$$\Lambda_1 = \int_{-\infty}^{\infty} \left[A(k) e^{(k^2 - k_{s1}^2)^{\frac{1}{2}} z} + B(k) e^{-(k^2 - k_{s1}^2)^{\frac{1}{2}} z} \right] e^{i(kx - \omega t)} dk, \tag{16-1}$$

and

$$\Lambda_2 = \int_{-\infty}^{\infty} \left[C(k) e^{-(k^2 - k_{s2}^2)^{\frac{1}{2}} z} \right] e^{i(kx - \omega t)} dk. \tag{16-2}$$

The coefficients, A(k), B(k), and C(k) are determined by applying the boundary conditions:

$$t_{yz} \Big|_{z=0} = \frac{-Y}{2\pi} \int_{-\infty}^{\infty} e^{i(kx - \omega t)} dk, \tag{16-3}$$

which is the inverse transform of a line force in the Y-direction acting with time dependence $e^{-i\omega t}$, and the continuity of shear-stress and tangential displacement at the surface z = H, i.e.,

$$t_{yz}^{(1)} \Big|_{z=H} = t_{yz}^{(2)} \Big|_{z=H},$$
$$u_y^{(1)} \Big|_{z=H} = u_y^{(2)} \Big|_{z=H}. \tag{16-4}$$

The values determined are:

$$A = \frac{Y}{2\pi i k \mu_1 \Delta_L} \left[\mu_2 (k^2 - k_{s2}^2)^{\frac{1}{2}} - \mu_1 (k^2 - k_{s1}^2)^{\frac{1}{2}} \right] e^{-(k^2 - k_{s1}^2)^{\frac{1}{2}} H},$$

$$B = \frac{-Y}{2\pi i k \mu_1 \Delta_L} \left[\mu_2 (k^2 - k_{s2}^2)^{\frac{1}{2}} + \mu_2 (k^2 - k_{s1}^2)^{\frac{1}{2}} \right] e^{(k^2 - k_{s1}^2)^{\frac{1}{2}} H}, \tag{16-5}$$

$$C = \frac{-Y}{ik\Delta_L} (k^2 - k_{s1}^2)^{\frac{1}{2}} e^{(k^2 - k_{s2}^2)^{\frac{1}{2}} H},$$

where

$$\Delta_L = 2(k^2 - k_{s1}^2)^{\frac{1}{2}} \Big[\mu_1 (k^2 - k_{s1}^2)^{\frac{1}{2}} \mathrm{Sinh}\{(k^2 - k_{s1}^2)^{\frac{1}{2}} H\}$$
$$+ \mu_2 (k^2 - k_{s2}^2)^{\frac{1}{2}} \mathrm{Cosh}\{(k^2 - k_{s1}^2)^{\frac{1}{2}} H\} \Big]. \tag{16-6}$$

Substituting into equations (16-1) and (16-2), we have

$$\Lambda_1 = \frac{-Y}{\pi i \mu_1} \int_{-\infty}^{\infty} \Big[\mu_2 (k^2 - k_{s2}^2)^{\frac{1}{2}} \mathrm{Sinh}\{(k^2 - k_{s1}^2)^{\frac{1}{2}} (H-z)\}$$
$$+ \mu_1 (k^2 - k_{s1}^2)^{\frac{1}{2}} \mathrm{Cosh}\{(k^2 - k_{s1}^2)^{\frac{1}{2}} (H-z)\} \Big] \frac{e^{i(kx - \omega t)} dk}{k\Delta_L}, \tag{16-7}$$

and

$$\Lambda_2 = \frac{-Y}{\pi i} \int_{-\infty}^{\infty} \left[(k^2 - k_{s1}^2)^{\frac{1}{2}} e^{\{(k^2 - k_{s2}^2)^{\frac{1}{2}}(H-z)\}} \right] \frac{e^{i(kx - \omega t)} dk}{k\Delta_L} . \qquad (16\text{-}8)$$

The contour of integration and the singularities of the integrands must be specified in (16-7) and (16-8). They are as in Figure 16-2. There are branch cuts

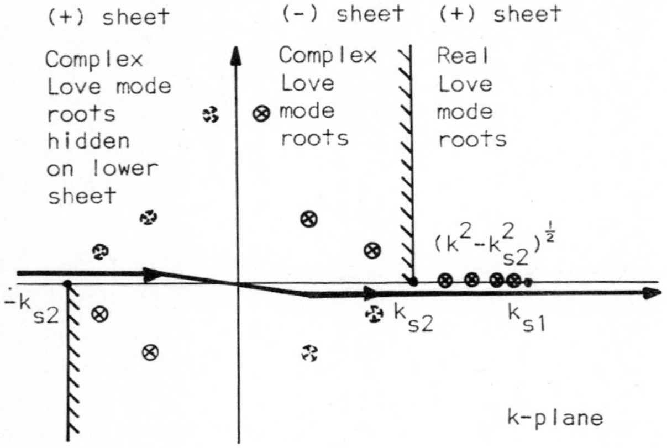

Figure 16-2. Branch cuts and singularities for equations (16-7) and (16-8).

starting at $\pm k_{s2}$ as shown, but because the factor $(k^2 - k_{s1}^2)^{\frac{1}{2}}$ always occurs in pairs, there is no branch cut associated with the point k_{s1}. We shall show later, that in a "layer cake" geometry, the only branch cuts are those associated with the half-space. The path goes as indicated below the branch point at k_{s2} and above the branch point at $-k_{s2}$ (to insure outward traveling waves for positive $\underline{x}$) and it will be deformed upward to pick up the pole contributions plus an additional factor due to the branch line integral around the cut from k_{s2}.

Let us now look at (16-6). This will have zeros when

$$\mu_1(k^2 - k_{s1}^2)^{\frac{1}{2}} H \,\text{Sinh}\{(k^2 - k_{s1}^2)^{\frac{1}{2}} H\} + \mu_2(k^2 - k_{s2}^2)^{\frac{1}{2}} H \,\text{Cosh}\{(k^2 - k_{s1}^2)^{\frac{1}{2}} H\} = 0. \quad (16\text{-}9)$$

For $k > k_{s1} > k_{s2}$, there will be no positive real roots. (When $k_{s2} < k_{s1}$, there will be no positive real roots at all. This corresponds to the case where the half-space is a low speed material and no critical refraction occurs from a source on the surface.) For $k_{s1} > k > k_{s2}$, (16-7) becomes

$$-\mu_1(k_{s1}^2 - k^2)^{\frac{1}{2}} H \,\text{Sin}\{(k_{s1}^2 - k^2)^{\frac{1}{2}} H\} + \mu_2(k^2 - k_{s2}^2)^{\frac{1}{2}} H \,\text{Cos}\{(k_{s1}^2 - k^2)^{\frac{1}{2}} H\} = 0,$$

or

$$\text{Tan}\left[\frac{\omega H}{v_{s1}} \left(1 - \frac{v_{s1}^2}{c^2}\right)^{\frac{1}{2}}\right] = \frac{\mu_2(1 - c^2/v_{s2}^2)^{\frac{1}{2}}}{\mu_1(c^2/v_{s1}^2 - 1)^{\frac{1}{2}}} . \qquad (16\text{-}10)$$

Real roots of this equation represent waves propagating in the plus x-direction with velocities between v_{s1} and v_{s2}, and which diminish exponentially as they penetrate into the half-space. The wave motion associated with a real root is known as a propagating mode. Their nature will be discussed later. These roots of (16-10) can be obtained by a graphical construction.

It can be noted from (16-10) that if $c = v_{s2}$, the right-hand side is zero and the left-hand side then determines certain characteristic frequencies. They are:

$$\frac{\omega_n^c H}{v_{s1}} = n\pi(1 - v_{s1}^2/v_{s2}^2)^{-\frac{1}{2}}, \tag{16-11}$$

where n = 0 is included. This makes a convenient number to designate the propagating modes by, and, as we shall see later, it also represents the number of nodal surfaces in the layer. For frequencies less than ω_n^c in a given mode, the root values k_n become complex. The associated modes are called leaking modes because they continually radiate energy into the lower half-space and thus attenuate as they travel in the x-direction.

As the root reaches the branch point k_{s2}, it goes onto the lower sheet reversing direction along the real axis. It then collides with a root already on the real axis of the lower sheet and two complex conjugate roots are formed. One travels into the first quadrant and moves to the left slightly above the real axis. As the root moves along, it passes near the origin, and we can investigate its position quantitatively. For $k_{s1} > k_{s2} > \mathrm{Re}\{k\}$, (16-9) becomes

$$\mu_1(k_{s1}^2 - k^2)^{\frac{1}{2}} H \,\mathrm{Sin}\{(k_{s1}^2 - k^2)^{\frac{1}{2}} H\} + i\mu_2(k_{s2}^2 - k^2)^{\frac{1}{2}} H \,\mathrm{Cos}\{(k_{s1}^2 - k^2)^{\frac{1}{2}} H\} = 0. \tag{16-12}$$

It is apparent that the complex roots k_n occur in fours, i.e., if k_n is a root so are $-k_n$, k_n^*, and $-k_n^*$. We now rewrite (16-12) in the form

$$\frac{\mathrm{Sin}\{k_{s1}H(1 + \varepsilon)\}}{\mathrm{Cos}\{k_{s1}H(1 + \varepsilon)\}} = -i\,\frac{\mu_2 v_{s1}(1 + \delta)}{\mu_1 v_{s2}(1 + \varepsilon)}, \tag{16-13}$$

where we have made the substitutions (assuming that kH is small with respect to $k_{s1}H$)

$$(k_{s1}^2 - k^2)^{\frac{1}{2}} H = k_{s1}H(1 + \varepsilon), \tag{16-14}$$

$$(k_{s2}^2 - k^2)^{\frac{1}{2}} H = k_{s2}H(1 + \delta). \tag{16-15}$$

If the right-hand side of (16-13) is large (e.g., $\mu_2 >> 1$, tending toward a rigid half-space) then we would seek a root where the cosine factor was small, i.e., near $k_{s1}H = (2n-1)\pi/2$. This gives, assuming ε and δ are both small,

$$-\frac{1}{\mathrm{Tan}(k_{s1}H\varepsilon)} \simeq -i\,\frac{\mu_2 v_{s1}}{\mu_1 v_{s2}},$$

or

$$\varepsilon \simeq -i\,\frac{\mu_1 v_{s2}}{\mu_2 v_{s1}}\,\frac{2}{(2n - 1)\pi}. \tag{16-16}$$

For small $\underline{k}$, we have from (16-14)

$$kH \simeq k_{s1}H(-2\varepsilon)^{\frac{1}{2}},$$

and from (16-16)

$$kH \simeq \left[\frac{i\mu_1 v_{s2}}{\mu_2 v_{s1}}(2n - 1)\pi\right]^{\frac{1}{2}} = \left[\frac{i(\mu_1\rho_1)^{\frac{1}{2}}}{(\mu_2\rho_2)^{\frac{1}{2}}}(2n - 1)\pi\right]^{\frac{1}{2}}; \tag{16-17}$$

where the second form of the right-hand side is written in terms of the acoustic impedances, $(\mu_i\rho_i)^{\frac{1}{2}}$, etc. This will be small compared to $k_{s1}H$ for large n or for a large impedance contrast. On the other hand, if the right-hand side of (16-13) is small (e.g., $\mu_2 << \mu_1$, tending towards a stress-free bottom), we look for roots where the sine factor is small, i.e., near $k_{s1}H = n\pi$, (n = 0 is excluded). Equation (16-13) then approximates to

$$\text{Tan}(k_{s1}H\varepsilon) \simeq -\frac{i\mu_2 v_{s1}}{\mu_1 v_{s2}}.$$

Expanding the tangent for small arguments, we obtain

$$\varepsilon \simeq -\frac{i\mu_2 v_{s1}}{\mu_1 v_{s2}}\frac{1}{n\pi}; \tag{16-18}$$

and

$$kH \simeq \left[\frac{i\mu_2 v_{s1}}{\mu_1 v_{s2}}2n\pi\right]^{\frac{1}{2}} = \left[\frac{i(\mu_2\rho_2)^{\frac{1}{2}}}{(\mu_1\rho_1)^{\frac{1}{2}}}2n\pi\right]^{\frac{1}{2}}. \tag{16-19}$$

Again this is small compared to $k_{s1}H$ for large n or large impedance contrast. In the first case the root locus crosses the 45° line in the first quadrant of the k-plane at $\omega H/v_{s1} = (2n - 1)\pi/2$ (zero excluded) and in the second case for $\omega H/v_{s1} = n\pi$ (zero excluded). When the impedance contrast is high, both values of kH are very small, and we have the organ-pipe condition with very high phase velocity, low group velocity, and low attenuation. That is, all the energy is trapped as standing waves in the layer. The associated modes are known as Organ-Pipe modes or singing modes.

Finally, for $\omega \to 0$, we seek roots where k remains finite. Remembering that the root has passed to the left of k_{s2} as $k_{s2} \to 0$, and that it lies on the lower sheet, equation (16-9) can be written as

$$\mu_1 kH\,\text{Sinh}(kH) - \mu_2 kH\,\text{Cosh}(kH) \simeq 0. \tag{16-20}$$

Writing $kH = \delta + i\varepsilon$, this can be rewritten as $\text{Tanh}(\delta + i\varepsilon) = \mu_2/\mu_1$. Separating into real and imaginary parts, we have

$$\frac{\text{Tanh}\delta(1 + \text{Tan}^2\varepsilon)}{1 + \text{Tanh}^2\delta\,\text{Tan}^2\varepsilon} = \mu_2/\mu_1,$$

and

$$\frac{\text{Tan}\varepsilon(1 - \text{Tanh}^2\delta)}{1 + \text{Tanh}^2\delta\,\text{Tan}^2\varepsilon} = 0.$$

The second part can be satisfied by either $\text{Tan}\varepsilon = 0$, or by $\text{Tan}\varepsilon = \infty$. In the former case we have

$$kH = \text{Tanh}^{-1}(\mu_2/\mu_1) + in\pi. \tag{16-21}$$

In the latter, we find

$$kH = \mathrm{Tanh}^{-1}(\mu_1/\mu_2) + i\left(\frac{2n-1}{2}\right)\pi. \tag{16-22}$$

Inasmuch as μ_2 is generally greater than μ_1, (16-22) is the usual case.

Although the zero-frequency roots are all ($n \geq 1$) located on the lower sheet, they can make a contribution to the seismic response in the layer. When the integral (16-7) is evaluated by sweeping the contour up around the branch cut into the first quadrant as in Figure 16-3, the contributions from the lower sheets combine in such

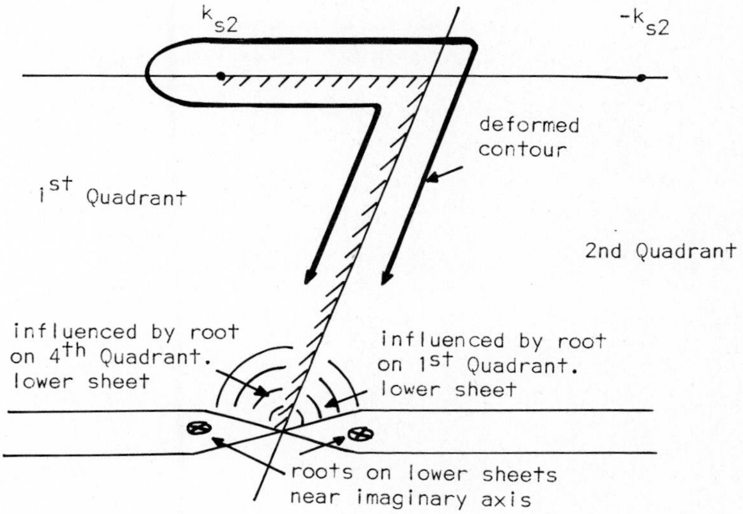

Figure 16-3. Contribution of lower sheet roots to Love-mode propagation in a layer overlying a half-space.

a way as to give a standing wave pattern that is attenuated rapidly with increasing $\underline{x}$. Such standing modes would be important near sources or vertical discontinuities in the layer. This will be discussed in more detail in the Rayleigh-wave case. The standing modes can be distinguished from organ-pipe modes because of this standing-wave condition. Organ-pipe modes reverberate but also propagate in the x-direction.

Summarizing the behavior of the complex Love-wave roots in the k-plane as a function of frequency, we find the following. As ω increases from zero, the branch point k_{s2} leaves the vicinity of the origin and advances along the Re{k}-axis. The root k_n (and its complex conjugate k_n^*) moves in the same direction as the branch point, but more slowly. With the branch line drawn vertically (as in Figure 16-2), the portion of the (-)-sheet containing k_n is uncovered as k_{s2} becomes greater than $\mathrm{Re}\{k_n\}$. The contribution of the root is negligible, however, until it approaches the real axis where the attenuation factor $e^{-\mathrm{Im}\{k_n x\}}$ becomes small. Here k_n accelerates relative to k_{s2} and crosses the vertical branch cut again ($\mathrm{Re}\{k_n\} = k_{s2}$), reaching the real axis at a point ahead of k_{s2}. It collides with k_n^* to form a double root which immediately separates into two real roots. One of the roots moves along the Real{k}-axis in the positive direction while remaining on the (-)-sheet. The other root is overtaken by the branch point at cutoff frequency and emerges on the (+)-sheet pulling ahead of the branch point. In this range, it represents a normal Love higher mode propagating without attenuation.

If we go back and evaluate (16-7) by deforming the contour upward, we have

$$\Lambda_1 = -\frac{2Y}{\mu_1}\sum\frac{\text{modes}}{k_n} + \text{branch line integral.}$$

The branch line integral makes little contribution at great distance (except when a root of the denominator is near the branch point) and will be ignored. Then

$$u_y = -\frac{\partial\Lambda}{\partial x} = \frac{2iY}{\mu_1}\sum\text{modes} + \text{another branch line integral,}$$

where

$$\text{modes} = \sum_0^\infty \left[\begin{array}{l} \mu_2(k_n^2-k_{s2}^2)^{\frac{1}{2}}\text{Sinh}\{(k_n^2-k_{s1}^2)^{\frac{1}{2}}(H-z)\} \\ +\mu_1(k_n^2-k_{s1}^2)^{\frac{1}{2}}\text{Cosh}\{(k_n^2-k_{s1}^2)^{\frac{1}{2}}(H-z)\}\end{array}\right]\frac{e^{i(k_nx-\omega t)}}{\Delta'_{L_n}}, \qquad (16\text{-}23)$$

and

$$\Delta'_{L_n} = \left.\frac{\partial\Delta_L}{\partial k}\right|_{k=k_n}.$$

However, at the roots of Δ_L, equation (16-9) holds. Expanding the hyperbolic functions in the numerator of (16-23), we obtain:

$$\begin{aligned}&\mu_2(k_n^2-k_{s2}^2)^{\frac{1}{2}}\text{Sinh}\{(k_n^2-k_{s1}^2)^{\frac{1}{2}}H\}\text{Cosh}\{(k_n^2-k_{s1}^2)^{\frac{1}{2}}z\}\\ &\quad-\mu_2(k_n^2-k_{s2}^2)^{\frac{1}{2}}\text{Cosh}\{(k_n^2-k_{s1}^2)^{\frac{1}{2}}H\}\text{Sinh}\{(k_n^2-k_{s1}^2)^{\frac{1}{2}}z\}\\ &\quad+\mu_1(k_n^2-k_{s1}^2)^{\frac{1}{2}}\text{Cosh}\{(k_n^2-k_{s1}^2)^{\frac{1}{2}}H\}\text{Cosh}\{(k_n^2-k_{s1}^2)^{\frac{1}{2}}z\}\\ &\quad-\mu_1(k_n^2-k_{s1}^2)^{\frac{1}{2}}\text{Sinh}\{(k_n^2-k_{s1}^2)^{\frac{1}{2}}H\}\text{Sinh}\{(k_n^2-k_{s1}^2)^{\frac{1}{2}}z\}.\end{aligned}$$

The second and fourth terms cancel because of (16-9). Then we have

$$\sum\text{modes} = \sum_0^\infty\left[\frac{\mu_2(k_n^2-k_{s2}^2)^{\frac{1}{2}}\text{Sinh}\{(k_n^2-k_{s1}^2)^{\frac{1}{2}}H\}+\mu_1(k_n^2-k_{s1}^2)^{\frac{1}{2}}\text{Cosh}\{(k_n^2-k_{s1}^2)^{\frac{1}{2}}H\}}{\Delta'_{L_n}}\right] \qquad (16\text{-}24)$$

$$\cdot\text{Cosh}\{(k_n^2-k_{s1}^2)^{\frac{1}{2}}z\}e^{i(k_nx-\omega t)}.$$

The factor

$$\text{Cosh}\{(k_n^2-k_{s1}^2)^{\frac{1}{2}}z\}e^{i(k_nx-\omega t)} = \text{Cos}\{(k_{s1}^2-k_n^2)^{\frac{1}{2}}z\}e^{i(k_nx-\omega t)} \qquad (16\text{-}25)$$

describes the spatial dependence of the nth mode and the remainder in brackets is called the <u>excitation coefficient</u> or <u>excitation function</u>. Now at cutoff, the cosine function reads (from (16-11)),

$$\text{Cos}(n\pi z/H).$$

This function starts at 1 for $z = 0$, equals $(-1)^n$ at $z = H$, and has <u>n</u> nodes in between. There are n-1 modes preceeding it regardless of frequency. Consequently there will always be n nodal planes for the nth real mode. For complex k_n the number of nodes will depend on the magnitude of its imaginary part. For example, if k were pure imaginary, say $k = i\xi$, then the z-dependence would read

$$\mathrm{Cos}\{(k_{s1}^2 + \xi^2)^{\frac{1}{2}} z\}.$$

16.3. Rayleigh Modes (P-SV).

We will now look at the Rayleigh type of surface motion in a layer over a half-space. The geometry is given in Figure 16-4. The analysis will be very much similar to the Love-wave case, except that because of P-SV coupling, there will be the additional complications introduced by the extra potentials necessary to satisfy

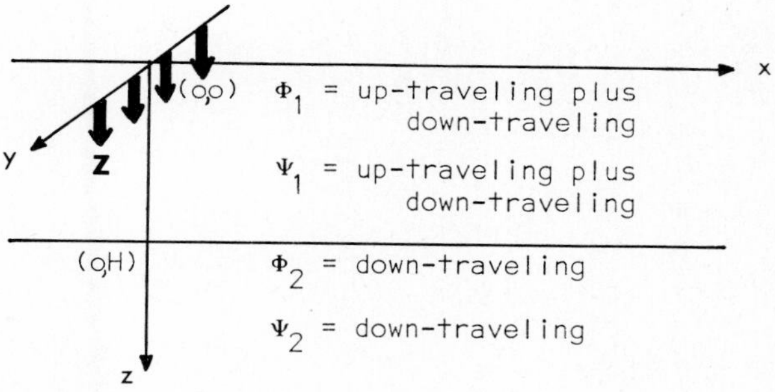

Figure 16-4. Geometrical relationships for Rayleigh-waves in a layer overlying a half-space.

the boundary conditions of this problem. We can set up our mathematical model by letting

$$u_x = \frac{\partial \Phi}{\partial x} - \frac{\partial \Psi^{\dagger}}{\partial z}, \qquad (6\text{-}25)$$

$$u_z = \frac{\partial \Phi}{\partial z} + \frac{\partial \Psi^{\dagger}}{\partial x}, \qquad (6\text{-}27)$$

$$\frac{t_{zz}}{\mu} = -k_s^2 \Phi + 2\left(\frac{\partial^2 \Psi^{\dagger}}{\partial x \partial z} - \frac{\partial^2 \Phi}{\partial x^2}\right), \qquad (6\text{-}31)$$

$$\frac{t_{xz}}{\mu} = 2\frac{\partial^2 \Phi}{\partial x \partial z} + \frac{\partial^2 \Psi^{\dagger}}{\partial x^2} - \frac{\partial^2 \Psi^{\dagger}}{\partial z^2}. \qquad (6\text{-}33)$$

As in Chapter 11, we can write

$$\Phi_1 = \int_{-\infty}^{\infty}\left[A(k)e^{(k^2-k_{p1}^2)^{\frac{1}{2}}z} + B(k)e^{-(k^2-k_{p1}^2)^{\frac{1}{2}}z}\right]e^{i(kx-\omega t)}dk,$$

$$\Phi_2 = \int_{-\infty}^{\infty}\left[C(k)e^{-(k^2-k_{p2}^2)^{\frac{1}{2}}z}\right]e^{i(kx-\omega t)}dk, \qquad (16\text{-}26)$$

$$\Psi_1^{\dagger} = \int_{-\infty}^{\infty}\left[D(k)e^{(k^2-k_{s1}^2)^{\frac{1}{2}}z} + E(k)e^{-(k^2-k_{s1}^2)^{\frac{1}{2}}z}\right]e^{i(kx-\omega t)}dk,$$

$$\Psi_2^{\dagger} = \int_{-\infty}^{\infty}\left[F(k)e^{-(k^2-k_{s2}^2)^{\frac{1}{2}}z}\right]e^{i(kx-\omega t)}dk.$$

The coefficients A(k) through F(k) are determined by applying the boundary conditions. These are

$$t_{zz}\Big|_{z=0} = -\frac{Z}{2\pi}\int_{-\infty}^{\infty}e^{i(kx-\omega t)}dk = -Z\delta(x)e^{-i\omega t},$$

$$t_{zx}\Big|_{z=0} = 0,$$

$$t_{zz}\Big|_{z=H} \text{ is continuous ,}$$

$$t_{zx}\Big|_{z=H} \text{ is continuous ,} \qquad (16\text{-}27)$$

$$u_z\Big|_{z=H} \text{ is continuous ,}$$

$$u_x\Big|_{z=H} \text{ is continuous.}$$

The first condition represents a line force in the positive z-direction (downwards). On substitution into (6-25) - (6-33), we find that A(k) through F(k) are determined by the following equations:

$$\gamma_1^2 A + \gamma_1^2 B + 2ik\beta_1 D - 2ik\beta_1 E = -Z/(2\pi\mu_1),$$

$$2ik\alpha_1 A - 2ik\alpha_1 B - \gamma_1^2 D - \gamma_1^2 E = 0 ,$$

$$\gamma_1^2 e^{\alpha_1 H}A + \gamma_1^2 e^{-\alpha_1 H}B - \gamma_2^2 e^{-\alpha_2 H}C + 2ik\beta_1 e^{\beta_1 H}D - 2ik\beta_1 e^{-\beta_1 H}E + 2ik\beta_2 e^{-\beta_2 H}F = 0 ,$$

$$2ik\alpha_1 e^{\alpha_1 H} - 2ik\alpha_1 e^{-\alpha_1 H}B + 2ik\alpha_2 e^{-\alpha_2 H}C - \gamma_1^2 e^{\beta_1 H}D - \gamma_1^2 e^{-\beta_1 H}E + \gamma_2^2 e^{-\beta_2 H}F = 0, \qquad (16\text{-}28)$$

$$ike^{\alpha_1 H}A + ike^{-\alpha_1 H}B - ike^{-\alpha_2 H}C - \beta_1 e^{\beta_1 H}D + \beta_1 e^{-\beta_1 H}E - \beta_2 e^{-\beta_2 H}F = 0,$$

$$\alpha_1 e^{\alpha_1 H}A - \alpha_1 e^{-\alpha_1 H}B + \alpha_2 e^{-\alpha_2 H}C + ike^{\beta_1 H}D + ike^{-\beta_1 H}E - ike^{-\beta_2 H}F = 0,$$

where

$$\alpha = (k^2-k_p^2)^{\frac{1}{2}}; \qquad \beta = (k^2-k_s^2)^{\frac{1}{2}}; \qquad \gamma^2 = 2k^2-k_s^2. \qquad (16\text{-}29)$$

Solving, we find

$$A = \frac{\Delta_1}{\Delta_R}, \quad B = \frac{\Delta_2}{\Delta_R}, \quad C = \frac{\Delta_3}{\Delta_R}, \quad D = \frac{\Delta_4}{\Delta_R}, \quad E = \frac{\Delta_5}{\Delta_R}, \quad F = \frac{\Delta_6}{\Delta_R},$$

where the determinant Δ_R is given by:

$$\Delta_R = \tag{16-30}$$

$$\begin{vmatrix} \gamma_1^2 & \gamma_1^2 & 0 & 2ik\beta_1 & -2ik\beta_1 & 0 \\ 2ik\alpha_1 & -2ik\alpha_1 & 0 & -\gamma_1^2 & -\gamma_1^2 & 0 \\ \gamma_1^2 e^{\alpha_1 H} & \gamma_1^2 e^{-\alpha_1 H} & -\gamma_2^2 e^{-\alpha_2 H} & 2ik\beta_1 e^{\beta_1 H} & -2ik\beta_1 e^{-\beta_1 H} & 2ik\beta_2 e^{-\beta_2 H} \\ 2ik\alpha_1 e^{\alpha_1 H} & -2ik\alpha_1 e^{-\alpha_1 H} & 2ik\alpha_2 e^{-\alpha_2 H} & -\gamma_1^2 e^{\beta_1 H} & -\gamma_1^2 e^{-\beta_1 H} & \gamma_2^2 e^{-\beta_2 H} \\ ike^{\alpha_1 H} & ike^{-\alpha_1 H} & -ike^{-\alpha_2 H} & -\beta_1 e^{\beta_1 H} & \beta_1 e^{-\beta_1 H} & -\beta_2 e^{-\beta_2 H} \\ \alpha_1 e^{\alpha_1 H} & -\alpha_1 e^{-\alpha_1 H} & \alpha_2 e^{-\alpha_2 H} & ike^{\beta_1 H} & ike^{-\beta_1 H} & -ike^{-\beta_2 H} \end{vmatrix},$$

and $\Delta_1, \ldots, \Delta_6$ are the various determinants formed by substituting the column

$$\begin{bmatrix} z/(2\pi\mu_1) \\ 0 \\ 0 \\ 0 \\ 0 \\ 0 \end{bmatrix}$$

for the first through the sixth columns of the determinant.

It serves no useful purpose to substitute the values of A, ... , F into (16-26) in that the level of intuition is exceeded and one can't make any additional general conclusions on inspection. It should be noted that each additional interface will add an additional four rows and four columns to the Rayleigh determinant, and two rows and two columns to the Love determinant.

The contour of integration runs just as in the Love case. However, the disposition of the singularities is different. It is not obvious from the nature of (16-26) and (16-30) that there will be only one set of branch cuts; those associated with the half-space. We shall prove this in a later section however. The location of the roots of (16-30) has to be found numerically. It is found that one real root $(\overline{S})$ always lies between k_{R1} and k_{R2}, i.e., the wave velocity lies between the Rayleigh velocities of the half-space and the layer. This is physically reasonable inasmuch as a long period wave would not notice the layer, and a short period wave would not penetrate the half-space. The other real roots lie between k_{s1} and k_{s2}, the number depending upon the frequency of excitation. At a particular frequency, the cut-off frequency, a given root passes to the left of k_{s2}. It then starts upon a path similar to the Love-wave case.

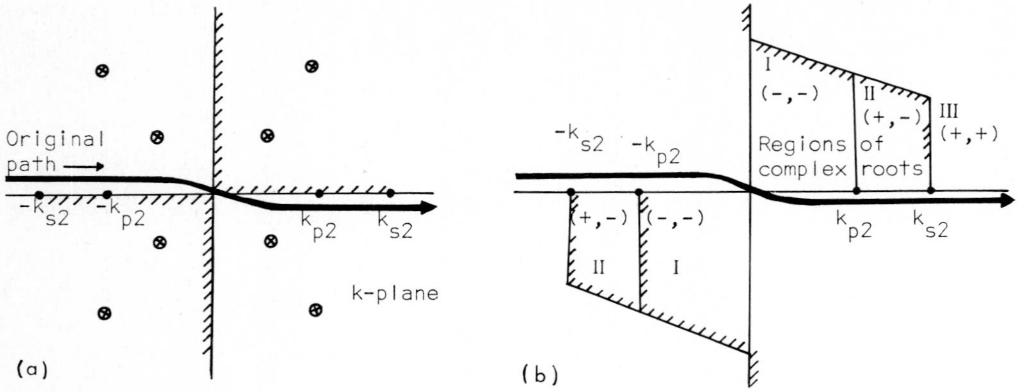

Figure 16-5(a). Zero frequency complex roots on the (+,+)-sheet of integration. Branch cuts are positioned so that all square roots have positive real parts. (Not to scale.)

(b). Branch cuts relocated so as to expose portions of lower Riemann sheets. The signs associated with the sheets are indicated. Roots on the (+,-)-sheet between k_{p2} and k_{s2} affect the value of the integrand. Roots on the (-,-)-sheet between 0 and k_{p2} do also. Likewise on the negative k-axis.

Having discussed the real roots of (16-30), we now have to look at complex roots as well as roots on other Riemann surfaces. We shall adopt a commonly used notation to indicate the Riemann sheet under consideration. That is

$$(+,+) \Longrightarrow \mathrm{Re}\{(k^2 - k_p^2)^{\frac{1}{2}}\} \geq 0, \quad \mathrm{Re}\{(k^2 - k_s^2)^{\frac{1}{2}}\} \geq 0;$$

$$(+,-) \Longrightarrow \quad '' \quad \geq 0, \quad '' \quad < 0;$$

$$(-,+) \Longrightarrow \quad '' \quad < 0, \quad '' \quad \geq 0;$$

$$(-,-) \Longrightarrow \quad '' \quad < 0, \quad '' \quad < 0.$$

The branch cuts that naturally achieve this over the whole plane are "L"-shaped as in Figure 16-5(a). If we deform these cuts slightly as in Figure 16-5(b), we expose portions of the lower sheets. That part of the (-,-)-sheet is called region I, the part of the (+,-)-sheet is called region II, and for completeness the (+,+)-sheet in the vicinity of k_{s2} will be called region III.

Now let us look at the singularities. Besides the real roots on the (+,+)-sheet, we note that as $\omega \to 0$, Watson (1972) found an infinite number of complex roots in each of the four quadrants. The real part of these roots increases logarithmically, and the imaginary part increases arithmetically. Values of these roots are given by Watson (1972, Table 3). He also found a similar distribution of roots on the three lower sheets. These complex roots need a little explanation. It would seem at first glance that such roots would be forbidden. However, since the roots occur in fours, we find (not proved) on sweeping the contour upwards, that we pick up the residues in pairs with the factors multiplying the terms e^{ikx} having the same magnitude and differing only in phase. That is, we pick up terms like

$$Me^{i\phi_1}e^{i(\xi + i\eta)x} + Me^{i\phi_2}e^{i(-\xi + i\eta)x} ,$$

where $\xi = \mathrm{Re}\{k\}$, $\eta = \mathrm{Im}\{k\}$. This can be rearranged in the form

$$2Me^{-\eta x} \mathrm{Cos}\left[\xi x + (\phi_1 - \phi_2)/2\right] e^{i(\phi_1 + \phi_2)/2} ,$$

which is immediately recognized as an attenuated standing wave in the vicinity of $x = 0$. Such waves are necessary to write solutions in the vicinity of sources and boundary discontinuities.

Other roots of importance are those found in region II. They are associated with PL, shear-coupled PL, and OP modes. The notation PL comes from the Long-period P-waves observed on long-period instruments. OP is Watson's (1970) notation, resulting from the steep portions of the phase-velocity curves above the velocity v_{p2}, typical of the resonant vibrations of an Organ Pipe.

These roots arise from zero frequency roots on the (+,-)- and (-,+)-sheets. There are several cases which are diagrammed in Figure 16-6. The K_n^{+-} roots in general follow a direct path into region II as OP modes. Their initial phase velocity tends to zero in such a way that K_n^{+-} has a finite value. However, a few go down

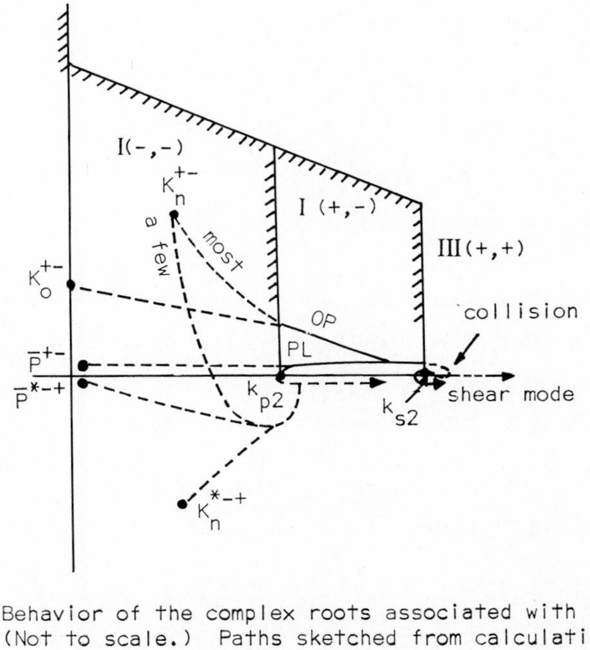

Figure 16-6. Behavior of the complex roots associated with PL and OP modes. (Not to scale.) Paths sketched from calculations by Watson (1970). Paths of $\bar{P}^{*-+}$ and K_n^{*-+} are shown to avoid confusion.

into the fourth quadrant (on the (-,+)-sheet) and enter region II as PL modes. All the K_n^{*-+} roots move in the fourth quadrant until they approach the branch point k_{p2}. They then enter region II as PL modes. We consider the path of K_n^{*-+} to avoid overlapping paths in Figure 16-6. The path of K_n^{-+} is mirrored in the Re{k}-axis. The initial phase velocity is likewise zero. There are three exceptional roots however. K_o^{+-} starts on the imaginary axis ($c \to \infty$) and becomes an OP mode. $\overline{P}^{+-}$ (one of the Lamb complex roots - see Section 11-4) starts with a phase velocity of v_{p2} and follows the path of the PL modes. $\overline{P}^{*-+}$ (the other Lamb complex root) starts with a phase velocity of v_{p2}, wanders in the fourth quadrant and becomes a PL root. Again, we have diagrammed the conjugate root to avoid confusion. As can be seen from the figure, the major difference between between OP and PL is the high attenuation of OP as it passes the branch point k_{p2}. Both types have low attenuation in the region of k_{s2}. This is associated with shear-coupled PL.

A summary of root behavior (according to Watson) is given below:

$\overline{S}$	$\longrightarrow$				$\longrightarrow$	RM_{10}		Fundamental
K_o^{+-}	$\longrightarrow$	Σ_1^{+-}	$\longrightarrow$	OP_{11}	$\longrightarrow$	RM_{21}		1st Shear
$\overline{P}^{+-}$	$\longrightarrow$		$\longrightarrow$	PL_{21}	$\longrightarrow$	RM_{11}		2nd Shear
$\overline{P}^{-+}$	$\longrightarrow$		$\longrightarrow$	PL_{22}	$\longrightarrow$	RM_{22}		etc.
K_1^{+-}	$\longrightarrow$	Σ_2^{+-}	$\longrightarrow$	OP_{12}	$\longrightarrow$	RM_{12}		
K_1^{-+}	$\longrightarrow$	Π_1^{-+}	$\longrightarrow$	PL_{23}	$\longrightarrow$	RM_{23}		
K_2^{+-}	$\longrightarrow$	Σ_3^{+-}	$\longrightarrow$	OP_{13}	$\longrightarrow$	RM_{13}		(Real-ω analysis)
K_2^{-+}	$\longrightarrow$	Π_2^{-+}	$\longrightarrow$	PL_{24}	$\longrightarrow$	RM_{24}		
K_4^{+-}	$\longrightarrow$	Σ_5^{+-}	$\longrightarrow$	OP_{14}	$\longrightarrow$	RM_{14}		
K_3^{+-}	$\longrightarrow$	Σ_4^{+-}	$\longrightarrow$	PL_{25}	$\longrightarrow$	RM_{25}		
K_5^{+-}	$\longrightarrow$	Σ_6^{+-}	$\longrightarrow$	OP_{15}	$\longrightarrow$	RM_{15}		

The notation Σ, Π is due to Gilbert who used a real-k rather than a real-ω approach. As $k \to 0$, one obtains the organ-pipe resonant frequencies (a complex frequency corresponding to attenuated standing waves) associated with pure P(Π) and S(Σ) resonances in a scalar wave-guide. The notation RM_{ij}, RM_{2j} is a slight modification of the notation of Tolstoy and Usdin (1953). I felt that the change from <u>11</u> to <u>10</u> was necessary because of the unique character of the fundamental Rayleigh mode. It is the only one without a cutoff frequency; all frequencies from zero to infinity being propagated. Also the decrement of "1" in the RM_1-branch makes for a better merging of the RM_{1j} with the OP_{1j}. However, in a two-layered model, Watson (1970) finds that $OP_{1j} \to RM_{2j}$ and $PL_{2j} \to RM_{1j}$. In still another one-layered model the transition of Σ's and Π's to PL's and OP's is different. Thus, a uniformly consistent notation is impossible. For example, even in the above transitions, $PL_{21} \rightarrow RM_{11}$, with 21→11 the inconsistency. The notation 1st shear, 2nd shear, etc. has to do with the limits of the pure-real normal modes which have a phase velocity equal to the half-space shear velocity at cutoff. This velocity then decreases to the layer shear velocity as frequency becomes very large.

Lastly, we find that region I roots contribute some reverberatory behavior and are identified with the Σ^{--} and Π^{--} roots of Gilbert (1964). The K_n^{--} roots become Σ^{--} roots, and then leave region I staying on a lower sheet as in Figure 16-7. The $-K_n^{*++}$ roots start in the second quadrant, migrate to the (-,-)-sheet in the first

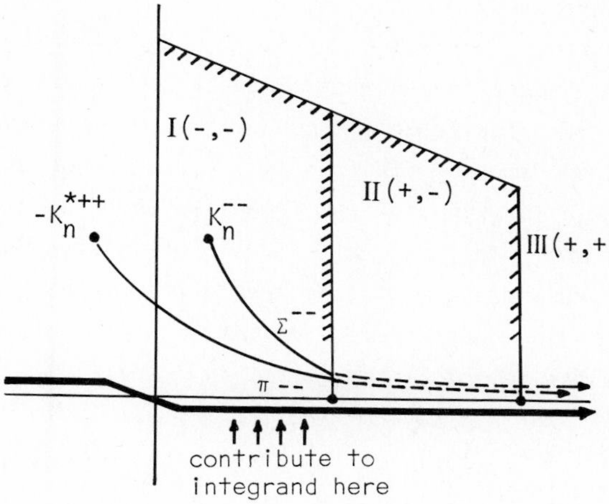

Figure 16-7. Behavior of the complex roots associated with Gilbert's (1964) Σ^{--} and Π^{--} roots. (Not to scale.) Paths sketched from calculation by Watson (1970).

quadrant, become Π^{--} roots, and then leave region I staying on a lower sheet. These roots generally have higher attenuation inasmuch as they leak both P and S energy into the half-space, whereas the Σ^{+-} and Π^{-+} modes leak only S.

On substitution of the values of A — F into (16-26), then substituting (16-26) into (6-25) and (6-27), we would finally obtain

$$u_x = \sum{}^{(1)} \text{modes} + \text{branch line integral},$$

$$u_z = \sum{}^{(2)} \text{modes} + \text{another branch line integral}.$$

The simplification of the $\sum$ modes term by cancelling out factors of Δ_R becomes prohibitive. However, by direct evaluation (Kovach and Anderson, 1964) it appears as if the higher modes have n nodal surfaces for the z-component of displacement and n + 1 nodal surfaces for the x-component. Despite the complexity of the factors multiplying $e^{i(k_n x - \omega t)}$, one can evaluate the excitation function at the surface (z = 0). Laster et al (1965) have made such calculations and they are shown in Figure 16-14. In such a calculation, the factor exp $\left[\mathrm{Im}\{k_n x\}\right]$ must be included as part of the excitation function. Therefore, the relative excitation between propagating and leaking modes will be a function of the source-receiver distance, expressed in terms of the layer thickness H.

16.4. Impulsive Response of the Surface Wave Modes — Phase and Group Velocity.

We have seen in Sections 16.2 and 16.3 that for two-dimensional surface waves, the time-harmonic form of the solution was given by

$$M_j = B_j\left[k_j(\omega)\right] e^{i(k_j x - \omega t)}, \qquad (16\text{-}31)$$

where M_j is some quantity we wish to determine, e.g., P-potential, tangential displacement, normal stress, etc. The factor $B_j\left[k_j(\omega)\right]$ is a function of the jth root, k_j, of a particular dispersion relation given by setting either Δ_L or Δ_R equal to zero. This dispersion relation is determined by the layer parameters and the frequency. That is, for each value of the frequency ω, we can determine a number of roots of the dispersion relation, and for the moment we shall consider the jth one. Equation (16-31) is valid when the time excitation is of the form $e^{-i\omega t}$. When an impulsive time source is present, we must take the Fourier transform of the time function and multiply M_j by the Fourier spectrum of the source. We must also include the frequency response of the seismic instruments. In general, we will not know the source spectrum, so we will want to determine as much as possible about a layered structure from an integral of the type

$$I_j(x,t) = \frac{1}{2\pi}\int_{-\infty}^{\infty} A\{k_j(\omega)\}e^{i\{\phi(\omega) + k_j x - \omega t\}}\,d\omega. \qquad (16\text{-}32)$$

Here in general, $A(\omega)e^{i\phi(\omega)}$ is unknown and represents the total frequency response of source, layering excitation coefficient, and instrument. For convenience in later manipulations, we have separated this factor into an amplitude part and a phase part.

In most physically applicable situations, the factors $A\{k_j(\omega)\}$ and $\phi(\omega)$ will be rather slowly varying functions of ω, but for large $\underline{x}$ ($k_j x \gg 1$) the phase factor ($k_j x - \omega t$) will be a rapidly varying function of frequency. This will lead to both constructive and destructive interference, as a function of frequency, at a given large value of $\underline{x}$ and as a result the integral (16-32) will assume a rather simple form. This simplified form can be obtained mathematically by making two approximate evaluations of (16-32). The first approximation is known as the Stationary Phase Approximation, and the second is known as the Airy Phase Approximation.

Mathematical Approximation and Physical Interpretation. In the first approximation, we seek a point (or points) of stationary phase, i.e., a value of the frequency ω where the quantity

$$\Phi = \phi + kx - \omega t, \qquad (16\text{-}33)$$

in the integral (16-32) has a stationary value, i.e., where $d\Phi/d\omega = 0$. For convenience, we will drop the mode subscript $\underline{j}$. At this point of stationary phase, the factor $A\{k(\omega)\}$ will make a large contribution, while over other intervals of frequency the contribution of this factor will be minimized by the rapid variation of the phase factor. About such a point of stationary phase we can then approximate the phase factor Φ of the integral (16-32) by

$$\Phi(\omega) \simeq \phi_o(\omega_o) + k_o(\omega_o)x - \omega_o t + \tfrac{1}{2}\left.\frac{\partial^2\Phi}{\partial\omega^2}\right|_{\omega=\omega_o}(\omega-\omega_o)^2 + O\left[(\omega-\omega_o)^3\right]; \qquad (16\text{-}34)$$

where $k_o = k_o(\omega_o)$ is determined by the equation

$$\left.\frac{\partial\Phi}{\partial\omega}\right|_{\omega=\omega_o} = \left.\frac{\partial\phi}{\partial\omega} + x\frac{\partial k}{\partial\omega} - t\right|_{\omega=\omega_o} = 0, \qquad (16\text{-}35)$$

and the relation of $\underline{k}$ to ω is given by the dispersion relation of the particular mode under discussion. We then have

$$I(x,t) \simeq \frac{A\{k_o(\omega_o)\}}{2\pi} e^{i(\phi_o + k_o x - \omega_o t)} \int_{-\infty}^{\infty} e^{i\frac{1}{2}(\partial^2\Phi/\partial\omega^2)\big|_{\omega=\omega_o}(\omega-\omega_o)^2} d(\omega-\omega_o),$$

where the rapid cancellation of A outside the vicinity of ω_o allows us to replace it with its value at the stationary point over the entire range of integration. This constant value may then be removed from under the integral sign. For all real time functions represented by integrals such as (16-32), we will have

$$\Phi(-\omega) = -\Phi(\omega),$$

and consequently there will be a similar contribution from a point of stationary phase at $-\omega_o$. However, for real time functions A will be even. The two contributions then combine giving

$$I(x,t) \simeq \frac{A\{k_o(\omega_o)\}}{\pi} \operatorname{Re}\left[e^{i(\phi_o + k_o x - \omega_o t)} \int_{-\infty}^{\infty} e^{i\frac{1}{2}(\partial^2\Phi/\partial\omega^2)\big|_{\omega=\omega_o}(\omega-\omega_o)^2} d(\omega-\omega_o)\right].$$

The above integral is of the standard form

$$\int_{-\infty}^{\infty} e^{-\frac{1}{2}\alpha^2\xi^2} d\xi = 2\pi^{\frac{1}{2}}/\alpha.$$

In our particular case, we have

$$\alpha = \left(-i\left.\frac{\partial^2\Phi}{\partial\omega^2}\right|_{\omega=\omega_o}\right)^{\frac{1}{2}} \begin{cases} = \left(\left|\dfrac{\partial^2\phi}{\partial\omega^2} + x\dfrac{\partial^2 k}{\partial\omega^2}\right|\right)^{\frac{1}{2}} e^{-i\pi/4}, & \dfrac{\partial^2\Phi}{\partial\omega^2} > 0; \\ = \left(\left|\dfrac{\partial^2\phi}{\partial\omega^2} + x\dfrac{\partial^2 k}{\partial\omega^2}\right|\right)^{\frac{1}{2}} e^{+i\pi/4}, & \dfrac{\partial^2\Phi}{\partial\omega^2} < 0. \end{cases}$$

Combining all these results, we have

$$I(x,t) \simeq (2/\pi)^{\frac{1}{2}} \frac{A\{k(\omega_o)\}}{\left(\left|\dfrac{\partial^2\phi}{\partial\omega^2} + x\dfrac{\partial^2 k}{\partial\omega^2}\right|\right)^{\frac{1}{2}}} \operatorname{Cos}(\phi_o + k_o x - \omega_o t \pm \pi/4), \qquad (16\text{-}36)$$

where the plus sign is to be taken if $\left(\frac{\partial^2\phi}{\partial\omega^2} + x\frac{\partial^2 k}{\partial\omega^2}\right)$ is positive, and the minus sign if this factor is negative. This approximation will be valid as long as the higher order terms in the expansion (16-34) can be ignored with respect to $(\partial^2\phi/\partial\omega^2)(\omega-\omega_o)^2$. This obviously fails when $\partial^2\phi/\partial\omega^2 = 0$, and we shall develop the second approximation for this region. The finer methematical details of this stationary phase approximation are discussed by Pekeris (1948) and by Erdelyi (1956).

Now that we have this first mathematical approximation to (16-32), let us examine the physical content. First of all, from (16-35), we see that ω_o and k_o will be functions of the particular observation variables, i.e., $\underline{x}$ and $\underline{t}$. The value of (16-36) must be calculated for each value of $\underline{x}$ and $\underline{t}$ of interest, i.e., it is a <u>point approximation</u>. Once having obtained this expression, many authors feel that it is obvious that the angular frequency and wave-number at a particular point in space and time are given by ω_o and k_o respectively. This is right, but it is not obvious. In a form such as (16-36), the <u>instantaneous frequency</u> is defined by

$$\begin{aligned}\omega_{inst} &= -\frac{\partial}{\partial t}(k_o x + \phi_o - \omega_o t \pm \pi/4)\\ &= -\left(x\frac{\partial k_o}{\partial\omega} - t + \frac{\partial\phi}{\partial\omega}\right)_{\omega=\omega_o}\frac{\partial\omega_o}{\partial t} + \omega_o \qquad (16\text{-}37)\\ &= \omega_o,\end{aligned}$$

and the <u>local wave-number</u> is defined by

$$\begin{aligned}k_{local} &= \frac{\partial}{\partial x}(k_o x + \phi_o - \omega_o t \pm \pi/4)\\ &= k_o + \left(x\frac{\partial k}{\partial\omega} - t + \frac{\partial\phi}{\partial\omega}\right)_{\omega=\omega_o}\frac{\partial\omega_o}{\partial x} \qquad (16\text{-}38)\\ &= k_o.\end{aligned}$$

That these two expressions achieve this simple relationship is because (16-35) holds.

Equations (16-37) and (16-38) tell us that both the instantaneous frequency and the local wave-number travel with the velocity U, where

$$\frac{1}{U(\omega)} = \frac{t}{x} = \left.\frac{\partial k}{\partial\omega}\right|_{\omega=\omega_o}. \qquad (16\text{-}39)$$

The velocity U is called the <u>group velocity</u>, for it is the velocity with which the maximum of a wave packet with a Gaussian frequency spectrum will travel in a dispersive medium (Bohm, 1951, p. 65). The concept of group velocity is much more appropriate to the "dots and dashes" of wireless telegraphy or to light quanta than to dispersive surface waves. In the case of concern to us, we find that the instantaneous frequency of the seismic signal changes appreciably from cycle to cycle. Very seldom can it be considered a narrow band-width wave-packet. In many special cases, but not in general, it can also be shown that U is the velocity of energy propagation (Biot, 1957).

If we consider the points of constant phase in (16-36), we find that the variation of

$$\Phi = \phi_o + k_o x - \omega_o t \pm \pi/4,$$

i.e.,

$$\delta\Phi = \delta\phi_o + k_o\delta x + x\delta k_o - \omega_o\delta t - t\delta\omega_o,$$

must be zero. But from (16-35) we have that

$$\delta\Phi_o + x\delta k_o - t\delta\omega_o = 0,$$

and consequently we find that the points of constant phase travel with the phase velocity given by

$$\frac{\delta x}{\delta t} = \frac{\omega_o}{k_o} = c(\omega_o). \tag{16-40}$$

However, since ω_o is a function of both x and t, the instantaneous frequency and local wave-number both change as this point of constant phase moves out from the origin.

Lastly, the term involving distance in the denominator of (16-36) is not a geometrical divergence factor, but represents the dispersion of energy as the impulsive disturbance propagates. This may most easily be understood by looking at a small range of frequencies propagating with a small range of group velocities. After one arbitrary unit of time, this range of frequencies will be spread out over a range of distance

$$dx = tdU.$$

After two units of time the spread in distance will be

$$dx = 2tdU.$$

The fact that x appears as a square root takes care of the fact that it is energy, not amplitude that is dispersed.

An Example. Consider the case where the wave-number k is related to the frequency by the relation

$$k = \frac{\omega}{c} = \frac{\omega}{c_o}\ (1 + a\omega^2). \tag{16-41}$$

In this case we have

$$c \simeq c_o(1 - a\omega^2). \tag{16-42}$$

From (16-39) we have

$$\frac{dk}{d\omega} = \frac{1}{c_o}\ (1 + 3a\omega^2) = \frac{1}{U}\ ,$$

or

$$U \simeq c_o(1 - 3a\omega^2); \tag{16-43}$$

i.e., the group velocity decreases three times as fast as the phase velocity. These relationships are sketched in Figure 16-8.

Setting

$$\frac{c_o}{1 + 3a\omega_o^2} = \frac{x}{t},$$

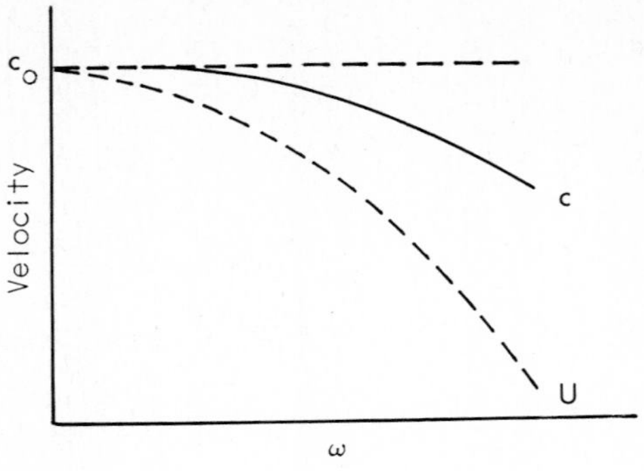

Figure 16-8. Relationships between phase and group velocities.

we find that

$$\omega_o = \left[\frac{1}{3a} \left(\frac{c_o t}{x} - 1 \right) \right]^{\frac{1}{2}} = \left[\frac{1}{3a} \frac{c_o \tau}{x} \right]^{\frac{1}{2}}, \tag{16-44}$$

where we have set

$$t = x/c_o + \tau,$$

and τ is the local time measured from the first wave arrival at $t_o = x/c_o$. It is clear from (16-44) that the statements about ω_o and k_o being functions of $\underline{t}$ and $\underline{x}$ is true. The variation of frequency with time is sketched in Figure 16-9.

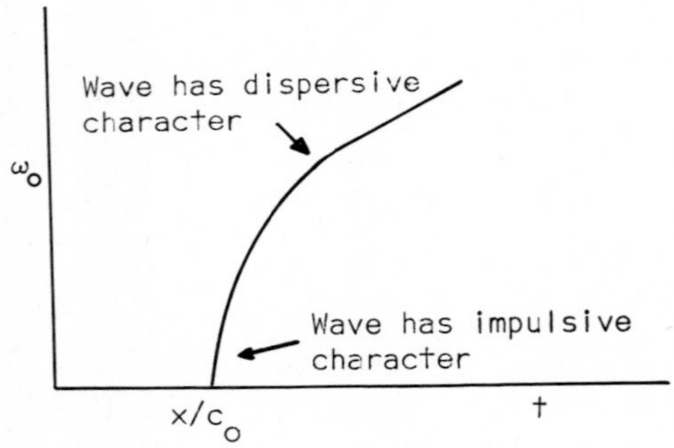

Figure 16-9. Frequency-Time diagram (Sonogram presentation) for a dispersive wave with dispersion relation as in (16-41).

In those cases where the first approximation breaks down because of the nulling of the factor $\partial^2\phi/\partial\omega^2$, there is a different expansion of the phase factor about this particular point. That is, we write

$$\Phi = \phi_a + k_a x - \omega_a t + \left[\left(\frac{\partial \phi}{\partial \omega} + x \frac{\partial k}{\partial \omega} \right)_{\omega=\omega_a} - t \right] (\omega - \omega_a)$$
$$+ \frac{1}{6} \left(\frac{\partial^3 \phi}{\partial \omega^3} + x \frac{\partial^3 k}{\partial \omega^3} \right)_{\omega=\omega_a} (\omega - \omega_a)^3 + \ldots \tag{16-45}$$

where ω_a is determined by the equation

$$\left. \frac{\partial^2 \Phi}{\partial \omega^2} \right|_{\omega=\omega_a} = 0, \tag{16-46}$$

and is not a function of t. The integral of (16-32) can then be approximated by (again assuming A(k) = A(-k)):

$$I \simeq \frac{2A(k_a)}{\pi} \operatorname{Cos}(k_a x - \omega_a t + \phi_a) \int_0^\infty \operatorname{Cos}(Pu + \frac{Qu^3}{3})du, \tag{16-47}$$

where

$$u = \omega - \omega_a,$$

$$P = \left(\frac{\partial \phi}{\partial \omega} + x \frac{\partial k}{\partial \omega} \right)_{\omega=\omega_a} - t, \tag{16-48}$$

and

$$Q = \tfrac{1}{2} \left(\frac{\partial^3 \phi}{\partial \omega^3} + x \frac{\partial^3 k}{\partial \omega^3} \right)_{\omega=\omega_a}. \tag{16-49}$$

The integral in (16-47) is a tabulated function known as the Airy function. Combining results, we have

$$I(x,t) \simeq \frac{A(k_a)\operatorname{Cos}(k_a x - \omega_a t + \phi_a)}{|Q|^{1/3}} \operatorname{Ai}\left[\frac{P \operatorname{Sgn}\{Q\}}{|Q|^{1/3}} \right]. \tag{16-50}$$

The Airy function Ai(z) is sketched in Figure 16-10. For positive arguments it decreases rapidly, while for negative arguments it oscillates with decreasing

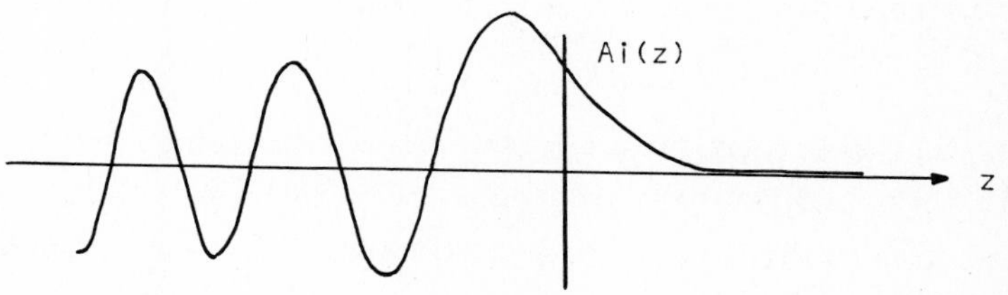

Figure 16-10. A sketch of the Airy function, Ai(z).

amplitude and increasing frequency. If $\omega_a \neq 0$, the Airy function acts as an envelope for a wave of that frequency, while if $\omega_a = 0$, the Airy function acts as the wave motion itself. For a phase and group velocity relationship as in Figure 16-11(a), the wave motion is given in Figure 16-11(b), with corresponding points labeled. It is assumed here that the amplitude spectrum is relatively uniform.

Returning to the Love-modes of Section 16.2, we can use the above analysis to plot the new quantities c (defined by (16-40)) and U (defined by (16-39)) against the angular frequency ω for each of the modes. Such a plot is known as a dispersion curve. In the case where k is complex, we will define

$$c = \omega/\mathrm{Re}\{k\}, \tag{16-51}$$

and

$$U = d\omega/d\ \mathrm{Re}\{k\}. \tag{16-52}$$

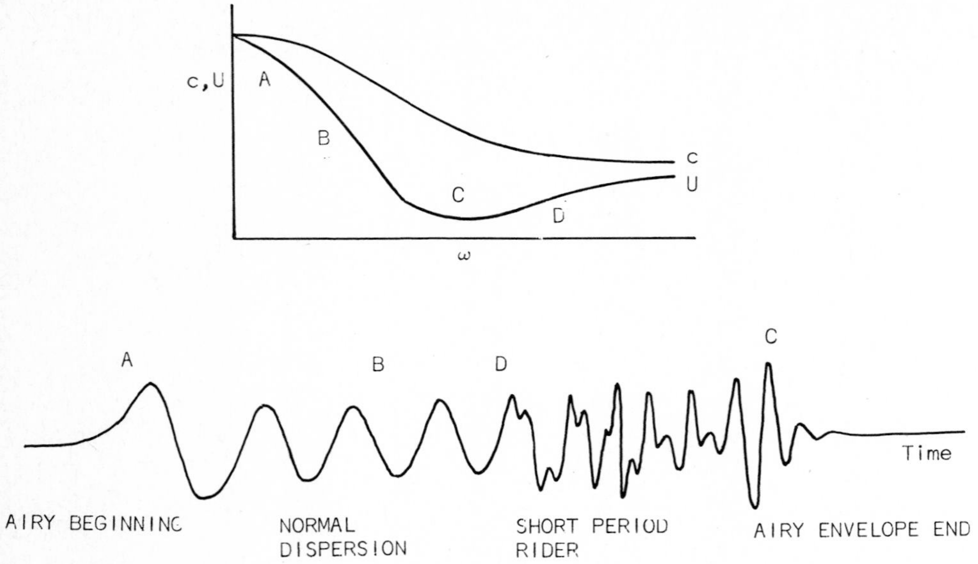

Figure 16-11(a). Simple dispersion curve corresponding to one layer overlying a half-space.

(b). Corresponding impulse response showing dispersed wave-train with Airy Phase at its tail.

Although others define these quantities (in the case of complex $\underline{k}$) as $c = \mathrm{Re}\{\omega/k\}$ and $U = \mathrm{Re}\{d\omega/dk\}$, this then requires both complex ω and complex $\underline{k}$. In our case, we are requiring ω to be strictly real. Formulas (16-51 and (16-52) seem to make the most sense to me in terms of the stationary phase analysis given above. The imaginary part of $\underline{k}$ is combined into the term $A\{k_j(\omega)\}$ in (16-32). We shall have a bit more to say about this at the end of this section. For Love-modes, we then have dispersion curves as in Figure 16-12. An alternative formulation, based on requiring U to be real, is given by Radovich and DeBremaecker (1974). However, their figure corresponding to Figure 16-12 below bears little resemblance to what we have shown here. My preference lies with Figure 16-12 and the real-ω analysis.

An extensive literature exists for real Love-waves; much of this can be found referenced in Chapter 4 of Ewing, Jardetzky and Press (1957). However, the work of Watson (1970) and Radovich and Debremaecker (1974) seem to be the only ones concerned with complex Love roots.

One extensively studied model for Rayleigh-waves in a layer over a half-space has been that of a brass layer over a steel half-space. This was easily studied in

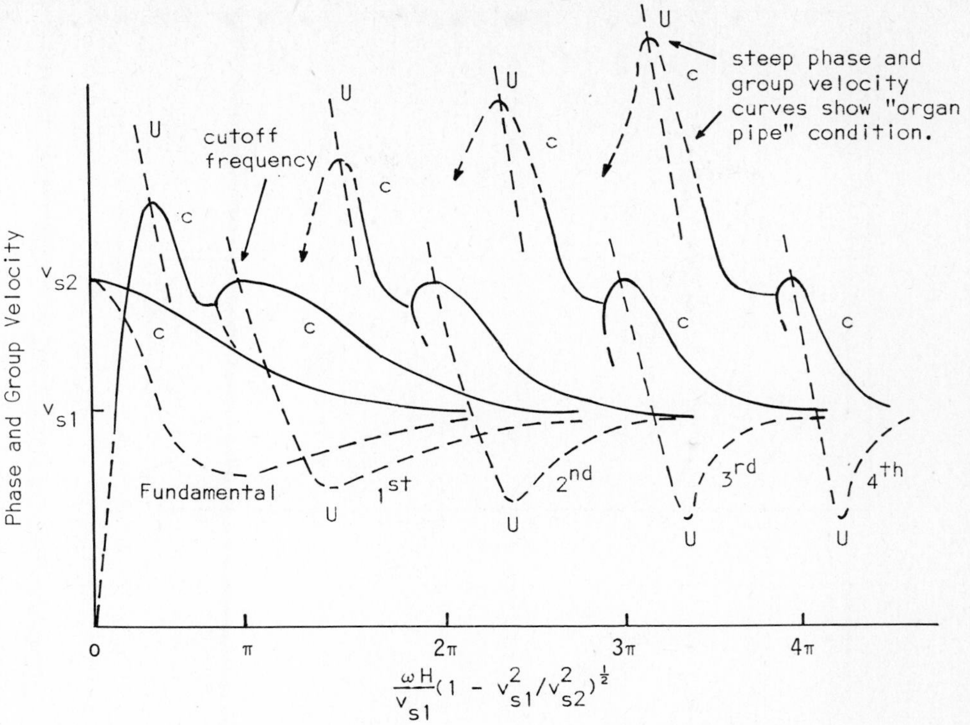

Figure 16-12. Dispersion curves for Love-waves in a simple layer over a half-space (schematic). Solid lines based on computations by Watson (1970), dotted lines based on discussion in the text.

the laboratory and was the first instance (Knopoff et al, 1960) of laboratory observation of the reverberatory PL behavior, though not so identified. Follow-up by Gilbert and Laster (1962a) using the same brass over steel model determined the behavior as that of PL. Laster et al (1965) worked out the dispersion curves for this model including the complex PL modes. These are presented in Figure 16-13(b). Finally Watson (1972) continued to explore the complex roots of this model. His results are presented in Figure 16-13(a). In addition to the dispersion curves, Laster et al (1965) calculated the relative excitation of propagating and PL modes. This is shown in Figure 16-14. Su and Dorman (1965) have a quite different approach to determining the PL modes. They analyze the excitation coefficient as a function of phase-velocity and period for a plane-wave incident from below the layer. This approach gives additional insight into the nature of surface-wave modes. As in the case of real Love-waves, Ewing, Jardetzky and Press (1957) has an extensive reference list to studies of real Rayleigh-wave modes.

We have already alluded to the real-k approach of Gilbert (1964). Abramovici (1968) following Gilbert's approach, noted that the succession from zero-frequency roots to the PL modes was model dependent just as Watson had noted in the real-ω

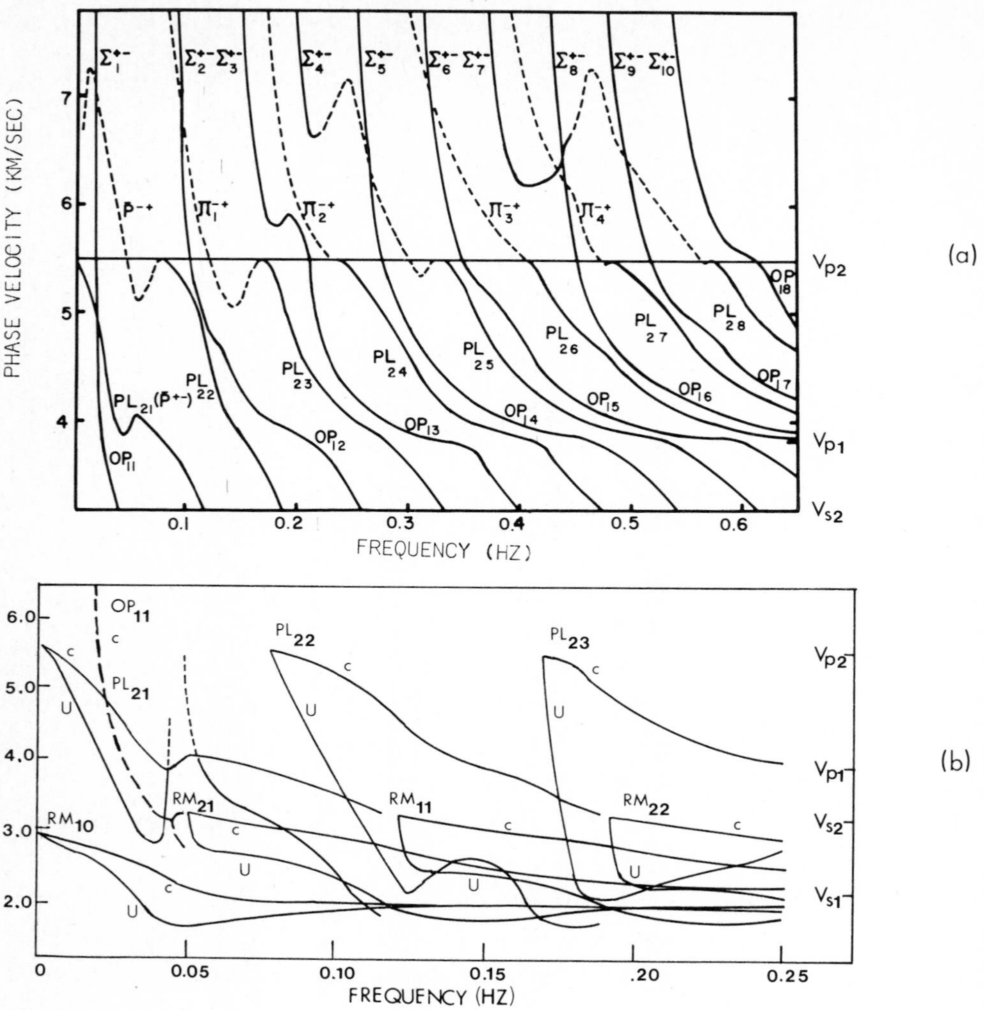

Figure 16-13(a). One layer over a half-space. Phase velocities for Σ, Π, PL and OP modes. (After Watson, 1972.)

(b). Phase and group velocities for PL and propagating Rayleigh modes. (After Laster et al, 1965.) Dashed line for OP_{11} taken from (a) above.

SCALED MODEL PROPERTIES (BRASS OVER STEEL): v_{p1} = 3.83 km/sec, v_{s1} = 2.14 km/sec, H = 20 km; v_{p2} = 5.49 km/sec, v_{s2} = 3.24 km/sec, ρ_2/ρ_1 = 1.0.

approach. A great deal of investigation into the nature of the complex roots for real-k has been carried out by DeBremaecker and his students (DeBremaecker 1967, 1968; Cochran, Woeber and DeBremaecker, 1970; Stalmach and DeBremaecker, 1973). The results for real propagating modes is the same as presented above. However, for the PL and OP modes there is a rather different structure. For the real-k method, the

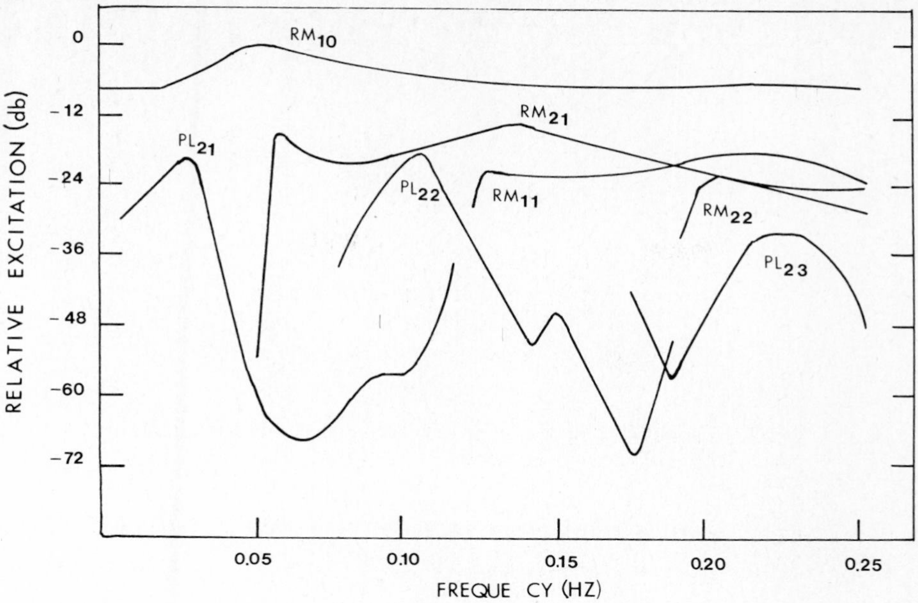

Figure 16-14. Relative excitation of propagating Rayleigh modes and PL modes in the brass over steel model of Figure 16-13. The curves are calculated at a distance of x = 25H. It should be noted that at this distance, the PL modes are comparable in power with the higher propagating modes. (After Laster et al, 1965.)

higher mode curves descend to the velocity v_{p2} in a stairstep manner similar to high-order modes in a thin plate (Tolstoy and Usdin, 1957). For the real-ω case, the higher order modes cross each other in a lattice arrangement. There is no real difference — essentially the same values of ω and c are covered by both nets of high-order dispersion curves.

16.5. Orthogonality of Modes in Two Dimensions.

The following demonstration of the orthogonality of surface-wave modes is essentially due to Herrera (1964) and slightly modified to take into account some points considered by McGarr and Alsop (1967). Alsop (1968) gives a more sophisticated analysis that takes into account the presence of body-waves. Some additional discussion is given by Pekeris (1948) in the case of fluid layers. Consider a region in two dimensions (Figure 16-15) bounded by lines at $x = -x_2$, $x = x_1$, $z = z_1$ and the free surface $z = 0$.

Consider Equation (8-8). If all sources were outside the shaded region of Figure (16-15) we would have

$$\int (W_j T_{ij},_i - U_j S_{ij},_i)\, dV = 0, \qquad (16\text{-}53)$$

where dV includes the shaded volume. In a similar manner, (16-53) may be shown to be a completely general relationship for any two solutions of the elastic wave equation

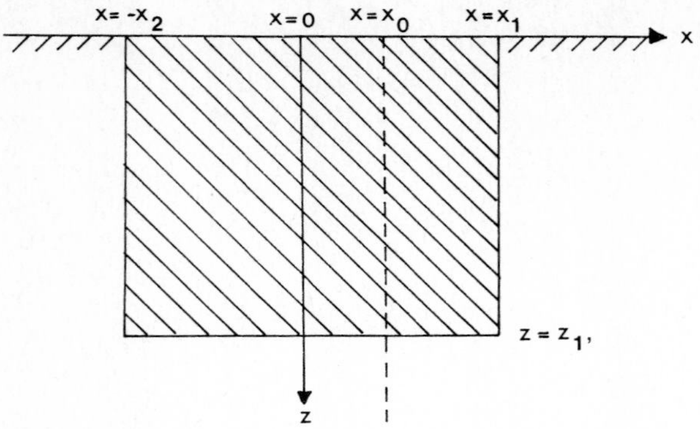

Figure 16-15. Geometric relationships for proof of orthogonality of surface-wave modes.

provided there are no sources within the volume of integration. In particular, we can use it for two surface-wave modes. We will also take the complex conjugate of one and obtain

$$\int [U_j^n T_{ij,i}^{m*} - U_j^{m*} T_{ij,i}] \, dV = 0. \tag{16-54}$$

Following the reasoning leading to (8-13), we can write

$$\int [U_j^n T_{ij}^{m*} - U_j^{m*} T_{ij}^n] \, n_i dS = 0. \tag{16-55}$$

Now in two dimensions the surface degenerates to a line, and with a stress-free upper surface the integral along the line z = 0 vanishes, leaving

$$- \int_0^{z_1} [U_j^n T_{xj}^{m*} - U_j^{m*} T_{xj}^n]_{x=-x_2} dz + \int_{-x_2}^{x_1} [\ \ ]_{z=z_1} dx + \int_0^{z_1} [\ \ ]_{x=x_1} dz = 0.$$

The negative sign in the first term comes from the negative x-direction of the normal. As z_1 tends to infinity, the second integral vanishes. We then see that

$$\int_0^{\infty} [U_j^n T_{xj}^{m*} - U_j^{m*} T_{xj}^n]_{x=x_0} dz = \text{constant}. \tag{16-56}$$

Now in two dimensions, the nth mode will look like

$$U_j^n(z) e^{ik_n x},$$

and the complex conjugate of the mth mode will look like

$$U_j^{m*}(z) e^{-ik_m x}.$$

The stresses will look like

$$T_{xj}^n(z) e^{ik_n x}, \quad T_{xj}^{m*}(z) e^{-ik_m x}.$$

Substituting into (16-56), one obtains

$$e^{i(k_n - k_m)x_o} \int_o^\infty [U_j^n T_{xj}^{m*} - U_j^{m*} T_{xj}^n] \, dz = \text{constant}. \qquad (16\text{-}57)$$

Now the only way that this can happen is for the integral to vanish unless m = n. Thus

$$\int_o^\infty [U_j^n T_{xj}^{m*} - U_j^{m*} T_{xj}^n] \, dz = C_n \delta_{mn}. \qquad (16\text{-}58)$$

The modes can be normalized so that $C_n = 1$. For normalized Rayleigh modes, this would then reduce to

$$\int_o^\infty [U_x^n T_{xx}^{m*} + U_z^n T_{xz}^{m*} - U_x^{m*} T_{xx}^n - U_z^{m*} T_{xz}^n] \, dz = R_n \delta_{mn}; \qquad (16\text{-}59)$$

and for Love modes, this would reduce to

$$\int_o^\infty [U_y^n T_{xy}^{m*} - U_y^{m*} T_{xy}^n] \, dz = L_n \delta_{mn}. \qquad (16\text{-}60)$$

I am quite sure that the same relations would hold for axially symmetric waves. However, for three-dimensional waves, SV and SH are not defined with respect to the plane x = 0, but only with respect to the plane z = 0. Hence P, SV, and SH all couple together in addition to the complications of multiple directions of propagation. Again, however, something like the above relations could probably be derived.

As an example of how all this doesn't do very much for us, we can take a look at the 90°-corner problem. This is described in Figure 16-16. Essentially we wish to prescribe the stresses on the left face while maintaining a stress-free upper surface.

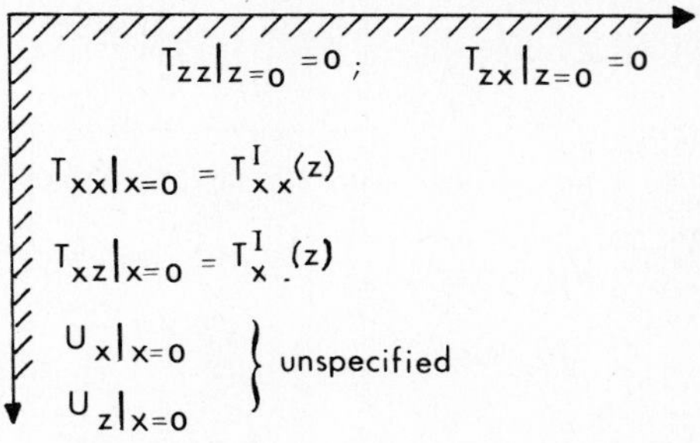

Figure 16-16. Prescribed stress distributions in the 90°-corner problem.

Here we can write:

$$T^I_{xx} = \Sigma a_m T^m_{xx} \quad \text{(given)},$$

$$T^I_{xz} = \Sigma a_m T^m_{xz} \quad \text{(given)},$$

$$U^I_x = \Sigma a_m U^m_x \quad \text{(unknown)},$$

$$U^I_z = \Sigma a_m U^m_z \quad \text{(unknown)},$$

where the a_m's are to be calculated. Applying (16-59), we find that

$$\int_0^\infty [U_x^n T_{xx}^{l*} + U_z^n T_{xz}^{l*} - U_x^{l*} T_{xx}^n - U_z^{l*} T_{xz}^n] dz = a_n R_n .$$

However, the last two terms of the integrand are unknown; if one substitutes the above expansions for them, it leads to an infinite set of equations for the a_n's. This may be further approximated by a finite number of equations in the same number of unknown a_n's.

Another way to approximate a solution to this problem is not to worry about using orthogonal properties, but to minimize the function

$$F = |T_{xx}^l - \sum_{n=1}^{N} a_n T_{xx}^n|^2 + |T_{xz}^l - \sum_{n=1}^{N} a_n T_{xz}^n|^2$$

by adjusting the parameters a_n. However, neither of these procedures is very good, having only limited or non-existent ranges of useful convergence. (Note: I write from terrible experience!) "Useful convergence" is defined as a sum which approaches a limit sufficiently rapidly that one does not have to use infinite amounts of computer time to evaluate it.

For more complicated problems such as the reflection of Rayleigh-waves from a vertical interface between two different media as in Figure 16-17, the orthogonality relationships help to reduce the number of unknowns, but do not solve the problem.

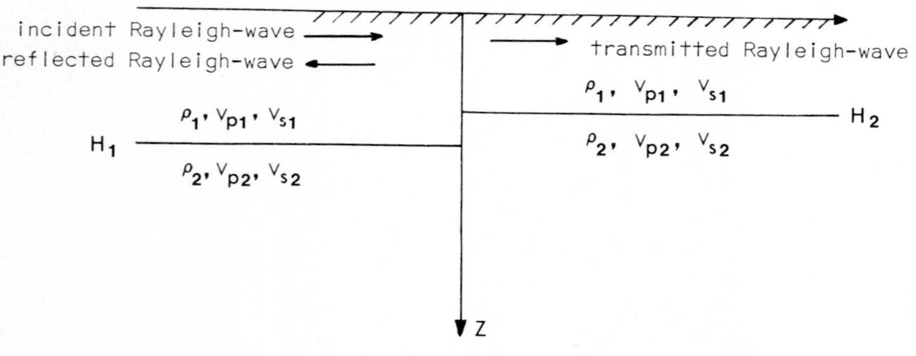

Figure 16-17. Scattering of Rayleigh-waves from a medium with a vertical step-discontinuity.

16.6. Some Relations Between Modes and Rays.

Pekeris (1950) found an equality between all the modes of a scalar wave-guide and all the rays emanating from a point source in a three-dimensional wave-guide. We shall show that the same relations hold in two dimensions using similar methods.

The Scalar Wave-guide — Mode Theory. The source function will be given by a slight modification of (7-11) as

$$\Phi_o = \frac{1}{2}\int_{-\infty}^{\infty} \frac{e^{-(k^2-k_o^2)^{\frac{1}{2}}|z-h|}\, e^{ikx}\, dk}{(k^2-k_o^2)^{\frac{1}{2}}} . \tag{16-61}$$

We introduce the scattered wave function as

$$\Phi_s = \frac{1}{2}\int_{-\infty}^{\infty} \left[A(k)e^{(k^2-k_o^2)^{\frac{1}{2}} z} + B(k)e^{-(k^2-k_o^2)^{\frac{1}{2}} z} \right] \frac{e^{ikx}\, dk}{(k^2-k_o^2)^{\frac{1}{2}}} . \tag{16-62}$$

The geometry is as in Figure 16-18. We shall put in very simple boundary conditions,

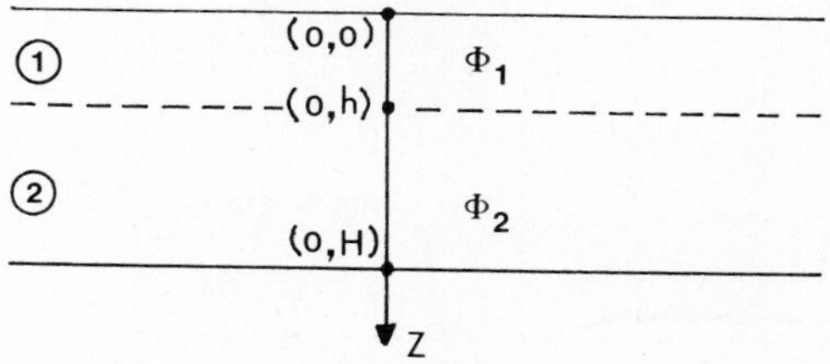

Figure 16-18. Geometrical relationships for the scalar wave-guide.

requiring that:

$$\Phi_{total} = \Phi_o + \Phi_s = 0 \quad \text{at } z = 0,\ z = H. \tag{16-63}$$

The satisfaction of (16-63) leads to the following values for the coefficients A(k) and B(k):

$$A = \frac{-e^{-(k^2-k_o^2)^{\frac{1}{2}}(H-h)} + e^{-(k^2-k_o^2)^{\frac{1}{2}}(H+h)}}{e^{(k^2-k_o^2)^{\frac{1}{2}} H} - e^{-(k^2-k_o^2)^{\frac{1}{2}} H}} ,$$

$$B = \frac{-e^{(k^2-k_o^2)^{\frac{1}{2}}(H-h)} + e^{-(k^2-k_o^2)^{\frac{1}{2}}(H-h)}}{e^{(k^2-k_o^2)^{\frac{1}{2}} H} - e^{-(k^2-k_o^2)^{\frac{1}{2}} H}} .$$

The wave functions in each of the two regions may be written (adding together Φ_o and Φ_s)

$$\Phi_1 = \int_{-\infty}^{\infty} \frac{Sh\{(k^2-k_o^2)^{\frac{1}{2}}(H-h)\}\, Sh\{(k^2-k_o^2)^{\frac{1}{2}} z\}\, e^{ikx}\, dk}{(k^2-k_o^2)^{\frac{1}{2}}\, Sh\{(k^2-k_o^2)^{\frac{1}{2}} H\}} , \tag{16-64}$$

$$\Phi_2 = \int_{-\infty}^{\infty} \frac{Sh\{(k^2-k_o^2)^{\frac{1}{2}} h\}\, Sh\{(k^2-k_o^2)^{\frac{1}{2}}(H-z)\}\, e^{ikx}\, dk}{(k^2-k_o^2)^{\frac{1}{2}}\, Sh\{(k^2-k_o^2)^{\frac{1}{2}} H\}} . \tag{16-65}$$

There are no branch cuts in the k-plane, but there are poles when $(k^2-k_o^2)^{\frac{1}{2}} H = in\pi$;

that is, when

$$k_n^2 = k_o^2 - \frac{n^2\pi^2}{H^2} \qquad [n = 1, 2, \ldots]. \tag{16-66}$$

If $k_o > n\pi/H$, k_n is real and we have a propagating mode. If $k_o < n\pi/H$, k_n is imaginary and we have a standing mode. The distribution of poles is shown in Figure 16-19. Applying the residue theorem to (16-64) and (16-65), we have

$$\Phi_{1 \text{ or } 2} = \frac{2\pi i}{H} \sum_{n=1}^{\infty} \frac{e^{ik_n x}}{k_n} \operatorname{Sin}\left(\frac{n\pi h}{H}\right) \operatorname{Sin}\left(\frac{n\pi z}{H}\right). \tag{16-67}$$

That there is no discontinuity (at $z = h$) in the modes is a general result, whether in scalar or elastic layered media. The artificial discontinuity at $z = h$ disappears

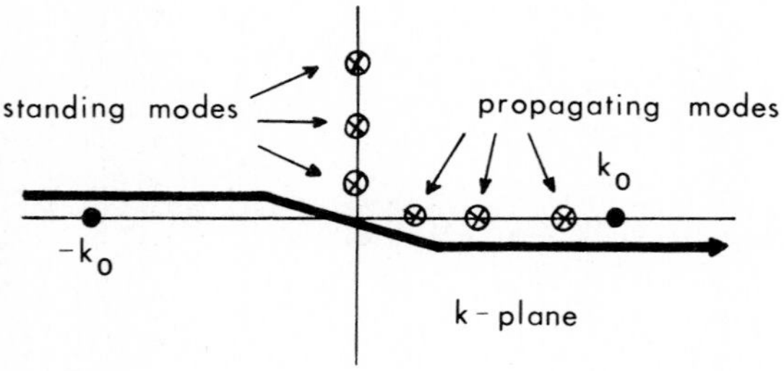

Figure 16-19. Poles for the scalar wave-guide problem.

upon evaluation of the contour integrals.

The Scalar Wave-guide — Ray Theory. Since the reflection coefficient at both upper and lower surfaces equals minus one for the boundary conditions specified, we can also formulate a solution to the problem using a system of images along a vertical plane through the source line as in Figure 16-20. The system consists of a source at $z = h$, its images at $z = -h$ and at $z = 2H - h$, their images at $z = 2H + h$ and $z = -(2H - h)$, etc. Each image has a sign opposite that of its two nearest neighbors. Let us now see how our integral representation (16-64) relates to this formulation.

Expanding the integrand of Φ_1, we have that

$$\frac{\operatorname{Sh}\{(k^2-k_o^2)^{\frac{1}{2}} z\}\operatorname{Sh}\{(k^2-k_o^2)^{\frac{1}{2}}(H-h)\}}{\operatorname{Sh}\{(k^2-k_o^2)^{\frac{1}{2}} H\}} =$$

$$\tfrac{1}{2}\left(1 + e^{-2(k^2-k_o^2)^{\frac{1}{2}} H} + e^{-4(k^2-k_o^2)^{\frac{1}{2}} H} + \ldots\right) \begin{bmatrix} e^{-(k^2-k_o^2)^{\frac{1}{2}}(h-z)} - e^{-(k^2-k_o^2)^{\frac{1}{2}}(h+z)} \\ + e^{-(k^2-k_o^2)^{\frac{1}{2}}(2H-h+z)} - e^{-(k^2-k_o^2)^{\frac{1}{2}}(2H-h-z)} \end{bmatrix}.$$

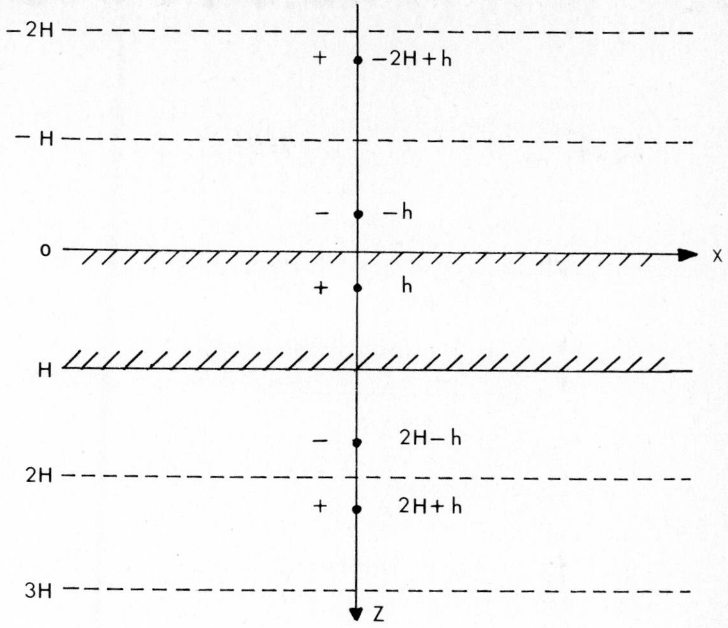

Figure 16-20. Image sources for the solution of the scalar wave-guide problem.

Using (7-11) and (7-12) we see that

$$\Phi_1 = G\{x^2+(h-z)^2\}^{\frac{1}{2}} - G\{x^2+(h+z)^2\}^{\frac{1}{2}}$$

$$G\{x^2+(2H+h-z)^2\}^{\frac{1}{2}} - G\{x^2+(2H+h+z)^2\}^{\frac{1}{2}} + G\{x^2+(2H-h+z)^2\}^{\frac{1}{2}} - G\{x^2+(2H-h-z)^2\}^{\frac{1}{2}}$$

$$G\{x^2+(4H-h+z)^2\}^{\frac{1}{2}} - G\{x^2+(4H-h-z)^2\}^{\frac{1}{2}}$$

or

$$\Phi_1 = G\{x^2 + (h - z)^2\}^{\frac{1}{2}} - G\{x^2 + (h + z)^2\}^{\frac{1}{2}}$$

$$+ \sum_{n=1}^{\infty} \left[\begin{array}{l} G\{x^2 + (2nH + h - z)^2\}^{\frac{1}{2}} - G\{x^2 + (2nH + h + z)^2\}^{\frac{1}{2}} \\ + G\{x^2 + (2nH - h + z)^2\}^{\frac{1}{2}} - G\{x^2 + (2nH - h - z)^2\}^{\frac{1}{2}} \end{array} \right] ; \qquad (16\text{-}68)$$

where

$$G\{x^2 + (h - z)^2\}^{\frac{1}{2}} = \frac{i\pi}{2} H_o^{(1)} \left[k\{x^2 + (h - z)^2\} \right] . \qquad (16\text{-}69)$$

The same result follows from (16-65).

The Equality of the Mode Sum and the Ray Sum. An interesting relationship was used by Pekeris (1950) that shows the mathematical equality of (16-67) and (16-68). This is Poisson's Formula (Titchmarsh, 1937, p. 60) which reads

$$\beta^{\frac{1}{2}} \left[\tfrac{1}{2} F_c(0) + \sum_{n=1}^{\infty} F_c(n\beta) \right] = \alpha^{\frac{1}{2}} \left[\tfrac{1}{2} f(0) + \sum_{n=0}^{\infty} f(n\alpha) \right] , \qquad (16\text{-}70)$$

where

$$F_c(p) = (2/\pi)^{\frac{1}{2}} \int_0^\infty f(t)\mathrm{Cos}(pt)dt \quad \text{and } \alpha\beta = 2\pi.$$

The conditions for the applicability of (16-70) are that f(t) is of bounded variation and that f(t) tends to zero as $\pm$ tends towards infinity. These are just the standard conditions on Fourier transforms. If we identify the mode terms as our f's, where $\alpha = 1$, then we have

$$f(n) = \frac{2\pi i}{H} \frac{e^{i(k_o^2 - n^2\pi^2/H^2)^{\frac{1}{2}} x}}{(k_o^2 - n^2\pi^2/H^2)^{\frac{1}{2}}} \mathrm{Sin}(\frac{n\pi h}{H})\mathrm{Sin}(\frac{n\pi z}{H}).$$

Then

$$F_c(p) = \frac{2\pi i}{H} (2/\pi)^{\frac{1}{2}} \int_0^\infty \mathrm{Cos}(np) \frac{e^{i(k_o^2 - n^2\pi^2/H^2)^{\frac{1}{2}} x}}{(k_o^2 - n^2\pi^2/H^2)^{\frac{1}{2}}} \mathrm{Sin}(\frac{n\pi h}{H})\mathrm{Sin}(\frac{n\pi z}{H})\, dn.$$

If we define $k = n\pi/H$, then $dn = Hdk/\pi$, and

$$F_c(p) = 2i(2/\pi)^{\frac{1}{2}} \int_0^\infty \mathrm{Cos}(Hpk/\pi) \frac{e^{i(k_o^2 - k^2)^{\frac{1}{2}} x}}{(k_o^2 - k^2)^{\frac{1}{2}}} \mathrm{Sin}(kn)\mathrm{Sin}(kz)\, dk.$$

Now let $p = 2n\pi$ (then $\beta = 2\pi$), and $F_c(p)$ becomes

$$F_c(2n\pi) = 2i(2/\pi)^{\frac{1}{2}} \int_0^\infty \mathrm{Cos}(2nHk) \frac{e^{i(k_o^2 - k^2)^{\frac{1}{2}} x}}{(k_o^2 - k^2)^{\frac{1}{2}}} \mathrm{Sin}(kh)\mathrm{Sin}(kz)\, dk.$$

But

$$\mathrm{Cos}(2nHk)\mathrm{Sin}(kh)\mathrm{Sin}(kz) = \frac{1}{4}\begin{bmatrix} \mathrm{Cos}\{k(2nH + z - h)\} + \mathrm{Cos}\{k(2nH - z + h)\} \\ -\mathrm{Cos}\{k(2nH + z + h)\} - \mathrm{Cos}\{k(2nH - z - h)\} \end{bmatrix}.$$

Using (7-11) and (16-69) we have

$$F_c(2n\pi) = (2\pi)^{-\frac{1}{2}} \begin{bmatrix} G\{x^2 + (2nH + z - h)^2\}^{\frac{1}{2}} + G\{x^2 + (2nH - z + h)^2\}^{\frac{1}{2}} \\ -G\{x^2 + (2nH + z + h)^2\}^{\frac{1}{2}} - G\{x^2 + (2nH - z - h)^2\}^{\frac{1}{2}} \end{bmatrix}.$$

Now

$$\tfrac{1}{2} F_c(0) = (2\pi)^{-\frac{1}{2}} \left[G\{x^2 + (z - h)^2\}^{\frac{1}{2}} - G\{x^2 + (z + h)^2\}^{\frac{1}{2}} \right],$$

and $\frac{1}{2}f(0) = 0$, hence

$$\sum_{n=1}^{\infty} \text{modes} = \sum_{0}^{\infty} \text{rays}. \qquad (16\text{-}71)$$

Following a suggestion of Pekeris (1950), McCracken (1957) found another way of proving the equality of (16-71) using the Euler Sum Formula (Whittaker and Watson, 1935, p. 127).

Despite the obvious success of the above analysis, the mode-ray relationship (16-71) has not been proved when there are branch-cuts present, as in the case of a layer over a half-space. Knopoff (1958a) shows how the formal Love-wave solution can be expanded into a series of images in the layer over a half-space problem. For P- and SV-waves, the image expansion gets increasingly more complicated, but Pekeris et al (1965) were able to do it. Since the formal relation could not be established, several people then tried to use powerful computers to sum up the ray expansions to determine their relationship to modes. Pekeris and Longman (1958) added up some 400 rays to show the transition of rays to modes in the case of a liquid layer over a liquid half-space. Pekeris et al (1965) added up some 320 rays to show the same transition for the case of an elastic layer over an elastic half-space. In this second paper, the body phases and the Rayleigh and Shear modes are developed quite well, but the PL modes do not show up. This is not suprising, since they are quite small experimentally. Dainty and Dampney (1972) added up 340 rays to look particularly at the PL modes.

One final remark on the mode-ray relationship (16-71). As H gets very large, we can set $n\pi/H = k$ and $\pi/H = dk$. The mode sum (16-67) then makes the following formal transformation:

$$\frac{2\pi i}{H}\sum\frac{e^{ik_n x}}{k_n}\mathrm{Sin}\left(\frac{n\pi h}{H}\right)\mathrm{Sin}\left(\frac{n\pi z}{H}\right)\xrightarrow[H\to\infty]{}$$

$$\int_0^\infty \frac{e^{-(k_o^2-k^2)^{\frac{1}{2}}x}}{(k^2-k_o^2)^{\frac{1}{2}}}\left[\mathrm{Cos}\{k(z-h)\}-\mathrm{Cos}\{k(z+h)\}\right]dk$$

$$= G\{x^2+(z-h)^2\}^{\frac{1}{2}} - G\{x^2+(z+h)^2\}^{\frac{1}{2}}.$$

Thus, the solution to the problem is the sum of the source plus its negative image in the plane $z = 0$.

Modes — A Constructive Interface Between Rays. Tolstoy and Usdin (1953) have worked out an extensive set of relationships between modes and rays using the requirement of constructive interference to determine the dispersion relation for the propagating modes. Before we consider the more complicated case of an elastic wave-guide, let us consider the two-dimensional scalar wave-guide. A plane wave propagating in such a wave-guide may be written (modifying (9-4) slightly):

$$e^{ik_o(x\mathrm{Sin}\theta \pm z\mathrm{Cos}\theta)} \qquad (16\text{-}72)$$

We can interpret the factor $k_o\mathrm{Sin}\theta$ as the wave-number in the x-direction, and $k_o\mathrm{Cos}\theta$ as the wave-number in the z-direction. We may then write

$$k_o\mathrm{Sin}\theta = k = \omega/c, \qquad (16\text{-}73)$$

$$k_o\mathrm{Cos}\theta = k\mathrm{Cot}\theta = (k_o^2 - k^2)^{\frac{1}{2}}. \qquad (16\text{-}74)$$

The rays representing this plane wave reflecting from the top and bottom of such a wave-guide are shown in Figure 16-21. To achieve a modal pattern (i.e., a standing wave in the z-direction), the vertical phase change on traveling the path $\overline{ABCD}$ must be a multiple of 2π. The horizontal phase will take care of itself due to Snell's

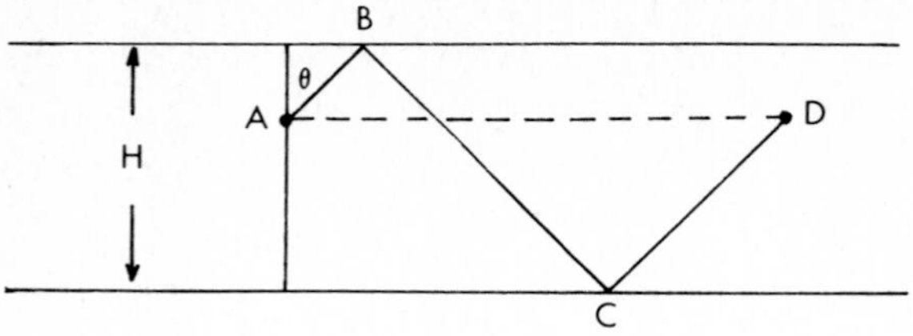

Figure 16-21. Plane-waves propagating in a scalar wave-guide.

Law. The vertical distance along $\overline{ABCD}$ is 2H, hence we must have

$$2kH\cot\theta + \text{phase change on reflection} = 2n\pi. \tag{16-75}$$

For Dirichlet boundary conditions, i.e., $\Phi = 0$, the reflection coefficient is equal to -1. This means a phase change of $-\pi$ on each reflection. We then have

$$kH(c^2/v^2 - 1)^{\frac{1}{2}} = (n + 1)\pi \quad [n = 0,1,2,\dots], \tag{16-76}$$

but this is just the same as (16-66). No D.C. mode is possible. In the case of Neumann boundary conditions, that is, $\partial\Phi/\partial n = 0$, the reflection coefficient is equal to +1 and we have

$$kH(c^2/v^2 - 1)^{\frac{1}{2}} = n\pi \quad [n = 0,1,2,\dots], \tag{16-77}$$

and there can be a D.C. mode.

Following Tolstoy and Usdin (1953), we now consider the <u>elastic wave-guide</u>. The geometrical relationships are as in Figure 16-22. Consider all the waves arriving at point E as P-waves. The conditions for constructive interference are now more

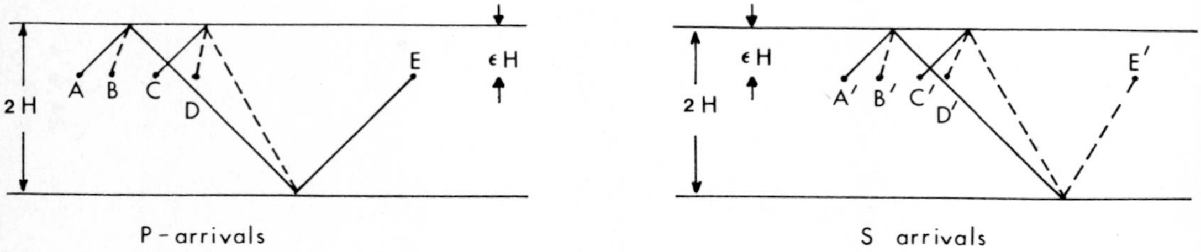

Figure 16-22. P- and S-waves in an elastic wave-guide. Solid lines are P, dashed lines are S. Phases PPP (starting at A), SPP (B), PSP (C) and SSP (D) must arrive at point E in phase and with an amplitude equal to that of the P phase going through A. Similarly for S phases arriving at E'.

complicated. Not only must the vertical phase difference of the sum be a multiple of 2π, but the amplitude of the sum arriving at E must be equal to the amplitude of the sum arriving at A. The wave must be identical at the two points. We will normalize to the S-ray and call the amplitude (which may be complex) of the P-ray K. The condition of equality of all vertically traveling P-rays is

$$K\left[R_{pp}^2 e^{i4\alpha H} + R_{ps}R_{sp} e^{i2(\alpha+\beta)H}\right] + 1\left[R_{sp}R_{pp} e^{i\{(4-\varepsilon)\alpha+\varepsilon\beta\}H} + R_{ss}R_{sp}e^{i\{(2-\varepsilon)\alpha+(2+\varepsilon)\beta\}H}\right] = K. \quad (16\text{-}78)$$

The condition on the S-rays is

$$K\left[R_{pp}R_{ps} e^{i\{(2+\varepsilon)\alpha+(2-\varepsilon)\beta\}H} + R_{ps}R_{ss} e^{i\{\varepsilon\alpha+(4-\varepsilon)\beta\}H}\right] + 1\left[R_{sp}R_{ps}e^{i2(\alpha+\beta)H} + R_{ss}^2 e^{i4\beta H}\right] = 1, \quad (16\text{-}79)$$

where

$$\alpha = (k_p^2 - k^2)^{\frac{1}{2}}; \quad \beta = (k_s^2 - k^2)^{\frac{1}{2}}. \quad (16\text{-}80)$$

This is an over-determined system, so it puts a constraint upon $\underline{k}$ which is

$$\frac{1 - R_{pp}^2 e^{i4\alpha H} - R_{ps}R_{sp} e^{i2(\alpha+\beta)H}}{R_{sp}R_{pp} e^{i\{(4-\varepsilon)\alpha+\varepsilon\beta\}H} + R_{ss}R_{sp}e^{i\{(2-\varepsilon)\alpha+(2+\varepsilon)\beta\}H}}$$

$$= \frac{R_{pp}R_{ps}e^{i\{(2+\varepsilon)\alpha+(2-\varepsilon)\beta\}H} + R_{ps}R_{ss}e^{i\{\varepsilon\alpha+(4-\varepsilon)\beta\}H}}{1 - R_{sp}R_{ps} e^{i2(\alpha+\beta)H} - R_{ss}^2 e^{i4\beta H}}. \quad (16\text{-}81)$$

Using the relationships

$$R_{pp} = -R_{ss} \quad \text{and} \quad (R_{pp}R_{ss} - R_{ps}R_{sp}) = 1, \quad (16\text{-}82)$$

which the student can show are true using (9-16), (9-17), (9-23), (9-24) and the relation $\text{Sin}\theta_s = a\text{Sin}\theta_p$; (16-81) reduces to

$$2e^{i2(\alpha+\beta)H}\left[\text{Cos}\{2(\alpha+\beta)H\} - R_{pp}^2\text{Cos}\{2(\alpha-\beta)H\} + 1 - R_{pp}^2\right] = 0. \quad (16\text{-}83)$$

The condition on $\underline{k}$ then becomes

$$R_{pp}(k) = \left[\frac{1 + \text{Cos } 2(\alpha+\beta)H}{1 + \text{Cos } 2(\alpha-\beta)H}\right]^{\frac{1}{2}} = \frac{\text{Sin}(\alpha+\beta)}{\text{Sin}(\alpha-\beta)}. \quad (16\text{-}84)$$

To get this into a recognizable form, we must evaluate the quantity

$$\frac{1 - R_{pp}}{1 + R_{pp}} = \frac{\text{Sin}(\alpha-\beta)H - \text{Sin}(\alpha+\beta)H}{\text{Sin}(\alpha-\beta)H + \text{Sin}(\alpha+\beta)H} = -\frac{\text{Tan}\beta H}{\text{Tan}\alpha H} \quad (16\text{-}85)$$

$$= \frac{(1 - 2a^2\text{Sin}^2\theta_p)^2}{4a^3\text{Sin}^3\theta_p\text{Cos}\theta_p(1 - a^2\text{Sin}^2\theta_p)^{\frac{1}{2}}} = \frac{(2k^2 - k_s^2)^2}{4k^2(k_p^2 - k^2)^{\frac{1}{2}}(k_s^2 - k^2)^{\frac{1}{2}}}.$$

But (16-83) is symmetric in α and β, so (16-85) can be rewritten as

$$\frac{(2k^2 - k_s^2)^2}{4k^2(k_p^2 - k^2)^{\frac{1}{2}}(k_s^2 - k^2)^{\frac{1}{2}}} = -\left[\frac{\text{Tan}\{(k_s^2 - k^2)^{\frac{1}{2}}\}H}{\text{Tan}\{(k_p^2 - k^2)^{\frac{1}{2}}\}H}\right]^{\pm 1}. \quad (16\text{-}86)$$

This is the dispersion relation for the elastic plate with the plus exponent giving rise to the antisymmetric modes, and the minus sign to the symmetric modes.

Tolstoy and Usdin (1953) have worked out several other problems. These are:

Elastic Layer with rigid boundaries.
Elastic Layer - one rigid, one free boundary.

Elastic Layer over liquid half-space (Ice sheet).
Elastic Layer in an infinite liquid.
Elastic Layer between two different liquid half-spaces.

Elastic Layer over elastic half-space.
Elastic Layer in an infinite elastic medium.
Elastic Layer between two different elastic half-spaces.

Elastic Layer between liquid half-space and elastic half-space.
(Air-coupled Rayleigh-wave)
Elastic Layer over liquid layer over elastic half-space.
Liquid Layer over elastic layer over elastic half-space.
(Rayleigh-waves under the ocean)

Alsop (1970) has related modes and rays in the case of leaky-modes in a liquid layer over an elastic half-space. It is a somewhat more sophisticated analysis than the above.

MANY LAYERS OVER A HALF-SPACE – RAY THEORY

17.1. Travel-Time Curves For Reflected Rays.

Inasmuch as the conversion of P to S and S to P is minimized at vertical incidence (Chapter 12), seismograms obtained in the vicinity of the source will be much less complicated than those in which the rays travel at greater angles. Also, if the angle of incidence upon the layered structure (Figure 17-1) is small, there will be no added complication from refracted arrivals. Let us first consider the travel-time for a ray traveling vertically beneath the source. In this case

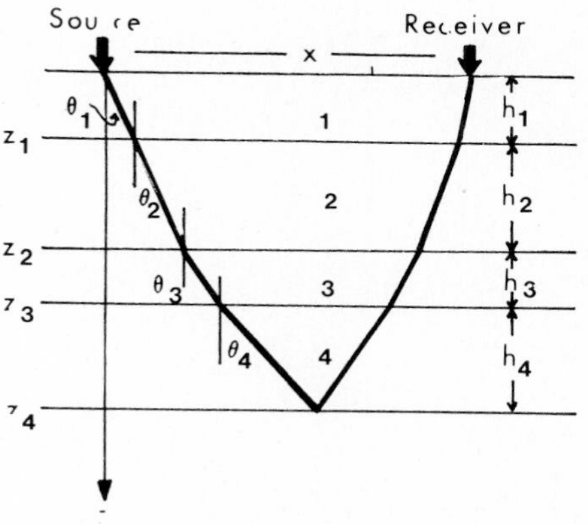

Figure 17-1. Ray path of a body-phase in multilayered, isotropic, homogeneous medium.

we have for a ray penetrating to the nth interface:

$$T_n^o = 2 \sum_{k=1}^{n} h_k/v_k, \qquad (17\text{-}1)$$

where h_k is the thickness of the kth layer and v_k is the velocity of P-waves in the kth layer. The true average velocity for such a medium is then seen to be

$$[\bar{v}_n]_{true} = \sum_{k=1}^{n} h_k \Big/ \sum_{k=1}^{n} t_k = \sum_{k=1}^{n} h_k \Big/ \sum_{k=1}^{n} (h_k/v_k). \qquad (17\text{-}2)$$

That is, the reciprocal of the true average velocity is the space-weighted average of reciprocal velocities. For convenience, we will temporarily drop the summation limits and use them implicitly when $\sum$ appears.

Now consider what happens in such a medium when off-vertical rays are examined. For a receiver at a point $\underline{x}$ removed from the source, we have

$$T_n = 2 \sum h_k/(v_k \text{Cos}\theta_k). \qquad (17\text{-}3)$$

Considering the difference in time as one moves out from the source, we find

$$\Delta T_n = T_n - T_n^o \simeq \sum h_k \theta_k^2 / v_k \simeq \delta^2 \sum h_k v_k, \tag{17-4}$$

where we have used the following small angle approximation for Snell's Law:

$$\frac{\theta_1}{v_1} \simeq \frac{\theta_2}{v_2} \simeq \quad \ldots \quad \simeq \frac{\theta_n}{v_n} \simeq \delta. \tag{17-5}$$

Since

$$x = 2 \sum h_k \mathrm{Tan}\theta_k \simeq 2\delta \sum h_k v_k, \tag{17-6}$$

we can combine (17-4) and (17-6) and obtain

$$\Delta T_n \simeq x^2/(4 \sum h_k v_k). \tag{17-7}$$

It should be noted that this expression for ΔT_n has been derived from exact physics (insofar as plane-waves are applicable) using numerical approximations that become more accurate as the angles θ_k get smaller.

Rewriting (17-7) slightly, we have

$$\sum h_k v_k \simeq x^2/(4\Delta T_n).$$

Then

$$h_n v_n = \frac{x^2}{4} \left(\frac{1}{\Delta T_n} - \frac{1}{\Delta T_{n-1}} \right).$$

Using

$$h_n = \tfrac{1}{2} v_n (T_n^o - T_{n-1}^o),$$

we finally obtain

$$v_n^2 = \frac{x^2}{2(T_n^o - T_{n-1}^o)} \left[\frac{1}{\Delta T_n(x)} - \frac{1}{\Delta T_{n-1}(x)} \right]. \tag{17-8}$$

From the differential nature of the factors in (17-8) it is immediately obvious that poor data will give even poorer estimates of v_n^2.

An alternative formulation almost equivalent to (17-8) can be obtained when one considers the observation that for moveout distances of the order of the maximum depth of penetration, the curve of T^2 vs. x^2 is almost a straight line. This is clearly shown in Grant and West (1965, Fig. 5-11, 5-12). This means that we can write the travel-time relation as

$$T_n^2 = (T_n^o)^2 + x^2/[\bar{v}_n]^2_{app}, \tag{17-9}$$

where $[\bar{v}_n]_{app}$ is the apparent average velocity of the section between the surface and the maximum depth of penetration. Logically, this is equivalent to saying that there is no refraction at the layer interfaces and that the travel-time in each layer is inversely proportional to the P-wave velocity in that layer, with the constant of proportionality (which varies as a function of angle) being the same for all layers. This statement is not the same as the relationship expressed by equation (17-4). Writing $T_n = T_n^o + \Delta T_n$, we have approximately

$$\Delta T_n \simeq x^2 / \left\{ 2T_n^o [\bar{v}_n]^2_{app} \right\}. \tag{17-10}$$

Comparing the two forms of ΔT_n, (17-10) and (17-7), we find that

$$[\bar{v}_n]^2_{app} \simeq (4\sum h_k v_k)/2T^o_n = \sum h_k v_k / \sum t_k = \sum t_k v_k^2 / \sum t_k = [\bar{v}_n]^2_{rms}. \quad (17\text{-}11)$$

That the RMS velocity should be a meaningful average may seem a little strange. However to the same order of approximation that we have been using, we find that

$$[\bar{v}_n]_{rms} = \left\{ \sum h_k v_k / \sum (h_k/v_k) \right\}^{\frac{1}{2}} = \sum h_k / \sum (h_k/v_k) + O\left[\left(\frac{v_i - v_j}{v_i}\right)^2 \right] \simeq [\bar{v}_n]_{true}.$$

This approximation is made by noting that the average reciprocal is approximately equal to the reciprocal of the average:

$$\sum (h_k/v_k) / \sum h_k \simeq \sum h_k / \sum (h_k v_k).$$

For a 20% range of velocities in layers of equal thickness, this approximation is good to better than 1%.

Evaluating (17-11) for the n-1st and n[th] layer, we obtain:

$$T^o_{n-1}[\bar{v}_{n-1}]^2_{app} = \sum_{k=1}^{n-1} t_k v_k^2,$$

$$T^o_n \; [\bar{v}_n]^2_{app} = \sum_{k=1}^{n-1} t_k v_k^2 = T^o_{n-1}[\bar{v}_{n-1}]^2_{app} + (T^o_n - T^o_{n-1})v_n^2.$$

Combining, we can write another expression for v_n^2 in the form

$$v_n^2 = \frac{T^o_n[\bar{v}_n]^2_{app} - T^o_{n-1}[\bar{v}_{n-1}]^2_{app}}{T^o_n - T^o_{n-1}}. \quad (17\text{-}12)$$

This is known as the Dix Equation. Dix (1955) points out that the squares of the apparent velocities $[\bar{v}_n]^2_{app}$ are best computed from the $T^2 - x^2$ curve (17-9), with x-values near the origin, that is, for small deviations from the normal. It is again obvious that the differential nature of this expression will magnify any errors in the raw data. The effects of dipping interfaces add considerable difficulty to the problem of determining accurate interval velocities v_n. Dix (ibid) suggests ways to overcome this problem.

Once we have obtained the layer velocities v_k, we can then calculate the velocity-depth function using the following relationship:

$$z_n = \sum_{k=1}^{n} h_k = \tfrac{1}{2} \sum_{k=1}^{n} v_k(T^o_k - T^o_{k-1}), \quad (17\text{-}13)$$

where z_n is the depth to the nth interface. Between the various approximations made in obtaining (17-8) and (17-12) and their inherent instability to data containing errors, one cannot obtain very precise velocity-depth curves using them. However, the real value of <u>reflection seismology</u> lies in its ability to delineate <u>structure</u>. One is able to follow the contours (even if only roughly) of certain layers in the earth's crust.

The great majority of effort in exploration geophysics is devoted to reflection seismology. The reader is referred to standard texts for further details of theory, interpretation, and data processing techniques. Particularly useful are Grant and West (1965), Dobrin (1976), and Telford et al (1976).

17.2. Travel-Time Curves for Refracted Rays.

For flat-layered structures, the velocity-depth curves can also be constructed using the travel-time curves obtained from refracted rays. This method eliminates the problem associated with the differential coefficients given above, but it introduces a new difficulty in some cases. This new difficulty will be discussed shortly.

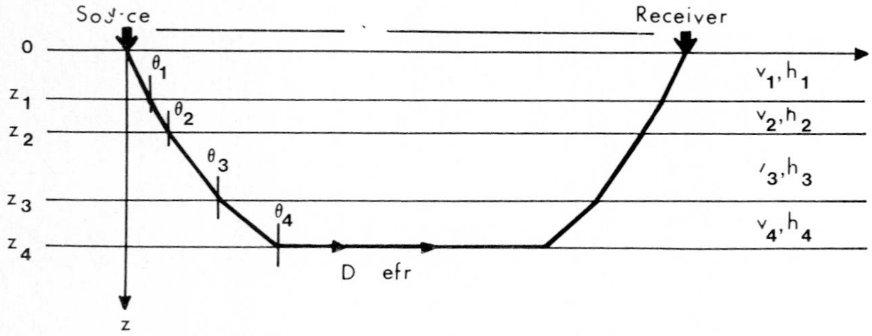

Figure 17-2. Ray path of a critically refracted body-phase in a medium composed of many homogeneous isotropic layers.

Consider the travel-time for the critically refracted ray shown in Figure 17-2. This is given by

$$T_n = \sum \frac{2h_k}{v_k \mathrm{Cos}\theta_k} + \frac{D_{refr}}{v_{n+1}}, \qquad (17\text{-}14)$$

where D_{refr} is the distance traveled by the ray at the interface between the nth layer and the $n{+}1^{st}$ layer. From Snell's Law, we may obtain the following relations:

$$\mathrm{Sin}\theta_k = v_k/(v_{n+1}), \qquad \mathrm{Cos}\theta_k = (1 - v_k^2/v_{n+1}^2)^{\frac{1}{2}}, \qquad (17\text{-}15)$$

$$\mathrm{Tan}\theta_k = v_k/(v_{n+1}^2 - v_k^2)^{\frac{1}{2}}.$$

The distance traveled by the critically refracted ray is given by

$$D_{refr} = x - \sum 2h_k \mathrm{Tan}\theta_k$$

$$= x - \sum \frac{2h_k v_k}{(v_{n+1}^2 - v_k^2)^{\frac{1}{2}}}. \qquad (17\text{-}16)$$

Substituting into (17-14) and using (17-15), we finally get

$$T_n = \frac{x}{v_{n+1}} + 2 \sum \frac{h_k (v_{n+1}^2 - v_k^2)^{\frac{1}{2}}}{v_{n+1} v_k}. \qquad (17\text{-}17)$$

The intercept time of the nth refracted ray is given by

$$T_n^o = 2 \sum \frac{h_k (v_{n+1}^2 - v_k^2)^{\frac{1}{2}}}{v_{n+1} v_k} . \qquad (17\text{-}18)$$

In the case where the velocities are monotonically increasing, each layer is represented by a straight line segment as in Figure 17-3. The slope of each line segment is equal to the inverse of the velocity in the next layer below. These

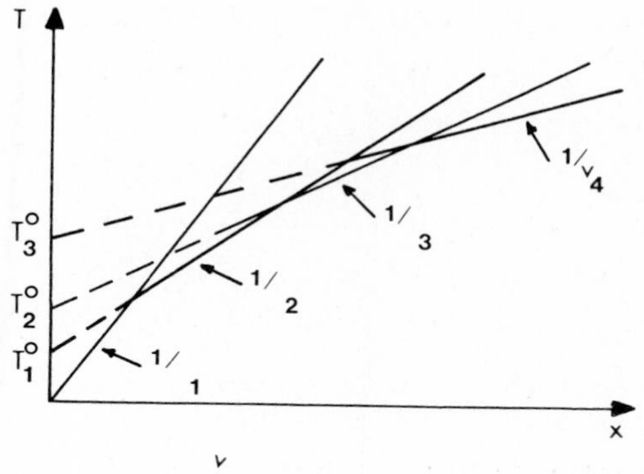

Figure 17-3. Travel-time segments corresponding to the individual layers in a layered medium with a monotonically increasing velocity.

values may then be substituted into (17-18) to solve for the $\underline{n}$ values of h_k. This set of $\underline{n}$ equations in $\underline{n}$ unknowns is particularly easy to solve, since it is triangular in form; one new unknown, h_k, is added with each additional equation. Even with an always increasing velocity, complications arise. For certain values of the parameters, one can find the <u>hidden-layer problem</u>. The travel-time diagram would appear as in Figure 17-4. In this case, the events on the line segment corresponding to the second layer would never appear as <u>first-arrivals</u>. If later arrivals were not

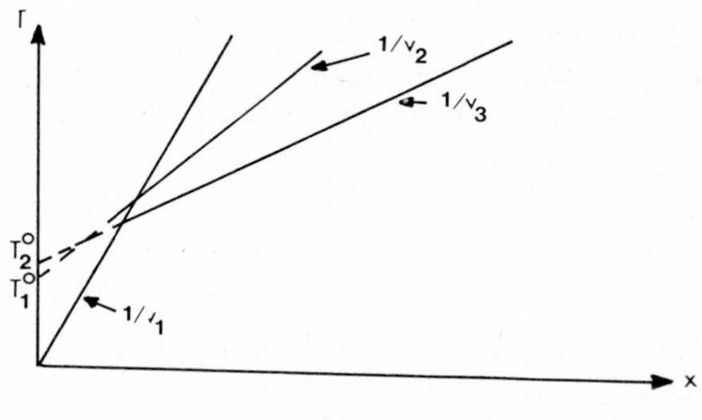

Figure 17-4. Travel-time segments corresponding to the "hidden-layer" problem. Events along the line with slope $1/v_2$ are not first arrivals.

recognized as defining the line as in the figure, the presence of the second layer would go undetected. An attempt to use (17-18) in the normal manner would give an erroneous thickness for the sum of layers One and Two compensating for the non-observation of v_2. Green (1962) presents rather clearly the conditions under which this hidden-layer phenomenon may occur.

An even more insidious case occurs when the layer velocities do not always increase with depth. Let us examine a case where a low-velocity layer is present. Equation (17-18) is still valid for those layers where a refracted arrival is present; the low-velocity layer is accounted for in the sum of (17-16). However, no refracted ray will be present in this layer. In Figure 17-5 we see the ray being turned back toward the normal as it travels from a layer of higher velocity into a layer of relatively lower velocity. Now we will observe no straight line segment in Figure 17-3 corresponding to the low-velocity layer. Consequently, we will think that there are less layers than there really are. The depths calculated from (17-18) will then be all wrong for layers lying below the low-velocity layer.

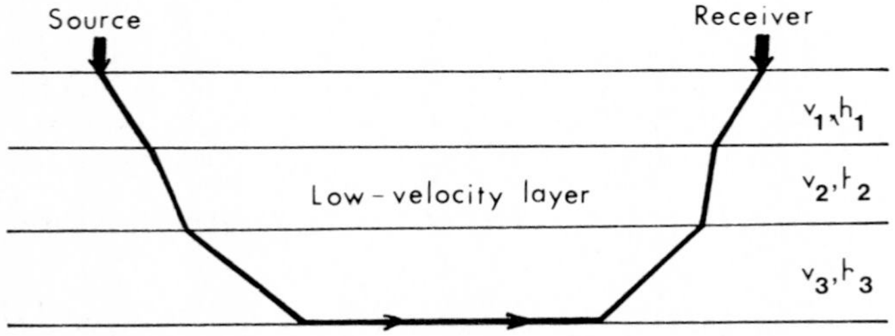

Figure 17-5. Ray path for the critically refracted ray in a medium with a low-velocity layer.

However, this low-velocity layer can still be detected by the reflection method if the velocity contrast is high enough.

One smaller problem of some interest has been the transmission of energy by a thin high-velocity layer imbedded within a layered medium of lower velocity. When the high-velocity layer is smaller than one wavelength, one would expect complications to arise such as constructive and destructive interference between waves reflected at the top and bottom of the layer. Spencer (1965a) and Rosenbaum (1965) have considered the problem of an elastic layer in an elastic medium. Rosenbaum (1961) has investigated arrivals from an elastic layer within a fluid medium. Additional considerations concerning the attenuation of the seismic signal have been given by Mellman and Helmberger (1974).

17.3. Rays Incident Upon a Plane-Layered Structure From Below.

When a seismic wave from a distant earthquake travels deep into the earth as in

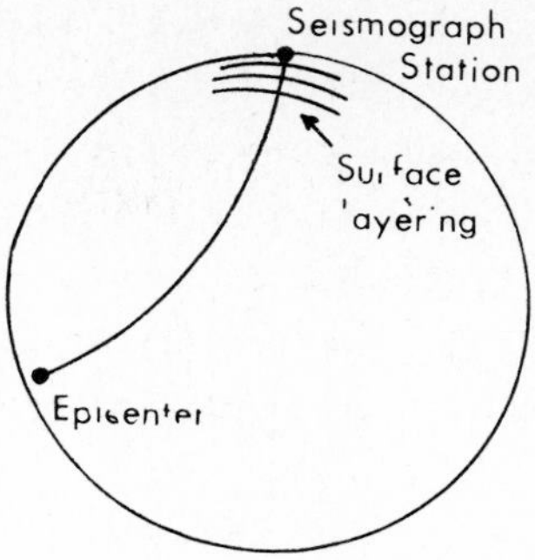

Figure 17-6. Ray from a distant earthquake arriving at a seismograph station with nearly vertical incidence.

Figure 17-6, it emerges at the seismograph station at the opposite end of its path at nearly vertical incidence. We shall discuss these matters at much greater length in Chapters 22 and 23. As the wave passes through the surface layers immediately beneath the seismograph station, it undergoes multiple reflections and transmissions giving rise to constructive and destructive interference. Inasmuch as there are characteristic lengths involved, properties of the observed waves will be wavelength or frequency dependent. One such property, wave amplitude, will be directly involved in the determination of amplitude vs. distance curves. This frequency dependence may not always be a hinderance. A frequency analysis of seismograms recorded from distant earthquakes can lead to an interpretation of the crustal structure beneath a seismograph station.

<u>Matrix Relations For P- and SV-Waves</u>. First of all, we shall ignore the slight curvature of the upper layers of the earth. Our model is illustrated in Figure 17-7. The half-space will be considered as the nth layer. The following is based largely

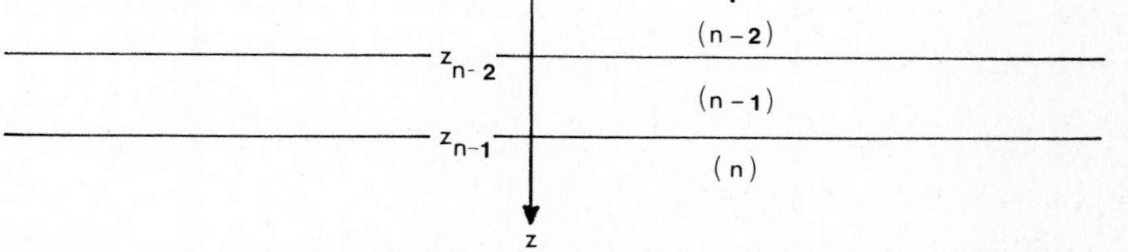

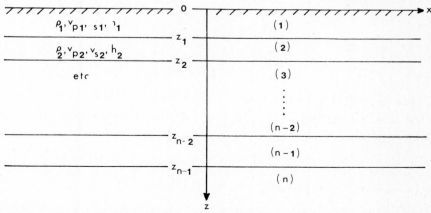

Figure 17-7. Plane-parallel model of the earth's surface. Each layer is homogeneous and isotropic.

on the formulation of Harkrider (1964); however, some of the notation has been altered to conform to that of previous chapters in this work. Although his paper was devoted to surface-waves, we have seen in Chapter 16 that there is a close relationship between propagating modes and multiply reflected body-waves in a multi-layered structure. Consequently there will be a close connection between what we outline here and what we do in Chapter 20 for modes in identical structures.

The potentials for the mth layer can be written:

$$\Phi^{(m)} = \left[A^{(m)}e^{\alpha^{(m)}z'^{(m)}} + B^{(m)}e^{-\alpha^{(m)}z'^{(m)}}\right]e^{i(kx - \omega t)},$$
$$\Psi^{\dagger(m)} = \left[C^{(m)}e^{\beta^{(m)}z'^{(m)}} + D^{(m)}e^{-\beta^{(m)}z'^{(m)}}\right]e^{i(kx - \omega t)}; \qquad (17\text{-}19)$$

where

$$\alpha^{(m)} = [k^2 - k^2_{p(m)}]^{\frac{1}{2}}, \qquad \beta^{(m)} = [k^2 - k^2_{s(m)}]^{\frac{1}{2}},$$

as in (16-80), and

$$z'^{(m)} = z - z^{(m-1)}$$

is a local measure of vertical distance. The displacements and stresses are related to the potentials as on page 167:

$$u_x = \frac{\partial\Phi}{\partial x} - \frac{\partial\Psi^\dagger}{\partial z}, \qquad u_z = \frac{\partial\Phi}{\partial z} + \frac{\partial\Psi^\dagger}{\partial x},$$
$$\frac{t_{zz}}{\mu} = -k_s^2\Phi + 2\left(\frac{\partial^2\Psi^\dagger}{\partial x\partial z} - \frac{\partial^2\Phi}{\partial x^2}\right), \qquad (17\text{-}20)$$
$$\frac{t_{xz}}{\mu} = \frac{\partial^2\Psi^\dagger}{\partial x^2} + 2\frac{\partial^2\Phi}{\partial x\partial z} - \frac{\partial^2\Psi^\dagger}{\partial z^2}.$$

Substituting equations (17-19) into (17-20) we find that

$$ku_x = ik^2(Ae^{\alpha z'} + Be^{-\alpha z'}) - k\beta(Ce^{\beta z'} - De^{-\beta z'}),$$
$$ku_z = k\alpha(Ae^{\alpha z'} - Be^{-\alpha z'}) + ik^2(Ce^{\beta z'} + De^{-\beta z'}),$$
$$t_{zz}/\mu = (2k^2 - k_s^2)(Ae^{\alpha z'} + Be^{-\alpha z'}) + 2ik\beta(Ce^{\beta z'} - De^{-\beta z'}),$$
$$t_{xz}/\mu = 2ik\alpha(Ae^{\alpha z'} - Be^{-\alpha z'}) - (2k^2 - k_s^2)(Ce^{\beta z'} + De^{-\beta z'});$$

where we have dropped the layering subscript (m) for the time being, and have suppressed the common factor $e^{i(kx - \omega t)}$. To simplify the resulting matrix, we want to rewrite the four above relations in the following form:

$$ku_x = ik^2\left[(A+B)\text{Ch}\alpha z' + (A-B)\text{Sh}\alpha z'\right] - k\beta\left[(C-D)\text{Ch}\beta z' + (C+D)\text{Sh}\beta z'\right],$$
$$ku_z = k\alpha\left[(A-B)\text{Ch}\alpha z' + (A+B)\text{Sh}\alpha z'\right] + ik^2\left[(C+D)\text{Ch}\beta z' + (C-D)\text{Sh}\beta z'\right], \qquad (17\text{-}21)$$
$$t_{zz}/\mu = (2k^2 - k_s^2)\left[(A+B)\text{Ch}\alpha z' + (A-B)\text{Sh}\alpha z'\right] + 2ik\beta\left[(C-D)\text{Ch}\beta z' + (C+D)\text{Sh}\beta z'\right],$$
$$t_{zx}/\mu = 2ik\alpha\left[(A-B)\text{Ch}\alpha z' + (A+B)\text{Sh}\alpha z'\right] - (2k^2 - k_s^2)\left[(C+D)\text{Ch}\beta z' + (C-D)\text{Sh}\beta z'\right].$$

Evaluating these equations at $z' = 0$ (the top of the layer) and using the summation convention for repeated subscripts, we have

$$U_i(0) = E_{ij}(0)A_j; \qquad (17\text{-}22)$$

where

$$U_i(0) = \begin{pmatrix} ku_x \\ ku_z \\ t_{zz} \\ t_{zx} \end{pmatrix}, \qquad (17\text{-}23)$$

$$A_j = \begin{pmatrix} A+B \\ A-B \\ C+D \\ C-D \end{pmatrix}, \qquad (17\text{-}24)$$

and

$$E_{ij}(0) = \begin{pmatrix} ik^2 & 0 & 0 & -k\beta \\ 0 & k\alpha & ik^2 & 0 \\ (2k^2 - k_s^2)\mu & 0 & 0 & 2ik\beta\mu \\ 0 & 2ik\alpha\mu & -(2k^2 - k_s^2)\mu & 0 \end{pmatrix}. \qquad (17\text{-}25)$$

We will need the inverse of this matrix later, so we must solve (17-22) for the A_i in terms of $U_j(0)$. In subscript notation we have

$$A_i = \{E(0)\}^{-1}_{ij}\, U_j(0); \qquad (17\text{-}26)$$

where

$$\{E(0)\}^{-1}_{ij} = \frac{1}{k_s^2\mu} \begin{pmatrix} -2i\mu & 0 & -1 & 0 \\ 0 & -\dfrac{(2k^2 - k_s^2)\mu}{k\alpha} & 0 & -\dfrac{ik}{\alpha} \\ 0 & -2i\mu & 0 & 1 \\ \dfrac{(2k^2 - k_s^2)\mu}{k\beta} & 0 & -\dfrac{ik}{\beta} & 0 \end{pmatrix}. \qquad (17\text{-}27)$$

That this is the inverse of (17-25) can be verified by evaluating the product

$$E_{ij}(0)\{E(0)\}^{-1}_{jk} = \delta_{ik}.$$

Despite their complication, all that is said by the relationships (17-22) and (17-26) is that the potential amplitudes A_i are linear functions of the displacements and stresses and vice-versa. That is, if one is given the potential amplitudes A_i, then by use of (17-22) one may calculate the displacements and stresses at $z' = 0$. Likewise, given the displacements and stresses, the potential amplitudes A_i may be calculated from (17-26). In a similar manner, we may evaluate (17-21) at $z'^{(m)} = h^{(m)} = z^{(m)} - z^{(m-1)}$ and find that

$$U_i(h) = E_{ij}(h)A_j; \qquad (17\text{-}28)$$

where the matrix $E_{ij}(h)$ is given by (17-29)

$$E_{ij}(h) = \begin{pmatrix} ik^2 \mathrm{Ch}\alpha h & ik^2 \mathrm{Sh}\alpha h & -k\beta \mathrm{Sh}\beta h & -k\beta \mathrm{Ch}\beta h \\ k\alpha \mathrm{Sh}\alpha h & k\alpha \mathrm{Ch}\alpha h & ik^2 \mathrm{Ch}\beta h & ik^2 \mathrm{Sh}\beta h \\ (2k^2 - k_s^2)\mu \mathrm{Ch}\alpha h & (2k^2 - k_s^2)\mu \mathrm{Sh}\alpha h & 2ik\beta\mu \mathrm{Sh}\beta h & 2ik\beta\mu \mathrm{Ch}\beta h \\ 2ik\alpha\mu \mathrm{Sh}\alpha h & 2ik\alpha\mu \mathrm{Ch}\alpha h & -(2k^2 - k_s^2)\mu \mathrm{Ch}\beta h & -(2k^2 - k_s^2)\mu \mathrm{Sh}\beta h \end{pmatrix}.$$

The inverse of this matrix will not be needed.

If we substitute (17-26) into (17-28), we obtain

$$U_i(h) = E_{ij}(h)\{E(0)\}^{-1}_{jk} U_k(0) = F_{ik}(h)U_k(0), \qquad (17\text{-}30)$$

where

$$F_{11} = F_{44} = \left[\gamma^2 \mathrm{Ch}\alpha h - (\gamma^2 - 1)\mathrm{Ch}\beta h\right],$$

$$F_{12} = F_{34} = -i\left[(\gamma^2 - 1)\frac{k}{\alpha}\mathrm{Sh}\alpha h - \gamma^2 \frac{\beta}{k}\mathrm{Sh}\beta h\right],$$

$$F_{13} = F_{24} = -\frac{i\gamma^2}{2\mu}\left[\mathrm{Ch}\alpha h - \mathrm{Ch}\beta h\right],$$

$$F_{14} = \frac{\gamma^2}{2\mu}\left[\frac{k}{\alpha}\mathrm{Sh}\alpha h - \frac{\beta}{k}\mathrm{Sh}\beta h\right],$$

$$F_{21} = F_{43} = -i\left[\gamma^2 \frac{\alpha}{k}\mathrm{Sh}\alpha h - (\gamma^2 - 1)\frac{k}{\beta}\mathrm{Sh}\beta h\right],$$

$$F_{22} = F_{33} = -\left[(\gamma^2 - 1)\mathrm{Ch}\alpha h - \gamma^2 \mathrm{Ch}\beta h\right], \qquad (17\text{-}31)$$

$$F_{23} = -\frac{\gamma^2}{2\mu}\left[\frac{\alpha}{k}\mathrm{Sh}\alpha h - \frac{k}{\beta}\mathrm{Sh}\beta h\right],$$

$$F_{31} = F_{42} = -2i\mu(\gamma^2 - 1)\left[\mathrm{Ch}\alpha h - \mathrm{Ch}\beta h\right],$$

$$F_{32} = -\frac{2\mu}{\gamma^2}\left[(\gamma^2 - 1)^2 \frac{k}{\alpha}\mathrm{Sh}\alpha h - \gamma^4 \frac{\beta}{k}\mathrm{Sh}\beta h\right],$$

$$F_{41} = \frac{2\mu}{\gamma^2}\left[\gamma^4 \frac{\alpha}{k}\mathrm{Sh}\alpha h - (\gamma^2 - 1)^2 \frac{k}{\beta}\mathrm{Sh}\beta h\right],$$

$$\gamma^2 = 2k^2/k_s^2.$$

It can be seen that these elements form a matrix with a kind of "reverse symmetry", i.e., a symmetry about the "other diagonal". It should also be noted that for real-k, each of the elements F_{ij} is either pure real or pure imaginary regardless of whether _k_ is greater or less than k_p or k_s. Again, despite its complication, all that is said by (17-30) is that the displacement and stress components on one side of the layer are linearly related to the displacement and stress components on the other side of the layer.

Although we have been following rather closely the analysis of Harkrider (1964), these matrix relationships were first noted by Thomson (1950) and given general application to the propagation of elastic waves in the earth by Haskell (1953). Consequently, this matrix formulation is known as the Thomson-Haskell method. Some additional theoretical considerations applicable to these relationships are given by Gilbert and Backus (1966) and Kennett (1972). Richards (1971) develops an alternative scheme for numerically continuing P- and SV-waves through a multilayered medium. But now, back to our exposition.

By direct computation, it may be verified that

$$F_{ij}^{(m)}(h)F_{jk}^{(m)}(h') = F_{ik}^{(m)}(h + h'). \tag{17-32}$$

The physical content of this equation is that a single layer of thickness h + h' must be equivalent in physical properties to two adjacent layers of thicknesses h and h'. By letting h' = -h, we have

$$F_{ij}^{(m)}(h)\ F_{jk}^{(m)}(-h) = F_{ik}^{(m)}(0) = \delta_{ij}. \tag{17-33}$$

Therefore we must have

$$\{F^{(m)}(h)\}_{jk}^{-1} = F_{jk}^{(m)}(-h). \tag{17-34}$$

The reciprocal of the matrix $F_{ij}^{(m)}$ is obtained by changing the sign of h. This reciprocal matrix will be needed later.

Let us now impose the interfacial boundary conditions. These may be written as

$$U_i^{(m)}\{z^{(m-1)}\} = U_i^{(m-1)}\{z^{(m-1)}\}; \tag{17-35}$$

which says that there is continuity of displacement and stress across the interface between the m-1st and the mth layers. Starting with the free surface, we may apply (17-35) to successive layers obtaining:

$$\begin{aligned}
U_i^{(2)}\{z^{(1)}\} &= U_i^{(1)}\{z^{(1)}\} = F_{ij}^{(1)}U_j^{(1)}(0),\\
U_i^{(3)}\{z^{(2)}\} &= U_i^{(2)}\{z^{(2)}\} = F_{ij}^{(2)}U_j^{(2)}\{z^{(1)}\} = F_{ij}^{(2)}F_{jk}^{(1)}U_k^{(1)}(0),\\
&\vdots\\
U_i^{(n)}\{z^{(n-1)}\} &= U_i^{(n-1)}\{z^{(n-1)}\}\\
&= F_{ij}^{(n-1)}F_{jk}^{(n-2)}\ \cdots\cdots\ F_{lm}^{(2)}F_{mn}^{(1)}U_n^{(1)}(0)\\
&= \mathcal{F}_{in}U_n^{(1)}(0).
\end{aligned} \tag{17-36}$$

Again, all that we are showing is that there is a linear relationship between the displacement and stress components in the half-space and these components on the free surface.

From the form of $F_{ij}^{(m)}$ and its inverse, it can be seen that it is of the class of matrices whose inverse is given by

$$\{F^{(m)}\}^{-1}_{ij} = (-1)^{i+j}\{F^{(m)}\}_{p+1-j,p+1-i}, \tag{17-37}$$

where $F^{(m)}$ is a p x p matrix. This fact was first noted by Harkrider (1964). In our case this means that

$$\{F^{(m)}\}^{-1} = \begin{pmatrix} F_{44} & -F_{34} & F_{24} & -F_{14} \\ -F_{43} & F_{33} & -F_{23} & F_{13} \\ F_{42} & -F_{32} & F_{22} & -F_{12} \\ -F_{41} & F_{31} & -F_{21} & F_{11} \end{pmatrix},$$

since $F_{44} = F_{11}$, $F_{34} = F_{12}$, etc. The reader can then show that for this specific class of matrices, the inverse of the product matrix is also given by (17-37),

$$\{\mathcal{F}\}^{-1} = \begin{pmatrix} \mathcal{F}_{44} & -\mathcal{F}_{34} & \mathcal{F}_{24} & -\mathcal{F}_{14} \\ -\mathcal{F}_{43} & \mathcal{F}_{33} & -\mathcal{F}_{23} & \mathcal{F}_{13} \\ \mathcal{F}_{42} & -\mathcal{F}_{32} & \mathcal{F}_{22} & -\mathcal{F}_{12} \\ -\mathcal{F}_{41} & \mathcal{F}_{31} & -\mathcal{F}_{21} & \mathcal{F}_{11} \end{pmatrix}. \tag{17-38}$$

However, in this case, $\mathcal{F}_{44} \neq \mathcal{F}_{11}$, etc. Because of the form of the reciprocal matrix, we have a set of relationships between the elements of the layer product matrix, $\mathcal{F}$. These are:

$$(-1)^{j+k}\mathcal{F}_{ij}\mathcal{F}_{5-k,5-j} = \delta_{ik}. \tag{17-39}$$

For example, we have

$$\mathcal{F}_{11}\mathcal{F}_{24} - \mathcal{F}_{12}\mathcal{F}_{23} + \mathcal{F}_{13}\mathcal{F}_{22} - \mathcal{F}_{14}\mathcal{F}_{21} = \delta_{13} = 0.$$

Such relationships will be useful in future analysis.

Going back to (17-26), we can write

$$A_i^{(m)} = \{E^{(n)}(0)\}^{-1}_{ij}\mathcal{F}_{jk}U_k^{(1)}(0) = J_{ik}U_k^{(1)}(0). \tag{17-40}$$

This is the final general relationship we need. It turns out that many useful specific relationships can be derived from this equation. Let us write it out in full:

$$\begin{aligned} A^{(n)} + B^{(n)} &= J_{11}ku_x^{(1)}(0) + J_{12}ku_z^{(1)}(0) + J_{13}t_{zz}^{(1)}(0) + J_{14}t_{zx}^{(1)}(0), \\ A^{(n)} - B^{(n)} &= J_{21}ku_x^{(1)}(0) + J_{22}ku_z^{(1)}(0) + J_{23}t_{zz}^{(1)}(0) + J_{24}t_{zx}^{(1)}(0), \\ C^{(n)} + D^{(n)} &= J_{31}ku_x^{(1)}(0) + J_{32}ku_z^{(1)}(0) + J_{33}t_{zz}^{(1)}(0) + J_{34}t_{zx}^{(1)}(0), \\ C^{(n)} - D^{(n)} &= J_{41}ku_x^{(1)}(0) + J_{42}ku_z^{(1)}(0) + J_{43}t_{zz}^{(1)}(0) + J_{44}t_{zx}^{(1)}(0). \end{aligned} \tag{17-41}$$

The following combinations of coefficients will occur often enough to warrant giving them a name.

$$K = J_{11} + J_{21}, \qquad P = J_{31} + J_{41},$$
$$L = J_{12} + J_{22}, \qquad Q = J_{32} + J_{42}, \qquad (17\text{-}42)$$
$$M = J_{13} + J_{23}, \qquad R = J_{33} + J_{43},$$
$$N = J_{14} + J_{24}, \qquad S = J_{34} + J_{44}.$$

Let us now consider the case of P- and SV-waves incident from below, where the surface, $z = 0$, is prescribed to be free of stress. This means that the last two terms in each of equations (17-41) vanish. We then obtain

$$A^{(n)} + B^{(n)} = J_{11}ku_x^{(1)}(0) + J_{12}ku_z^{(1)}(0),$$
$$A^{(n)} - B^{(n)} = J_{21}ku_x^{(1)}(0) + J_{22}ku_z^{(1)}(0),$$

and

$$C^{(n)} + D^{(n)} = J_{31}ku_x^{(1)}(0) + J_{32}ku_z^{(1)}(0),$$
$$C^{(n)} - D^{(n)} = J_{41}ku_x^{(1)}(0) + J_{42}ku_z^{(1)}(0).$$

Adding each pair of equations and solving for the surface displacements, we find that

$$u_x^{(1)}(0) = \frac{2}{k}\frac{AQ - CL}{KQ - LP}; \qquad u_z^{(1)}(0) = -\frac{2}{k}\frac{AP - CK}{KQ - LP}. \qquad (17\text{-}43)$$

For incident P-waves, we have that $|u^p| = k_pA$, and for incident S-waves we have that $|u^s| = k_sC$, when only real angles of incidence are considered, i.e., $k^2 < k^2_{p,s}$. The factor 2 in (17-43) comes from the doubling of particle motion at a free surface. Using this normalization, we then have four <u>crustal-response functions</u>:

$$T_x^p = \frac{1}{k_pk}\left|\frac{Q}{KQ - LP}\right|, \qquad T_z^p = \frac{1}{k_pk}\left|\frac{P}{KQ - LP}\right|,$$
$$T_x^s = \frac{1}{k_sk}\left|\frac{L}{KQ - LP}\right|, \qquad T_z^s = \frac{1}{k_sk}\left|\frac{K}{KQ - LP}\right|. \qquad (17\text{-}44)$$

These are analogous to the bracketed terms in (9-41), (9-42), (9-44), and (9-45). Haskell (1962) has prepared contour diagrams for the two displacement amplitudes for both P and SV as a function of period and angle of incidence upon a simple crustal section. In addition, he presents contour diagrams of the phase differences between vertical and horizontal components of displacement.

The reflected potential amplitudes $B^{(n)}$ and $D^{(n)}$ are given by

$$B^{(n)} = \tfrac{1}{2}\{K^\dagger ku_x^{(1)}(0) + L^\dagger ku_z^{(1)}(0)\}$$
$$D^{(n)} = \tfrac{1}{2}\{P^\dagger ku_x^{(1)}(0) + Q^\dagger ku_z^{(1)}(0)\}; \qquad (17\text{-}45)$$

where

$$K^\dagger = J_{11} - J_{21} \qquad P^\dagger = J_{31} - J_{41},$$
$$L^\dagger = J_{12} - J_{22}, \qquad Q^\dagger = J_{32} - J_{42}. \qquad (17\text{-}46)$$

Values of $B^{(n)}$ and $D^{(n)}$ may thus be computed from (17-45) after solving for $u_x^{(1)}(0)$

and $U_z^{(1)}(0)$. These may be looked at as generalized reflection coefficients from the earth's structured surface. In his analysis, Haskell (ibid) has also worked out the partition of energy between reflected P- and SV-waves for incoming P-waves with varying angles of incidence and frequency of excitation. Again he has plotted up a contour diagram.

Matrix Relationships for SH-Waves. The geometry of Figure 17-7 will still be applicable, and we will have

$$u_y^{(m)} = -\frac{\partial\Lambda^{(m)}}{\partial x}, \tag{6-26}$$

$$t_{zy}^{(m)} = -\mu^{(m)}\frac{\partial^2\Lambda^{(m)}}{\partial x\partial z}; \tag{6-34}$$

where

$$\Lambda^{(m)} = \left[A^{(m)}e^{\beta^{(m)}z'^{(m)}} + B^{(m)}e^{-\beta^{(m)}z'^{(m)}}\right]e^{i(kx - \omega t)}. \tag{17-47}$$

Here $\beta^{(m)}$ and $z'^{(m)}$ are the same as in (17-19). Substituting (17-47) into (6-26) and (6-34) gives the following relationships

$$ku_y = -ik^2(Ae^{\beta z'} + Be^{-\beta z'}),$$

$$t_{zy}/\mu = -ik\beta(Ae^{\beta z'} - Be^{-\beta z'});$$

where we have again dropped the layer superscript and the common factor $e^{i(kx-\omega t)}$. This pair can be symmetrized as in (17-23) giving

$$\left.\begin{aligned} ku_y &= -ik^2\left[(A+B)\mathrm{Ch}\beta z' + (A-B)\mathrm{Sh}\beta z'\right], \\ t_{zy}/\mu &= -ik\beta\left[(A-B)\mathrm{Ch}\beta z' + (A+B)\mathrm{Sh}\beta z'\right]. \end{aligned}\right. \tag{17-48}$$

Evaluating this pair at the top of the mth layer, we have

$$U_i(0) = E_{ij}(0)A_j; \tag{17-49}$$

where

$$U_i(0) = \begin{pmatrix} ku_y(0) \\ t_{zy}(0) \end{pmatrix}, \tag{17-50}$$

$$A_j = \begin{pmatrix} A+B \\ A-B \end{pmatrix}, \tag{17-51}$$

and

$$E_{ij}(0) = \begin{pmatrix} -ik^2 & 0 \\ 0 & -ik\beta\mu \end{pmatrix}. \tag{17-52}$$

We can also write A_i in terms of $U_j(0)$, obtaining

$$A_i = \{E(0)\}_{ij}^{-1}U_j(0); \tag{17-53}$$

where

$$\{E(0)\}^{-1}_{ij} = \begin{pmatrix} i/k^2 & 0 \\ 0 & i/(k\beta\mu) \end{pmatrix}. \qquad (17\text{-}54)$$

This matrix is reciprocal to $E_{ij}(0)$, as may be calculated directly;

$$E_{ij}(0)\{E(0)\}^{-1}_{jk} = \delta_{ik}. \qquad (17\text{-}55)$$

Equations (17-48) may be evaluated at $z' = h$, giving

$$E_{ij}(h) = \begin{pmatrix} -ik^2\mathrm{Ch}\beta h & -ik^2\mathrm{Sh}\beta h \\ -ik\beta\mu\mathrm{Sh}\beta h & -ik\beta\mu\mathrm{Ch}\beta h \end{pmatrix}. \qquad (17\text{-}56)$$

As in the case of P- and SV-waves, we have

$$U_i(h) = E_{ij}(h)\{E(0)\}^{-1}_{jk}U_k(0) = F_{ik}U_k(0); \qquad (17\text{-}57)$$

where

$$\begin{aligned} F_{11} &= F_{22} = \mathrm{Ch}\beta h, \\ F_{12} &= k\mathrm{Sh}\beta h/(\beta\mu), \\ F_{21} &= \beta\mu\mathrm{Sh}\beta h/k. \end{aligned} \qquad (17\text{-}58)$$

The reciprocal matrix is given by

$$\{F(h)\}^{-1}_{jk} = F_{jk}(-h), \qquad (17\text{-}59)$$

and the reciprocal product matrix is given by

$$\{\mathcal{F}\}^{-1}_{ij} = \begin{pmatrix} \mathcal{F}_{22} & -\mathcal{F}_{12} \\ -\mathcal{F}_{21} & \mathcal{F}_{11} \end{pmatrix}. \qquad (17\text{-}60)$$

The interfacial boundary conditions of continuity of tangential displacement and tangential stress give us the relationship between the displacements and stresses at the top of the half-space and those at the surface. Finally, using (17-53) we can evaluate the potential amplitudes in the half-space, obtaining

$$A^{(n)}_i = \{E^{(n)}(0)\}^{-1}_{ij}\mathcal{F}_{jk}U^{(1)}_k(0) = J_{ik}U^{(1)}_k(0). \qquad (17\text{-}61)$$

Inasmuch as there has been little use of crustal transfer coefficients for either type of S-wave, we have not worked out relations similar to (17-44). However, Haskell (1960) has worked out surface amplitude values, presenting them in the form of a contour diagram as a function of frequency and angle of incidence. In this case there is only one component of surface displacement, so phase differences as with P-SV are not appropriate. Likewise, all the energy is reflected at the free surface and back through the stack of layers.

<u>Matrix Relationships When Liquid Layers are Present</u>. For the liquid layer, we need only one potential which is given by

$$\Phi = (Ae^{\alpha z'} + Be^{-\alpha z'}). \quad (17\text{-}62)$$

The displacements and stresses are given by

$$ku_z = k\frac{\partial\Phi}{\partial z} = k(\alpha Ae^{\alpha z'} - \alpha Be^{-\alpha z'}), \quad (17\text{-}63)$$

$$t_{zz} = -\lambda k_p^2(Ae^{\alpha z'} + Be^{-\alpha z'}). \quad (17\text{-}64)$$

Symmetrizing, we have

$$\begin{aligned} ku_z &= k\alpha\left[(A-B)\mathrm{Ch}\alpha z' + (A+B)\mathrm{Sh}\alpha z'\right], \\ t_{zz} &= -\lambda k_p^2\left[(A+B)\mathrm{Ch}\alpha z' + (A-B)\mathrm{Sh}\alpha z'\right]. \end{aligned} \quad (17\text{-}65)$$

At $z' = 0$, we have

$$V_i(0) = D_{ij}(0)A_j, \quad (17\text{-}66)$$

where V_i represents the normal displacement and stress in the liquid. These two quantities are enough to specify the motion in the liquid layer. The components of V_i may be written:

$$V_i = \begin{pmatrix} ku_z \\ t_{zz} \end{pmatrix}. \quad (17\text{-}67)$$

The components of A_j are

$$A_j = \begin{pmatrix} A+B \\ A-B \end{pmatrix}, \quad (17\text{-}68)$$

and the matrix $D_{ij}(0)$ is given by

$$D_{ij}(0) = \begin{pmatrix} 0 & k\alpha \\ -\lambda k_p^2 & 0 \end{pmatrix}, \quad (17\text{-}69)$$

with inverse

$$\{D(0)\}^{-1}_{ij} = \begin{pmatrix} 0 & -1/(\lambda k_p^2) \\ 1/(k\alpha) & 0 \end{pmatrix}. \quad (17\text{-}70)$$

At $z' = h$, we have

$$V_i = D_{ij}(h)A_j; \quad (17\text{-}71)$$

where

$$D_{ij}(h) = \begin{pmatrix} k\alpha\mathrm{Sh}\alpha h & k\alpha\mathrm{Ch}\alpha h \\ -\lambda k_p^2\mathrm{Ch}\alpha h & -\lambda k_p^2\mathrm{Sh}\alpha h \end{pmatrix}. \quad (17\text{-}72)$$

The layer matrix relating the vector V_i at the top and the bottom of the liquid layer is given by

$$G_{ij} = \begin{pmatrix} \mathrm{Ch}\alpha h & -k\alpha \mathrm{Sh}\alpha h/\lambda k_p^2 \\ -\lambda k_p^2 \mathrm{Sh}\alpha h/k\alpha & \mathrm{Ch}\alpha h \end{pmatrix}. \qquad (17\text{-}73)$$

Then,

$$V_i(h) = G_{ij}V_j(0). \qquad (17\text{-}74)$$

If we wanted to find surface amplitudes and reflection coefficients for a layered structure of various liquids, we would proceed as in the previous discussions leading to (17-44) and (17-45).

Let us now see what happens when we intermix solid and liquid layers. A solid layer between two liquid layers may be generalized to cover all relevant cases. Now this solid layer may be a sandwich of several solid layers, in which case the matrix elements F_{ij} will refer to the product matrix for all the solid layers bounded by the liquid layers, one on the top and one on the bottom. The geometry and relevant equations are diagramed in Figure 17-8. The vanishing of the tangential stress on the bottom of the solid layer gives the following relation between the vector components at the top of the layer:

$$F_{41}ku_x^{(m)} + F_{42}ku_z^{(m)} + F_{43}t_{zz}^{(m)} = 0. \qquad (17\text{-}75)$$

This in turn may be solved for $ku_x^{(m)}$ giving

$$ku_x^{(m)} = -(F_{42}/F_{41})ku_z^{(m)} - (F_{43}/F_{41})t_{zz}^{(m)}. \qquad (17\text{-}76)$$

This value may then be substituted into the vector components at the bottom of the layer, giving rise to

$$\begin{aligned} ku_z^{(m+1)} &= ku_z^{(m)\prime} = (F_{22} - F_{21}F_{42}/F_{41})ku_z^{(m)} + (F_{23} - F_{21}F_{43}/F_{41})t_{zz}^{(m)}, \\ t_{zz}^{(m+1)} &= t_{zz}^{(m)\prime} = (F_{32} - F_{31}F_{42}/F_{41})ku_z^{(m)} + (F_{33} - F_{31}F_{43}/F_{41})t_{zz}^{(m)}; \end{aligned} \qquad (17\text{-}77)$$

where the primed notation indicates evaluation at the bottom of the layer. This implies that going from the top of the solid to the top of the under-lying liquid may be represented by the matrix relation

$$V_i^{(m+1)} = F_{ij}^{(m)\dagger}V_j^{(m)}; \qquad (17\text{-}78)$$

where

$$F_{ij}^{\dagger} = \begin{pmatrix} F_{22} - F_{21}F_{42}/F_{41} & F_{23} - F_{21}F_{43}/F_{41} \\ F_{32} - F_{31}F_{42}/F_{41} & F_{33} - F_{31}F_{43}/F_{41} \end{pmatrix}. \qquad (17\text{-}79)$$

It should be remembered that if the solid layer is composite, the elements of $F^{\dagger}$ are to be formed from the appropriate elements of the product matrix for all the solid layers making up the composite.

LIQUID

$$ku_z^{(m-1)'} = G_{11}ku_z^{(m-1)} + G_{12}t_{zz}^{(m-1)}$$
$$t_{zz}^{(m-1)'} = G_{21}ku_z^{(m-1)} + G_{22}t_{zz}^{(m-1)}$$

$m-1^{st}$ layer

SOLID

$$ku_x^{(m)} = -(F_{42}/F_{41})ku_z^{(m)} - (F_{43}/F_{41})t_{zz}^{(m)}$$
$$ku_z^{(m)} = G_{11}ku_z^{(m-1)} + G_{12}t_{zz}^{(m-1)}$$
$$t_{zz}^{(m)} = G_{21}ku_z^{(m-1)} + G_{22}t_{zz}^{(m-1)}$$
$$t_{zx}^{(m)} = 0$$

m^{th} layer

$$ku_x^{(m)'} = F_{11}ku_x^{(m)} + F_{12}ku_z^{(m)} + F_{13}t_{zz}^{(m)}$$
$$ku_z^{(m)'} = F_{21}ku_x^{(m)} + F_{22}ku_z^{(m)} + F_{23}t_{zz}^{(m)}$$
$$t_{zz}^{(m)'} = F_{31}ku_x^{(m)} + F_{32}ku_z^{(m)} + F_{33}t_{zz}^{(m)}$$
$$t_{zx}^{(m)'} = F_{41}ku_x^{(m)} + F_{42}ku_z^{(m)} + F_{43}t_{zz}^{(m)} = 0$$

LIQUID

$$ku_z^{(m+1)} = (F_{22} - F_{21}F_{42}/F_{41})ku_z^{(m)} + (F_{23} - F_{21}F_{43}/F_{41})t_{zz}^{(m)}$$
$$t_{zz}^{(m+1)} = (F_{32} - F_{31}F_{42}/F_{41})ku_z^{(m)} + (F_{33} - F_{31}F_{43}/F_{41})t_{zz}^{(m)}$$

$m+1^{st}$ layer

Figure 17-8. Solid (possibly composite) layer between two liquid layers, with conditions imposed by the vanishing of the tangential stresses.

Going from the top of a liquid layer to the top of a solid layer (which may be composite) is much easier. In this case, one only need use the matrix relations given by

$$V_i^{(m+1)} = G_{ij}^{(m)}V_j^{(m)}. \tag{17-80}$$

The starting relationships need a little discussion. If the topmost layer is a solid with a free surface, we will have the following relationships:

$$ku_x^{(1)'} = F_{11}ku_x^{(1)} + F_{12}ku_z^{(1)},$$

$$ku_z^{(1)'} = F_{21}ku_x^{(1)} + F_{22}ku_z^{(1)},$$

$$t_{zz}^{(1)'} = F_{31}ku_x^{(1)} + F_{32}ku_z^{(1)},$$

$$0 = F_{41}ku_x^{(1)} + F_{42}ku_z^{(1)}.$$

The last equation may be used to solve for $ku_x^{(1)}$. This value is then substituted into $ku_z^{(1)'}$ and $t_{zz}^{(1)'}$ giving

$$ku_z^{(2)} = ku_z^{(1)'} = (F_{22} - F_{21}F_{42}/F_{41})ku_z^{(1)},$$

$$t_{zz}^{(2)} = t_{zz}^{(1)'} = (F_{32} - F_{31}F_{42}/F_{41})ku_z^{(1)};$$

which is the same relation as (17-78) with $t_{zz}^{(1)}$ set equal to zero. The starting relations when the top layer is a liquid are given by (17-80) with $t_{zz}^{(1)}$ set equal to zero.

Thus the two-component vector V_i serves to represent the dynamics of an intermixture of liquid and solid layers, the matrix $F_{ij}^{\dagger}$ taking care of the transition from solid to liquid, and the matrix G_{ij} taking care of the transition from liquid to solid. However, the elements of the full 4 x 4 matrix F_{ij} (or the product matrix if the layer is composite) must be used in constructing the matrix $F_{ij}^{\dagger}$.

When we come to the half-space, we must again look at the situation. In the case of a solid (possibly composite) half-space we may write as in (17-41):

$$\begin{aligned}
A^{(n)} + B^{(n)} &= J_{11}^{\S}u_x^{(n)\S} + J_{12}^{\S}u_z^{(n)\S} + J_{13}^{\S}t_{zz}^{(n)\S}, \\
A^{(n)} - B^{(n)} &= J_{21}^{\S}u_x^{(n)\S} + J_{22}^{\S}u_z^{(n)\S} + J_{23}^{\S}t_{zz}^{(n)\S}, \\
C^{(n)} + D^{(n)} &= J_{31}^{\S}u_x^{(n)\S} + J_{32}^{\S}u_z^{(n)\S} + J_{33}^{\S}t_{zz}^{(n)\S}, \\
C^{(n)} - D^{(n)} &= J_{41}^{\S}u_x^{(n)\S} + J_{42}^{\S}u_z^{(n)\S} + J_{43}^{\S}t_{zz}^{(n)\S},
\end{aligned} \qquad (17\text{-}81)$$

where the $n^{\S th}$ layer is the topmost solid layer of the half-space. $J_{11}^{\S}$, $J_{12}^{\S}$, etc., are calculated for all the layers below the last liquid layer. Inasmuch as the half-space is a solid, then the layer immediately above must be liquid, which is why the fourth term does not appear in each of the equations (17-81). Before equations (17-81) can be used, however, we need additional information. For one example, see Section 19.4.

If the half-space is liquid (represented by the superscript n) and composite (with the uppermost liquid layer being the $n^{\S th}$ layer), we have

$$A^{(n)} + B^{(n)} = \{D^{(n)}(0)\}^{-1}_{1j} G^{\S}_{jk} V^{(n)\S}_{k},$$

$$A^{(n)} - B^{(n)} = \{D^{(n)}(0)\}^{-1}_{2j} G^{\S}_{jk} V^{(n)\S}_{k}, \qquad (17\text{-}82)$$

where $G^{\S}_{ij}$ is the layer matrix of the composite.

An alternative formulation for the inclusion of liquid layers is given by Dorman (1962).

CHAPTER 18

MANY LAYERS OVER A HALF-SPACE — BODY-WAVE OBSERVATIONS

18.1. General.

The analysis of body-wave observations in an essentially flat-layered earth would seem to be a rather simple process corresponding to the uncomplicated formulation of Chapter 15. However, a number of hampering factors creep in. In reality, the earth is neither homogeneous nor isotropic, the layers are neither flat (disregarding the earth's curvature) nor of constant thickness, and the observing surface is neither flat nor composed of competent rock to which good source-receiver coupling may be made.

The source of body-wave energy can be either man-made, giving rise to explosion seismology, or from a natural disturbance, to earthquake seismology. (At extremely short ranges, a number of other sources may also be used.) When contemplating a detailed examination of crustal layers and sometimes the upper mantle as well, explosion seismology is almost always used. This is because the source parameters, coordinates and origin time, are known precisely. However, the range is limited to a few thousand kilometers and depth of penetration to a few hundred kilometers. For greater penetration, one generally uses earthquake sources. But, at depths of penetration greater than a few hundred kilometers, one then has to take into account the curvature of a spherically-layered earth. Consequently, explosion seismology is generally associated with reflection and refraction in a flat-layered earth model, while earthquake seismology is identified with a spherical model. The discussion of body-waves in the latter model will be put off to Chapter 23.

Nonetheless, there is one case where waves from distant sources are analyzed in conjunction with a flat model. This is the case of body waves of nearly vertical incidence from below a flat-layered structure as in Section 17.3. These observations will be discussed in Section 18.5.

18.2. Reflection Seismology.

Reflection seismology is largely the domain of the exploration geophysicist. His region of interest is usually a volume a few kilometers on a side. In a simple layered model, he has to be able to distinguish primary reflections at each interface from those which have been multiply reflected between interfaces. In addition to this problem, seismic energy can be scattered from breaks in the layered structure (faults) as well as from small inhomogeneous regions. Arrivals from these virtual sources often interfere with the desired primary reflections. In recent years, powerful data-processing techniques have been developed which allow for the correction and correlation of literally hundreds of thousands of seismic wavelets. It is not in the province of this book to discuss these matters. The reader is referred

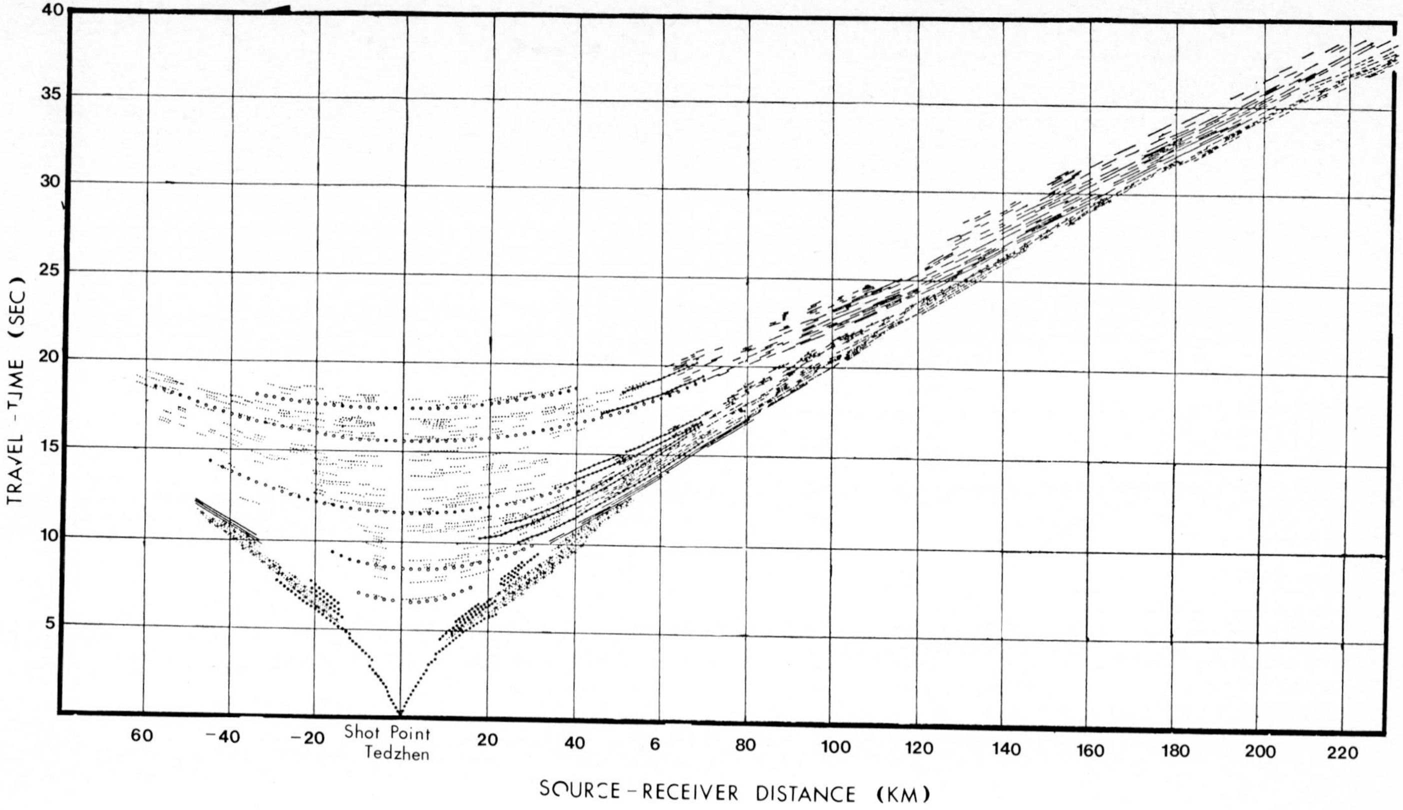

Figure 18-1. Example of Deep Seismic Sounding work in the USSR. Travel-time curves for reflected and refracted waves. Recorded in Turkmen SSR. (After Beloussov et al, 1962.)

to standard texts, e.g., Grant and West (1965), Dobrin (1976), and Telford et al (1976) for an elementary treatment; and to Robinson (1967), Kanasewich (1973), and Claerbout (1976) for more advanced discussions.

Observations. In regions where seismic noise is at a low level, reflected waves can be defined as coming from interfaces deep within the crust and sometimes into the upper mantle. Particularly successful in this regard have been seismologists from the U.S.S.R. They have been involved with an extensive program known as deep seismic sounding (DSS) starting in the late 1940's. A comprehensive discussion of techniques and results may be found in Kosminskaya (1971). Figure 18-1 shows some of the deeper penetrations achieved by Beloussov et al (1962). Reflections may be seen at times close to twenty seconds after the initial shot. Similar work, though not on as large a scale, has been carried out in Canada (Hall and Hajnal, 1973) and in Europe (Dohr and Meissner, 1975).

As can be seen from Figure 18-1, the coherence between seismic wavelets generally does not persist beyond a few kilometers. This is not surprising since geological maps indicate that the earth's crust is quite heterogeneous on this scale. The predominant frequencies involved with DSS are 5-15 HZ. With velocities of 5-8 Km/Sec, this indicates a resolution of a few hundred meters. At these frequencies, the nature of seismic wavelets will be determined by an integrated effect over all the lithologic variation within this few hundred meter interval. This type of response amounts to a constructive and destructive inteference phenomena. Strong reflected wavelets can arise in one place while at another with only slightly different lithologic character, the reflection will be much weaker. Despite these difficulties, the deepest coherent reflections seen on Figure 18-1 came from the vicinity of the Mohorovicic discontinuity (the boundary between the crust and the mantle) which in this region is at a depth of slightly more than 50 Km.

Interpretation. Standard techniques involve the use of (17-8) or (17-12) to determine the velocities corresponding to the medium above the nth reflecting interface. Once the layer velocities have been established, the use of (17-13) allows the determination of a velocity-depth structure. The use of apparent velocities allows some discrimination of multiply reflected events whose travel-paths remain higher in the layered structure than a primary reflection arriving at the same time. The situation is sketched in Figure 18-2. On the other hand, multiples occurring near the primary reflector are much more difficult to identify because their apparent velocity is almost the same as that of the primary reflection.

For scattered (diffracted) events, there is also some velocity discrimination. For the one layer case shown in Figure 18-3 we find that the travel-time can be written as

$$T_p^D = \{H + (x^2 + H^2)^{\frac{1}{2}}\}/v_p = \frac{2H}{v_p}(1 + \frac{x^2}{4H^2} + \dots). \quad (18\text{-}1)$$

Then

$$T_p^D - T_p^o \simeq x^2/(2v_pH) = x^2/(v_p^2\, T_p^o). \qquad (18\text{-}2)$$

A similar analysis of the travel-time of the primary reflected arrival, equation (15-1), would lead to

$$T_p^R - T_p^o \simeq x^2/(4v_pH) = x^2/(2v_p^2\, T_p^o). \qquad (18\text{-}3)$$

Thus the apparent velocity of the scattered event is less than that of the primary reflection. Note that for other orientations of source, scatterer and receiver, the results will be somewhat different.

Additional problems in intrepretation occur because both primary reflections and scattered arrivals come from points not on the vertical plane between the source and receiver. That is, the three-dimensional nature of the earth must be considered. Additional information must be obtained by parallel and crossing seismic reflection lines. The large-scale computer processing of the vast amount of data obtained in three-dimensional interpretation is in its developmental stages at the present time.

18.3. Refraction Seismology.

Refraction seismology is used much less frequently by the exploration geophysicist. However, it is the primary means of investigation for looking at the earth's crust and upper mantle. Because of environmental problems, large-yield sources have generally been set off in large bodies of water. Use has been made also of large explosions such as quarry blasts; sometimes an occasional nuclear test. Consequently, the coverage is not uniform and one has to make interpretations with rather less data than one would like. The regions considered are generally areas one or two hundred kilometers deep and of the order of a thousand kilometers in length. Regional

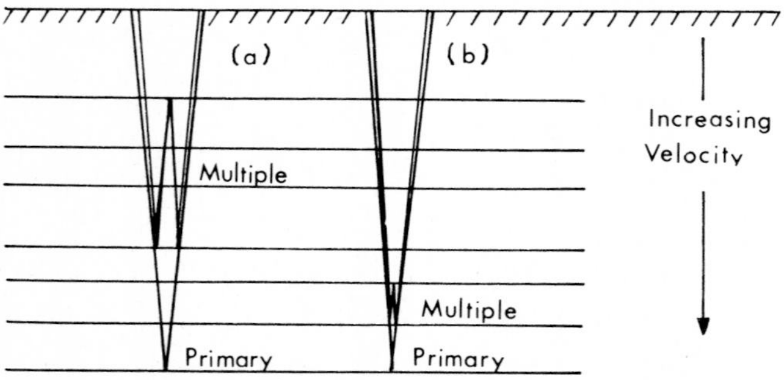

Figure 18-2. Multiply reflected seismic events. Case (a) can be distinguished rather easily by velocity analysis since its travel-path remains in the shallower portion of the section where the velocity is slower. Case (b) is more difficult to discriminate against in that both primary and multiple events have similar velocities.

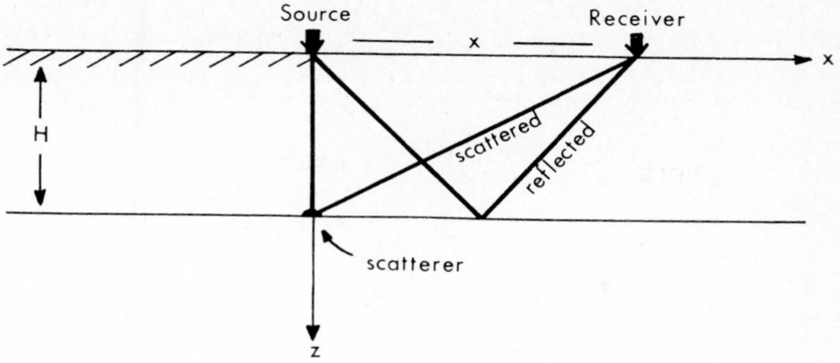

Figure 18-3. Geometrical relationships for scattered (diffracted) events

coverage is attained by connecting segments. Several collections of observations and interpretation are available. Three American Geophysical Union Monographs (#10, Steinhart and Smith, 1966; #12, Knopoff, Drake and Hart, 1968; and #13, Hart, 1969) as well as one special volume of Tectonophysics (Mueller, 1973) are particularly useful. A slightly older volume (Steinhart and Meyer, 1961) describes the work of a group at the University of Wisconsin during 1956 - 1959. This latter work is particularly valuable because of a unified presentation plus a review of earlier work. We might note here that sophisticated data-processing is generally not applicable because of the fairly wide separation of receiving units in a refraction line. However, there has been considerable interest in amplitudes in order to determine detailed structure as well as the lossy properties of crustal and upper mantle materials.

Observations. In a series of papers (Massé et al, 1972; Massé, 1973, 1974, and 1975; and Massé and Alexander, 1974) Massé and others have demonstrated the power of refraction seismology to delineate the structure of the upper part of the earth. Figure 18-4 is taken from Massé (1973) and is a refraction profile across the Canadian Shield. The scale on the left is what is known as reduced travel-time, $T - \Delta/v_o$. This new variable allows the use of an expanded time scale for the figure in that the large time interval between the shot and the arrival of P can be subtracted out by a judicious choice of the velocity v_o.

Interpretation. In the analysis of refraction data, one uses the slopes of the straight line-segments of Figure 17-3 to determine the layer velocities. Then equation (17-18) is solved to find the successive layer thicknesses. Assuming that there is no ambiguity in interpreting the line-segments and intercept times, Steinhart and Meyer (1961) present the necessary statistical formulation to determine slopes, intercept times, and crossover distances. They also discuss the effect of dipping layers on these calculations. Borcherdt and Healy (1968) also discuss the statistical problem and apply the Steinhart and Meyer (ibid) formulation to 18 seismic refraction profiles in the Basin and Range province of the Western United States. This paper

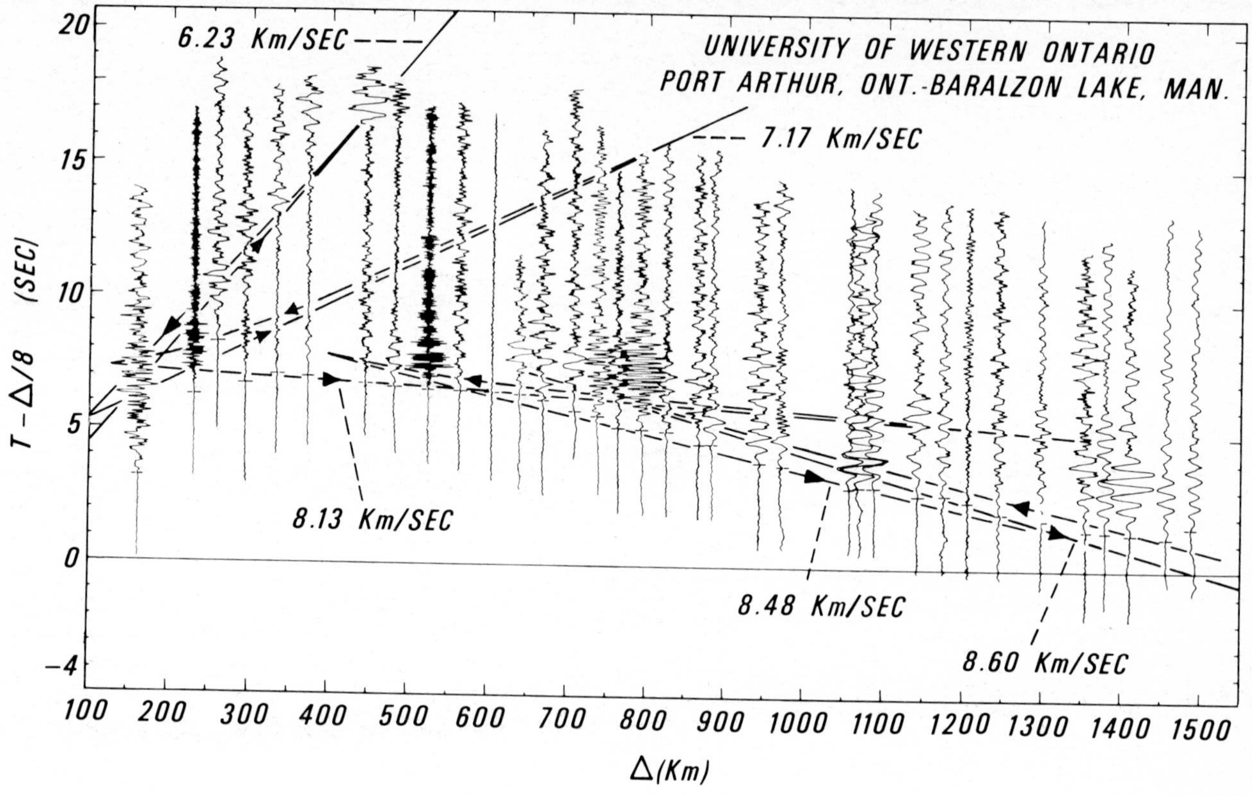

Figure 18-4. Example of explosion seismic traces in the Canadian Shield. Refraction and reflection lines superimposed on a record section adapted from Warren et al (1968). (After Massé, 1973.)

gives a good idea of the spread in data over a single physiographic province. Dowling (1970) gives some additional considerations regarding dipping layers. Berry (1971) considers the hidden-layer problem and the more general problem of unresolved structure in the lower part of each refracting layer.

A word of caution should now be introduced. In using real refraction travel-time data as in Figure 18-5, one can fit the observed points by a number of straight-line combinations — possibly even a curved line segment. The number of lines, and thus layers, is quite often left up to the person making the interpretation of the data. This number may or may not have real significance. Continuously varying

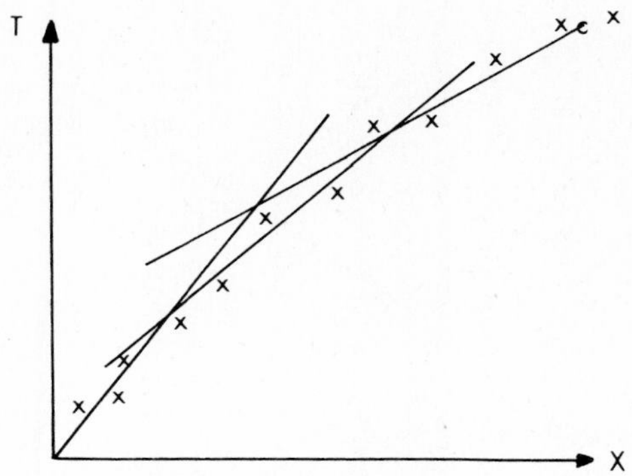

Figure 18-5. Hypothetical travel-time diagram of first arrivals from a refraction profile. Either two or three line-segments may be fitted to the data.

velocity structures give continuously varying travel-time curves. Fitting them by straight lines may be entirely artificial, the resulting layered structure having no physical meaning. It may only be a discrete approximation to a continuously varying structure. A paper by Knopoff and Teng (1965) discusses some points in the fitting of ambiguous travel-time data.

<u>Models</u>. Those areas of the earth's surface most densely covered by refraction profiles include Western Europe, the United States and the Union of Soviet Socialist Republics. One method of presentation for crustal models is the <u>fence-diagram</u> where the velocity sections are isometrically projected above a regional map. A number of such diagrams are given by Healy and Warren (1969), one of which is reproduced as Figure 18-6. The figure shows both velocities and crustal thicknesses in a combined presentation with the linkages between sections clearly shown. One may get a range of variability of continental crustal velocity sections from this diagram, although the Western United States is hardly typical continental crust and upper mantle.

"Typical" continental and oceanic crustal models are shown in Figure 18-7. One must bear in mind that the actual structures are quite variable. The continental crust can thicken considerably to 70 km or so under the highest mountain ranges and

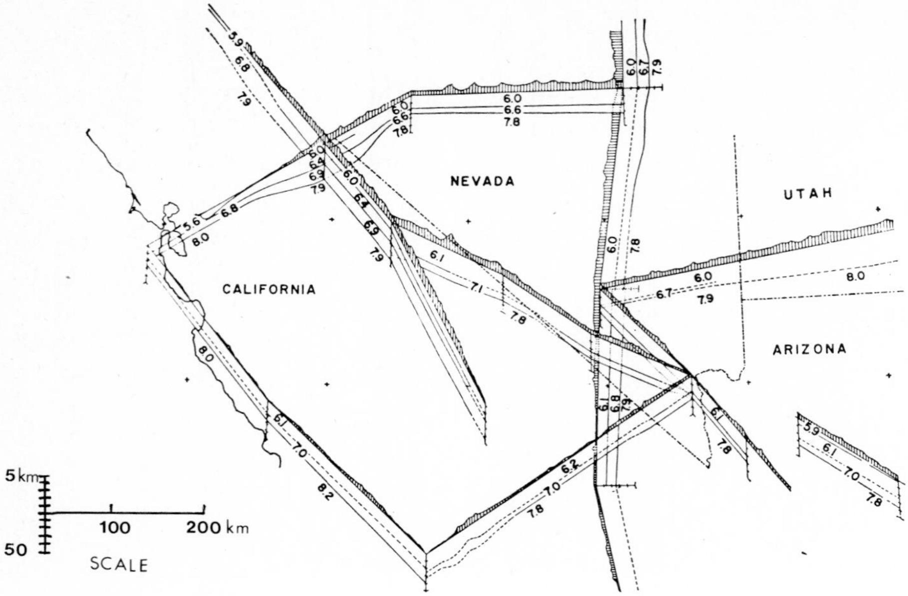

Figure 18-6. Fence diagram in the western portion of the United States. Numbers refer to the average velocity (km/sec) of compressional waves in the crustal and upper mantle layers; layer thicknesses show no vertical exaggeration. Altitude above sea level is shaded and has a 5x vertical exaggeration. (After Healy and Warren, 1969.) Copyrighted by AGU.

to less than 30 km in tectonic regions. The oceanic crust is more uniform; however, anomoulous regions occur at both oceanic ridges and at active continental margins where the deep trenches lie. At the former, there appears to be an intermixing of mantle and oceanic layers. At the latter there seems to be a mixing of mantle, oceanic layer, thick terrigenous sediments and some of the continental granitic layer. Seismic profiles for both ridges and active continental margins abound, but generally the well defined layer boundaries become dotted lines interspersed with question marks.

Rays refracted through the granitic layer are known as P_g, through the basaltic layer as P^*, and through the upper mantle as P_n. A similar notation holds for the more difficult to identify S-phases. The transition defined by P^* between the granitic and basaltic layers seems to be rather sharp in some places (particularly in Western Europe) and diffuses to a broad transition zone elsewhere. A multi-layered crust seems appropriate at other places. Closs (1969) reviews the evidence concerning this. The Mohorovicic discontinuity between crust and mantle appears to be

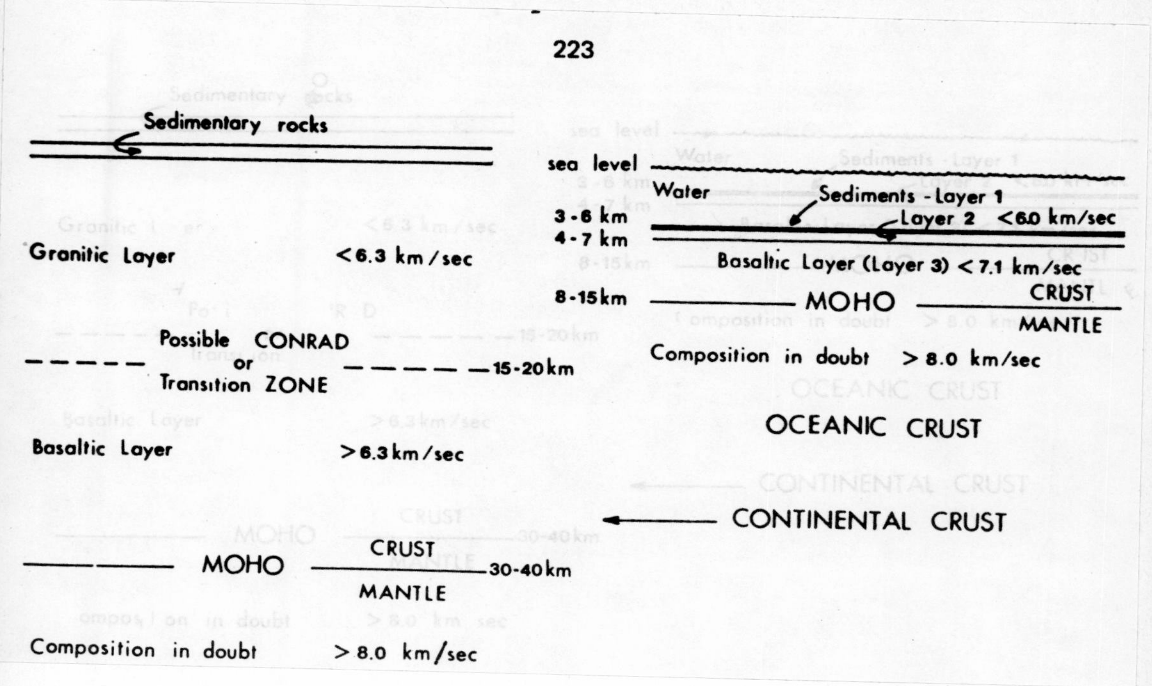

Figure 18-7. Schematic crustal sections for continental and oceanic regions of the world.

almost everywhere, both under oceans and continents. Exceptions seem to be under tectonic regions — rift zones on continents, ridges in oceans, and the deep trench areas of active continental margins. In addition to these widespread features, Landisman et al (1971) give evidence in many places for a low-velocity layer within the continental crust.

A summary of refraction profile data appears in McConnell et al (1966). Cummings and Shiller (1971) present a list of references including much refraction data since 1950, which they used to prepare a world-wide map of crustal thickness. Toksöz et al (1969) give a spherical harmonic synthesis of sixth order and degree showing a smoothed world-wide crustal thickness. Correlating the large amount of data available for the United States, Herrin and Taggart (1968) show the regional variation of P_n velocity; here presented as Figure 8-8. Note the large number of areas in the Western Third of the United States where $P_n < 8.0$ km/sec. This region is one of the possible places that the Moho may not exist. If the 7.5 - 7.8 material represents mantle rocks, it is most unusual. More likely, this region represents a mixture of crustal and mantle material. This widespread area, the Basin and Range Province, has been the subject of much intensive study in order to define its anomalous character.

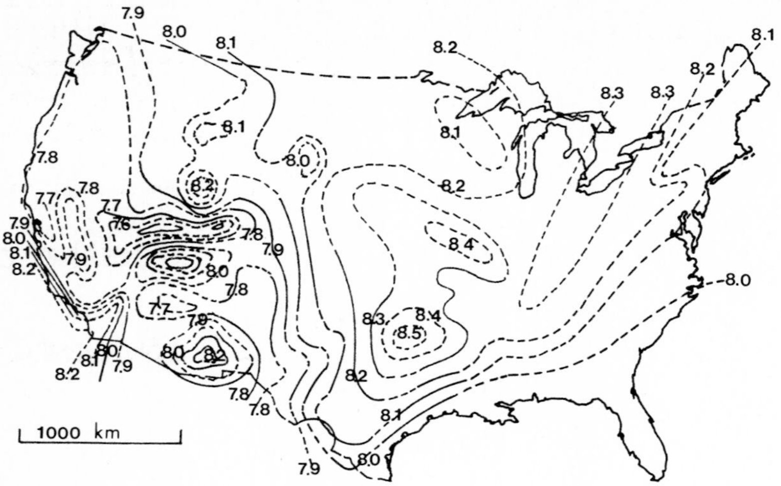

Figure 18-8. Apparent P_n velocities in the United States. Velocities in km/sec. Regions with $P_n < 8.0$ km/sec represent tectonically active areas. (After Herrin and Taggert, 1962.)

18.4. Time-Term Analysis.

An alternative method of relating large quantities of refraction data over regions greater than 100 kilometers square has been the systematic use of time-terms. These are quantities associated with that part of the travel path above the critically refracting interface. We have seen (pp. 158-159) that their value can depend on structure, for we can rewrite (15-12) as

$$T_d = \frac{H_d \mathrm{Cos}\theta_c}{v_{p1}} + \frac{H_u \mathrm{Cos}\theta_c}{v_{p1}} + \frac{x\mathrm{Cos}\alpha}{v_{p1}} ; \tag{18-4}$$

where $H_u = H_d + x\mathrm{Sin}\alpha$, α is the dip of the refracting interface, and x is the source-receiver distance. If the seismometers are laid out in a linear array over several values of x, one can compute values for the individual parameters. On the other hand, if we have many source locations, each accompanied by several receiver locations, a different analysis is needed. Time-term analysis is associated with a system of equations which represents the network of shot-points and seismographs:

$$T_{ij} = a_i + b_j + \Delta_{ij}/v. \tag{18-5}$$

Here, a_i and b_j are time-terms associated with the source and receiver points respectively, Δ_{ij} is the map distance between them and v is the refractor velocity.

The resemblance to (18-4) is obvious. Approximations made are that H_d and H_u are slant heights while a_i and b_j are referred to the vertical and that $\cos\alpha$ is approximated by unity. If the refractor has much topographic relief, these approximations can break down and require a second approximation to obtain a meaningful result.

Early work using time-terms seems to have centered in geophysical exploration in the 1930's. A brief account of its historical development can be found in Wilmore and Bancroft (1960). This paper plus an earlier one by Scheidegger and Wilmore (1957) seem to have been the earliest attempts at systematically interpreting large amounts of such data. The Lake Superior Seismic Experiment was the first large scale application of the time-term method. Berry and West (1966), and Smith, Steinhart, and Aldrich (1966) continue along lines suggested by the earlier work cited above, but using a more modern matrix notation. These two early interpretations were subject to disagreement by some. O'Brien (1968), Morris (1972), and Bamford (1973) have given alternative interpretations and methods of analyzing the original data.

If we had only three points, each of which served as source and receiver, the observations would look like:

$$\begin{aligned} T_{12} &= a_1 + a_2 + \Delta_{12}/v, \\ T_{23} &= a_2 + a_3 + \Delta_{23}/v, \\ T_{31} &= a_3 + a_1 + \Delta_{31}/v. \end{aligned} \qquad (18\text{-}6)$$

Assuming that <u>v</u> was determined by a linear array of receivers at one of the three points, we could then solve the three equations (18-6) for the three unknowns, a_1, a_2, and a_3. If there is a large collection of data, we find that there are two approaches to treating a system like (18-5). The first considers all observations as independent and uses a "least squares" method to solve for the unknown values of a_i and b_j. This least squares technique has a number of problems associated with it including large amounts of data and a rather unstable inverse. To deal with these problems, Mereu (1966) and Bamford (1973, 1976) suggest an iterative technique. This seems to work out well. An alternative method of achieving stability is the introduction of a mathematical representation for the refracting interface of variable depth. This surface may be given by either polynomials or Fourier series. The continuously variable time-terms associated with this surface have been called <u>delay-time-functions</u>. These concepts were introduced by Raitt et al (1969). Bamford (1976) contrasts the delay-time-function approach with a grouped-data method he proposes, called <u>MOZAIC</u>. This latter method reduces the independence of the individual observations by grouping them in a systematic manner. Additional comments on the errors involved have been given by Reiter (1970) and Barr (1971).

Once the time-terms are obtained by one of the above methods, one then has to make an additional interpretation. Recalling the multi-layer form of the intercept time (17-18), we would have

$$a_i^{(n)} = \sum_{k=1}^{n} \frac{h_k(v_{n+1}^2 - v_k^2)}{v_{n+1} v_k} ; \tag{18-7}$$

where _i_ stands for the ith station and _k_ for the kth layer in a stack above the nth interface. Most times a further analysis for velocity-depth structure calls for data that is not available. There are two alternatives. One is to convert the time-terms to thicknesses by using an average crustal velocity or some other simple model. The other is to let the ambiguity remain and look only for general relationships. The reader may mentally make the conversion to a variable crustal thickness. Figure 18-9 presents results obtained from Western Germany by Bamford (1976). If we assume an average time-term of 3.0 sec., then the variability is ± 20%. Unless there are large compensating irregularities in crustal velocities, crustal thicknesses should vary by approximately the same amount. Similar variability is reported for the Pacific Ocean floor (Morris et al, 1969).

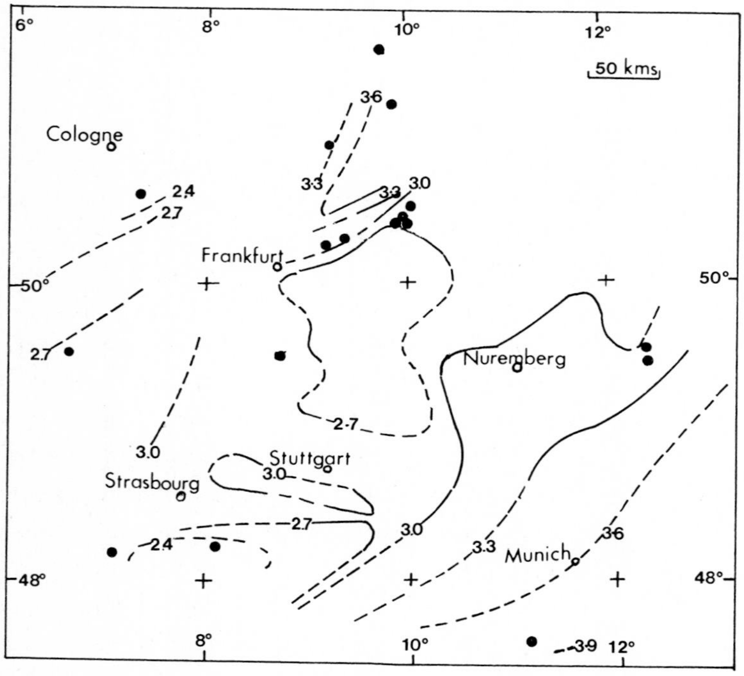

Figure 18-9. Time-term contours for P_n constructed from a MOSAIC analysis (in seconds). ● Shotpoints; o Cities. (After Bamford, 1976.)

This may be contrasted to the variability in Figure 18-8, where we have a P_n velocity of 8.0 ± 0.5 km/sec or a little more than 6%. Variations of similar magnitude in the structure of the crust and upper mantle have been noted on even larger scale figures in conjunction with studies of the earth's properties under large seismic arrays. We shall have more to say on this in Chapter 23. Thus simple crustal models — as in Figure 18-7 — have only a "conceptual" reality. In applications to general geophysics such models have numerous useful applications that are not significantly harmed by variations such as we have seen in Figure 18-8 and 18-9.

18.5. Crustal Transfer Functions

In Section 17.3 we had developed the P-SV crustal response functions, T_x^P etc. However, the actual frequency-domain response at the free surface will be a product of the source spectrum, medium transmission characteristics, and the crustal response. Haskell (1962) recognized this and calculated phase differences between vertical and horizontal components of surface displacement. However, it is well known (possibly not so well known then) that phase-spectra are considerably less stable than amplitude spectra in the presence of random components such as noise. Consequently, phase differences have not seen much use in determining crustal structure. Phinney (1964) proposed to use the amplitude ratio u_z/u_x as an indicator of crustal structure. In terms of our previous notation, this can be written as

$$\frac{|u_z|}{|u_x|} = \frac{|T_z^P|}{|T_x^P|} = \frac{|P|}{|Q|} = \frac{|J_{31} + J_{41}|}{|J_{32} + J_{42}|} ; \qquad (18\text{-}9)$$

where we have used (17-44) and (17-42). He then worked out a number of simple crustal models as well as compared observed data with theoretical models. This involved a fit to "oceanic" structure at Bermuda and to "continental" structure at Alburquerque. Recognizing the possibilities in this analysis, Fernandez (1967) constructed "master curves" for the response and transfer functions of one- and two-layered systems as well as phase differences between vertical and horizontal components of motion. Hannon (1964) has constructed theoretical seismograms showing the reverberatory character of the seismic record for a relatively simple seismic pulse. Rogers and Kisslinger (1972) used theoretical seismograms to evaluate crustal response functions for models with dipping interfaces. They found that errors in crustal thickness were less than 5% for dips up to 25%. McMechan (1976a) used theoretical seismograms to determine crustal transfer functions for angles of incidence deviating considerably from the vertical. Such calculations are appropriate for body-waves arising from epicentral distances of 10°-35°.

Probably the most systematic study and use of crustal transfer functions has been by Kurita. Starting in 1969 he has considered many possibilities to determine crustal structure. He has examined amplitude ratios for P, SV, and $(SV/SH)_{horizontal}$ as well as phase differences. In his later papers (Kurita; 1973a, 1973b, 1974, and

1976) he has reccommended the combined use of crustal transfer functions, surface-wave dispersion, travel-time residuals, and synthetic seismograms to determine a unambiguous crustal structure. Figure (18-10) shows a comparison of model transfer functions with observed spectral ratios at three locations in the Southeastern United States. Additional references to the usage of crustal transfer functions can be found in Båth (1974, p. 294).

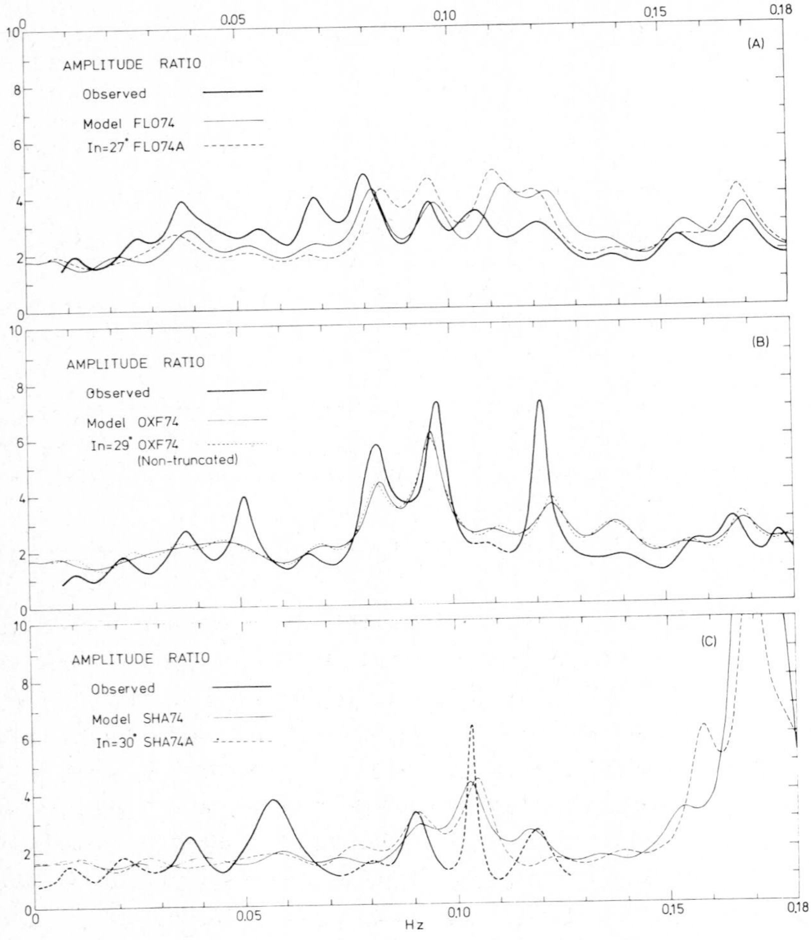

Figure 18-10. Crustal transfer functions determined at three locations in the Southeastern United States. Truncated CTF's (Kurita, 1973a) have been used in all cases but one. (After Kurita, 1976.)

18.6. Advanced Inversion Techniques.

Synthetic Seismograms. With the advent of large computers, the way was opened for a better comparison of theory and observation. The computations necessary to plot the displacements corresponding to Lamb's problem (Figures 11-6 and 11-9) would have been prohibitive on a mechanical desk calculator of 1950's vintage. However, on 1970's electronic computers they are trivial. Not only does the computer do the calculations, but it also oversees the plotting. While the seismograms corresponding to Lamb's problem might be called synthetic, I should prefer to use the term (as in previous chapters) theoretical. That is, a theoretical seismogram would be the computed displacements (strains, stresses, etc.) from a "type" problem such as Lamb treated. This then would allow the term synthetic seismogram to be used for constructs of a more investigative nature. There seem to be three major uses for synthetic seismograms:

(a). The construction of a more refined model.

(b). The determination of the relative importance of various phases resulting from a complicated model.

(c). The determination of what portions of the observations are due to a given partial model.

We will now consider these uses in a bit more detail.

Use (a) can best be illustrated by the work of Helmberger and his colleagues who have used synthetic seismograms to refine the structure of the crust and upper mantle. Of particular relevance to flat-layered media are a study of the Bering Sea area (Helmberger, 1968) and an investigation of the lateral variations in the Western United States and Northern Mexico (Helmberger, 1972). Layer properties were varied until a good agreement with observations was obtained. In the first paper, the interaction of reflected and refracted energy to form head waves was found important. In the second, the interaction of head waves and PL phases was determined by an appropriate model.

There are two cases of importance under item (b). We have already mentioned the synthesis of modes from rays (p. 191). This is an an example of considering the relative importance of the many individual reflections contributing to the surface wave modes. Continuing the work of Pekeris et al (1965), Abramovici (1970) constructs seismograms for layered media giving both vertical and horizontal components of motion. The second area of importance occurs in exploration seismology. Here the problem lies in the identification of "dependable" primary reflections. That is, reflections which can be followed over several kilometers and also be identified with particular reflecting horizons. The reflection seismogram taken at vertical incidence is the sum of primary events, multiple reflections, and scattered energy. By constructing models based on measured properties (well-log data) of the few thousands of feet of section of interest and then calculating synthetic seismograms, the major reflectors can be correlated with the appropriate primary reflections. Examples of this usage can be found in Peterson et al (1955), Wuenschel (1960), and Trorey (1962).

Category (c) is most applicable to geometries that cannot be solved exactly. One example is the interaction of seismic waves with edges, corners, and points. One can generally treat all but the diffraction effects. By comparing a synthetic seismogram with a record obtained from a physical model, the relative nature and importance of the diffraction effects can be measured. Pertinent to layered media are Kane (1966), Ishii and Ellis (1970a and 1970b), Rogers and Kisslinger (1972), Hong and Helmberger (1977), and Langston (1977); all of whom were considering multiple reflections in media with dipping layers.

It should be obvious that computing synthetic seismograms will take large amount of computer time. First to evaluate the impulse response at a useful number of discrete time points, and second to repeat this for many hundreds of wave arrivals. Early investigators used a ray expansion of exact solutions (similar to that found on p. 188), evaluating a few hundred rays. One formulation used by a group at the Weizmann Institute in Israel is given by Abramovici and Alterman (1965). Müller (1970) analyzes a number of approximations and finds that the most difficult task is to accurately model continuously varying transition zones with a small number of homogeneous layers. Fuchs and Müller (1971) show that this problem can be attacked by using matrix-methods (with time-harmonic excitation) on a large number of layers to get frequency dependent reflection coefficients. They then perform a Fourier synthesis to return to the time domain. This is known as the reflectivity method. Alterman and Loewenthal (1972) give a general treatment of generating synthetic seismograms, but put most emphasis upon finite-difference methods. Wiggins and Helmberger (1974) present methods used by Helmberger and his colleagues in constructing refined crustal models. These have been called the generalized ray method. In this paper, some consideration was given to minimizing the number of arrivals actually computed. Hron and others (Hron, 1972; Kanasewich et al, 1973; Hron et al, 1974) as well as Kennett (1974b) have put a great deal of effort into solving this problem.

Inversion of Reflection Seismograms. If a high-quality seismogram were obtained from normally incident reflections in a flat-layered medium, Robinson and Treitel (1978) show how to procees the record to obtain all the reflection coefficients as a time sequence. (Note that the velocity — layer thickness ambiguity still remains.) This work was based on earlier work by these two authors. Relevant earlier studies are given by Goupillaud (1961), Kunetz and D'Erceville (1962), Sherwood and Trorey (1965), and Claerbout (1968). An alternative treatment is given by Hron and Razavy (1977) who used methods from theoretical physics for the solution of scattering problems. These sophisticated techniques have a big "IF", however, in that the real earth generally fails to meet the specifications listed above. One informative sidelight of the Robinson and Treitel (ibid) analysis is that the authors were able to show that in a randomized sequence of reflecting layers (for a precise definition, see their paper) the multiple reflections mutually tend to cancel while the primaries tend to amplify. If this had not been the case in nature, exploration seismology would have had a different history.

Inversion of Refraction Seismograms. Attempts to improve the determination of structure have proceeded along a number of lines. Berry and West (1966a) have derived relations for head-wave amplitudes and have given a graphical presentation for the results for multilayered media. Cerveny and Ravindra (1971, pp. 188-224) have devoted a whole chapter to the head-wave problem in a multilayered earth. Mueller and Landisman (1971) have advocated the use of first and later arrivals from both reflected and refracted arrivals in a "Unified Method" of crustal interpretation. Braile and Smith (1975) have added the use of synthetic seismograms to this combined method. Braile (1973) has constructed a linearized inversion scheme for determining crustal structure from a combination of reflected and refracted travel-time curves.

MANY LAYERS OVER A HALF-SPACE – MODE THEORY

19.1. Rayleigh-Waves From Surficial Stress Distributions.

As in Chapter 17, we shall follow the exposition of Harkrider (1964) with a few changes in notation necessary for compatability with previous chapters. The reader should not try to interchange formulas here with those of Harkrider. The relations derived previously for rays in a layered medium (17-41) give us all we need to solve this problem. The surficial sources are prescribed and given by

$$t_{zz}^{(1)}(0) = t_{zz}^{o}(k)e^{i(kx - \omega t)}, \qquad t_{zx}^{(1)}(0) = t_{zx}^{o}(k)e^{i(kx - \omega t)}. \tag{19-1}$$

For convenience, we will assume that the stresses are δ-functions in x so that $t_{zz}^{o}(k) = -Z/2\pi$ and $t_{zx}^{o}(k) = -X/2\pi$. Inasmuch as there will be only downward traveling waves in the half-space, $A^{(n)}$ and $C^{(n)}$ will be zero. We can then eliminate $B^{(n)}$ and $D^{(n)}$ by adding equations (17-41) in pairs and obtain

$$0 = Kku_x^{(1)}(0) + Lku_z^{(1)}(0) + Mt_{zz}^{o} + Nt_{zx}^{o}, \qquad 0 = Pku_x^{(1)}(0) + Qku_z^{(1)}(0) + Rt_{zz}^{o} + St_{zx}^{o}. \tag{19-2}$$

Solving for $u_x^{(1)}(0)$ and $u_z^{(1)}(0)$, we find that

$$ku_x^{(1)}(0) = -\frac{(MQ - LR)t_{zz}^{o} + (NQ - LS)t_{zx}^{o}}{KQ - LP},$$

and

$$ku_z^{(1)}(0) = -\frac{(KR - MP)t_{zz}^{o} + (KS - NP)t_{zx}^{o}}{KQ - LP}. \tag{19-3}$$

Restoring the spatial dependence by integrating over all wave-numbers k, we obtain

$$u_x(x,0,t) = \frac{1}{2\pi}\int_{-\infty}^{\infty} \frac{(MQ - LR)Z + (NQ - LS)X}{KQ - LP} e^{i(kx - \omega t)} \frac{dk}{k},$$

and

$$u_z(x,0,t) = \frac{1}{2\pi}\int_{-\infty}^{\infty} \frac{(KR - MP)Z + (KS - NP)X}{KQ - LP} e^{i(kx - \omega t)} \frac{dk}{k}. \tag{19-4}$$

Remembering that

$$J_{ik} = \{E^{(n)}(0)\}_{ij}^{-1}F_{jk},$$

and that K, L, ... R, S are combinations of the J_{ij}'s, we see that branch cuts to maintain single valuedness of the integrand will only be necessary for the square roots associated with the half-space, $\alpha^{(n)}$ and $\beta^{(n)}$. This is because each element

F_{pq} of the product matrix F_{jk} is even in both α and β. The only multivalued terms are in the factor

$$\{E^{(n)}(0)\}^{-1}_{ij}.$$

Consequently the statement of Chapter 16 to the effect that the only branch cuts needed for integrals like (19-4) are those associated with the half-space has been proved.

The roots of KQ - LP are the Rayleigh roots as in the single layer case, but they exhibit a more complicated dispersion. Haskell (1953) has shown in general that for long wavelengths, the ordering of the layers is immaterial, and that for short wavelengths, the layering matrix F_{ij} can be factored so as to display the factors for interface (generalized Stoneley) waves at each interface. The reader is referred to his paper for specifics.

With these considerations out of the way, we can then write

$$u_x(x,0,t) = \text{X-branch line integral} + i \sum_{j=0}^{m} \left. \frac{(MQ - LR)Z + (NQ - LS)X}{k\partial(KQ - LP)/\partial k} \right|_{k=k_j} e^{i(k_j x-\omega t)}, \tag{19-5}$$

$$u_z(x,0,t) = \text{Z-branch line integral} + i \sum_{j=0}^{m} \left. \frac{(KR - MP)Z + (KS - NP)X}{k\partial(KQ - LP)/\partial k} \right|_{k=k_j} e^{i(k_j x-\omega t)}.$$

The number m can vary depending upon the path chosen for the branch line integrals, and the propagation can include normal, leaking, or standing modes according to the locations of the roots k_j.

At the roots of the generalized Rayleigh denominator we have that

$$KQ\Big|_{k=k_j} = LP\Big|_{k=k_j}. \tag{19-6}$$

With judicious use of this relation, equations (17-39), and the definitions (17-40) and (17-42), it can be shown that

$$(K/L)\ (LS - NQ) = (KS - NP) = (LR - MQ) = (L/K)\ (KR - MP), \tag{19-7}$$

all terms being evaluated at $k = k_j$.

Using these relationships, we can rewrite (19-5) somewhat more simply:

$$u_x(x,0,t) = \text{X-branch line integral} - i \sum_{j=0}^{m} \left. \frac{L}{K} \frac{(KR - MP)(Z + LX/K)}{k\partial(KQ - LP)/\partial k} \right|_{k=k_j} e^{i(k_j x-\omega t)}, \tag{19-8}$$

$$u_z(x,0,t) = \text{Z-branch line integral} + i \sum_{j=0}^{m} \left. \frac{(KR - MP)(Z + LX/K)}{k\partial(KQ - LP)/\partial k} \right|_{k=k_j} e^{i(k_j x-\omega t)}.$$

The common factor,

$$E^R_j = \left. \frac{i(KR - MP)}{k\partial(KQ - LP)/\partial k} \right|_{k=k_j}, \tag{19-9}$$

can be looked at as a modal excitation coefficient for the jth normal or leaking mode. It is a characteristic of the layer cake structure, and as we shall see later, occurs regardless of the source type. For the standing modes, one must pick up roots in pairs and formulate an analogous expression. The ratio L/K represents the relative efficiencies of vertical and horizontal forces in exciting Rayleigh-waves. It also appears as -L/K in the ratio of $u_{x_j}(x,0,t)/u_{z_j}(x,0,t)$, i.e., the ratio of x- and z-displacements due to either vertical or horizontal excitation. The negative sign comes from the not-quite reciprocal nature of the interchange as discussed on page 123. See Harkrider and Anderson (1966) for an alternative evaluation of (19-9) in terms of energy integrals evaluated over the layers.

19.2. The Rayleigh-Wave Problem for Sources at Depth.

If the sources are not stress distributions on the surface, then the problem becomes more complicated but still tractable. The analysis below is also adapted from Harkrider (1964), and reformulated in our notation. First of all, we will diagram the problem. The source distribution is to lie on the plane z = D. We will introduce a fictitious interface at this level in the mth layer. The relationships are diagramed in Figure 19-1. The source distribution on the plane z = D will be

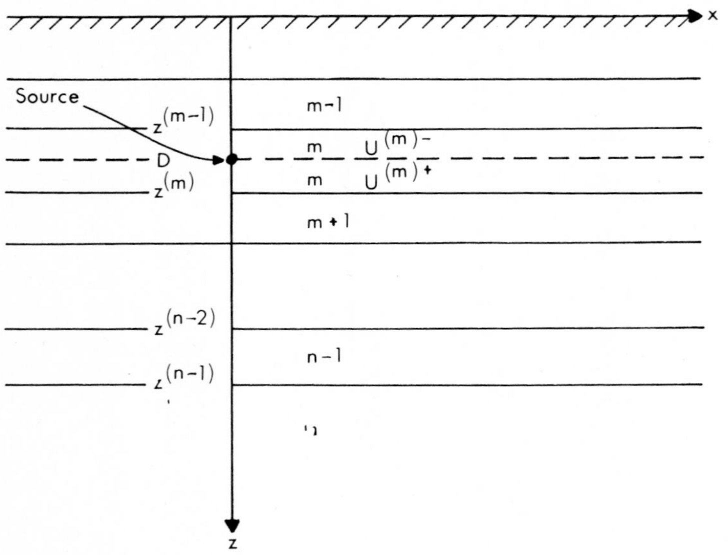

Figure 19-1. Layer relationships for a buried source. Source is at depth D in the mth layer.

defined as a jump in the vector $U_i^{(m)}$, where for the time being, we will again suppress the common factor $e^{i(kx - \omega t)}$. We write

$$U_i^{(m)+} = U_i^{(m)-} + \delta U_i^{(m)}. \qquad (19\text{-}10)$$

Now the vector $U_i^{(m)+}$ can be continued downward by the following equation:

$$U_i^{(n-1)}\{z^{(n-1)}\} = \mathcal{F}^{+}_{ij} U_j^{(m)+}(D), \tag{19-11}$$

where

$$\mathcal{F}^{+}_{im} = F_{ij}^{(n-1)} F_{jk}^{(n-2)} \;\cdots\cdots\; F_{lm}^{(m)}\{z^{(m)} - D\}. \tag{19-12}$$

In a similar way the surface values of displacement and stress can be carried downward giving

$$U_i^{(m)-}(D) = \mathcal{F}^{-}_{ij} U_j^{(1)}(0); \tag{19-13}$$

where

$$\mathcal{F}^{-}_{im} = F_{ij}^{(m)}\{D - z^{(m-1)}\} F_{jk}^{(m-1)} \;\cdots\cdots\; F_{lm}^{(1)}. \tag{19-14}$$

From (17-32) we see that

$$\mathcal{F}^{+}_{ij} \mathcal{F}^{-}_{jk} = \mathcal{F}_{ik}. \tag{19-15}$$

Let us now try to evaluate the half-space potential amplitudes as in (17-40). We see that

$$\begin{aligned} A_i^{(n)} &= \{E^{(n)}(0)\}^{-1}_{ij} \mathcal{F}^{+}_{jk} U_k^{(m)+}(D) \\ &= \{E^{(n)}(0)\}^{-1}_{ij} \mathcal{F}^{+}_{jk} \mathcal{F}^{-}_{kl} \left[\{\mathcal{F}^{-}\}^{-1}_{lm} U_m^{(m)-}(D) + \{\mathcal{F}^{-}\}^{-1}_{lm} \delta U_m^{(m)} \right], \end{aligned} \tag{19-16}$$

where we have used (19-10). Now using (19-13) and (19-15) we obtain

$$\begin{aligned} A_i^{(n)} &= \{E^{(n)}(0)\}^{-1}_{ij} \mathcal{F}_{jl} \left[U_l^{(1)}(0) + \{\mathcal{F}^{-}\}^{-1}_{lm} \delta U_m^{(m)} \right] \\ &= J_{il} W_l, \end{aligned} \tag{19-17}$$

where J_{il} is as in (17-40). The vector

$$W_i = \begin{pmatrix} k u_x^{(1)}(0) + \Pi_1 \\ k u_z^{(1)}(0) + \Pi_2 \\ \Pi_3 \\ \Pi_4 \end{pmatrix} \tag{19-18}$$

represents the resulting displacements and stresses at the surface, and

$$\Pi_i = \{\mathcal{F}^{-}\}^{-1}_{ij} \delta U_j, \tag{19-19}$$

are known quantities calculated from the jump discontinuities in $U_i^{(m)}$. The vector Π_i gives the stresses and displacements as seen at the interface $z = 0$ as though it were just another interface. If we have no incident waves from the half-space, we can combine pairs of (19-17) obtaining

$$Kku_x^{(1)}(0) + Lku_z^{(1)}(0) = -(K\Pi_1 + L\Pi_2 + M\Pi_3 + N\Pi_4) = -\Theta_1,$$
$$Pku_x^{(1)}(0) + Qku_z^{(1)}(0) = -(P\Pi_1 + Q\Pi_2 + R\Pi_3 + S\Pi_4) = -\Theta_2. \tag{19-20}$$

We can solve this pair for $u_x^{(1)}(0)$ and $u_z^{(1)}(0)$, obtaining

$$u_x^{(1)}(0) = \frac{1}{k}\left(\frac{L\Theta_2 - Q\Theta_1}{KQ - LP}\right),$$
$$u_z^{(1)}(0) = \frac{1}{k}\left(\frac{P\Theta_1 - K\Theta_2}{KQ - LP}\right). \tag{19-21}$$

One should remember that these results are valid only when the factor $e^{i(kx-\omega t)}$ is restored and a wave-number synthesis is performed to give a real source distribution. This will be done shortly with a specific example.

We can now show that in this formulation, reciprocity is explicitly demonstrated. To this end, let us rewrite (19-17) slightly:

$$A_i^{(n)} = J_{il}\left[U_l^{\text{surface}} + \{\mathcal{F}(0 \to D)\}_{lm}^{-1}\delta U_m^D\right], \tag{19-22}$$

where we have put in an explicit representation of the superscripts which should be self-evident. Now, as in (19-13) we can write

$$U_i^{D'} = \mathcal{F}_{ij}(0 \to D')U_j^{\text{surface}}. \tag{19-23}$$

This implies

$$U_i^{\text{surface}} = \{\mathcal{F}(0 \to D')\}_{ij}^{-1}U_j^{D'}; \tag{19-24}$$

where D' will be our point of observation at some level beneath the surface. Substituting into (19-22), we have

$$A_i^{(n)} = J_{il}\left[\{\mathcal{F}(0 \to D')\}_{lm}^{-1}U_m^{D'} + \{\mathcal{F}(0 \to D)\}_{lm}^{-1}\delta U_m^D\right]. \tag{19-25}$$

Now in (19-25) the excitation is given as δU_m^D and the resultant displacements and stresses are given by $U_m^{D'}$. But what do we mean by resultant displacement? We mean that vector which if sustained as a jump to zero at D' would maintain the same displacements and stresses between the surface and the fictitious interface D'. If D is greater than D', this must be a conceptual relationship. Hence by altering the roles of $U_m^{D'}$ and δU_m^D we do not change the values of $A_i^{(n)}$ and reciprocity is formally established. This is just what we said in Section 8.4: "Keep the same diagram, relabel source and receiver."

An example: Vertical Line-Stress Discontinuity in a Single Layer Over a Half-Space. The problem is sketched in Figure 19-2. In this simple case (19-17) reduces to

$$A_i^{(2)} = \{E^{(2)}(0)\}_{ij}^{-1}F_{jl}^{(1)}(h)W_l,$$

where W_l is given by (19-18). For this type of stress discontinuity, we have

$$(t_{zz})e^{-i\omega t} = \frac{-Z}{2\pi}\int_{-\infty}^{\infty} e^{i(kx - \omega t)}\,dk.$$

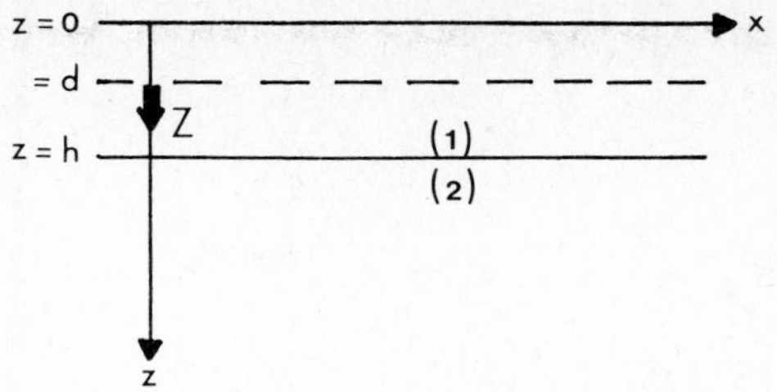

Figure 19-2. Geometrical relationships for a point force in a layer overlying a half-space.

Consequently

$$\delta U_i = (0,0,-Z/2\pi,0),$$

and

$$\Pi_i = -\frac{Z}{2\pi}\{F^{(1)}(d)\}^{-1}_{i3} = -\frac{Z}{2\pi}F^{(1)}_{i3}(-d),$$

where we have used (17-34). We can then write, as in (19-4),

$$u_x(x,0,t) = \int_{-\infty}^{\infty} \frac{L\Theta_2 - Q\Theta_1}{KQ - LP} e^{i(kx - \omega t)} \frac{dk}{k},$$

and

$$u_z(x,0,t) = \int_{-\infty}^{\infty} \frac{P\Theta_1 - K\Theta_2}{KQ - LP} e^{i(kx - \omega t)} \frac{dk}{k}.$$

Each of these integrals can be evaluated as a sum of two branch line integrals, a contribution due to the residue at the pole $k = 0$, and a sum of modes due to the residues of the poles at the roots of the generalized Rayleigh denominator $KQ - LP$. The first three items generally do not contribute anything of interest, so we will not consider them here, but write

$$u_x(x,0,t) \simeq 2\pi i \sum_{j=0}^{N} \frac{L\Theta_2 - Q\Theta_1}{k\partial(KQ - LP)/\partial k}\Bigg|_{k=k_j} e^{i(k_j x - \omega t)}, \tag{19-26}$$

$$u_z(x,0,t) \simeq 2\pi i \sum_{j=0}^{N} \frac{P\Theta_1 - K\Theta_2}{k\partial(KQ - LP)/\partial k}\Bigg|_{k=k_j} e^{i(k_j x - \omega t)}; \tag{19-27}$$

where k_j are the N roots of $KQ - LP$. Equations (19-6) and (19-7) make possible some simplification of the terms in (19-26) and (19-27). Expanding the factors containing Θ_1 and Θ_2, we have

$$\begin{aligned} L\Theta_2 - Q\Theta_1 &= LP\Pi_1 + LQ\Pi_2 + LR\Pi_3 + LS\Pi_4 - QK\Pi_1 - QL\Pi_2 - QM\Pi_3 - QN\Pi_4 \\ &= (LR - QM)\Pi_3 + (LS - QN)\Pi_4 \\ &= -\frac{L}{K}(MP - KR)(\Pi_3 + L\Pi_4/K), \end{aligned} \tag{19-28}$$

$$\begin{aligned} P\Theta_1 - K\Theta_2 &= PK\Pi_1 + PL\Pi_2 + PM\Pi_3 + PN\Pi_4 - KP\Pi_1 - KQ\Pi_2 - KR\Pi_3 - KS\Pi_4 \\ &= (PM - KR)\Pi_3 + (PN - KS)\Pi_4 \\ &= (MP - KR)(\Pi_3 + L\Pi_4/K). \end{aligned} \tag{19-29}$$

That only Π_3 and Π_4 show up in the mode sum is a general property of the layered geometry and the surface boundary conditions, not of the source. The composition of $\delta U_i^{(m)}$ is immaterial. We can then write

$$u_x(x,0,t) \simeq -2\pi i \cdot \sum_{j=0}^{N} \left. \frac{\frac{L}{K}(MP - KR)(\Pi_3 + \frac{L}{K}\Pi_4)}{k\partial(KQ - LP)/\partial k} \right|_{k=k_j} e^{i(k_j x - \omega t)}, \tag{19-30}$$

and

$$u_z(x,0,t) \simeq 2\pi i \sum_{j=0}^{N} \left. \frac{(MP - KR)(\Pi_3 + \frac{L}{K}\Pi_4)}{k\partial(KQ - LP)/\partial k} \right|_{k=k_j} e^{i(k_j x - \omega t)}.$$

The factor $(\Pi_3 + L\Pi_4/K)$ takes care of the depth dependence of the source. The depth dependence of the point of observation would show up (if one wanted it) through the multiplication of the layer matrix to the point of observation.

19.3. Love-Waves from Surficial Stress Distributions.

We now look to equation (17-61). Writing out this relationship in full, we have

$$A^{(n)} + B^{(n)} = J_{11}ku_y^{(1)}(0) + J_{12}t_{zy}^{(1)}(0),$$

$$A^{(n)} - B^{(n)} = J_{21}ku_y^{(1)}(0) + J_{22}t_{zy}^{(1)}(0). \tag{19-31}$$

For our surface stress distribution, we prescribe

$$t_{zy}^{(1)}(0) = -\frac{Y}{2\pi} e^{i(kx - \omega t)}. \tag{19-32}$$

Now for surface sources, the coefficient $A^{(n)}$ must be identically zero, and then adding equations (19-31) leads to:

$$0 = (J_{11} + J_{21})ku_y^{(1)}(0) + (J_{12} + J_{22})t_{zy}^{(1)}(0). \tag{19-33}$$

Solving, we find that

$$u_y^{(1)}(0) = -\frac{1}{k}\frac{(J_{12} + J_{22})}{(J_{11} + J_{21})} t_{zy}^{(1)}(0). \tag{19-34}$$

Restoring the spatial dependence gives

$$u_y(x,0,t) = \frac{Y}{2\pi} \int_{-\infty}^{\infty} \frac{(J_{12} + J_{22})}{(J_{11} + J_{21})} e^{i(kx - \omega t)} \frac{dk}{k} \tag{19-35}$$

$$= \text{Y-branch line integral} + iY \sum_{j=0}^{N} \left. \frac{1}{k} \frac{(J_{12} + J_{22})}{\partial(J_{11} + J_{21})/\partial k} \right|_{k=k_j} e^{i(k_j x - \omega t)}.$$

This equation can be simplified. Multiplying the layer product matrix by its reciprocal (17-60) gives us

$$\mathcal{F}_{11}\mathcal{F}_{22} - \mathcal{F}_{12}\mathcal{F}_{21} = 1. \tag{19-36}$$

At the zeros of the denominator, we have

$$\Delta_L = (J_{11} + J_{21}) = \{E^{(n)}\}_{11}^{-1}\mathcal{F}_{11} + \{E^{(n)}\}_{22}^{-1}\mathcal{F}_{21} = 0. \tag{19-37}$$

Likewise, we have that

$$\begin{aligned}(J_{12} + J_{22}) &= \{E^{(n)}\}^{-1}_{11}F_{12} + \{E^{(n)}\}^{-1}_{22}F_{22} \\ &= \{E^{(n)}\}^{-1}_{22}\left[-(F_{21}F_{12}/F_{11}) + F_{22}\right] \\ &= \{E^{(n)}\}^{-1}_{22}/F_{11} = i/\{k\beta^{(n)}\mu^{(n)}F_{11}\}.\end{aligned} \tag{19-38}$$

Combining, we rewrite (19-35) as

$$u_y(x,0,t) = \text{Y-branch line integral} - \frac{Y}{\mu^{(n)}}\sum_{j=0}^{N}\frac{1}{k^2\beta^{(n)}F_{11}\partial(\Delta_k)/\partial k}\Bigg|_{k=k_j} e^{i(k_jx-\omega t)}. \tag{19-39}$$

The factor

$$E^L_j \equiv \frac{1}{\mu^{(n)}\beta^{(n)}k^2F_{11}\partial(\Delta_L)/\partial k}\Bigg|_{k=k_j}, \tag{19-40}$$

is the modal excitation coefficient for Love-waves and is characteristic of the layered sequence. See Harkrider and Anderson (1966) for an alternative evaluation of (19-40) in terms of energy integrals. Anderson and Harkrider (1968) give excitation partials for small changes in model parameters.

For the simple case of a layer over a half-space, we have

$$\Delta_L = \frac{i}{k^2\beta^{(2)}\mu^{(2)}}\left[\mu^{(2)}\beta^{(2)}\mathrm{Ch}\beta^{(1)}h + \mu^{(1)}\beta^{(1)}\mathrm{Sh}\beta^{(1)}h\right]. \tag{19-41}$$

The bracketed portion of this expression is the same as in (16-6) and is the Love-wave dispersion relation for a single layer.

Again, sources at depth may be included as discussed in the case of Rayleigh-waves. See Harkrider (1964) for details.

19.4. Surface Waves When Liquid Layers Are Present.

Two important cases can be developed from the results of Section 17.3. The first is the case of propagation of

Rayleigh Waves in Oceanic Basins. If we represent the continuation of the surface values through all layers, both liquid and solid, including the last liquid layer above a solid composite half-space, by $G^{\dagger}_{ij}$, we have at the bottom of this composite liquid:

$$\begin{aligned}ku_z^{(n)\S} &= G^{\dagger}_{11}ku_z^{(1)}(0) + G^{\dagger}_{12}t_{zz}^{(1)}(0), \\ t_{zz}^{(n)\S} &= G^{\dagger}_{21}ku_z^{(1)}(0) + G^{\dagger}_{22}t_{zz}^{(1)}(0).\end{aligned} \tag{19-42}$$

Since $t_{zz}^{(1)} = 0$ at the surface of the ocean, these equations may be solved for $t_{zz}^{(n)\S}$, giving

$$t_{zz}^{(n)\S} = G^{\dagger}_{21}ku_z^{(n)\S}/G^{\dagger}_{11}. \tag{19-43}$$

Again $A^{(n)}$ and $C^{(n)}$ in (17-81) are identically zero, and so adding in pairs, we obtain (on using (19-43)):

$$0 = K^{\S} k u_x^{(n)\S} + (L^{\S} + M^{\S} G^{\dagger}_{21}/G^{\dagger}_{11}) k u_z^{(n)\S}, \qquad (19\text{-}44)$$

$$0 = P^{\S} k u_x^{(n)\S} + (Q^{\S} + R^{\S} G^{\dagger}_{21}/G^{\dagger}_{11}) k u_z^{(n)\S};$$

where $K^{\S}$, $L^{\S}$, etc., are calculated for all the solid layers below the last liquid layer. If there is to be a non-trivial solution of this pair, we must have

$$(K^{\S}Q^{\S} - L^{\S}P^{\S}) + (K^{\S}R^{\S} - M^{\S}P^{\S}) G^{\dagger}_{21}/G^{\dagger}_{11} = 0. \qquad (19\text{-}45)$$

This is the required dispersion relation for Rayleigh-waves in a (composite) half-space overlain by an intermixture of liquid and solid layers. The second term, in the case of the ocean and its unconsolidated sedimentary bottom, acts as a perturbation on the first term which may be identified as the Rayleigh dispersion relation if the ocean were not present.

As an example, let us assume that the liquid layer is uniform. In this case, the $G^{\dagger}$ matrix is the simple G matrix for the liquid layer and the matrix for the half-space is what we would calculate with the water absent. From (17-73) we have

$$G^{\dagger}_{21} = -\frac{\lambda k_p^2 \mathrm{Sh}\alpha h}{k\alpha}, \qquad G^{\dagger}_{11} = \mathrm{Ch}\alpha h;$$

and the dispersion relation (19-45) becomes

$$(K^{\S}Q^{\S} - L^{\S}P^{\S}) - (K^{\S}R^{\S} - M^{\S}P^{\S}) \frac{\lambda k_p^2 \mathrm{Tanh}\alpha h}{k\alpha} = 0. \qquad (19\text{-}46)$$

It should be noted here that <u>Love-waves are unaffected</u> by the presence of an oceanic water layer.

<u>Waves in a Floating Ice Sheet</u>. Here we have two additional conditions. The condition of no energy incident from below requires that $A^{(n)} = 0$. The pair (17-82) may then be added, giving

$$0 = Y^{\S} k u_z^{(n)\S} + Z^{\S} t_{zz}^{(n)\S}; \qquad (19\text{-}47)$$

where

$$Y^{\S} = \{D^{(n)}(0)\}^{-1}_{1j} G^{\S}_{j1} + \{D^{(n)}(0)\}^{-1}_{2j} G^{\S}_{j1},$$
$$Z^{\S} = \{D^{(n)}(0)\}^{-1}_{1j} G^{\S}_{j2} + \{D^{(n)}(0)\}^{-1}_{2j} G^{\S}_{j2}. \qquad (19\text{-}48)$$

Using (19-43), we have from (19-47)

$$\left[Y^{\S} + Z^{\S}\, G^{\dagger}_{21}/G^{\dagger}_{11} \right] k u_z^{(n)\S} = 0. \qquad (19\text{-}49)$$

For a non-trivial solution we must then have

$$Y^{\S} + Z^{\S} G^{\dagger}_{21}/G^{\dagger}_{11} = 0; \qquad (19\text{-}50)$$

which is the dispersion relation for mixed layers above a liquid half-space.

<u>Sources</u>. By being careful, particularly about inverse matrices, one could probably introduce sources into a complete formulation including liquid layers. As

long as the sources are in the solid layers, and only a composite liquid layer is above, one definitely can — see Harkrider (1964).

19.5. Other Considerations.

Three-Dimensional Problems. The dispersion relations derived in this chapter are valid for three dimensions; however, the layering matrices and the excitation coefficients derived from them will be slightly different. Also, the distinction between SH and SV becomes more complex: both types of motion have an azimuthal dependence about the source, and the type of motion will have to be carefully watched. Harkrider (1964 and 1970) does a rather complete analysis of the three-dimensional problem. Much the same ground is covered in a review paper by Takeuchi and Saito (1972). However, these authors also relate surface-wave theory to normal mode theory for a spherical earth. We shall consider this in Chapter 24.

Alternative Formulations. Knopoff (1964a), Dunkin (1965), and Thrower (1965), have presented alternative formulations using augmented 6 x 6 matrices formed by 2 x 2 combinations of the Thompson-Haskell matrices. These formulations are not quite as physically meaningful as that given above. The latter two were devised explicity to overcome loss of precision difficulties encountered when using the Thompson-Haskell matrices directly. Harkrider (1970) has shown the relationship between the 6 x 6 method and the Thompson-Haskell 4 x 4 method. Watson (1970a) has shown how the 6 x 6 method may be reduced to 5 x 5. Schwab and Knopoff have presented a series of papers in which they closely examined all the above methods with the aim of producing the most efficient computational algorithm for determining surface-wave dispersion relations. Much of this work has been summarized in a review paper by Schwab and Knopoff (1972). For actual computations, their formulations are recommended.

The above methods have all been exact. In a number of cases, though, one would like approximate methods that would be of more general applicability. A special need arises when there is some variability in either the lateral or vertical directions that can't be matched by the plane-layered model. Variational Methods are very powerful in this regard. Kennett (1974a), Jobert et al (1975), Jobert (1976), and Wiggins (1976) treat these methods and their relationship to the matrix methods outlined above. Another pair of approximate analyses are those of finite element methods (Lysmer and Drake, 1972) and of finite difference methods (Boore, 1972). Lysmer (1970) derives dispersion relations for a number of cases using the finite element method.

Oceanic and Atmospheric Waveguides. Both the ocean and the atmosphere are vertically stratified by temperature differences. This stratification can be modeled by plane-layered media as above, and all the power of the Thompson-Haskell matrix techniques can be applied. Certain aspects of propagation are also simplified in that both are fluid rather than solid media. An excellent theoretical treatment of the oceanic waveguide is given by Pekeris (1948). This work is accompanied by two observational papers by Worzel and Ewing (1948) and Ewing and Worzel (1948). More modern treatments can be found in Officer (1958) and Tolstoy and Clay(1966). The usefulness of these matrix techniques in working with the atmospheric waveguides was noted early by Press and Harkrider (1962). Acoustic and gravitational waves in the atmosphere are treated by Gossard and Hooke (1975). Atmospheric tides are considered by Chapman and Lindzen (1970). The interaction of such waves with the ionosphere was reviewed by Yeh and Liu (1974). Two useful bibliographies are given by Thomas et al (1971 and 1972). Despite the way shown by Press and Harkrider (ibid) not much of the work in the atmospheric waveguide has made use of the matrix formulation.

CHAPTER 20

MANY LAYERS OVER A HALF-SPACE – SURFACE-WAVE OBSERVATIONS

20.1. General.

The analysis of surface-waves in multilayered media rests on two modern developments, the long-period seismograph and the electronic computer. Early instrumentation recorded surface-waves as "Long Period" waves. Rayleigh-waves were given the shorthand notation LR while Love-waves were denoted by LQ (Q from the German Querwellen - transverse waves). However, the frequency band recorded was so narrow little interpretation was possible. Despite these limitations Ewing and Press (1959) made a phase-velocity study of Middle North America at a period of 20 seconds. Instruments of various vintages were responsive at this particular period. We have seen in Chapter 10 that the depth of penetration depends upon the period of the surface-wave. It was found that for sufficiently long periods, the flat-layer approximation breaks down, and sphericity has to be considered. This occurs at about 60 seconds for Rayleigh-waves, but for Love-waves as low as 15 seconds (Anderson and Töksoz, 1963). Thus modern instrumentation forces the consideration of spherical-earth models for an adequate description of surface-wave propagation. Modern computers have allowed the use of powerful data processing methods for the analysis of observations. The study of dispersive surface-waves requires a special technique which we shall consider in Section 20.2. Observations are voluminous; representative ones will be given in Section 20.3. Surface-wave observations taken over the interval 1950-1970 extend from periods of a few seconds to over 300 seconds. At less than 15 seconds, the observations are very dependent upon the thickness of the sedimentary rock section which ranges from zero to occasionally as much as ten kilometers. The band from 15 to 300 seconds is the usual range considered for determining crustal and upper mantle models. With much quality data available, time and effort could be expended upon obtaining a reliable means of inversion. Generalized inversion methods, considered in Section 20.4, allow a discussion of uniqueness and resolution. In Section 20.5, some unsophisticated models are given as representing changes in crust and upper mantle structure.

In his second general surface-wave paper, Harkrider has computed dispersion curves from flat-earth models for both Rayleigh- and Love-waves. These curves (Figure 20-1) should not be used for interpretation, but are presented to give an indication of the variation with different types of structure, in this case, Shield and Oceanic. Sphericity corrections need to be applied before comparison with observed data. Figure 20-2 shows the relative excitation of the fundamental and higher modes for each type of surface-wave.

Kovach (1965 and 1978) has given two excellent review papers on the whole subject of surface-waves. Each contains many references.

244

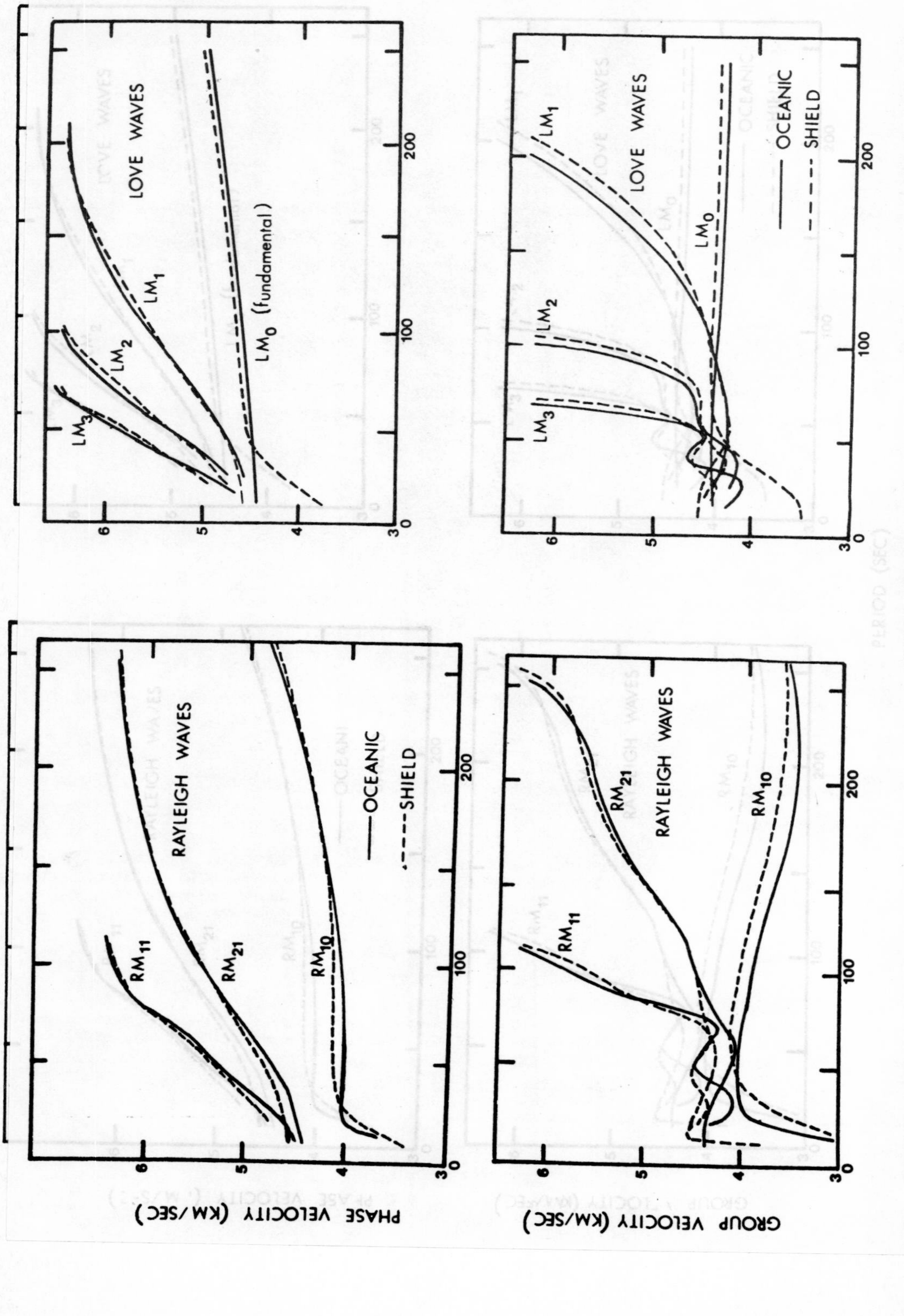
PHASE VELOCITY (KM/SEC)
GROUP VELOCITY (KM/SEC)
PERIOD (SEC)
RAYLEIGH WAVES
LOVE WAVES
RM_{11}
RM_{21}
RM_{10}
LM_0 (fundamental)
LM_1
LM_2
LM_3
LM_0
OCEANIC
SHIELD
0
100
200
3
4
5
6

Figure 20-1. Phase and group velocities for Rayleigh and Love modes. Curves are shown for shield and oceanic structures. Flat-earth model parameters may be found in the original paper. After Harkrider (1970); notation changed for Rayleigh and Love modes.

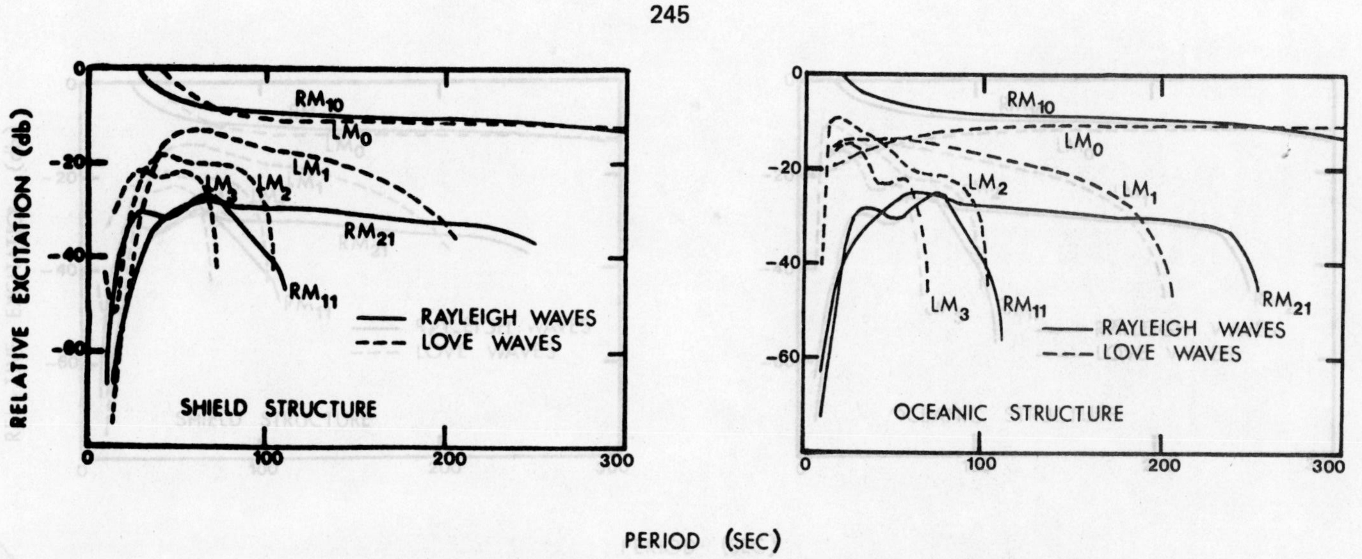

Figure 20-2. Relative Excitation of Rayleigh and Love Modes. After Harkrider (1970); notation changed for Rayleigh and Love modes.

20.2. The Determination of Phase and Group Velocity.

Data Analysis — Peak and Trough Method. Consider the dispersive sinusoidal wave-train of Figure 20-3. By numbering the peaks, troughs, and zero crossings

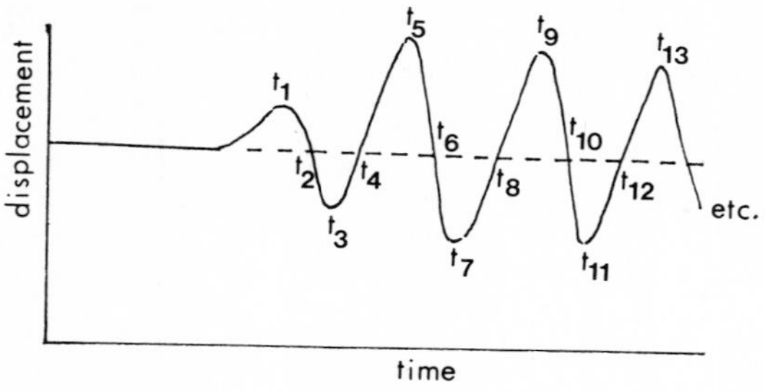

Figure 20-3. Peak and Trough analysis of a single dispersive wave-train.

(a phase interval of $\pi/2$) we can compute the instantaneous frequency at each time point t_i as:

$$\omega_i^{inst} = - \left.\frac{d\Phi_S}{dt}\right|_{\Phi_S=\phi_i} \simeq \frac{6\pi}{8(t_{i+1} - t_{i-1}) - (t_{i+2} - t_{i-2})} ; \qquad (20\text{-}1)$$

where

$$\Phi_S = (k_o x - \omega_o t + \phi_o \pm \pi/4) \qquad (20\text{-}2)$$

is the apparent phase in those situations where the stationary phase approximation is valid. The form of the denominator comes from a standard numerical derivative formula. Once we have computed the instantaneous frequency ω_i^{inst} at each phase point ϕ_i, we can then eliminate the parameter i and obtain a plot of frequency vs. time similar to Figure 16-9. The validity of this peak and trough method depends upon two conditions being fulfilled. First of all, we must be dealing with the written "signature" of only one mode, and secondly the train must be well dispersed. The latter requirement is usually fulfilled by observing the seismic disturbance at sufficient distance so that the various frequencies have enough time to become well separated. If this is true, then the separation of modes (should two or more be present) can be accomplished quite often by judicious filtering in the time domain. This method has largely been supplanted by the Fourier analysis of digital data. It is still useful however, when only paper recordings are available. Details may be found in Brune et al (1960).

Data Analysis — Fourier Analysis Method. If we take the Fourier transform of (16-32), we find that

$$\overline{\Phi}(x,\omega) = \int_{-\infty}^{\infty} e^{i\omega\tau} d\tau \frac{1}{2\pi} \int_{-\infty}^{\infty} A\{k(\omega')\} e^{i\{\phi(\omega') + k(\omega')x - \omega'(t_o+\tau)\}} d\omega'$$

$$= A\{k(\omega)\} e^{i\{\phi(\omega) + k(\omega)x - \omega t_o\}} = A\{k(\omega)\} e^{i(\Phi_H - 2\pi N)} , \qquad (20\text{-}3)$$

where t_o is some convenient fiducial origin for the local time variable τ. The phase ambiguity of $2\pi N$ is explicitly included for later use.

Fourier analysis thus removes the restriction of large distance required by the stationary phase approximation. That is, the amplitude and phase are calculable at all distances without having been separated into resolvable peaks and troughs. All that is required is that only one mode be present, and even that may be accomplished by judicious filtering as before. Additional considerations can be found in Y. Sato (1955) who introduced these methods into the determination of phase and group velocities.

Another problem, quite serious, arises where there is a fair amount of contamination by noise or other interfering signals. From the nature of (20-3), the amplitude and phase spectra at a given angular frequency ω have contributions from wherever it appears, not just from the one spot it would be in a pure dispersed signal. Pilant and Knopoff (1964) introduced the group-delay filter (a time-variable digital band-pass filter) to pass only those frequencies appropriate to a smoothed group-velocity curve constructed from the raw data. This process could be applied iteratively to obtain a new group-velocity curve, etc. One can obtain good amplitude and phase spectra except when the group-delay varies rapidly. Large changes in the filter necessary to follow these variations introduce erroneous results. This problem can be largely eliminated by the residual dispersion method discussed below, and one need not worry about constructing the "best possible" filter.

Data Analysis.— Residual Dispersion Method. For two "nice" time-functions, $f_1(t)$ and $f_2(t)$, each square integrable, we have from Fourier theory that:

$$\int f_1(t + \tau) f_2^*(t)dt \iff \overline{f}_1(\omega)\overline{f}_2^*(\omega). \qquad (20\text{-}4)$$

The time-function on the left, suitably normalized, is recognized as the cross-correlation between f_1 and f_2. Writing $\overline{f}_1(\omega)$ and $\overline{f}_2^*(\omega)$ in terms of amplitude and phase spectra we have

$$\int f_1(t + \tau) f_2^*(t)dt \iff A_1(\omega)A_2(\omega)e^{i\{\Phi_1(\omega) - \Phi_2(\omega)\}}. \qquad (20\text{-}5)$$

Thus the Fourier analysis of the cross-correlation between $f_1(t)$ and $f_2(t)$ gives the phase difference between the two spectra. Dziewonski et al (1972) found that if $f_1(t)$ were the observed seismogram and $f_2(t)$ were a reference synthetic seismogram matching $f_1(t)$ fairly closely, then the phase difference could be calculated very accurately. For contaminated signals, the effect of misplaced energy could be greatly reduced by limiting the time-lag τ by a suitable "window" function (Båth, 1974, Sec. 4.4). It might be noted here that if both $f_1(t)$ and $f_2(t)$ are observations, then phase differences between two stations can be computed directly by (20-5). Again, only one modal "signature" should be present.

Data Analysis — Other Considerations. In an extensive review article, Dziewonski and Hales (1972) have discussed the above techniques as well as several other useful methods including the construction of sonogram displays of group- and phase-delay. An excellent treatment of the whole subject of spectral analysis in a geophysical context is given by Båth (1974). He gives a voluminous list of references. Section 7.2 refers particularly to the analysis of dispersive surface-waves.

It can be seen from Figure 20-1 that in the period range 25-100 seconds, the group-velocity curves for Love-waves are intertwined. This means that two (or more) dispersed wave-trains can be occupying the same time-window. This was noted observationally by Thatcher and Brune (1969, 1973), and James (1971). Boore (1969) seemed to feel that such interference would lead to scatter, but produce no uniform bias. Forsyth (1975a) used a least squares technique to reduce this scatter and produce Love-wave phase-velocity curves from 33 to 167 seconds period over a number of regions in the Southeast Pacific ocean.

Single Station Method — Group-Velocity. From (20-3) we have that

$$d\Phi_H/d\omega = d\phi/d\omega + x/U - t_o, \tag{20-6}$$

where we have used (16-39). Then

$$U = x/[t_o + d\Phi_H/d\omega - d\phi/d\omega] . \tag{20-7}$$

The term $d\phi/d\omega$ is composed of two parts, one due to phase variation in the source, the other due to the phase response of the recording instrumentation. This latter is usually calculable (Chapter 33) while the former causes some problems particularly at distances less than 1000 km. We shall comment upon source phase shortly. With this exception, then, the group-velocity U measured at one station is inherently absolute. It represents the velocity of travel for the energy at a particular frequency, and is determined by the average structure between source and station.

Single Station Method — Phase-Velocity. Again from (20-3) we have that

$$\Phi_H/\omega = \phi/\omega + x/c - t_o + 2\pi N/\omega , \tag{20-8}$$

where $2\pi N$ is a constant independent of frequency. Rewriting,

$$c = x/ [t_o + \{\Phi_H - \phi - 2\pi N\}/\omega] . \tag{20-9}$$

Besides the source phase portion of ϕ, we now have to determine the unknown integer N. If the distance is small enough that one unit either side of N gives absurd values for the phase-velocity c, then N can be chosen correctly. But at such short distances the source phase must be known, not guessed at; for then it only contributes to the ambiguity in N. For this reason, single station determinations of phase-velocity are not common.

Two Station Method — Phase-Velocity. Here we set

$$\delta\Phi_H = \Phi_H(x + \delta x) - \Phi_H(x) + 2\pi\delta N = k\delta x - \omega\delta t_o, \tag{20-10}$$

where δN is a possible integer ambiguity between the two observations at x and $x + \delta x$. We then have

$$c = \delta x/ [\delta t_o + (\delta\Phi_H - 2\pi\delta N)/\omega] . \tag{20-11}$$

Here the partially unknown term $\phi(\omega)$ cancels out if identical instrumentation is used at two stations on the same azimuth from the source. One tries to use station separations and frequencies such that δN is not ambiguous. If δx is larger than 10 wavelengths, one may have difficulty in determining δN. Thus the determination of phase-velocity of surface-waves between two stations is inherently differential in nature. However, the differencing of phase-delay times (Φ_H/ω) between two stations necessitates extremely good time data.

Two Station Method — Group-Velocity. From (20-11) we can write

$$d(\delta\Phi_H)/d\omega = \delta x/U - \delta t_o \ . \tag{20-12}$$

This leads to

$$U = \delta x/\left[\delta t_o + d(\delta\Phi_H)/d\omega\right] . \tag{20-13}$$

Here the term $d\phi/d\omega$ cancels out, but we are left with a frequency derivative of a differential quantity $\delta\Phi_H$. This measurement is decidedly unstable in the presence of noise. Consequently, two-station determinations of group-velocity are rare.

Relations Between Phase-and Group-Velocity. From (16-39) we have

$$U = \frac{d\omega}{dk} = \frac{d(kc)}{dk} = c + k\frac{dc}{dk} = c - L\frac{dc}{dL} , \tag{20-14}$$

and

$$\frac{1}{U} = \frac{dk}{d\omega} = \frac{d(\omega/c)}{d\omega} = \frac{1}{c}\left[1 - \frac{\omega}{c}\frac{dc}{d\omega}\right] = \frac{1}{c}\left[1 + \frac{T}{c}\frac{dc}{dT}\right] . \tag{20-15}$$

One solution of the second differential equation in terms of ω and $\underline{c}$ is

$$c = \frac{\omega\cdot\delta x}{\int \frac{\delta x d\omega}{U} + \text{const}} = \frac{\omega\cdot\delta x}{\int \frac{d\Phi_H}{d\omega} d\omega + \text{const}} = \frac{\omega\cdot\delta x}{\delta\Phi_H + \text{const}} . \tag{20-16}$$

Here we have assumed that $\phi(\omega)$ is negligible. If we have a certain $\delta\Phi_H$, an additive constant does not change the group-velocity on differentiation. On the other hand, the group-velocity can be integrated to give a family of phase-velocity curves with an arbitrary constant of integration. Solanki (1974) has done this using already obtained two-station phase-velocity data at longer periods to obtain the constant. This circumvents the loss of peak-to-peak coherence at short periods.

From the relations above, we see that the phase-velocity is the more fundamental, and that new information can be added from the group-velocity curve only when phase cannot be accurately determined. The group-velocity curve does, however, serve as a check on our data (Nyman et al, 1977). Also, because of the ambiguity in integrating the group-velocity to get a phase-velocity, structures cannot be uniquely determined by group-velocity curves alone (Pilant and Knopoff, 1970).

The Phase at the Source. That part of ϕ due to the source presents several problems. It depends both on the time history of the source as well as upon its spatial orientation and location.

For either impulse- or step-function time histories, there will be a constant phase (0 or $\pi/2$ respectively) for the time part. These two cases correspond to the observation of waves long compared with the duration of rupture or to waves which are so short that the effects of rupture termination do not have to be considered. The former is most applicable to surface-wave studies and the latter to body-wave studies. For more complicated time histories, exact results are difficult to obtain without a more representative theory of source mechanism.

Knopoff and Schwab (1968) have shown that the source orientation, unless purely vertical or purely horizontal, may contribute a portion of $\phi(\omega)$ which varies with frequency. This may be easily seen from equations (19-30) where both components depend upon the source term $\Pi_3 + L\Pi_4/K$. From the defining relation (19-19), we have that

$$\Pi_3 = \Pi_3^z + \Pi_3^x = \{F^-\}_{33}^{-1}\delta U_3 + \{F^-\}_{34}^{-1}\delta U_4, \tag{20-17}$$

with a similar relationship for Π_4. In the one-layer case, $F_{ij}^- \to F_{ij}^{(1)}(d)$. But from (17-31) we see that F_{33} is real and F_{34} is imaginary. From (17-34) we see that this is also true for the inverses $\{F\}_{33}^{-1}$ and $\{F\}_{34}^{-1}$. Consequently the phase of Π_3 (and also Π_4) will depend upon the source orientation as expressed in the relative proportions of δU_3 and δU_4. A similar analysis holds for the three-dimensional problem.

Frez and Schwab (1976) have investigated the importance of source depth upon initial phase considering several source models. The reason for this interest is that with shallow sources the free surface gives rise to an echo effect with attendent phase-shifts.

The Multistation Method. Three stations were used to determine phase-velocities of Rayleigh-waves by Evernden (1953, 1954) and Press (1956) and the method was extended to more than three stations by Aki (1961). In Figure 20-4 we sketch the

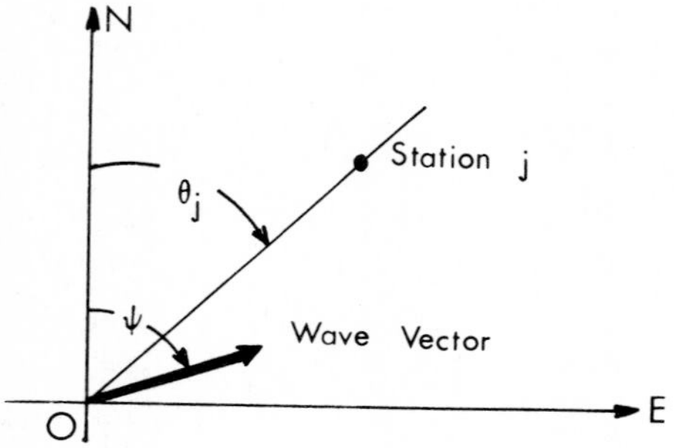

Figure 20-4. Geometrical relationships for the multistation method.

relationships for any one of the observing stations. Consider the arrival of a particular event at station j in a local network centered at O. We have

$$\begin{aligned} T_{ij} &= T_{io} + r_j \mathrm{Cos}(\psi_i - \theta_j)/v_i \\ &= T_{io} + r_j \mathrm{Cos}\theta_j \mathrm{Cos}\psi_i/v_i + r_j \mathrm{Sin}\theta_j \mathrm{Sin}\psi_i/v_i, \end{aligned} \tag{20-18}$$

where T_{io} is the time that the particular event crosses the origin O. If we have N stations, i.e., j = 1,2,...,N, there will be N equations for the three unknowns

$$T_{io},\ \mathrm{Cos}\psi_i/v_i,\ \mathrm{Sin}\psi_i/v_i\ .$$

These unknowns will be functions of the parameter i, which may be discrete for "peak and trough" methods, or tabular values for the Fourier Analysis method. If N is greater than three, we can solve the system by least squares and get a measure of our error as well as the three unknowns. If N equals three, we can still solve the system (20-18) but obtain no knowledge of the error.

In the case of three stations (tripartite method), an analysis of the relationships will aid in interpretation. The apparent velocity of a particular event as it travels down each of the two legs of triangle ABC (Figure 20-5) is given by

$$1/v_{AB} = (T_B - T_A)/\overline{AB} = \text{Cos}\psi\,/v,$$
$$1/v_{AC} = (T_C - T_A)/\overline{AC} = \text{Cos}(\theta - \psi)/v. \tag{20-19}$$

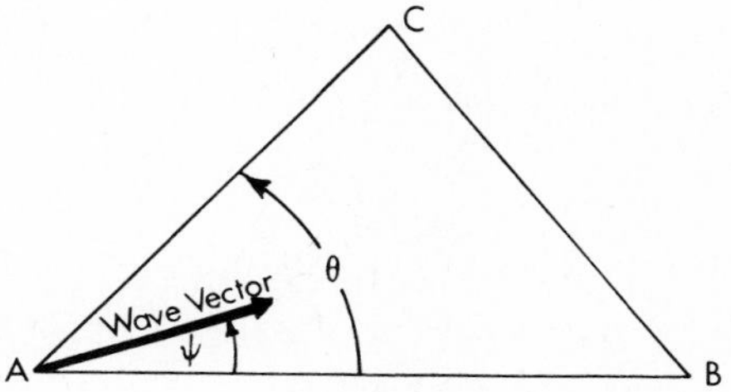

Figure 20-5. Geometrical relationships for the tripartite method.

These can be combined into

$$\frac{1}{v_{AB}^2} + \frac{1}{v_{AC}^2} - \frac{2\text{Cos}\theta}{v_{AB}v_{AC}} = \frac{\text{Sin}^2\theta}{v^2}. \tag{20-20}$$

On differentiating, we find that

$$\text{Sin}^2\theta\,\frac{dv}{v^3} = \left(\frac{1}{v_{AB}} - \frac{\text{Cos}\theta}{v_{AC}}\right)\frac{dv_{AB}}{v_{AB}^2} + \left(\frac{1}{v_{AC}} - \frac{\text{Cos}\theta}{v_{AB}}\right)\frac{dv_{AC}}{v_{AC}^2}.$$

But

$$\frac{1}{v_{AB}} = \frac{\text{Cos}\psi}{v}, \qquad \frac{1}{v_{AC}} = \frac{\text{Cos}(\theta - \psi)}{v};$$

and we finally obtain

$$\text{Sin}\theta\,\frac{dv}{v} = \text{Cos}\psi\text{Sin}(\theta - \psi)\,\frac{dv_{AB}}{v_{AB}} + \text{Sin}\psi\text{Cos}(\theta - \psi)\,\frac{dv_{AC}}{v_{AC}}. \tag{20-21}$$

If we look at dv/v as a fractional error, we see that if the wave comes down one leg of the triangle, then the error introduced by an apparent velocity error down the other leg will be minimized. Such errors could be introduced into the tripartite system by an anomalous region outside the triangle as shown in Figure 20-6. Knopoff et al (1966) advocated choosing great circle paths lying close to one leg of the triangle as the paths through A and B in the figure. Additional considerations are given by Knopoff et al (1967). Schwab and Kausel (1976) add an additional station and are able to consider circular wave fronts. They show that lateral heterogeneity can be investigated by the addition of two stations.

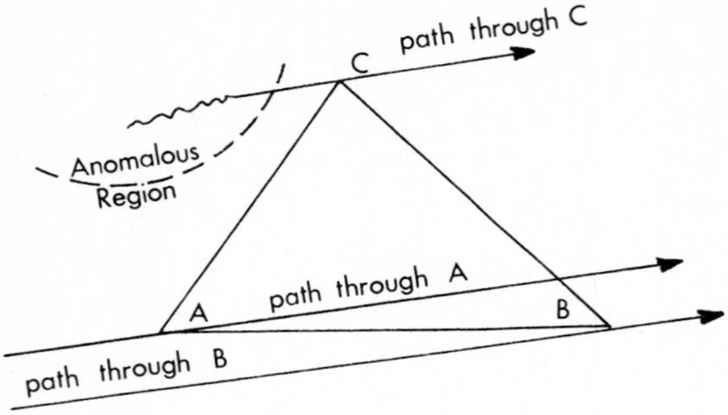

Figure 20-6. An anomalous region outside the tripartite network can cause erroreous phase-velocity determinations unless the great circle paths lie along one of the legs of the triangle.

Multipath Propagation. In his work analyzing Rayleigh-waves in the San Francisco region, Evernden (1953) noted a "beaty" character to the surface-waves crossing his array as well as a variation in apparent azimuth of propagation. He attributed this to energy arriving along non-great-circle paths. Pilant and Knopoff (1964) attributed sharp minima in surface-wave spectra to multiple arrivals. These could be due to temporal and/or spatial distribution at the source or due to multipath propagation. In either study, the tripartite network was insufficient to resolve the question of where the multipathing might occur. Using the much greater resolution of the large aperture seismic array (LASA) in Montana, Capon (1970, 1971) was able to demonstrate that both Rayleigh- and Love-wave multipathing was associated with refractions and reflections of the wave-trains at continental margins. Capon and Evernden (1971) identified multipath events as a part of the Rayleigh-wave coda (an extended portion of the surface-wave signal lasting several times the duration of the high-energy portion described by the dispersion analysis of Chapter 16). Michaels (1977) has suggested the demodulation of "beaty" signals caused by any sort of interference to obtain each of the two interferring signals, but his results consider only synthetic data.

20.3. Observations.

Observations of surface-wave dispersion at periods of less than 100 seconds are appropriate to a flat-layered model. Waves at periods between 100 and 500 seconds are referred to as mantle waves and need calculations applicable to a spherical model and will be referred to in Chapter 25. For periods over 200 seconds, there is constructive interference on multiple circlings of the globe leading to resonance. The individual peaks can be resolved and are known as the free periods of oscilliation of the earth. Observations of the free periods will be considered in Chapter 25. An early summary covering all three ranges is given by Oliver (1962). In this summary, the tremendous variability of velocity at periods less than 10 seconds period has been observed and associated with the thickness of sedimentary rock, unconsolidated sediments, and water layers. We finally note that because of the Love-wave interferences mentioned in the previous section, and because Rayleigh-waves are isolated from Love-waves on vertical component seismic instruments; most recent observations have been of Rayleigh-waves. Other summaries of observational data on surface-waves are given by Brune (1969) and Dorman (1969).

The most complete set of observations of Rayleigh-wave phase velocity data (mostly taken by the two-station method) is given by Knopoff (1972a). We have chosen to present his data for shields, aseismic platforms, rift zones and mountains. A later study on the Pacific Ocean (Leeds, 1973) is given to complete the set shown in Figure 20-7. (Additional details on the Pacific Ocean study are given by Kausel et al, 1974.) All these data were processed in essentially the same way, which means that biasing (if any) should not vary from figure to figure. Three letter combinations are adopted station codes. Region 1 — Region 8 correspond to derived data for paleomagnetically dated age divisions of the Pacific Ocean. Region 1 to 150 million years; 3 to 100 million years; 6 to 30 million years; and 8 to 5 million years. GUA 1 and GIE 10 refer to single station phase-velocity measurements. The path GUA 1 lies mostly in Region 1 while GIE 10 lies largely along the East Pacific Rise.

It can be seen that there is rather good uniformity for shields, platforms, and rift zones while mountains and oceans show great variability. In the case of the mountains, you are looking at two different types with two different elevations. The Alps are fold mountains and the Andes are uplifted. The Andes, with their greater elevation, have deeper roots and lower phase velocities between 20 and 50 seconds period. The variability of the Oceans is explained by Leeds et al (1974) on the basis of the age of the ocean floor as determined from paleomagnetic data. Since there is a correlation of depth with age, the younger ocean floor over the ridges and the older floor in the basins, the regional differences can also be correlated with ocean structure should the age interpretation be modified. Although only the ocean has been illustrated as continuously varying in character, there is also a gradation going from shields to rift zones.

As an example of the continuous variation from one province to another, results from two regional studies are given in Figure 20-8. These diagrams represent the phase-velocity of Rayleigh-waves crossing North America and western Eurasia. Low-velocity regions are seen in the Basin and Range province, the St. Lawrence River area, the northern Mid-Atlantic Ridge, and in the central Mediterranean. The former three are proposed rift zones, while the central Mediterranean is a proposed

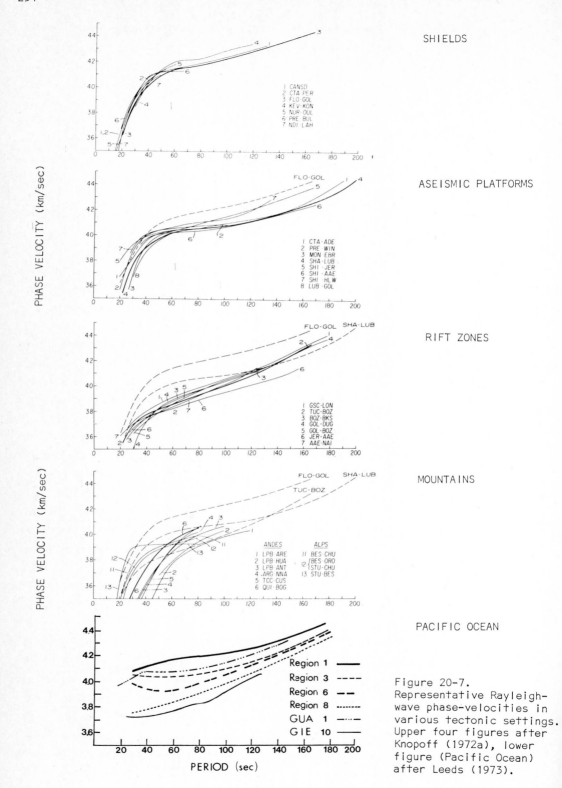

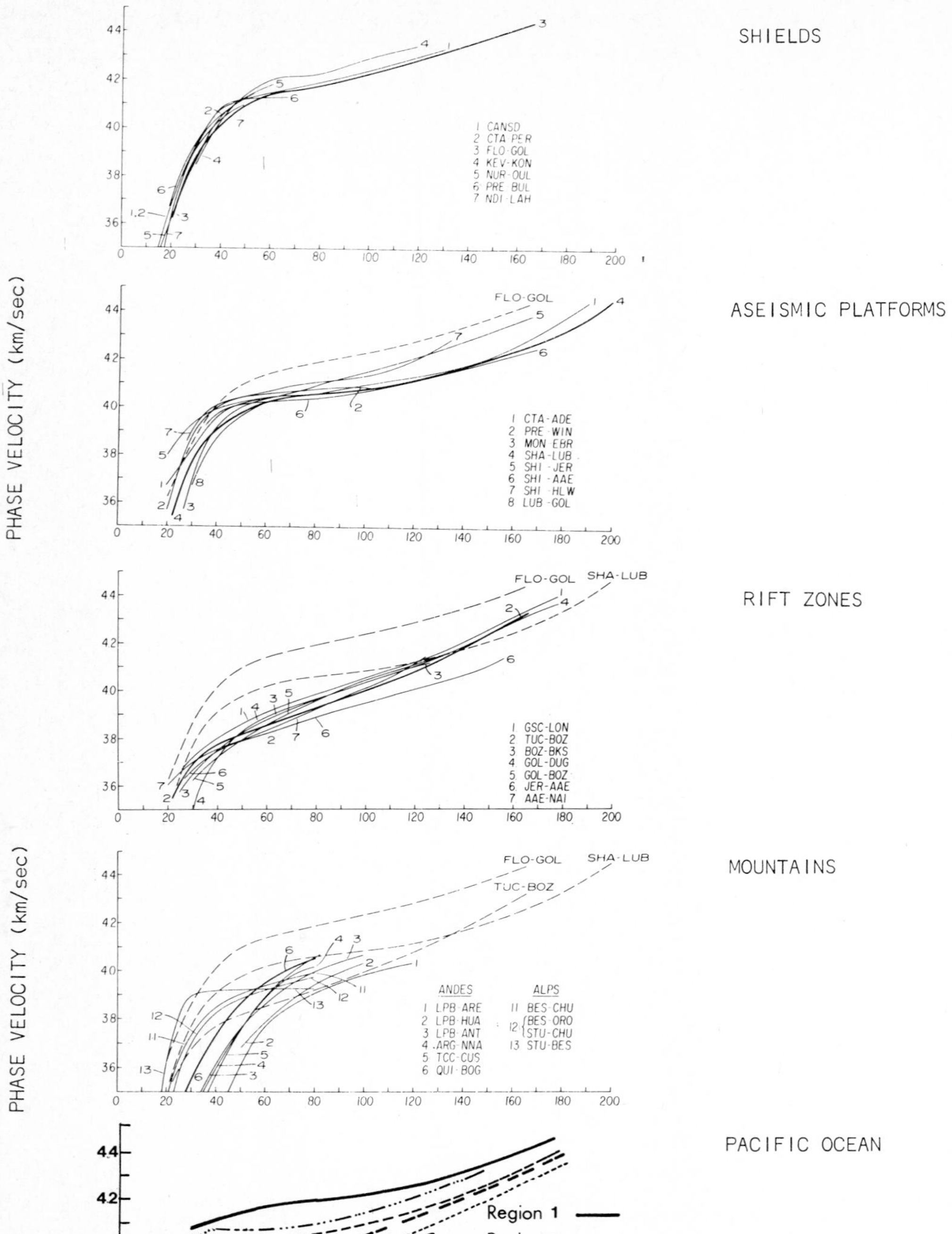

Figure 20-7. Representative Rayleigh-wave phase-velocities in various tectonic settings. Upper four figures after Knopoff (1972a), lower figure (Pacific Ocean) after Leeds (1973).

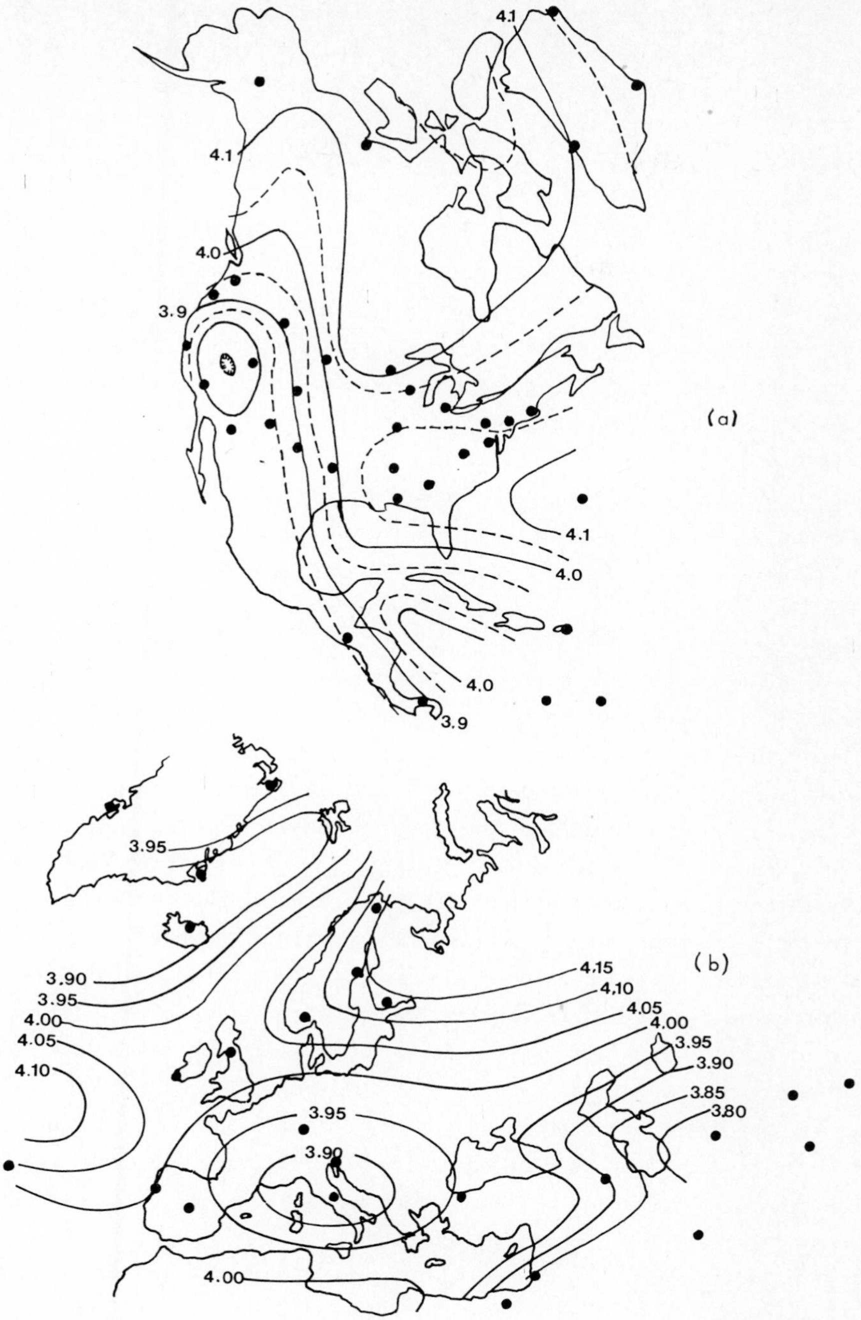

Figure 20-8. Rayleigh-wave phase-velocity contours at 51 seconds period. Part (a) from Pilant (unpublished data) and Part (b) after Lee (1975). Seismic stations are given by dots.

collision zone between Africa and Europe. High-velocity regions are found over the Canadian shield, the North American Basin, the European Basin, and the Baltic Shield.

Additional regional studies have been done in Canada by Wickens (1971) and Hashizume (1976), the United States by Biswas and Knopoff (1974), the North Atlantic by Tarr (1969) and Weidener (1974, 1975), and the southeast Pacific by Forsyth (1975, 1975a).

Higher Surface-Wave Modes. Crampin (1964a and b, 1966a and b) has made an extensive study of higher mode dispersion both for Love- as well as for Rayleigh-waves. Nolet (1975) gives curves for the fundamental Rayleigh and the first six shear modes crossing an array of seismic stations in western Europe. A number of impulsive type arrivals (Li, Lg, Rg, Sa, Pa) have been identified by many authors as higher modes of Love- and Rayleigh-waves. Kovach (1965, Table 5) gives a summary of such observations.

PL and Shear Coupled PL Modes. Observations of these reverberatory waves following P and S have been made by many. Some of the better observations are Oliver and Major (1960), Oliver (1964), Su and Dorman (1965) and Poupinet and Wright (1972).

20.4. Interpretation.

Three problems are presented in the interpretation of the observations in the previous sections: To sort out the various signals and identify them with corresponding dispersion curves, to identify the effects of various structures upon the observed wavetrains and lastly (and most difficult) to determine the actual subsurface structure lying between two locations on the earth. The first problem can be treated by solving what is known as the forward problem — the calculation of dispersion curves and excitation functions for given structures. From these two sets of data, synthetic seismograms may be produced to compare with observations. The second can be illuminated by the completion of "partials," i.e., the partial derivatives relating changes in dispersion and excitation characteristics given a small change in the layered structure. Again this is a forward problem. The third question is resolved by tackling the inverse problem — the determination of a model whose properties match a given set of observations. This latter problem may be investigated in an iterative trial-and-error way; by searching for all possible models, rejecting those which are not acceptable; or in a direct manner. In all three methods, a tabulation of "partials" is a necessary component.

Synthetic Seismograms. After having determined the best ways to compute dispersion curves (Schwab and Knopoff, 1972), it was only natural to continue with the construction of dispersion curves and synthetic seismograms. Further developing a method introduced by Aki (1960a), Calcagnile et al (1976) present an efficient method of constructing synthetic seismograms composed of dispersive wavetrains. They accomplished this with a piecewise approximation of the inverse Fourier integral by quadratic and linear functions in amplitude and phase respectively. This allows an analytic evaluation of each piece. They also worked out the conditions so that the pieces were no longer required to lie between equal intervals of frequency. An excellent example of synthetic seismograms produced in this manner is given by Liao et al (1978). In a series of papers (Knopoff et al, 1973; Schwab et al 1974; Nakanishi et al, 1976; Kausel et al, 1977; Mantovani et al, 1977), the UCLA group made interpretations of Lg, Sa and some of the oscillating behavior of Sn using the higher mode dispersion curves of Love-waves. Calcagnile and Panza (1974) and Panza and Calcagnile (1975) looked at the contributions of higher Rayleigh-wave modes to Sa, Lg, Li and Rg. Some earlier work in this regard was given by Kovach and Anderson (1964).

Partial Derivatives. The contruction of partial derivatives relating changes in phase- and group-velocities is of interest for two reasons. First of all, they tell us what the effect of changes in the parameters of a given layer will do to a particular dispersion curve. Secondly, they are crucial in the computation of models. Figure 20-9 shows partial derivatives of Rayleigh-wave phase-velocities for a Shield Model with respect to layer shear velocities. In this model, layer thicknesses have been chosen in a logarithmically increasing manner. This is because deep layers have wide partials and thinner layers would not increase the resolution much.

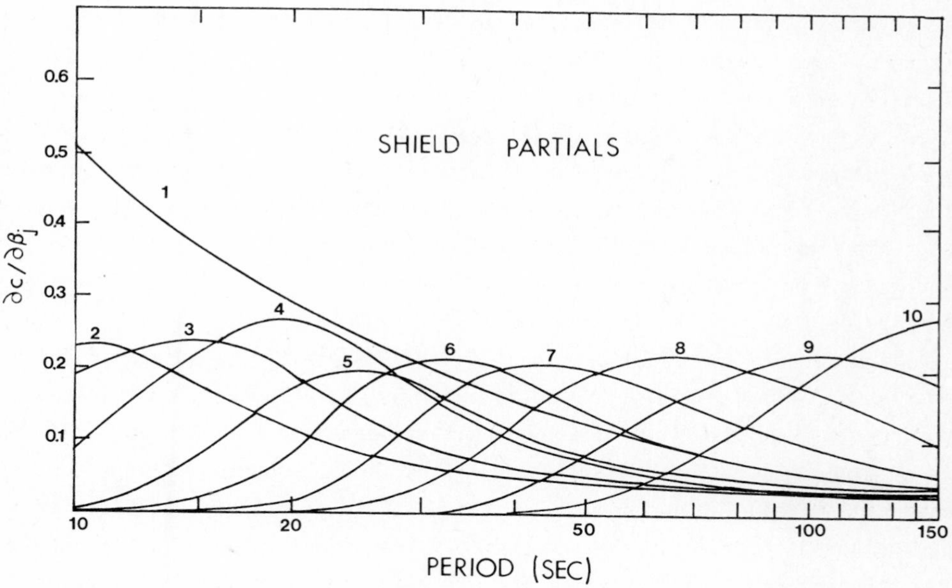

Figure 20-9. Rayleigh-wave phase-velocity partials for a model shield structure. Layer thicknesses: 1 (10 km), 2 (5km), 3 (7 km), 4 (10 km), 5 (14 km), 6 (22 km), 7 (32 km), 8 (50 km), 9 (70 km), 10 (100 km), 11 (140 km), 12 (∞). c is the Rayleigh-wave phase-velocity and v_j is the shear velocity in layer j.

The construction of partial derivative curves is one of the more time consuming tasks in theoretical seismology. However, once done, they do not have to be continuously regenerated. The simplest way to compute them is to perturb the parameters of each layer by a small amount and compute a new dispersion curve (Brune and Dorman, 1963); this was also done in Figure 20-9. The difference between the new curve and the old curve divided by the parameter change gives an approximation of the partial derivative as a function of period. For Rayleigh-waves there are three parameters (λ_j, μ_j, ρ_j) and 3N new dispersion curves must be computed. For Love-waves, 2N new curves have to be calculated. For simple three- and four-layer models this is not prohibitive, but when the number of layers passes ten, a new method is needed.

Using Rayleigh's principle, Jeffreys (1961) suggested that changes in dispersion curves could be computed from ratios of energy integrals. Takeuchi et al (1962) following up this suggestion produced partial derivative curves for the Canadian Shield model of Brune and Dorman (ibid). Both Takeuchi et al (ibid) and Anderson (1964) have worked out partials for a spherical earth; the former using radial integration of the differential equations for both Rayleigh- and Love-waves and the latter using an earth stretching transformation (Anderson and Töksoz, 1963) to obtain correct partials for Love-waves. But there were difficulties with group-velocity calculations in that these involved derivatives of the changes in phase-velocities. Accuracy of the results and the speed of computation were not pleasing. Harkrider (1968) derived the variational equations in detail pointing out that certain assumptions made by some investigators were not necessary nor were they true in many cases. The use of these assumptions had led to errors in the calculation of group-velocity partials. Anderson and Harkrider (1968) then worked out partial derivatives for Love-wave phase-velocity, group-velocity, and excitation coefficients. The problem relating to the length of computation pemained. Kosloff (1975) and Rodi et al (1975) have suggested improvements in the speed of computation.

Inversion. As an example of inversion technique, we will use a rather simple model — that of inverting fundamental mode Rayleigh-waves to determine the shear velocity structure in the crust and upper mantle. There is not enough information here to also model compressional velocity and density. (Burkhard and Jackson (1976) note some of the dangers in not considering these parameters.) One may use rather general relationships linking P and S velocities (e.g., assuming Poisson's ratio σ = constant), P velocity and density (Wollard, 1959, Fig. 7) to constuct the model. However, only the shear velocity is reported. One normally starts with some initial model (abstracted from the literature for a similar tectonic setting) and proceeds to refine it. The simplest refinement is to use a linearized model (which may then have to be tested and iterated if necessary) such as

$$\Delta c_k = \sum_j (\partial c_k/\partial v_j)\Delta v_j = \sum_j P_{kj}\Delta v_j \qquad (k = 1,2,\dots M;\ j = 1,2,\dots N); \qquad (20\text{-}22)$$

where

$$\Delta c_k = c_k^{obs} - c_k^{model} ,$$

c_k is the phase-velocity at frequency point $\underline{k}$, v_j is the shear-wave velocity in layer $\underline{j}$, $\partial c_k/\partial v_j$ are computed for small changes of the model in question, and Δv_j are the changes necessary to correct the model. Summations will be explicitly written out for the next few pages.

If N = M and the matrix of coefficients P_{ij} is non-singular, we can write

$$\Delta v_j = \sum_k (\mathbf{P})^{-1}_{jk} \Delta c_k. \qquad (20\text{-}23)$$

This rarely works unless a good deal of care is taken. One reason is that along a smoothly varying phase-velocity curve, the quantities Δc_k will be independent only if the points are separated widely enough. That is, the true number of independent observations quite often turns out less than the number of layers. One has generally chosen too many layers in an attempt to achieve maximum resolution. From a graphical presentation of partial derivatives (Figure 20-9) it can be seen that layer 6 affects periods between 15 and 150 seconds. The layers chosen for this shield model were of logarithmically increasing thickness. Layer 1 was arbitrarily chosen to be 10 km thick since fundamental mode Rayleigh-waves at teleseismic distances do not have the short-period components to resolve the top 10 km of the earth's crust. If the layers had been chosen half as thick, there would have been twice as many curves in Figure 20-9 but their individual spreads would have not decreased significantly. Even the choice here of six layers per ten-fold increase in thickness seems to have been high. Thus with a real lack of information, $\{\Delta c_k\}_{effective}$, and possibly too many layers, it is no wonder that (20-23) does not work. Even if you can perform the calculations implicit (using elimination methods on (20-22) rather than direct inversion) in (20-23), $\mathbf{P}$ will be nearly singular and some additional means of insuring stability is needed.

The next logical step is to consciously overdetermine the problem by getting data from as wide a frequency-band as possible, hopefully at least three octaves, 16-128 sec, for example. Then choose a small number of layers (the number will depend upon the quality of the data), say five. One can then use least squares techniques to obtain

$$\Delta v_j = \sum_{q,k} (\mathbf{P}^T\mathbf{P})^{-1}_{jq} P_{kq} \Delta c_k. \qquad (20\text{-}24)$$

This result will be somewhat more stable, but $\mathbf{P}^T\mathbf{P}$ may still be nearly singular and give extreme changes in Δv_j for noisy Δc_k. A successful example of the use of the least squares technique was that of Brune and Dorman (1963) where they constructed an eight-layer model of the Canadian Shield using five octaves of Rayleigh-wave data, plus additional Love-wave and body-wave information.

Given a set of partials, as in Figure 20-9, one can go through a "trial and error" sequence, stopping when the residuals Δc_k are below some acceptable level. In this process, solutions are successively modified. This is somewhat tedious but requires no sophisticated programming. However, one does not know how many other "acceptable" models there are or what they might be. An organized approach called the "hedgehog" method has been used by Knopoff (1972). Using considerable computer time, this method finds a lattice-work of acceptable models. This method gives a general feeling for the uncertainties in both depth and velocities.

In none of the above methods has anything been done to try and improve the resolution on a systematic basis. Backus and Gilbert (1967, 1968, 1970) and Gilbert (1971a) have developed a powerful formalism known as linear inverse theory. Condensed summaries are given by Parker (1970) and Gilbert (1972a). The Backus-Gilbert papers were largely concerned with free oscillation data, while Parker was inverting electrical conductivity data. A related study by Der et al (1970) was designed to treat surface wave data and is particularly straightforward for a first encounter with linear inverse theory. For our brief treatment, we shall generally follow their exposition.

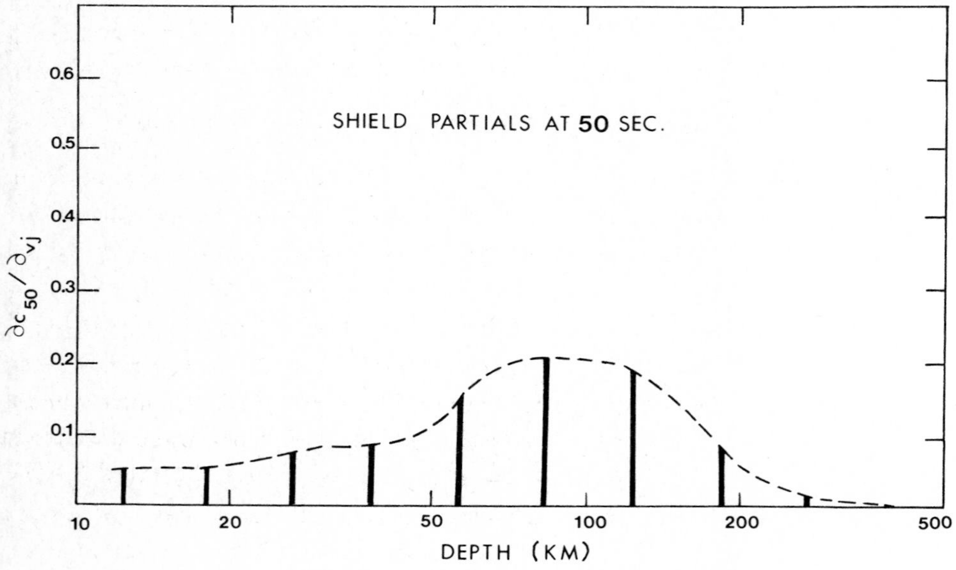

Figure 20-10. Rayleigh-wave phase-velocity partials at constant period (50 sec) plotted vs. depth. The bars are placed at the geometric mean depth of each layer.

An alternative solution of (20-22) can be obtained by trying to group the observations in such a way as to concentrate the resultant as much as possible around a given depth. That is, we create new data D_i such that

$$D_i \equiv \sum_k w_{ik} \Delta c_k = \sum_{k,j} w_{ik} P_{kj} \Delta v_j . \qquad (20\text{-}25)$$

But we can choose the weights w_{ik} rather arbitrarily. If we can choose them so that

$$\sum_k w_{ik} P_{kj} \simeq \delta_{ij}, \qquad (20\text{-}26)$$

where δ_{ij} is the Kronecker-delta, then from (20-22) we can write

$$D_i \equiv \sum_k w_{ik} \Delta c_k \simeq \Delta v_i .$$

But this is a first approximation to the solution for which we are looking. Here, then, is a third solution. In Figure 20-10, you can see that wave propagation at 50 sec period is affected by layers 2 through 10. The new grouping $\sum_k w_{ik} P_{kj}$ (the sum is over all partials affected by j^{th} layer) will be concentrated around the jth layer. Again, however, there are instability problems. The values of Δv_j so calculated may be large enough to violate both local linearity and common sense. To achieve some stability, Der et al (ibid) make the following adjustment.

If the data points are sufficiently far apart, small observational errors should not correlate and one can write

$$\mathrm{Var}\{D_i\} = \sum_k w_{ik}^2 \mathrm{Var}\{\Delta c_k\} \quad . \qquad (20\text{-}27)$$

For a reasonable inverse, these small observational errors should not produce wild changes in D_i, so we can require

$$\sum_k \ w_{ik}^2 \mathrm{Var}\{\Delta c_k\} = \text{minimum}. \qquad (20\text{-}28)$$

Finally, we come to the calculation of w_{ik}. We require that

$$\sum_k \ (w_{ik} P_{kj} - \delta_{ij})^2 = \text{minimum}. \qquad (20\text{-}29)$$

We can couple equations (20-28) and (20-29) together requiring a joint minimum:

$$\sum_k (w_{ik} P_{kj} - \delta_{ij})^2 + \beta^{(j)} \sum_k w_{jk}^2 \mathrm{Var}\{\Delta c_k\} = \text{minimum}. \qquad (20\text{-}30)$$

where $\beta^{(j)}$ is a constant for the jth layer which relates a "trade-off" between <u>resolution</u> (best approximation to δ_{ij}) and minimal variance. Small $\beta^{(j)}$ maximizes resolution at the expense of variance. Performing the required minimization, we find that the M equations of condition for the jth layer are:

$$\sum_{m,n} w_{jm} P_{mn} P_{kn} + \beta^{(j)} w_{jk} \mathrm{Var}\{\Delta c_k\} = P_{kj} . \qquad (20\text{-}31)$$

Der et al (1970) give a number of additional considerations necessary to the complete solution of the problem as well as resolution diagrams constructed according to (20-26). By adding in Love-waves and the first higher modes of both Rayleigh- and Love-waves, these authors show rather dramatically the increased resolution that may be obtained.

The story does not stop here; the problem posed by the set of equations (20-22) is a very general one; that of solving the matrix equation

$$\mathbf{A}\mathbf{x} = \mathbf{y} , \qquad (20\text{-}32)$$

where $\mathbf{A}$ is a rectangular (rather than square) matrix. A number of authors have written on its solution; a particularly lucid account is given by Lanczos (1961, Sec 3.7 ff). The stability of the inverse is related to the eigenvalues of $\mathbf{A}$ and its transpose. Smith and Franklin (1969) and Wiggins (1972) have cast the geophysical inverse problem in terms of this matrix formulation. Jackson (1972) gives a clear analysis of this method in terms of a "best" answer for the inverse problem.

A review summarizing inverse theory has been given by Parker (1977). Other aspects are treated by Parker (1972), Jackson (1973, 1976), Sabatier (1977), and Kennett and Nolet (1978).

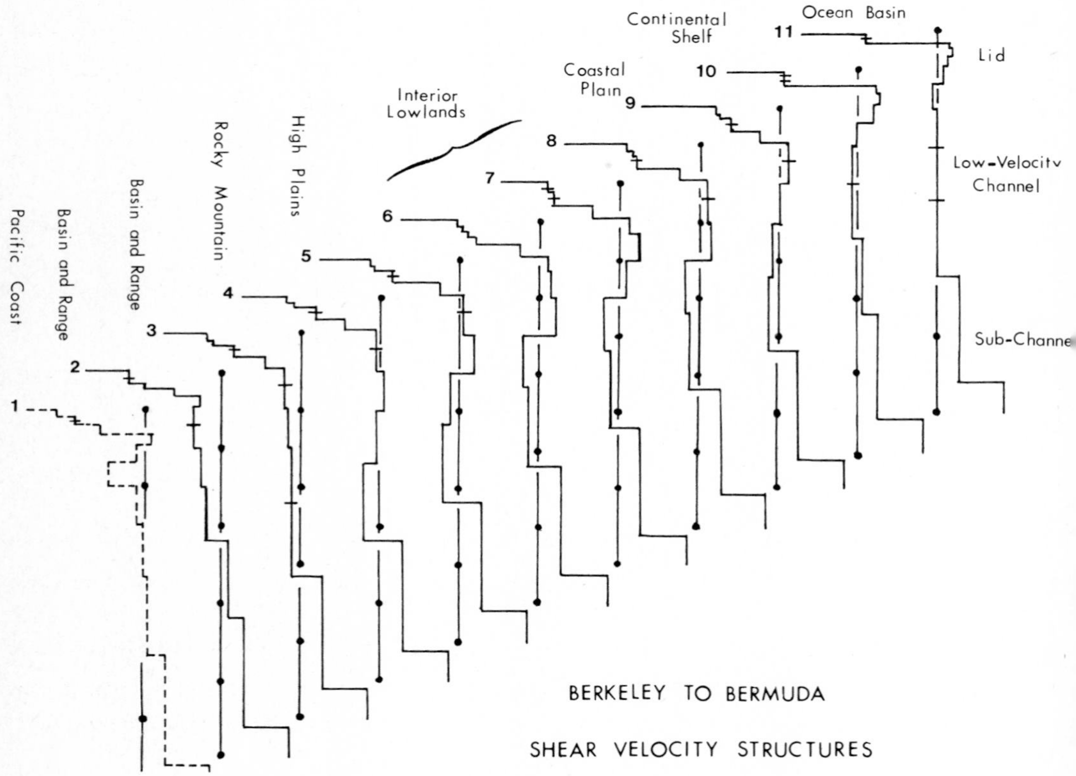

Figure 20-11. Models constructed to match Rayleigh-wave phase-velocity dispersion across the central United States. • mark 100 km depth intervals; vertical lines are at a velocity of 4.5 km/sec with 1.0 km/sec separation between the lines.

20.5. Models.

Because of their relative simplicity, most models of the crust and upper mantle have been constructed by one of the simpler schemes, mostly by the trial-and-error iterative method. We have already mentioned the least squares solution of Brune and Dorman (1963) and the "hedgehog" method (Knopoff, 1972a). Using a sequence of phase-velocity maps for 16 to 128 seconds period, I have constructed a series of models for shear velocity structure going from San Francisco to Bermuda, across the central United States. These are shown in Figure 20-11. All eleven models were constructed by the trial-and-error method, so no "best fit" is claimed. However, models departing very far from those shown would not fit the observations. Most structures except shields are shown. The figure shows crustal layering, the Moho transition (except for the Basin and Range where it might not be present), the lithospheric lid (again, except for the Basin and Range), the low velocity channel (asthenosphere) and a rather uniform sub-channel in the mantle. The reason for this uniformity is artificial, in that I did not have data to interpret structure this deeply and a model was used that matched world-wide averages. Where the fixed layers of the model do not fit a discontinuity, a two-step change is often shown; this is not necessarily real.

CHAPTER 21

ASYMPTOTIC RAY THEORY

21.1. Some Properties of the Scalar Wave Equation in Inhomogeneous Media.

Consider the scalar wave equation:

$$\nabla^2\Phi = \frac{1}{c^2}\frac{\partial^2\Phi}{\partial t^2}, \qquad (21\text{-}1)$$

where c is a variable wave velocity whose values depend upon position. If we assume a time dependence of $e^{-i\omega t}$, (21-1) becomes

$$\nabla^2\Phi + k_m^2\Phi = 0, \qquad (21\text{-}2)$$

where $k_m = \omega/c$ and is also a function of position. The solution of this equation is very difficult, even in special cases, but we can demonstrate several general features of ray theory. Analogous to the plane-wave solutions of Chapter 9, let us write

$$\Phi = A(x,y,z)e^{ik_oS(x,y,z)} \qquad (21\text{-}3)$$

as a general solution of (21-2). The factor $k_o = \omega/c_o$ is a wave-number associated with a reference wave velocity c_o. This reference velocity is introduced to conform to traditional nomenclature developed for optics where c_o is the velocity of light in free space. We then have

$$\frac{\partial\Phi}{\partial x} = ik_o\Phi\frac{\partial S}{\partial x} + \frac{\Phi}{A}\frac{\partial A}{\partial x} = ik_o\Phi\frac{\partial S}{\partial x} + \Phi\frac{\partial}{\partial x}(\ln A),$$

and

$$\frac{\partial^2\Phi}{\partial x^2} = -k_o^2\Phi\left(\frac{\partial S}{\partial x}\right)^2 + 2ik_o\Phi\left[\frac{1}{2}\frac{\partial^2 S}{\partial x^2} + \frac{\partial}{\partial x}(\ln A)\frac{\partial S}{\partial x}\right] + \Phi\left[\frac{\partial}{\partial x}(\ln A)\right]^2 + \Phi\frac{\partial^2}{\partial x^2}(\ln A).$$

Hence on substitution of (21-3) into (21-2) we obtain

$$\nabla^2\Phi + k_m^2\Phi = -k_o^2\Phi\left[\left(\frac{\partial S}{\partial x}\right)^2 + \left(\frac{\partial S}{\partial y}\right)^2 + \left(\frac{\partial S}{\partial z}\right)^2 - \frac{k_m^2}{k_o^2}\right] + 2ik_o\Phi\left[\tfrac{1}{2}\nabla^2 S + \nabla(\ln A)\cdot\nabla S\right]$$
$$+ \Phi\nabla(\ln A)\cdot\nabla(\ln A) + \Phi\nabla^2(\ln A). \qquad (21\text{-}4)$$

At high frequencies, the right hand side terms containing k_o will be dominant, and we find that equality can be approximated if we have

$$\left(\frac{\partial S}{\partial x}\right)^2 + \left(\frac{\partial S}{\partial y}\right)^2 + \left(\frac{\partial S}{\partial z}\right)^2 - n^2 = 0; \qquad (21\text{-}5)$$

where

$$n^2 = k_m^2/k_o^2,$$

and

$$\tfrac{1}{2}\nabla^2 S + \nabla(\ln A)\cdot\nabla S = 0. \qquad (21\text{-}6)$$

Equation (21-5) is known as the Eikonal Equation. The quantity n is known as the index of refraction. Equation (21-6) gives the geometrical spreading effect on the amplitude A. The terms of (21-4) which do not involve k_o have to be taken into account when we consider critically refracted rays and diffractive effects. We will have a bit more to say on this shortly. The surfaces S = constant are the wave surfaces on which all motion will be in phase, and the normals to this surface are the ray directions.

The normal direction is defined by the gradient of S, and a differential vector in the normal direction is given by

$$d\mathbf{x} = a'\nabla S, \tag{21-7}$$

where $\underline{a'}$ is a constant of proportionality. Since

$$(dx)^2 + (dy)^2 + (dz)^2 = (ds)^2 , \tag{21-8}$$

where $\underline{ds}$ is the element of arc length in the normal direction, we can also write

$$d\mathbf{x}/ds = a\nabla S, \tag{21-9}$$

where $\underline{a}$ is another constant to be determined. Evaluating, we should have

$$\left| \frac{d\mathbf{x}}{ds} \right|^2 = a^2 \left[\left(\frac{\partial S}{\partial x} \right)^2 + \left(\frac{\partial S}{\partial y} \right)^2 + \left(\frac{\partial S}{\partial z} \right)^2 \right] = a^2 n^2 = 1,$$

or

$$a = 1/n. \tag{21-10}$$

Then (21-9) becomes

$$n d\mathbf{x}/ds = \nabla S. \tag{21-11}$$

Taking a derivative of (21-11) with respect to arc length, we have

$$\frac{d}{ds}\left(n \frac{d\mathbf{x}}{ds}\right) = \frac{d}{ds}(\nabla S) = \nabla\left(\nabla S \cdot \frac{d\mathbf{x}}{ds}\right) = \nabla n. \tag{21-12}$$

This equation determines the ray paths as a function of the index of refraction.

<u>SH-Waves in a Vertically Inhomogeneous Medium</u>. From (5-2) and (5-7) we have

$$(\lambda u_{k},{}_{k}\delta_{ij}),{}_{i} + \{\mu(u_{i},{}_{j} + u_{j},{}_{i})\},{}_{i} = \rho \ddot{u}_j . \tag{21-13}$$

For SH-Waves in such a medium, we have

$$u_x = 0, \qquad u_y = u_y(x,z), \qquad u_z = 0, \qquad \mu = \mu(z), \qquad \rho = \rho(z).$$

This gives on substitution,

$$\frac{\partial}{\partial x}\left(\mu \frac{\partial u_y}{\partial x}\right) + \frac{\partial}{\partial z}\left(\mu \frac{\partial u_y}{\partial z}\right) = \rho \ddot{u}_y . \tag{21-14}$$

Performing the indicated operations, we have

$$\mu\nabla^2 u_y + \frac{\partial \mu}{\partial z}\frac{\partial u_y}{\partial z} = \rho \ddot{u}_y . \tag{21-15}$$

If we identify our variable velocity as

$$c = \left[\frac{\mu(z)}{\rho(z)} \right]^{\frac{1}{2}}, \tag{21-16}$$

then we will also have to know that $\partial\mu/\partial z$ is also very small before (21-15) can be put into the form of (21-1). Similar statements hold for P- and SV-Waves.

In an inhomogeneous earth, we see that the approximations of ray theory are many, and we should be glad that they are generally valid, for without them, we would lose one of the most valuable tools in our investigation of the earth.

21.2. Fermat's Principle.

This principle says that the time for a ray to travel between two points must be stationary with respect to small variations of the path. That is, the travel-time must be a minimum or a maximum, most often a minimum. We want to find stationary values of the integral

$$I = c_o \int_{P_1}^{P_2} dt = c_o \int_{P_1}^{P_2} \frac{ds}{c} = \int_{P_1}^{P_2} n ds. \tag{21-17}$$

Now, we can represent the arc length ds as a function of a dummy parameter dσ by

$$ds = \left[\left(\frac{\partial x}{\partial \sigma}\right)^2 + \left(\frac{\partial y}{\partial \sigma}\right)^2 + \left(\frac{\partial z}{\partial \sigma}\right)^2 \right]^{\frac{1}{2}} d\sigma. \tag{21-18}$$

Substitution into (21-17) gives

$$I = \int_{P_1}^{P_2} n(x,y,z) \left[\left(\frac{\partial x}{\partial \sigma}\right)^2 + \left(\frac{\partial y}{\partial \sigma}\right)^2 + \left(\frac{\partial z}{\partial \sigma}\right)^2 \right]^{\frac{1}{2}} d\sigma$$

$$= \int_{P_1}^{P_2} F(x,y,z,\frac{\partial x}{\partial \sigma},\frac{\partial y}{\partial \sigma},\frac{\partial z}{\partial \sigma}) d\sigma. \tag{21-19}$$

It is sufficient for a stationary value of I that Euler's Equations (Morse and Feschbach, 1953, p.277) be satisfied, i.e.,

$$\frac{\partial F}{\partial x_i} - \frac{d}{d\sigma} \left[\frac{\partial F}{\partial \left(\frac{\partial x_i}{\partial \sigma}\right)} \right] = 0, \qquad i = 1,2,3. \tag{21-20}$$

Carrying out the indicated operations on (21-19), we obtain

$$\left[\left(\frac{\partial x}{\partial \sigma}\right)^2 + \left(\frac{\partial y}{\partial \sigma}\right)^2 + \left(\frac{\partial z}{\partial \sigma}\right)^2 \right]^{\frac{1}{2}} \nabla n - \frac{d}{d\sigma} \left\{ \frac{n\left(\frac{\partial \mathbf{x}}{\partial \sigma}\right)}{\left[\left(\frac{\partial x}{\partial \sigma}\right)^2 + \left(\frac{\partial y}{\partial \sigma}\right)^2 + \left(\frac{\partial z}{\partial \sigma}\right)^2 \right]^{\frac{1}{2}}} \right\} = 0.$$

But using (21-18), this becomes

$$\frac{d}{ds}\left(n \frac{d\mathbf{x}}{ds}\right) = \nabla n. \tag{21-21}$$

Thus Fermat's principle tells us that the stationary time path is the ray path given by the Eikonal Equation. This gives us a somewhat simpler means of computation, for we no longer have to solve the non-linear Eikonal Equation to get the ray paths. However, we do not get this for free. Without a solution to the Eikonal Equation, equation (21-6) cannot be solved for the amplitude over the wave surface. It turns out, though, that sometimes rather simple geometrical considerations will give us the amplitude relations we need.

21.3. Rays in a Vertically Inhomogeneous Medium.

In such a medium, the parameters describing its physical properties vary as a function of depth only. They may vary either continuously or in a jump-like manner. We can compute the time necessary for a disturbance to go from point P_o to point P along the path $\overline{P_oQ}$ in Figure 21-1. This time can be written as

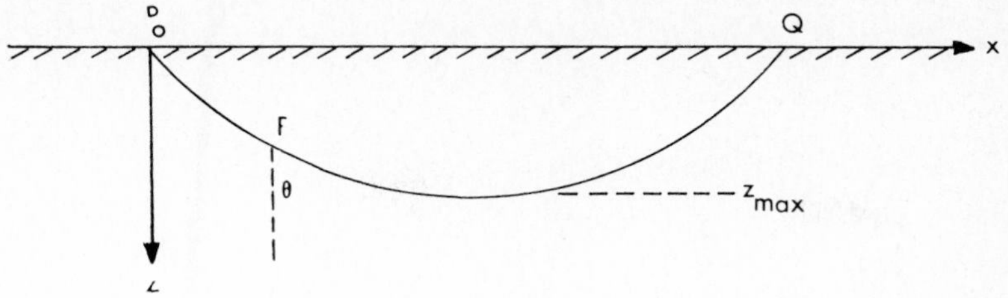

Figure 21-1. Ray path geometry in a vertically inhomogenous medium.

$$T(P) = \int_{P_o}^{P} c^{-1} ds = \int_{P_o}^{P} c^{-1} \left[(dz)^2 + (dx)^2 \right]^{\frac{1}{2}} = \int_{P_o}^{P} c^{-1} \left[1 + (dz/dx)^2 \right]^{\frac{1}{2}} dx. \quad (21\text{-}22)$$

If we write $z' = dz/dx$, and set

$$c^{-1} \left[1 + (dz/dx)^2 \right]^{\frac{1}{2}} = F(z,z'),$$

the calculus of variations tells us that in order to have a stationary value of the travel-time T(P), in this case a minimum time, we must have

$$\frac{d}{dx} \left(\frac{\partial F}{\partial z'} \right) - \frac{\partial F}{\partial z} = 0.$$

But

$$\frac{d}{dx} \left[F(z,z') - z' \frac{\partial F}{\partial z} \right] = z' \left[\frac{\partial F}{\partial z} - z' \frac{d}{dx} \left(\frac{\partial F}{\partial z'} \right) \right] = 0,$$

by the stationarity condition. Consequently, the bracketed expression is a first integral of the ray path. Now

$$z' \partial F/\partial z' = c^{-1} (z')^2 \left[1 + (z')^2 \right]^{-\frac{1}{2}} .$$

Combining, we have that

$$F(z,z') - z' \partial F/\partial z' = c^{-1} \left[1 + (z')^2 \right]^{-\frac{1}{2}} = p;$$

where $\underline{p}$ is a constant known as the ray parameter. But $z' = \mathrm{Cot}\theta$, so we can write our first integral as

$$\frac{\mathrm{Sin}\theta(z)}{c(z)} = p. \quad (21\text{-}23)$$

This is just the generalized form of Snell's Law for such a medium.

We can relate the ray parameter $\underline{p}$ to two physically observable quantitites. At the surface, we have

$$p = \mathrm{Sin}\theta_s / c_s = (v_{apparent})^{-1} = dT/dX , \quad (21\text{-}24)$$

where T is the travel-time to a surface distance X. The second is that at the maximum depth of penetration of the ray, $\theta = 90^o$ and

$$p = 1/c_{max}. \quad (21\text{-}25)$$

Eliminating the parameter $\underline{p}$ from these two equations gives one further relation,

$$dT/dX = 1/c_{max}; \quad (21\text{-}26)$$

i.e., the apparent surface velocity is equal to the medium velocity at the depth of greatest penetration.

From (21-23) the following may be derived on differentiation with respect to an element of arc length along the ray:

$$- c^{-2} \operatorname{Sin}\theta \; dc/ds + c^{-1} \operatorname{Cos}\theta \; d\theta/ds = 0.$$

But

$$dc/ds = dc/dz \; dz/ds = dc/dz \operatorname{Cos}\theta;$$

and on substitution, we have

$$d\theta/ds = p \; dc/dz. \tag{21-27}$$

This says that the curvature of a ray in a vertically inhomogeneous medium is directly proportional to the velocity gradient.

We can now proceed to look at the variation of energy flux as rays travel away from the source as in Figure 21-2. We will assume that the source is uniformly radiating power at the rate of P ergs/sec/steradian. Then the intensity at any

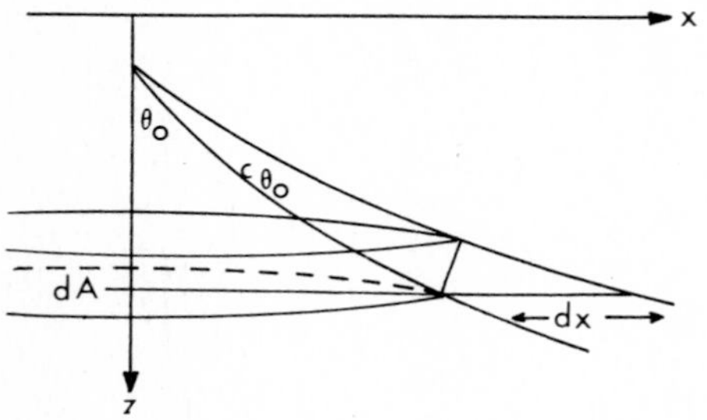

Figure 21-2. Ray geometry for energy flux calculations.

distance will be given by

$$I dA = P d\Omega_o \; ; \tag{21-28}$$

where dA is the area swept out by a ray bundle between θ_o and $\theta_o + d\theta_o$ as it is rotated about the z-axis, and $d\Omega_o$ is the solid angle at the source subtended by the bundle. We have then

$$I = P \frac{d\Omega_o}{dA} = P \frac{2\pi \operatorname{Sin}\theta_o d\theta_o}{2\pi x |\operatorname{Cos}\theta| dx} = P \frac{\operatorname{Sin}\theta_o d\theta_o}{x |\operatorname{Cos}\theta| dx} \; . \tag{21-29}$$

Now $\underline{x}$ is a function of both the depth of penetration and of the original angle with which the bundle of rays left the source, but at constant depth $\underline{z}$, we have

$$dx = \frac{\partial x}{\partial \theta_o} d\theta_o \; . \tag{21-30}$$

We can then rewrite (21-29) as

$$I(x,\theta_o) = \frac{P \operatorname{Sin}\theta_o}{x |\operatorname{Cos}\theta \, \partial x/\partial\theta_o|} \; . \tag{21-31}$$

If the velocity increases monotonically with depth, then all rays leaving the source will strike the surface. If we let the surface values of $\underline{x}$ be given by X, then on the surface we will have

$$I(X,\theta_o) = \frac{P \operatorname{Sin}\theta_o}{X |\operatorname{Cos}\theta_s \, \partial X/\partial\theta_o|} \; . \tag{21-32}$$

From (21-23) and (21-24) we have that

$$\frac{d}{dX}\left(\frac{\text{Sin}\theta_o}{c_o}\right) = \frac{\text{Cos}\theta_o}{c_o}\frac{d\theta_o}{dX} = \frac{d}{dX}\left(\frac{dT}{dX}\right) = \frac{d^2T}{dX^2} ;$$

since c_o is the velocity at the source and is not a function of θ_o. Upon substitution, (21-32) then becomes

$$I(X,\theta_o) = \frac{Pc_o\text{Tan}\theta_o}{X}\left|\frac{1}{\text{Cos}\theta_s}\frac{d^2T}{dX^2}\right| . \qquad (21\text{-}33)$$

The surface amplitude is thus related to the second derivative of the travel-time curve. The factor X takes care of the cylindrical spreading of the energy over all depths from 0 to ∞.

Finally, we wish to look at the travel-time itself. For convenience we will consider a surface source. In this case

$$T(p) = 2\int_0^{z_{max}} \frac{ds}{c} = 2\int_0^{z_{max}} \frac{dz}{c\ \text{Cos}\theta} = 2\int_0^{z_{max}} \frac{dz}{c(1 - p^2c^2)^{\frac{1}{2}}} , \qquad (21\text{-}34)$$

and

$$X(p) = 2\int_0^{z_{max}} \text{Tan}\theta dz = 2p\int_0^{z_{max}} \frac{c\ dz}{(1 - p^2c^2)^{\frac{1}{2}}} . \qquad (21\text{-}35)$$

If we define the inverse velocity as a new variable, i.e.,

$$q = 1/c, \qquad (21\text{-}36)$$

the two previous equations may be rewritten:

$$T(p) = 2\int_0^{z_{max}} \frac{q^2dz}{(q^2 - p^2)^{\frac{1}{2}}} , \qquad (21\text{-}37)$$

and

$$X(p) = 2p\int_0^{z_{max}} \frac{dz}{(q^2 - p^2)^{\frac{1}{2}}} . \qquad (21\text{-}38)$$

Either pair of equations, (21-34) and (21-35) or (21-37) and (21-38), parametrically determine the travel-time curve T = T(X). This will be true regardless of low-velocity regions as long as we stay on the ray. Problems with such regions will arise when we try to invert, however. A new variable τ defined by

$$\tau(p) = T(p) - pX(p) , \qquad (21\text{-}39)$$

has been defined by Gerver and Markushevich (1966). As we shall see, this variable has a number of useful properties and can be helpful in the inverse problem. From (21-37) and (21-38), we have that

$$\tau(p) = 2\int_0^{z_{max}} (q^2 - p^2)^{\frac{1}{2}}dz. \qquad (21\text{-}40)$$

Geometrically, τ is the intercept to the travel-time curve since p = dT/dX. Likewise from (21-24) we see that

$$\frac{d\tau}{dp} = \frac{d\tau}{dX}\frac{dX}{dp} - X - p\frac{dX}{dp} = -X. \tag{21-41}$$

We can get some information about the general behavior of the travel-time curve due to a surface source, by recasting (21-38). First of all, taking the derivative with respect to the ray parameter p we have

$$\frac{dX}{dp} = 2\int_0^{z_{max}} \frac{dz}{(q^2 - p^2)^{\frac{1}{2}}} + 2p\frac{d}{dp}\int_0^{z_{max}} \frac{dz}{(q^2 - p^2)^{\frac{1}{2}}}. \tag{21-42}$$

But

$$\int_0^{z_{max}} \frac{dz}{(q^2 - p^2)^{\frac{1}{2}}} = \int_0^{z_{max}} \frac{dz}{dq}\frac{dq/dz}{(q^2 - p^2)^{\frac{1}{2}}}\,dz = \int_0^{z_{max}} f(q)\frac{dq/dz}{(q^2 - p^2)^{\frac{1}{2}}}\,dz$$

$$= f(q)\mathrm{Cosh}^{-1}\left(\frac{q}{p}\right)\Big|_{q_s}^{p} - \int_0^{z_{max}} f'(q)\mathrm{Cosh}^{-1}\left(\frac{q}{p}\right)\frac{dq}{dz}\,dz$$

$$= -f(q_s)\mathrm{Cosh}^{-1}\left(\frac{q_s}{p}\right) - \int_0^{z_{max}} f'(q)\mathrm{Cosh}^{-1}\left(\frac{q}{p}\right)\frac{dq}{dz}\,dz;$$

where we have written $f(q)$ for dz/dq. Differentiating this expression with respect to p, we can substitute into (21-42) obtaining

$$\frac{dX}{dp} = \frac{2q_s f(q_s)}{(q_s^2 - p^2)^{\frac{1}{2}}} + 2\int_0^{z_{max}} \left[f(q) + qf'(q)\right]\frac{dq/dz}{(q^2 - p^2)^{\frac{1}{2}}}\,dz. \tag{21-43}$$

Now we make another change of variables, i.e.,

$$\zeta(z) = c^{-1}dc/dz = d(\ln c)/dz. \tag{21-44}$$

Then

$$qf(q) = q\,dz/dq = -c\,dz/dc = -\zeta^{-1}.$$

But

$$\frac{d}{dz}\left[qf(q)\right] = \left[f(q) + qf'(q)\right]\frac{dq}{dz} = \frac{1}{\zeta^2}\frac{d\zeta}{dz}.$$

Substituting these values into (21-43) leads to

$$\frac{dX}{dp} = -\frac{2}{\zeta_s(q_s^2 - p^2)^{\frac{1}{2}}} + 2\int_0^{z_{max}} \frac{d\zeta/dz}{\zeta^2(q^2 - p^2)^{\frac{1}{2}}}\,dz. \tag{21-45}$$

This expression gives the slope of the p vs. X curve. Since $p = dT/dX$, we can now get a qualitative picture of the behavior of the T vs. X curve. The two terms of (21-45) give us the information we need. First of all, we see that

$$(q_s^2 - p^2)^{\frac{1}{2}} = q_s\mathrm{Cos}\theta, \tag{21-46}$$

which says that the first term will be large and negative for rays that just penetrate the surface of the earth. This term will then decrease in magnitude as p decreases and the ray penetrates more deeply. (See Fig. 21-2.) The second term will cancel the first one only when $d\zeta/dz$ is sufficiently large to give a significant positive contribution to the integrand. Normally, the factor $d\zeta/dz$ will be very small; it is zero for $c^{-1}dc/dz$ = constant, i.e., for velocities such as

$$c = c_s e^{\alpha z} \qquad (\alpha > o).$$

A general analysis of (21-45) can be found in Officer (1958, p. 56 ff). We shall look at a few specific cases involving media where the velocities change uniformly with depth.

21.4. Rays in a Medium With Uniform Gradients.

Taking the derivative of Snell's Law (21-23) with respect to z we find that

$$\text{Cos}\theta \frac{d\theta}{dz} = p \frac{dc}{dz} = pc', \tag{21-47}$$

where c' is a constant expressing the velocity gradient in the medium. Then

$$X = 2 \int_{\theta_s}^{\pi/2} \text{Tan}\theta dz(\theta) = 2 \int_{\theta_s}^{\pi/2} \frac{\text{Sin}\theta}{\text{Cos}\theta} dz(\theta) = \frac{2}{c'p} \int_{\theta_s}^{\pi/2} \text{Sin}\theta d\theta = \frac{2\text{Cos}\theta_s}{c'p}. \tag{21-48}$$

But

$$\text{Cos}\theta_s = (1 - p^2c_s^2)^{\frac{1}{2}}.$$

Hence

$$X(p) = \frac{2(1 - p^2c_s^2)^{\frac{1}{2}}}{c'p} = \frac{2c_s}{c'} \text{Cot}\theta_s, \tag{21-49}$$

and

$$\frac{dX}{dp} = - \frac{2}{c'p^2(1 - p^2c_s^2)^{\frac{1}{2}}}. \tag{21-50}$$

These two relationships are sketched in Figure 21-3. Likewise

$$T = 2 \int_0^z \frac{dz}{c\ \text{Cos}\theta} = \frac{2}{c'} \int_{\theta_s}^{\pi/2} \frac{d\theta}{\text{Sin}\theta} = \frac{2}{c'} \ln\left[\text{Cot}(\theta_s/2)\right]. \tag{21-51}$$

We then find that

$$\text{Sinh}(\frac{c'T}{2}) = \tfrac{1}{2}\left[e^{\frac{1}{2}c'T} - e^{-\frac{1}{2}c'T}\right] = \tfrac{1}{2}\left[\text{Cot}(\theta_s/2) - \text{Tan}(\theta_s/2)\right] = \text{Cot}\theta_s.$$

But from (21-51) we find that

$$\text{Cot}\theta_s = \frac{c'X}{2c_s},$$

and we have

$$T = \frac{2}{c'} \text{Sinh}^{-1}(\frac{c'X}{2c_s}). \tag{21-52}$$

The travel-time curve (21-52) and the ray paths are also sketched in Figure 21-3.

To evaluate the intensity, we need to calculate

$$\frac{\partial X}{\partial \theta} = - \frac{2c_s}{c'\text{Sin}^2\theta_s} = - \frac{X}{\text{Cos}\theta_s \text{Sin}\theta_s},$$

and from (21-31) we have

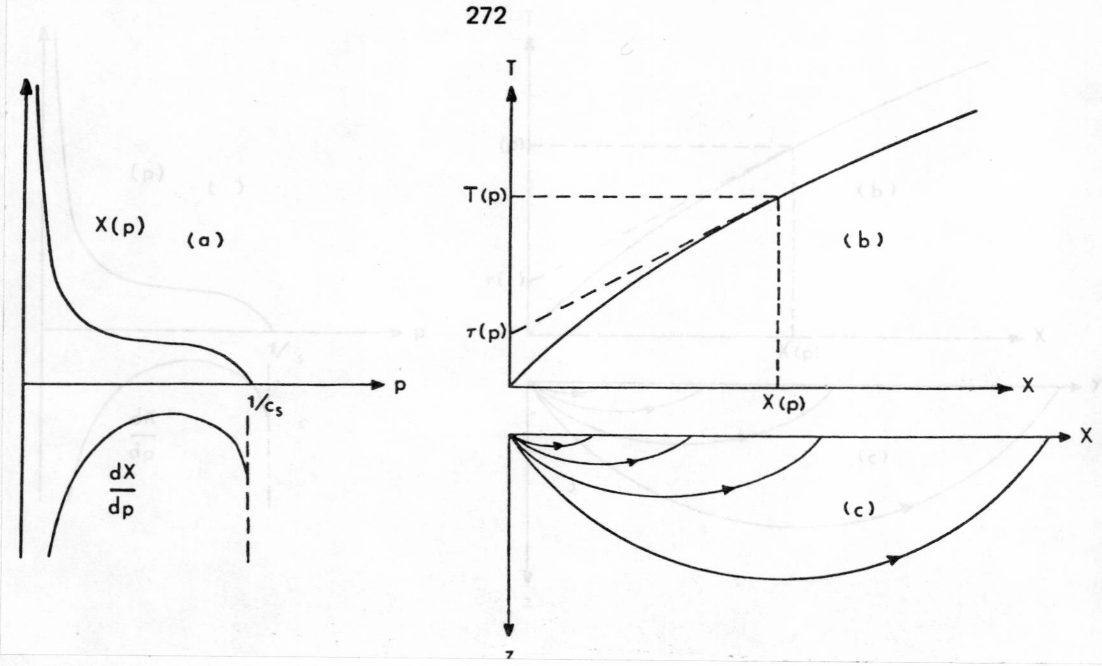

Figure 21-3. X(p), dX/dp, travel-time curve, τ(p), and ray paths in a medium with steadily increasing velocity.

$$I = \frac{PSin^2\theta_s}{X^2} = \frac{P}{X^2}\left[1 + \frac{c'^2X^2}{4c_s^2}\right]^{-1} . \tag{21-53}$$

Inasmuch as X increases as p decreases, i.e., as θ decreases, this is a faster rate of dropoff than normal geometric divergence.

Let us now calculate the moveout X and the travel-time T in a surficial layer of material of uniform velocity gradient as in Figure 21-4. Following the same line of reasoning as above, we find that

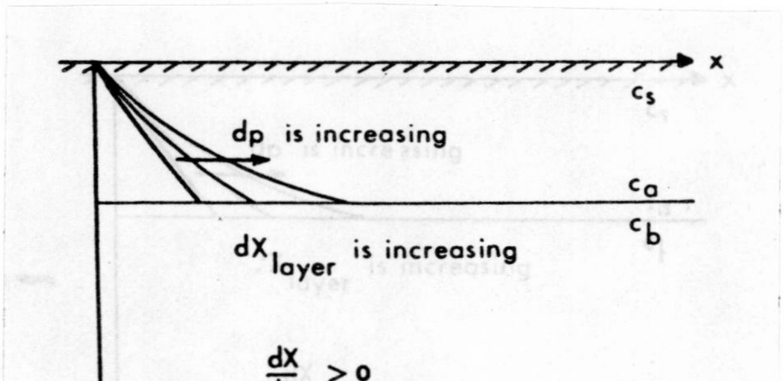

Figure 21-4. A surficial layer with uniform velocity gradients. Here dX/dp is positive.

$$X_{layer} = \frac{2c_s}{c_s'}\left[\frac{\cos\theta_s - \cos\theta_a}{\sin\theta_s}\right], \quad (21\text{-}54)$$

$$T_{layer} = \frac{2}{c_s'}\ln\left[\frac{\tan(\theta_a/2)}{\tan(\theta_s/2)}\right]. \quad (21\text{-}55)$$

This represents that portion of the travel-time and that portion of the moveout distance actually spent by the ray in the surficial layer. Since all rays are portions of circles, the geometry of Figure 21-4 shows that dX/dp will be positive for all rays which intersect the interface at angles between 0° and 90°. Since this is just opposite of the behavior of dX/dp when the entire ray path lies in a medium with a uniform velocity gradient, we see that there exists a possibility for a reversal of sign depending upon what the ray does as it enters the second medium. With this in mind, we can discuss the effect of a jump increase or a jump decrease in a medium with an otherwise uniform velocity gradient.

Jump Increase in Velocity. Figure 21-5 shows the various relationships when a low-speed layer overlies a high-speed half-space. Ray A lies entirely within the

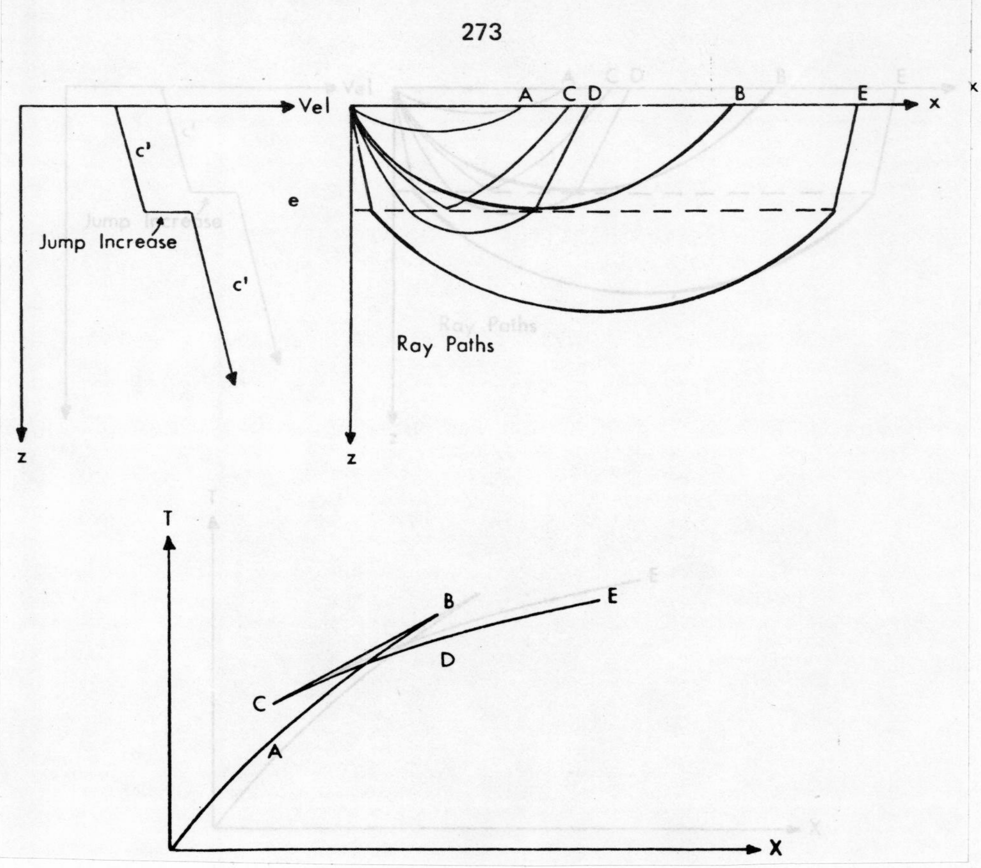

Figure 21-5. Relationships for a medium in which there is a jump increase in velocity with otherwise uniform gradients.

surficial layer. Ray B is a limiting ray; rays whose ray parameter is less than that of Ray B are reflected at the interface. Ray C is the ray which strikes the discontinuity at the critical angle. Ray D, on refraction, has upward curvature and consequently its total distance is less than that of ray B. As the rays penetrate more deeply, they again move out in distance as in ray E. The end result is a triplication in the travel-time curve. At the ends of the travel-time branches (points B and C) there will be amplification of the wave-form due to simultaneous arrival of energy with two ray-parameters.

Jump Decrease in Velocity. In this case we have a low-velocity channel. Again ray A propagates wholly in the layer. Ray B is the limiting ray; rays with ray parameter less than ray B are refracted downwards into the lower medium as well as reflected. However, there is no critical angle in this case. Ray C represents a ray with an infinitesimal decrease in ray parameter from ray B. Ray D travels a longer distance in the lower medium than ray C, but as we have discussed above, it travels much less in the layer and the total result is a decrease in moveout over ray C. Again as the rays penetrate more deeply, we get a normal moveout as in ray E. Between

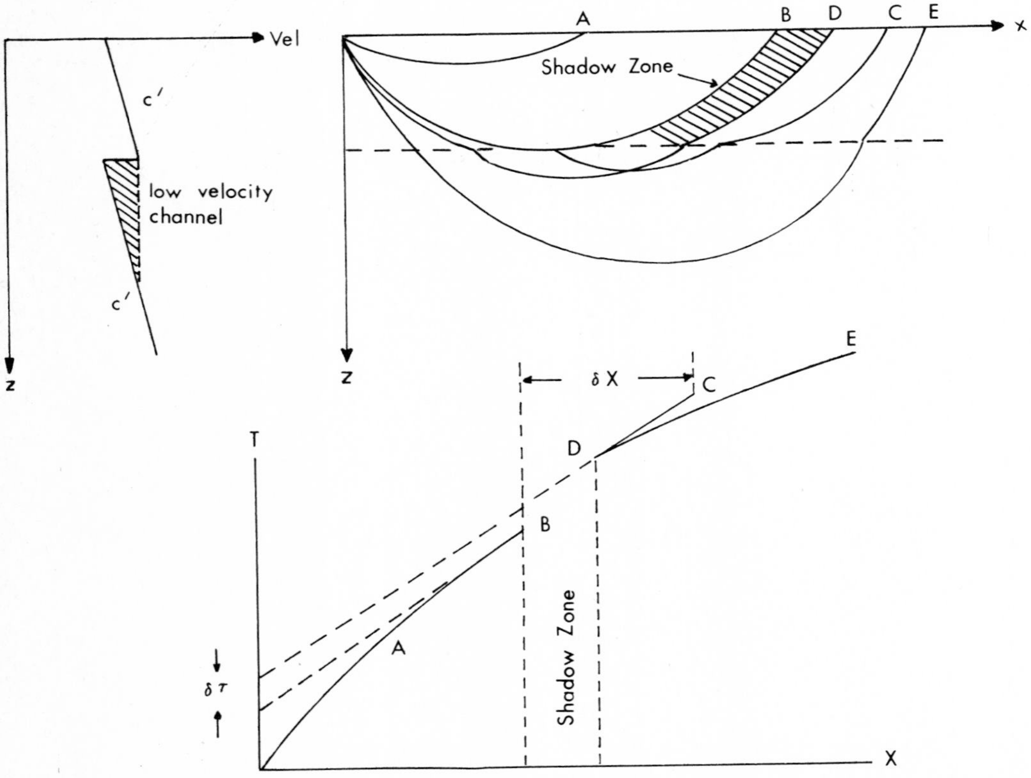

Figure 21-6. Relationships for a medium in which there is a jump decrease in velocity with otherwise uniform gradients. Rays B and C have the same slope, dT/dX, but there is a jump in both T and X values.

the points of emergence for rays B and D we have a <u>shadow zone</u>. Also, the travel-time curve from C to E shows a cusp at D. The relationships are diagrammed in Figure 21-6.

In media where the velocity gradient is not uniform, the effects described above are still true, although the analysis becomes more complicated. Thus in a vertically inhomogeneous medium, large increases (including discontinuous ones) lead to a triplication of the travel-time curve and all decreases lead to a shadow zone.

A large number of other cases are treated by Kauffman (1953) in which he works out temporal and spatial coordinates for any point on the ray path as well as surface to surface travel-time curves.

21.5. Inversion for a Vertically Inhomogeneous Medium.

We will start with the equation

$$X(p) = 2p \int_0^{z_{max}} \frac{dz}{(q^2 - p^2)^{\frac{1}{2}}} = 2p \int_{q_s}^{p} \frac{dz/dq}{(q^2 - p^2)^{\frac{1}{2}} dq} . \qquad (21\text{-}38)$$

If we now integrate both sides of this equation with respect to p, including all rays from $\theta = 90^\circ$ to the ray which becomes horizontal at depth z_1 where $q_1 = (c_1)^{-1}$, we have

$$\int_{q_s}^{q_1} \frac{X(p)dp}{(p^2 - q_1^2)^{\frac{1}{2}}} = 2 \int_{q_s}^{q_1} dp \int_{q_s}^{p} \frac{p \; dz/dq \; dq}{(p^2 - q_1^2)^{\frac{1}{2}}(q^2 - p^2)^{\frac{1}{2}}} .$$

Inverting the order of integration, we have

$$\int_{q_1}^{q_s} \frac{X(p)dp}{(p^2 - q_1^2)^{\frac{1}{2}}} = 2 \int_{q_s}^{q_1} \frac{dz}{dq} dq \int_{q_1}^{q} \frac{pdp}{(p^2 - q_1^2)^{\frac{1}{2}} (q^2 - p^2)^{\frac{1}{2}}} \qquad (21\text{-}56)$$

$$= \pi \int_{q_s}^{q_1} \frac{dz}{dq} dq = \pi \int_0^{z_1} dz = \pi z_1 .$$

The region of integration is shown in Figure 21-7 to clearly demonstrate the new

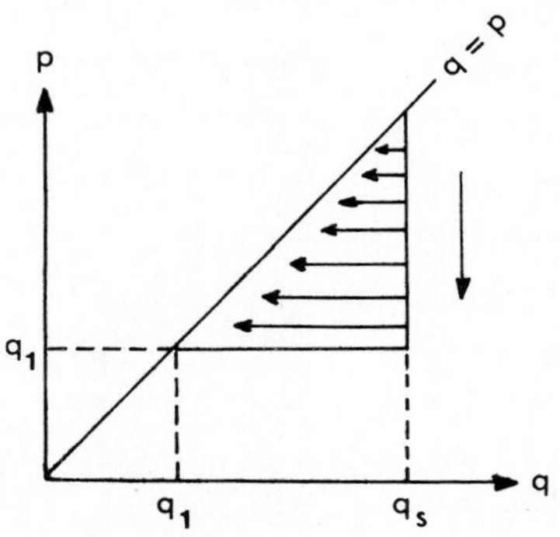

Figure 21-7. Limits of integration in the inversion problem.

limits on the integrals. Now the left hand side can be integrated by parts, giving

$$X(p)\mathrm{Cosh}^{-1}\left(\frac{p}{q_1}\right)\Bigg|_{q_1}^{q_s} - \int_{q_1}^{q_s} \frac{dX(p)}{dp}\mathrm{Cosh}^{-1}\left(\frac{p}{q_1}\right)dp.$$

But $X = 0$ when $p = q_s$, so that the first term vanishes. Changing the limits of integration in the second term, we find that

$$\int_0^{X_1} \mathrm{Cosh}^{-1}\left(\frac{p}{q_1}\right)dX = \pi z_1. \qquad (21\text{-}57)$$

But $p = dT/dX$, and $q_1 = dT/dX\,|_{X=X_1}$. Hence we have

$$z_1 = \frac{1}{\pi}\int_0^{X_1} \mathrm{Cosh}^{-1}\left[\frac{dT/dX}{(dT/dX)_{X=X_1}}\right] dX. \qquad (21\text{-}58)$$

The integral in (21-58) must be done numerically from the experimental travel-time curve. Remembering that at the deepest point of the ray's penetration we have

$$c(z_1) = \left[\frac{dT}{dX}\Big|_{X=X_1}\right]^{-1},$$

we can then construct a velocity-depth curve. It is obvious from the derivation that $X(p)$ must be continuous and single-valued. This means that all decreases in velocity (low-velocity zones) invalidate (21-58).

The above method of derivation of the inversion formula was due to Rasch (see Jeffreys, 1970, p.50), and will be used again in the radially inhomogeneous case. However, we jumped rather quickly past one point (for the sake of continuity) in which we gave the value of

$$I = \int_a^b \frac{p\,dp}{(p^2 - a^2)^{\frac{1}{2}}(b^2 - p^2)^{\frac{1}{2}}} = \pi/2. \qquad (21\text{-}59)$$

We can evaluate this integral as follows. First of all, change to a new variable, $\xi = 1/p$. Then we have

$$I = \int_{1/b}^{1/a} \frac{d\xi}{\xi(1 - a^2\xi^2)^{\frac{1}{2}}(b^2\xi^2 - 1)^{\frac{1}{2}}},$$

where the contour of integration runs as in Figure 21-8. The integrand dies away sufficiently rapidly so that we may close the contour at infinity. By Cauchy's theorem, we have that

$$4I = 2\pi i(\text{Sum residues enclosed}) = 2\pi i\left(\frac{1}{i}\right),$$

or that $I = \pi/2$ as we have stated in (21-59).

Finally we note that the variable τ may also be inverted to determine a velocity depth function (Bessonova et al, 1976). From (21-56) we have that

$$z_1 = \frac{1}{\pi}\int_{q_1}^{q_s} \frac{X(p)dp}{(p^2 - q_1^2)^{\frac{1}{2}}}.$$

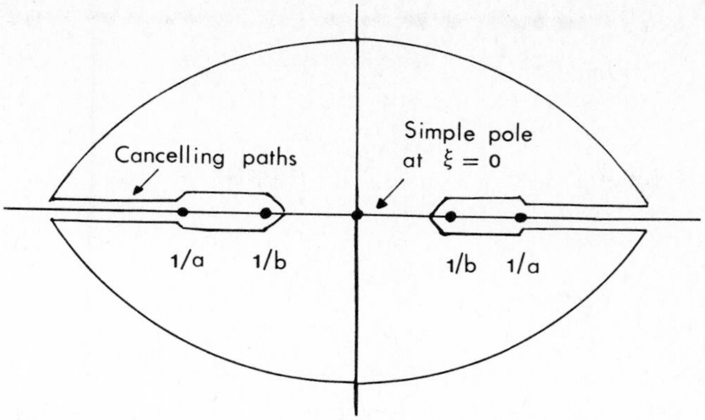

Figure 21-8. Contour of integration for evaluation of (21-59).

Then from (21-41) we have that

$$d\tau = - X(p)dp ,$$

so we can finally write

$$z_1 = \frac{1}{\pi} \int_0^{\tau(q_1)} \frac{d\tau}{\{p^2(\tau) - q_1^2\}^{\frac{1}{2}}} . \tag{21-60}$$

The above analytic methods of inversion have a number of difficulties. Besides the obvious one of failing to work in the presence of a low-velocity zone, it also requires that the whole travel-time curve be known including the exact nature of all triplications. Now the travel-time curve as given by (21-37) and (21-38) is a ray-parametric curve and not a first-arrival curve. That is, the triplications represent second and third arrivals. These later arrivals are difficult to observe and seismologists do not always agree on the number and nature of the branches of a seismic travel-time curve.

The problem of the low-velocity zone was considered by Slichter (1932) where he examined some minimum and maximum properties of a single zone which could be deduced from the two travel-time discontinuities $\delta\tau$ and δX (Figure 21-6). Gerver and Markushevich (1966, 1967) made an extensive study of the properties of realizable travel-time curves. They showed that the inversion could be carried out while retaining an ambiguity due to the unknown details of the velocity structure in the channels. They further indicated that if sources were located between each of the channels, then the channel positions could be determined, the between-channel velocity could be calculated, and that certain limits could be placed on the indeterminate material velocity within the channel.

Following up on these studies, Bessonova and others (Bessonova et al; 1970, 1974, 1976) developed the τ-method for the inversion of travel-times. This involved calculating limits for the variable $\tau(p)$ (21-39) from the envelope of observed travel-time data. They then transformed these limits into an envelope for the velocity-depth function. In their 1976 paper, they not only suggested inversion according to (21-60) but also made the necessary modifications to include low-velocity channels. An alternative formulation of this type scheme has been given by McMechan and Wiggins (1972). A rather different approach has been taken by Davies and Chapman (1975).

L. E. Johnson and F. Gilbert (1972a and b) have made extensive use of the variable

$$\tau = T - pX$$

in a lineared inversion technique for rays in a spherical earth. We shall show in Chapter 22 that for travel-time calculations, the spherical-earth equivalents of T(p) and X(p) can be reduced to a flat-earth model by the appropriate transformation. Kennett (1976) works out the details for a flat-earth model and applies both the Johnson-Gilbert linear inversion scheme and the Bessonova extremal-bound procedure to invert a long-range refraction profile in France. A similar dual application to a marine profile is made by Kennett and Orcutt (1976).

Returning to the problem of the details of the velocity-structure in the low-velocity channel, McMechan (1977) works out some necessary relations for determining this enigmatic variable. He notes that the quantity of detailed information necessary precludes extensive application.

21.6. Other Considerations.

Waves at Transition Zones. The reflection and refraction of scalar waves in general inhomogeneous media is complex, but there exists a small class of regions described by hypergeometric regions that can be treated exactly (Phinney, 1970). M. Hron and C. H. Chapman (1974a) study the "shadow-zone" phenonomena from such a region and the same authors (1974b) study reflection and head-waves from transition regions. Although P-SV waves cannot be treated in this manner, the scalar case provides some additional understanding of the processes involved.

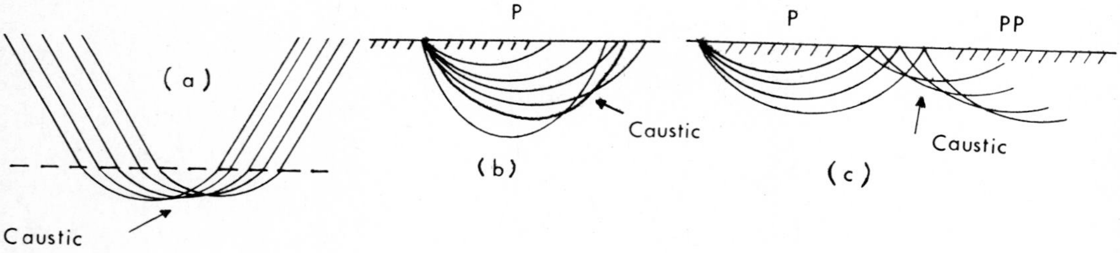

Figure 21-9a. A caustic formed at the turning point of plane waves entering an inhomogeneous region.

b. A caustic formed with a rapid increase in velocity with depth.

c. A caustic formed by reflection from the surface of an inhomogeneous medium.

Phase Changes at Caustics. Tolstoy (1968) cleared up a misunderstanding that had existed concerning the turning points of seismic rays. He pointed out that phase-shifts were associated not with the turning point but with the fact that a ray had touched a caustic. The error is easy to comprehend, for in the simple case of plane waves in inhomogeneous media treated by the WKBJ method of approximation, there is a phase shift of $\pi/2$ associated with the turning point. The situation is illustrated in Figure 21-9a. In this particular case, however, it is a coincidence that the turning points and the caustic surface coincide. In cases (b) and (c) this is not the case. Besides the examples shown in Figure 21-9, caustics are found associated with low-velocity zones and with crossing the axial line in a spherically symmetric body. Hill (1974) and Choy and Richards (1975) point out that caustic surfaces are associated with non-minimum time-paths and that the phase-shift of $\pi/2$ is a high-frequency phenomenon. They also give a number of examples of what phase-shifts do to pulse shapes. We have already discussed shape changes caused by phase-shifts in Section 9.5.

Ray Tracing. In simple media which vary only with the vertical (or radial direction), ray paths can be traced out in an approximate model constructed by straight-line segments (homogeneous layers) or by segments of constant curvature (dc/dz = constant for vertical media and $c(r) = a - br^2$ for spherical media). No recourse is needed to the integration of (21-21). However, for those seismic rays generated from earthquakes a problem arises. In the source region, the crustal structure can only be adequately represented by a three-dimensional model. One then needs to numerically integrate (21-21). Jacob (1970), Sorrells et al (1971), Cerveny et al (1974), and Julian and Gubbins (1977) all discuss aspects of this problem.

Ray Series. In the first part of the chapter, we wrote out the requirement that $A\exp\{ik_oS\}$ be a solution of the wave equation in inhomogeneous scalar media as equation (21-4). We noted that A and S were coupled together through the solution

$$\tfrac{1}{2}\nabla^2 S + \nabla(\ln A)\cdot\nabla S = 0. \qquad (21\text{-}6)$$

This was valid to first order, but we have already seen that higher order terms exist in the case of head waves. To take full account of these higher terms, we must write

$$2ik_o\left[\tfrac{1}{2}\nabla^2 S + \nabla(\ln A)\cdot\nabla S\right] + \nabla(\ln A)\cdot\nabla(\ln A) + \nabla^2(\ln A) = 0. \qquad (21\text{-}61)$$

One means of solution is to write

$$A(x,y,z) = \sum_{n=0}^{\infty} A_n(x,y,z)(ik_o)^{-n}. \qquad (21\text{-}62)$$

On substitution into (21-61), and equating terms with equal powers of k_o, we have that

$$2\left[\tfrac{1}{2}\nabla^2 S + \nabla(\ln A_n)\right] + \nabla(\ln A_{n-1})\cdot\nabla(\ln A_{n-1}) + \nabla^2(\ln A_{n-1}). \qquad (21\text{-}63)$$

This gives a scheme for determining each A_n from the previous one (A_{-1} is defined to be zero). Even here, $A_o(x,y,z)$ has normally to be determined by some means other than as a solution of (21-6).

In an elastic medium, the motion is governed by the vector equation (5-8). The solution in terms of a vector equivalent of (21-3) is much more complicated and leads to a three-term recurrence relation between vector amplitude coefficients. Relevant details of the theory are given by Karal and Keller (1959), F. Hron and Kanasewich (1971) and by Cerveny and Ravindra (1971, Chapter 2). The second reference is particularly aimed at a method of producing synthetic seismograms while the third is directed towards the evaluation of amplitude coefficients for head waves in multi-layered media.

CHAPTER 22

A RADIALLY INHOMOGENEOUS EARTH – RAY THEORY

22.1. General.

The complications of a spherical earth are many. Plane-waves are no longer appropriate so we lose their relative simplicity of interpretation. However, many occasions arise in which the asymptotic ray theory developed in Chapter 21 can be used to get ray paths and intensities for a wave approaching an interface. If curvature can be neglected, plane-wave reflection and transmission coefficients may be used to determine the partition of energy at the interface and then asymptotic theory may be applied to carry the ray onward. Cerveny and Ravindra (1971, Sec. 2.3) give the necessary details if the curvature of the wavefront and/or the interface need to be taken into consideration. Some of the simpler parts of asymptotic ray theory for radially inhomogeneous media will be developed in Section 22.2.

A rather simple change of variables exists which will transform a spherical-earth into a flat-earth model. Although we have mentioned this in the preceeding chapter with regard to more elaborate means of solving the inverse problem, we shall also write out the exact inversion formula for radially inhomogeneous media insofar as it is valid. This equation derived in Section 22.3 is extremely simple to use and can be applied directly to a spherical-earth travel-time curve if no low-velocity layers are present.

When one comes to the core of the earth, diffraction phenomena have to be considered. These are caused by the velocity drop at the core-mantle boundary. The full power of complicated mathematical techniques is necessary for an analysis of this problem. When there is some structure to the core-mantle boundary region, one has to employ a spherical version of the Thompson-Haskell matrix technique, a radial formulation of the WKBJ approximation, a direct numerical integration of the equations of motion, or some combination of the three. Some aspects of core diffraction will be considered in Section 22.4.

22.2. Rays in a Radially Inhomogeneous Earth.

In our simplified earth model, we will allow the parameters to vary as a function of radius only. They may vary either continuously or in a jump-like manner. For this model, the time necessary for a seismic disturbance to travel from P_o to P is given by

$$T(p) = \int_{P_o}^{P} \frac{ds}{c(r)} = \int_{P_o}^{P} \frac{1}{c(r)} \left[dr^2 + r^2 d\phi^2\right]^{\frac{1}{2}}$$

$$= \int_{P_o}^{P} \frac{1}{c(r)} \left[r^2 + \left(\frac{dr}{d\phi}\right)^2\right]^{\frac{1}{2}} d\phi. \qquad (22\text{-}1)$$

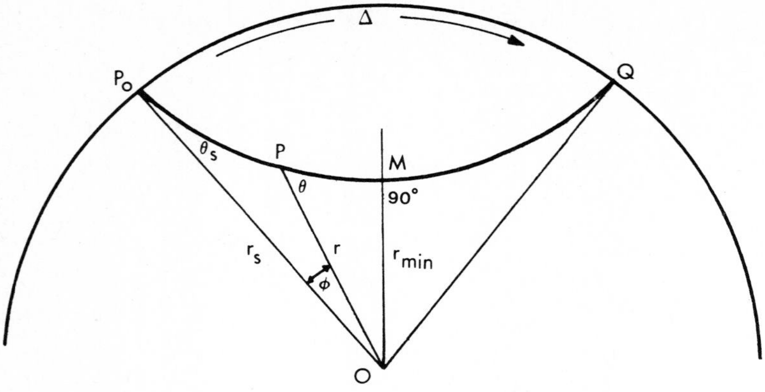

Figure 22-1. Geometric relationships for a ray traveling in a radially inhomogeneous earth model.

If we write $r' = dr/d\phi$, and set

$$\frac{1}{c(r)}\left[r^2 + \left(\frac{dr}{d\phi}\right)^2\right]^{\frac{1}{2}} = F(r,r'),$$

the calculus of variations tells us that in order to have a stationary value of the travel-time T(p), in this case a minimum time, we must have

$$\frac{d}{d\phi}\left(\frac{\partial F}{\partial r'}\right) - \frac{\partial F}{\partial r} = 0.$$

On the other hand, we see that

$$\frac{d}{d\phi}\left[F(r,r') - r'\frac{\partial F}{\partial r'}\right] = r'\left[\frac{\partial F}{\partial r} - \frac{d}{d\phi}\left(\frac{\partial F}{\partial r'}\right)\right] = 0,$$

and consequently the first bracketed expression is a first integral of the condition for a stationary value. Carrying out the indicated computations, we find that

$$r^2c(r) = p(r^2 + r'^2)^{\frac{1}{2}}, \tag{22-2}$$

where p is a constant.

If we introduce the quantity

$$q = \frac{r}{c(r)}, \tag{22-3}$$

and solve (22-2) for r' ($= dr/d\phi$), we have

$$d\phi = - dr(p/r)(q^2 - p^2)^{-\frac{1}{2}}. \tag{22-4}$$

Since the path is symmetric, the epicentral distance from P_o to Q can be written

$$\Delta(p) = 2\int_{r_{min}}^{r_s} \frac{p\,dr}{r(q^2 - p^2)^{\frac{1}{2}}}. \tag{22-5}$$

Using this equation, we can find Δ(p) for various values of the parameter p if we know the values of r, c(r), and hence of q(r). On the other hand, this equation may be regarded as an integral equation to be solved for q if we know Δ(p) for all p. Before we solve the integral equation, we shall derive an expression for the travel-time as a function of p, and we shall identify the ray parameter p in this case.

Returning to equation (22-1) we find that

$$T(p) = 2\int_{P_o}^{M} \frac{1}{c(r)}\left[1 + r^2\left(\frac{d\phi}{dr}\right)^2\right]^{\frac{1}{2}}dr = 2\int_{r_{min}}^{r_s} \frac{1}{c(r)}\left[1 + \frac{p^2}{q^2 - p^2}\right]^{\frac{1}{2}}dr,$$

or

$$T(p) = 2\int_{r_{min}}^{r_s} \frac{q^2 dr}{r(q^2 - p^2)^{\frac{1}{2}}}. \tag{22-6}$$

From this equation we can find T(p) if we know q(r). In the spherical case the variable τ can be calculated from $\tau(p) = T(p) - p\Delta(p)$, or

$$\tau(p) = 2\int_{r_{min}}^{r_s} (q^2 - p^2)^{\frac{1}{2}} dr/r. \tag{22-7}$$

Equations (22-5), (22-6), and (22-7) are similar to (21-38), (21-39), and (21-40) respectively. In fact, the following change of variables (Gerver and Markushevich, 1966) connects one set to the other:

$$X = R\Delta, \quad Z = R\ln(R/r), \quad c(z) = (R/r)c(r). \tag{22-8}$$

This transformation allows all travel-time and velocity-inversion problems in a spherical earth to be transformed exactly to a flat-earth case. Such problems are known as kinematic problems. It should be remembered that asymptotic ray theory is still only a high-frequency approximation, so this tranformation is only valid in the high-frequency limit in either case. The calculation of reflection and transmission coefficients at discontinuities are dynamic problems; for this type of problem the transformation given by (22-7) is sometimes a useful approximation to go from a spherical to a flat-earth model.

The ray parameter is similar to that in a horizontally stratified medium, but the functional form varies. In Figure 22-2, we see that

$$r d\phi = ds\,\mathrm{Sin}\theta = (r^2 + r'^2)^{\frac{1}{2}} d\phi\,\mathrm{Sin}\theta,$$

or

$$r = \mathrm{Sin}\theta(r^2 + r'^2)^{\frac{1}{2}}.$$

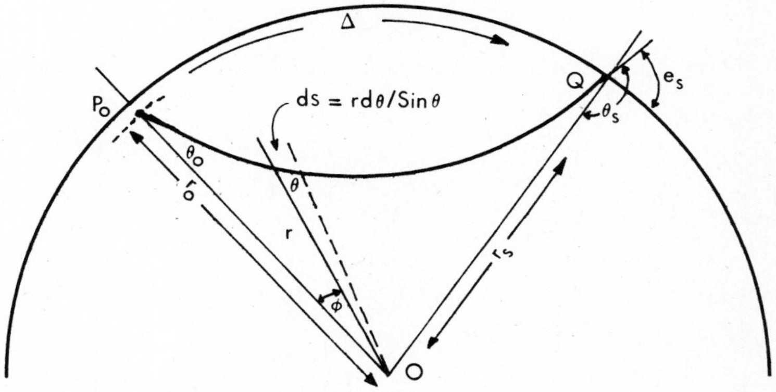

Figure 22-2. Geometrical relationships for ray paths.

But from (22-2), we have

$$(r^2 + r'^2)^{\frac{1}{2}} = r^2/\{p\ c(r)\}.$$

Hence

$$p = r\mathrm{Sin}\theta/c(r) = \text{constant}. \tag{22-9}$$

This is the generalized form of Snell's law for a radially inhomogeneous medium. At the lowest point on the path, i.e., $r = r_{min}$ we have $\theta = 90^{\circ}$, and

$$p = r_{min}/c(r_{min}) = q_{min}. \tag{22-10}$$

At the surface, the wavefront goes by with an apparent velocity

$$c(r_s)/\mathrm{Sin}\theta_s = \text{apparent velocity} = r d\Delta/dT.$$

Hence

$$p = r_s \mathrm{Sin}\theta_s/c(r_s) = dT/d\Delta. \tag{22-11}$$

We see that p is a function of Δ, but as in the flat-earth case it is constant for all values of r along any given ray.

Using the geometrical considerations leading to (21-33), we find that the intensity of the waves can be written

$$I = P\,\frac{d\Omega}{dA} = \frac{P\ \mathrm{Sin}\theta_o}{r_s^2 \mathrm{Sin}\Delta} \left| \frac{1}{\mathrm{Cos}\theta_s} \frac{d\theta_o}{d\Delta} \right| . \tag{22-12}$$

From (22-11) we have

$$\frac{d\theta_o}{d\Delta} = \frac{c_o}{r_o \mathrm{Cos}\theta_o} \frac{d^2T}{d\Delta^2} ;$$

whence

$$I = \frac{Pc_o\ \mathrm{Sin}\theta_o}{r_s^2 r_o \mathrm{Sin}\Delta\mathrm{Cos}\theta_o} \left| \frac{1}{\mathrm{Cos}\theta_s} \frac{d^2T}{d\Delta^2} \right| = \frac{Pc_o\ \mathrm{Tan}\theta_o}{r_s^2 r_o \mathrm{Sin}\Delta} \left| \frac{1}{\mathrm{Cos}\theta_s} \frac{d^2T}{d\Delta^2} \right| . \tag{22-13}$$

Using

$$r_s \mathrm{Sin}\theta_s/c_s = r_o \mathrm{Sin}\theta_o/c_o$$

to rewrite this expression in terms of θ_s, we finally obtain

$$I = \frac{P}{r_s^2 \mathrm{Sin}\Delta} \frac{q_s}{q_o} (q_o^2 \mathrm{Cot}^2\theta_s - q_s^2 \mathrm{Cos}^2\theta_s)^{-\frac{1}{2}} \left| \frac{d^2T}{d\Delta^2} \right| . \tag{22-14}$$

Traditionally, seismologists use the angle of emergence e_s instead of θ_s, where

$$e_s = \theta_s - 90^{\circ}. \tag{22-15}$$

The intensity formula can then be written

$$I = \frac{P}{r_s^2 \mathrm{Sin}\Delta} \frac{q_s}{q_o} (q_o^2 \mathrm{Tan}^2 e_s - q_s^2 \mathrm{Sin}^2 e_s)^{-\frac{1}{2}} \left| \frac{d^2T}{d\Delta^2} \right| . \tag{22-16}$$

We can get an expression for $d\Delta/dp$ in much the same manner as dX/dp in the flat-layer case. Starting with (22-5), we have

$$\frac{d\Delta}{dp} = 2 \int_{r_{min}}^{r_s} \frac{dr}{r(q^2 - p^2)^{\frac{1}{2}}} + 2p\,\frac{d}{dp} \int_{r_{min}}^{r_s} \frac{dr}{r(q^2 - p^2)^{\frac{1}{2}}} . \tag{22-17}$$

If we now set

$$f(q) = \frac{1}{r}\frac{dr}{dq}, \tag{22-18}$$

(22-17) can be written as

$$\frac{d\Delta}{dp} = 2\int_{r_{min}}^{r_s} \frac{f(q)dq/dr}{(q^2 - p^2)^{\frac{1}{2}}} dr + 2p\frac{d}{dp}\int_{r_{min}}^{r_s} \frac{f(q)dq/dr}{(q^2 - p^2)^{\frac{1}{2}}} dr. \tag{22-19}$$

On integrating the second term by parts, we have

$$\int_{r_{min}}^{r_s} \frac{f(q)dq/dr}{(q^2 - p^2)^{\frac{1}{2}}} dr = f(q)\text{Cosh}^{-1}(\frac{q}{p}) \Big|_p^{q_s} - \int_{r_{min}}^{r_s} f'(q)\text{Cosh}^{-1}(\frac{q}{p}) \frac{dq}{dr} dr$$

$$= f(q_s)\text{Cosh}^{-1}(\frac{q_s}{p}) - \int_{r_{min}}^{r_s} f'(q)\text{Cosh}^{-1}(\frac{q}{p}) \frac{dq}{dr} dr.$$

Performing the indicated differentiation in (22-19), we have

$$\frac{d\Delta}{dp} = -\frac{2q_s f(q_s)}{(q_s^2 - p^2)^{\frac{1}{2}}} + 2\int_{r_{min}}^{r_s} \frac{\{f(q) + qf'(q)\}\, dq/dr}{(q^2 - p^2)^{\frac{1}{2}}} dr. \tag{22-20}$$

Introducing another variable $\zeta(r)$, where

$$\zeta(r) = \frac{r}{c}(\frac{dc}{dr}) = \frac{d(\ln c)}{d(\ln r)}, \qquad \text{(normally negative)} \tag{22-21}$$

we find that

$$\frac{1}{qf(q)} = \frac{r}{q}\frac{dq}{dr} = \frac{d(\ln q)}{d(\ln r)} = 1 - \frac{d(\ln c)}{d(\ln r)} = 1 - \zeta.$$

Then

$$\frac{d}{dr}\{qf(q)\} = (f + qf')\frac{dq}{dr} = \frac{d}{dr}(\frac{1}{1-\zeta}) = \frac{1}{(1-\zeta)^2}\frac{d\zeta}{dr}, \tag{22-22}$$

and we have

$$\frac{d\Delta}{dp} = \frac{-2}{(q_s^2 - p^2)^{\frac{1}{2}}(1 - \zeta_s)} + 2\int_{r_{min}}^{r_s} \frac{d\zeta/dr}{(q^2 - p^2)^{\frac{1}{2}}(1 - \zeta)^2} dr. \tag{22-23}$$

This is similar to the expression derived for the vertically inhomogeneous medium (21-45), but differs in one major respect. Low-velocity regions are allowed as long as ζ remains less than +1. Otherwise, the travel-time curves behave in much the same manner as discussed previously. A comprehensive analysis of travel-time curves is given by Bullen (1963, Sec. 7.3).

Special Solutions: Three special solutions are of interest in that the time-distance relationships can be rather easily obtained. The first is the travel-time curve for a homogeneous sphere (Figure 22-3).

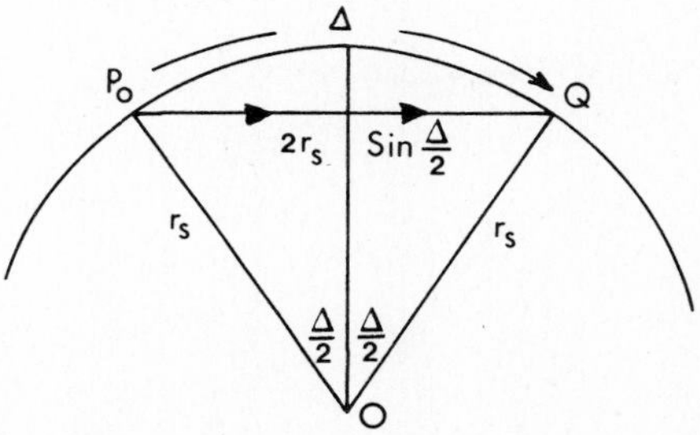

Figure 22-3. Ray path in a homogeneous sphere.

The required relationship is given easily by trigonometry:

$$T = (2r_s/c_s)\mathrm{Sin}(\Delta/2). \tag{22-24}$$

In the second case we want a solution where the integral in (22-23) vanishes, i.e., ζ = constant. The velocity appropriate to this case is

$$c(r) = c_s\left(\frac{r_s}{r}\right)^k, \qquad (k > -1). \tag{22-25}$$

Then

$$\zeta = \frac{r}{c}\left(\frac{dc}{dr}\right) = -k, \tag{22-26}$$

$d\zeta/dr$ vanishes, and

$$\frac{d\Delta}{dp} = \frac{-2}{k+1}\,\frac{1}{(q_s^2 - p^2)^{\frac{1}{2}}}\,.$$

Integrating we have that

$$\Delta = -\frac{2}{k+1}\int_{q_s}^{p}\frac{dp}{(q_s^2 - p^2)^{\frac{1}{2}}} = \frac{2}{k+1}\,\mathrm{Cos}^{-1}\left(\frac{p}{q_s}\right), \tag{22-27}$$

or

$$p = q_s\mathrm{Cos}\{\tfrac{1}{2}(k+1)\Delta\} = \frac{dT}{d\Delta}\,.$$

Whence

$$T = \frac{2q_s}{k+1}\,\mathrm{Sin}\{\tfrac{1}{2}(k+1)\Delta\} = \frac{2r_s}{(k+1)c_s}\,\mathrm{Sin}\{\tfrac{1}{2}(k+1)\Delta\}. \tag{22-28}$$

The third solution can be found by looking for ray paths of <u>constant</u> <u>curvature</u>. From the geometry of Figure 22-4, we find that

$$ds = \rho d\psi \quad \text{or} \quad \rho = ds/d\psi.$$

But

$$dr = -ds\mathrm{Cos}\theta \qquad \text{and} \qquad dw = r\mathrm{Cos}\theta d\psi.$$

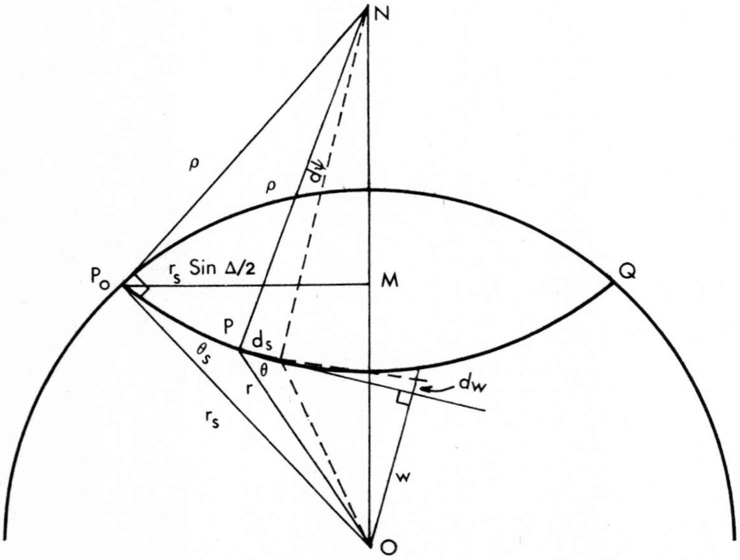

Figure 22-4. Geometrical relationships for determining rays of constant curvature.

Also

$$w = r\mathrm{Sin}\theta = pc \quad \text{and} \quad dw = pdc.$$

Hence

$$\rho = -\frac{dr}{\mathrm{Cos}\theta}\frac{r\mathrm{Cos}\theta}{pdc} = -\frac{r}{p}\frac{dr}{dc},$$

i.e.,

$$\frac{1}{\rho} = -\frac{p}{r}\frac{dc}{dr}. \tag{22-29}$$

For rays of constant curvature, we can integrate this equation obtaining

$$\tfrac{1}{2}r^2 + \text{constant} = -p\rho c.$$

Working in reverse, we see that a velocity distribution of the form

$$c(r) = a - br^2 \tag{22-30}$$

will give rise to circular rays of radius $(2pb)^{-1}$. Jeffreys (1970, p. 62-63) gives the distance and travel-time relation as follows:

$$\Delta = 2\mathrm{Tan}^{-1}\{\lambda^{-1}\mathrm{Cot}\theta_s\}, \tag{22-31}$$

and

$$T = \frac{2r_s}{c_s(\lambda^2 - 1)^{\frac{1}{2}}}\,\mathrm{Sinh}^{-1}\{(\lambda^2 - 1)^{\frac{1}{2}}\mathrm{Sin}\tfrac{1}{2}\Delta\}, \tag{22-32}$$

where

$$\lambda = 1 + 2br_s^2/c_s. \tag{22-33}$$

Using (22-21), we can rewrite (22-29) as

$$1/\rho = -\zeta\mathrm{Sin}\theta/r. \tag{22-34}$$

This says that $|\rho| > r$ if $0 < \zeta < 1$, and the ray will bend toward the center of the earth. $|\rho|$ will be smallest when $\mathrm{Sin}\theta = 1$ and at that point we have $|\rho| = r/\zeta$. If $\zeta = +1$, we see that a ray perpendicular to the radius vector $\underline{r}$ will stay at constant depth. If $\zeta > +1$, $|\rho| < r$ and the ray will spiral inward.

Odegard and Fryer (1977) work out similar relations for seismic rays in a spherical medium whose velocity varies linearly with $\underline{r}$. The resulting expressions are somewhat more complex than the above.

A power series expansion of any of the forms (22-24), (22-28), (22-32), and (21-52) is informative. They all take the form

$$T \simeq a\Delta - b\Delta^3 + \ldots \tag{22-35}$$

As long as Δ is sufficiently small, we cannot tell the difference between a curved earth with any of the three velocity variations above and a flat earth with a linear variation. One must have a sufficient portion of the travel-time curve to distinguish between the models.

22.3. Inversion for a Radially Inhomogeneous Earth.

Here again, we integrate both sides of (22-5) with respect to $\underline{p}$ while weighting with the factor $(p^2 - q_1^2)^{\frac{1}{2}}$, i.e.,

$$\int_{q_1}^{q_s} \frac{\Delta(p)dp}{(p^2 - q_1^2)^{\frac{1}{2}}} = \int_{q_1}^{q_s} dp \int_p^{q_s} \frac{2p\ dr/dq}{r(q^2 - p^2)^{\frac{1}{2}}(p^2 - q_1^2)^{\frac{1}{2}}} dq$$

$$= 2\int_{q_1}^{q_s} \frac{dr}{dq}\frac{dq}{r} \int_{q_1}^{q} \frac{pdp}{(q^2 - p^2)^{\frac{1}{2}}(p^2 - q_1^2)^{\frac{1}{2}}}$$

$$= \pi \int_{r_1}^{r_s} \frac{dr}{r} = \pi \ln\left(\frac{r_s}{r_1} \right).$$

As in Chapter 21, the left-hand side can be integrated by parts giving:

$$\int_{q_1}^{q_s} \frac{\Delta(p)dp}{(p^2 - q_1^2)^{\frac{1}{2}}} = \Delta \mathrm{Cosh}^{-1}\left(\frac{p}{q_1} \right) \Bigg|_{q_1}^{q_s} - \int_{q_1}^{q_s} \frac{d\Delta(p)}{dp} \mathrm{Cosh}^{-1}\left(\frac{p}{q_1} \right)dp$$

$$= \int_0^{\Delta_1} \mathrm{Cosh}^{-1}\left\{ \frac{p(\Delta)}{q_1} \right\}d\Delta.$$

So, in total we have

$$\pi \ln\left(\frac{r_s}{r_1} \right) = \int_0^{\Delta_1} \mathrm{Cosh}^{-1}\left(\frac{p}{q_1} \right)d\Delta = \int_0^{\Delta_1} \mathrm{Cosh}^{-1}\left[\frac{dT/d\Delta}{(dT/d\Delta)_{\Delta=\Delta_1}} \right] d\Delta. \qquad (22\text{-}36)$$

In this case too, we must numerically integrate the experimental travel-time curve to obtain the maximum depth of any ray, r_1. From (22-10) and (22-11) we then obtain the final result

$$c(r_1) = \frac{r_1}{(dT/d\Delta)_{\Delta=\Delta_1}}. \qquad (22\text{-}37)$$

Again it is obvious that $\Delta(p)$ must be continuous and single-valued. However, as we have seen in the previous section, small decreases in velocity are permissable as long as ζ remains less than +1.

As we have mentioned previously, the methods of Bessonova et al (1970, 1974, 1976) can be used by making the earth-flattening transformation (22-8). However, L. E. Johnson and F. Gilbert (1972a and b) have worked directly in the spherical domain.

22.4. Diffraction by the Core of the Earth.

The velocity in the earth's core at the core-mantle interface is a good deal lower (∿8km/sec) than the velocity of the mantle material (∿14km/sec) so P-waves striking the core are both reflected and refracted inwards as in Figure 22-5. However, the ray just grazing the core follows around the core boundary emanating energy as it travels along the interface. These diffracted waves are much weaker than direct P-waves but they can be observed in the shadow zone. Information about the core-mantle boundary can be obtained from reflected rays and diffracted rays. The latter are very sensitive to details of the interfacial structure, but their mathematical treatment is complicated indeed.

289

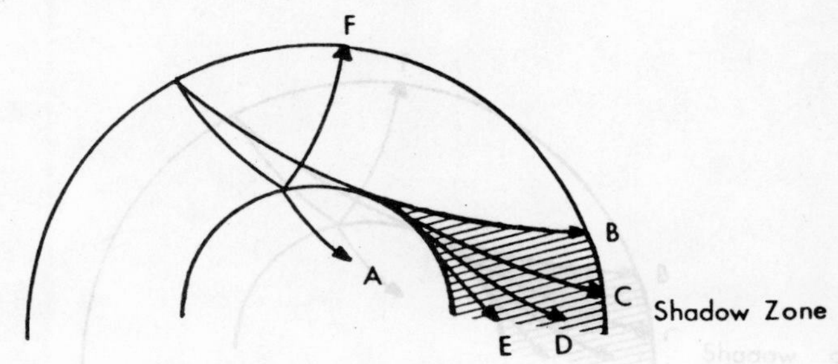

Figure 22-5. P-wave ray paths in the vicinity of the core. A - ray refracted inwards, B - grazing ray, C,D,E - diffracted rays, F - reflected ray.

One reason for this complication lies in the form of the solution of the wave equation in scalar coordinates. For axially symmetric surficial source and receiver, this can be written as

$$\Phi = e^{-i\omega t} \sum_n (n + \tfrac{1}{2}) A_n j_n(kr) P_n(\mathrm{Cos}\theta), \tag{22-38}$$

where the coefficients A_n are to be evaluated from the boundary conditions. In this form, Φ represents a sum of normal modes for the sphere. If the source were buried, the form of Φ would be more complicated, but most of the ideas to follow would carry through. An extremely large number of these normal modes must be summed to approximate a ray response. An alternative method is given by the <u>Watson Transformation</u> (Chapman and Phinney, 1972). This allows one to write

$$\begin{aligned} \Phi &= e^{-i\omega t} \sum_n (n + \tfrac{1}{2}) A_n j_n(kr) P_n(\mathrm{Cos}\theta) \\ &= e^{-i\omega t} \frac{1}{2i} \int_\Gamma \frac{\nu d\nu}{\mathrm{Cos}(\nu\pi)} P_{\nu-\frac{1}{2}}\{\mathrm{Cos}(\pi - \theta)\} A_\nu j_\nu(kr), \end{aligned} \tag{22-39}$$

Where Γ is a suitable contour. By an appropriate deformation of Γ, one can find a new set of singularities in the ν-plane some of which are related to the diffracted arrivals. One has to make a number of simplifications in the real problem to obtain the desired singularities.

Duwalo and Jacobs (1959) applied these methods to determine the diffraction (time-harmonic) by a fluid core in an otherwise homogeneous earth model. Knopoff and Gilbert (1961) considered the problem of diffraction (impulsive) by a fluid sphere in a homogeneous elastic medium. In the impulsive case, additional approximations have to be made to facilitate use of the Cagniard-deHoop method to obtain seismic wavelets. However wide-band seismic recordings of the 1960s made it possible to investigate the spectral response as an additional aid to delineating structure at the core-mantle boundary. Phinney and Alexander (1966) used a spherically layered transition region to represent structure and the spherical form of the Thompson-Haskell matrices to attack this problem. The matrix continuation of the displacement-stress vector works in well with the modal representation (22-38), after which the Watson transformation is used to calculate the spectra of diffracted waves. Chapman and Phinney (1970) used a combination of the WKBJ approximation and numerical integration to obtain diffraction spectra. Chapman and Phinney (1972) give an excellent review and comparison of these various techniques. Richards (1973), Chapman (1974a), and Richards (1976) have all suggested improvements in the WKBJ method for use in diffraction problems.

CHAPTER 23

A RADIALLY INHOMOGENEOUS EARTH – BODY-WAVE OBSERVATIONS

23.1. Seismic Phases and Travel-Time Curves.

First of all, there is a bit of nomenclature to set forth. The P- and S-waves that travel through the various regions of the earth (mantle, core, inner core) have been assigned one letter abbreviations as follows:

P = P-wave in mantle, S = S-wave in mantle,

c = reflection from core-mantle boundary,

K = P-wave in core, No S-waves in core (fluid),

i = reflection from inner core boundary,

I = P-wave in inner core, J = S-wave in inner core,

p = short leg P-wave in mantle, s = short leg S-wave in mantle.

These may be illustrated best by Figure 23-1.

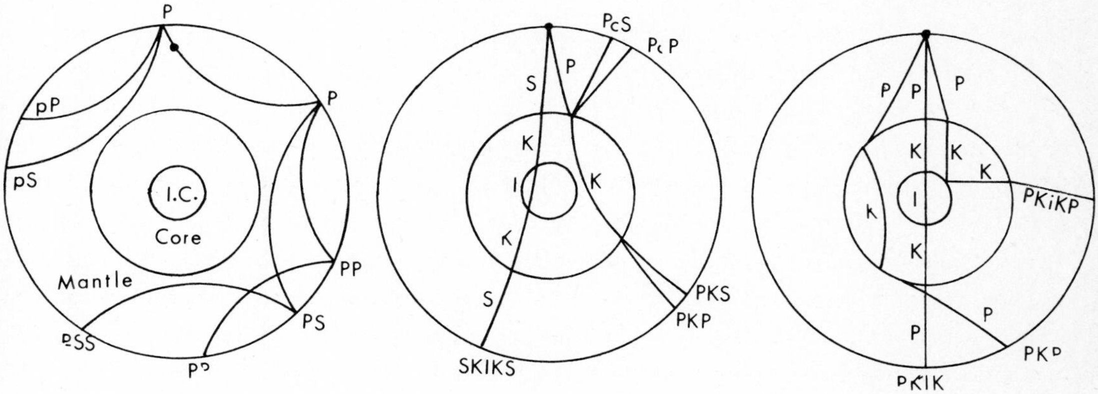

Figure 23-1. Seismic phases in the earth (schematic).

To facilitate explanation, we shall present two diagrams without prior discussion. The first (Figure 23-2) is an accurate construction by Gutenberg (1959) showing the various P-waves as they travel through the earth, together with the time of transit to any given point. The numerical subscripts associated with PKP_1 and PKP_2 refer to the two branches of the PKP travel-time curve. The lines of equal transit time also trace out the wave-fronts as the waves pass through the earth.

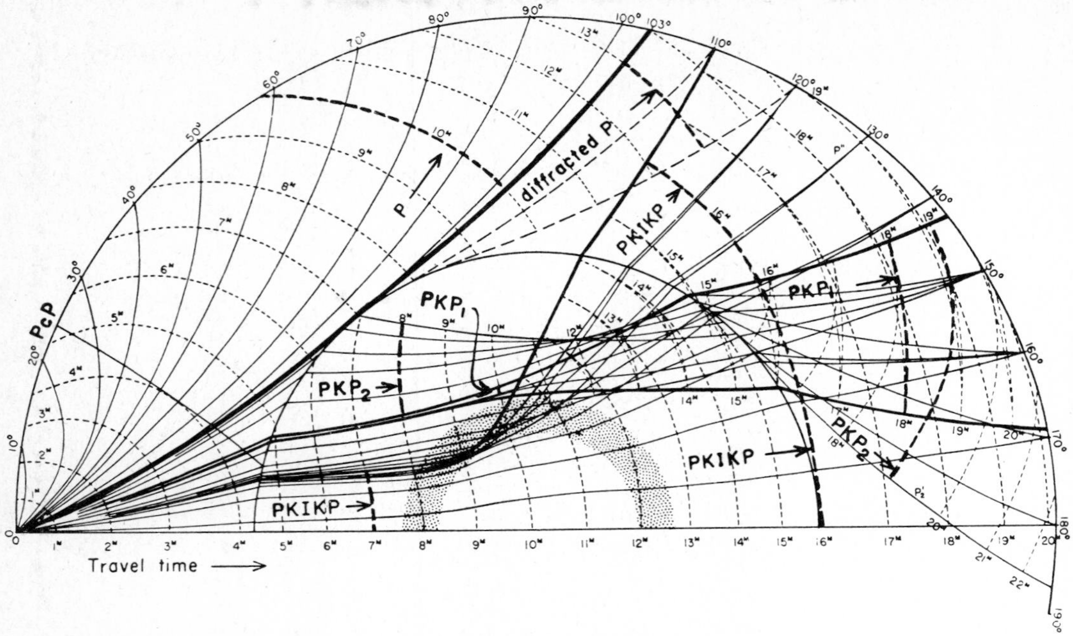

Figure 23-2. Various P-phases through the mantle, core, and inner core. After Gutenberg (1959).

The second figure gives the travel-time curves as constructed by Jeffreys and Bullen (1940). These curves represent something rather special in that they are still a reference standard, even though there has been universal agreement that they contain some errors. These errors were discovered when arrivals from accurately timed nuclear blasts in the South Pacific were compared with the values given by the J-B curves. A number of different investigators set about revising portions of them and found that there were significant regional variations that we shall consider later. These regional differences have made the compilation of a new set of comprehensive travel-time curves very difficult. A working group under the chairmanship of E. Herrin (1968) constructed a revised set of travel-time tables for P, PcP, and PKP. Using the same selected earthquakes as the Herrin group, Randall (1971a, 1971b) constructed a travel-time table for S. When you consider that later workers had the benefit of electronic computers, more modern instrumentation and almost thirty years of additional thought, it is testimony to the monumental work of Jeffreys and Bullen that it still remains a standard of comparison. The Jeffreys-Bullen travel-time curves are given in Figure 23-3.

OPPOSITE PAGE

Figure 23-3. Jeffreys-Bullen travel-time curves for a surface focus. LR and LQ are surface-waves. (After Jeffreys, 1970.)

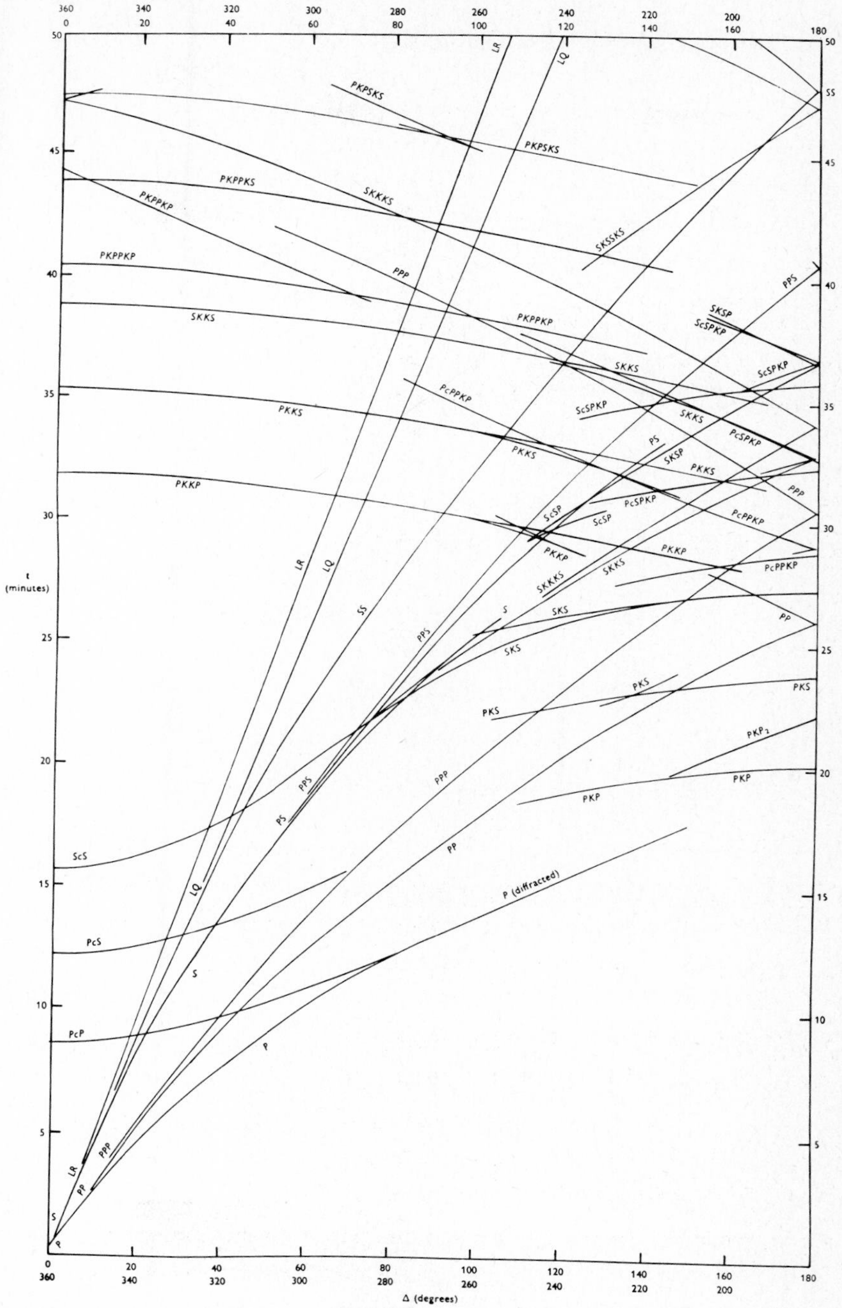

t
(minutes)
Δ (degrees)
P
S
PP
LR
PPP
PcP
PcS
ScS
LQ
PS
PPS
SS
SKS
SKKS
PKS
PKP
PKP₂
P (diffracted)
PKKP
PKKS
SKKS
PKPPKP
PKPPKS
PKPSKS
SKSSKS
ScSP
PcSPKP
ScSPKP
SKSP
PcPPKP

The reading of the various phases on the seismogram is an art. Some help in this regard is given by Simon (1969) and Lehmann (1970). Müller and Kind (1976) lay out a whole suite of records side by side over the range $0^\circ < \Delta < 180^\circ$. Although only a few phases are marked, their set of figures gives an overview of the relationship between the same event recorded at several distances.

23.2. The Construction of Travel-Time Curves.

If we are starting from the very beginning, we have to construct a preliminary travel-time curve using eyewitness accounts to locate the epicenter and origin times of the first few events that we start with. After that, we have to solve (by the method of least squares) a system of the type

$$\delta\tau + \varepsilon(\Delta') + \frac{1}{r_e}\frac{\partial T^*}{\partial \Delta'}(\delta x \operatorname{Sin} Az' + \delta y \operatorname{Cos} Az') + \frac{\partial T^*}{\partial h}\delta h = T_\Delta - T_o' - T^*(\Delta'), \qquad (23\text{-}1)$$

where

$\delta\tau$ = correction in source time (sec),

δx = correction to the east (km),

δy = correction to the north (km),

δh = correction to the depth (km),

$\varepsilon(\Delta')$ = travel-time correction at distance Δ' due to using a preliminary travel-time curve,

$\frac{\partial T^*}{\partial \Delta'}$ = slope in trial travel-time curve,

$\frac{\partial T^*}{\partial h}$ = change in trial travel-time with source depth,

$T^*(\Delta')$ = calculated travel-time using trial curve for preliminary distance,

Δ' = preliminary distance (geocentric coordinates — degrees),

Az' = preliminary azimuth of seismic ray as it leaves the focus,

T_Δ = P-arrival time at true distance (sec) — observed,

T_o' = preliminary origin time (sec),

r_e = radius of the earth (km).

The unknown correction to the preliminary travel-time curve $\varepsilon(\Delta')$ has a different value for each observing station. One cannot solve the system (23-1) unless some further restriction is put on the values that may be taken by $\varepsilon(\Delta')$. Such a restriction may be to require that $\varepsilon(\Delta')$ be linear over small intervals so that several equations of the form (23-1) lie within any one interval. The method of least squares may then be used provided due caution is taken to insure the stability of the solution.

If the errors in the travel-time curve are small, a process of iteration may also be used as suggested by Bullen (1963, Sec 10.4). First of all, one sets $\varepsilon = 0$ and solves by least squares for δx, δy, δh, and $\delta\tau$ in a given earthquake. Then one solves each equation for $\varepsilon(\Delta')$, and combines results in an appropriate manner. A new travel-time curve is constructed and you go through the process again.

The travel-time curve thus obtained may be smoothed by various methods outlined in the articles accompanying the 1968 Seismological Tables. A number of statistical considerations are also included among the papers presented in the Volume 58, Number 4 issue of the Bulletin of the Seismological Society of America.

There is an additional constraint on these calculated curves. With the advent of large seismic arrays, the quantity $dT/d\Delta$ may be observed directly. These observations may then be compared with the values of $dT/d\Delta$ obtained by differentiation of the constructed curves. In many cases the two results seemed to be incompatible. This has led to slight revisions of the travel-time curves. However, there are still meaningful differences which have been attributed to regional variations in upper mantle velocity distribution and/or anomalous structures beneath the arrays or source regions. The ultimate degree of refinement possible is still being worked out. Some considerations in this respect are given by Lomnitz (1971).

Some useful formulas from spherical trigonometry are given below. The geometry is as in Figure 23-4. With the advent of the electronic computer, no changes need be made in order to facilitate computation, and the formulas may be used directly.

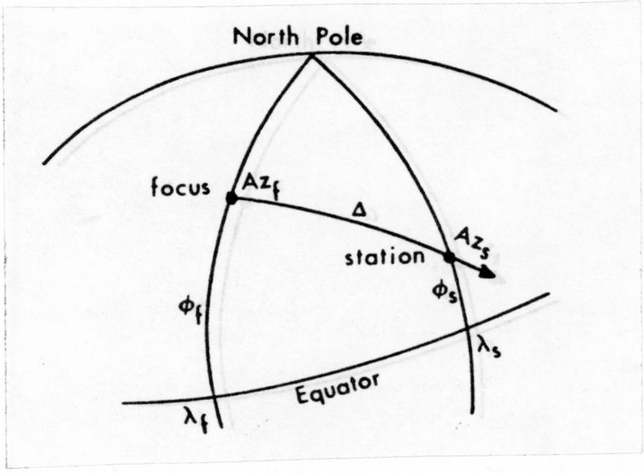

Figure 23-4. Spherical relationships between geographic coordinates of focus and station.

First of all, the epicentral distance is given by

$$\text{Cos}\Delta = \text{Sin}\phi'_s\ \text{Sin}\phi'_f + \text{Cos}\phi'_s\ \text{Cos}\phi'_f\ \text{Cos}(\lambda_s - \lambda_f); \qquad (23\text{-}2)$$

where ϕ'_s and ϕ'_f are geocentric latitudes related to the geodetic latitudes of station and focus by

$$\text{Tan}\phi'_{geocentric} = 0.993277\ \text{Tan}\phi_{geodetic}. \qquad (23\text{-}3)$$

This change is necessitated by the ellipticity of the earth. A more comprehensive study of the effect of ellipticity is given by Dziewonski and Gilbert (1976). Care should be taken in using the longitude if the surface trace of the ray path crosses the 180° meridian. The azimuth of the ray as it leaves the focus can be calculated from

$$Az_f = Sin^{-1}\left[Sin(\lambda_s - \lambda_f)Cos\phi_s'/Sin\Delta\right] \quad , \tag{23-4}$$

and the azimuth of the ray trace as it goes through the observation station is given by

$$Az_s = 180^\circ - Sin^{-1}\left[Sin(\lambda_s - \lambda_f)Cos\phi_f'/Sin\Delta\right] \quad . \tag{23-5}$$

It should be noted that these two angles are not the same.

23.3. The Location of the Source Region.

If we assume that the travel-time curves for P and S are known, then these curves can supply us with the data we need to determine the source location. The source epicenter is the point on the earth's surface directly above the focus or hypocenter. We can also obtain a measure of the origin time. The source distance can be read from a graph of $T_s - T_p$ vs. Δ (Figure 23-5). This figure has been constructed from the data of Figure 23-3.

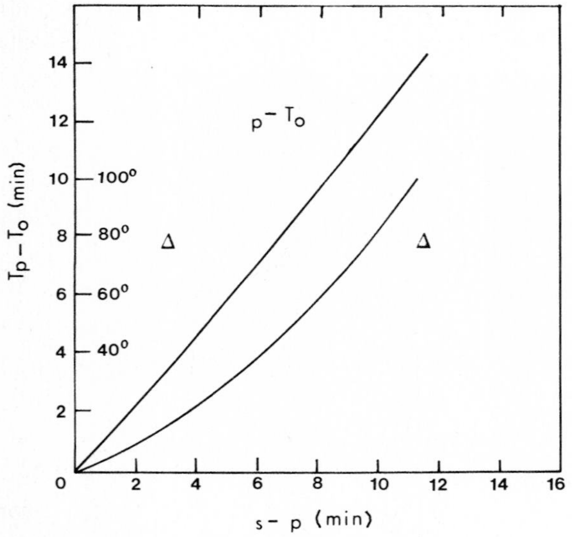

Figure 23-5. Chart for determining Δ and T_o given the arrival time of P and S, T_p and T_s.

Three such distances will suffice to give a unique preliminary position — obtained as the intersection of three circles of the proper radius about three observing stations. Two distances will reduce the problem to a choice of the proper one of two intersections. This choice can be made using either the polarity of particle motion or the relative frequency of seismic activity at the two possible sites. If only one circle is available, particle motion may be used to give a choice of azimuth. The ambiguity between the forward and backward direction can then (sometimes) be removed by noting the relative seismic activity at the two possible sites.

Once the preliminary epicenter and origin time are known, data from all reporting stations may be used to refine the location of the epicenter, depth of focus, and time of origin. For each station we may write

$$\delta\tau + \frac{1}{r_e}\frac{\partial T}{\partial \Delta'}(\delta x \mathrm{Sin}Az' + \delta y \mathrm{Cos}Az') + \frac{\partial T}{\partial h}\delta h = T_\Delta - T_o - T(\Delta'), \qquad (23\text{-}6)$$

where $T(\Delta')$ represents an adopted travel-time curve. The system of equations (23-6) may then be solved by the method of least squares to give values for the four corrective unknowns $\delta\tau$, δx, δy, δh.

The problems in designing computer programs to calculate source parameters are many and varied. Representative papers that treat various aspects are Bolt (1960), Flinn (1965), Buland (1976), E. G. C. Smith (1976), Crossan (1976), and Lomnitz (1977a). The Buland paper discusses the best ways of treating the problem by ordinary least squares techniques (with additional comments by Smith), Crossan shows how to use the generalized inverse matrix (p.262) in its solution. The Lomnitz paper is rather unusual in that it uses distance residuals rather than travel-time residuals. The author says that his new method is "absolutely convergent" and is much faster than standard programs.

The problems involved in determining individual epicenters are compounded by the anomalous character in the source region and in the station vicinity. Douglas (1967) and Freedman (1967a) suggest grouping the data together in order to determine more accurate epicentral location and/or station corrections. Lilwall and Douglas (1970) used the joint epicenter method to determine station corrections at a large number of points over the globe. They then compared these corrections with ones obtained by somewhat different means by Herrin and Taggert (1968). Dewey (1972) uses the joint hypocenter determination to relocate a large number of earthquakes using a master event. Bolt (1973) proposed that a number of master earthquakes be universally adopted to improve the precision of seismicity studies. The master event idea has also been suggested by Evernden (1969a) with regard to improving the depth capabilities of hypocentral location. A master event has also been used by von Seggern (1972) and Crough and Van der Voo (1973) who felt that the use of surface waves would eliminate some of the tectonic complications in source regions because such seismic waves travel on the surface of the earth. Detracting from this advantage, the timing accuracy of surface waves is not as great. However, there seems to be some overall improvement in structurally complicated regions when using surface waves for relocation of the source.

Depth Determination. The problem of depth determination using a least squares solution to equation (23-6) lies in the fact that at large teleseismic distances, the partial derivatives with respect to depth are extremely flat. Although this has been known for a long time, Lomnitz (1977b) gives a figure for these curves at various depths and shows that a change of depth is almost equivalent to a change of origin time. This requires additional means for the direct determination of focal depth.

The use of the time difference between the seismic phases pP and P has been used for a long time as one evidence of depth. An indication of focal depth is given by the general character of surface waves which are excited relatively less in deep focus earthquakes. However, discrimination between earthquakes and explosions requires rather precise depth estimates. Cohen (1970) has used frequency domain methods to determine the time interval between pP and P. This study was based upon earlier work by Bogert et al (1963). Frequency domain analysis works best with short-period body-waves. Herrmann (1976) has used synthetic seismograms to model the character of P-waves recorded on long-period instruments with some success in the determination of focal depth. The Fourier spectra of surface waves is also an indication of focal depth, and may be used in its determination. Tsai and Aki (1970) have given some good examples of such spectra. Canitez and Toksöz (1971) have used the spectral ratio L/R as a measure of focal depth as well as of source characteristics. Massé et al (1973) present a large number of figures and tables relating to the determination of focal depth from surface wave spectra.

Precision of Source Location. The imprecision of source location consists of two parts; a random error and a systemic mislocation. Flinn (1965), Freedman (1967b), and Evernden (1969b and 1971a) discuss many aspects of the accuracy of source location. Current precision runs to a few kilometers in relative location and a few tens of kilometers in absolute location. The precision of source location is closely tied into the precision of travel-time curves. A number of papers previously indicated in this respect are also relevant. The systematic mislocation of hypocenters has been discussed by Underwoood and Lilwall (1969) and Lilwall and Underwood (1970). For the ultimate treatment of this problem one has to resort to the construction of models in the source region and then use the seismic ray methods mentioned in Section 21.6. However, the computer processing necessary to this on a routine basis is prohibitive and maps such as presented by Lilwall and Underwood (ibid) should be a help in refining travel-time curves.

23.4. Observations.

The data relating to P- and S-waves can be divided into three groups. The first group are those taken at distances less than 30° and whose travel path lies generally in the crust and upper mantle. Hales (1972) reviews P-wave data and Helmberger and Engen (1974) consider S-wave data. There appears to be a greater degree of uniformity to body-wave data recorded in the second group (30° — 100°+). Hales and Herrin (1972) give an excellent summary of data in this range. The third group includes core-mantle data and observations of core phases. The former is reviewed by Cleary (1974) and the latter by Qamar (1973). The reader is referred to these sources for details of the observed data. The citations in the remainder of this Section will largely be of special interest, or to data not referenced in the above.

Crust and Upper Mantle. The upper 300 km or so of the earth's surface seems to have not only most of the regional variation, but it also includes the low-velocity zone for S. A similar zone for P is not as well documented. Helmberger (1973a) has found that synthetic seismograms are quite useful in determining the properties of the zone, in that shadow zone effects and attenuation are intermixed. His best model for P structure has a high-velocity lid above a thin low-velocity layer. Synthetic seismograms were similarly used in the Helmberger and Engen (ibid) paper.

A graphic picture of the regional differences is given by Herrin and Taggert (1968) for the United States where there is a very high density of stations. These station corrections have been computed relative to the 1968 Travel-Time Tables. In

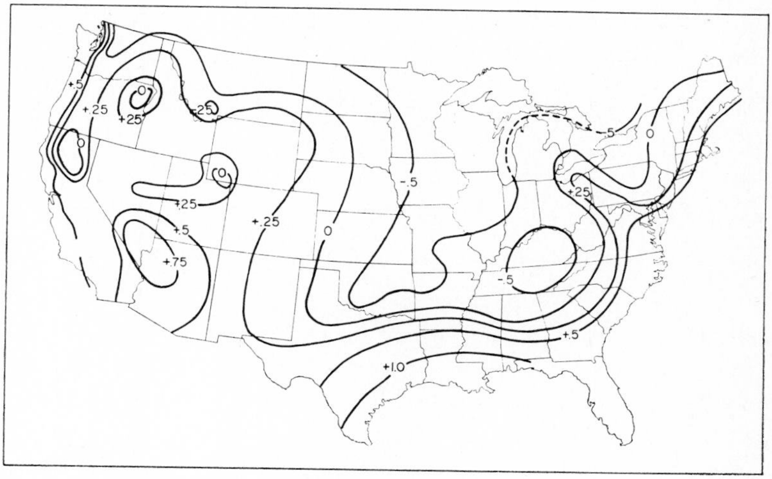

Figure 23-6. Mean-station corrections for the United States. Data relative to the 1968 Travel-Time Tables. (After Herrin and Taggert, 1968.)

addition to the U.S. data shown in the figure, Herrin and Taggert (1968) also give world-wide station anomalies. A comparison is also made with an earlier study made by Cleary and Hales (1966). Using the joint-eipcenter method, Lilwall and Douglas (1970) compute a world-wide set of station anomalies and find a good correlation with those of Herrin and Taggert in the U.S., but the correlation is not as good for European stations. Sengupta and Julian (1976) used about 3300 arrival times from deep-focus earthquakes to eliminate the upper mantle inhomogeneities as much as possible. They present a comparison of the three sets of data above with their results. Although most of the variation in a diagram such as Figure 23-6 is ascribed to differences in the crust and upper mantle, Sengupta and Julian (ibid) find that there is significant lateral variation in the mantle below 2000 km deep.

A number of investigators have looked into the regional variation of S. Hart (1975) makes a study of the lower mantle using explosion data. Hales and Roberts

(1970) present a diagram for S (their Figure 10) similar to that of Herrin and Taggert in Figure 23-6 above. They also compare their S station anomalies with the P anomalies of Cleary and Hales (1966) finding a ratio of about 4. Hales and Herrin (1972) consider various implications of this ratio. Duschenes and Solomon (1977) find a correlation of S anomalies with age of the oceanic lithosphere. Sipkin and Jordan (1975, 1976) and Okal and Anderson (1975) look into S anomalies as determined from ScS and multiple ScS phases.

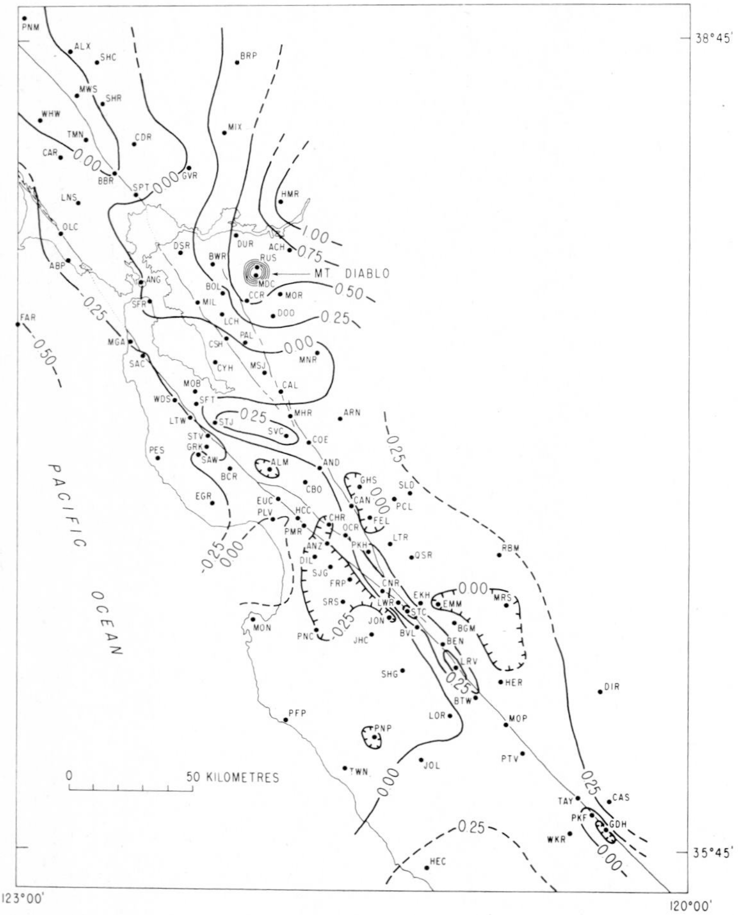

Figure 23-7. Contour map of average station residuals. The contour interval is 0.25 sec. The residuals on Mt. Diablo are -0.67 and -0.50 sec. Dots are station locations. (After Robinson and Iyer, 1976.)

P anomalies occur on an even smaller scale than in Figure 23-6. Figure 23-7 gives station residuals for a small region in the vicinity of San Francisco. The total variation is of the order of 1.5 sec over a small region. The length of this array is slightly more than 400 km. Iyer and Healy (1972) find a variation of 0.6

sec over a 200 km diameter area under the LASA observatory in Montana. Using methods outlined by Aki (1977), Husebye et al (1976) inverted about 1500 teleseismic residuals to obtain a three-dimensional model of P-velocity structure for the region shown in Figure 23-7. Using similar methods, but with local earthquakes, Aki and Lee (1976) constructed a three-dimensional velocity model in a highly instrumented portion of the region covered in Figure 23-7.

The Mantle. Some papers more recent than the ones listed in Hales and Herrin (1972) are Mitchell and Helmberger (1973), S and ScS; Gogna (1973), S, PcP and ScS; Sengupta and Julian (1976), P from deep earthquakes; and Jeffreys (1977), P and S for $\Delta > 95^{\circ}$. A comprehensive study of precursors to P'P' (the core phase PKPPKP) by Whitcomb and Anderson (1970) has shed a good deal of light upon regions of high-velocity gradient in the mantle.

A large number of $dT/d\Delta$ observations are also listed in Hales and Herrin (ibid). Two more recent ones are Wright and Cleary (1972) — $(dT/d\Delta)_P$ and Robinson and Kovach (1972) — $(dT/d\Delta)_S$. During the 1970s much interest has been centered in the analysis of the capability of large and medium size seismic arrays to determine $dT/d\Delta$ as an

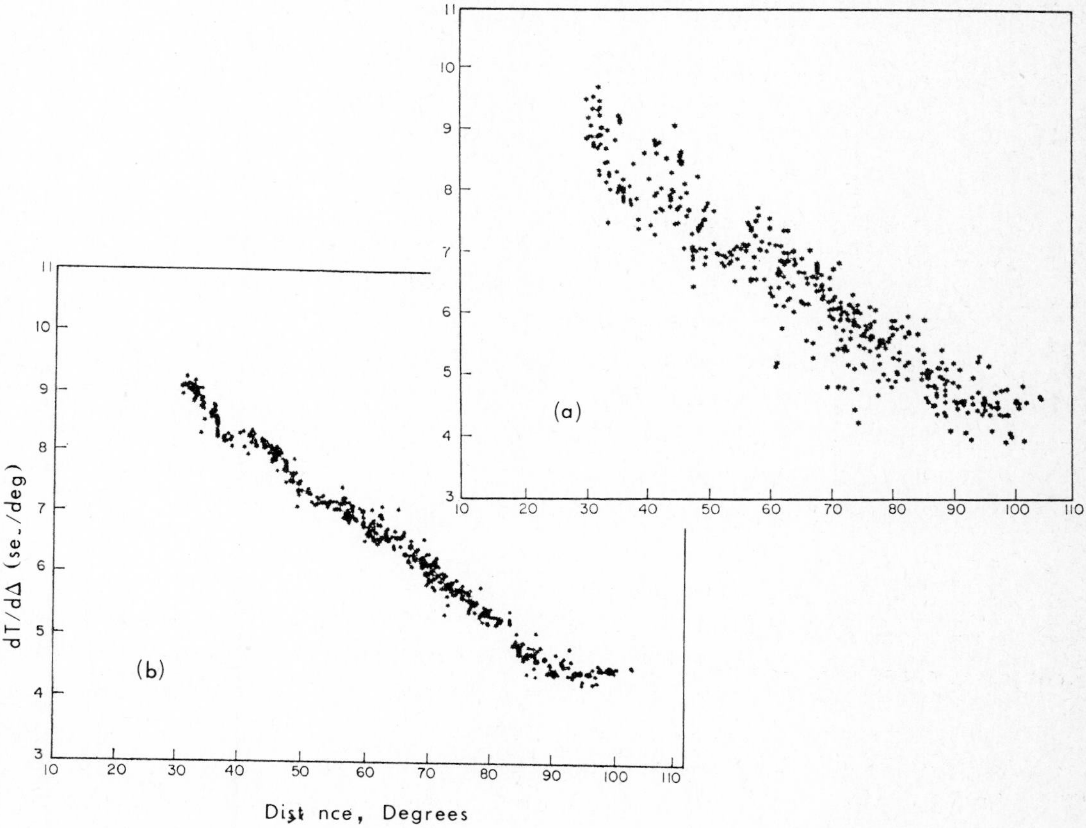

Figure 23-8(a). Uncorrected values of $dT/d\Delta$ measured at four arrays.
(b). Corrected values of $dT/d\Delta$ using site corrections containing a two-term azimuthal component. (After Corbishley, 1970).

independent variable. Figure 23-8a shows uncorrected values of dT/dΔ obtained by Corbishley (1970) at four medium size arrays in Australia, Canada, India, and the United Kingdom. Figure 23-8b shows the corrected data. Corbishley (ibid) shows how the data may be combined in a meaningful manner so that variations from the linear trend are not obscured by the regional variations. It is certainly evident that values of dT/dΔ obtained from a single array are not going to be very useful. Berteussen (1975 and 1976) even shows that anomalies in dT/dΔ and apparent azimuth of arrival are dependent upon the array configuration. To investigate the causes of this sort of variation, a number of studies have been made of two large arrays, LASA and NORSAR. Aki, Christofferson, and Husebye (1976, 1977) have made three-dimensional velocity structures for both arrays. Many of the earlier studies are referenced by these authors and in the two Berteussen papers mentioned above.

The Core and the Core-Mantle Boundary. Some papers not referenced in Qamar (1973) are Julian et al (1972), array studies of PKJKP; Müller (1973a), amplitude studies of PKP; Masse et al (1974), PKIIKP; Engdahl et al (1974), PKiKP and the radius of the inner core; Stamou and Båth (1974), SKP; Kind and Müller (1977), amplitudes and travel-times of SKS; and Agrawal (1978), PKP (GH branch).

The nature of the core-mantle boundry has been studied by both reflected and diffracted waves. Some recent papers using reflected phases are Buchbinder and Poupinet (1973), PcP study; Engdahl and Johnson (1974), differential PcP travel-times and core radius; Dziewonski and Haddon (1974), a critical review of the core-mantle radius; Müller et al (1977), amplitudes of PcP. The value of the core-mantle radius given by Dziewonski and Haddon is 3485 ± 3km, somewhat larger than the Jeffreys (1939) estimate of 3473.1 ± 2.5km. This small difference is remarkable, given almost forty years and demonstrates the refining process that earth models are undergoing.

Studies of diffracted events from the core-mantle boundary have been given by Alexander and Phinney (1966), Phinney and Cathless (1969), and Phinney and Alexander (1969). A number of authors have proposed a model composed of relatively small, but strong, inhomogeneities which act as scattering centers. Doornbos (1976) summarizes much of this work.

23.5. Advanced Inversion Techniques.

Synthetic Seismograms. The production of synthetic seismograms in spherically-layered media is much the same as with flat-layered media (Sec. 8.6) with due regard taken for the nature of the eigenfunctions. One may use asymptotic ray theory, generalized ray theory, or sum up large numbers of normal modes. The first is well suited for inhomogeneous layers, while the second is most useful with a large number of thin homogeneous layers. The mode-summing procedure has proven rather inefficient and is useful only for long-period response where the previous two are inadequate. Examples of summing large numbers of spherical normal modes have been given by Landisman et al (1970) and by Alterman and Loewenthal (1972).

Gilbert and Helmberger (1972) have adopted a generalized ray theory to the spherical case giving a number of examples. Helmberger (1973b) gives additional examples computing synthetic seismograms from seventeen to forty degrees. Chapman (1974b) notes that for the radially inhomogeneous case, the results given by the generalized ray method are band-limited in frequency. Low frequencies are excluded by keeping only a finite set of rays and by the nature of the asymptotic expansion of the Bessel and Legendre functions in the solution. On the other hand, high frequencies are not available because all wavelengths have to be long compared to the discretized layer thicknesses. To eliminate these problems, Chapman (1974b, 1976a, and 1976b) adopts propagator matrices (Gilbert and Backus, 1966) to spherically-layered media. Additional comments on the construction of such matrices are given by Cisternas et al (1973) and Jobert (1974).

The construction of synthetic seismograms using generalized ray theory is much easier in a flat-layered medium. This has lead to an investigation of the dynamic problem. Müller (1971), Hill (1972), and Chapman (1973), have all studied earth-flattening approximations. For fluids and SH-waves, exact transformations can be found, while for P-SV, compromises have to be made.

A novel method called the quantized ray theory has been introduced by Wiggins and Madrid (1974). This scheme has the simplicity of asymptotic ray theory while maintaining much of the accuracy of generalized ray theory. The fundamental idea is to break up a p-Δ curve into a number of segments, empirically determining the impulse response due to each segment of the curve. Additional considerations are given by Wiggins (1976a), McMechan (1974), and McMechan and Dey-Sarkar (1976). Synthetic seismograms have been computed for mantle phases by Müller and Kind (1976) using the reflectivity method while Choy (1977) has used a frequency-dependent full wave theory (improved WKBJ approximation) to synthesize core-phases.

Inversion. Keilis-Borok and Yanovskaja (1967) give a summary of methods of inversion prior to the Backus-Gilbert linear inversion and the Bessonova extremal methods. Although not as efficient as the extremal method, Monte Carlo techniques have been used for body-waves (Wiggins, 1969). This paper is particularly interesting in that it shows the relationships between velocity structure, dT/dΔ curves, and travel-time curves for many models. L. E. Johnson and F. Gilbert (1972b) construct a spherical-earth model using linearized inversion. Wiggins et al (1973) and Bessanova et al (1976) construct envelopes of possible models using the extremal procedure.

Models. Because of the ambiguities introduced by possible low-velocity zones, presentations of models will be postponed to Chapter 25, where they have been constructed using additional information from free-oscillations. However, models constructed from only body-wave data are given in the last three papers above.

A RADIALLY INHOMOGENEOUS EARTH – MODE THEORY

24.1. General.

The problem of the free oscillations of a homogeneous sphere is an old one. Love (1911) in a monumental work, summarized previous studies as well as solved the problem of the free oscillations of a gravitating compressible sphere with applications to the earth. However, it was not until the advent of large scale computers in the late 1950s that the problem of an inhomogeneous gravitating compressible earth could be attempted. Eringen and Suhubi (1975, Section 8.13ff) give a more modern presentation, considering both the vibrations of a homogeneous sphere and the motions of a spherical cavity in a homogeneous medium. Some of their material is derived from an extensive study of the homogeneous sphere by Y. Sato, T. Usami and others, much of which is not readily accessible. Three papers relevant to the free oscillations of the earth are Y. Sato and T. Usami (1962), Y. Sato, T. Usami and M. Ewing (1962) and T. Usami and Y. Sato (1964).

It might be well to mention here that the radially inhomogeneous spherical body is the only finite elastic body that has proved amenable to exact solution. Exact solutions are not obtainable for the cube or the rectangular cylinder due to interactions at the edges. The usefulness of approximate solutions for these two cases is only marginal. The vibrating ellipsoid also has to be solved by approximate methods. However, these approximations are much more useful and will be considered more in Section 24.2.

The full theory is not necessary where most of the particle motion is confined to the surface of the earth and a number of approximations are useful in calculating the motions of mantle waves (periods of 100 to 500 seconds). These will be outlined in Section 24.3. Section 24.4 will be given to a short discussion of modes and rays in a spherical earth.

24.2. The Nature of Spherical Wave Motion.

In Section 6.4 we saw that the solutions of the elastic wave equation appropriate to spherical coordinates were

$$= \mathbf{e}_r \frac{\partial\Phi}{\partial r} + \mathbf{e}_\theta \frac{1}{r}\frac{\partial\Phi}{\partial\theta} + \mathbf{e}_\phi \frac{1}{r\text{Sin}\theta}\frac{\partial\Phi}{\partial\phi}, \qquad \text{(P)} \quad (24\text{-}1)$$

$$\mathbf{N} = -\mathbf{e}_r \frac{n(n+1)\ell\Psi}{r} - \mathbf{e}_\theta \frac{\ell}{r}\frac{\partial}{\partial r}(r\frac{\partial\Psi}{\partial\theta}) - \mathbf{e}_\phi \frac{\ell}{r\text{Sin}\theta}\frac{\partial}{\partial r}(r\frac{\partial\Psi}{\partial\phi}), \qquad \text{(SV)} \quad (24\text{-}2)$$

and

$$\mathbf{M} = \mathbf{e}_\theta \frac{1}{\text{Sin}\theta}\frac{\partial\Lambda}{\partial\phi} - \mathbf{e}_\phi \frac{\partial\Lambda}{\partial\theta}; \qquad \text{(SH)} \quad (24\text{-}3)$$

where Φ is a solution of

$$\nabla^2\Phi - (1/v_p^2)\partial^2\Phi/\partial t^2 = 0, \qquad (6\text{-}16)$$

Ψ is a solution of

$$\nabla^2\Psi - (1/v_s^2)\partial^2\Psi/\partial t^2 = 0, \tag{6-17}$$

and Λ is a solution of

$$\nabla^2\Lambda - (1/v_s^2)\partial^2\Lambda/\partial t^2 = 0. \tag{6-18}$$

These equations are appropriate to a homogeneous sphere or spherical shell. One can combine the above components of motion and separate out the spherical harmonic part $Y_n^m(\theta,\phi)$ from each of the potential functions Φ, Ψ, and Λ, obtaining (in the case of periodic excitation $e^{-i\omega t}$):

$$u_r = U(r)Y_n^m(\theta,\phi)e^{-i\omega t},$$

$$u_\theta = V(r)\frac{\partial}{\partial\theta}Y_n^m(\theta,\phi)e^{-i\omega t}, \qquad \text{SPHEROIDAL OSCILLATIONS} \quad (24\text{-}4)$$

$$u_\phi = \frac{V(r)}{\mathrm{Sin}\theta}\frac{\partial}{\partial\phi}Y_n^m(\theta,\phi)e^{-i\omega t};$$

and

$$u_r \equiv 0,$$

$$u_\theta = \frac{W(r)}{\mathrm{Sin}\theta}\frac{\partial}{\partial\phi}Y_n^m(\theta,\phi)e^{-i\omega t}, \qquad \text{TOROIDAL OSCILLATIONS} \quad (24\text{-}5)$$

$$u_\phi = -W(r)\frac{\partial}{\partial\theta}Y_n^m(\theta,\phi)e^{-i\omega t};$$

where

$$Y_n^m(\theta,\phi) = P_n^m(\mathrm{Cos}\theta)e^{im\phi}. \tag{24-6}$$

In a uniform spherical earth, U, V, and W are appropriate combinations of spherical Bessel functions and their derivatives. However, for the real earth, one must substitute equations of the form (24-4) and (24-5) into the elastic equations of motion including gravity. Alterman et al (1959) have shown that this leads to a set of three second-order differential equations in the radius variable $\underline{r}$ for the spheroidal oscillations, and to one second-order differential equation in $\underline{r}$ (which doesn't involve gravity) for the toroidal oscillations. Not much insight is gained by working out the derivation of these equations, or by exhibiting them. The reader is refered to Alterman et al (ibid). These equations have then to be integrated numerically to obtain appropriate solutions whose associated stresses vanish on the free surface of the earth. For this numerical work, the second-order equations are converted to a system of six first-order equations for spheroidal oscillations and two first-order equations for toroidal oscillations. These first-order equations are also set out in the above paper. Wiggins (1976) has developed a Rayleigh-Ritz variational method of solving for the free modes of oscillation. His method is both stable and efficient.

An alternative formulation was given by Gilbert and MacDonald (1960) for the case of toroidal oscillations. They used a modified Thompson-Haskell matrix method which assumed a sequence of uniform spherical layers in the earth. The evaluation of the spherical layer-matrices at each interface is much less efficient than the numerical integration outlined above. Consequently most free oscillation calculations use numerical integration. However, the layer-matrices are appropriate for displacement calculations in source studies. P-SV matrices are given by Ben-Menahem (1964a) and Phinney and Alexander (1966). Teng (1970) gives the inverse of the spherical layer-matrix; Frazer (1977) works out P-SV matrices for inhomogeneous layers with the velocity variation $v = v_o(r/r_o)^{-b}$.

The spheroidal oscillations are designated by ${}_pS_n^m$ where p is the number of nodes along a radius, n is the number of nodes in latitude, and m is the number of nodes in longitude. The radial order number p in the case of spheroidal modes, is model dependent (Dziewonski and Gilbert, 1973) and some ambiguity occurs in comparison with observation. (See page 172 for a similar phenomenon in the case of surface-waves in a flat earth.) In a similar manner, the toroidal oscillations are labeled by ${}_pT_n^m$. The radial order number p has less ambiguity here. For a layered spherical earth, the periods of oscillation are found to be degenerate in the longitude parameter m. However this degeneracy is broken up by both the rotation of the earth, by departures from sphericity, and by inhomogeneities. We shall have more to say on this shortly.

To get some insight into the nature of the free oscillations, we shall look first at the distribution of u_r for the spheroidal modes. We shall assume m = 0. Figure 24-1 shows the radial displacement for the modes, ${}_0S_0$, ${}_0S_1$, and ${}_0S_2$. The second, ${}_0S_1$, corresponds to a rigid body translation. The plus sign indicates outward motion while the minus sign indicates inward motion. For the toroidal modes,

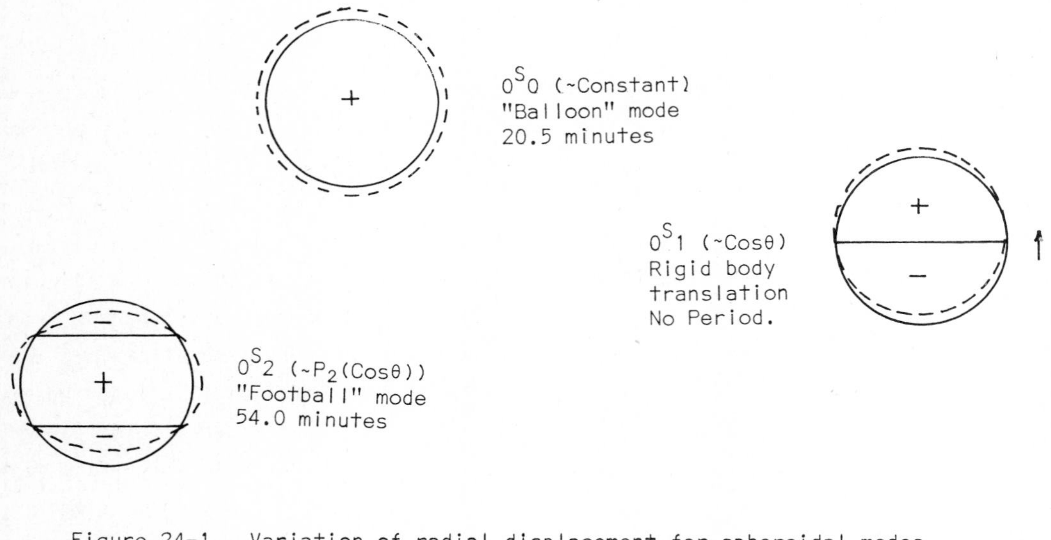

Figure 24-1. Variation of radial displacement for spheroidal modes.

we shall look at the tangential component of motion u_ϕ. In this case we must use the surface distrubution $\frac{\partial}{\partial\theta} Y_n^o(\theta,\phi)$ where we have again assumed no longitudinal dependence. Because of this derivative relationship, the mode, ${}_0T_0$ does not exist.

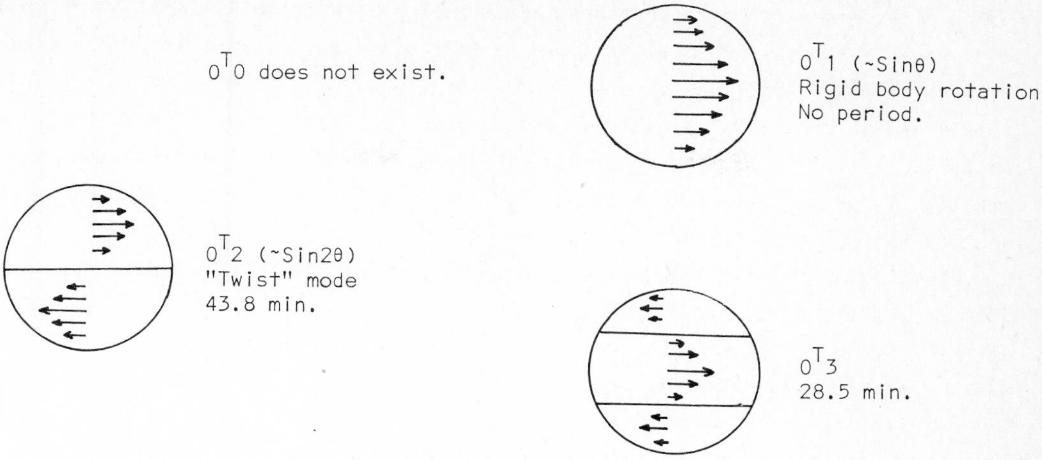

Figure 24-2. Variation of azimuthal displacement for toroidal modes.

It can be seen easily that modes such as $_pS_0$, $_pS_2$, etc., and $_pT_2$ etc., can have overtones with nodal surfaces within the body of the sphere. However, the two special cases of rigid body motion only occur for p = 0, i.e., for $_0S_1$ and $_0T_1$. For non-zero values of p, these modes can have overtones. For example, $_1S_1$ represents the outside of the sphere moving outward, the inside inward. Similarly, $_1T_1$ represents the outside of the sphere in a clockwise rotation, the inner portion in a counter-clockwise rotation.

Core Modes. Slichter (1961) found a peak at 86 minutes on the tidal gravimeter's reading of the great Chilean earthquake of 1960 and postulated that it might be due to a translational motion of the solid inner core. Although much numerical and theoretical work has been done since that time, the parameters are many and numerous approximations have to be made. Regardless of the model, the results seem to show that such modes would be barely detectable at the earth's surface. Two recent papers which will lead back to much of the earlier work are: M. L. Smith (1976), inner core oscillations; and P. Olsen (1977), fluid core waves.

An Aspherical Earth. Backus and Gilbert (1961) and Gilbert and Backus (1965) first studied the rotational splitting of the degenerate lines in the earth's free oscillation spectra. Dahlen (1968, 1969) added the effects of ellipticity. Gilbert (1971b) pursued the theory of splitting even further, and derived the diagonal sum rule for the earth's eigenfrequencies which says that the average over the multiplet is equal to the degenerate eigenfrequency. M. L. Smith (1974) worked out a new way of calculating free modes in the rotating, slightly elliptical case. Dahlen and Smith (1975) give an advanced treatment of the effect of rotation.

Dziewonski and Sailor (1976) point out that Dahlen (1975) had used Rayleigh's principle improperly and that numerical calculations involving the splitting were wrong. In his reply (Dahlen, 1976), and in papers by Woodhouse (1976) and Woodhouse and Dahlen (1978) the error is found to include other papers including the Dahlen (1968) study and a revised formulation is given.

The effects of lateral heterogeneities have been studied by Madariaga (1972), Luh (1973, 1974) and Luh and Dziewonski (1976). Moon and Wiggins (1977) propose a variational type finite element method for the solution of problems involving lateral inhomogeneities in the earth.

Stein and Geller (1977) examine the multiplet structure as a function of source type.

24.3. Surface-Waves on a Spherical Earth.

When the order number of the free oscillations becomes twenty or so, most of the wave energy is carried along the surface of the earth. In this case the effects of curvature dominate; gravity and the effects of the core are much less. The standing waves comprising the free modes can be considered as a superposition of traveling surface-waves of long wavelength. Surface-waves in the period range 100 - 500 sec. are referred to as mantle waves. Takeuchi and Saito (1972) give a comprehensive review of the theory and calculations for both free modes and surface-waves. We shall touch on a few relevant points.

Consider the eigenfunction expansion associated with a homogeneous spherical earth, i.e.,

$$\Phi = \sum_{k,m,n} a_{mn}(k)j_n(kr)P_n^m(\cos\theta)e^{i(m\phi - \omega t)} . \tag{24-7}$$

Let us relate these functions to the ones we have seen in Chapter 16.

First of all, we will look at the defining equation for the spherical Bessel function $j_n(kr)$. This is

$$\frac{d^2R}{dr^2} + \frac{2}{r}\frac{dR}{dr} + \left[k^2 - \frac{n(n+1)}{r^2}\right]R = 0, \tag{24-8}$$

with solution $R = j_n(kr)$. By a change in variable, $R = V/r$, equation (24-8) can be reduced to the following form:

$$\frac{d^2V}{dr^2} + \left[k^2 - \frac{n(n+1)}{r^2}\right]V = 0. \tag{24-9}$$

An inspection of this form shows that there are two types of solution. Where $k^2r^2 > n(n+1)$ we will have an oscillatory solution, and where $k^2r^2 < n(n+1)$ we will have an exponential solution. For proper behavior at the origin, the exponentially decreasing solution must be used. Thus the exponential drop-off towards the center of the sphere corresponds to the exponential drop-off of the surface-waves with depth, and the oscillatory part corresponds to the oscillations between nodal planes described previously.

Consider now the spherical harmonics with m = 0, i.e., having longitudinal symmetry. We have

$$P_n(\cos\theta) \simeq \left[\frac{2}{n\pi \sin\theta}\right]^{\frac{1}{2}} \left[(1 - \frac{1}{4n})\sin\phi - \frac{1}{8n}\cot\theta\cos\phi\right], \qquad (24\text{-}10)$$

where $\phi = (n + \frac{1}{2})\theta + \pi/4$, $n >> 1$, $\varepsilon < \theta < \pi - \varepsilon$, and ε is a small number. (See Jahnke and Emde, 1938, p. 117.) This can be further approximated by

$$P_n(\cos\theta) \simeq \left[\frac{2}{\pi n \sin\theta}\right]^{\frac{1}{2}} \cos\{(n + \tfrac{1}{2})\theta - \pi/4\}$$

$$\simeq \left[\frac{1}{2\pi n \sin\theta}\right]^{\frac{1}{2}} \left[e^{i\{(n + \frac{1}{2})\theta - \pi/4\}} + e^{-i\{(n + \frac{1}{2})\theta - \pi/4\}}\right], \qquad (24\text{-}11)$$

where we have neglected all additive terms containing n. This approximation says that in the center portion of the interval $0 < \theta < \pi$ (and also $\pi < \theta < 2\pi$) the Legendre function is oscillating at a rate of $(n + \frac{1}{2})$ cycles per 2π interval. As it passes the singular points 0 and π it stretches out a bit to have only n cycles in the 2π interval. This can be sketched out for $P_5(\cos\theta)$ as in Figure 24-3. The solid line depicts $P_5(\cos\theta)$ while the dotted line depicts $0.34 \cos(5.5\theta - \pi/4)$. First of all we note that the central portion of the interval $\pi < \theta < 2\pi$ lags the central portion of the interval $0 < \theta < \pi$ by $\pi/2$ i.e., there is a $\pi/2$ phase shift (or one-quarter wavelength) as the pole is crossed.

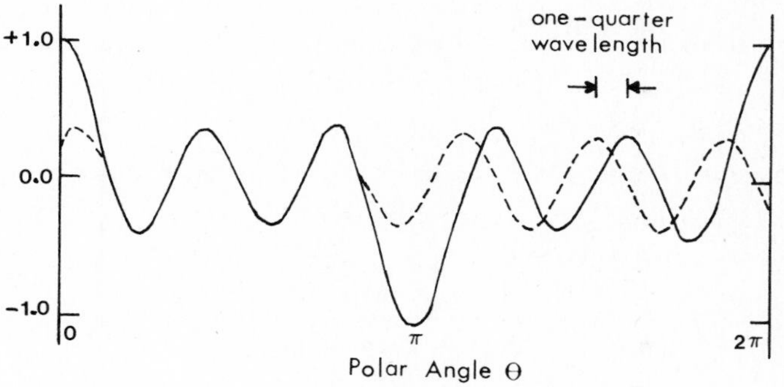

Figure 24-3. A comparison of $P_5(\cos\theta)$ and the function $0.34 \cos(5.5\theta - \pi/4)$.

Brune et al (1961) first observed and explained this polar phase shift. Each successive polar crossing adds an additional phase shift. (NOTE: This same effect due to the singularity of the coordinate system manifests itself in the far field representation of the outgoing wave solution in cylindrical coordinates as

$$H_0^{(1)}(kr) \simeq (kr)^{-\frac{1}{2}} e^{i(kr - \pi/4)},$$

where again we have the $\pi/4$ phase shift.)

We can also use the asymptotic form of $P_n(\cos\theta)$ to relate the apparent phase-velocity c of waves in the equatorial region to the order number n. Here we have

$$\theta = L/a = 2\pi/(n + \tfrac{1}{2}),$$

where "L" is the wavelength. But $c = Lf$, and we finally obtain

$$c \simeq \frac{2\pi a}{(n + \frac{1}{2})T_n}, \qquad (24\text{-}12)$$

where T_n is the period of oscillation of the nth mode. Using this relation, we can take the free periods of oscillation and construct equivalent phase-velocity curves for the mantle waves. Lastly we might note that the factor $\mathrm{Sin}^{-\frac{1}{2}}\theta$ in (24-10) and (24-11) represents the geometrical divergence factor necessary to maintain a constant energy flux across any ring of radius $a\mathrm{Sin}\theta$.

A more precise discussion of the transition from standing waves to traveling waves is given by Gilbert (1976a). Schwab and Kausel (1976a) note that as surface-wave observations are pushed towards longer wavelengths, a true determination of phase-velocity will depend upon the relative location of source and receiver. The polar-focusing effect both amplitude- and phase-modulates the waveforms and these modulations have to be taken into account. Likewise, rotation and ellipticity will contribute small corrections to the calculation of phase-velocity.

Surface-Waves on a Homogeneous Sphere. We have seen in Figure 22-3 that body-waves take a "short-cut" in a homogeneous sphere arriving more quickly than if they had traveled a circumferential path. For surface-waves, a similar effect holds. The more deeply penetrating waves have a circle of smaller radius to travel and apparently "speed-up". This effective velocity increase not only gives dispersion of Rayleigh-waves, but allows the existence of Love-waves in such a geometry even though composed of a homogeneous material. Rather exotic approximations are necessary to obtain the dispersion relations from the solutions to the spherical equations of motion. However, by the use of (24-12) and the free periods of oscillation calculated by Y. Sato and T. Usami (1962), one can show the effect very clearly. In Figure 24-4 we see the equivalent phase-velocities for both fundamental toroidal and

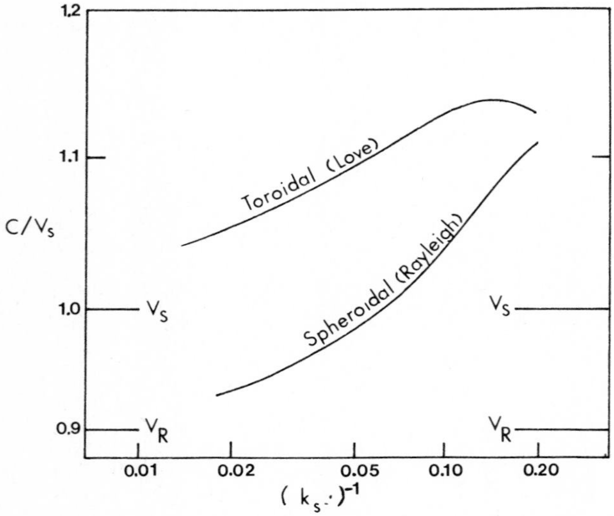

Figure 24-4. Dispersion curves for fundamental mode surface-waves on a homogeneous sphere.

spheroidal modes. The velocity associated with the fundamental toroidal mode approaches the shear velocity of the homogeneous sphere while that associated with the fundamental spheriodal mode approaches the Rayleigh velocity in a half-space composed of the same material.

Propagation of Love-Waves on a Radially Inhomogeneous Sphere. This problem obviously has great importance for the propagation of Love-waves on the earth's surface. For sufficiently short wavelengths, the flat-earth techniques developed previously may be used. For long wavelengths, interferences will occur as the waves travel around the earth and the complete theory for the torsional oscillations of a sphere must be used. For in-between wavelengths an earth stretching transformation developed by Biswas and Knopoff (1970) may be used. This approximation can be obtained in the following way.

We start with the equation of motion governing the azimuthally symmetric torsional oscillations of a radially inhomogeneous sphere (slightly modified from those given by Biswas and Knopoff):

$$\frac{1}{r^2}\frac{\partial}{\partial r}\left(r^2\frac{\partial\Lambda}{\partial r}\right)+\frac{1}{\mu}\frac{\partial\mu}{\partial r}\left(\frac{\partial\Lambda}{\partial r}-\frac{\Lambda}{r}\right)-\frac{\rho}{\mu}\ddot{\Lambda}+\frac{1}{r^2\operatorname{Sin}\theta}\frac{\partial}{\partial\theta}\left(\operatorname{Sin}\theta\frac{\partial\Lambda}{\partial\theta}\right)=0; \qquad (24\text{-}13)$$

where Λ is related to the displacement component u_ϕ by (24-3), i.e.,

$$u_\phi = -(\partial\Lambda/\partial\theta)e^{-i\omega t} .$$

Only the radial derivative operates on $\mu(r)$. The solution of (24-13) is given by

$$\Lambda = R(r)P_n(\operatorname{Cos}\theta)e^{-i\omega t} ;$$

where $R(r)$ satisfies the radial differential equation

$$\frac{1}{r^2}\frac{d}{dr}\left(r^2\frac{dR}{dr}\right)+\frac{1}{\mu}\frac{d\mu}{dr}\left(\frac{dR}{dr}-\frac{R}{r}\right)+\left[\frac{\omega^2\rho}{\mu}-\frac{n(n+1)}{r^2}\right]R=0. \qquad (24\text{-}14)$$

Making the substitutions

$$R = rV, \quad \frac{dR}{dr} = r\frac{dV}{dr}+V, \quad \frac{dR}{dr}-\frac{R}{r} = r\frac{dV}{dr} ; \qquad (24\text{-}15)$$

we obtain

$$\frac{1}{r^2}\frac{d}{dr}\left(r^3\frac{dV}{dr}\right)+\frac{1}{r^2}\frac{d}{dr}(r^2V)+\frac{1}{\mu}\frac{d\mu}{dr}\left(r\frac{dV}{dr}\right)+\left[\frac{\omega^2\rho}{\mu}-\frac{n(n+1)}{r^2}\right]rV=0.$$

Making the further change of variable

$$z = a\ln\left(\frac{a}{r}\right), \quad \frac{d}{dr} = -\frac{a}{r}\frac{d}{dz}, \quad \frac{dr}{dz} = -\frac{r}{a} ; \qquad (24\text{-}16)$$

and multiplying by μ/r gives

$$\mu\frac{a^2}{r^2}\frac{d^2V}{dz^2}-\frac{3a\mu}{r^2}\frac{dV}{dz}+\frac{a^2}{r^2}\frac{d\mu}{dz}\frac{dV}{dz}+\left[\omega^2\rho-\frac{\mu}{r^2}\{n(n+1)-2\}\right]V=0.$$

Regrouping, we have

$$\left[\frac{a^2}{r^2}\frac{d}{dz}\left(\mu\frac{dV}{dz}\right)-\frac{3\mu a}{r^2}\frac{dV}{dz}\right]+\left[\omega^2\rho-\frac{\mu}{r^2}\{n(n+1)-2\}\right]V=0.$$

A final regrouping, and multiplication by r^5/a^5 leaves us with

$$\frac{d}{dz}\left[\mu\left(\frac{r}{a}\right)^3\frac{dV}{dz}\right] + \left[\omega^2\rho\left(\frac{r}{a}\right)^5 - \mu\left(\frac{r}{a}\right)^3\left\{\frac{n(n+1)-2}{a^2}\right\}\right]V = 0. \qquad (24\text{-}17)$$

Let us now look at the equation for Love-waves in a vertically inhomogeneous medium (21-15), and write

$$u_y(x,z,t) = V(z)e^{i(kx - \omega t)}; \qquad (24\text{-}18)$$

where V(z) contains the complexity due to the inhomogeneous nature of the medium. Then we have as an equation determining V(z):

$$\frac{d}{dz}\left(\mu\frac{dV}{dz}\right) - k_f^2\mu V + \omega^2\rho V = 0. \qquad (24\text{-}19)$$

Comparing with (24-17), we can make the following identifications:

$$\begin{aligned} z &= a\ \ln(a/r), \\ \mu_f(z) &= \mu_s(r)(r/a)^3, \\ \rho_f(z) &= \rho_s(r)(r/a)^5, \\ (v_s)_f &= (v_s)_s(a/r), \\ k_f^2 &= \frac{1}{a^2}\{n(n+1) - 2\}; \end{aligned} \qquad (24\text{-}20)$$

where $\underline{a}$ is the radius of the sphere.

At this point the transformations are exact but not very useful. From (24-12) we see that

$$k_s a = 2\pi a/L \simeq n + \tfrac{1}{2},$$

whence

$$\frac{1}{a^2}\{n(n+1) - 2\} \simeq k_s^2 - \frac{9}{(2a)^2},$$

and the last of the relations (24-20) becomes

$$k_f^2 \simeq k_s^2 - \frac{9}{(2a)^2}. \qquad (24\text{-}21)$$

A second approximation comes in reducing the inhomogeneous medium to a series of flat homogeneous layers of varying properties according to (24-20). Then one can apply the standard flat-earth Thompson-Haskell matrix techniques. After solving the matrix equation for k_f, k_s is then computed from (24-21).

Anderson and Toksöz (1963) have used an alternative approach, which leads to an equivalent anisotropic flat model. The Universal Dispersion Tables computed by Anderson (1964) and Anderson and Harkrider (1968) were computed using this approximation and hence are essentially correct for a spherical earth. These Tables were mentioned on page 258. In a small discussion paper (Hill and Anderson, 1977) it was noted that Anderson and Toksöz had incorrectly used $k_s^2a^2 \simeq n(n+1)$ rather than $k_s^2a^2 \simeq n(n+1) + \frac{1}{4}$ as in (24-12). In this paper a number of other points relating the Anderson and Toksöz transformation to that of Biswas and Knopoff are discussed.

It might be noted that the first and fourth transformations of (24-20) are identical to the second and third of (22-8) for body-waves. These two transformations are necessary that velocities be correct (the <u>kinematic</u> problem). The second and third of (24-20) are necessary that the stresses be correct (the <u>dynamic</u> problem). Chapman (1973) shows that the Biswas-Knopoff transformation (24-20) is exact for SH and approximate for P-SV body-waves propagating in a sphere.

A similar <u>exact</u> earth-flattening transformation is not available for Rayleigh-waves. However Biswas (1972) having neglected some terms in the differential equation finds an approximate transformation as follows:

$$\begin{aligned} &z = a \ln(a/r), \\ &\lambda_f = \lambda_s, \qquad (v_p)_f = (v_p)_s(a/r), \\ &\mu_f = \mu_s, \qquad (v_s)_f = (v_s)_s(a/r), \\ &\rho_f = \rho_s(r/a)^2, \qquad k_f^2 = k_s^2 - \frac{1}{4a^2} . \end{aligned} \tag{24-22}$$

Additional improvement is found empirically by setting $\rho_f = \rho_s(r/a)^{2.275}$. This transformation was further refined by Schwab and Knopoff (1972, p. 141 ff.) by empirical adjustment of additional parameters. By considering the propagation of fundamental mode Rayleigh-waves in a number of models, Bolt and Dorman (1961) have worked out an empirical relation for periods less than 300 sec. This is

$$c_s = c_f(1 + 0.00016T); \tag{24-23}$$

where c_s is the true phase velocity in a curved earth, c_f is that calculated by the Thompson-Haskell matrix method for a flat earth and T is the period in seconds. North and Dziewonski (1976) have examined a number of models not included in the Bolt and Dorman study and propose that a more accurate form to use is:

$$c_s = c_f + 0.00016U_fT; \tag{24-24}$$

where U_f is the group-velocity corresponding to the flat-earth model. Gaulon et al (1970) have worked directly with the spherical layer-matrices (retaining the spherical Bessel functions) to calculate directly the disperion relations. Bhattacharya (1976) also works with spherical layer-matrices, but makes the layers inhomogeneous in such a way that the spherical Bessel functions can be replaced by exponentials as in the flat-earth matrix formulation. This speeds up the computer processing greatly.

24.4. <u>Modes and Rays in a Sphere</u>.

In Section 16.6 we discussed a number of relationships between modes and rays in a flat-layered medium. For a spherical medium, the transformations are much more complicated. Basically, one must use the Watson transformation (22-39) and appropriate expansions of the radial and Legendre functions. The necessary mathematical detail is worked out by Ben-Menahem (1964), and by Nolet and Kennett (1978). One of the simpler results can be worked out rather simply in a heuristic manner (Brune, 1964). This derivation preceded the mathematical verification by Ben-Menahem (ibid).

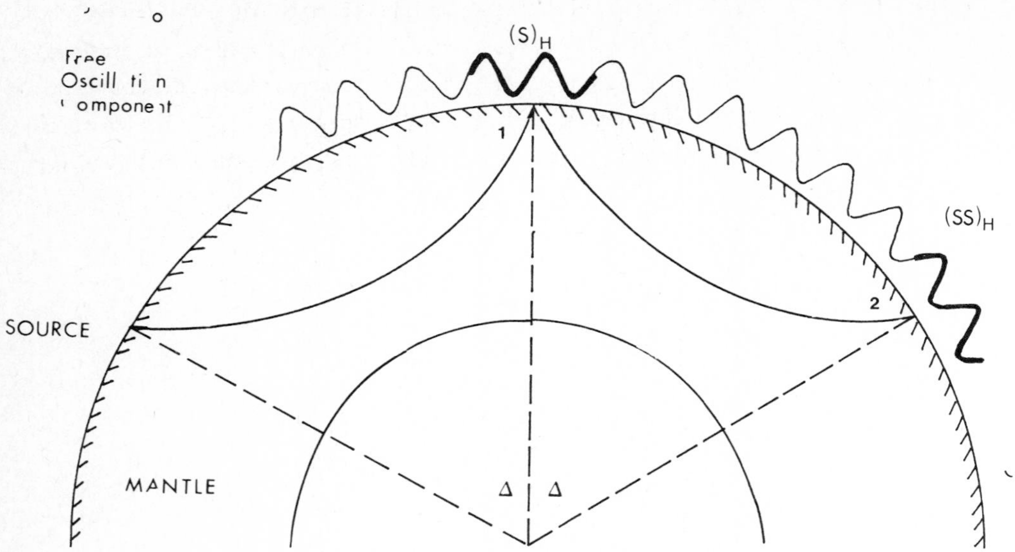

Figure 24-5. Geometric relationships between toroidal mode components and the body-wave pulses $(S)_H$ and $(SS)_H$.

Consider the phase difference between points 1 and 2 in Figure 24-5, where we denote the phases at the two points by ϕ_1 and ϕ_2. This phase difference will be that between the phase accumulated as the energy travels along the ray-path and the phase accumulated as the mode propagates along the surface. One can write

$$\phi_2 - \phi_1 + 2p\pi = 2\pi t/T - 2\pi a\Delta/cT, \qquad (24\text{-}25)$$

where $\underline{p}$ is the phase integer ambiguity, $\underline{t}$ is the travel-time for the distance Δ (radians), $\underline{a}$ is the radius of the sphere, and $\underline{c}$ is the modal phase-velocity along the surface. Now the phase-velocity $\underline{c}$ can be identified with apparent surface-velocity of the pulse, $ad\Delta/dt$. Making the substitutions and regrouping, one obtains

$$p = (t - \Delta dt/d\Delta)/T - (\phi_2 - \phi_1)/2\pi. \qquad (24\text{-}26)$$

The integer was identified as the radial order number of the torsional mode by Brune (1964) and subsequently verified theoretically by Ben-Menahem (1964). In practice, two stations are chosen along a great circle at distances such that the phase-velocities of the observed pulses are the same. For example, the phase-velocity of the S pulse at station one is the same as the phase-velocity of the SS pulse at station two which is twice as far away. A similar equivalency would occur for S at Δ and SSS at 3Δ. A Fourier spectrum is taken of each pulse and the phases ϕ_2 and ϕ_1 computed as a function of the period T. One then computes the function defined by (24-26) and interpolates along the T-axis to find integer values of $\underline{p}$. Using (24-12) one can identify the polar order number $\underline{n}$ from

$$n + \tfrac{1}{2} = \{2\pi/T(p)\}dt/d\Delta. \qquad (24\text{-}27)$$

Brune (ibid) and Brune and Gilbert (1974) have identified more than one hundred torsional overtones using the methods outlined above. Some additional considerations concerning P- and SV-waves are given by Brune (1966). From these relationships, we see that overtones of large angular order n correspond to higher modes of surface-waves which themselves may be looked on as interferences of body-waves of the type S, SS, ... $(S)_n$.

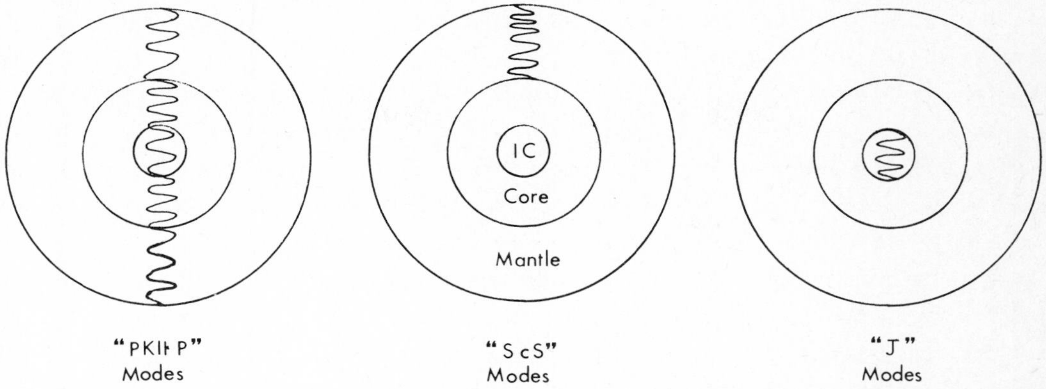

Figure 24-6. Schematic diagram of large p, low n modes in the earth. Spheroidal modes correspond to P-waves resonating through the whole earth ("PKIKP"). Toroidal modes correspond to SH-waves in the mantle ("ScS") and to SH-waves in the inner core ("J").

On the other hand, small values of polar order number n correspond to surface-waves of extremely high phase-velocity which are composed of body-waves at nearly vertical incidence. From energy considerations, Dziewonski and Gilbert (1973) classified such modes into three types. The spheroidal modes corresponding to P-waves passing through mantle, core, and inner core (PKIKP); and the toroidal modes broke into two groups — SH-waves in the mantle (ScS) and SH-waves in the inner core (J). Such waves are sketched in Figure 24-6. It might be noted that excitation of "J"-modes is unlikely.

It is not difficult to see that for body-waves of relatively high frequency the travel-time through any region should be relatively independent of frequency and consequently there should be an asymptotic behavior for the magnitude of frequency jumps as radial order number is increased. Anderssen and others in a series of papers (a recent one is Anderssen, 1977) have studied this asymptotic behavior. Lapwood (1975) noted that discontinuities in the layers would lead an oscillation about this asymptotic behavior. Gilbert (1975) looked at the high-frequency behavior of the differential equations involved to derive the asymptotic spacings.

CHAPTER 25

A RADIALLY INHOMOGENEOUS EARTH – FREE OSCILLATIONS AND MANTLE WAVES

25.1. The Periods of Free Oscillation.

Mode Identification. In Figures 24-1 and 24-2 we have indicated that the gravest periods of oscillation for the earth approach one hour. A comparison of the spectra of two very large earthquakes is given in Figure 25-1. Here it can be seen that one can identify the lower order fundamental modes quite well with trouble appearing only when one has to separate almost coincident pairs as at T_{17} - S_{16}. When the problem involves only the separation of toroidal from spheroidal modes, one can use particle motion as a simple discriminant. However, in these spectra from two relatively shallow earthquakes, higher overtones are little excited. The real problem then is to separate various orders and overtones from each other, all belonging to either the toroidal or spheroidal class. In addition to particle motion (polarization), one can use relative attenuation, group-velocity, spectral line-width, and a gradually changing line spacing to identify doubtful lines.

However, the big breakthrough came when it was found possible to use the differential excitation of the various lines as a means of line identification. Depending on the source mechanism, the spectral line amplitude and phase varies. Mendiguren (1973) used theoretical calculations from a body-wave derived source mechanism to adjust the phase of a particular spectral peak so that in a sum these peaks (from many records) would add in-phase while nearby peaks would add randomly. This stacking procedure was a tremendous success. A further improvement is called stripping. The stripping process (Gilbert and Dziewonski, 1975, Section 2.2) uses the orthogonal character of the vector spherical harmonics to separate out orders of various n into groups. Within each group it was not too difficult to identify the different radial overtones. An extended discussion of both the stacking and stripping procedures may be found in Nyman (1975).

The use of source-mechanism to interpret observation has been extended to multiplet-splitting by Geller and Stein (1977) and Stein and Geller (1978). Bolt and Currie (1975) have suggested the application of the maximum entropy method (Burg, 1972) to the obtaining of spectra of improved quality. This suggestion has not been widely applied, most spectra being obtained by power spectral methods (Båth, 1974, Chapter 3).

Observations. A list of observations containing 36 papers prior to 1973 may be found in Båth (1974, pp 328-329) and weighted average values have been calculated for observations through 1968 by Derr (1969a). With this early data, there was a problem of mis-identification at higher n values, consequently the more precise technique of stacking and stripping is now desirable. This additional sophisticated processing

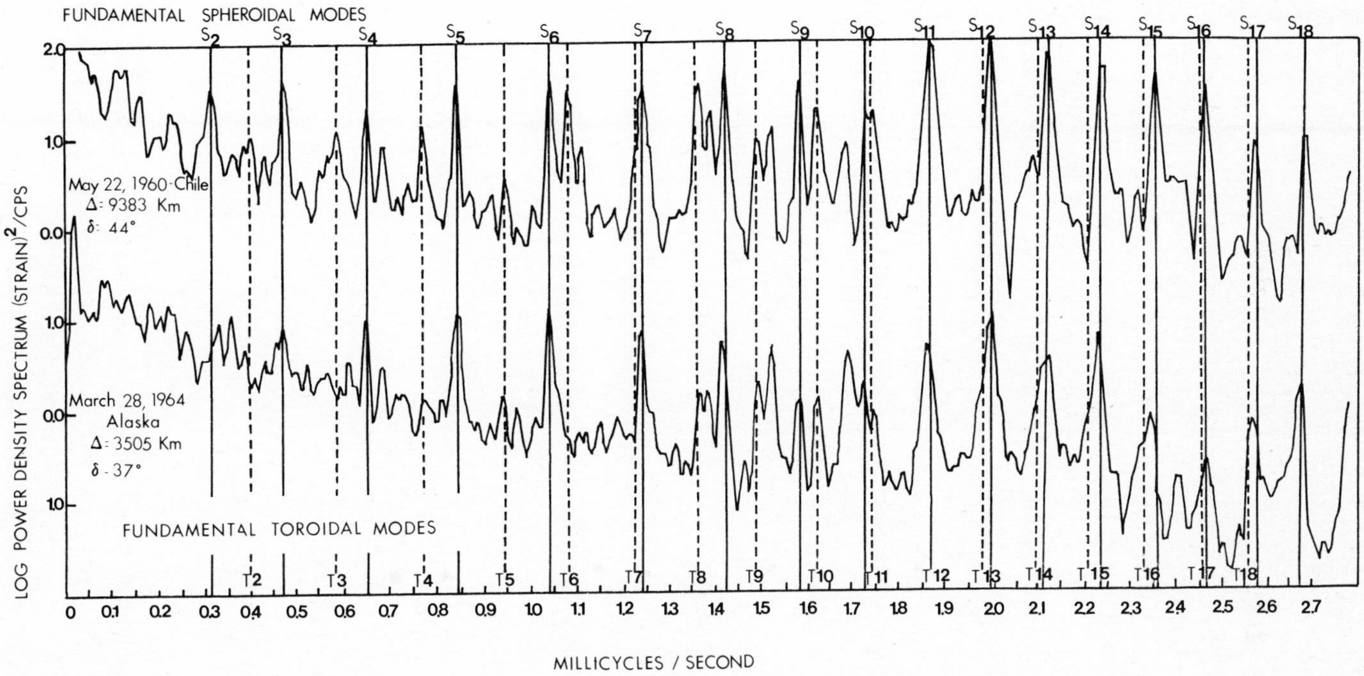

Figure 25-1. A comparison of spectra from the Chile 1960 and Alaska 1964 earthquakes. (After S. W. Smith, 1966.) Copyrighted by AGU.

has greatly lowered the output of observations. Four large works are relevant: Mendiguren (1973), Dziewonski and Gilbert (1972, 1973), and Gilbert and Dziewonski (1975).

25.2. Mantle Waves.

At periods between 100 and 500 seconds, surface-waves take on a unique character. The Rayleigh-waves have an Airy phase (Figure 16-11b) centered about a period of 240 seconds (R-phase) while the Love-waves take a rather impulsive shape known as the G-phase. The reason for this behavior is not difficult to understand in the light of the group-velocity character exhibited by Figure 20-1. A subscript is added to this notation to denote the travel path: R_1 and G_1 are direct from the epicenter, R_2 and G_2 come from the opposite direction along the great circle path, R_3 and G_3 travel the short leg plus once around, R_4 and G_4 travel the long leg plus once around, etc. These particular signals are generally of better quality than the normal 10 - 100 sec dispersive wavetrains associated with normal long-period instrumentation and rather accurate phase-velocities may be determined. In particular, if one takes ratios of the pairs $R_1 - R_3$, $G_2 - G_4$, $R_4 - R_6$, etc, the effects of source and receiver cancel out. A synopsis of great-circle mantle-wave phase-velocities is given in Table 25-1. Data are derived from Toksöz and Anderson (1966), Dziewonski and Landisman (1970), Kanamori (1970) and Nakanishi et al (1976a).

TABLE 25-1

Period	Rayleigh	Love
100 sec	4.09 km/sec	4.64 km/sec
150	4.30	4.78
200	4.58	4.92
250	4.92	5.07
300	5.29	5.23
350	5.62	5.40
400	5.94	5.55
450	6.17	5.70
500	6.40	5.84

Problems arise when one tries to take these great-circle path data, determine how much of a certain tectonic type region the great circle has crossed and then try to ascertain the phase-velocity associated with that type region. Kanamori (ibid) and Dziewonski (1970a) have differences between their regionalized phase-velocities that are greater than one would expect considering the accuracy of their starting data. Wu (1972) attempts to reconcile the differences by making an improved regionalization program. Madariaga and Aki (1972) show that "ray-theory" for surface-waves may not be applicable if the lateral heterogeneities are too great. Finally, it may be necessary to take ellipticity and rotation into effect (Dahlen, 1976).

Finally, using (24-12), one can get equivalent phase-velocities for the free modes of oscillation. Combining the regional data, the mantle wave data, and the equivalent phase-velocities, one can obtain a composite picture. Figure 25-2 shows the data for Rayleigh modes and Figure 25-3 presents the values for Love modes. It might be noted here that there are no cut-off frequencies for the higher modes as in Figures 16-12 and 16-13. This corresponds to the fact that energy leaking out of the uppermost layers passes through to the other side of the earth. It is not lost, as in the flat-layer case, but only undergoes a group-delay as it rejoins its parent mode.

From the data given in Figure 25-2 it seems that stable regions are quite similar (whether oceanic or continental), at least above 50 seconds period. Likewise tectonic regions. Older literature attempted to make the division on the basis of continents versus oceans. However it was soon realized that there was a great variety amongst continental types: shields, platforms, mountains, rift zones. Only more recently have the various oceanic types of young (ridge) and old (basin) been separated. Arcs and trenches occupy such a small portion of the ocean area that their properties are difficult to ascertain. The spread below 50 seconds period is largely due to the effect of water upon the Rayleigh-wave dispersion curve.

The unreliability of Love-wave measurements between 20 and 100 sec periods has been noted (p. 248); however, the data used in Figure 25-3 was especially processed by Forsyth (1975a) to minimize any problems. No such problems exist for analysis of the G-wave (Toksöz and Anderson, 1966). It can be seen that the fit between the two sets of data is fairly good, a small discrepancy existing at 150 seconds. Part of this may be due to difficulties with separating out "pure-path" data and part may be a residual of the Love-wave group-velocity interferences. It might be noted that very old ocean (100 - 200 million years) was not included in the area investigated by Forsyth (ibid). However from the trend of his curves, values for such ancient sea-floor would not be expected to lie much above the 10 - 50 million year curve.

25.3. Inversion.

It is obvious from the nature of the equations governing the free modes, that density will play a part. Consequently, density as well as velocity can be obtained on inverting the free mode data. This is in contrast to surface-waves which are relatively insensitive to density models. We have already considered (Section 20.4) the Backus-Gilbert linearized inversion method in the case of using surface-wave data to determine shear-velocity structure in a flat-earth model. The method is complicated now by having to invert for v_p, v_s and ρ. Most of the details for the spherical case are given by Backus and Gilbert (1967, 1968, 1970) and Gilbert (1971a). Gilbert (1972a) summarizes the technique as well as constructs a simple model from

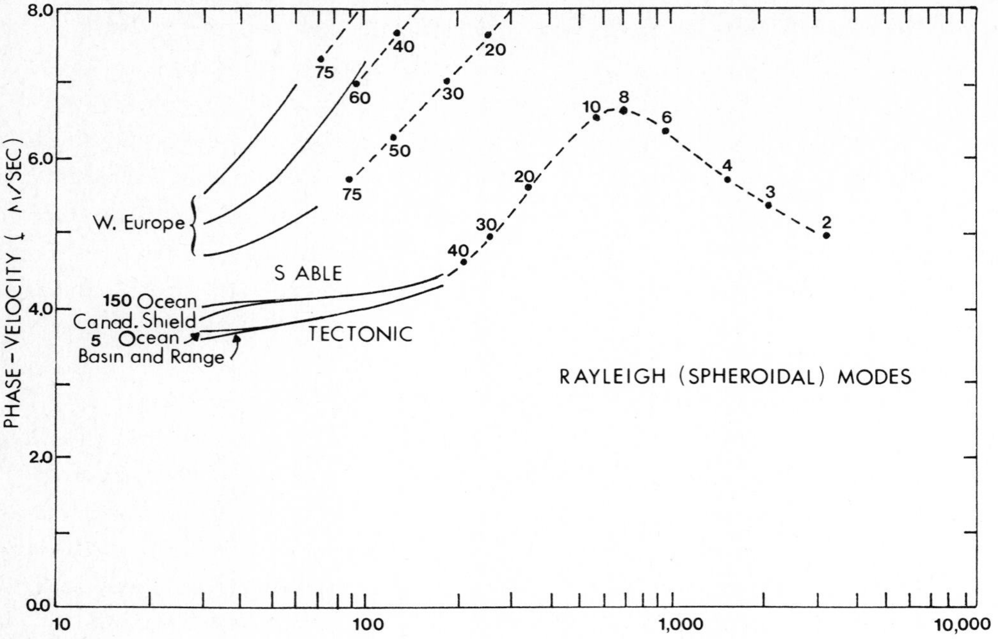

Figure 25-2. Composite Rayleigh-wave phase-velocities. Sources of data are:
5 Ocean, 150 Ocean; Leeds (1973).
Canadian Shield, Basin and Range; Pilant (unpublished).
W. Europe (higher modes); Nolet (1975).
Free modes; Gilbert and Dziewonski (1975).

15 free periods, mass, and moment of inertia. Gilbert (1976b) shows how group-velocity data might be included in the inversion.

Crucial to the linearized inversion scheme is the availability of partial derivatives with respect to the three parameters in a spherical earth. Published partials include Takeuchi and Sudo (1968), Wiggins (1968), Anderson and Kovach (1969) and Derr (1969a). These partials are only for the fundamental and first two radial overtones and additional partials are needed to work with data sets such as that of Gilbert and Dziewonski (1975). Dziewonski (1970b) shows that there are high correlations between some partials leading to non-uniqueness or a possible interchange between core radius and velocity or density distributions in the lower mantle.

In a significant paper, Gilbert, Dziewonski, and Brune (1973) show how great quantities of data do not provide as much information as one would think. For their model B497 they use the mass, moment of inertia, 368 modal frequencies and 127 ray travel-time data for a total of 497 gross earth data (GED). After applying rather stringent (but not unreasonable) requirements to check for independence, they find that the mass and moment of inertia each provide one significant earth datum (SED),

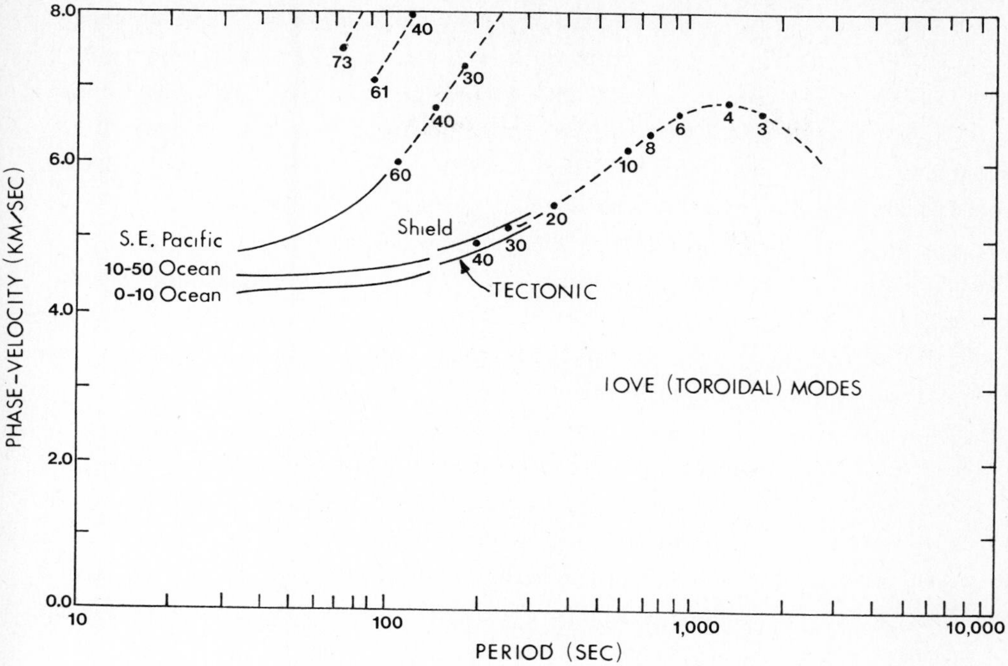

Figure 25-3. Composite Love-wave phase-velocities. Sources of data are: 0 - 10 Ocean, 10 - 50 Ocean; Forsyth (1975a). S. E. Pacific (1st higher Love); Forsyth (1975a). Shield, Tectonic; Toksöz and Anderson (1966). Free modes; Gilbert and Dziewonski (1975).

the modes provide 41 SED, and the rays provide 19 SED. However in combination a total of only 45 is achieved. The modes and rays contain much overlapping information. In the much larger set of data contained in the Gilbert and Dziewonski (1975) paper, a set of 1066 GED reduce to 57 SED. That is, a doubling of the quantity of gross earth data only achieved a 27% increase in significant earth data. Gilbert and Dziewonski (ibid) note that what is needed is "better" not "more" data. However, many factors work against attaining this goal. Attenuation limits the length of time that a signal stays above the noise level and consequently limits the resolution of the record. Lateral inhomogeneities in the earth add unknown biases to the observations — one trivial example is that most data is recorded on land. Perhaps another "breakthrough" will be forthcoming!

Synthetic Seismograms. Because of the complex nature of models of the radially inhomogeneous earth, the construction of synthetic seismograms from modal contributions is much less frequent than for those using one ray theory or another. Some examples are Landisman et al (1970), Alterman and Loewenthal (1972), Luh and Dziewonski (1975), Nakanishi et al (1977), and Mantovani et al (1977a).

25.4. Realistic Earth Models.

In the previous section, we discussed the difficulty of obtaining a large number of significant earth data. We have to outline v_p, v_s, ρ, and the radii of major discontinuities with 40 - 60 SED. Gilbert and Dziewonski took their large set of data and produced two models. The first, 1066a, is rather smooth in the upper mantle. This smoothness is artificial in that discontinuities were not allowed to develop. Their second model, 1066b, started with a model containing major discontinuities in the upper mantle whose nature was indicated by body-wave precursors (p. 301), synthetic seismogram studies of the upper mantle (p. 299), short-period surface-wave data (p. 262), and plausibility arguments relating v_p, v_s, and ρ. The refined model contained these discontinuities with modifications. Both models, 1066a and b, were constructed from free modes alone, without including ray data.

Because I rather favor the above arguments for discontinuities in the upper mantle (even though the evidence is not overwhelming), I have chosen Anderson and Hart's (1976) Model C2 for inclusion. It also turns out that the effects of attenuation strongly affect the model parameters in the upper mantle. The changes necessary have been made to this Model C2, and so its inclusion serves a dual purpose. The reader can decide for himself which is the better model. The data set used by Anderson and Hart (ibid) is a somewhat abbreviated portion of that set forth by Gilbert and Dziewonski (ibid), but heavily constrained in the inversion by both starting model and by reference to travel-time observations. The parameters for this model are given in Figure 25-4. Some other models constructed in such a fashion are the CAL5 series (Bolt and Uhrhammer, 1975) which use slightly different values for the free periods.

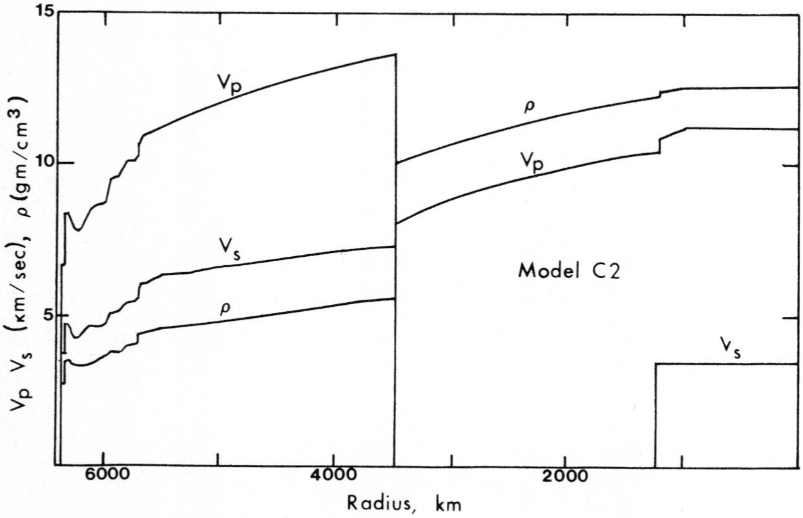

Figure 25-4. Model parameters for a realistic earth.
(After Anderson and Hart, 1976.) Copyrighted by AGU.

It might be noted that in most recent models, the shear-wave velocity in the inner core has been given values around 3.5 km/sec. The evidence from the free oscillation data requires the presence of a solid-inner core.

A number of earlier models have also received a great deal of attention. These are HB_1 (Haddon and Bullen, 1969), DI-11 (Derr, 1969b), and HB_2 (Bullen and Haddon, 1970). Although Monte Carlo methods are quite inefficient, they do give an idea of the possibilities of models which will fit the data. Press (1970a), gives a goodly number of examples.

Finally, Press (1970b) used Monte Carlo models to show the differences between regional structures. Dziewonski et al (1975) constructed continental and oceanic models by varying the crust and upper mantle above 420km. The lower part of the earth's structure was held constant.

ELASTIC WAVE DISSIPATION – THEORY

26.1. The Constant "Q" Model.

The reason for the introduction of the constant Q model is that it is found experimentally that for a period range of 10^{-5} to 10^{3} seconds, Q for most earth materials (though not over the whole range for any single one) lies between 100 and 1000. That is, for an eight-fold order of magnitude change in frequency, we have something less than an order of magnitude change in Q. However, a rigorously constant Q model gives trouble, both mathematical and physical. We shall have more to say about this later. Knopoff (1964b) gives an excellent review of both theory and observations. The notion of Q (Quality factor) can be developed in a number of ways, some of which follow.

Attentuation is traditionally thought of in terms of the damping of a simple harmonic oscillator. The equation of motion for such an oscillator is given by

$$\ddot{x} + 2\lambda\omega_o\dot{x} + \omega_o^2 x = 0, \qquad (26\text{-}1)$$

which has as its solution

$$x(t) = Ae^{-\lambda\omega_o t}\,\mathrm{Cos}\left[\omega_o(1 - \lambda^2)^{\frac{1}{2}}t - \phi\right]. \qquad (26\text{-}2)$$

Here A and ϕ are arbitrary constants chosen to fit the initial conditions. We will define (for a simple harmonic oscillator):

$$\frac{2\pi}{Q} \equiv -\left.\frac{\Delta E}{E}\right|_{1\text{ cycle}} \equiv -b \qquad \text{(Tabulated in Bradley and Fort, 1966)} \qquad (26\text{-}3)$$

$$\equiv -\left.\frac{2\Delta A}{A}\right|_{1\text{ cycle}} \equiv -2\delta \qquad (\delta = \text{logarithmic decrement}) \qquad (26\text{-}4)$$

$$\simeq 2\lambda\omega_o T = \frac{4\pi\lambda}{(1 - \lambda^2)^{\frac{1}{2}}} \qquad (\lambda << 1); \qquad (26\text{-}5)$$

where E is the peak energy stored in the oscillator and T is the period. We can also look at the quantity Q by considering a forced oscillator governed by the equation of motion:

$$\ddot{x} + 2\lambda\omega_o\dot{x} + \omega_o^2 x = (F/m)\mathrm{Cos}\omega t. \qquad (26\text{-}6)$$

The solution of this equation is given by

$$x(t) = \frac{(F/m)\mathrm{Cos}(\omega t - \varepsilon)}{\{(\omega_o^2 - \omega^2)^2 + 4\lambda^2\omega_o^2\omega^2\}^{\frac{1}{2}}}; \qquad (26\text{-}7)$$

where

$$\mathrm{Tan}\varepsilon = \frac{2\lambda\omega_o\omega}{\omega_o^2 - \omega^2}. \qquad (26\text{-}8)$$

This response has half-power points where the denominator is equal to $2^{\frac{1}{2}}$ times its minimum value, i.e., when

$$(\omega_o^2 - \omega^2)^2 + 4\lambda^2\omega_o^2\omega^2 = 8\lambda^2\omega_o^4.$$

Setting

$$\omega^2 = \omega_o^2 + 2\omega_o\delta\omega + (\delta\omega)^2,$$

we get

$$\delta\omega/\omega_o \simeq \lambda. \tag{26-9}$$

For small λ we then have (using 26-5)

$$1/Q = 2\delta\omega/\omega_o = 2\delta f/f_o; \tag{26-10}$$

i.e., Q is a measure of the quality factor of a resonant harmonic oscillator.

Another way of looking at Q comes from a consideration of the energy flux involved at the point of force application. Now

$$\langle -\dot{E} \rangle = \langle F\mathrm{Cos}\omega t \cdot \dot{x}(t) \rangle = \frac{\langle -F^2\omega \mathrm{Cos}\omega t(\mathrm{Sin}\omega t \mathrm{Cos}\varepsilon - \mathrm{Cos}\omega t \mathrm{Sin}\varepsilon)\rangle}{m\{(\omega_o^2 - \omega^2)^2 + 4\lambda^2\omega_o^2\omega^2\}^{\frac{1}{2}}}$$

$$= \frac{F^2\omega \mathrm{Sin}\varepsilon}{2m\{\quad\}^{\frac{1}{2}}}. \tag{26-11}$$

The total energy in the system is equal to the K.E. at the equilibrium position (i.e. x = 0) or to the P.E. at maximum excursion. The former is easier to evaluate, and we have for the total energy

$$E = \tfrac{1}{2} m\dot{x}^2 \Big|_{x=0} = \tfrac{1}{2}\frac{\omega^2F^2}{m\{\quad\}}. \tag{26-12}$$

Then, taking the ratio, we have

$$\frac{\langle -\dot{E}\rangle}{E} = \frac{\mathrm{Sin}\varepsilon}{\omega}\{\quad\}^{\frac{1}{2}} = 2\lambda\omega_o.$$

Multiplying by T, we have

$$\frac{\langle -\dot{E}\rangle T}{E} = -\frac{\Delta E}{E}\Big|_{1\ \text{cycle}} = \frac{2\pi}{Q} = 2\lambda\omega_o T;$$

as before in (26-5). Consequently Q is a measure of the rate of power absorption. Defined in this way, $2\pi/Q$ eliminates the power series expansion of the exponential in (26-2) and the approximation made thereby.

When one turns to an elastic medium, the simple harmonic oscillator model is no longer appropriate, particularly if anelastic effects are present. A more representative model can be constructed as follows. Consider the case of one-dimensional wave propagation in an anelastic medium — such propagation will be governed by:

$$\partial\sigma(x,t)/\partial x = \rho\partial^2u/\partial t^2, \tag{26-13}$$

$$\sigma(t) = \int_{-\infty}^{t} M(t - t')d\varepsilon(t'), \tag{26-14}$$

$$\varepsilon(x,t) = \partial u(x,t)/\partial x; \tag{26-15}$$

where σ = stress, ε = strain, u = displacement, and M is the stress response (generalized modulus function) to a step-function input of strain. An inverse to (26-14) can be written

$$\varepsilon(t) = \int_{-\infty}^{t} J(t - t')d\sigma(t'), \tag{26-16}$$

where J is the strain response (generalized compliance function) to a step-function input of stress. If we restrict ourselves to the time-harmonic case

$$\varepsilon(t) = \varepsilon_o e^{i\omega t}, \quad \sigma(t) = \sigma_o e^{i\omega t}; \tag{26-17}$$

we have

$$\varepsilon_o = J(\omega)\sigma_o, \quad \sigma_o = M(\omega)\varepsilon_o; \tag{26-18}$$

where $J(\omega)$ is the complex compliance and $M(\omega)$ is the complex modulus. They are given by

$$M(\omega) = i\omega \int_o^\infty M(t)e^{-i\omega t}, \tag{26-19}$$

$$J(\omega) = i\omega \int_o^\infty J(t)e^{-i\omega t}. \tag{26-20}$$

From (26-18) we see that

$$M(\omega)J(\omega) \equiv 1. \tag{26-21}$$

Traditionally, one writes

$$M(\omega) = M_1(\omega) + iM_2(\omega), \quad J(\omega) = J_1(\omega) - J_2(\omega). \tag{26-22}$$

Finally, putting all this back into (26-13), we have

$$M(\omega)\partial^2 u(x,\omega)/\partial x^2 = -\rho\omega^2 u(x,\omega). \tag{26-23}$$

An inspection of this form leads to an idea of a complex velocity $v(\omega)$ given by

$$v(\omega) = \{M(\omega)/\rho\}^{\frac{1}{2}} = \{\rho J(\omega)\}^{-\frac{1}{2}} = v_1 + iv_2. \tag{26-24}$$

Now Bland (1960, Chap. 2) finds that the stored energy in a system composed of springs and dashpots is a function not only of the complex compliance, but also of its first frequency derivative. O'Connell and Budiansky (1978) investigate this a bit further and find that although the maximum stored energy involves the first frequency derivative, the average stored energy does not. Therefore, they propose the following definition:

$$\frac{2\pi}{Q} = \frac{-\Delta E}{2E_{avg}}. \tag{26-25}$$

In the simple harmonic oscillator case, we have $E_{max} = 2E_{avg}$ and so the definition is consistent. In terms of this new definition, O'Connell and Budiansky (ibid) derive the following relations for Q.

$$Q = M_1/M_2 = J_1/J_2. \tag{26-26}$$

This particular relation has certain advantages when one is working with models of the complex modulus or the complex compliance. These authors go on to show that for high values of Q, the above observational relationships are applicable to its determination. Thus we still can use the following models to calculate Q in most cases.

Traveling Waves Decaying with Distance. Here we will write

$$\frac{2\pi}{Q} = -\left.\frac{2\Delta A}{A}\right|_L, \tag{26-27}$$

where ΔA is the change in amplitude over one wavelength "L". In this case, the appropriate form of solution is

$$f(x,t) = e^{-\alpha x} e^{-i(kx - \omega t - \phi)}, \tag{26-28}$$

where $k = 2\pi/L$. In one wavelength, $f(x,t)$ decreases to $e^{-\alpha L} = e^{-\alpha cT}$. Consequently,

$$-\Delta A/A \simeq \alpha cT,$$

and

$$\alpha = \frac{\pi}{QcT} = \frac{\omega}{2cQ} .$$

Inversely, we have

$$2\pi/Q = 2c\alpha/f. \tag{26-29}$$

In terms of the complex velocity $\underline{v}$ (26-24), one would find (using (26-26)) that

$$Q = \tfrac{1}{2}(v_1/v_2 - v_2/v_1). \tag{26-30}$$

The leading term of this expression is equivalent to the previous calculation (26-29).

<u>Standing Waves Decaying with Time</u>. Here one would like to write (in analogy with the simple harmonic oscillation)

$$\frac{2\pi}{Q_T} = -\left.\frac{2\Delta A}{A}\right|_T ,$$

where ΔA is the change in amplitude during one period "T". The form of the solution in the standing wave case can be written

$$f(x,t) = e^{-\beta t} e^{i(\omega t - \phi)} \operatorname{Sin}(kx); \tag{26-31}$$

where, for simplicity, we have assumed zero displacement boundary conditions. After one cycle, the amplitude will have decayed to $e^{-\beta T}$. For small decrements

$$-\Delta A/A \simeq \beta T;$$

whence

$$\beta = \pi/(Q_T T) = \omega/(2Q_T).$$

In terms of β, we have

$$2\pi/Q_T = 2\beta/f, \tag{26-32}$$

where $\underline{f}$ is the frequency. These relations would be fine if the waves were non-dispersive, but we shall see that all attenuation is accompanied by dispersion. To see the problems this leads to, we shall go back and consider (26-13) — (26-15).

On substitution of the form (26-31) into this triplet of equations, one finds

$$M(\omega + i\beta)k^2 = \rho(\omega + i\beta)^2. \tag{26-33}$$

Using (26-21) and simplifying, we have

$$k = (\omega + i\beta) \left[\rho J(\omega + i\beta)\right]^{\frac{1}{2}}. \tag{26-34}$$

In the standing-wave case, $\underline{k}$ is to be considered strictly real (and equal to $2\pi/L$) and (26-34) gives an equation to be solved for the appropriate value of $\omega + i\beta$. In the traveling-wave case we have that the frequency is real, and we must determine the real (k) and imaginary (α) parts of the wave-number from

$$(k - i\alpha) = \omega \left[\rho J(\omega)\right]^{\frac{1}{2}}. \tag{26-35}$$

Following a proof given by Knopoff et al (1964), we can expand (26-35) about the lossless condition where p_i are the parameters in J causing loss. To show this explicitly, we shall write $J = J(\omega,p_i)$. Then for traveling-waves we will have

$$k - i\alpha = k_o + \omega_o \sum \frac{\partial}{\partial p_i} \left[\rho J(\omega,p_i)\right]^{\frac{1}{2}}_{\omega_o,0} \delta p_i .$$

For most reasonable cases, the second term on the right is largely imaginary. We can therefore identify the imaginary portion of the wave-number as

$$\alpha \simeq -\mathrm{Imag}\{\omega_o \sum \frac{\partial}{\partial p_i} \left[\rho J(\omega,p_i)\right]^{\frac{1}{2}}_{\omega_o,0} \delta p_i\}. \qquad (26\text{-}36)$$

A similar expansion in the standing-wave case (26-34) leads to (since $\underline{k}$ remains fixed and the complex frequency $\omega + i\beta$ is to be found)

$$k = k_o + \frac{\partial k}{\partial \omega}\bigg|_{\omega_o,0} (\delta\omega + i\beta) + \omega_o \sum \frac{\partial}{\partial p_i} \left[\rho J(\omega,p_i)\right]^{\frac{1}{2}}_{\omega_o,0} \delta p_i .$$

Here $\partial k/\partial\omega$ is real (for propagating modes) and so we must have

$$\frac{\partial k}{\partial \omega}\bigg|_{\omega_o,0} \beta + \mathrm{Imag}\{\omega_o \sum \frac{\partial}{\partial p_i} \left[\rho J(\omega,p_i)\right]^{\frac{1}{2}}_{\omega_o,0} \delta p_i\} = 0.$$

On using (16-39), we have that

$$\beta = U\alpha, \qquad (26\text{-}37)$$

the basic result obtained by Knopoff et al (ibid). What this means is that the values of Q_T determined from (26-32) will be related to Q determined from (26-29) in the following way:

$$UQ_T = cQ. \qquad (26\text{-}38)$$

This relationship was noted by Brune (1962) when he analyzed apparent discrepancies between Q obtained from mantle waves and from free modes. However, in his paper it was assumed that the Q's determined by (26-32) were correct; those given by (26-29) should be altered. This tradition has been followed to the present and the figures of Chapter 27 conform to this decision. At the time, this choice seemed reasonable, but in light of the material constitutive relations (26-14) and (26-16) it seems that the alternative is more appropriate.

As we have mentioned, the constant Q model has difficulties. Physical difficulties arise when one tries to conceive of a model with no characteristic frequency (as required by a constant Q model). Also such a model would not be "causal". Knopoff and MacDonald (1958) suggested that a linear theory of attenuation predicts dissipation proportional to an even power of the frequency. This is inconsistent with the observations and consequently led them to investigate a number of non-linear models. There are two ways out of this problem. The first is to assume constant Q only over a finite band of frequencies, and that outside that band Q varies differently (Futterman, 1962). The second alternative is to assume:

26.2. An Almost Constant Q Model.

One model with almost constant Q was given by Lomnitz (1957). This was based on creep studies of igneous rocks (Lomnitz, 1956). Another instructive model based on analytical relationships was given by Strick (1967). Consider the transform pair:

$$f(t) \Longleftrightarrow F(\omega) = R(\omega) - iX(\omega).$$

First of all, we note that on separating into even and odd parts

$$f(t) = f_e(t) + f_o(t),$$

we have

$$F(\omega) = \int_{-\infty}^{\infty} f(t)e^{-i\omega t} dt = \int_{-\infty}^{\infty} f_e(t)\mathrm{Cos}\omega t - i \int_{-\infty}^{\infty} f_o(t)\mathrm{Sin}\omega t,$$

and

$$F(-\omega) = \int_{-\infty}^{\infty} f(t)e^{i\omega t} dt = \int_{-\infty}^{\infty} f_e(t)\mathrm{Cos}\omega t + i \int_{-\infty}^{\infty} f_o(t)\mathrm{Sin}\omega t.$$

That is, for a real time function, we have that

$$R(\omega) = R(-\omega),$$
$$X(\omega) = -X(-\omega).$$

Now for a causal time function, i.e., $f(t) \equiv 0$ for $t < 0$, we have that

$$f_o(t) = f_e(t)\mathrm{Sgn}\{t\},$$
$$f_e(t) = f_o(t)\mathrm{Sgn}\{t\}. \tag{26-39}$$

But

$$f_e(t) \Longleftrightarrow R(\omega),$$
$$f_o(t) \Longleftrightarrow -iX(\omega),$$
$$\mathrm{Sgn}\{t\} \Longleftrightarrow \frac{2}{i\omega} ;$$

whence (on forming the convolution integral)

$$X(\omega) = \frac{1}{\pi} \int_{-\infty}^{\infty} \frac{R(\omega')d\omega'}{\omega - \omega'} ; \tag{26-40}$$

and

$$R(\omega) = -\frac{1}{\pi} \int_{-\infty}^{\infty} \frac{X(\omega')d\omega'}{\omega - \omega'} . \tag{26-41}$$

We note however, that a factor $a_o\delta(t)$ gets lost in f_o when we use equation (26-39). Now $a_o\delta(t) \Longleftrightarrow a_o$. For reasonable $f(t)$, the integral in (26-41) vanishes as ω tends to infinity. Hence the pair should be written:

HILBERT TRANSFORM PAIR

$$X(\omega) = \frac{1}{\pi} \int_{-\infty}^{\infty} \frac{R(\omega')d\omega'}{\omega - \omega'} \quad \text{(odd in } \omega) \tag{26-42}$$

$$R(\omega) = R_\infty - \frac{1}{\pi} \int_{-\infty}^{\infty} \frac{X(\omega')d\omega'}{\omega - \omega'} \quad \text{(even in } \omega) \tag{26-43}$$

where $R_\infty = a_o$. A necessary and sufficient condition for the validity of this pair is the Paley-Wiener condition

$$\int_{-\infty}^{\infty} \frac{|\ln F(\omega)| d\omega}{1 + \omega^2} < \infty. \tag{26-44}$$

We can even go further. Writing

$$F(\omega) = A(\omega)e^{-i\phi(\omega)}, \tag{26-45}$$

we have

$$\ln F(\omega) = \ln A(\omega) - i\phi(\omega). \tag{26-46}$$

We can apply the Hilbert Transform Pair to this and obtain

$$\phi(\omega) = \frac{1}{\pi} \int_{-\infty}^{\infty} \frac{\ln A(\omega')d\omega'}{\omega - \omega'} , \tag{26-47}$$

and

$$\ln A(\omega) = \ln A_\infty - \frac{1}{\pi} \int_{-\infty}^{\infty} \frac{\phi(\omega')d\omega'}{\omega - \omega'} . \tag{26-48}$$

Finally, for the special function

$$F(\omega) = e^{-\alpha(\omega) - i\theta(\omega)} , \tag{26-49}$$

we have

$$\theta(\omega) = -\frac{1}{\pi} \int_{-\infty}^{\infty} \frac{\alpha(\omega')d\omega'}{\omega - \omega'} = -\frac{2\omega}{\pi} \int_{0}^{\infty} \frac{\alpha(\omega')d\omega'}{\omega^2 - \omega'^2} , \tag{26-50}$$

and

$$\alpha(\omega) = \alpha_c + \frac{1}{\pi} \int_{-\infty}^{\infty} \frac{\theta(\omega')d\omega'}{\omega - \omega'} = \alpha_c + \frac{2}{\pi} \int_{0}^{\infty} \frac{\theta(\omega')\omega'd\omega'}{\omega^2 - \omega'^2} ; \tag{26-51}$$

where we have made explicit use of the fact that $\alpha(\omega) = \alpha(-\omega)$ and $\theta(\omega) = -\theta(-\omega)$. Note that (26-50) says that if there is any attenuation, there will be accompanying dispersion. This has been verified experimentally by Wuenschel (1965).

Now Strick writes his plane wave as

$$e^{-\gamma(\omega)x + i\omega t} , \tag{26-52}$$

where

$$\gamma(\omega) = \alpha(\omega) + i\theta(\omega). \tag{26-53}$$

Then

$$c = \frac{\omega}{\theta} , \; Q = \frac{\omega}{2c\alpha} = \frac{\theta(\omega)}{2\alpha(\omega)} . \tag{26-54}$$

If we look at the most general form that $\theta(\omega)$ can take in a non-dispersive system, we find that

$$\theta(\omega) = \hat{\alpha}(\omega) + \omega\tau, \tag{26-55}$$

where $\tau = c_\infty^{-1}$ (representing a pure time delay) and $\hat{\alpha}$ is the Hilbert transform of $\alpha(\omega)$ given by (9-38). Thus we have

$$Q = \frac{\hat{\alpha}(\omega) + \omega\tau}{2\alpha(\omega)} . \tag{26-56}$$

For Q to be rigorously constant, we would have to require that $\hat{\alpha} \sim \omega$, and $\alpha \sim \omega$. But this is impossible for a causal time function. The closest one can come is

$$\alpha(\omega) = k_o|\omega|^s, \quad 0 < s < 1. \tag{26-57}$$

Then

$$\hat{\alpha}(\omega) = \mathrm{Tan}\left(\frac{s\pi}{2}\right)\mathrm{Sgn}\{\omega\}\alpha(\omega), \tag{26-58}$$

whence

$$\theta(\omega) = \mathrm{Tan}\left(\frac{s\pi}{2}\right)\mathrm{Sgn}\{\omega\}k_o|\omega|^s + \omega\tau, \tag{26-59}$$

and

$$\gamma = k_o|\omega|^s + i\left[\mathrm{Tan}\left(\frac{s\pi}{2}\right)\mathrm{Sgn}\{\omega\}k_o|\omega|^s + \omega\tau\right], \tag{26-60}$$

which are analytic except for s = 1.0. For physically interpretable quantities, we have (from (26-54))

$$\frac{1}{c} = \frac{1}{c_\infty} + \mathrm{Tan}\left(\frac{s\pi}{2}\right)k_o\omega^{s-1} \xrightarrow[\omega\to 0]{} \infty, \tag{26-61}$$

and

$$Q = \frac{1}{2}\,\mathrm{Tan}\left(\frac{s\pi}{2}\right) + \frac{\tau}{2k_o}\,\omega^{1-s} \xrightarrow[\omega\to\infty]{} \infty. \qquad (26\text{-}62)$$

For Q to be almost constant, we find that s must be near 1.0.

Replotting Wuenschel's (1965) data on a log-log scale, Strick (1967) finds that s = 0.9227 and everything is fine. However, another carefully conducted experiment in the Pierre Shale (McDonal et al, 1958) finds that s = 1.1. This is outside the theory developed here which requires $0 < s < 1$. Another defect in the theory as presented here is that a physical model which leads to the complex propagation function $\gamma(\omega)$ given by (26-60) could not be found. This led Strick to investigate other alternatives to such a promising representation for attenuation.

Now a great deal of work in modeling elastic modulii (and/or compliances) in polymers has been carried out where the models have consisted of a network of springs and dashpots. Much of this is summarized in Gross (1953) and Gross and Braga (1961). In particular, Pelzer (1957) constructed three models which have slowly varying Q's as a function of frequency. It was in this direction that further advances were to be made. If one substitutes (26-52) into (26-14), one obtains

$$\gamma^2(\omega) = -\rho\omega^2/M(\omega) = -\rho\omega^2 J(\omega). \qquad (26\text{-}63)$$

If Q is to remain almost constant, then the function γ will have to be a slowly varying function of frequency.

Strick has investigated a number of models and one that meets the requirements very satisfactorily is (Strick, 1978)

$$J = \frac{1}{\mu_\infty} + \frac{1}{\mu'}\ln\left(1 + \frac{\mu'}{i\omega\eta}\right). \qquad (26\text{-}64)$$

For frequencies such that $\exp(-\mu'/\mu_\infty) << \omega\eta/\mu' << 1$, we can write

$$\gamma = i\omega(\rho J)^{\frac{1}{2}} \simeq \frac{i\omega}{v_\infty}\left[1 - \frac{\mu_\infty}{\mu'}\ln(\omega\eta/\mu') - \frac{i\pi}{2}\frac{\mu_\infty}{\mu'}\right]^{\frac{1}{2}},$$

where $v_\infty = (\mu_\infty/\rho)^{\frac{1}{2}}$. Looking at the real and imaginary parts, we find that

$$\alpha(\omega) \simeq \frac{\omega\pi}{4v_\infty}\frac{\mu_\infty}{\mu'}\left[1 + \frac{\mu_\infty}{2\mu'}\ln(\omega\eta/\mu') + \ldots\right],$$

$$\theta(\omega) \simeq \frac{\omega}{v_\infty}\left[1 - \frac{\mu_\infty}{2\mu'}\ln(\omega\eta/\mu') + \ldots\right]. \qquad (26\text{-}65)$$

From (26-54), we have that

$$c(\omega) \simeq v_\infty\left[1 - \frac{\mu_\infty}{2\mu'}\ln(\omega\eta/\mu') + \ldots\right]^{-1}, \qquad (26\text{-}66)$$

and from (26-26) or (26-54) we have

$$Q(\omega) \simeq \frac{2}{\pi}\frac{\mu'}{\mu_\infty}\left[1 - \frac{\mu_\infty}{\mu'}\ln(\omega\eta/\mu')\right]. \qquad (26\text{-}67)$$

Strick (1978) designed this form specifically to have an $\alpha(\omega) \simeq \omega^s$ with s = 1. Generalizations to $0 < s < 2$ are forthcoming. Although the generalized models cover a wide range of s, most experimental data on earth materials have s-values in the vicinity of 1. We have previously mentioned two examples where s = 0.9227 and s = 1.1.

26.3. Mechanisms of Attenuation.

The other aim in mind with the development of the form (26-64) was the construction of a realizable physical model. Strick (1976) exhibited the infinite spring-dashpot network corresponding to (26-64). This can be done as follows. Consider an infinite network as in Figure 26-1. We can examine this network pair

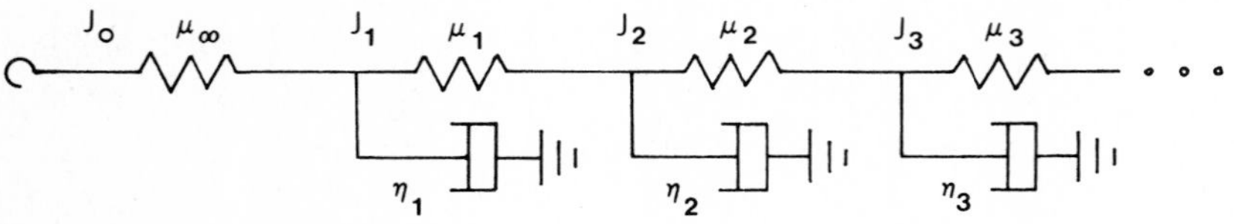

Figure 26-1. Infinite spring-dashpot network.

by pair and find that

$$J_n = \cfrac{1}{i\omega\eta_n + \cfrac{1}{\cfrac{1}{\mu_n} + J_{n+1}}} . \tag{26-68}$$

Starting at the beginning, we have

$$\begin{aligned} J_o &= \frac{1}{\mu_\infty} + J_1 \\ &= \frac{1}{\mu_\infty} + \cfrac{1}{i\omega\eta_1 + \cfrac{1}{\cfrac{1}{\mu_1} + \cfrac{1}{i\omega\eta_2 + \cfrac{1}{\cfrac{1}{\mu_2} + \cfrac{1}{i\omega\eta_3 + \cdots}}}}} \qquad (26\text{-}69) \\ &= \frac{1}{\mu_\infty} + \frac{1}{i\omega\eta_1+}\;\frac{1}{1/\mu_1+}\;\frac{1}{i\omega\eta_2+}\;\frac{1}{1/\mu_2+}\;\frac{1}{i\omega\eta_3+}\cdots \end{aligned}$$

The last line is a notation for the continued fraction expressed above. Now using relation 4.1.39 from Abramovitz and Stegun (1965, p. 68) one can show that

$$A\ \ln\left[1 + B\right] = \frac{1}{1/AB+}\;\frac{1}{2A/1^2+}\;\frac{1}{3/AB+}\;\frac{1}{4A/2^2+}\;\frac{1}{5/AB+}\;\frac{1}{6A/3^2+}\cdots$$

Expanding (26-64) by the continued fraction above, we have that

$$J = \frac{1}{\mu_\infty} + \frac{1}{i\omega\eta+}\;\frac{1}{2/\mu'+}\;\frac{1}{3i\omega\eta+}\;\frac{1}{1/\mu'+}\;\frac{1}{5i\omega\eta+}\;\frac{1}{2/3\mu'+}\cdots \tag{26-70}$$

Making the identification with (26-69) we have

$$\eta_n = (2n + 1)\eta, \qquad \mu_n = \frac{n}{2}\,\mu'. \tag{26-71}$$

That is, the dashpots and spring constants increase away from the point of stress application. It might be noted here that μ' is an effective spring constant which is the ratio of the real internal spring constant and an interaction parameter. Only the effective constant μ' may be determined by fitting curves to the observations.

This model then gives a good fit to experimental data, plus it has a physical realization consisting of an infinite spring-dashpot network. This latter property allows for interaction at all distances within the solid, something which earlier finite configurations failed to do.

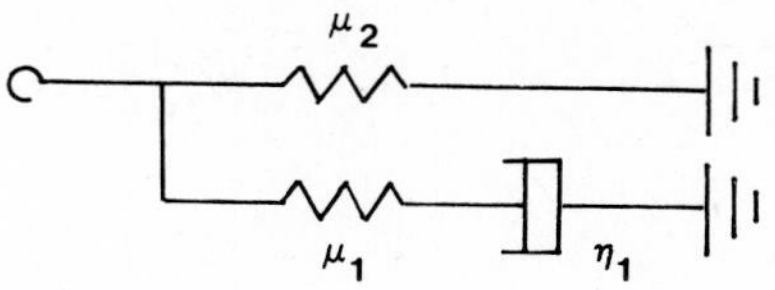

Figure 26-2. Three-element loss model.

One such combination is given in Figure 26-2. Here we have

$$J = J_1 - iJ_2 = \frac{1}{\mu_2 + \dfrac{i\omega\eta_1\mu_1}{\mu_1 + i\omega\eta_1}} = \frac{\mu_2 + (\mu_1 + \mu_2)\omega^2\tau^2 - i\mu_1\omega\tau}{\mu_2^2 + (\mu_1 + \mu_2)^2\omega^2\tau^2}; \qquad (26\text{-}72)$$

where we have introduced $\tau = \eta_1/\mu_1$. Then

$$Q = J_1/J_2 = (\mu_2/\mu_1)\left[1 + \{(\mu_1 + \mu_2)/\mu_2\}\omega^2\tau^2\right]/\omega\tau$$
$$\simeq (\mu_2/\mu_1)(1 + \omega^2\tau^2)/\omega\tau, \qquad (26\text{-}73)$$

for large μ_2/μ_1 (as is generally the case). Equations (26-73) represent a rather narrow absorption band as opposed to the rather broad minimum given by (26-67). Liu et al (1976) have proposed an attenuation model consisting of a spectrum of such three-element models to more accurately represent the constant Q data. A more extensive treatment is given by Kanamori and Anderson (1977). We see that two lines of reasoning (one mathematical and one physical) seem to converge upon similar physical models of attenuation.

Jackson and Anderson (1970) have published an extensive review of possible mechanisms of attenuation together with their relation to data available at the time. In this article, they favored grain-boundary relaxation, a high-temperature internal-friction background attributed to vacancy creation and diffusion, with losses due to partial melting possibly predominating in the low-velocity zone of the upper mantle. Anderson and Hart (1978a) feel that grain-boundary relaxation spread over a spectrum of relaxation times can explain most of the data.

26.4. The Dissipation of Surface-Waves, Rod and Plate Waves.

Rayleigh Waves. Press and Healy (1957) showed that if one sets up complex velocities of the form

$$v_p \longrightarrow v_p^o(1 + i/Q_p),$$

$$v_s \longrightarrow v_s^o(1 + i/Q_s), \qquad (26\text{-}74)$$

$$v_R \longrightarrow v_R^o(1 + i/Q_R);$$

corresponding to a constant Q model if Q is large, one can determine a relationship between the absorption coefficients for body- and Rayleigh-waves. Macdonald (1959) simplifies their expressions and gives the following relationship between the various Q's.

$$\frac{1}{Q_R} \simeq m\frac{1}{Q_p} + (1 - m)\frac{1}{Q_s} \qquad (Q_s > 10); \qquad (26\text{-}75)$$

where

$$m = \frac{a(2 - b)(1 - b)}{a(2 - b)(1 - b) - b(1 - a)(2 - 3b)} \left[= 0.134 \text{ for } \sigma = \tfrac{1}{4} \right], \qquad (26\text{-}76)$$

and

$$a = (v_R/v_p)^2, \quad b = (v_R/v_s)^2. \qquad (26\text{-}77)$$

<u>Some Other Relations</u>. White (1965, Chapter 3) has given the following. He starts by writing the Lamé parameters in complex form:

$$\lambda + i\,\mathrm{Sgn}\{\omega\}\lambda^\dagger, \qquad (26\text{-}78)$$
$$\mu + i\,\mathrm{Sgn}\{\omega\}\mu^\dagger.$$

Using a similar notation, he finds:

$$\left.\begin{aligned} \rho v_p^2 &= \lambda + 2\mu, \\ \rho v_p^{\dagger 2} &= \lambda^\dagger + 2\mu^\dagger, \\ \rho v_s^2 &= \mu, \\ \rho v_s^{\dagger 2} &= \mu^\dagger, \end{aligned}\right\} \quad \text{BULK MEDIA} \qquad (26\text{-}79)$$

$$\left.\begin{aligned} \rho v_{Rod}^2 &= \frac{\mu(3\lambda + 2\mu)}{\lambda + \mu}, \\ \rho v_{Rod}^{\dagger 2} &= \frac{\mu^2\lambda^\dagger + (3\lambda^2 + 4\lambda\mu + 2\mu^2)\mu^\dagger}{(\lambda + \mu)^2}, \\ \rho v_s^2 &= \mu, \\ \rho v_s^{\dagger 2} &= \mu^\dagger, \end{aligned}\right\} \quad \text{RODS} \qquad (26\text{-}80)$$

$$\left.\begin{aligned} \rho v_{plate}^2 &= \frac{4\mu(\lambda + \mu)}{\lambda + 2\mu}, \\ \rho v_{plate}^{\dagger 2} &= \frac{4\mu^2\lambda^\dagger + 4(\lambda^2 + 2\lambda\mu + 2\mu^2)\mu^\dagger}{(\lambda + \mu)^2}, \\ \rho v_s^2 &= \mu, \\ \rho v_s^{\dagger 2} &= \mu^\dagger. \end{aligned}\right\} \quad \text{PLATES} \qquad (26\text{-}81)$$

ELASTIC WAVE DISSIPATION – OBSERVATIONS

27.1. General.

There are a large number of mathematical models to represent the almost constant Q behavior of the earth. However, as we shall see, there is a great deal of scatter in the data and, at the present time, one cannot really distinguish between models. Certain materials have attenuations varying significantly from the linear and careful experiments may help in the elimination of certain models. Measurements made on the real earth have a number of problems which prevent the taking of very good data and so a certain amount of scientific intuition has to be exercised in the selection of data. We shall meet some of these difficulties shortly.

Details of experimental methods and data treatment can be found in White (1965, Chapter 3), Smith (1972), and Båth (1974, Sec. 7.4). Summaries of data can be found in the above as well as in Knopoff (1964b), Bradley and Fort (1966), Jackson and Anderson (1970), and Anderson and Hart (1978a and b).

27.2. Attenuation of Body-Waves.

The attenuation of body-waves (P and S) has been measured in at least four quite different situations. We shall discuss each in turn. Attenuation as used here refers to dissipative losses, not reduced amplitudes due to geometrical spreading. The measurement of attenuation is extremely difficult technically because all sorts of interferences take place. These interferences are generally unknown and random, so they can increase or decrease the amplitudes of seismic pulses. If losses are low and the Q of the material under test is high, these interference effects can easily give large errors to the measured Q's. It might also be pointed out that digital records and computers are almost a requirement, inasmuch as pulse height and width analysis is very innaccurate and Fourier methods should be used.

Using the notation of page 334, we can look at the relative dissipation of P- and S-waves. We have that

$$\rho\omega^2 k_p^{-2} = (k + 4/3\ \mu)(1 + \frac{k^\dagger + 4/3\ \mu^\dagger}{k + 4/3\ \mu}), \qquad (27\text{-}1)$$

where $\underline{k}$ is the bulk modulus. Keeping first order terms, we have

$$k_p \simeq \frac{\omega}{v_p}(1 - \tfrac{1}{2}\frac{k^\dagger + 4/3\ \mu^\dagger}{k + 4/3\ \mu} + \ldots),$$

and

$$Q_p^{-1} \simeq \frac{2v_p}{\omega} k_p^\dagger = \frac{k^\dagger + 4/3\ \mu^\dagger}{k + 4/3\ \mu}. \qquad (27\text{-}2)$$

Similarly

$$Q_s^{-1} \simeq \frac{\mu^\dagger}{\mu}; \qquad (27\text{-}3)$$

and finally

$$\frac{Q_p}{Q_s} \simeq \frac{\mu^\dagger}{k^\dagger + 4/3\ \mu^\dagger} \left(\frac{v_p^2}{v_s^2} \right). \qquad (27\text{-}4)$$

For a Poisson solid ($\lambda = \mu$), we have

$$Q_p/Q_s = 9/4 \Longrightarrow k^\dagger = 0.$$

$$= 18/11 \Longrightarrow k^\dagger = \mu^\dagger/2.$$

$$= 9/7 \Longrightarrow k^\dagger = \mu.$$

Laboratory Measurements. If it is possible to construct cylindrical samples of great precision, then the usual method is to measure the amplitude decrements of successive reflected pulse bursts as in Figure 27-1. One may then use (26-4) directly to relate the "Q" obtained to the carrier frequency of the pulse bursts. With

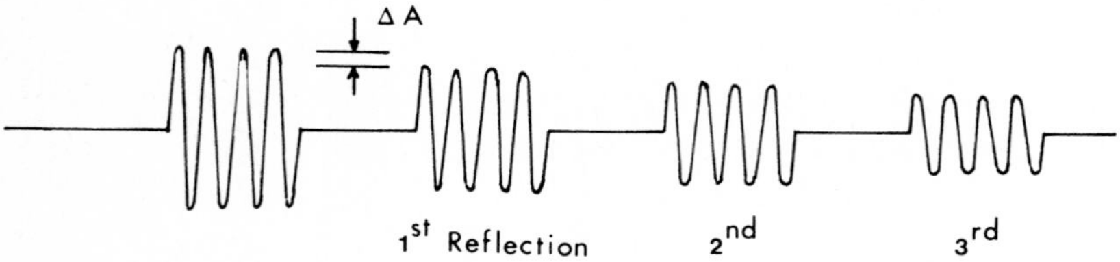

Figure 27-1. Pulse burst method of determining Q in laboratory materials.

samples of a few centimeters in characteristic dimension, this method is useful in the range 10khz - 10mhz. All sorts of materials may be measured. Knopoff (1964b) gives many details and results, and Bradley and Fort (1966) give a compilation of experimental data.

An alternative method that is useful when machined samples are not available is that known as the spectral ratio method. This method is also generally applicable to body-waves in the earth. In the case of laboratory models, it is particularly simple and the assumptions are most probably realized. In this method, one compares the spectra of individual seismic pulses at varying distances. In the time domain, we might be looking at simple pulses like those in Figure 9-9. For a pulse defined by $f(t)$, we then have the relation

$$f(t) \Longleftrightarrow F(\omega). \qquad (27\text{-}5)$$

For the pupose of measuring attenuation we write

$$|F(\omega)| = |F_o(\omega)| G(x) e^{-\alpha(\omega)x}, \qquad (27\text{-}6)$$

where $F_o(\omega)$ represents the combined effects of source, receiver, and associated coupling effects. This function is assumed to remain the same as the pulse travel-distance is varied, and thus is independent of the distance x. $G(x)$ is to represent the geometrical divergence, and it should be independent of frequency. The exponential factor represents the dissipative component. If we take the logarithms of the spectral ratio for pulses at two stations, we have

$$\ln\left\{\frac{|F_2(\omega)|}{|F_1(\omega)|}\right\} = \ln\left\{\frac{G(x_1)}{G(x_2)}\right\} - \alpha(\omega)(x_2 - x_1). \qquad (27\text{-}7)$$

Thus, if the function on the left-hand side of (27-7) is plotted against frequency as in Figure 27-2, the general level is determined by the geometric divergence, but the slope of the curve is a measure of Q^{-1}. If a straight line is not obtained, the data are too noisy, or a constant Q model is not appropriate.

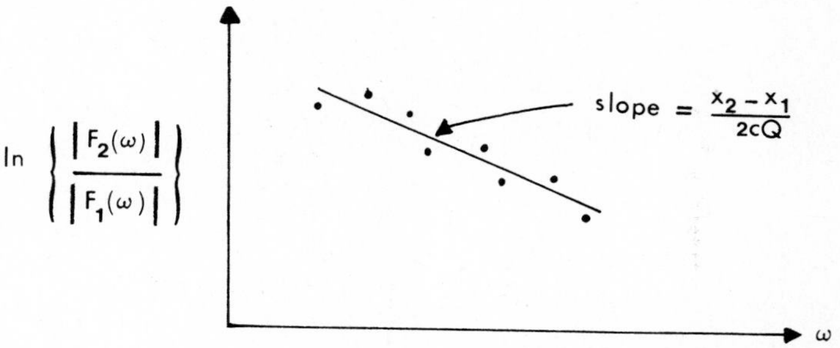

Figure 27-2. Spectral ratio method for determining attenuation in an elastic solid.

The Q obtained in laboratory measurements is greatly affected by minute cracks and pore spaces which generally close up when the lithostatic pressure exceeds a few kilobars. Consequently, the relation of laboratory measurements to <u>in situ</u> properties is a bit tenuous. However, elastic wave measurements under high confining pressures are much more difficult and few have been made relative to atmospheric pressure measurements. Simmons and Nur (1968) have also noted that rocks once buried deep in the earth do not seem to develop these cracks until they are very near the surface. That is, it requires large pressures to close these cracks, once opened, but the voids are not created until the confining pressure drops almost to zero.

<u>Small Field Experiments</u> are very difficult to carry out and hence are not very numerous. White (1965, Chapter 3) describes a few, and DeBremaecker et al (1966) give details of one since. There are also several papers dealing with the attenuation of seismic waves with distance as observed from nuclear explosions. The experiments described by White are for body-wave propagation in a single rock unit, while that of DeBremaecker et al looks at refracted arrivals. The latter assumes an amplitude dependence of

$$A = A_o x^{-2} e^{-\pi x f/(Q_p v_p)},$$

and finds Q values of 10 - 100 at frequencies of 15 - 50 hz. However, the data are not very good and only serve to show the difficulty of such measurements.

Part of the difficulty seems to be from scattering or from reinforcement within the refracting horizon because of velocity gradients. Figure 27-3 shows some of the possibilities. A discussion of these points, together with considerable analysis of much data is given by Hill (1971). In a later paper (Hill, 1973) he treats the mathematical theory. Additional mathematical discussion can be found in Cerveny and Ravindra (1971, Chapter 6).

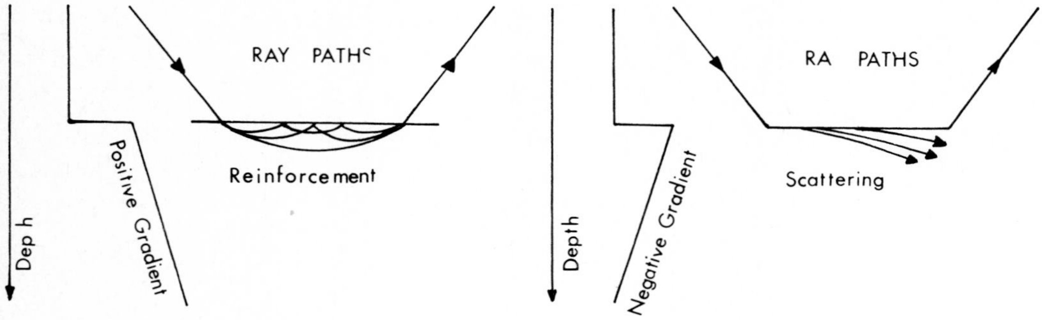

Figure 27-3. The effect of velocity gradients upon seismic head-waves complicating the determination of attenuation.

Teleseismic P and S. Teng (1968) has published a very comprehensive analysis of teleseismic body-waves. He assumes an amplitude function of the form

$$A(\omega,\Delta,\theta,\phi) = S_s(\theta,\phi)S_t(\omega)R(\omega)CR(\Delta,\omega)G(\Delta)e^{-ft^*}, \qquad (27\text{-}8)$$

where

$$t^* = \pi\int_{path} \frac{ds}{Qv} . \qquad (27\text{-}9)$$

The function $S_s(\theta,\phi)$ represents the spatial directivity function for the source and is assumed to be independent of frequency. That is, all the frequency dependence of the source is represented by $S_t(\omega)$ and is the same regardless of the direction of propagation. For point sources, this is generally true (for example, the sources of Section 7.4), but when the source has a finite dimension and/or the depth of the source is small, there will be a spatially directive frequency dependence. If the wavelengths are long with respect to source dimensions and small with respect to source depth (i.e., P and pP, etc., do not interfere), then such an assumption is not unwarranted. The best chances occur for intermediate and deep-focus earthquakes of small magnitude (4-6) measured on long-period seismographs (10 - 100 sec period). $R(\omega)$ represents the instrumental response and it is assumed to be calculable as in Chapter 33. The factor $CR(\Delta,\omega)$ is the crustal response function of Section 17.3 and is assumed to be calculable. $G(\Delta)$ represents the geometric divergence and is assumed to be frequency independent. This is valid as long as seismic ray theory is valid. The quantity t^* represents the averaging effect of the ray traveling through inhomogeneous dissipative media.

Taking logarithms again, we have

$$\ln\left\{\frac{A_2(\omega)R_1(\omega)CR(\Delta_1,\omega)}{A_1(\omega)R_2(\omega)CR(\Delta_2,\omega)}\right\} = \ln\left\{\frac{S_s(\theta_2,\phi_2)G(\Delta_2)}{S_s(\theta_1,\phi_1)G(\Delta_1)}\right\} - f(t_2^* - t_1^*). \qquad (27\text{-}10)$$

The left-hand side is to be calculated, the first term on the right-hand side gives the general level of the curve, and the second term is used to find the slope

$$\delta A_i = t_i^* - t_o^*, \qquad (27\text{-}11)$$

where the subscript i represents a general station and o represents a reference station. It should be obvious that only differences in the quantity t^* are significant, given the formulation of (27-10).

The next problem is to invert the integral of (27-9) using data like (27-11). This is most easily done using a spherically layered earth where the properties, both elastic and anelastic, of each layer are constant. Equation (27-11) can then be rewritten

$$\delta A_i = \pi \sum_{k=0}^{M} \frac{\Delta t_k^i - \Delta t_k^o}{Q_k} \qquad i = 1,2,\ldots N; \qquad (27\text{-}12)$$

where there are N+1 observations including the reference station. The quantity Δt_k^i represents the travel-time in the kth layer on the way to the ith station. This has to be computed from the postulated structure of the layered earth model. Δt_k^o is the travel-time for the reference ray and Q_k is the unknown Q of each layer. Teng chooses to solve the system (27-12) of N equations in the M unknown Q's by relaxation

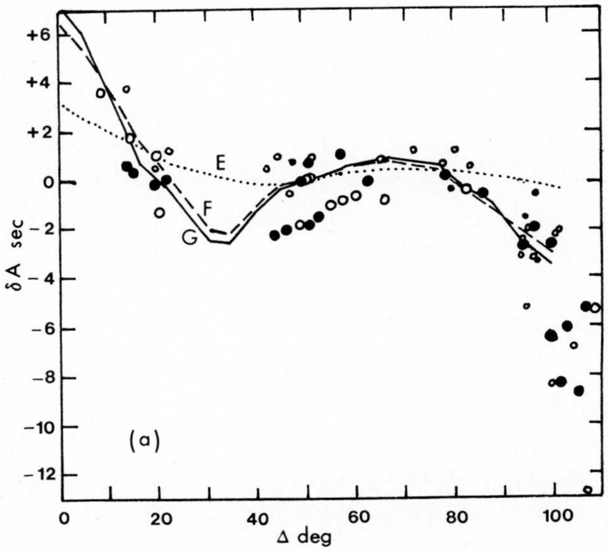

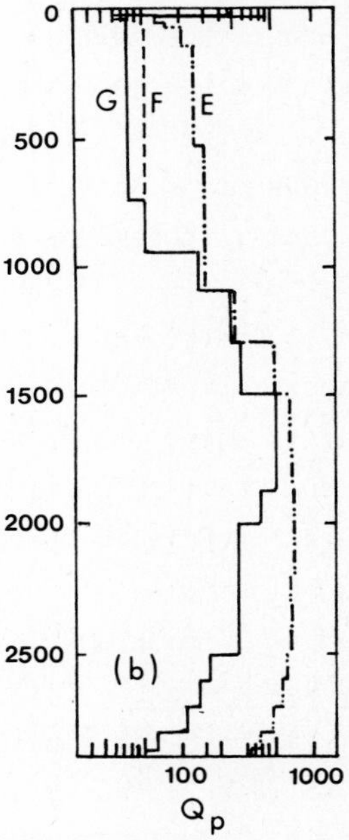

Figure 27-4(a). Differential attenuation data and theoretical curves from the accompanying models E, F and G.

(b). Three Q_p models for the mantle.

(After Teng, 1968.) Copyrighted by AGU.

methods. (An alternative method of inversion using least squares is given by L. M. Dorman (1968)). The raw data are given in Figure 27-4a together with theoretical curves taken from the Q models presented in Figure 27-4b.

The differential attenuation data (Figure 27-4a) is quite scattered, and one has problems in fitting a Q model to it. The data for $\Delta < 20^{\circ}$ are all taken from stations in tectonic regions of South America plus the surrounding ridges and island arcs and therefore may have a regional (tectonic) bias. The data after $\Delta \approx 90^{\circ}$ may be interfered with by core-mantle boundary interactions. Consequently, the weight to be assigned in the end regions is problematic. Secondly, as all stations are referred to a station at approximately 50°, all model curves will go through this point. Teng preferred model F whose slopes matched the data at both ends. However, the Q_p's given by this model are extremely low. In light of Q_s data to be presented later, these low values seem somewhat unreasonable. Model E, which does not fit the end data as well, has a Q structure with values about twice as great as Model F. If the Q values in the lowermost part of the mantle had been reduced as in models F and G, it might have been more compatible with the Q_s data.

Solomon and Toksöz (1970) determine regional values of δA_i over the continental United States for both P and S attenuation. They find that attenuation is relatively greater in the western United States with the exception of the Pacific Coast. There is also a hint of slightly higher attenuation in the extreme northeastern United States. Solomon (1973) uses this technique to find a narrow band of low Q material centered about the Mid-Atlantic Ridge.

Now matter how one struggles to find a model which will fit the data found by the above, there is still a further problem. The data used by these investigations were all taken from long-period seismograph records. A number of investigators (Dorman (1968), Archambeau et al (1969), Berzon et al (1974), Helmberger and Engen (1974), Der et al (1975), and Der and McElfresh (1977)) have looked at records from short-period instruments. They find Q values much higher than for the long-period data. These contradictory results constitute one of the strongest arguments that Q may actually depend on frequency since the propagation paths are much the same. Anderson and Hart (1978b) in their Q model choose to use the short-period rather than the long-period data in their inversion. The latter two of the papers listed above make regional studies of the U.S. Their overall patterns are much the same as those determined by Solomon and Toksöz (ibid), but at a much higher level of Q.

Core Reflections. Following a method initiated by Press (1956a), Anderson and Kovach (1964) used the body-waves $(ScS)_n$ and $(sScS)_n$ to determine average Q in the mantle. In their formulations, the final amplitude functions look like

$$A_n = A_o R_c^n R_s^{n-1} e^{-2n\gamma_2 h_2} e^{-(2n-1)\gamma_1 h_1} (2nH - h_1)^{-1}, \qquad (27\text{-}13)$$

and

$$_sA_n = A_o R_c^n R_s^n e^{-2n\gamma_2 h_2} e^{-(2n+1)\gamma_1 h_1} (2nH + h_1)^{-1}, \qquad (27\text{-}14)$$

for $(ScS)_n$ and $(sScS)_n$ respectively. The geometry of the problem is given in Figure 7-25. Here R_c and R_s are reflection coefficients at the core-mantle boundary and the

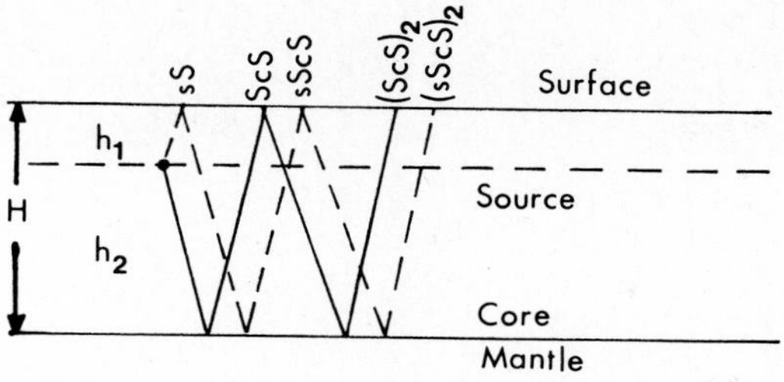

Figure 27-5. Geometrical relationships for the core-reflection problem.

surface respectively. For vertically incident shear-waves, we have for a reflection coefficient at the core-mantle interface:

$$R_c = \frac{(\mu_m\rho_m)^{\frac{1}{2}} - (\mu_c\rho_c)^{\frac{1}{2}}}{(\mu_m\rho_m)^{\frac{1}{2}} + (\mu_c\rho_c)^{\frac{1}{2}}} .$$

For shear-waves at the free surface we have $R_s = 1.0$. If one sets $\gamma_1 = 0$ and $\gamma_2 = 0$, one finds

$$A_{n+p}/A_n = R_c^p \frac{2nH - h_1}{2(n + p)H - h_1} . \tag{27-15}$$

For a deep-focus South American earthquake, Anderson and Kovach (ibid) found $A_3/A_2 = 0.52$ leading to

$$\mu_c < 0.013\ \mu_m\rho_m/\rho_c .$$

That is, the rigidity of the core has an upper bound of $2 \cdot 10^{10}$ dynes/cm^2. Any attenuation in the mantle material would reduce this value.

Alternatively, one can set both reflection coefficients equal to unity and ascribe all losses to attenuation. In this case

$$2pH\gamma_{avg} = \ln\left\{ \frac{2(n + p)H - h_1}{2nH - h_1} \frac{A_{n+p}}{A_n} \right\} , \tag{27-16}$$

or

$$2pH\gamma_{avg} = \ln\left\{ \frac{2(n + p)H + h_1}{2nH + h_1} \frac{{}_sA_{n+p}}{{}_sA_n} \right\} . \tag{27-17}$$

For the ratio $(ScS)_3/(ScS)_2$, $\gamma_{avg} = 4.0 \cdot 10^{-5}$, $Q_{25sec} = 508$. For $(sScS)_3/(sScS)_2$, $\gamma_{avg} = 4.7 \cdot 10^{-5}$, $Q_{25sec} = 440$. These are lower bounds for the average Q for the mantle.

Combining the two types of reflections, they found

$$2h_1\gamma_1 = \ln\left\{ \frac{2nH + h_1}{2nH - h_1} \frac{{}_sA_n}{A_n} \right\} , \tag{27-18}$$

with the ratio ${}_sA_2/A_2$ leading to $\gamma_1 = 1.42 \cdot 10^{-4}$ and a Q_{25sec} of 185. The ratio ${}_sA_3/A_3$ leads to $\gamma_1 = 1.75 \cdot 10^{-4}$ and a Q_{25sec} of 151. Again, these are lower bounds.

Once the Q of the upper mantle was known, that of the lower mantle could be calculated from

$$\gamma_2 h_2 = \gamma_{avg} H - \gamma_1 h_1 . \qquad (27\text{-}19)$$

This gave $\gamma_2 = 1.3 \cdot 10^{-5}$ and $Q_{25sec} = 1430$.

In their second paper, Kovach and Anderson (1964a) made additional corrections for the proper geometrical spreading using the factor:

$$\left(\frac{\mathrm{Sin}\theta_o}{r_e^2 \mathrm{Sin}\Delta \mathrm{Cos}\theta_s} \left| \frac{d\theta_o}{d\Delta} \right| \right)^{\frac{1}{2}} . \qquad (22\text{-}12)$$

Over a frequency range of 0.015 to 0.07hz, they found $Q_{avg} = 600$, $Q_{upper} = 200$, and $Q_{lower} = 2200$.

R. Sato and Espinoza (1967) took one of the earthquakes used above and extended the analysis to $\Delta = 80^o$. They found that $Q_{avg} = 581 \pm 33$, $\Delta \leqq 12^o$, 34-90 sec; that Q decreases with period (also found by Kovach and Anderson), and that Q decreases with epicentral distance. This last observation they used as evidence to say that partition of shear-wave energy upon reflection must be taken into account.

Kanamori (1967b) looks at the ratios ScS/ScP and PcS/PcP and finds (using earlier results on P and PcP (Kanamori (1967a))

$$\left. \begin{array}{l} Q_{s(avg)} = 230 \\ Q_{p(avg)} = 1.90\, Q_s \end{array} \right\} \text{1.5 - 5 sec period.}$$

These values are considerably less than the attenuation values cited previously.

Anderson and Hart (1978a) give a summary of reflected core phase Q measurements in the mantle adopting the value of $Q_s = 285$ for the period range 10 - 50 sec. They favor a mean value of $Q_s \approx 165$ for the mantle above 600 km depth. Anderson and Hart (1978b) also give a number of references to attenuation associated with core phases PnKP.

27.3 Attenuation of Surface-Waves and Free Oscillations.

Surface-wave attenuation has been measured by the traveling-wave method of Section 26.1 and then "corrected" by the ratio c/U to mesh with the free oscillation data measured by the standing-wave method. As we have noted before, page 328, the correction probably should have been the other way; free modes to surface-waves by the ratio U/c. Except for a percentage change of a few tens of points (which on a logarithmic plot doesn't amount to much) the relative apparent Q over the spectrum is given in Figures 27-6, -7, and -8. There is a lot of scatter which makes it difficult to invert the data. There is much additional data not incorporated in Figures 27-6 and 27-8; Anderson and Hart (1978b) give tables of all data available to them.

The data presented above is world-average in nature; however, some regional attenuation studies have been made. Most of these have been confined to middle North America where the density of seismic stations is high. We have already mentioned the

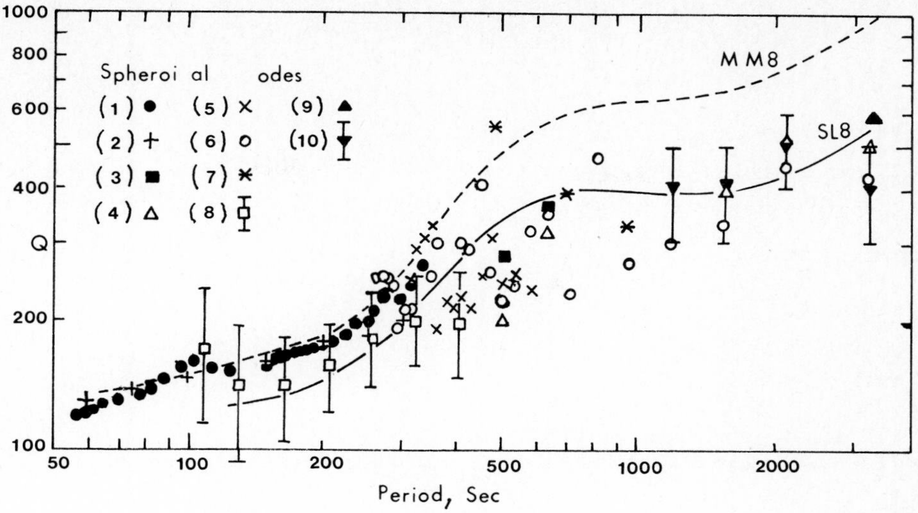

Figure 27-6. Q values for the spheroidal modes of oscillation. Superimposed on the data values are Q vs. Period curves MM8 (Anderson et al, 1965) and SL8 (Anderson and Hart, 1978b). Figure modified from Anderson and Hart (1978a). Sources of data are: (1) Anderson et al (1965); (2) Ben-Menahem (1965); (3) Ness et al (1961); (4) Slichter (1967); (5) Nowroozi (1968); (6) S.W. Smith (1972); (7) Bolt and Brillinger (1975); (8) Mills and Hales (1977); (9) A. Dziewonski (personal communication, 1976); (10) Stein and Geller (1978a), preliminary data.

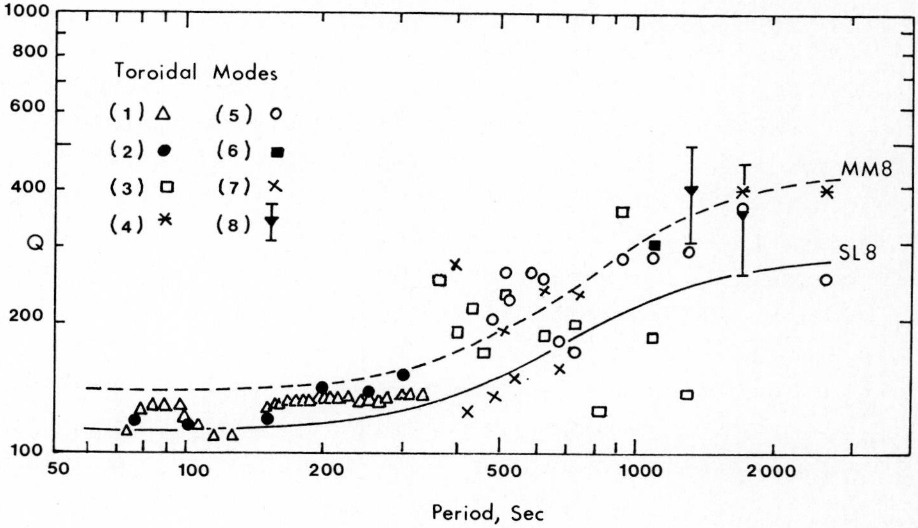

Figure 27-7. Q values for the toroidal modes of oscillation. MM8 and SL8 as in Figure 27-6. Figure modified from Anderson and Hart (1978a). Sources of data are: (1) Anderson et al (1965); (2) Ben-Menahem (1965), smoothed curve; (3) Bolt and Brillinger (1975); (4) S.W. Smith (1961); (5) S.W. Smith (1972); (6) Alsop et al (1961); (7) Nowroozi (1968); (8) Stein and Geller (1978a), preliminary data.

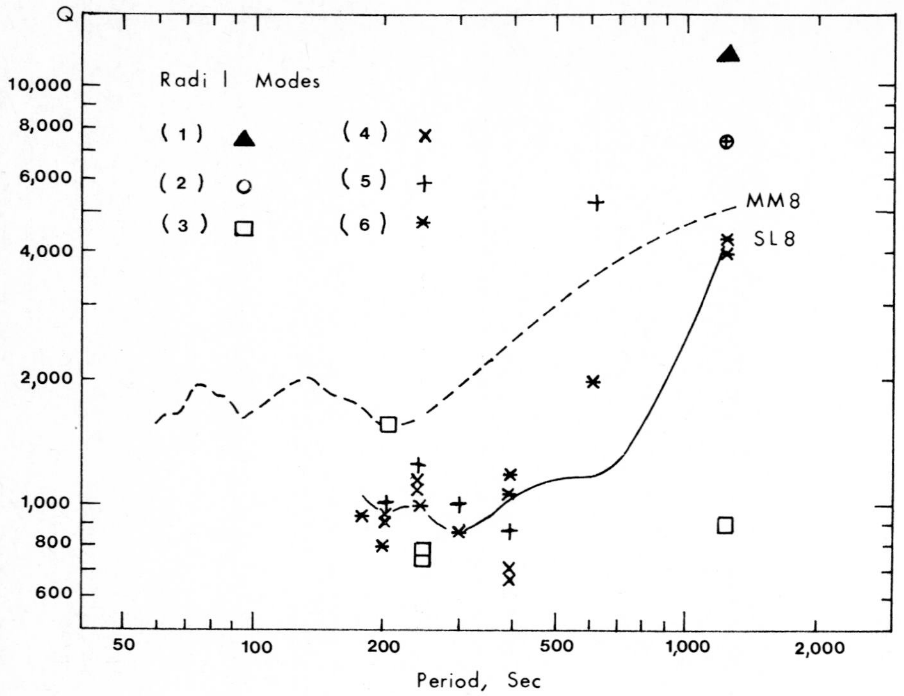

Figure 27-8. Q values for the radial modes of oscillation. MM8 and SL8 are as in Figure 27-6. Data compiled by Anderson and Hart (1978b). Sources are: (1) Slichter (1967); (2) Ness et al (1961); (3) S.W. Smith (1961, 1972); (4) Dratler et al (1971); (5) Buland and Gilbert (1978); (6) Sailor and Dziewonski (1978).

the east vs. west differences in regard to long- and short-period body-waves. Nuttli (1973) confirms this for Rayleigh-waves. In an extended series of studies, Mitchell and others (Hermann and Mitchell, 1975, North America; Mitchell et al, 1976, Pacific Ocean; Yacoub and Mitchell, 1977, Eurasia; plus earlier papers) have made detailed attenuation observations in a number of continental-sized regions. Nakanishi (1978) has observed path-dependent attenuation on great-circle paths.

27.4. Inversion.

Once Q data such as presented in the previous two sections has been obtained, one would then like to invert this data. One comprehensive inversion stood for ten years, that of Anderson et al (1965). Until recent Q data became available in the 1970s, there wasn't much opportunity for improving it. We shall look at their method of inversion, for it is instructive even though some criticism has been made recently (Lee and Solomon, 1978).

Anderson et al (ibid) approximate the complex wave-number by (notation has been changed somewhat):

$$k - i\alpha \simeq k + i \sum_{j=1}^{N} \left[\frac{\partial k}{\partial v_{p_j}} v^{\dagger}_{p_j} + \frac{\partial k}{\partial v_{s_j}} v^{\dagger}_{s_j} \right]$$

$$\simeq k + i \frac{k}{2} \sum_{j=1}^{N} \left[\left(\frac{v_p}{k} \frac{\partial k}{\partial v_p} \right)_j Q^{-1}_{p_j} + \left(\frac{v_s}{k} \frac{\partial k}{\partial v_s} \right)_j Q^{-1}_{s_j} \right]$$

$$\simeq k - i \frac{k}{2} \sum_{j=1}^{N} \left[\left(\frac{v_p}{c} \frac{\partial c}{\partial v_p} \right)_j Q^{-1}_{p_j} + \left(\frac{v_s}{c} \frac{\partial c}{\partial v_s} \right)_j Q^{-1}_{s_j} \right] \qquad (27\text{-}20)$$

$$\simeq k - i \frac{k}{2} Q^{-1}$$

$$\simeq k - i \frac{k}{2} \sum_{j=1}^{N} \left[\frac{\partial Q^{-1}}{\partial Q^{-1}_{p_j}} Q^{-1}_{p_j} + \frac{\partial Q^{-1}}{\partial Q^{-1}_{s_j}} Q^{-1}_{s_j} \right] ;$$

where the complex material velocities have been written $v_p + iv^{\dagger}_p$, $v_s + iv^{\dagger}_s$, $c + ic^{\dagger}$, j is the layer index, and

$$Q^{-1}_p \simeq 2v^{\dagger}_p/v_p, \; Q^{-1}_s \simeq 2v^{\dagger}_s/v_s, \; Q^{-1} \simeq 2\alpha/k. \qquad (27\text{-}21)$$

The latter relations have come from (26-29) and (26-30). This approximation allows the calculation of a modal Q from partial derivatives already computed. That is:

$$Q^{-1}_R = \sum_{j=1}^{N} \left[\left(\frac{v_p}{c_R} \frac{\partial c_R}{\partial v_p} \right)_j Q^{-1}_{p_j} + \left(\frac{v_s}{c_R} \frac{\partial c_R}{\partial v_s} \right)_j Q^{-1}_{s_j} \right] , \qquad (27\text{-}22)$$

and

$$Q^{-1}_L = \sum_{j=1}^{N} \left(\frac{v_s}{c_L} \frac{\partial c_L}{\partial v_s} \right)_j Q^{-1}_{s_j}. \qquad (27\text{-}23)$$

A fuller exposition of the physics leading to these results can be found in Anderson and Archambeau (1964). It might be noted here that even if the layer Q's are independent of frequency, the resultant Q_R and Q_L are not. Partials relating to the Q structure have been given by Anderson et al (ibid) and by Deschamps (1977).

Equations (27-22) and (27-23) are in the usual form for inversion, though in 1965 somewhat less sophisticated methods of inversion were available. These authors produced attenuation model MM8 for the mantle. Although we don't give the parameters associated with this model, the Q values associated with it are superimposed on Figures 27-6, -7, and -8. The need for improvement is apparent in the case of low-order spheroidal modes and for high-order radial modes. A number of simple mantle Q-models were given by Sailor and Dziewonski (1978) and the effect of Q changes in various regions is made clearly evident. Additional models were constructed by Deschamps (1977). Anderson and Hart (1978a and b) have developed a series of models for the whole earth of which we present the latest, SL8 in Figure 27-9. Q values associated with this model have been plotted with the data shown previously. The improvement is clear, but much better data need to be available. It might be pointed

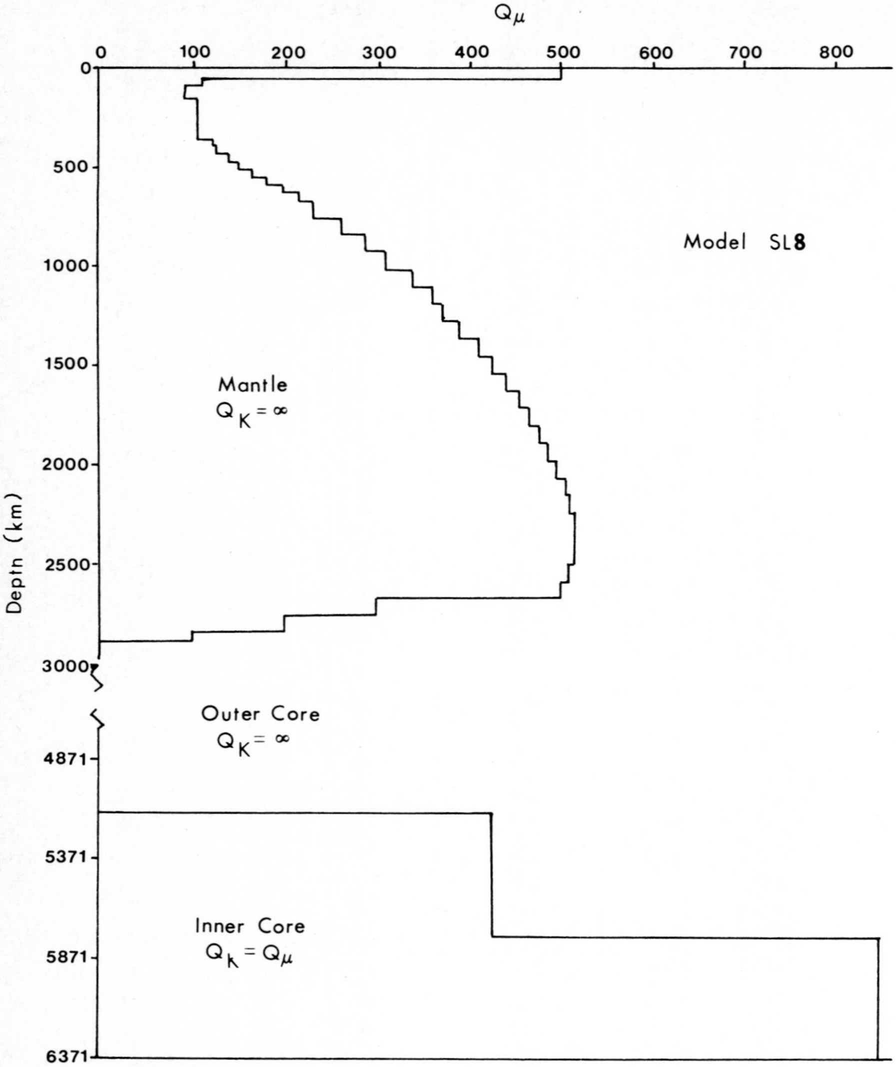

Figure 27-9. Q Model SL8 for the whole earth. (After Anderson and Hart, 1978b).

out that Model SL8 was heavily constrained by body-wave Q data to improve its overall reliability. Note that the attenuation due to the imaginary part of the bulk modulus (27-2) has been fitted best by zero in the mantle and core, having finite values only in the inner core.

A number of regional studies have been done on Q values obtained from surface-wave data. These give the Q structure for the crust and upper mantle. Some of the more recent ones are: Lee and Solomon (1975), western and eastern United States; Mitchell (1976), Pacific Ocean; Burton (1977), Eurasia; and Lee and Solomon (1978), western United States.

This latter paper was unique in that it accomplished a simultaneous inversion of both phase-velocity and Q. Some of Lee and Solomon's (1978) models allowed for a frequency-dependent Q. It is also in this paper that the inversion scheme given by (27-22) and (27-23) is criticized. Lee and Solomon note that the partial derivative expansion of Q^{-1} given there considers only the real part of the derivative of phase-velocity with respect to the real part of the layer velocity and point out that a full complex formulation should be used. They also feel that there may be problems in the two-step procedure of inverting for Q and then correcting the phase-velocities for accompanying dispersion (to be discussed shortly). They note that the Anderson-Hart two-step procedure leads to different results than the simultaneous procedure they advocate. A great deal of numerical experimentation will probably have to be done to find the "best" method in terms of stability, resolution, and accuracy. The trade-offs are not obvious, for Lee and Solomon find that the simultaneous inversion leads to better resolution for Q and poorer resolution for shear-velocity. In addition, a complex formulation will significantly increase computer time for inversion.

<u>The Effect of Q on Velocity Models</u>. If one considers the vibrations of a simple harmonic oscillator, one finds that the angular frequency is given from (26-2) as

$$\omega = \omega_o(1 - \lambda^2)^{\frac{1}{2}} \simeq \omega_o(1 - \frac{1}{4Q^2})^{\frac{1}{2}},$$

where we have used (26-5). We see that for large Q the resonant frequency is changed only slightly. Consequently most investigators have ignored what was assumed to be a second-order effect. During the years that the attenuation-dispersion pairs mentioned above were being developed, a number of authors noted that the dispersion accompanying attenuation ought to be considered more systematically. Randall (1976) demonstrated that the velocity changes were significant in calculations involving three quite different models. In the case of Strick's (1978) logarithmic model, we can write

$$c(\omega) \simeq v_\infty \left[1 + \frac{1}{\pi Q} \ln(\omega\eta/\mu') + \ldots \right] . \qquad (27\text{-}24)$$

Kanamori and Anderson (1977) find a similar form for their model, Savage and O'Neill (1975) compare Futterman's (1962) and Lomnitz's (1957) models again giving the same approximate relation for phase-velocity. They note that there is not much possibility to choose between the two models. Strick (personal communication) finds that quantities derived from (26-64) agree to four significant figures with calculated values given by Savage and O'Neill from the approximate Lomnitz relations. A similar agreement will probably be true with the Liu et al (1976) model. Large differences show up only outside the seismic band and models will probably have to be distinguished on the basis of creep studies. A discussion of experimental work in this area is outside the scope of this book. Suffice it to say that much creep behavior cannot be approximated by constant Q models and at least one extra parameter will have to be introduced.

Since v_∞ is hard to determine, Liu et al suggest the use of a reference frequency of 1 hz which is in the range of short period seismic records where most body-wave studies are carried out. Then one has (using (27-24))

$$\frac{c(\omega)}{c(2\pi)} \simeq 1 + \frac{1}{\pi Q} \ln\left(\frac{\omega}{2\pi} \right). \tag{27-25}$$

For body-wave velocities this reduces to

$$\frac{v_p(\omega)}{v_p(2\pi)} \simeq 1 + \frac{1}{\pi Q_p} \ln\left(\frac{\omega}{2\pi} \right),$$

$$\frac{v_s(\omega)}{v_s(2\pi)} \simeq 1 + \frac{1}{\pi Q_s} \ln\left(\frac{\omega}{2\pi} \right);$$

whence

$$\frac{\Delta v_p}{v_p(2\pi)} \simeq \frac{1}{\pi Q_p} \ln\left(\frac{\omega}{2\pi} \right), \quad \frac{\Delta v_s}{v_s(2\pi)} \simeq \frac{1}{\pi Q_s} \ln\left(\frac{\omega}{2\pi} \right). \tag{27-26}$$

Then looking at the changes in the Rayleigh- and Love-wave phase-velocities, one finds

$$\begin{aligned} \frac{\Delta C_R}{C_R} &= \sum_{j=1}^{N} \left[\left(\frac{\partial C_R}{\partial v_p} \right)_j \left(\frac{\Delta v_p}{C_R} \right)_j + \left(\frac{\partial C_R}{\partial v_s} \right)_j \left(\frac{\Delta v_s}{C_R} \right)_j \right] \\ &= \frac{1}{\pi} \ln\left(\frac{\omega}{2\pi} \right) \sum_{j=1}^{N} \left[\left(\frac{v_p}{C_R} \frac{\partial C_R}{\partial v_p} \right)_j Q_{p_j}^{-1} + \left(\frac{v_s}{C_R} \frac{\partial C_R}{\partial v_s} \right)_j Q_{s_j}^{-1} \right] \\ &= \frac{1}{\pi Q_R(\omega)} \ln\left(\frac{\omega}{2\pi} \right), \end{aligned} \tag{27-27}$$

and

$$\begin{aligned} \frac{\Delta C_L}{C_L} &= \sum_{j=1}^{N} \left(\frac{\partial C_L}{\partial v_s} \right)_j \left(\frac{\Delta v_s}{C_L} \right)_j \\ &= \frac{1}{\pi} \ln\left(\frac{\omega}{2\pi} \right) \sum \left(\frac{v_s}{C_L} \frac{\partial C_L}{\partial v_s} \right)_j Q_{s_j}^{-1} \\ &= \frac{1}{\pi Q_L(\omega)} \ln\left(\frac{\omega}{2\pi} \right). \end{aligned} \tag{27-28}$$

In the last step of these equations, we have used (26-23) and (26-24) respectively. $Q_R(\omega)$ and $Q_L(\omega)$ are given by data such as in Figures 27-6, -7, -8. Again the partial derivative expansions have been done in the real, not complex domains.

The latest in a series of corrected velocity models is given by Anderson et al (1977): this is model 4Q2 which is shown in Figure 27-10. The corrections to this velocity structure were accomplished with Q model MM8 (Anderson et al, 1965) not with one of the later SL series. The largest changes can be noted in the upper 1000 km of the earth where Q values are the lowest. Changes in the lowermost portion of the mantle due to the decreased Q there (Figure 27-9) are not seen, for MM8 does not have this particular low Q region.

The corrected velocity structure goes a long way toward resolving the discrepancy between travel-times computed from a model which gives correct free periods and

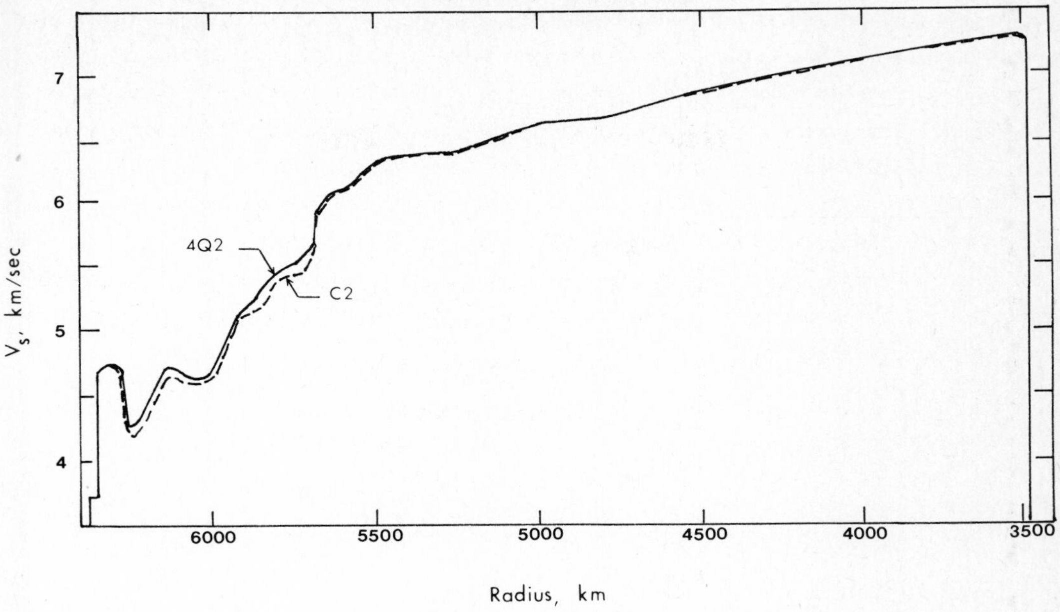

Figure 27-10. Shear-wave velocity structure plotted for models C2 (see also Figure 25-4) and 4Q2. C2 is a perfectly elastic model while 4Q2 has been corrected for dispersion due to attenuation. (After Anderson et al, 1977.) Copyrighted by AAAS.

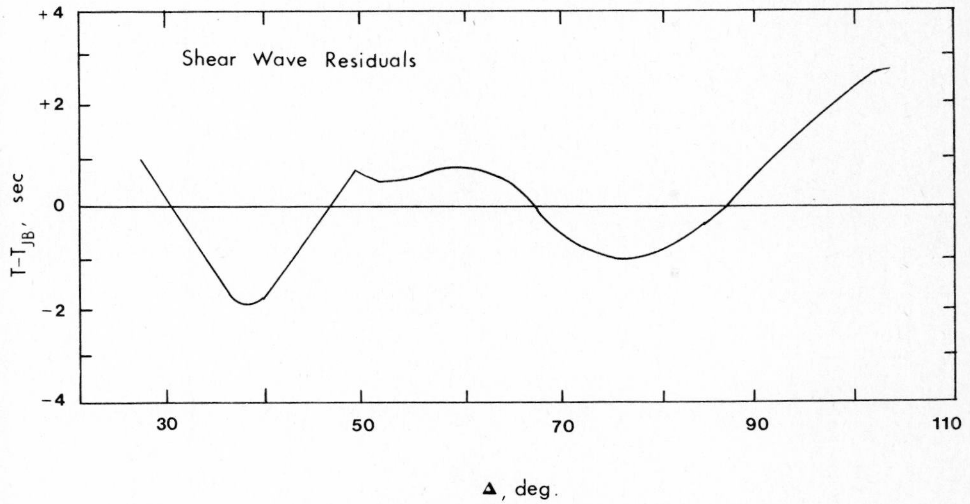

Figure 27-11. Shear-wave travel-time residuals calculated from Model 4Q2. Time differences relative to Jeffreys-Bullen travel-time curves. (After Anderson et al, 1977.) Copyrighted by AAAS.

travel-times as observed. Figure 27-11 shows the shear-wave residuals as calculated from model 4Q2 relative to the Jeffreys-Bullen travel-time curves. The "base Line" error has clearly been removed even though there is some coherent behavior left. Although this seems to solve the problem, there may still remain other causes of base line error. For example, average world models cannot adequately accurately depict both continental and oceanic structure at the same time: see Hales (1974).

Synthetic Seismograms have been calculated for body-wave wavelets in attenuating models by Helmberger (1973a), Helmberger and Engen (1974) and Kennett (1975). The latter author notes that much of the fine detail found in perfectly elastic models is smoothed out by attenuation and suggests that upper mantle transitions may be sharper than previously thought.

CHAPTER 28

THE SEISMIC SOURCE – THEORY

28.1. General.

The nature of the seismic source has been very elusive. Getting data for analyzing a single event required writing to dozens of seismic observatories to request copies of their records. The installation of a world-wide network of standardized stations in the early 1960's made microfilm records of over one hundred stations available to research centers on a regular basis. The installation of this network was one portion of a program for the detection of clandestine nuclear explosions. As the data came in, it became apparent that there was much that we did not know about discriminating between natural earthquakes and man-made explosions. The problem of accurately identifying most nuclear explosions accelerated studies into the nature of the seismic source and into methods of obtaining source parameters from far-field seismic observations.

The great mass of data that came from the world-wide network was available to verify certain hypotheses of the "New Tectonics" (Isacks et al, 1968). The hypothesized motion of rigid "plates" riding on a "plastic layer" was seen as a possible mechanism for stress buildup leading to rupture and the quaking of the earth. The late 1960's and early 1970's saw a tremendous increase in the number of papers on source mechanism and how the postulated plate motions might fit in with seismic observations.

The decade of the 1970's has seen the resources of the seismic community, both instrumental and theoretical, directed toward a further understanding of the source mechanism with a view toward predicting earthquakes and possibly even controlling them. To enable this closer look at the source, the nature of fracture itself became part of the study and seismology became wedded to fracture mechanics.

In this chapter, we shall look more closely into the theoretical nature of seismic sources, extending the material of Chapter 7. The literature is so vast, and much of it is so extremely complicated that we shall touch on only the major points. The analysis of observations will be found in the next chapter. Reviews of source theory can be found in Archambeau (1968), Randall (1972a), and Ben-Menahem and Singh (1972). A summary of later work in the United States is given by Archambeau (1975).

Section 28.2 is given to a further look at point sources, with particular emphasis upon the calculation of directivity patterns which will be useful in determining the orientation of the fracture plane. Section 28.3 looks at the complications introduced by having an extended source with a moving singularity. Section 28.4 is essentially a list of various geometries in which the motions have been related to source orientation and time history. The final section, 28.5, is a look at some of the facets to be considered in the nature of fracture mechanism itself.

28.2. Point Sources.

A point source is a useful representation when the characteristic periods observed in the seismic record are long compared to the transit time of material rupture across the source region. We shall look at some examples in order of increasing complexity. One should note that, in the earth, isolated forces and couples with moment are generally excluded by the conditions of mechanical equilibrium, so our first models are mostly for illustration. In model seismology, however, one can apply isolated forces and couples to the surface of a model, with the equilibriating forces or moments being taken up in some external system. It is in the area of model seismology that most of the experimental verification of simple problems, such as Lamb's, has been carried out. However, the synthesis of realistic seismograms for earth models requires the use of more complicated sources.

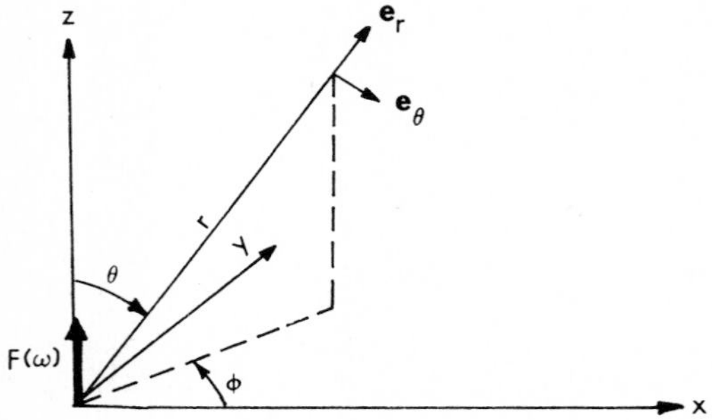

Figure 28-1. Geometrical relationships to describe the radiation pattern of an isolated force.

Forces. If we take our coordinate system as in Figure 28-1 with the isolated force directed along the z-axis, the elastodynamic field for a time-harmonic source is given by (7-89) and (7-90). In the far-field, the r^{-2} and r^{-3} terms die away rapidly and we have approximately:

$$\mathbf{w}^{(p)} \simeq \mathbf{e}_r \frac{F(\omega)\mathrm{Cos}\theta}{4\pi\rho v_p^2} \left(\frac{e^{ik_p r}}{r} \right), \tag{28-1}$$

and

$$\mathbf{w}^{(s)} \simeq -\mathbf{e}_\theta \frac{F(\omega)\mathrm{Sin}\theta}{4\pi\rho v_s^2} \left(\frac{e^{ik_s r}}{r} \right); \tag{28-2}$$

where $\mathbf{w}^{(p)}$ is the displacement associated with P-motion and $\mathbf{w}^{(s)}$ is the displacement associated with S-motion. The azimuthally symmetric directivity patterns are shown in Figure 28-2. Remember that these are isolated forces within the elastic body. Forces applied to a free surface would have patterns as determined from the solution of the three-dimensional Lamb's Problem. White (1965, pp 226-228) gives expressions and diagrams for these directivity patterns.

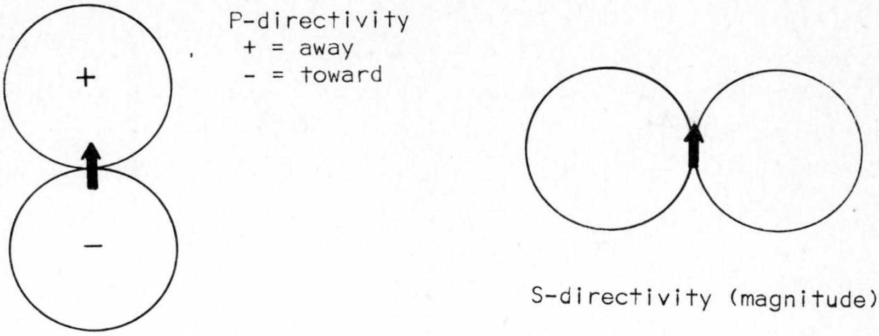

Figure 28-2. Directivity patterns for an isolated force.

Couples. The total motion due to an isolated time-harmonic force is given by

$$W_j = F_k \Gamma_{kj}(\mathbf{x}|\mathbf{x}^o;\omega); \qquad (8\text{-}19)$$

where $\mathbf{x}^o$ is the source point, $F_k(\omega)$ is the force-vector, and Γ_{kj} the infinite-space Green's function. If we add an oppositely directed force at a small distance $\delta\mathbf{x}$, we have

$$W_j = F_k \Gamma_{kj}(\mathbf{x}|\mathbf{x}^o + \delta\mathbf{x};\omega) - F_k \Gamma_{kj}(\mathbf{x}|\mathbf{x}^o;\omega) \simeq F_k \delta x_l \Gamma_{kj},_l, \qquad (28\text{-}3)$$

to first order. If we keep the product $F_k \delta x_l$ constant as δx_l goes to zero, i.e.,

$$M_{kl} \equiv F_k \delta x_l = \text{constant}; \qquad (28\text{-}4)$$

we then have

$$W_j = M_{kl} \Gamma_{kj},_l. \qquad (28\text{-}5)$$

We have thus shown that $\Gamma_{kj},_l$ is the motion due to an isolated couple M_{kl} as we noted on page 80 in regard to an interpretation of the representation theorem. Depending on the relative orientations of the force vector **F** and the relative displacement vector $\delta\mathbf{x}$, we can have couples with moment (**F** and $\delta\mathbf{x}$ not co-linear) and without moment (**F** and $\delta\mathbf{x}$ co-linear).

To examine the directivity patterns due to couples with moment, we consider a pair of oppositely directed forces in the z-direction separated by a small distance in the x-direction. Making the proper differentiations, we have:

$$\begin{aligned} \mathbf{W}_C^{(p)} &\simeq \frac{M_{zx}}{F}\frac{\partial}{\partial x}\mathbf{W}^{(p)} \simeq \mathbf{e}_r \frac{M_{zx}}{4\pi\rho v_p^2}\frac{\partial}{\partial x}\left(\text{Cos}\theta\ \frac{e^{ik_p r}}{r}\right) \\ &\simeq \mathbf{e}_r \frac{ik_p M_{zx}}{4\pi\rho v_p^2}\,\text{Sin}\theta\text{Cos}\theta\text{Cos}\phi\left(\frac{e^{ik_p r}}{r}\right); \end{aligned} \qquad (28\text{-}6)$$

and

$$\mathbf{W}_C^{(s)} \simeq -\mathbf{e}_\theta \frac{ik_s M_{zx}}{4\pi\rho v_s^2}\,\text{Sin}^2\theta\text{Cos}\phi\left(\frac{e^{ik_s r}}{r}\right). \qquad (28\text{-}7)$$

The directivity patterns for M_{zx} are shown in Figure 28-3.

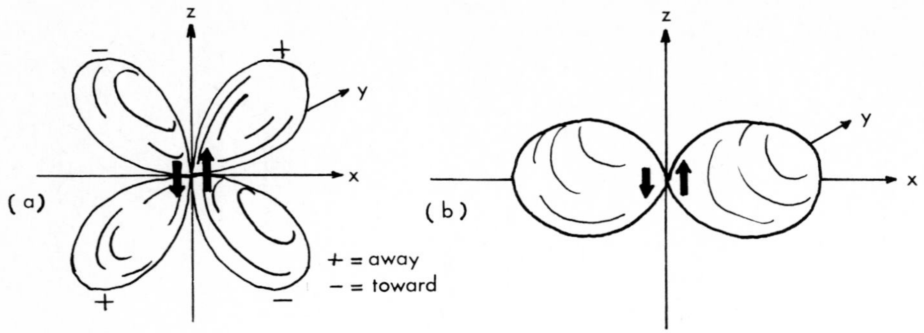

Figure 28-3. Directivity patterns for couples with moment.
(a) P-directivity for M_{zx}. (b) S-directivity for M_{zx}(magnitude).

On the other hand, if we had considered x-directed forces separated in the z-direction, that is, formed M_{xz}, we would have merely rotated the force arrows in Figure 28-3 by 90° counterclockwise and changed their signs. This would have left us with an unchanged pattern for P and a rotated pattern for S. We shall return to this point later.

For couples without moment, we take a pair of oppositely directed z-forces and separate them a short distance in the z-direction. This gives

$$\mathbf{w}^{(p)}_{LD} \simeq \frac{M_{zz}}{F}\frac{\partial}{\partial z}\mathbf{w}^{(p)} \simeq \mathbf{e}_r \frac{ik_p M_{zz}}{4\pi\rho v_p^2}\,\mathrm{Cos}^2\theta\left(\frac{e^{ik_p r}}{r}\right)$$

$$\simeq\left[\mathbf{e}_x \mathrm{Sin}\theta\mathrm{Cos}\phi\mathrm{Cos}^2\theta + \mathbf{e}_y\mathrm{Sin}\theta\mathrm{Sin}\phi\mathrm{Cos}^2\theta + \mathbf{e}_z\mathrm{Cos}^3\theta\right]\frac{ik_p M_{zz}}{4\pi\rho v_p^2}\left(\frac{e^{ik_p r}}{r}\right) \quad (28\text{-}8)$$

$$\simeq\left[\mathbf{e}_x\frac{xz^2}{r^3} + \mathbf{e}_y\frac{yz^2}{r^3} + \mathbf{e}_z\frac{z^3}{r^3}\right]\frac{ik_p M_{zz}}{4\pi\rho v_p^2}\left(\frac{e^{ik_p r}}{r}\right),$$

and

$$\mathbf{w}^{(s)}_{LD} \simeq \frac{M_{zz}}{F}\frac{\partial}{\partial z}\mathbf{w}^{(s)} \simeq -\mathbf{e}_\theta \frac{ik_s M_{zz}}{4\pi\rho v_s^2}\,\mathrm{Sin}\theta\mathrm{Cos}\theta\left(\frac{e^{ik_s r}}{r}\right)$$

$$\simeq\left[-\mathbf{e}_x\mathrm{Sin}\theta\mathrm{Cos}^2\theta\mathrm{Cos}\phi - \mathbf{e}_y\mathrm{Sin}\theta\mathrm{Cos}^2\theta\mathrm{Sin}\phi + \mathbf{e}_z\mathrm{Sin}^2\theta\mathrm{Cos}\theta\right]\frac{ik_s M_{zz}}{4\pi\rho v_s^2}\left(\frac{e^{ik_s r}}{r}\right) \quad (28\text{-}9)$$

$$\simeq\left[-\mathbf{e}_x\frac{xz^2}{r^3} - \mathbf{e}_y\frac{yz^2}{r^3} + \mathbf{e}_z\frac{(x^2+y^2)z}{r^3}\right]\frac{ik_s M_{zz}}{4\pi\rho v_s^2}\left(\frac{e^{ik_s r}}{r}\right).$$

The cartesian forms of these two vectors have been included for later use. The directivity pattern for a <u>linear doublet</u>, as such a couple is called, is given in Figure 28-4.

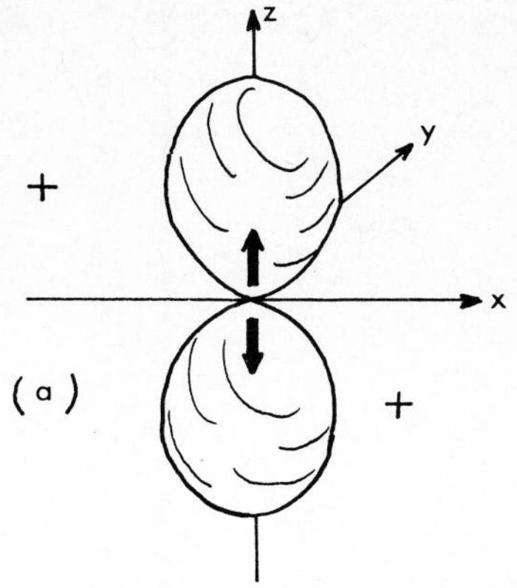

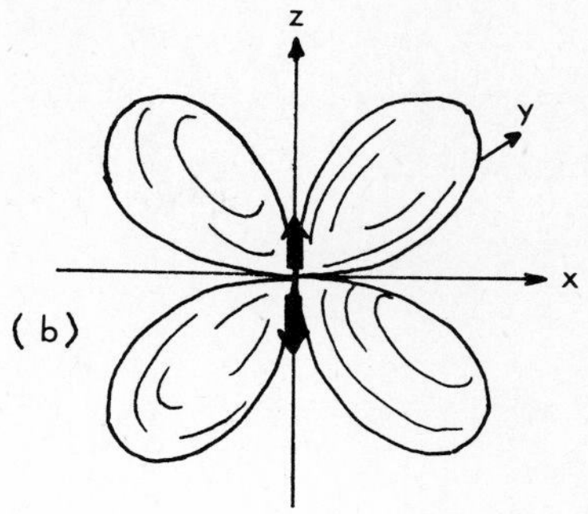

Figure 28-4. Directivity patterns for a linear doublet. (a) P-directivity for M_{zz}. (b) S-directivity for M_{zz}.

One final result can be obtained from (28-7) and (28-9). If we place one of three equal linear doublets along each of the cartesian axes at the origin (Figure 28-5), we find an interesting result. By permuting the coordinates in order we obtain (from (28-8) and (28-9)):

$$\mathbf{w}_D^{(p)} \simeq \mathbf{e}_r \frac{ik_p M}{4\pi\rho v_p^2} \left(\frac{e^{ik_p r}}{r} \right), \qquad (29\text{-}10)$$

and

$$\mathbf{w}_D^{(s)} \equiv 0;$$

where $M_{zz} = M_{yy} = M_{xx} = M$. This analysis shows the equivalency of three equal and mutually perpendicular linear doublets and a dilatational source. The directivity pattern is shown in Figure 28-5.

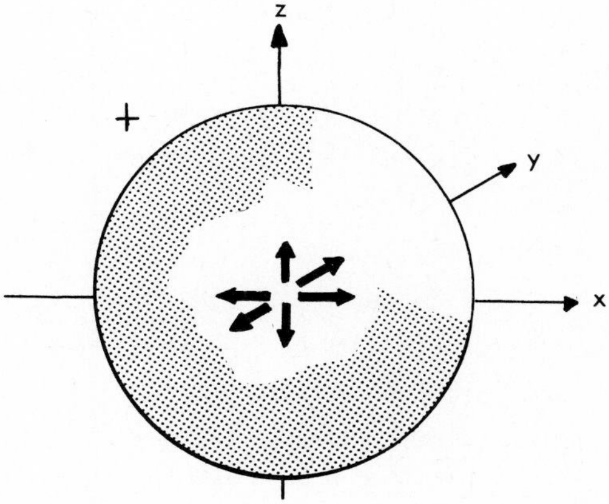

Center of Dilatation.

Figure 28-5. Directivity pattern for P-waves from three equal linear doublets. There are no S-waves. This is effectively a center of dilatation.

The effects of burying a couple directly beneath the free surface of an elastic half-space have been considered by Alterman and Aboudi (1970). Rayleigh-wave directivity patterns for a couple in a half-space have been given by Haskell (1963).

Double Couple. Here we have

$$\mathbf{w}_D^{(p)} \simeq \mathbf{e}_r \frac{ik_pM}{4\pi\rho v_p^2} 2\mathrm{Sin}\theta\mathrm{Cos}\theta\mathrm{Cos}\phi\left(\frac{e^{ik_pr}}{r}\right), \tag{28-11}$$

and

$$\mathbf{w}_{DC}^{(s)} \simeq -\mathbf{e}_\theta \frac{ik_sM}{4\pi\rho v_s^2} (\mathrm{Sin}^2\theta - \mathrm{Cos}^2\theta)\mathrm{Cos}\phi\left(\frac{e^{ik_sr}}{r}\right); \tag{28-12}$$

where we have required $M_{zx} = M_{xz} = M$. The directivity patterns are shown in Figure 28-6.

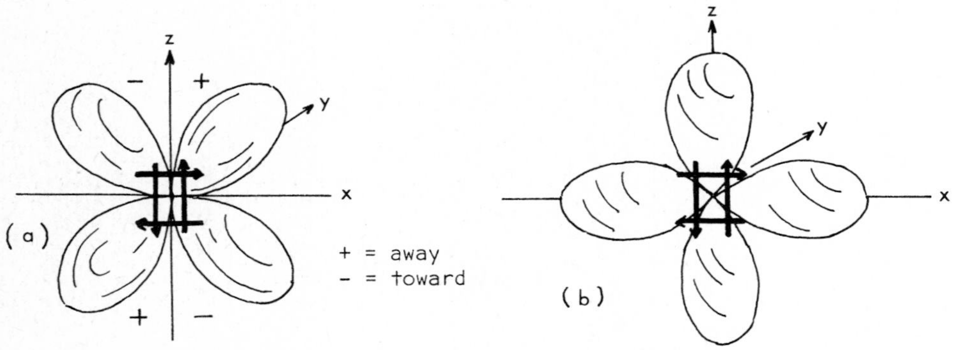

Figure 28-6. Directivity patterns for double couples without moment. (a) P-directivity. (b) S-directivity.

We have seen how, at large distance, a triple of linear doublets (Figure 28-5) can masquerade as a dilatational source. A similar effect can be seen in the directivity patterns for a double couple. The combination of two linear doublets, one positive and one negative as in Figure 28-7(a) gives the same directivity patterns for both P and S. That is, two oppositely signed linear doublets at right angles to each other behave (at distance) as a double couple.

This equivalence leads to some ambiguity in interpretation. In Figure 28-7(b), we can suitably combine a double couple, a center of dilatation, and a linear doublet to obtain a differently oriented double couple. We will again touch on this point in Chapter 29. Knopoff and Randall (1970) proposed the combination in Figure 28-7(c) as a third realistic candidate for a seismic source mechanism along with the double couple and center of dilatation. I have changed their terminology slightly to conform with White's (1965) usage of linear doublet and call the combination a compensated linear doublet. This combination does not produce volume change as does the linear doublet. Knopoff and Randall (ibid) go on to show that the directivity patterns for P-waves cannot choose between specially oriented combinations of double

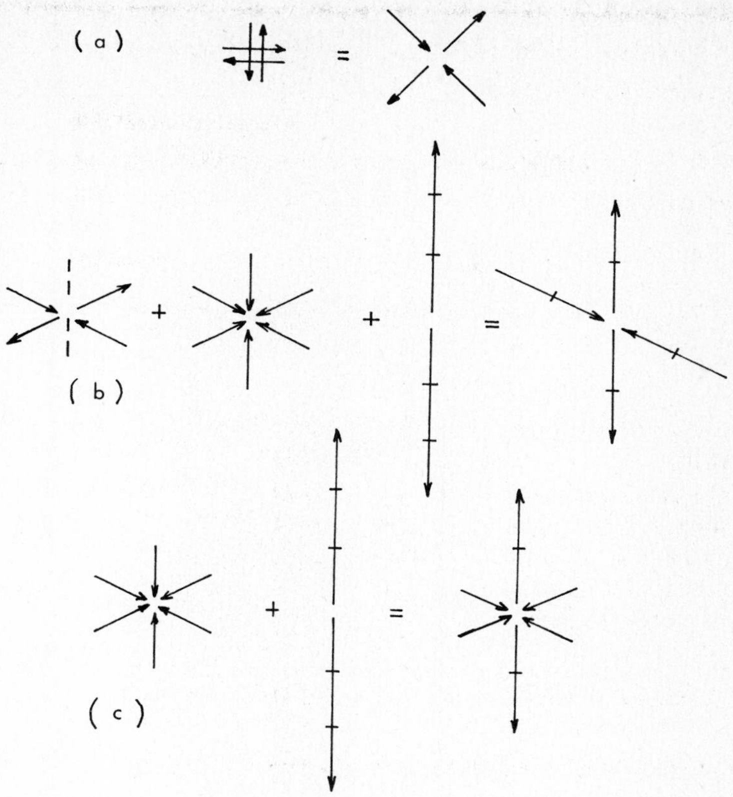

Figure 28-7. Equivalence of some simple source types.
(a) A double couple is equivalent to two oppositely signed linear doublets at right angles.
(b) A double couple equivalent in the x-y plane plus a center of dilatation plus a linear doublet in the z-direction equals a double couple equivalent in the y-z plane.
(c) A center of dilatation plus a linear doublet is equivalent to a compensated linear doublet.

couples and compensated linear doublets, and isolated double couples. Figure 28-7(b) is one example given by them. Patterns of S-waves are different, but difficult to observe experimentally.

The effects of burying a double couple directly beneath the free surface of an elastic half-space have been considered by Burridge et al (1964). Rayleigh-wave directivity patterns for a double couple in a half-space have been given by Haskell (1963).

Elastic Dislocations. Up to now, we have been constructing a few conceptual models and graphing their resulting directivity patterns. However, this must eventually be tied into the real earth by observational data. H. F. Reid (1910) plotted the observed displacements from the 1906 San Francisco earthquake. They were largest near the fault break and decreased with increasing distance from it as in Figure 28-7. Such patterns in displacement are called elastic dislocations. The first attempt to fit observations of this type surface breakage was with a single couple. This met with two objections, the aforementioned improbability of isolated couples and some S-wave data which were at variance with the single couple pattern of Figure 28-3.

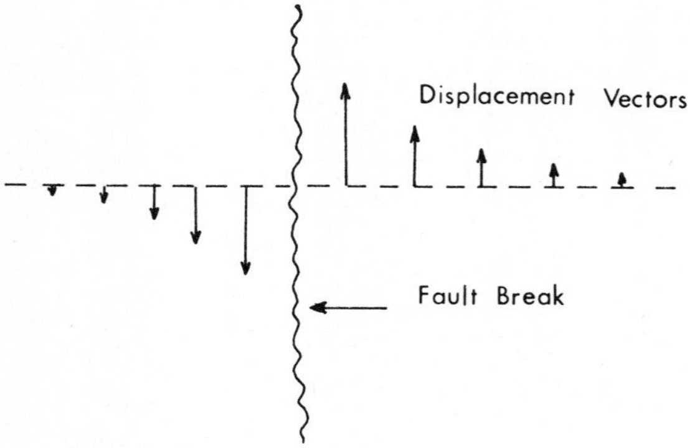

Figure 28-8. Schematic ground displacement pattern for horizontal rupture along a vertically dipping strike-slip fault.

Two attacks were mounted upon this problem. Stauder (1960) and a number of co-workers began a systematic study of S-wave directivity patterns. Because of the unquiet nature of the seismic record during the arrival of S, the results of this study were not unambiguous. However, they did lend support to a double-couple source. Theoreticians, knowing that the double-couple directivity pattern was four-lobed, then worked on this type of source mechanism. Steketee (1958) made an identification of elastic dislocations and double couples in the static case. His work was followed up by Maruyama (1963) who made a similar identification in the dynamic case. Knopoff and Gilbert (1959a, 1960), taking a somewhat different tack, investigated the directivity patterns of dislocation type sources working directly from a form of the representation theorem. In their 1960 paper, they found an equivalence between the directivity pattern from dislocation sources and from double couples. Shortly afterward, Burridge and Knopoff (1964) demonstrated a rigorous relationship between the two. Let us see how this equivalence occurs.

From (8-33) we have that the motion due to prescribed displacements is

$$U_j(\mathbf{x}^o,\omega) = -\int C_{kl}(\mathbf{x},\omega)G_{lj,k}(\mathbf{x}^o|\mathbf{x};\omega)dS; \qquad (28\text{-}13)$$

where in an isotropic homogeneous medium

$$C_{pq} = C_{qp} = \lambda U_m n_m \delta_{pq} + \mu(U_p n_q + U_q n_p). \qquad (8\text{-}32)$$

Equation (8-32) is a crucial regrouping, originally suggested by Burridge and Knopoff (ibid), that makes the following almost trivial. But, we have just seen that

$G_{lj,k}$ is the response to a force couple. We thus make an identification of $C_{kl}dS$ with M_{kl}. To keep with convention, let us reverse the normal so that it points into the medium from the crack (as in Figure 28-9), and we then see that

$$M_{kl} = \overline{C_{kl}dS} \equiv \left[C_{kl}^{(1)} - C_{kl}^{(2)}\right] dS, \qquad (28\text{-}14)$$

where the over-bar signifies a jump discontinuity.

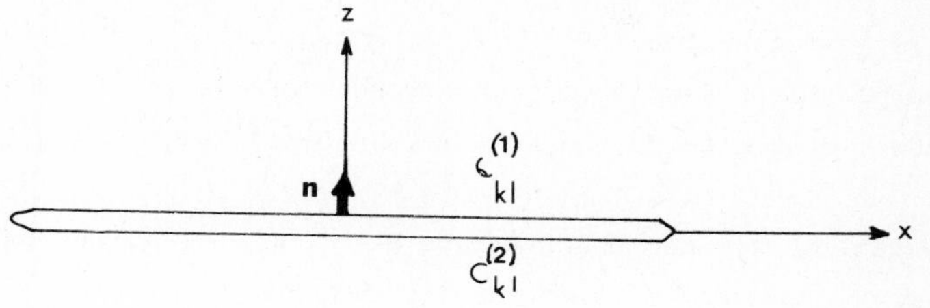

Figure 28-9. Geometrical relationships relating to the evaluation of the form (8-32).

Let us now make a few computations and see what we get. We choose z̲ to be the normal-direction and x̲ to be along the crack. For a jump in U_x, we have:

$$C_{xx} = 0,$$
$$C_{xz} = C_{zx} = \mu U_x n_z, \qquad \text{(terms containing } \underline{y} \text{ are zero)} \qquad (28\text{-}15)$$
$$C_{zz} = 0.$$

Hence, in the sum over l̲ and k̲ in the integral of (28-13) we pick up both C_{xz} and C_{zx}. Consequently, a displacement dislocation in the x-direction gives rise pointwise to a double couple without moment. Thus all mechanical objections were removed and double-couple radiation patterns (Figure 28-3) were shown to be compatible with displacement dislocation observations (Figure 28-8).

For a jump in U_y, we have:

$$C_{yy} = 0,$$
$$C_{yz} = C_{zy} = \mu U_y n_z, \qquad \text{(terms containing } \underline{x} \text{ are zero)} \qquad (28\text{-}16)$$
$$C_{zz} = 0.$$

Again we pick up the clockwise and counter-clockwise couples. Lastly, for a jump in U_z we have

$$C_{xx} = \lambda U_z n_z,$$
$$C_{yy} = \lambda U_z n_z, \qquad \text{(off-diagonal terms are zero)} \qquad (28\text{-}17)$$
$$C_{zz} = (\lambda + 2\mu) U_z n_z.$$

This can be represented by a set of three linear doublets, or a linear doublet in z̲ plus an equivalent source of discontinuous dilatation.

Multipole Expansions. The above analysis can be systematized by expanding the Green's function G_{ij} in vector spherical harmonics. Carrying out the indicated integration of (8-33) leaves the displacement represented as a summation over vector spherical harmonics — a multipole expansion. One then identifies the individual terms with simple forces, couples, quadrupoles (double couples), centers of dilatation, etc. Details are given by Archambeau (1968), Ben-Menahem and Singh (1972) and Randall (1972a).

A more sophisticated approach is given by Backus and Mulcahy (1976a and b), and Backus (1977a and b) who present a complete tensor formulation of the seismic source problem. Some additional aspects can be found in papers by McCowan (1976) and Geller (1976a).

Multipole expansion are necessarily made about some preferred axis related to the source geometry. However, this orientation may not be favorable with respect to boundaries within and on the surface of the material medium. Wason and Singh (1971a), Phinney and Burridge (1973), and Minster (1976) have given formulations to facilitate a change in coordinate systems.

The effects of burying multipolar sources beneath the free surface of an elastic half-space have been considered by Vered and Ben-Menahem (1976). They give a convenient table so that the effects upon simpler source types can be seen in relation to one another. Their formulation is obtained by taking the sum of direct and reflected fields, and is only approximate, but useful.

28.3. Extended Sources.

An extended source can be constructed in two different ways. One can create such a source as a distribution of point sources over a finite region acting simultaneously. An example would be a punch indenting the surface of a half-space. Alternatively, one can have a single point source trace out a moving pattern over a finite volume of material. The first case can be thought of as an infinite velocity limit of the second. In either example, one has to formulate the point source problem and then sum up the effects of the distribution by an integration over space and time. The two-dimensional case can be attacked in a number of ways (Eringen and Suhubi, 1975, Sec. 7.11, 7.12, and 7.13) but the results are not very applicable to realistic earthquake sources. The three-dimensional moving source is much more relevant; however, solutions are few. Exact treatments of the three-dimensional problem have been given by Payton (1964), Gakenheimer and Miklowitz (1969), and Roy (1974).

A useful method was developed by Ben-Menahem (1961) for surface-waves and later (Ben-Menahem (1962)) for body-waves. In his approximation, the period of the elastic wave signal is short with respect to the travel-time from the epicenter but of the order of the transit-time for the rupture to traverse the fracture zone. Although he gives a rigorous proof of the following results, they can also be derived from rather simple considerations.

Let us look at a simple source with time-history f(t). It will have a Fourier transform

$$F(\omega) = \int_{-\infty}^{\infty} f(t)e^{-i\omega t} dt. \quad (28\text{-}18)$$

Adding a simple complication, let us take a signal composed of f(t) plus a delayed and diminished signal af(t) where $a \leq 1$, i.e.,

$$g(t) = f(t) + af(t - \Delta t). \quad (28\text{-}19)$$

We then have

$$G(\omega) = (1 + ae^{-i\omega\Delta t})F(\omega). \quad (28\text{-}20)$$

The factor $(1 + ae^{-i\omega\Delta t})$ is a modulating factor, changing a rather smooth spectrum $F(\omega)$ into one with a large number of maxima and minima, Figure 28-10. The spacing and depth of these minima allow the recovery of the two parameters a and Δt.

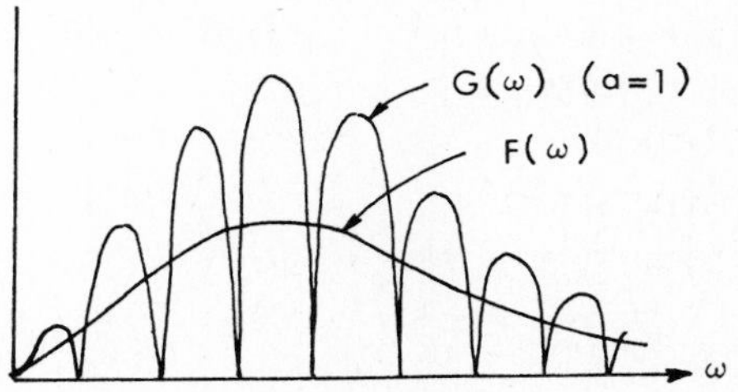

Figure 28-10. Modulation of signal spectrum because of finite duration.

If we write a more complicated time-function by superposing a number of signals proportional to f(t) together in a continuum, i.e.,

$$g(t) = \frac{1}{\tau_o}\int_o^{\tau_o} a(\tau)f(t - \tau)d\tau, \quad (28\text{-}21)$$

we have that

$$G(\omega) = \frac{1}{\tau_o} A(\omega)F(\omega). \quad (28\text{-}22)$$

In the particular case $a(\tau) \equiv 1$, i.e., all components are weighted equally, then

$$\frac{1}{\tau_o} A(\omega) = \frac{\mathrm{Sin}(\omega\tau_o/2)}{(\omega\tau_o/2)} . \quad (28\text{-}23)$$

The effect of this new modulation factor is similar to that in Figure 28-10.

Consider the situation described in Figure 28-11. We will have a disturbance propagating from 0 to b along the z-axis. The apparent time of rupture as seen at point P will be a function of the aximuth angle θ_o of the point of observation. Simple considerations give us

$$\tau_o = \frac{b}{v_r} - \frac{b\mathrm{Cos}\theta_o}{v_t} ; \quad (28\text{-}24)$$

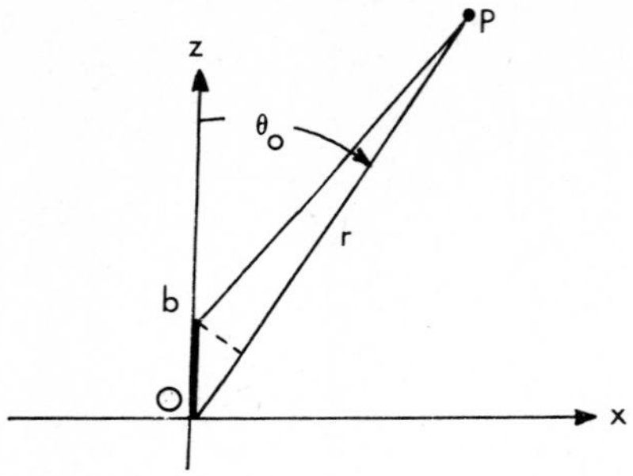

Figure 28-11. Geometrical relationships for the calculation of the directivity modulation factor due to a finite source.

where $\underline{b}$ is the length of rupture, v_r is the velocity of rupture, $b\text{Cos}\theta_o$ is the projection of $\underline{b}$ onto $\underline{r}$, and v_+ is the velocity of propagation of the particular seismic phase. Substituting this into (28-23), we obtain

$$A(\omega,\theta_o) \sim \frac{\text{Sin}\left\{\frac{\pi b}{Tv_+}\left[\frac{v_+}{v_r} - \text{Cos}\theta_o\right]\right\}}{\left\{\frac{\pi b}{Tv_+}\left[\frac{v_+}{v_r} - \text{Cos}\theta_o\right]\right\}} . \tag{28-25}$$

For many sources, we have that the spatial directivity is the same in the forward and backward directions. One then can take the ratio

$$\frac{A(\omega,\theta_o)}{A(\omega,\theta_o + \pi)} = \frac{\left[\frac{v_+}{v_r} + \text{Cos}\theta_o\right]}{\left[\frac{v_+}{v_r} - \text{Cos}\theta_o\right]} \; \frac{\text{Sin}\left\{\frac{\pi b}{Tv_+}\left[\frac{v_+}{v_r} - \text{Cos}\theta_o\right]\right\}}{\text{Sin}\left\{\frac{\pi b}{Tb_+}\left[\frac{v_+}{v_r} + \text{Cos}\theta_o\right]\right\}} . \tag{28-26}$$

Such ratios are useful in determining v_r and $\underline{b}$ if the orientation of the fault is known. A range of values for different θ allows the separation of $\underline{b}$ from v_r.

A somewhat different approximation is made by Hirasawa and Stauder (1965) and Savage (1965). The two papers are complementary. They consider times after the wave arrival short compared to the time for the rupture to traverse the source region. That is, they ignore the effects of the cessation of motion ("stopping phase"). They both find that the directivity pattern is modified by the propagating character. Savage explicitly gives this modulation factor as:

$$M_1(\theta) = \frac{v_r}{\{1 - (v_r/v_+)\text{Cos}\theta_o\}} , \tag{28-27}$$

for a singularity propagating in one direction (unilateral); and as

$$M_2(\theta) = \frac{2v_r}{\{1 - (v_r/v_+)^2\text{Cos}^2\theta_o\}} , \tag{28-28}$$

for a singularity propagating equally in opposite directions (bilateral). This means that the zeros of the original directivity patterns are unaffected; a point made by Hirasawa and Stauder.

Savage then goes on to show that the resulting response consists of roughly rectangular impulses whose height is given by the above factors but whose duration is proportional to $b/M_1(\theta)$. Thus the integrated area is not dependent upon direction. This is essentially what is said by equation (28-25) for small b or large T.

28.4. Applications.

At this point in the book, the complexity of elastic wave propagation almost becomes overwhelming. In the analysis of the elastodynamic transfer of energy from point A to point B, we could include:

a) The time-history of the rupture,
b) the multipolar nature of the source mechanism,
c) layering in the source regions,
d) layering and curvature along the propagation path,
e) layering in the observing region,
f) the multipolar nature of the observing instrumentation,
g) the impulse response of the recording instrument,
h) anisotropy, inhomogeneity, or anelasticity of materials at any point of the above.

Thus, to work out any given example, there are so many parameters that the only way to proceed is to set up the problem as a series of black boxes operating in sequence. Each item above, with the exception of a), is fairly well understood, and we have studied b) — f) and h) so far. Item f) is very similar to item b), and item g) will be covered in Chapter 33. However, the time-history of the source depends critically upon the properties of materials, which are only partially understood, and a definitive dynamical theory of rupture has yet to be devised. We shall discuss some aspects of this problem in the next section.

The review article of Ben-Menahem and Singh (1972) is a distillation of a tremendous amount of work done by these two scientists and some of their co-workers. It basically covers items b) through f) above in a perfectly elastic earth-model. However, even the seventy-six pages of their paper barely outlines hundreds of pages of their previous work. In this section, we shall briefly catalog a number of papers, particularly those with tables or diagrams, which should prove helpful. The list is by no means complete (most of the original work of Ben-Menahem and Singh is not included and references should be obtained from their review article), but it should indicate the scope of problems that can be solved.

Flat-layered Media — Body-Waves. T-C. Lee and T-L. Teng (1973) derive polar radiation patterns for P- and SV-waves from a number of source types. Ben-Menahem and Vered (1973) extend the Cagniard method for use with dislocation sources in flat-layered media. Helmberger (1974) works out generalized ray theory for shear dislocations in such media. Langston and Helmberger (1975) apply these techniques to shallow dislocation sources in a layered half-space. Jobert (1975) gives the form of propagator and Green matrices for use with forces and dislocations. Bache and Harkrider (1976) consider a general multipolar souce in a layered stack.

Flat-layered Media — Surface-Waves. Haskell (1964a) works out the directivity functions for a number of point sources from forces to double couples. Ben-Menahem and Harkrider (1964) provide both directivity patterns as well as spectra for a number of dipolar sources buried in a flat-layered earth. Saito (1967) derives directivity functions for a vertically inhomogeneous earth using variational methods without the necessity of a layered structure. Ben-Menahem et al (1970) work out spectral responses for three Rayleigh and four Love modes in a multilayered earth. Panza et al (1973, 1975a, 1975b) give many of the properties of Rayleigh-waves in such media.

Spherically-layered Media — Body-Waves. Ben-Menahem et al (1965) give a rather complete analysis of body-waves in a sphere with many tables and figures. Jobert (1977) derives approximate propagator and Green matrices for use with body-force and dislocation sources. In a series of papers (Singh et al, 1972; Shimshoni et al, 1973; Sylman et al, 1974), theoretical amplitudes of body-waves from dislocation sources have been calculated.

Spherically-layered Media — Surface-Waves and Free Oscillations. Saito (1967) works out modal excitation coefficients in a radially inhomogeneous sphere without a layered structure. Ben-Menahem et al (1971) give almost one hundred pages of tables and figures relating to the amplitudes of terrestrial line spectra resulting from excitation by a dislocation source. Doornbos (1977) computes the excitation of normal modes by sources with a volume change. Bhattacharya (1978) has worked out improved Thomson-Haskell matrices to be used in calculating the excitation of Rayleigh-waves in a spherically-layered earth.

Static Deformations and the Excitation of the Chandler Wobble. Following the great Alaskan earthquake of 1964, instrumentation was sufficiently well developed so that residual deformations were found (Press, 1965) at a number of points on the globe. This, in itself, warranted a theoretical study of such displacements, but a suggestion was put forward by Mansinha and Smylie (1967) that the Chandler Wobble was excited by large earthquakes. The Chandler Wobble is a free motion associated with the earth's rotation described by Officer (1974, Sec. 10.4 and 10.7). It has a period of approximately 430 days and would be damped out by losses in the earth if it were not continually excited. Since this was a resonable suggestion, literally dozens of papers were forthcoming in the next few years to verify this possibility.

A number of articles dealing with the pure formalism will first be mentioned. Ben-Menahem and Singh (1968), Singh (1970), and Yeatts (1973) all considered the problem of static multipoles in a flat-layered medium. Ben-Menahem et al (1969), Wason and Singh (1971b), and Wason and Singh (1972) considered the spherical-earth model.

The calculation of changes in the moment of inertia tensor was next, for it is these changes that drive the Chandler Wobble. Smylie and Mansinha (1971) calculated the effects of large earthquakes upon the moment of inertia. They felt there were no theoretical difficulties in accounting for the excitation. Dahlen (1973a) achieved different results which essentially agreed with those of Israel et al (1973). These two papers found that large earthquakes were capable of exciting only a small part of the observed wobble. Alternative methods of calculation were given by Rice and Chinnery (1972) and by Israel and Ben-Menahem (1975). The latter paper confirmed the earlier results of Israel et al (ibid). A number of papers have been concerned about the proper treatment of the fluid core. Chinnery (1975) reviews many of these problems and gives new results.

Prestressed Media. We have seen that the dilatational sources of Chapter 7 and their force equivalent, Figure 28-5, do not produce S in the vicinity of the source. However, the observed elastic wave fields from nuclear explosions (Archambeau and Sammis, 1970; Toksoz et al, 1971; Aki and Tsai, 1972) contained a large proportion of Love-waves in many cases. Asymmetry of the source, interaction with fractures (pre-existing or produced in the explosion), and the release of pre-existing tectonic stress have all been proposed as mechanisms for the production of SH energy. The first effect should be small except in extremely heterogeneous conditions, the second is largely intractable, but the third lends itself to theoretical analysis.

The three summary papers listed above contain dozens of references to both observational, experimental, and theoretical work. A number of recent investigations have been reported by Archambeau (1972), Burridge and Alterman (1972), Singh (1973) and Dahlen (1972a, 1972b, 1973b).

28.5. Models of Fracture.

In a model of fracture, there are a large number of features to explain. We want to determine why a fracture starts; that is, what are the conditions of stress, rock strength, and friction that come together to initiate fracture. Secondly we want to know how fast the rupture itself will propagate. Aside from its intrinsic interest, the rupture velocity should be able to give us information about the three quantities above. Thirdly, a model of fracture should tell us why a fracture, once started, ceases to run. At a first order of approximation, the inhomogeneities present in the earth's crust seem enough to cause local stress build-ups and variation in strength and friction. A stress concentration sufficient to overcome the resistance of the rock to fracture will start the rupture, and when the stress diminishes or the resistance increases the rupture will stop. However, one would like to do better than this. In addition to the details of each individual rupture, one would like a (possibly augmented) model to say something about the small fractures — foreshocks — that sometimes occur prior to the main rupture and about the diminishing series of fractures — aftershocks — that take place afterward.

The classic text of Richter (1958), particularly Part One, "Nature and Observation of Earthquakes", gives a wealth of information about the phenomena we would like to model. A review of effects associated with Japanese earthquakes is given by Kanamori (1973). A summary of seismic as well as associated effects is given by Rikitake (1976) with particular application to the prediction of earthquakes. The nature of the fracture process in earthquakes is reviewed by Dieterich (1974).

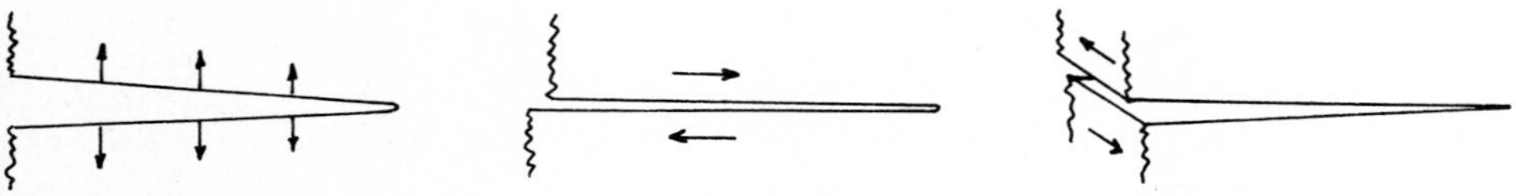

Figure 28-12. Three modes of crack propagation.

There are three modes of rupture at the tip of a crack and these are illustrated in Figure 28-12. The first, a tension crack, is appropriate to materials under tensile stress — a situation found only at the very surface of the earth. At depths below a few hundred meters, lithostatic pressure would keep the rock faces together. This leaves the latter two, in-plane shear cracks and anti-plane shear cracks, as suitable models for rupturing of the earth during an earthquake. The tensional model, with its applications to engineering, was the first to be solved analytically (Broberg, 1960 and Kostrov, 1964a). Analytical solutions to in-plane shear were given by Kostrov (1964b). The semi-infinite anti-plane shear crack was studied by Kostrov (1966, 1974). Burridge and Willis (1969) and Richards (1973a) considered the three-dimensional situation of elliptical cracks. In all these problems, a limiting velocity, v_S or v_R, was found because of energy absorption at the crack tip. This energy was absorbed in overcoming cohesion, the natural resistance of crystalline material to fracture. The cohesive forces must be broken for the crack to propagate.

However, most earthquakes take place along previously ruptured zones where the strength of the earth materials is much less than previously unbroken rock. The forces preventing movement may be largely frictional, although cementation by interstitial fluids may provide some "new" material. Burridge and others (Burridge and Halliday, 1971; Burridge, 1973; Burridge and Levy, 1974) have considered rupturing shear cracks where static friction had to be overcome to start rupture and dynamic friction provided the energy absorption mechanism. Without cohesion at the crack tip, the crack could propagate at v_P in some cases. In two and three dimensions, an analytical solution including both cohesion and dynamic friction has not been given, but Knopoff et al (1973a) and Knopoff and Mouton (1975) have given one-dimensional models combining both sources of energy absorption. In these models, rupture is subsonic.

Madariaga (1976) has used finite difference methods to examine the circular shear fault. His numerical results compare well with Kostrov's (1964b). Richards (1976a) extended his work with the elliptical crack to discuss conditions on the fracture surface and in the near-field. Modifying the tip conditions, Das and Aki (1977a) find that the P-velocity may be reached for low-strength materials. This modified tip condition then allows the introduction of "barriers" which may or may not be broken as the crack proceeds (Das and Aki, 1977b). It might be noted that there is some experimental evidence for rupture velocities exceeding v_s (T. L. Johnson and C. H. Scholz, 1976).

Prior to most of the work on rupture velocity, there was some interest in modeling the aftershock character of earthquakes. Burridge and Knopoff (1967) constructed an analog model consisting of spring-interconnected blocks sliding on a rough surface in response to a driving force applied to each block through an additional spring. Such a model generates a random sequence of "fractures" but no aftershocks. To obtain aftershocks, they introduce viscous elements under some of the masses instead of static friction. Looking for a more realistic model, Knopoff (1972b) suggests that the long-range nature of stresses at the crack tip may penetrate to other weaker areas. In addition he suggests a gradual diminishment of the critical stress by creep. Dieterich (1972) also suggests time-dependent friction, but finds in numerical experiments that it is still necessary to include viscoelasticity to generate aftershocks. The theoretical nature of the aftershock process is in need of improvement.

CHAPTER 29

THE SEISMIC SOURCE – OBSERVATIONS

29.1. General.

The observations that may be made at the surface of the earth are many and varied. In Chapter 23 we discussed the spatial location of the earthquake focus or hypocenter using either body- or surface-waves. Now we wish to look a little more closely at the details of the source. If the source can be represented as an elastic dislocation, we want to be able to find the orientation of the fault plane and the magnitude and direction of slip upon this plane. If the source can be represented as a phase change, then we would like to determine the variation in material properties at the source. We would also like to know the stresses acting, the volumes involved, and the amount of energy released. Lastly, in a comprehensive study of source mechanism, we may hope to determine regional and global stress patterns. These are important for an understanding of tectonic processes.

Extensive reviews have been given by Stauder (1962) and Khattri (1973). They make almost three hundred references to all aspects of the "source" problem. Ben-Menahem et al (1968) describe a computer program and give flow-charts for a large-scale processing of surface- and body-wave data to determine focal mechanism.

The material presented in the following two sections will assume a dislocation source rather than a phase-change, even though Knopoff and Randall (1970) show a complete equivalence in the directivity patterns. In particular, these authors show that a sudden change in shear modulus in material subjected to a plane shear field is equivalent to a double-couple dislocation. They also show that a sudden change in shear modulus in the presence of axial strain is equivalent to a linear doublet. The choice of model is easy for near-surface earthquakes since the rupture patterns can often be seen. On the other hand there has been much speculation as to the nature of deep earthquake sources. In earlier studies involving long-period P-waves (~10 sec period), it seemed that there might be a dilatational component to records from deep sources, but Randall (1972a) finds little evidence for this in a detailed study of six deep earthquakes. He cautions, however, that more work needs to be done before dilatational components at these periods can be dismissed. For longer periods (~100 sec), Gilbert and Dziewonski (1975) report dilatational components precursory to the main shock. They speculate on the possibility of "silent earthquakes" occuring at depth, noting that only very sensitive instruments could detect them.

The final section will be given to a brief summary of the nuclear detection and identification problem. Two books have been recently published on this critical subject; Bolt (1976) and Dahlman and Israelson (1977), so we will only give a few ideas and examples. This problem, which looks relatively easy given the quite different nature of explosive and earthquake sources, has turned out to have many unexpected problems.

29.2. Spatial Parameters at the Source.

Fault plane solutions for P- and S-waves. Consider a source of seismic energy located at 0 in Figure 29-1. First of all, one must compute the angle i_o at which a particular ray has left the source. From Snell's law for shperical rays (22-9) and (22-11) we have that

$$\text{Sin } i_o = \frac{v(r)}{r} \frac{dT}{d\Delta} \Big|_P ; \qquad (29\text{-}1)$$

where $dT/d\Delta$ is the apparent slowness (sec/radian) at point P. This calculation need only be made once for various values of r and Δ and the results tabulated. Tables for i_o are given by Pho and Behe (1972) for P and by Chandra (1972) for S. These are based upon the "1968 Seismological Tables" (Herrin, 1968) and upon Randall's (1971b) tables for S respectively. These tables cover $20^\circ \leq \Delta \leq 100^\circ$ and $0 \leq h \leq 700$km.

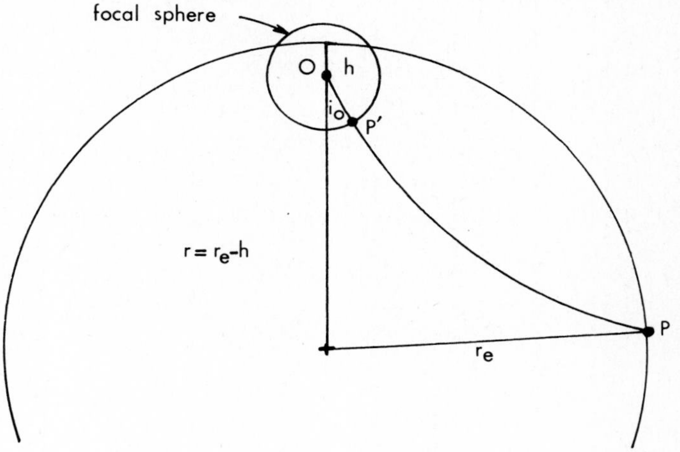

Figure 29-1. The relationships of the focal sphere to the seismic source and its emerging rays. The earthquake focus is at point 0.

The next step is the addition of a useful construct, the focal sphere. The seismic ray from 0 to P pierces the focal sphere at P'. When the records of a number of stations are treated in this way, the focal sphere will be marked with a number of points. By using standard projections, quantities associated with these points on the focal sphere can be mapped. The most commonly mapped quantities are the first motions of P. However one can also map first motions of S, and even amplitudes of P. It should be obvious that the XY and YZ planes of either Figure 28-3 or 28-6 cut the focal sphere in two great circles at right angles. For dislocation type sources, one of these corresponds to the fault plane while the other is called the auxiliary plane. There is an ambiguity here that cannot be resolved by first motion patterns of P alone; S must be used. However a series of ruptures along a fault zone will often line up in such a way that the choice can be made.

The most common projection is the stereographic projection, where P' is projected onto plane A as in Figure 29-2. Unless the great circles created where

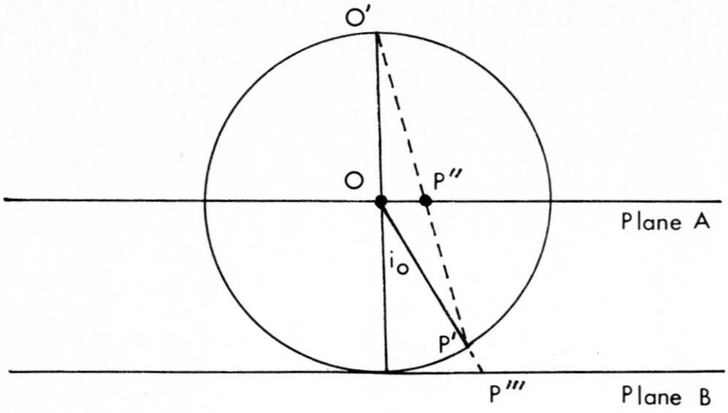

Figure 29-2. Projections of the focal sphere. P" is the stereographic projection and P" is the central projection of P'.

the fault plane and auxilliary plane pierce the focal sphere pass through O', these circles will project as circles at right angles onto plane A. Stauder (1962) has plotted compressions (away) and dilatations (toward) from the earthquake of 10 July 1958 in a number of projections. We first illustrate the stereographic projection in Figure 29-3. In this projection, distant stations appear at the center while nearer stations appear at the edges.

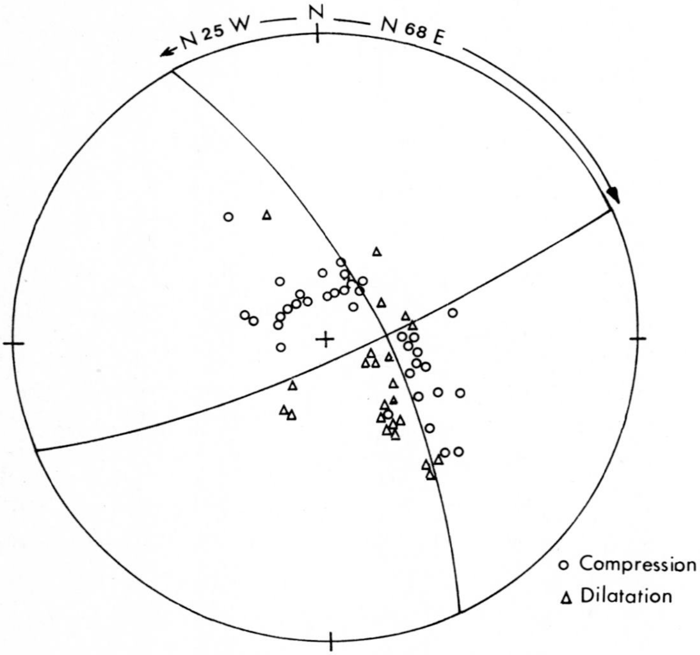

Figure 29-3. The stereographic projection of compressions and dilatations for the earthquake of 10 July 1958. The circular lines superimposed are the fault-plane solution. (After Stauder, 1962.)

After plotting all the compressions and dilatations, two circles at right angles are placed so as to divide the two types of first motions. This is known as a fault-plane solution. When this is completed, the strikes of the two planes may be read directly from the diagram. The maximum distance from the center to each circle is a measure of the dip. In this case

$$D = \frac{\pi}{2} - 2\mathrm{Tan}^{-1}\left(\frac{d}{r_f}\right);$$

where D is the dip angle, <u>d</u> is the radial distance to the circle, and r_f is the radius of the focal sphere. In the example here, Stauder (ibid) finds

Plane 1: STRIKE N25W; DIP 72°NE.

Plane 2: STRIKE N68E; DIP 82°SE.

Note that the projected circles cross at right angles, but that the sum of the two angles measured from North is greater than 90°.

It should be clear that the circles drawn in Figure 29-3 are not completely arbitrary. The radius of curvature must be greater than that of the focal sphere, and the circles have to meet at right angles. This merely reflects that two planes at right angles cannot have arbitrary strike and dip. Using direction cosines to work out the orthogonality condition, we have

$$\mathrm{Cos}S_1 \cdot \mathrm{Sin}D_1 \cdot \mathrm{Cos}S_2 \cdot \mathrm{Sin}D_2 + \mathrm{Sin}S_1 \cdot \mathrm{Sin}D_1 \cdot \mathrm{Sin}S_2 \cdot \mathrm{Sin}D_2 + \mathrm{Cos}D_1 \cdot \mathrm{Cos}D_2 = 0; \qquad (29\text{-}2)$$

where S is the strike and D is the dip respectively. For this to work out correctly, the strike must be taken with the origin on the right as you face the down-dip direction.

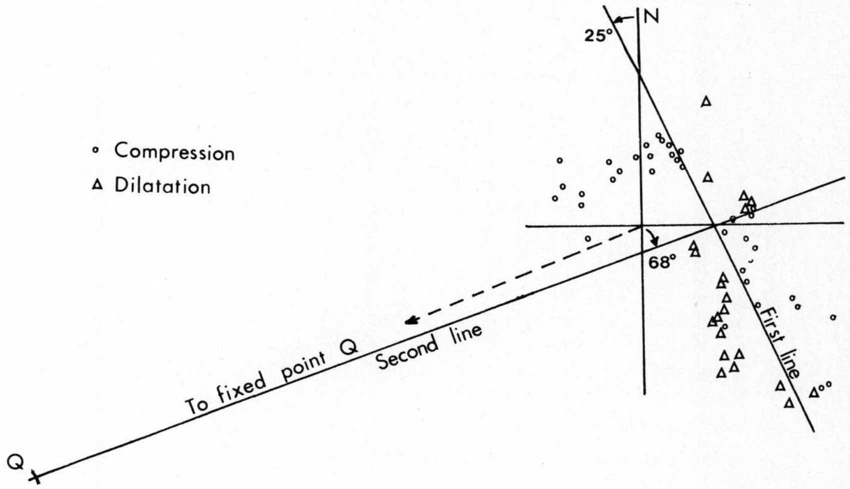

Figure 29-4. The central projection of the earthquake of 10 July 1958. The straight lines superimposed are the fault plane solution. More points are plotted here as the figure is more spread out. Point Q is described in the text. (Modified from Stauder, 1962).

The central projection has P' projected to P''' on Plane B (Figure 29-2). Since the origin of the focal sphere lies on the intersection of the two planes, their projections will be straight lines. This makes this projection particularly nice for use with digital computers. In this projection distant stations appear at the center. The strikes of the two planes may be read from the angles between the lines and the N - S coordinate axis. The dips may be calculated from

$$D = \mathrm{Cot}^{-1}(d/r_f),$$

where d is now the perpendicular distance from the origin to each of the lines. Figure 29-4 shows the same earthquake in the central projection.

Again, these lines are not independent and the calculated quantities have to satisfy (29-2). This can be translated into a geometrical construction similar to one given for the extended distance projection by DeBremaecker (1956). If the "first line" is considered good, it can be held fixed and the other line varied. Now this "second line" will have to contain the point lying on the normal to the first plane where it pierces the focal sphere. This point (Q) will plot at a distance $d' = r_f \mathrm{Tan} D$ starting from 0 in a direction perpendicular to the first line. All "second lines" will then have to pass through this point Q. One disadvantage of the central projection is that only one-half of the focal sphere can be used, as $i_o = 90^o$ projects to infinity.

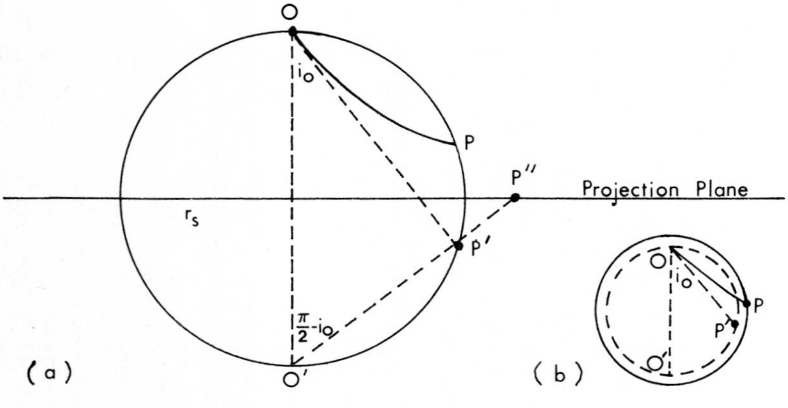

Figure 29-5(a). The extended distance projection. The event arriving at point P would arrive at P' (the extended position) in a homogeneous sphere. This then is projected stereographically to point P'' (extended distance). The radius r_s is a scaled radius for projection.

(b). Modification for deep earthquakes using a "stripped" earth.

An earlier, but much used, projection is the extended distance projection. Here, the whole earth is projected rather than the focal sphere. This limits its use to shallow earthquakes unless a "stripped" earth is used. That is, points 0, P' and O' all are set at the focal depth of the earthquake, while P remains on the earth's surface as in Figure 29-5(b). The previously mentioned tables are applicable to this case. In this representation, the two planes will generally intersect the earth's surface as two small circles. The extended distance projection will take the two small circles and normally project them into two circles on the projection plane. Except in special cases, the two small circles will not be orthogonal nor will be their projections. In the extended distance projection, near stations are close to the origin.

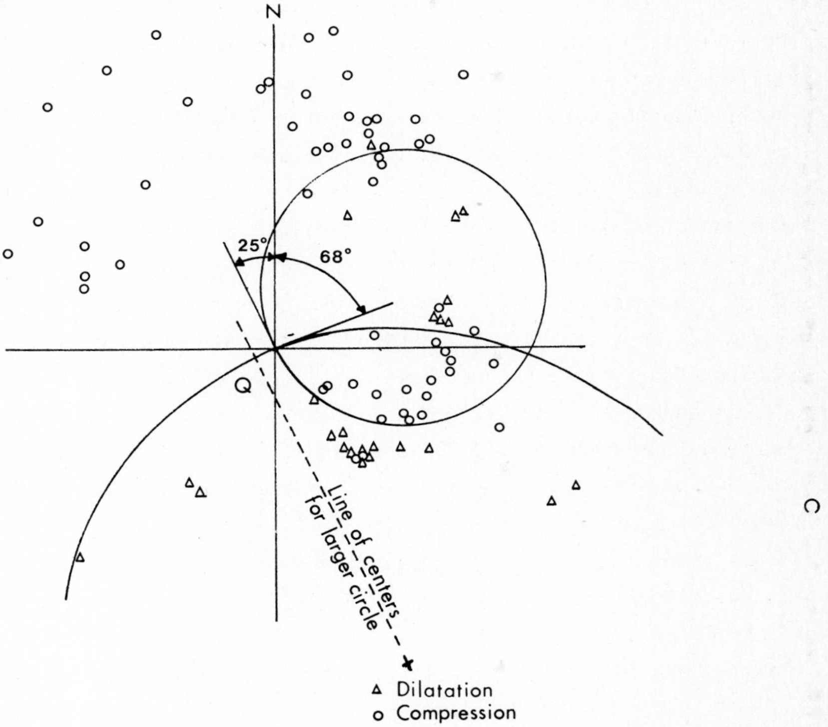

Figure 29-6. The extended distance projection of the earthquake of 10 July 1958. The circles superimposed here are the fault plane solution. Still more points are plotted here than in Figure 29-4. The point Q is described in the text. (Modified from Stauder, 1962).

After drawing the circles separating the four quadrants, one can read the strikes of the two planes as in the Figure. The radii of the two circles are a measure of the dips which are given by

$$D = \mathrm{Tan}^{-1}(d/r_s),$$

where d is the diameter of the circle and r_s is the radius of the model sphere. The interdependence of the two circles can be shown geometrically. Let us assume that the small circle is correct. Then the line normal to the plane represented by this circle will start at the origin perpendicular to the small circle. This normal line will pierce the model sphere at a point whose projection will be Q'. Its distance from the origin will be $d' = r_s \mathrm{Cot}D$. By construction, the larger projected circle will have to pass both through the origin O and the fixed point Q. Its center will then have to lie on the perpendicular bisector of $\overline{OQ}$. This construction was due to DeBremaecker (1956).

A word now on the reliability of the solutions obtained by computer. Knopoff (1961a and b) devised a scheme of weighting each point (compression or rarefaction) according to its distance from the orthogonal nodal planes. By small increments of motion of his projected lines, he was able to obtain the half-width of the probability maximum. This, he felt, should be a measure of the reliability of the solution. His method was modified somewhat by Wickens and Hodgson (1967). They re-examined 618 earthquakes; 70 had unique solutions in which none of the parameters vary by more than 10 degrees. The use of Fisher statistics has been suggested by Whitney and Merrill (1974).

The use of amplitudes has been limited for a number of reasons: a high signal to noise ratio must be maintained in order to recover signals near the nodal planes, deep focus shocks are generally used to avoid contamination with pP, and the data processing is considerably more involved in that a number of corrections due to propagation effects must be applied before directivity patterns can be fitted to the data. Teng and Ben-Menahem (1965) and Khattri (1969) give some examples of this usage. Randall and Knopoff (1970) and Randall (1972a) have used amplitude data to look for any compensated linear doublet component in the deep earthquake source.

The problem of S is much more difficult. A rather complete discussion is given by Stauder(1962). Numerical methods were given by Stevens (1964, 1967) and Udias (1964). Dillinger et al (1972) and Pope (1972) have worked out combined methods using both P and S data.

Fault-plane solutions have been instrumental in the development of the "New Tectonics". Sykes (1967) used long-period P-phases from worldwide network records to determine the sense of motion along the fracture zones perpendicular to the oceanic ridges. He was able to show that these fracture zones belonged to a tectonic setting called a transform fault (Wilson, 1965) rather than being a simple displacement of the ridge axis — see Figure 29-7. Isacks and Molnar (1971) made a comprehensive study of the stress patterns in those portions of the lithospheric plates descending into the asthenosphere and found that tensile stresses parallel to the dip were prevalent at shallow and intermediate depths. Below 300 km the predominant stress pattern was compressive and again parallel to the dip. The transition region between these two stress regimes was marked by a strong decrease in seismicity. This decrease will be discussed again in Chapter 30.

Source Parameters from Surface-wave Spectra. If the focal region is small, it behaves much like a point source and the method of phase equalization can be used to determine source mechanism. The method of phase equalization is a correction of the surface wavetrain for dispersion, i.e., a phase deconvolution. By removing the

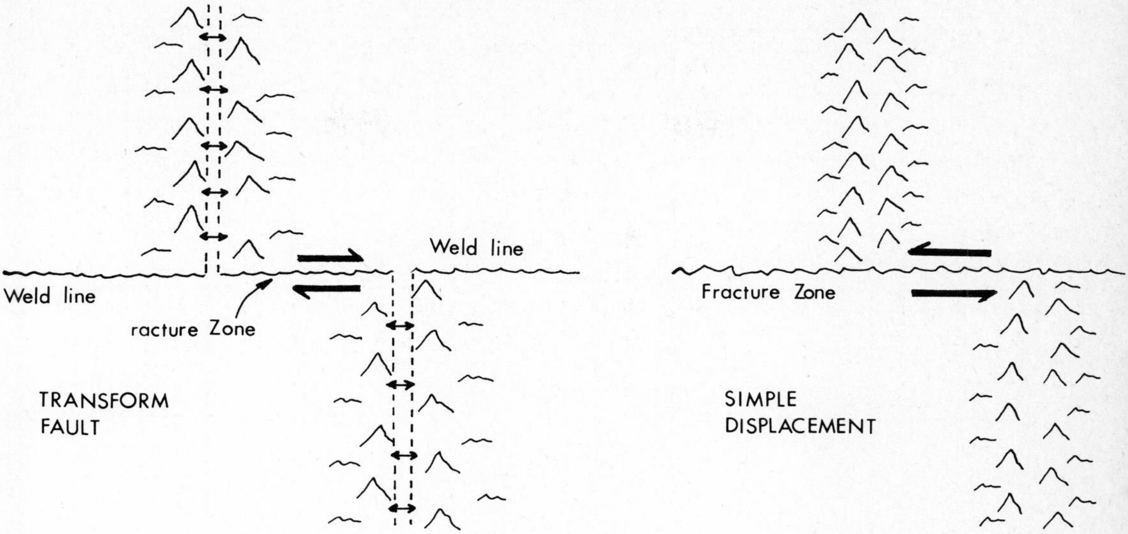

Figure 29-7. Transform faults and simple ridge-axis displacement.

dispersive effects, the wavetrain can be reconstituted at some standard distance so that directivity patterns can be observed. Sato (1956) and Aki (1960a) give alternative ways of computing the impulse at standard distance. These methods have been further extended by Aki (1960b and c) and used by Kanamori (1970a and b) to determine source directivity patterns.

Kanamori's results are shown in Figure 29-8 for the Kurile Island earthquake of 13 October 1963. The nodal lines are drawn for a 45° dip-slip source. It should be noted that phase equalization is a much more complicated procedure than reading compressions and dilatations. Consequently, this method will probably be applied only in special cases. Also, the dispersion must be known accurately requiring the compilation of regional phase velocity maps.

The theory of Section 28.3 has been applied by Press et al (1961) to determine the fault length and rupture velocity of large earthquakes. These studies were continued by Ben-Menahem and Toksöz (1962a, b, 1963) for some large earthquakes of the 1950's and 1960's. Figure 29-9(a) shows the "holes" (interferences) in the spectrum of the phases R_2 and R_3 due to finite length of rupture and finite velocity of rupture. In Figure 29-9(b), one can see the ratio of spectra for the phases R_2 and R_3. Above it is given an empirical curve corresponding to a given fault length and rupture velocity; the velocity of 3 km/sec is less than all but the shear-velocity in the uppermost crustal layers. Studies such as this give significant information to put back into the fracture process which was discussed in Section 28.5.

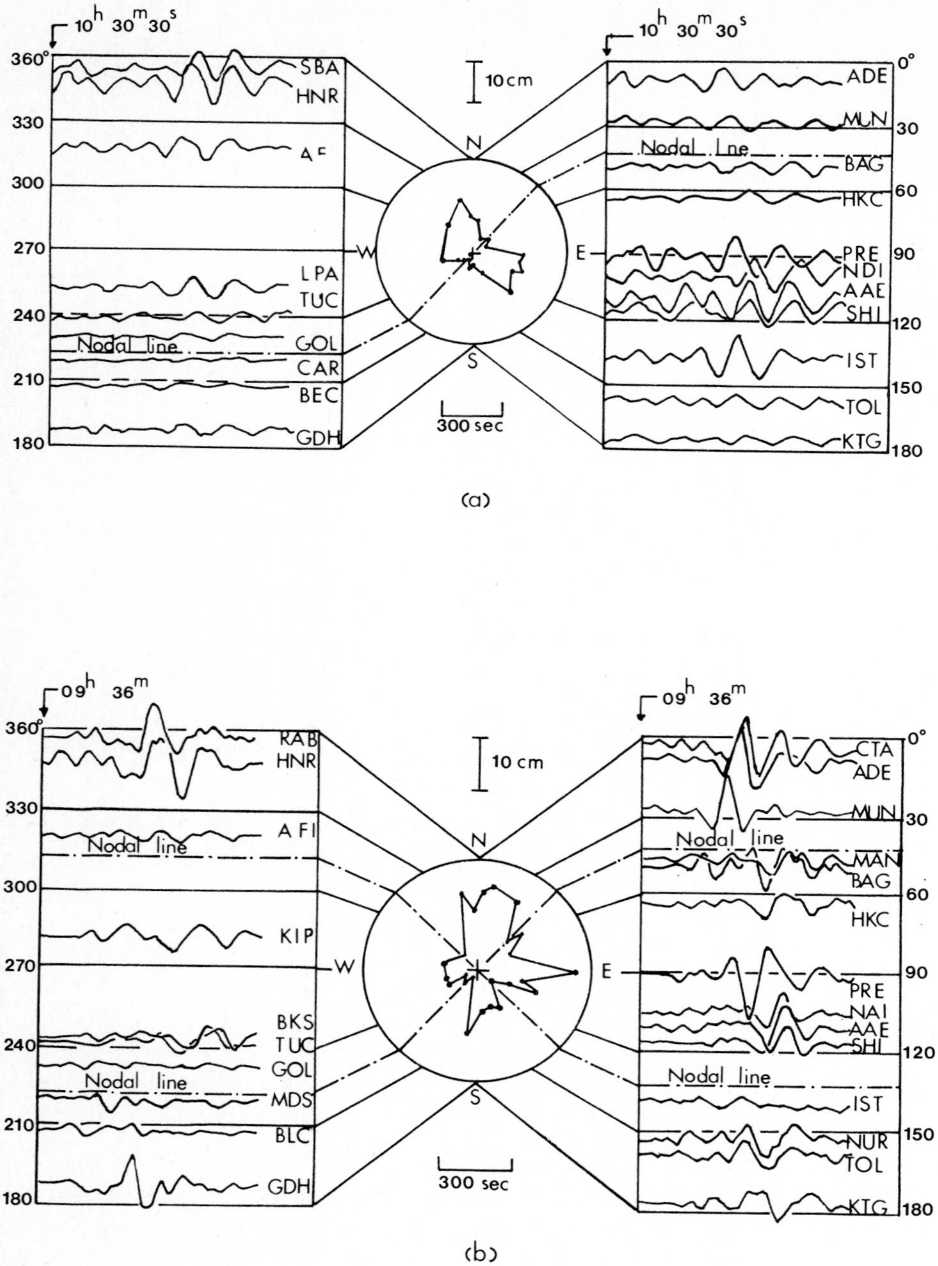

Figure 29-8. Equalized surface-waves for the Kurile Islands earthquake of 13 October 1963. All seismograms are equalized to a distance of $\pi/2$ (propagation distance of $7\pi/2$). The vertical scale gives the trace amplitude on the standard 30 - 100 long-period seismograms with a magnification of 1500.

(a). Rayleigh-waves — R_4. (b). Love-waves — G_4.

(After Kanamori, 1970a. Copyrighted by AGU.)

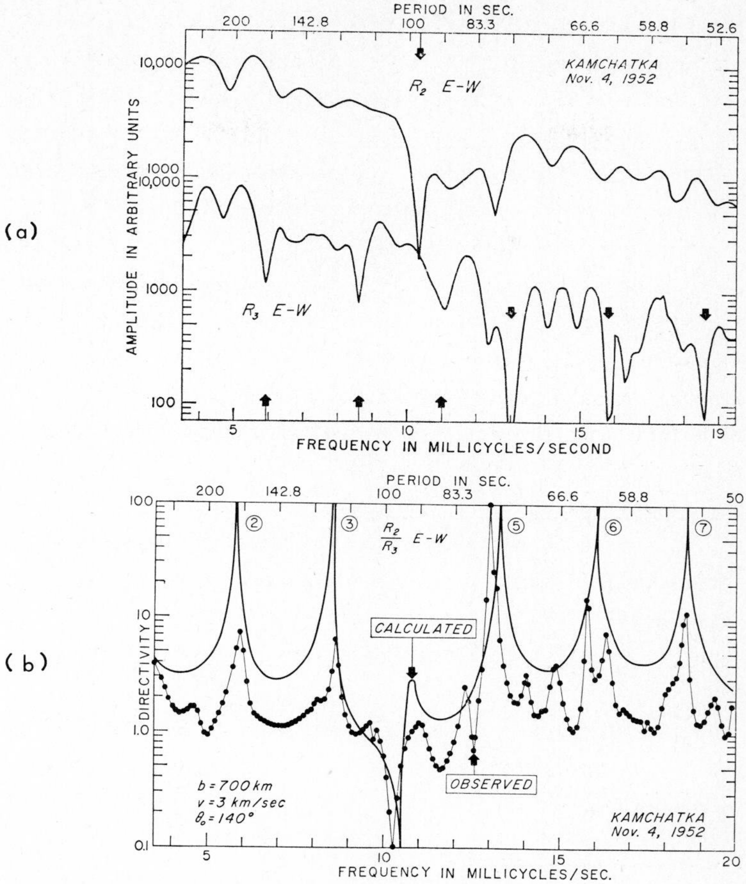

Figure 29-9. The analysis of seismic surface-waves from the 4 November 1952 Kamchatka earthquake.

(a). Amplitude spectra of mantle phases R_2 and R_3. Holes in spectra correspond to destructive interference due to rupture propagation across a finite source.

(b). Observed versus calculated directivities. (After Ben-Menahem and Toksöz, 1962b. Copyrighted by AGU.)

Canitez and Toksöz (1971) and Weidener and Aki (1973) used surface-waves to get focal mechanism as well as depth of focus. Panza (1974) has considered the use of higher modes in the determination of focal mechanism. Mendiguren (1977a) shows how to determine the seismic moment tensor components from surface-wave spectra.

<u>Source Parameters from Free Oscillations</u>. We have already mentioned the article by Ben-Menahem et al (1971) with its voluminous tables and charts. Ben-Menahem (1971) and Ben-Menahem et al (1972) applied these tables to the determination of focal parameters of the 1960 Chilean and the 1964 Alaskan earthquakes respectively. Gilbert and his colleagues (Gilbert 1971c, 1973a; Gilbert and Dziewonski, 1975; and

Gilbert and Buland, 1976) have published a series of papers giving methods designed to retrieve the seismic moment tensor from free oscillation data. The third paper of this group has applied these methods to two deep earthquakes. Some additional considerations along this line have been given by Mendiguren, (1977b).

Source Parameters from Static Stain Data. The nature of the fault plane may also be determined from near-field residual deformations and strain data. Chinnery (1961, 1963) developed the theory applicable to strike-slip faulting and has applied it to five different faults with surface breaks (Chinnery, 1964). In a later paper (Chinnery, 1965), he showed that there must be vertical displacements associated with strike-slip faulting. Using dislocation theory developed by Maruyama (1964), Savage and Hastie (1969) have worked out the parameters for a dip-slip earthquake. The above studies have been made on near-field data. Jovanovich (1975) has been able to obtain source parameters from two well documented earthquakes recorded on strain instruments up to a few hundred kilometers away.

29.3. Moment, Magnitude and Energy Release.

These quantities should be derivable from the material presented in Chapter 28 and in the previous section. Particulary useful papers which gather together the necessary ideas are Randall (1973), Kanamori and Anderson (1975) and Ohnaka (1976).

Moment and Source Dimensions. We define seismic moment M_o for a shear crack as

$$M_o = \mu \cdot U_{avg} \cdot S; \tag{29-3}$$

where μ is the shear modulus and U_{avg} is the average displacement jump over the area of the crack, S. This is consistent with (8-32). We then have from (28-11) and (28-12) and for a step-function in time:

$$|W_{DC}^{(p)}| = \frac{M_o}{4\pi\rho v_p^2} \frac{\mathrm{Sin}2\theta\mathrm{Cos}\phi}{r}; \tag{29-4}$$

i.e.,

$$M_o = \frac{4\pi\rho v_p^2 |W_{DC}^{(p)}| r}{\mathrm{Sin}2\theta\mathrm{Cos}\phi}. \tag{29-5}$$

As long as there is a permanent offset U_{avg}, the strain history is not important as one takes the zero-frequency limit of the double couple term due to such an offset. Traditionally, this has been written in the form

$$M_o^{(p)} = \frac{\Omega_o^{(p)}}{D^{(p)}(\theta,\phi)} 4\pi\rho v_p^3 r^\dagger; \tag{29-6}$$

where

$\Omega_o^{(p)}$ = long-period spectral level of the P-waves, (corrected for instrumental effects)

$D^{(p)}(\theta,\phi)$ = directivity pattern for P-waves.

ρ = density,

$r^\dagger$ = a corrected distance representing the geometrical divergence in a spherical earth,

v_p = compressional wave-velocity.

Randall (1971c) has generalized the above to the case where linear doublets are considered as well as double couples to get what he calls a shear invariant. The shear invariant reduces to M_o/μ when only plane shear is present.

Now the static problem of the elastic dislocation is considerably easier to solve than the kinematic or dynamic problem. Two solutions of interest are given by Knopoff (1958b) and Keilis-Borok (1959). The former considered the problem of a surface crack in a half-space while the latter considered the problem of a buried circular crack. We will look at the surface crack in more detail. The method of solution is relatively complex, so we will start with results given by Knopoff. The geometry is given in Figure 29-10. This is a useful approximation to that of strike-slip motion along an infinitely long vertical surface fault; however, end effects are neglected. See also Figure 28-8.

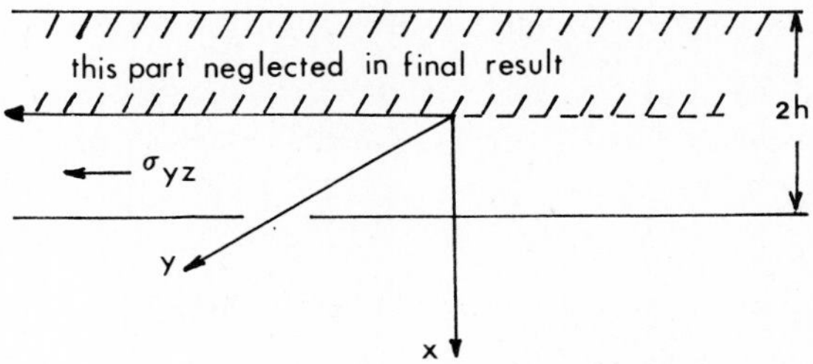

Figure 29-10. Geometrical relations for the infinite surface crack in a plane shear field.

Knopoff finds that

$$U_z \Big|_{x=0} = \pm A_o(y^2 + h^2)^{\frac{1}{2}} \xrightarrow[|y| \to \infty]{} A_o y, \qquad (29\text{-}7)$$

and

$$U_z \Big|_{y=0} = 0 \qquad |x| > h$$
$$= A_o(h^2 - x^2)^{\frac{1}{2}} \qquad |x| \le h; \qquad (29\text{-}8)$$

where

$$A_o = \sigma_{yz}/\mu. \qquad (29\text{-}9)$$

The first expression (29-7) can be used to find the half-width of the fault. One computes the displacement relative to the uncracked state as

$$\delta U_z = A_o\{(y^2 + h^2)^{\frac{1}{2}} - y\}. \qquad (29\text{-}10)$$

This reaches its half-amplitude point when

$$y_{\frac{1}{2}} = 3h/4. \qquad (29\text{-}11)$$

The geometry of relative displacement in the y-z plane is shown in Figure 29-11.

The maximum displacement is found in the face of the crack at x = 0, and is given by

$$U_{max} = 2A_o h. \qquad (29\text{-}12)$$

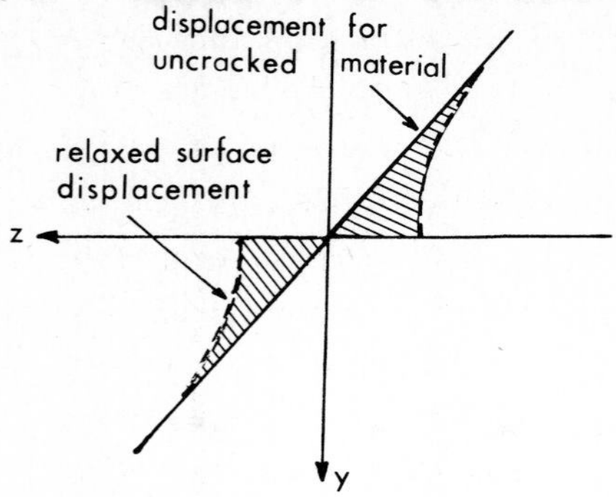

Figure 29-11. Displacements for the linear crack.

One can use expression (29-8) to find the average displacement on one side of the crack. Doubling this to obtain the total jump in displacement, one finds that

$$U_{avg} = \pi U_{max}/4. \tag{29-13}$$

From (29-3) then, the moment for a finite length L of such a crack would be calculated as

$$M_o = \mu \frac{\pi}{4} U_{max} hL. \tag{29-14}$$

Since the shear stress on the crack faces is zero, the <u>stress drop</u> is equal to the original shear stress σ_{yz}, i.e.,

$$\Delta\sigma = \mu A_o = \mu U_{max}/(2h) = 2\mu U_{max}/(\pi h). \tag{29-15}$$

One can eliminate U_{max} from (29-14) and (29-15) to obtain the static stress drop in terms of the moment:

$$\Delta\sigma = 2M_o/(\pi h^2 L). \tag{29-16}$$

Now equation (29-15) is of a general form

$$\Delta\sigma = C\mu(U_{avg}/\ell_c); \tag{29-17}$$

where U_{avg} is the average displacement and ℓ_c is a characteristic dimension. In the linear crack problem $C = 2/\pi$, $U_c = U_{avg}$, and ℓ_c equals the depth of the crack. If the crack were not to pierce the surface, then one could write a similar formula with $C = 4/\pi$ and ℓ_c equal to the full width of the crack. This is another instance of the free surface modifying the effects of an isolated singularity. In the case of the circular crack studied by Keilis-Borok (1959), one has

$$\Delta\sigma = 7\pi\mu U_{avg}/(16a), \tag{29-18}$$

$$M_o = \mu\pi a^2 U_{avg}, \tag{29-19}$$

and

$$\Delta\sigma = 7M_o/(16a^3). \tag{29-20}$$

Kanamori and Anderson (1975), in an extensive study of large earthquakes, have used (29-20) to examine the relation between moment and fault area. Now the area S of a circular crack is given by πa^2, so (29-20) can be rewitten as

$$M_o = \left(\frac{16\Delta\sigma}{7\pi^{3/2}} \right) S^{3/2}. \qquad (29\text{-}21)$$

The area of the fault plane can be estimated by either the aftershock area or from the methods of Ben-Menahem and Toksöz (pp. 375ff.) using directivity patterns. The results of Kanamori and Anderson are shown in Figure 29-12. This diagram shows a remarkable linearity between log (Moment) and log (Fault Area) indicating a constant stress drop for large earthquakes. Inter-plate earthquakes have stress drops of about 30 bars while intra-plate earthquakes have stress drops nearer 100 bars. Inter-plate earthquakes take place along the jointure between two moving lithospheric plates while the intra-plate earthquakes occur well away from the boundaries. We shall see shortly that for smaller earthquakes the stress drop increases.

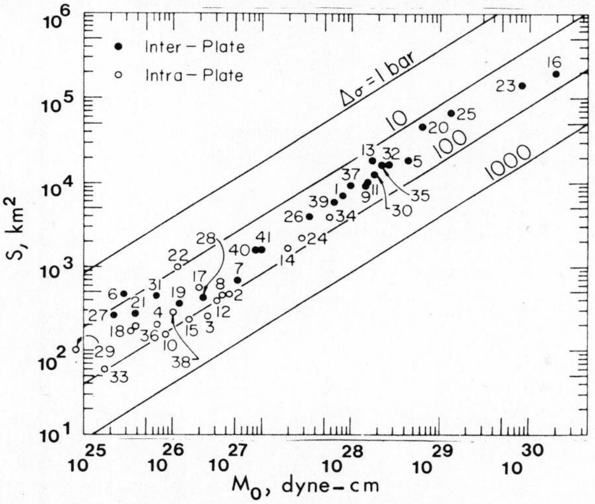

Figure 29-12. Relation between S (fault surface area) and M_o (seismic moment). The straight lines give the relation for circular cracks with constant Δσ (stress drop). The numbers with each event correspond to a table in the original paper. (After Kanamori and Anderson, 1975.)

By considering a number of kinematic models, several investigators have been able to relate a characteristic source dimension ℓ_c, to the spectra of the far-field seismic radiation. The most generally used model has been that of Brune (1970). By considering the effective stress available to accelerate two sides of the fault, he obtained a spectrum as in Figure 29-13. Here $\Omega_o^{(s)}$ is the low-frequency level related to the moment by (29-6), $f_o^{(s)}$ is a <u>corner frequency</u> given by

$$f_o^{(s)} = \frac{2.34 v_s}{2\pi a_o^{(s)}}, \qquad (29\text{-}22)$$

and the response has high-frequency <u>spectral roll-off</u> of ω^{-2}. These results are for the case where the stress drop is 100%. If the stress drop is less, there are modifications and the reader is referred to the original paper. Turning (29-22) around, we can find the characteristic dimension $a_o^{(s)}$ from the corner frequency by

$$a_o^{(s)} = \frac{2.34 v_s}{2\pi f_o^{(s)}} . \qquad (29\text{-}23)$$

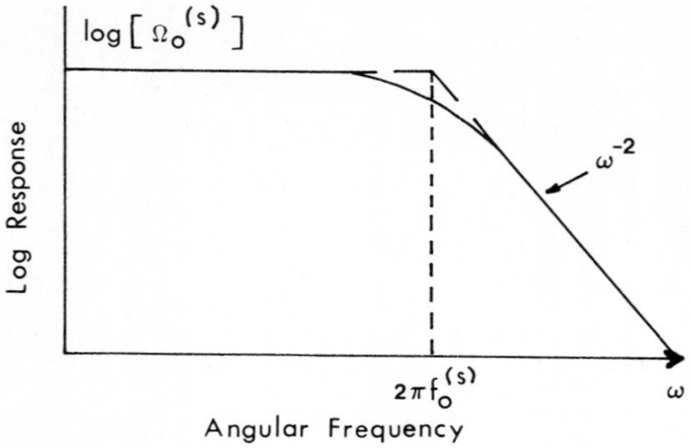

Figure 29-13. Idealized S-wave frequency spectrum for the circular dislocation model of Brune (1970).

Brune's solution was for S-waves, but a corresponding one for P-waves was not obtained. We have noted in a number of instances that S-wave data are hard to work with. Needing a corresponding result for P-waves, Hanks and Wyss (1972) generalized the results and wrote

$$a_o^{(p)} = \frac{2.34 v_p}{2\pi f_o^{(p)}} . \qquad (29\text{-}24)$$

It can be seen that in this generalization, $f_o^{(p)}$ will be greater than $f_o^{(s)}$ by the ratio v_p/v_s. This theoretically unjustified step caused a great deal of discussion in the literature. We shall come back to this point. Hanks and Wyss (ibid) worked out parameters for three large earthquakes. The data they used for one of the earthquakes is shown in Figure 29-14. The characteristic dimensions determined from (29-24) agreed well with surface faulting observed in the three cases.

Hanks and Thatcher (1972) worked up a number of earthquakes from the Gulf of California determing Ω_o and f_o; and calculating the corresponding moments, source dimensions, and stress drops. Their data is presented in Figure 29-15. Here the scatter is considerably greater and is generally attributed to an oversimplicity of the Brune model and consequent inaccuracies in the determination of a_o. The reported

OPPOSITE PAGE

Figure 29-14(a). Long-period P-waves. Turkey earthquake of 22 July 1967. Signals are presented at their proper azimuth. The fault plane solution is by McKenzie (personal communication).

(b). Corresponding P-wave spectra. (After Hanks and Wyss, 1972.)

(a)

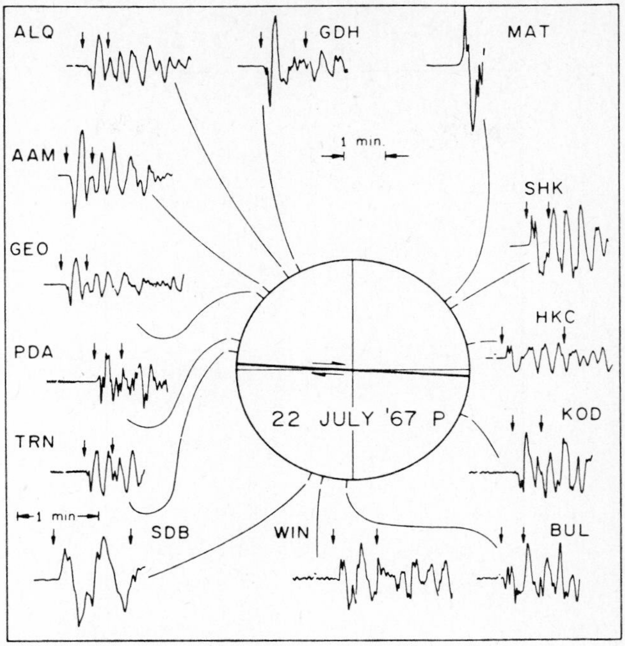
ALQ
GDH
MAT
AAM
1 min.
SHK
GEO
HKC
PDA
KOD
22 JULY '67 P
TRN
1 min
SDB
WIN
BUL

(b)

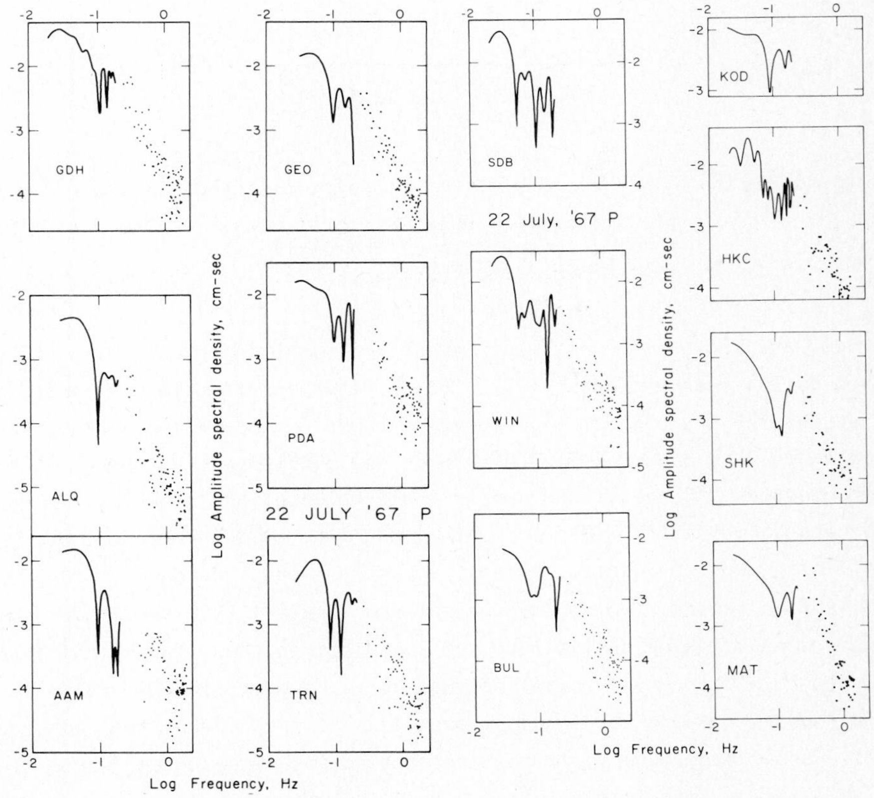
GDH
GEO
SDB
KOD
HKC
22 July, '67 P
ALQ
PDA
WIN
SHK
22 JULY '67 P
AAM
TRN
BUL
MAT
Log Amplitude spectral density, cm-sec
Log Frequency, Hz

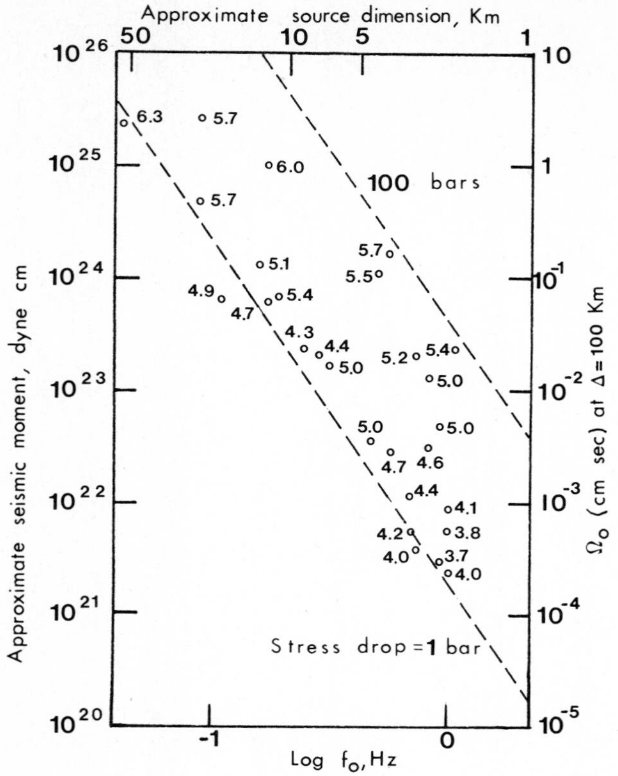

Figure 29-15. $\Omega_o - f_o$ representation of seismic sources in the northern Gulf of California — Baja California region. The number represents the local magnitude of the earthquake. (After Hanks and Thatcher, 1972. Copyrighted by AGU.)

magnitudes are shown along with each event. A much larger collection of $\Omega_o - f_o$ data is given by Thatcher and Hanks (1973) for Southern California earthquakes.

Besides the Brune model used above, a number of other models have been proposed. Haskell (1966) and Savage (1972) have proposed models leading to an ω^{-3} roll-off at high frequencies. Observations supporting this model are given by Evernden (1977b) and Frazier and North (1978). Haskell (1964b) and Aki (1967) propose other models with a ω^{-2} roll-off. Most evidence, however, supports the ω^{-2} dependence at high frequencies. In this regard, two studies by Aki and Chouet (1975) and Chouet et al (1978), involving sophisticated analysis of the seismic coda, have found that this roll-off can be determined from very small earthquakes. However, they found that the stress drop increases from one bar to about one kbar as the size of the earthquake increases. About magnitude 5, it then returns to the 10 - 100 bar range of Figure 29-12. A hint of this is given in Figure 29-15 where the smaller events hover around the one bar line for stress-drop. The reason for this sophisticated treatment of the data is that small events give signals that are of much higher frequency content and thus are subject to attenuation and scattering.

The problem of the corner frequencies is much more difficult to solve. Kinematical models proposed by Dahlen (1974), Savage (1974), and Burridge (1975) show that $f_o^{(p)}$ should be less than $f_o^{(s)}$ unless rupture takes place at speeds greater than v_s. Observations supporting this relation have been made by Bakun et al (1976). Madariaga (1976) has constructed a dynamic model in which $f_o^{(p)} \approx 1.5\ f_o^{(s)}$, but also varies as a function of azimuth. Molnar et al (1973) and Wyss and Shamey (1975) as well as the previously cited papers by Hanks and Wyss (1972), Hanks and Thatcher (1972), and Thatcher and Hanks (1973) all found that $f_o^{(p)} > f_o^{(s)}$. The paper by Wyss and Shamey (ibid) is particularly relevant in that the authors considered only deep earthquakes where possible free surface effects would not occur.

Now that we have considered the quantities, moment, stress drop, corner frequency, and rate of spectral roll-off, we have to take stock. Of these quantities, moment seems to be the best defined since it is a low-frequency limit and the static model is tractable in a number of cases. The other three quantities seem to be quite dependent upon the particular model. Knopoff and Mouton (1975) go so far as to ask the question: "Can one determine seismic focal parameters from the far-field radiation?" Their answer is not very optimistic based upon detailed numerical studies. Besides effects related to individual models of fracture, the interaction with the free surface has also to be considered. Two recent papers dealing with this problem are Boore and Dunbar (1977) and Langston (1978).

The rectangular dislocation seems to be one of the most easily treated kinematic models which is also realistic. This model was developed by Haskell (1964b), and extended by Haskell and a number of others (Haskell, 1969; Thomson and Haskell, 1972; Haskell and Thompson, 1972; Boatwright and Boore, 1975; and Madariaga, 1978). Haskell's model was the starting point for the spectral studies of Haskell (1966) and Aki (1967) mentioned previously.

Magnitude. Although each of the quantities discussed above may be considered as a measure of the size of an earthquake based upon one or more characteristics of the source region, its size has been traditionally determined in other ways. The earliest measure was the seismic intensity "I". This was a quantified description of the effects of an earthquake in a given locality. Although the maximum intensity is a useful number in designing structures, it depends strongly upon both the distance to and depth of the ruptured zone at the focus. Details of the intensity scale can be found in Richter (1958, Chapter 11). Another measure was needed; one that would reflect conditions at the source. This was the magnitude scale developed by Richter. This scale is based upon signal amplitude on the seismic record itself, corrected for distance and depth of source effects. The evolution of this scale, which started as a local measure of Southern California earthquakes, will be found in Richter (1958,

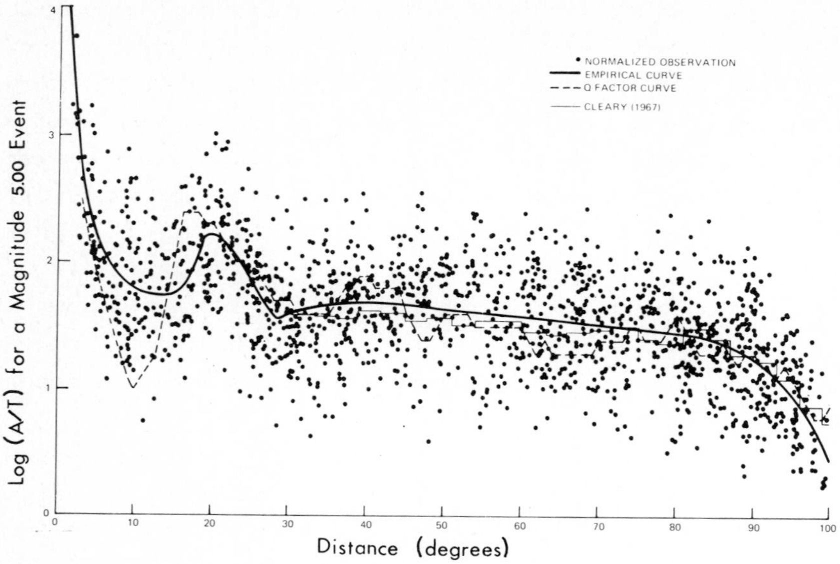

Figure 29-16. Surface empirical and Q factor amplitude-distance curves with normalized amplitude data. A given in millimicrons. (After Veith and Clawson, 1972.)

Chapter 22). Additional details may be found in Gutenberg and Richter (1956) and in Båth (1966).

Figure 29-16 shows the kind of data that one may expect in plotting values of $\log_{10}(A/T)$ for short-period body-waves from many events normalized to give a common amplitude level. There is considerable scatter in the data and an attempt to get a magnitude-distance relation from this is beset with many problems. For body-waves, magnitude is defined by

$$m_b = \log_{10}(A/T) + f_1(\Delta,h) + C, \qquad (29\text{-}25)$$

where A is the trace amplitude at the period T of maximum amplitude, $f_1(\Delta,h)$ is an empirically determined function of epicentral distance and focal depth, and C is a station constant. The function $f_1(\Delta,h)$ can also be computed for a theoretical model of the earth and compared with observation. The "Q Factor" curve of Figure 29-16 was developed by Gutenberg and Richter (ibid) and is essentially the negative of $f_1(\Delta,h)$. Another line in the figure represents results of Cleary (1967).

Veith and Clawson (1972) have developed a new set of correction factors which they call "P Factors". These new terms were derived from the analysis of some 2400 short-period P-wave amplitudes from 43 large explosions at 19 different sites. The Q Factors of Gutenberg and Richter (ibid) were derived from earthquake data. The P Factors give a standard deviation of 0.356 for magnitude, while the Q Factors give a standard deviation of 0.380. Although the reduction is not much, Veith and

Clawson state that it is statistically significant at the 95% level. It is pretty obvious that much of the remaining scatter is of a regional nature and that much more data will have to be gathered before we can write a correction factor as $f(\Delta;\phi_s,\lambda_s;\phi_r,\lambda_r)$, i.e., include the latitude and longitude of source and receiver. Veith and Clawson present their results graphically in Figure 29-17 and give a table of values by one-degree increments for various source depths. Similar curves can be found in Richter (1958, Appendix VIII) including magnitude factors for S- and PP-phases. Using the P Factors, one can write

$$m_b = \log_{10}(A/T) + P(\Delta,h) + C, \tag{29-26}$$

where A is the ground amplitude in millimicrons and T is the period in seconds.

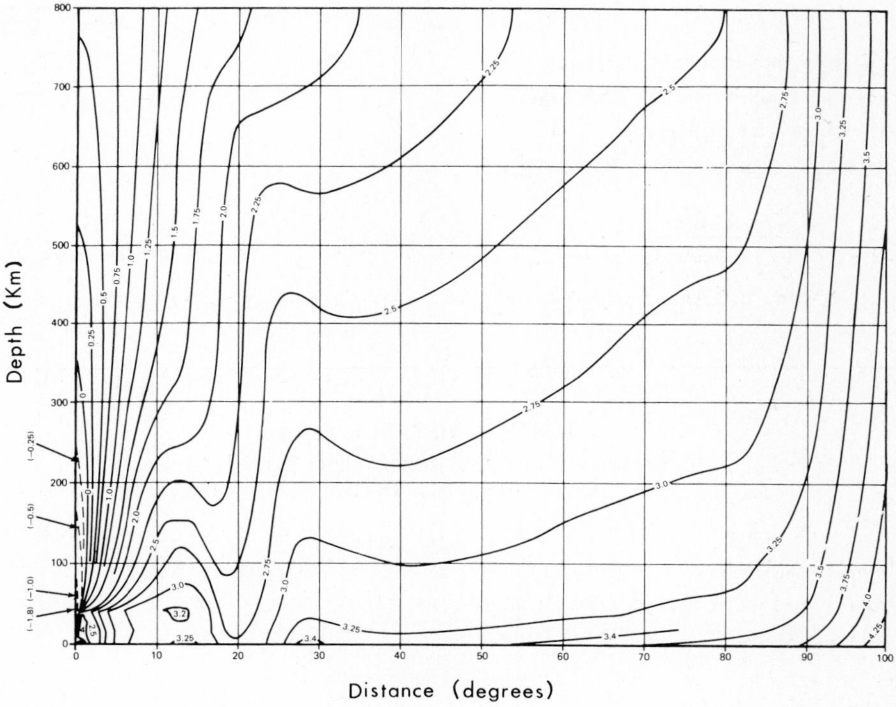

Figure 29-17. P Factors as a function of distance and depth. (After Veith and Clawson, 1972.)

In some other recent studies, McMechan and Workman (1974) have found that P-wave amplitude behavior between 10° and 35° is controlled primarily by the upper mantle structure. Booth et al (1974) and Sengupta and Toksöz (1977) found that long-period P and short-period P have somewhat different curves. The latter paper used records from deep-focus earthquakes to eliminate near-surface effects. Corrections for source and receiver heterogeneity were also applied. Evernden and Clark (1970) and Kaila and Sarkar (1975) have found that the correction factor curves seem to have

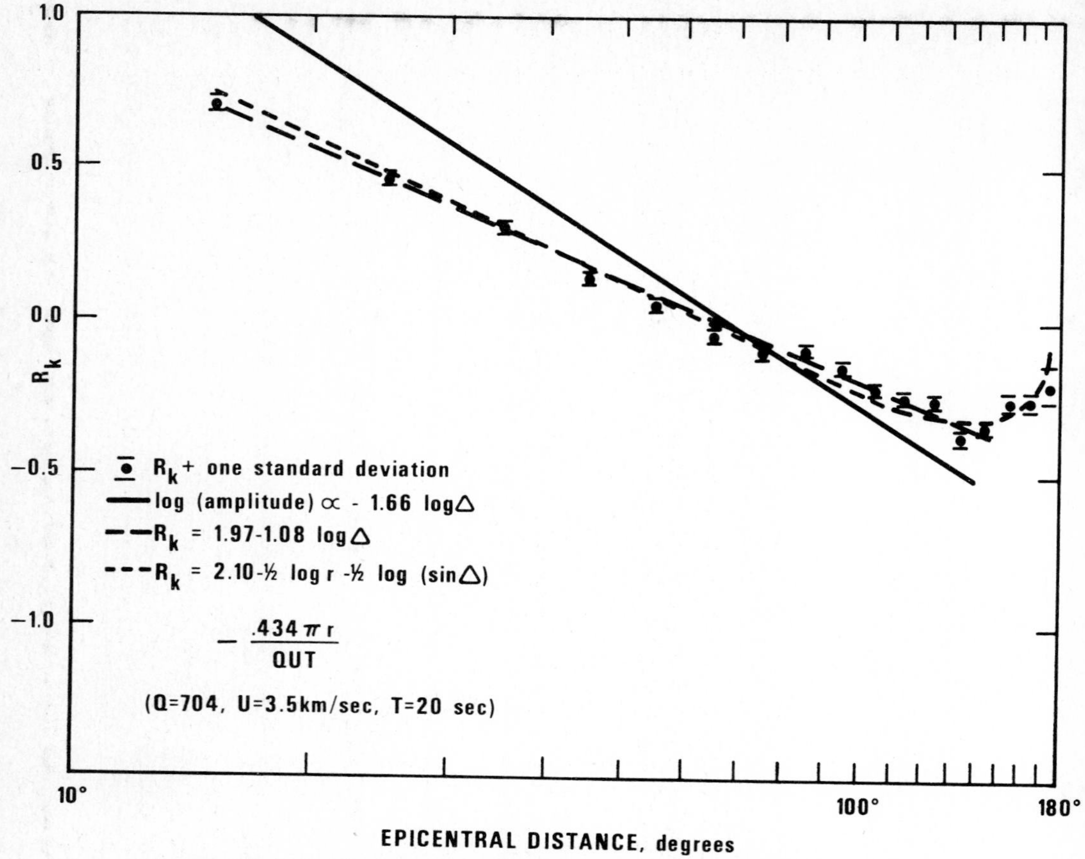

Figure 29-18. Distance effects for 20-sec Rayleigh-waves. (After Von Seggern, 1977.)

discontinuous segments. The curves in all these works vary slightly from each other. But, with all the regional corrections necessary, it may be some time before a standard curve is adopted.

Similar data can be used to determine magnitude from Rayleigh-waves. Love-waves are difficult to use because there is interference between Love and Rayleigh and between Love and Love higher modes. Here we can write

$$M = M_s = \log_{10} A_H + f_2(\Delta, h) + C \ , \qquad (29\text{-}27)$$

where A_H is the maximum zero-to-peak amplitude of the horizontal component of the fundamental-mode Rayleigh-wave having a period of about 20 sec, and $f_2(\Delta,h)$ is a correction factor. The effect of <u>h</u> is not too well known, and generally only surface sources (<50 km) are considered. It should be noted that deep-focus earthquakes produce small surface waves.

Figure 29-18 shows Rayleigh-wave amplitude data reduced from over 4000 observations of almost 700 events recorded on stations of the HGLP network (Sec. 33.5). Von Seggern (1977) finds that the reduction in amplitude with distance is considerably less than that observed in defining the currently used surface-wave magnitude formula (Gutenberg, 1945). Von Seggern's data were taken from vertical-component instruments, but he does not feel that this should be an important factor in the differences between the two studies. According to his study, a proposed formula for surface-wave magnitude is given by

$$M_s = \log_{10}A + 1.08 \log_{10}\Delta - 0.22; \qquad (29\text{-}27a)$$

where A is the amplitude in millimicrons peak-to-trough and Δ is the epicentral distance in degrees. As can be seen from the figure, this is quite different from the slope of -1.66 commonly used. A somewhat more complex formula based upon the theory of surface-waves traveling over an attenuating spherical earth is given in the figure. This expression takes care of the antipodal focusing of surface-waves. Similar results are obtained in another extensive study by Thomas et al (1978). They find the $\log_{10}\Delta$ coefficient to be 1.15.

Having now defined the magnitude scales for body-waves and surface-waves, it would seem that there should be no more problems. However, it was noted that magnitudes given by the two formulations were not the same. Gutenberg and Richter (1956) found that

$$M_s = 1.59\, m_b - 3.97, \qquad (29\text{-}28)$$

with equality at 6.75. A number of papers (Aki, 1967; Evernden, 1975; Geller, 1976b) have attempted to explain this relation on the basis of a scaling law for earthquake spectra. The scaling law was a combination of theoretical and empirical data leading to a calculable seismic spectrum for an earthquake of a given size. One such law was that of Brune's (1970) which we have discussed in some detail (p.381). Aki (1967) looked at both ω^{-2} and ω^{-3} scaling laws to find which was the better at explaining the observations. After an extensive study of magnitude data, Evernden (1975) found that the M_s:m_b curve was non-linear — see Figure 29-19. He proposed that this non-linear curve might best be fit by an ω^{-3} scaling law. With the spectral curves as drawn by Evernden, m_b (measured at one second period) saturates at something less than 7 relative to the increasing values of M_s (measured at 20 sec period). Geller (1976b) plots the scaling law as a function of moment, rather than magnitude, and found a saturation of M_s at something greater than 8 with increasing moment. We shall have more to say about this later. Having noted this saturation, Brune and Engen (1969) had earlier proposed that a "Mantle Wave Magnitude" be taken at 100 sec period.

A number of other schemes for determining magnitude have been proposed, mostly dealing with the way in which the amplitude of body- or surface-waves was to be determined, e.g., over a spectral band rather than at one frequency. Howell (1972) discusses some of these methods. A number of authors have proposed the use of the length of the seismic signal as a measure of magnitude. Two recent papers in this area are Real and Teng (1973) and Hermann (1975). The use of the LG phase to compute magnitude has been suggested by Baker (1970), Street et al (1975) and Street (1976).

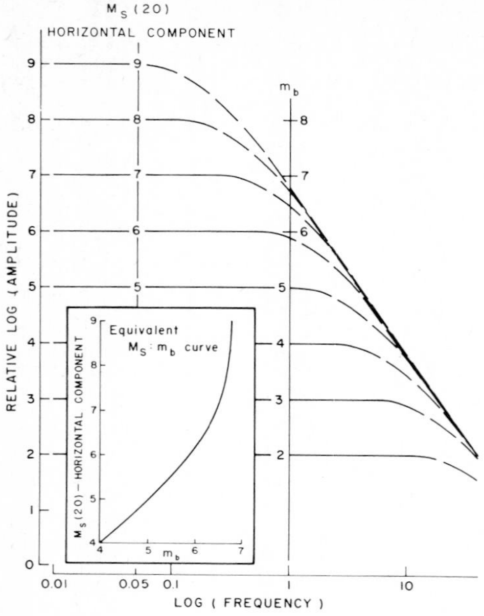

Figure 29-19. Scaling law of earthquake spectra proposed by Everdern. Measuring m_b at 1 second period gives the $M_s:m_b$ relation in the inset figure. (After Evernden, 1975.)

It really is amazing that magnitude calculations should work at all. The amplitudes of the various seismic phases are affected by source type, source orientation, source depth, and velocity structure in both source and observing regions. The logarithmic nature of the magnitude scale smooths out some of the variation; a large number of observations also helps. With a large data set, statistical considerations have to be applied. Three papers relating to the statistical estimation of seismic magnitude are Freedman (1967c), von Seggern (1973) and Ringdal (1976).

Magnitude vs. Moment. The magnitude of an earthquake has been defined as the log of the record amplitude at a certain characteristic period (1 sec or so for body-waves and 20 sec for surface-waves). Depending upon the scaling law associated with a particular model, one can calculate the signal amplitude at a particular period. After some extensive calculations based upon Haskell's (1964b) model, Kanamori and Anderson (1975) found that:

$$M_s \sim \log M_o, \qquad \text{small earthquakes, short risetimes}$$

$$M_s \sim \frac{2}{3} \log M_o, \qquad \text{most large earthquakes } (6<M_s<8) \tag{29-29}$$

$$M_s \sim \frac{1}{3} \log M_o, \qquad \text{great earthquakes, long risetimes.}$$

These variations are due to the positions of the corner frequencies relative to 20 sec. The corner frequencies reflect two times characteristic of the dislocation, the rise time and the rupture duration. Data supporting these calculations are given in Figure 29-20. Events (16) and (23), Chile 1960 and Alaska 1964 respectively, are clearly off the trend indicating the saturation of M_s with increasing moment mentioned previously.

Combining (29-21), which has $M_o \sim S^{3/2}$, with the middle relation of (29-29) gives

$$M_s \sim \frac{2}{3} \log S^{3/2} \sim \log S, \qquad (29\text{-}30)$$

a relation which had been noted by Utsu and Seki (1954).

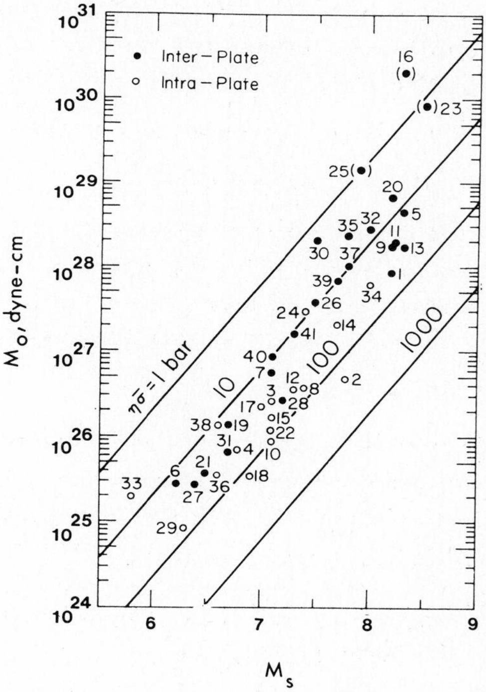

Figure 29-20. Relation between M_s (20 sec surface-wave magnitude and M_o. The straight lines are for constant apparent stress. $\mu = 3\times10^{11}$ dynes/cm^2. (After Kanamori and Anderson, 1975.)

<u>Magnitude</u> <u>vs</u>. <u>Energy</u>. If we return to the static models of fracture, some simple considerations can give estimates of the energy available to go into seismic waves. For an element of surface along the crack, the energy released can be equated to the work necessary to return the crack to its original state before rupture. This energy is given as

$$\delta W = U\sigma_{avg} = \tfrac{1}{2}U(\sigma_o + \sigma_1),$$

where U is the relative displacement between the two sides of the crack, σ_o is the initial shear stress, and σ_1 is the final shear stress along the crack. Since σ_o and σ_1 are the same over the whole surface of the crack, the total energy available would be

$$\Delta W = U_{avg} \cdot \sigma_{avg} \cdot S.$$

That available for seismic waves would be

$$E_s = \eta \Delta W = (M_o/\mu)\eta\sigma_{avg}, \qquad (29\text{-}31)$$

where η is the seismic efficiency, and we have used (29-3). The quantity $\eta\sigma_{avg}$ has been called the <u>apparent stress</u>.

Extensive studies have been carried out to determine the energy carried away in seismic surface-wave trains, and independent relations have been given by Gutenberg and Richter (1956) and Båth (1958). They are:

$$\log E_s = 1.5\, M_s + 11.8, \text{ (Gutenberg and Richter)} \qquad (29\text{-}32)$$

$$\log E_s = 1.44\, M_s + 12.24. \text{ (Båth)} \qquad (29\text{-}33)$$

Here M_s is the magnitude determined from 20 second Rayleigh-waves, and E_s is the energy over all frequencies available to the observers. These relationships were computed without the use of any scaling laws such as led to (29-29). If one substitutes (29-31) into (29-32), the resulting equation between moment and magnitude is:

$$\log M_o = 1.5\, M_s + 11.8 - \log(\eta\sigma_{avg}/\mu). \qquad (29\text{-}34)$$

The lines represented by this equation for several values of $\eta\sigma_{avg}$ are overlain on Figure 29-20. Kanamori and Anderson (1975) note that $\eta\sigma_{avg}$ for inter-plate earthquakes ~ 10-20 bars, while $\eta\sigma_{avg}$ ~ 50 bars for intra-plate earthquakes.

Comparing Figures 29-12 and 29-20 shows that

$$\eta\sigma_{avg} \sim \tfrac{1}{2}\Delta\sigma \qquad (29\text{-}35)$$

for both inter- and intra-plate earthquakes. If all energy loss in the fracture is assigned to either seismic waves or to friction along the crack, one can write:

$$\eta = (\sigma_{avg} - \sigma_f)/\sigma_{avg}\,. \qquad (29\text{-}36)$$

Rewriting (29-36), we have

$$\sigma_f - \sigma_1 = \tfrac{1}{2}(\sigma_o - \sigma_1) - \eta\sigma_{avg} = \tfrac{1}{2}\Delta\sigma - \eta\sigma_{avg}\,. \qquad (29\text{-}37)$$

If one substitutes relation (29-35) into (29-37), one finds that

$$\sigma_f \sim \sigma_1. \qquad (29\text{-}38)$$

This says that faulting stops when the stress drops to the level of the frictional stress. There may be some overshoot ($\sigma_1 < \sigma_f$) due to dynamic stress concentrations at the rupture front.

Now the <u>effective stress</u> to sustain fracture is given by

$$\sigma_{eff} = \sigma_o - \sigma_f, \qquad (29\text{-}39)$$

(Brune, 1970; Kanamori, 1972). Then if (29-35) is a realistic evaluation of the data, we have that

$$\sigma_{eff} = \sigma_o - \sigma_1 = \Delta\sigma; \qquad (29\text{-}40)$$

i.e., the <u>effective stress is approximately equal to the stress-drop</u>. Kanamori and Anderson (ibid) give a number of observational references to support (29-40). Since the stress-drop has been seen to be relatively constant for large earthquakes, then the effective stress must be also. This has allowed Kanamori and Anderson to insert a condition of dynamic similarity into the Haskell (1964b) model enabling them to integrate the square of the seismic spectrum. After integration they find that $E_s \sim L^3$. For the static model, constant stress drop (29-16) implies $M_o \sim h^2L$. Thus for faults of constant aspect ratio, h/L = constant, and (29-31) also says that $E_s \sim \eta\sigma_{avg}L^3$.

Because magnitude saturates as moment increases, m_b at 7^- and M_s at 8^+, there is a need for a new magnitude scale for very large earthquakes. Brune and Engen (1969) made a start with amplitude measurements made at 100 sec period. Noting that many investigators have observed the linear relation between moment and fault area (Figure 29-12), Kanamori (1977) makes another suggestion. First the quantity

$$W_o = \tfrac{1}{2}\Delta\sigma(M_o/\mu) \sim M_o/(2\times10^4) \qquad (29\text{-}41)$$

is computed. The quantity 2×10^4 is proposed as an empirical average value for $\Delta\sigma/(2\mu)$. Then a new magnitude, M_w, is calculated

$$\log W_o = 1.5M_w + 11.8. \qquad (29\text{-}42)$$

Since log (moment) ~ $1.5M_s$ for $M_s<8$ (Figure 29-20), this new scale fits onto the older scale given by (29-32). Kanamori (ibid) shows that W_o is a reasonable quantity to use for an energy estimate. For example, if $\sigma_1 = 0$, then $\Delta\sigma = 2\sigma_{avg}$ and W_o is the total energy release. If $\sigma_1 = \sigma_f$, then

$$\eta\sigma_{avg} - \sigma_f = \tfrac{1}{2}\Delta\sigma;$$

and $W_o = E_s$. If σ_1 is some arbitrary value between zero and σ_o, then

$$\Delta W = U_{avg}\cdot\sigma_{avg}\cdot S = W_o + U_{avg}\cdot\sigma_1\cdot S\ .$$

Thus W_o is a minimum estimate of seismic energy release.

29.4. Nuclear Detection and Identification.

The development of nuclear weapons has necessitated an extensive program of testing and refinement. Testing in the atmosphere or hydrosphere is rather easily detected. The fission products remaining after the explosion are a dangerous pollutant. The use of space for test purposes entails difficult problems in logistics and communications. This leaves underground testing as the most viable alternative and opens a connection with seismology. Nuclear explosions in alluvium, or in soft or hard rock create seismic waves of a character different from those leaving an earthquake epicenter. It is the purpose of this section to discuss the problem of the separation of earthquakes and underground nuclear explosions. Two monographs have been written recently by Bolt (1976) and by Dahlman and Israelson (1977). Neither of them is extremely technical, but intended for a fairly wide audience. Each however, has large numbers of references. Blandford (1977) gives a brief review article with a few additional references. Evernden (1977a) makes a critical evaluation of routinely available data for identifying earthquakes.

Source Characteristics of Explosions. The problem of a radially directed stress pulse inside a spherical cavity has been treated by many authors. The purely elastic case is quite straightforward and the details may be found in Eringen and Suhubi (1975 Sec. 6.8 — the cylindrical cavity is touched upon in Sec. 6.11). However, with large nuclear blasts, there is a considerable region about the source which responds in a non-elastic manner, vaporizing, fracturing, and deforming plastically. For this reason, seismologists have tried to match a theoretical function (satisfying the elastic wave equation) at the radius where the stresses begin to behave elastically. This matching requires the measurement of close-in pressures and displacements, a technical feat in itself. Let us look at the conventional treatment of the problem.

Similar to (6-56), one defines a reduced displacement potential ψ by

$$u_r(r,t) = -\frac{\partial}{\partial r}\left[\frac{\psi(t - r/v_p)}{r}\right] . \tag{29-43}$$

The radial stress from (4-14) and (2-54) is then

$$t_{rr}(r,t) = \rho v_p^2 \partial u_r(r,t)/\partial r + 2\rho(v_p^2 - 2v_s^2)u_r(r,t)/r \quad . \tag{29-44}$$

For a time harmonic source, we now have

$$U_r = -\frac{\partial}{\partial r}\left[\frac{\Psi(\omega)e^{ik_p r}}{r}\right] \longrightarrow ik_p\Psi(\omega)\ (\frac{e^{ik_p r}}{r}) .$$

On comparison with (29-10), we see that the equivalent moment of an explosive source is given by

$$M(t) = 4\pi\rho v_p^2 \psi(t). \tag{29-45}$$

For long times, $\psi(t)$ will approach a constant and the spatial derivative will not operate on $\psi(t - r/v_p)$; giving

$$u_r(r,\infty) = \psi(\infty)/r^2 , \tag{29-46}$$

$$t_{rr}(r,\infty) = -4\rho v_s^2 \psi(\infty)/r^3 . \tag{29-47}$$

A non-zero value of $\psi(\infty)$ corresponds to permanent deformation in the non-linear zone around the explosive source.

A number of investigators have used empirical models to construct reduced potential functions with free parameters to match close-in observations. Haskell (1967), Mueller and Murphy (1971), von Seggern and Blandford (1972), and Aki et al (1974) have all done considerable work in trying to fit the observed data. The initial rise of the potential function is still uncertain, theoretical considerations and observations

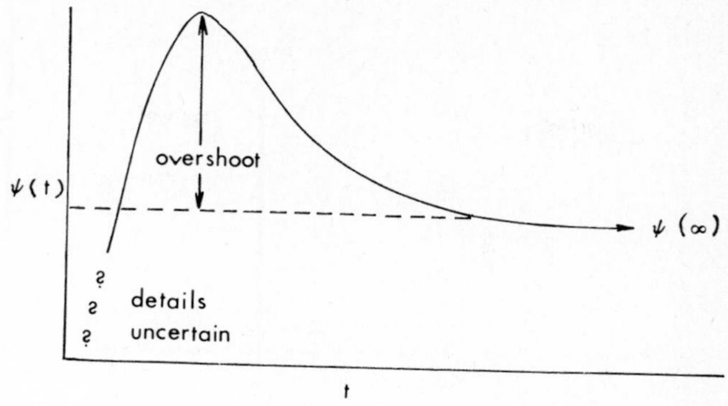

Figure 29-21. Reduced potential function used to fit observational data.

being somewhat in conflict. However, the overshoot in Figure 29-21 seems to be required.

The basic theory embodied in equations (29-43) through (29-47) immediately suggests two methods of discriminating explosions from earthquakes. A radially symetric source function implies that all observation points should see a compression as first motion. This, however, is not the case and a combination of factors have been suggested to explain why not. On page 365 we noted that source asymmetry, pre-stress in the source region, and wave interaction with fractures all play their part in the region near the source — a low signal to noise ratio makes difficult a positive decision for compression at teleseismic distances. Secondly, a source function like that in Figure 29-21 leads to a relatively simple displacement function (29-43). Consequently, discrimination using signal complexity has been seriously considered. However, conditions near the source — layering, burial depth, fracturing, spalling, and cavity collapse — contribute to a more complex signal in a number of cases. Such a discriminant has not proved too reliable.

Spectral Characteristics of the Explosive Source. When one considers that explosive sources develop extremely high stresses compared with earthquakes, it would seem that explosions should be associated with more rapid risetimes and shortened durations (the characteristic dimension should be smaller) than earthquakes of equal seismic moment. Both von Seggern and Blandford (1972) and Aki et al (1974) found that the farfield spectrum has a finite low-frequency level, a peaking at the corner frequency, and an ω^{-2} roll-off with increasing frequency. A sketch is given in Figure 29-22. The size of the high-frequency peak is related to the amount of overshoot in Figure 29-21. One final note. Even though discriminants based on signal complexity have not been too reliable, Evernden (1977b) reopens the question with some success using sliding windows to compute time-dependent spectra of the signal coda.

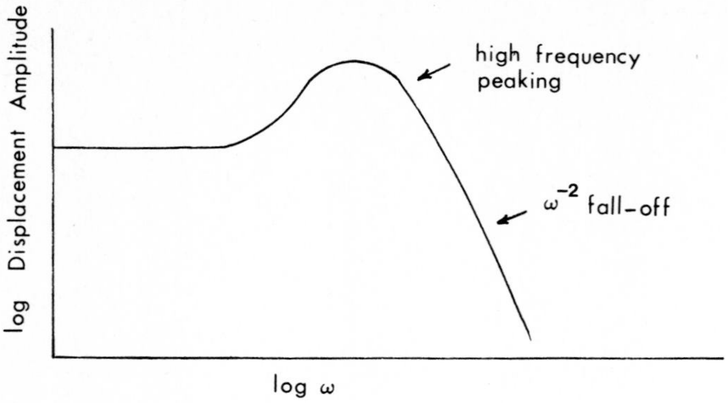

Figure 29-22. Far-field displacement spectrum.

The M_s : m_b Discriminant Because of the difference in far-field displacement spectra (compare Figure 29-22 with Figure 29-19) one reason for the success of the difference $M_s - m_b$ as an earthquake discriminant is apparent. However, there are even more reasons that it is successful. Gilbert (1973b) gives some general considerations showing that $M_s - m_b$ for earthquakes should be about 0.5 magnitude units greater than for explosions. In his derivation, size, duration, burial depth, and radiation pattern were not incorporated. Adding in some of these details, Douglas et al (1971 and 1973) have constructed models showing that $M_s - m_b \sim -1$ for explosions and $M_s - m_b \sim 0$ for earthquakes. Certain fault orientations, however, give very small surface waves at 20 sec period and reduce the value of the M_s : m_b discriminant. Other criteria can be called into play in these cases.

It must be pointed out that the spectral determinations of von Seggern and Blandford (1972) and of Aki et al (1974) depend upon the nature of the explosive source. Similar reduced displacement potentials have been incorporated in the two Douglas et al papers. Thus the general level effects deduced by Gilbert (ibid), and the relative spectral level of the two observational papers are combined in the theoretical seismograms produced by Douglas et al (1973). Many details must also be put into the calculation of earthquake spectra. Blandford (1975) considers many factors and finds that displacement spectra for earthquakes can assume many different shapes, but with an ultimate ω^{-3} roll-off at high frequencies. Figure 29-23, below, shows a typical separation of earthquakes and explosions. Evernden (1977a) shows how good this separation really is in a study of about 1000 earthquakes with $m_b \geq 4.5$.

Discrimination by Accurate Location. If an event can be located at depths greater than 30 km, in relatively deep oceans, or in a populated area, one can be quite certain that it is not an explosion. This puts a great premium upon precision methods of location.

Evasion. A number of schemes have been proposed to conceal, or at least

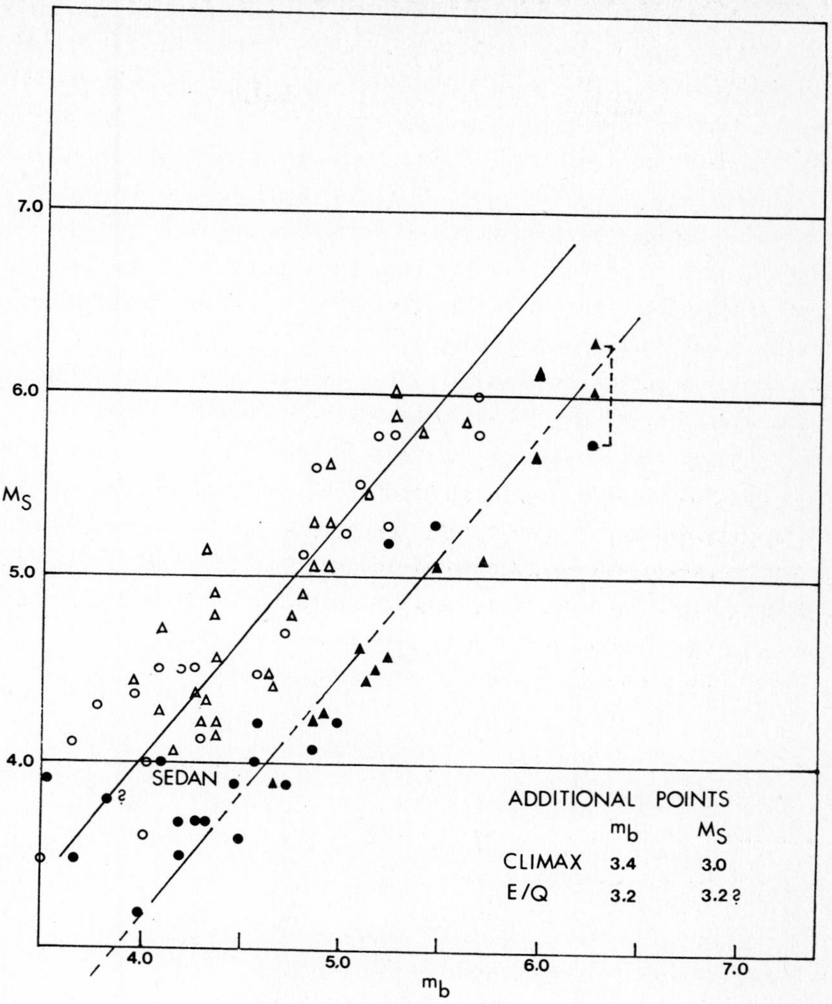

Figure 29-23. M_s vs. m_b for U.S. earthquakes and explosions. Open symbols are earthquakes, closed symbols are explosions. Circles are from Evernden et al, triangles are from Basham (1969). (After Evernden et al, 1971. Copyrighted by AGU.)

downgrade the yield of, underground nuclear explosions. In a series of three large papers, Evernden (1976a, b, c) has made a detailed study of all facets of evasion schemes, both from the point of view of the concealer and of the monitor. We can only sketch some of these proposals.

It is obvious that an explosive shot which is below noise level at all observing stations will not be detected. Consequently, there is a level at which concealment is possible. The development of seismic stations with improved capabilities (Sec 33.5) has largely been directed to this area. However, the placement of the shot in alluvium or a cavity tends to muffle the explosion. This could reduce marginally detectable events to below noise level. However with large devices, alluvium is not generally deep enough to keep the radioactive debris from venting and large cavities are extremely expensive to construct. Thus, there are practical limits to this method.

Firing several nuclear devices in a predetermined sequence can increase the low-frequency signal content and increase the signal's complexity. However, at sufficiently high frequencies, the levels are increased also, and sophisticated methods are available to break this evasion scheme.

Another promising scheme is that of hiding an explosion in an earthquake — eithe triggered by a natural event at the explosion site or with its signal below the coda level of some large earthquake at some distance. The former is logistically difficul because one is never sure when a natural earthquake will happen, and maintaining a nuclear testing site (temporary) would be expensive. The latter method of concealmen can be overcome by selective positioning of observation sites and by sophisticated signal processing. This is possible since the character of coda and primary signal do differ.

CHAPTER 30

SEISMICITY

30.1. Global Extent.

The earthquakes of the world generally occur in three belts: the Circum-Pacific belt ("Rim of Fire"), the Alpine-Himalayan Belt, and the active seismic ridges in the oceanic basins. These belts can be seen clearly in Figure 30-1. This represents earthquakes located by the (U.S.) National Ocean Survey during the year 1970. Because of a non-uniform coverage of reporting stations, somewhat smaller events are included in the northern hemisphere. Recent compilations of a longer duration have been given by Barazangi and Dorman (1969) for latitudes between 70°N and 70°S and Tarr (1970) for the polar regions. Figure 30-1 and the two studies mentioned used data reported by a modern network of instruments which were able to locate events larger than magnitude 4 in many regions.

A classic study in the distribution of earthquakes throughout the world was made by Gutenberg and Richter in 1949 and later updated (Gutenberg and Richter, 1954). A compilation of data in the same style was given by Rothé (1969) covering some 5000 earthquakes during the period 1953 - 1965. A special study of the Circum-Pacific Belt has been published by Duda (1965) listing details of 1263 earthquakes throughout the years 1897 - 1964 with magnitudes greater than 7.0. According to Duda, 76% of the seismic energy release occurs on this belt. A regional study of the European area has been given by Karnik in two volumes (1969, 1971), while government agencies put out a frequently updated publication, Earthquake History of the United States. Two particularly valuable resources are extended catalogs of seismicity in Southern California (Hileman et al, 1973) and in Northern California (Bolt and Miller, 1975). These catalogs have proved valuable in relating earthquake occurances to known faulting and in statistical studies of earthquake time sequences.

The seismic belts separate relatively aseismic regions which have been characterized (Le Pichon et al, 1973, have written an excellent text giving full details of the new Plate Tectonics) as rigid plates (lithosphere) riding over the soft low-velocity zone (asthenosphere). Names of the larger plates are superimposed upon the seismicity map of Figure 30-1. The rigid plates move away from the active seismic ridges in the ocean basins with new oceanic crust being formed behind them. Concurrent seismic activity is generally small (Mag. <7.0) and shallow. Since new material is being created at the ridges, it must be taken up elsewhere. This occurs in regions called subduction zones where one plate rides under another. Most of these are found around the rim of the Pacific Ocean, although there is much evidence that the Alpine-Himalayan belt is the expression of a collision between the African and Australian plates with the Eurasian plate. These subduction zones have associated with them almost planar (deformed slightly to match an arcuate surface expression) regions dipping deeply under the continents where most of the earthquakes take place. These planar

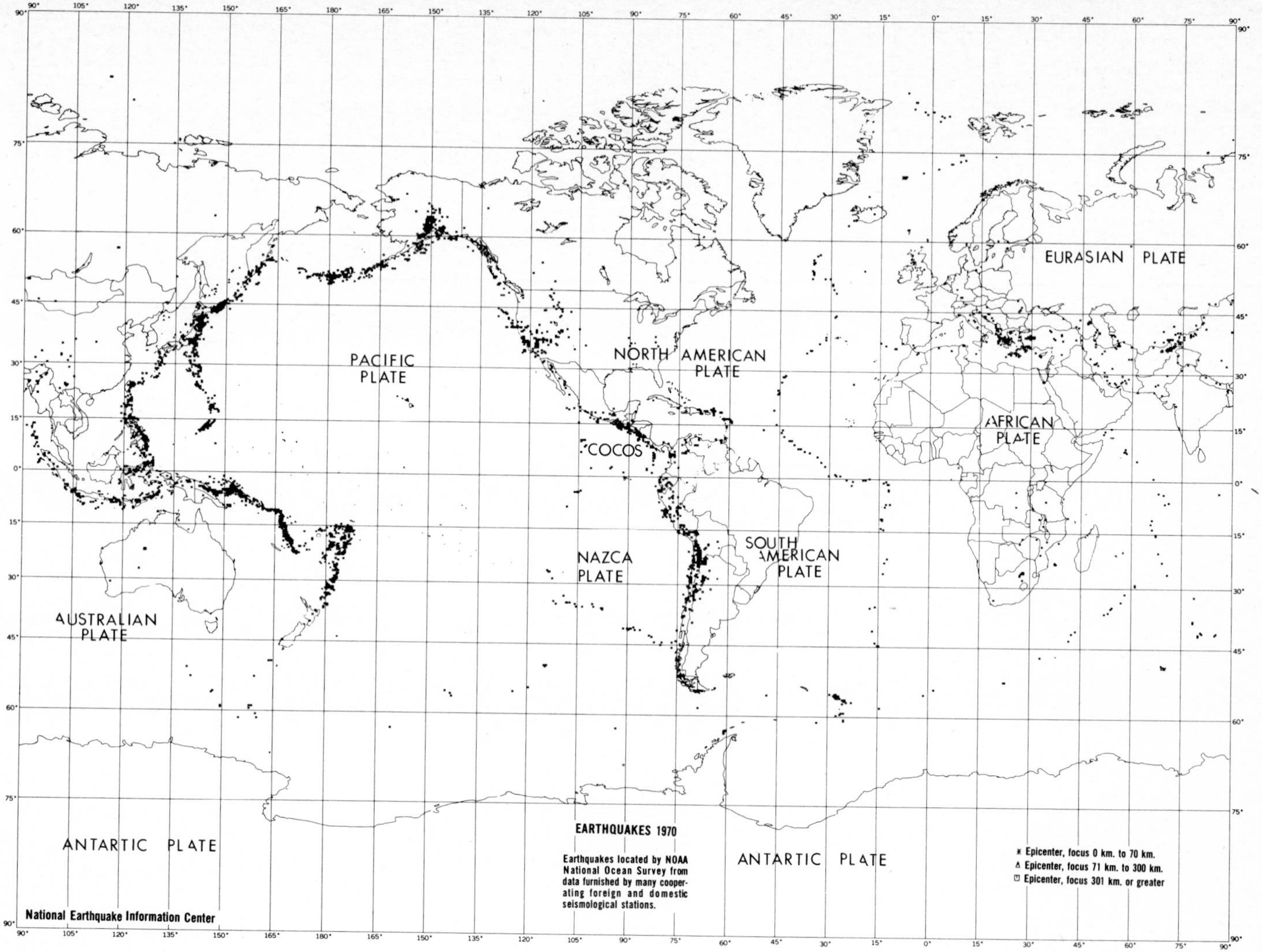
EURASIAN PLATE
NORTH AMERICAN PLATE
PACIFIC PLATE
AFRICAN PLATE
COCOS
NAZCA PLATE
SOUTH AMERICAN PLATE
AUSTRALIAN PLATE
ANTARTIC PLATE
ANTARTIC PLATE
EARTHQUAKES 1970
Earthquakes located by NOAA National Ocean Survey from data furnished by many cooperating foreign and domestic seismological stations.
* Epicenter, focus 0 km. to 70 km.
△ Epicenter, focus 71 km. to 300 km.
□ Epicenter, focus 301 km. or greater
National Earthquake Information Center

regions have been named Benioff Zones after extensive study by their namesake (see Benioff, 1954, for example). We shall return to them in the next section. Good expressions of the Benioff Zone are lacking in the Alpine-Himalayan belt, although deep earthquakes occasionally happen there. The Pacific rim is made up largely of oceanic-continental and oceanic-oceanic collisions while the Alpine-Himalayan belt is a continental-continental collision. The increased combined crustal thickness in the last case may be related to the different tectonic style. It is in these collision zones that the largest earthquakes take place, with events found at all depths to something over 700 km. A third boundary type occurs when two plates slide past each other. The best studied example is the San Andreas Fault where the Pacific plate moves past the North American plate. Earthquakes along here can be large, but all are shallow.

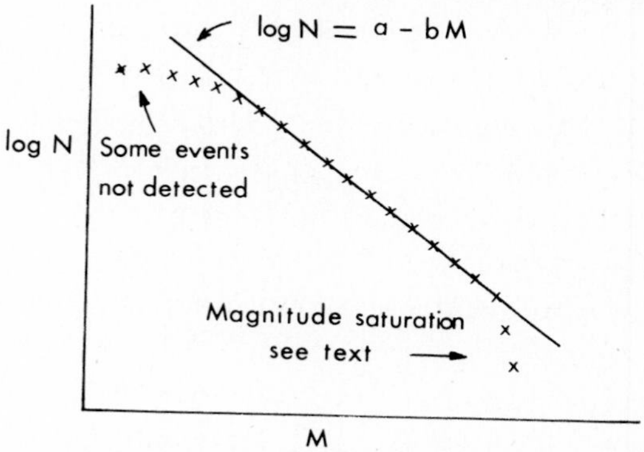

Figure 30-2. Cumulative number of earthquakes exceeding a given magnitude (Schematic).

Number of Earthquakes vs. Magnitude. Many investigators have compiled statistics relating the yearly number of earthquakes in a given region as a function of magnitude. A summary of many such investigations plus additional specially compiled data has been put together by Evernden (1970). The results are generally as in Figure 30-2. Over a middle range, the data can be fitted with a line of the form

$$\log N = a - bM, \tag{30-1}$$

where $\log = \log_{10}$, N = the number of earthquakes with magnitudes greater than M and a and b are constants. Various studies use a number of magnitude definitions; M here will be a generalized magnitude. If, on the other hand, one wishes to describe the number of earthquakes in a given magnitude interval, a simple calculation will show that a line of the same slope, but with different constant value, will fit the data.

OPPOSITE PAGE

Figure 30-1. World Seismicity, 1970. Names of the larger lithospheric plates have been added. Seismicity map courtesy National Oceanic and Atmospheric Administration/Environmental Data Service, Boulder, Colorado.

In the Evernden study, b-values are found to range between 0.6 and 1.5. In a more localized study, Kaila and Rao (1975) have produced maps of Europe showing a similar range of values. This study, in addition, shows a strong correlation of a with b.

Two problems arise in obtaining adequate data for fitting a straight line. At the low magnitude end, there will be a level below which seismic instruments will detect a decreasing fraction of the total events occuring. On a world wide basis, this takes place at magnitudes of 4.5 - 5.0. Regionally, however, earthquakes with negative magnitudes (microearthquakes) can all be recorded near a sensitive seismometer. This decreased sensitivity causes a flattening of the log N vs. M curve. At the high magnitude end, two effects cause a steepening of the curve. Most obvious is the fact that at large magnitudes (>7.0) one must wait quite a long time for sufficient data to accumulate. For example, there are approximately 17 earthquakes per year (Kanamori, 1977) over the entire world whose size exceeds magnitude 7.0. Secondly, we have seen that very large earthquakes are associated with large rupture zones, not large stress buildups (Figures 29-12 and 29-20). Thus, there is a certain saturation of Magnitude with increasing size which we have noted previously on page 393. Given the rather random boundaries of the lithospheric plates and the dimensions of geologic inhomogeneities such as mountain ranges, stable continental platforms, etc; it is not unreasonable to assume a limit to the length of an incipient rupture zone where nearly uniform stress buildup could occur without rupture. According to Geller's (1976b) Table 1, twelve large earthquakes, all with moment greater than 10^{28} dyne-cm, have lengths (and/or widths) lying between 10^2 and 10^3 km. This group includes Chile 1960, and Alaska 1964.

Kanamori and Anderson (1975) have introduced a simple argument to show that this range of b-values is quite reasonable. They write

$$n(M_s)S(M_s) \sim \Sigma = \text{constant}; \qquad (30\text{-}2)$$

where n is the incremental number of earthquakes within a magnitude range ΔM_s, S is the area of the dislocation, and Σ is the total area of the highly stressed zone subject to fracture. This stress may be relieved by a large number of small earthquakes or by a few large ones. Once relieved, buildup would have to start all over. Taking logarithms, one finds

$$\log n \sim -\log S \sim -M_s; \qquad (30\text{-}3)$$

where we have used (29-30). The coefficient in 30-3 is thus -1.0, in agreement with the observed average. Considerably more detailed discussions of the relation (30-1) have been given by Wyss (1973)and Caputo (1977).

30.2. Variation with Depth.

The world's earthquakes have been divided into three groupings on the basis of depth: Shallow, <70 km; Intermediate, 70-300 km; Deep, >300 km. Now that the nature of the earthquake producing zones is better understood, it makes more sense to only consider the depth variations along the Benioff zones associated with converging lithospheric plates. The earthquakes associated with the seismic ridges are largely confined to the fracture zone region (Figure 29-7) of the Transform fault. This fracture zone is very similar to the boundary between the two litospheric plates sliding past each other. Most of the friction will take place in the lithosphere which in these regions is less than 100 km thick. Consequently, adding the generally shallow earthquakes produced in these two regions into the whole distribution should lessen, not increase, an understanding of seismicity.

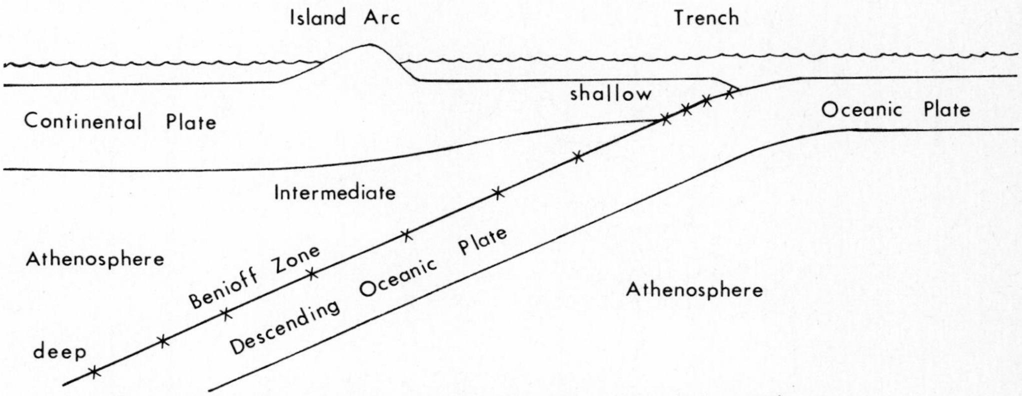

Figure 30-3. Schematic of an active continental margin — including trench, island arc, and Benioff zone.

Figure 30-3 shows a schematic model of the descending oceanic slab under a continental margin. A number of Benioff zones associated with several island arcs have been studied by Isacks et al (1968) and the distribution of earthquakes with depth is given in Figure 30-4. In all arcs, the number of earthquakes decreases rapidly through the top 100 km. In the intermediate depth range there is a leveling out in three cases and in the deep zone there is an increase in number of earthquakes per unit interval. In additional studies, Isacks et al (1969) examined the focal mechanisms of the Tonga - Kermadec region and found that the shallow events are quite different from the deep ones. Isacks and Molnar (1971) in a world wide study of mantle earthquakes have found that the descending lithospheric slab behaves as a stress guide aligning earthquake producing stresses parallel to the inclined seismic zones.

30.3. Variarion with Time.

We have already discussed one aspect of the variation of earthquakes with time, i.e., the number of earthquakes per year with greater than a given magnitude. Many studies have been carried out with the purpose of providing more detail to the time behavior of earthquakes. In examining such behavior, one first looks for any periodic components and then at the nature of the random sequence remaining.

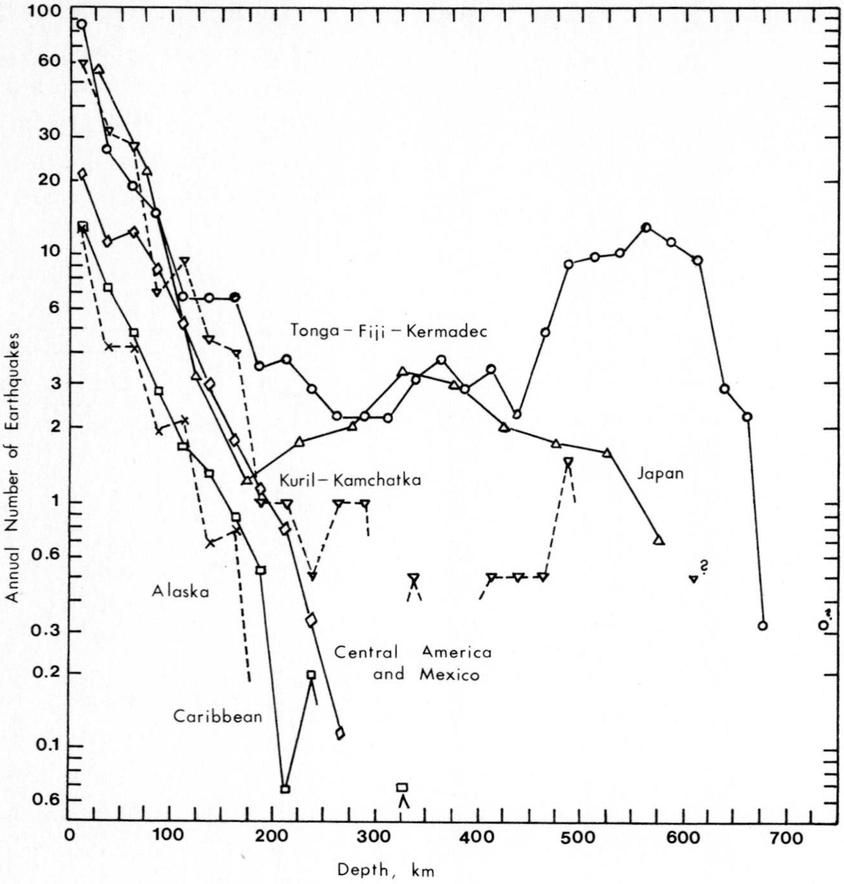

Figure 30-4. Number of earthquakes per 25 km interval (except Japan, which is in percent). Curves are not normalized for sample length and only shape is significant. (After Isacks et al, 1968. Copyrighted by AGU.)

The largest periodic stresses acting at the earth's surface are those due to the attraction of the sun and the moon. This attraction gives rise to the solid earth tides. Catalogs of earthquake sequences have been searched unsuccessfully for a tidal effect — worldwide, Shlien (1972); Southern California, Knopoff (1964c); two small regions in Central California, Shudde and Barr (1977). Tests on sequences more localized in both space and time have been somewhat more successful. Ryall et al (1968) found tidal correlations with microaftershock sequences and Klein (1976) found a relation between earthquake swarms and the earth tide. Heaton (1975) found tidal triggering of shallow large magnitude oblique-slip and dip-slip earthquakes. Both the Klein and Heaton papers have extensive references to earlier work with the tidal correlation problem.

The analysis of the remaining random catalog is greatly complicated by the presence of aftershock sequences. Aftershocks are events that follow closely in time and nearby in space to a major seismic event. To first order, it is easy enough to see such a sequence on a space-time diagram. However, a precise measure of whether a given earthquake is an aftershock is difficult to formulate. Also, it is not clear whether all earthquakes in a given space-time region should be considered aftershocks. That is, some of the events following a large earthquake might have occurred even if an aftershock sequence had not been present.

The simplest random sequence is the Poisson process. Here the probability of occurence is small and constant over time. The probability of $\underline{n}$ earthquakes per unit time would be given by

$$P_1(n) = \lambda^n e^{-\lambda}/n! \quad , \tag{30-4}$$

where λ is the average number per unit time. Also, in such a process, the probability of any time interval "T" between earthquakes is given by

$$P_2(T) = \lambda e^{-\lambda T} \quad . \tag{30-5}$$

Equation (30-5) indicates that there will be some clustering of events, in that small intervals are much more probable than large ones. By direct computation however, the reader can show that 50% of the intervals are shorter than λ^{-1}, while 50% are greater; i.e., the average interval is λ^{-1}. This clustering effect of the Poisson process makes it difficult to separate out aftershocks from a purely random sequence obeying Poisson statistics. An especially readable derivation of equations of the type (30-4) and (30-5) is given by Meyer (1975, p. 203 and p. 216). A number of possible statistical models applicable to earthquake sequences are given by Lomnitz (1974, Chap. 5-8). Additional discussion can be found in Vere-Jones (1975).

An analysis of the USCGS catalog (Jan. 1961 - Aug. 1968) was made by Shlein and Toksöz (1970) to see if worldwide earthquakes obeyed Poisson statistics. They found that the Poisson process was inappropriate and developed another model. A similar result was obtained on the Southern California catalog (1934-1957) by Knopoff (1964d). In both cases, the occurrence of aftershocks kept the observations from fitting a Poisson distribution. The former paper tried to incorporate them in their new model while the latter tried unsuccessfully to remove them. A new method was developed by Knopoff and Gardner (1972) to remove aftershocks by using windows with variable intervals of space and time (as a function of magnitude). Those techniques were applied to the Southern California catalog (now 1932-1971) to show that the sequence was Poissonian (Gardner and Knopoff, 1974). One interesting result from their study was that two-thirds of the listed events were found to be aftershocks. Of the remaining events, the Poissonian character is attributed to the much more numerous smaller events.

The Knopoff-Gardner method of removing aftershock sequences has been found to be applicable to many non-Poissonian events and thus improves the "Poissonicity" of a catalog (Rundle and Jackson, 1977). Rundle and Jackson (ibid) felt that this could lead to an unwarranted identification of an earthquake sequence as Poissonian. Such a danger, together with the great preponderance of aftershocks found by Gardner and Knopoff (ibid) suggest that more sophisticated models be considered. Some recent alternative models have been provided by Knopoff (1971), Lomnitz (1974), Shlien and Toksöz (1975), and Kagan and Knopoff (1976).

The statistical description of earthquake processes also has to take account of changing patterns of seismicity. This has been studied by Mogi (1974, several earlier papers starting in 1968 are referenced) and Kelleher (1970, 1972). These authors noted several seismicity gaps in which there is general agreement that strain energy is being stored rather than being released in a series of small earthquakes. These gaps, then, would be the site of future major earthquakes. Two prominent such gaps appear in Figure 30-1 along the Alaskan Peninsula (155°-162°W) and between the Aleutian Islands and the Kamchatka Peninsula (165°-170°E). Statistical models describing the time-rate of changes in seismicity associated with seismic gaps have been proposed in a series of papers by Kagan and Knopoff (1976, 1977, 1978).

30.4. Microearthquakes.

The level of earth noise drops rapidly as the frequency of observation increases (see Chapter 32). This means that at frequencies of 1-100 hz, seismometers of increased gain ($>10^6$) may be used to detect ever smaller earthquakes near to the point of observation. Events with magnitudes less than 2 or 3 have been called microearthquakes. Early investigations over limited areas were carried out by Asada and others (Asada, 1957; Asada et al, 1958). These early studies indicated that these microearthquakes were very similar to their bigger brothers. For example, b-values (equation (30-1)) are similar; Isacks and Oliver (1964) summarize several results showing that b is slightly less than 1.0. However, extensive studies were possible only after the development of portable high-gain seismographic instruments. Currently microearthquakes are being studied to aid in the interpretation of regional tectonics, to help with the analysis of the source mechanism, and as a possible predictor of the occurrence of larger shocks with destructive potential.

Extensive studies of regional tectonic character have been carried out in Nevada (Oliver et al, 1966; Stauder and Ryall, 1967, Ryall and Malone, 1971) and in Southern California (Brune and Allen, 1967). The major effort here is the accurate location o events in three-dimentional space for possible correlation with fault planes, both known and unknown. Arrays of instruments are set out with the S-P times used to compute range. Because of the high frequencies associated with microearthquakes, the seismic wave trains attenuate rapidly and most events recorded by any one instrument lie within a few tens of kilometers. This means that for regional studies, arrays of a few instruments must be moved many times to cover a large area. In these studies, high levels of microearthquake activity were generally associated with areas of recen earthquake activity or fault creep. This is consistent with the extrapolation of the N vs. m_b curves to microearthquake magnitudes. In the Brune and Allen (1967) study, certain sections of the San Andreas Fault were found to be "locked." That is, there was a minimal level of microseismic activity. This was correlative with a low-level of shocks with magnitudes >3.0. Locked segments of major transcurrent faults are generally assumed to be accumulating strain energy with release by a destructive event some time in the future. Additionally, this study found no definite changes in level of microseismic activity preceding major shocks.

Studies of source mechanisms of these extremely small events have been carried out by a number of investigators. Some examples of fault-plane solutions (p. 369) are given by Stauder and Ryall (1967) and Ryall and Malone (1971). Studies of their spectral characteristics have been carried out by Douglas and others (B. M. Douglas et al, 1970; B. M. Douglas and A. Ryall, 1972). They have computed stress drops of 0.04 to 0.60 bars, but feel that these may be only partial stress drops and if so, the interpretation of the spectra must be modified.

Lastly, Udias and Rice (1975) have made a statistical study of a 4 year record of microearthquake activity taken at the San Andreas Geophysical Observatory near Hollister, California. Their work showed a strong tendency toward clustering, leading to large deviation from a Poisson distribution. They also noted a long-term variation in the activity, suggesting that conclusions drawn from short-term studies may not be accurate.

30.5. Aftershocks and Foreshocks.

Aftershocks are earthquakes of lesser magnitude following closely in time and nearby in space to a "main shock." These three descriptors are difficult to define precisely. In an area without much seismic activity, such a cluster of events is rather easy to spot, but in a seismically active area, there is much difficulty in separating main and aftershocks. We have already mentioned the Southern California study (Gardner and Knopoff, 1974) in which two-thirds of the events were classed as aftershocks according to their time-space criterion. The term "main shock" is also difficult to define, because as size gets larger there are fewer and fewer large shocks in any given region and the appropriate time-space windows are more difficult to describe. Aftershocks even appear to occur after some larger microearthquakes.

Aftershocks have been used to delimit the area of rupture in a major earthquake prior to determining seismic moment (p. 381). However, only early aftershocks must be used as there is a tendency for the region defined by the aftershock zone to increase in area with time. The temporal history of aftershocks is also open to question. All agree that the number of aftershocks decreases, but the exact manner is not certain. Many investigators (e.g. Lomnitz, 1966a; but also see reference list in Lomnitz, 1974, p.88) find that average magnitude remains constant as their number decreases. However, not all earthquakes of a given magnitude are followed by aftershocks (Mogi; 1963; Udias and Rice, 1975). A comprehensive study of aftershocks and microaftershocks following the 1964 Alaskan earthquake has been carried out by Page (1968), the relationship between stress drop and b-values has been considered by Gibowicz (1973). Recent investigations by Kagan and Knopoff (1976, 1977, 1978) have applied sophisticated statistical analysis to a number of earthquake catalogs to delineate the relationship between main and aftershocks.

Foreshocks are those earthquakes which immediately precede and occur in the vicinity of a main shock. Again the time and space relationships are difficult to define precisely. In a normally quiet area, foreshocks stand out, but a relatively high seismic background disguises their identity. At the present time, most events are recognized as foreshocks only in retrospect, that is, after the main shock has occured. However, there is some indication (Lindh et al, 1978a) that foreshocks may be distinguished by a consistant change in focal mechanism from a regional pattern; Lindh et al (ibid) give a number of references to other recent papers where such changes have been noted.

The presence of foreshocks suggests their use as predictors of the main shock which is to come. However, in the Mogi (1963) study, only 4% of some 1500 earthquakes in and near Japan were preceded by identifiable foreshocks. Rikitake (1976, Sec. 8.3) gives a comprehensive summary of many aspects of foreshocks. The Kagan and Knopoff (1976, 1977, 1978) studies of earthquake catalogs describe many features of foreshock sequences.

Earthquake Swarms represent an intermediate state in which the earthquake activity builds to a maximum and then dies away without there being a main shock which stands out. Mogi (1963) describes their properties, listing 133 swarms between 1926 and 1961. The most studied earthquake swarm was that which took place at Matsushiro, Japan (Hagiwara and Iwata, 1968) during a two year period starting in October 1965. Mogi (ibid) feels that this type of phenomenon takes place in heterogeneous and/or fractured regions in which large strains are not able to build up.

30.6. The Problem of Prediction.

The loss of life and the property damage associated with large earthquakes demands that efforts be made to predict them so that human suffering may be minimized and that property losses may be lessened. The problem of prediction can be divided into two areas which have considerable overlap. The first is time scale: long range, months-years; intermediate range, days-weeks; short range, minutes-hours. The principal considerations here are socio-economic and relate to society's response to a warning (with associated probability) that an earthquake is predicted. Different responses are required depending upon the time available for reaction. The second area is that of the means of prediction. This may be split into two catagories: statistical methods of prediction and phenomenological methods of prediction. The U.S. National Academy of Sciences (1975) has published a 142 page study of the implication of public policy relating to the first area. The second area is covered by two recent monographs (Lomnitz, 1974; Rikitake, 1976). The former has strongly emphasized the statistical aspects and the latter the phenomenological aspects. The reader is referred to these three sources for more detailed descriptions. Public policy is outside the scope of this book, but we will briefly summarize the main points of statistical and phenomenological prediction below.

Statistical Prediction. The temporal and spatial properties of earthquake sequences have been outlined above. We have indicated that deterministic components of such sequences are at best only marginally defined; they are certainly not obvious. The presence of aftershocks is partially deterministic in that they do follow a main shock. However, the aftershock sequence itself seems to be random. The main sequence appears to be a random process though its Poissonicity is not definitely established. The main thrust of statistical prediction is the establishment of a probability value for the occurance of an earthquake of a given magnitude, for the occurance of a certain value of ground acceleration (here the geological conditions at the site enter), or for the occurance of a certain amount of damage (here site conditions and building design have to be incorporated). Once these probabilities have been determined, contour maps of the various levels of hazard can be plotted. Hopefully, then, these factors are taken into account in the siting of new construction and in the design of structures of sufficient strength to withstand the expected groundshaking.

Phenomenological Prediction. For structures and communities that have been built without due regard to the hazard level, the ability of predicting statistically can only set insurance rates; it is not very satisfactory from a personal point of view. One would like to be able to move people, dangerous materials, and valuable items prior to a destructive earthquake. To do this, some deterministic predictor is necessary.

The earliest candidates were changes in seismicity. We have discussed foreshocks above and have noted that they precede only a small proportion of mainshocks. Studies continue in this area to see if this non-occurance is real or only a limitation of our present means of identifying foreshocks against a seismic background. Changes in the rate of earthquake occurrence, as expressed by the b-value, have also been noted before large earthquakes and numerous investigations by regional seismic networks are underway to determine the inter-relationship. In this regard, studies of microearthquakes have not been able to determine if they are aftershocks of a previous event, foreshocks of a forthcoming major shock, or some combination. If microearthquakes can be definitely linked to the main N vs. M curve, they will be of a substantial use in determining changes in the level of seismicity.

If ground rupture is the result of stress accumulation exceeding some critical value, then careful monitoring of this buildup could also be a predictor. To this end there are extensive programs of ground measurement: leveling surveys to determine uplift or downwarping, determination of tilt, trilateration surveys to measure regional compression or extension, as well as local observations of creep along active faults. Anomalous measurements of all of these have preceeded some major earthquakes, while in other cases anomalies have been followed by no event and in still others, large earthquakes were preceded by no anomalous events. Both ground measurements and the seismicity studies above need a more definitive theoretical basis for the initiation of fracture with a given stress accumulation and a given tectonic setting.

Laboratory studies on the fracture of rocks (Brace et al, 1966) gave rise to a model of the processes taking place in the earth near the earthquake source (Nur, 1972; Aggarwal et al, 1973; and Scholz et al, 1973). This construct is called the dilatancy-diffusion model, and the reader is referred to the above papers for details. The main points of this model are the following. Microcracks are created as the effective stress (difference between confining stress and pore pressure) rises. This increased open space then lessens the pore pressure, increasing the effective stress. The result is a strengthening of the rock; no more cracks are formed. As water flows into the newly created pore space, the pore pressure drops until saturation is reached then it begins to rise again. All this time, the tectonic stress has been rising and at a certain level the increasing pore pressure triggers the earthquake. While present, the unsaturated pore space reduces the effective bulk modules of the rock and lowers the P-wave velocity, while the flow of fluids changes the electrical properties of the rock.

This change in V_p (while V_s remains relatively constant) has launched dozens of monitoring projects in which one monitors the ratio T_s/T_p for waves from a given source region which pass through a suspected zone of incipient fracture. In the Garm region of Russia, results have been spectacularly successful (Semenov, 1969). In New York state, Aggarwal et al (1975) correctly predicted an earthquake of magnitude 2.6 in mid-1973. This success created a feeling of optimism among earthquake seismologists that a useful means of prediction was in sight. This early optimism has been lessened considerably by careful work along the San Andreas Fault in California. McEvilly and Johnson (1974) monitored T_s/T_p for 70 quarry blasts in central California recorded at seven stations in the area, and found that 97% of the readings lay within 1% of the mean values. They concluded that the dilatant volume must be less than 5 km in lateral dimension for earthquakes of magnitude 4.5 - 5.4. Lindh et al (1978b) showed that travel-time anomalies preceding two magnitude 5 earthquakes were related to variations in source location and not due to material property changes. These and other studies have shown that an increased density of monitoring stations will be necessary to define what seem to be rather small volumes affected by the dilatancy phenomenon.

In addition to the velocity anomalies, the movement of fluids should make changes in the apparent resistivity of the dilatant region and if earth currents are flowing they should show changes also. Such variations have been noted in several cases. Also the microfracturing process should release the radioactive decay product Radon into the underground water system. Anomalous values of Radon content in well water have been noted in several instances preceding earthquakes. As with the T_s/T_p anomalies, monitoring networks have been set up in seismic zones to observe those quantities.

Rikitake (1976, pp. 272-288) gives an extensive list of premonitory phenomena in the case of many world-wide earthquakes. There are almost three hundred precursors listed, but when one considers the large number of potentially destructive earthquakes over a comparable period of time, we see that much more data needs to be collected; much more needs to be understood.

CHAPTER 31

SEISMIC NOISE — THEORY

31.1. Wave-Generated Microseisms.

It has been known for several decades (Gutenberg (1958) and Darbyshire (1962) give a summary of older work) that there was a large peak in the spectrum of earth noise at periods of 6-8 seconds. These signals, called microseisms, are quite noticeable on the records produced by long-period instruments as can be seen in Figure 31-1. Because they are most prominent at coastal stations, it was natural to associate this earth noise with wave action. However, on measurement, it was found that the predominant period in ocean swell was on the order of 12-16 seconds. Consequently there existed a real problem as to the origin of the 6-8 second microseismic background. (It might be noted here that there is also a peak in the earth noise spectrum which corresponds to the 12-16 second swell — see Figure 32-1.) However, theory developed by Longuet-Higgins (1950) and modern array "beam-forming" measurements (see Sec. 33.5) have definitely associated the microseismic peak with ocean swell through a non-linear frequency doubling process to be outlined below. A thorough statistical treatment of the problem has been given by Hasselmann (1963). Additional considerations have been given by Darbyshire and Okeke (1969) and Okeke (1972). Microseisms in the 12-16 second band are called primary frequency microseisms, while those in the 6-8 second band are called double frequency microseisms.

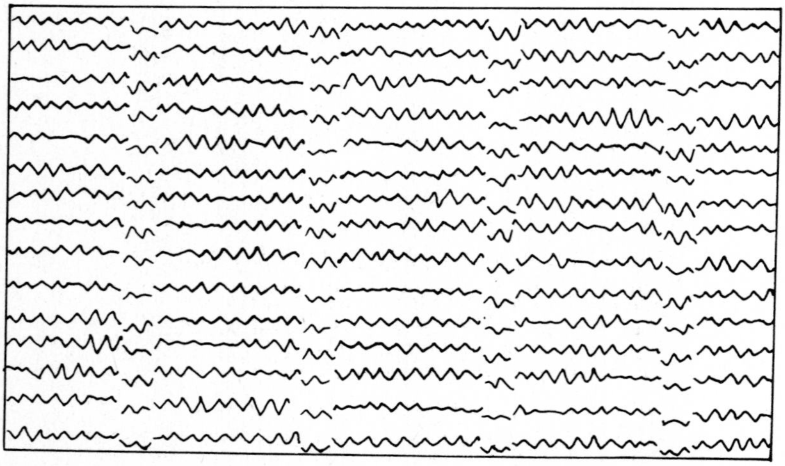

Figure 31-1. Double frequency microseisms recorded on the long-period N-S component at Pittsburgh, Pennsylvania.

To work out this frequency doubling, we must start with the equations of fluid dynamics:

$$\frac{\partial \mathbf{v}}{\partial t} + \mathbf{v}\cdot\nabla\mathbf{v} = \mathbf{f} - \frac{1}{\rho}\nabla p \quad . \tag{31-1}$$

We then introduce the velocity potential by

$$\mathbf{v} = -\nabla\Phi \ , \tag{31-2}$$

and furthermore assume that = - ∇U, i.e., forces are derivable from a potential; and $\rho = \rho(p)$, i.e., temperature effects are not important. We can write the equation of motion as

$$\nabla\left\{-\frac{\partial\Phi}{\partial t} + \frac{v^2}{2} + U\right\} + \frac{1}{\rho}\nabla p = 0. \tag{31-3}$$

If we dot this equation with d**r** and integrate along a path from one point to another, we find that

$$\int_{\text{path}} \frac{1}{\rho}\,dp - \frac{\partial\Phi}{\partial t} + \frac{v^2}{2} + U = \Theta(t). \tag{31-4}$$

This equation is known as the generalized Bernoulli equation. If we consider an incompressible medium, then from the equation of continuity,

$$\nabla\cdot\rho\mathbf{v} + \frac{\partial\rho}{\partial t} = 0; \tag{31-5}$$

we have that

$$\nabla\cdot\mathbf{v} = \nabla^2\Phi = 0. \tag{31-6}$$

Then we can write (assuming $U = -gz$)

$$\frac{p-p_s}{\rho} - gz = \frac{\partial\Phi}{\partial t} - \frac{v^2}{2} + \Theta(t). \tag{31-7}$$

Now Φ has as a general solution, $\Phi = A(k)e^{i(kx-kz)}$, hence at great depth we find that

$$\frac{p-p_s}{\rho} - gz = \Theta(t); \tag{31-8}$$

the exponentially decaying term reducing $\partial\Phi/\partial t$ and $\frac{1}{2}v^2$ to a negligible contribution. Thus it is possible to have a pressure variation that consists of a hydrostatic increase with depth plus a time-varying term that is constant in space.

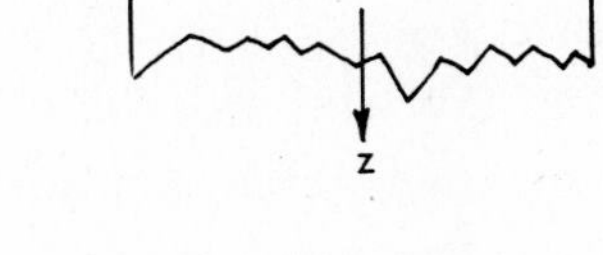

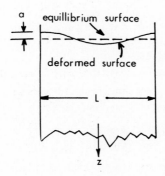

Figure 31-2. Geometric relations for a standing-wave deformation of the oceanic surface.

Let us consider a one wavelength portion of a standing-wave pattern in the ocea (Figure 31-2). From the equilibrium position we have an increase in the center of gravity twice each cycle as water flows from the troughs to the ridges and back agai We can compute

$$\text{Mass} * \delta\,(\text{Center of Gravity}) = \int_0^L \tfrac{1}{2}h(x)\rho h(x)dx$$

$$= \int_0^L \tfrac{1}{2}a^2\rho\ \text{Cos}^2\omega t\text{Cos}^2 kxdx \qquad (31\text{-}9)$$

$$= \frac{1}{8}\ La^2\rho(1 + \text{Cos}2\omega t).$$

Now the changing pressure at the bottom can be written as

$$p = \frac{\text{Mass}}{L}\ \frac{\partial^2\delta(\text{C.G.})}{\partial t^2} = -\ \tfrac{1}{2}\rho a^2\omega^2\text{Cos}2\omega t. \qquad (31\text{-}10)$$

That is, the change in the center of gravity is reflected in a time-varying pressure on the bottom having a doubled frequency with respect to the surface time-variation.

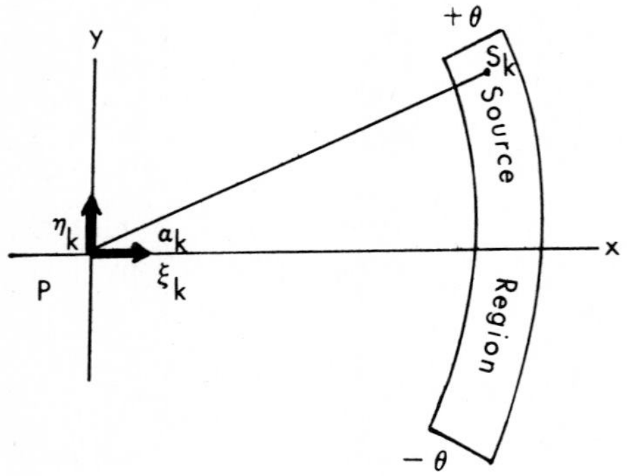

Figure 31-3. Components of particle motion for elementary waves at the point P. The component ζ_k is positive upward. Sources are uniformly distributed over the sector from -θ to +θ with random phases.

31.2. Some Properties of Random Signals.

Although Hasselmann (1963) gives a description of the power spectral properties of primary and double frequency microseisms, this is not well suited to the description of the seismic record as depicted in Figure 31-1. Strobach (1965) has given a statistical description of the amplitude distribution on the seismic record that agrees well with observation. We shall only set down his principal results, referrin the reader to the original paper for the derivations. Strobach (ibid) models the observed signal as a superposition of Rayleigh-waves originating in the sector from -θ to +θ and passing through the point P with random phases. In the source region, all angles α_k are equally probable as are all source phases ϕ_k. Thus from each source point, S_k, we have that corresponding displacement components are:

$$\xi_k = e^{i\omega t}(e^{-i\phi_k}\text{Cos}\alpha_k)\ ,$$

$$\eta_k = e^{i\omega t}(e^{-i\phi_k}\text{Sin}\alpha_k)\ , \qquad (31\text{-}11)$$

$$\zeta_k = e^{i\omega t}\ iR\ e^{-i\phi_k}\ ;$$

where Cos α_k and Sinα_k give the orthoginal separation of the horizontal components. R is the ratio of vertical to horizontal components for Rayleigh-waves (p. 101). To find the three components of motion, we must sum the contributions from the individual oscillators. Strobach writes:

$$\sum \xi_k = xe^{-i\phi_x} = x_1 + ix_2\ ,$$

$$\sum \eta_k = ye^{-i\phi_y} = y_1 + iy_2\ ,$$

$$\sum \zeta_k = iRze^{-i\phi_z} = iR(z_1 + iz_2).$$

Since the phases ϕ_k are random, these sums give a description of the ground motion in terms of a Gaussian probability distribution with appropriate means and variances. Strobach finds:

$$P(x) = \frac{x}{\sigma_x^2}\, e^{-|x|^2/2\sigma_x^2}\ ,$$

$$P(y) = \frac{y}{\sigma_y^2}\, e^{-|y|^2/2\sigma_y^2}\ , \qquad (31\text{-}12)$$

$$P(z) = \frac{z}{\sigma_z^2}\, e^{-|z|^2/2\sigma_z^2}\ ;$$

where

$$\sigma_x^2 = \frac{N}{4\theta}\left\{\theta + \tfrac{1}{2}\,\text{Sin}2\theta\right\}\ ,$$

$$\sigma_y^2 = \frac{N}{4\theta}\left\{\theta - \tfrac{1}{2}\,\text{Sin}2\theta\right\}\ , \qquad (31\text{-}13)$$

$$\sigma_z^2 = \frac{N}{2}\ .$$

He then proceeds to calculate the horizontal amplitudes "ρ" as:

$$P(\rho)\big|_{\sigma_x \neq \sigma_y} = \frac{\rho}{\sigma_x^2 - \sigma_y^2}\left\{e^{-\rho^2/2\sigma_x^2} - e^{-\rho^2/2\sigma_y^2}\right\}\ ,$$

$$(31\text{-}14)$$

$$P(\rho)\big|_{\sigma_x = \sigma_y} = \frac{\rho^3}{2\sigma_x^4}\, e^{-\rho^2/2\sigma_x^2}\ .$$

Observations are plotted in Figure 31-4. In the case of the vertical component,

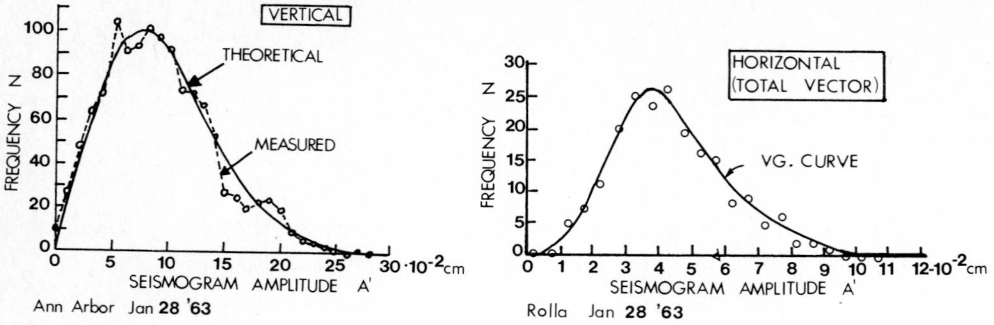

Figure 31-4. Frequency distribution of microseismic amplitudes. (a) Vertical amplitudes plotted against the theoretical curve. (b) Horizontal amplitudes. Only an average curve is plotted because the theoretical curve has an unknown parameter θ; see text. (After Strobach, 1965.)

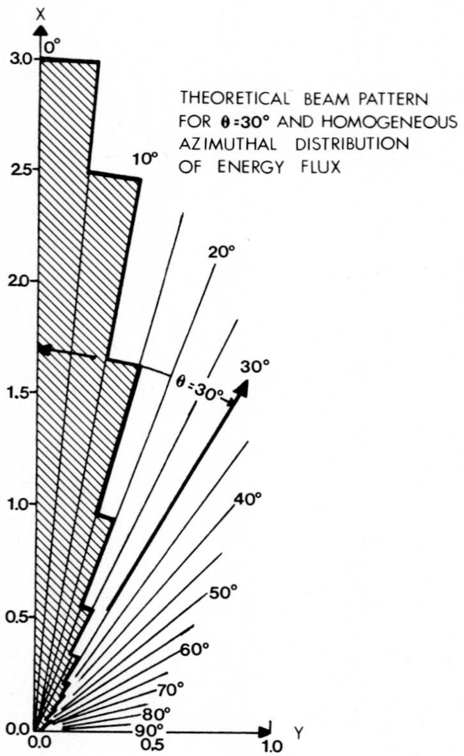

Figure 31-5. Theoretical beam pattern for a homogeneous azimuthal distribution of energy flux. The incoming waves are from a 60° sector ($-30° \leq \theta \leq 30°$). Only the right hand side is plotted. (After Strobach, 1965.)

the theoretical curve is given by the last of (31-12) and can be explicitly shown since σ_z is independent of θ. For the horizontal component, the shape of the curves given by (31-14) is dependent upon the unknown θ, consequently a theoretical curve cannot be given in this case. However, Strobach (ibid) notes that the average of the observed curve is very similar to that given by the second of (31-14) where $\theta = 90^{\circ}$ and $\sigma_x = \sigma_y$.

Strobach then proceeds to work out the beam patterns for the horizontal component. This is shown in Figure 31-5.

His final task is to determine the relationship between horizontal and vertical amplitudes. He finds that if the vertical amplitude is a maximum, then the horizontal amplitude has a great probability of being a maximum also, and the horizontal vector will point in the general direction of approach of the waves. Conversely, if the vertical amplitude is zero, then the x-component of the horizontal will probably be very small. The y-component will remain, even though small, and might be misinterpreted as a Love-wave.

CHAPTER 32

SEISMIC NOISE — OBSERVATIONS

32.1. The Level of Seismic Noise.

Brune and Oliver (1959) gave a summary of measurements of the seismic noise background between 0.01 - 20 seconds period. However, extensions of their curves to longer periods awaited improved instrumentation. Frantti et al (1962) made additional measurements at 16 sites covering the period range 0.03 - 2 seconds. By the early 1970's, the needed instrumentation was available. An extensive study was done by Savino et al (1972a)for the range 8 - 120 seconds. Their data was taken from a pressure sealed chamber deep (543 meters) in a New Jersey mine. This study noted a minimum in the earth noise between 30 and 40 seconds period that the authors attributed to a transition between swell-generated microseisms and atmospherically generated ground motion at periods greater than 40 seconds. A second study (Fix, 1972) covers the period range 0.1 - 2560 seconds and is representative of the lowest noise levels found. As with the previous experiment, the apparatus here was located in an Arizona mine (130 meters deep) with pressure-sealing doors. It is data from this location which is presented in Figure 32-1. The levels of noise between 20 and 200 seconds period at eleven High-Gain, Long-Period (HGLP — see Sec. 33.5) seismograph stations around the world, with varying depths of burial, have been given by Murphy and Savino (1975). Although the minimum levels of some of the HGLP sites are not as low as those in deep mines, the shapes of the noise spectra are similar. A somewhat different experiment (Berger and Levine, 1974) involving laser strainmeters (Sec. 33.4) has produced a noise-level spectrum of earth strain from 0.01 seconds to greater than 10^7 seconds period. The figure in their paper does not have as much detail as the Fix (ibid) data. In particular, the microseismic peak and 30-40 second spectral windows are not seen.

For short-period noise of an intermittent nature, two papers are of interest. Walker et al (1964) present some beautiful sonograms in which they recognized line spectra due to machinery as well as the continuous background. Robertson (1965) made a study of physical and topographic factors as related to short-period wind noise. He noted that topography rather than lithology contributed most to short-period wind noise.

OPPOSITE PAGE

Figure 32-1. Earth motion amplitude spectral densities. (a) Vertical component. (b) Horizontal (55° azimuth) component. (After Fix, 1972.)

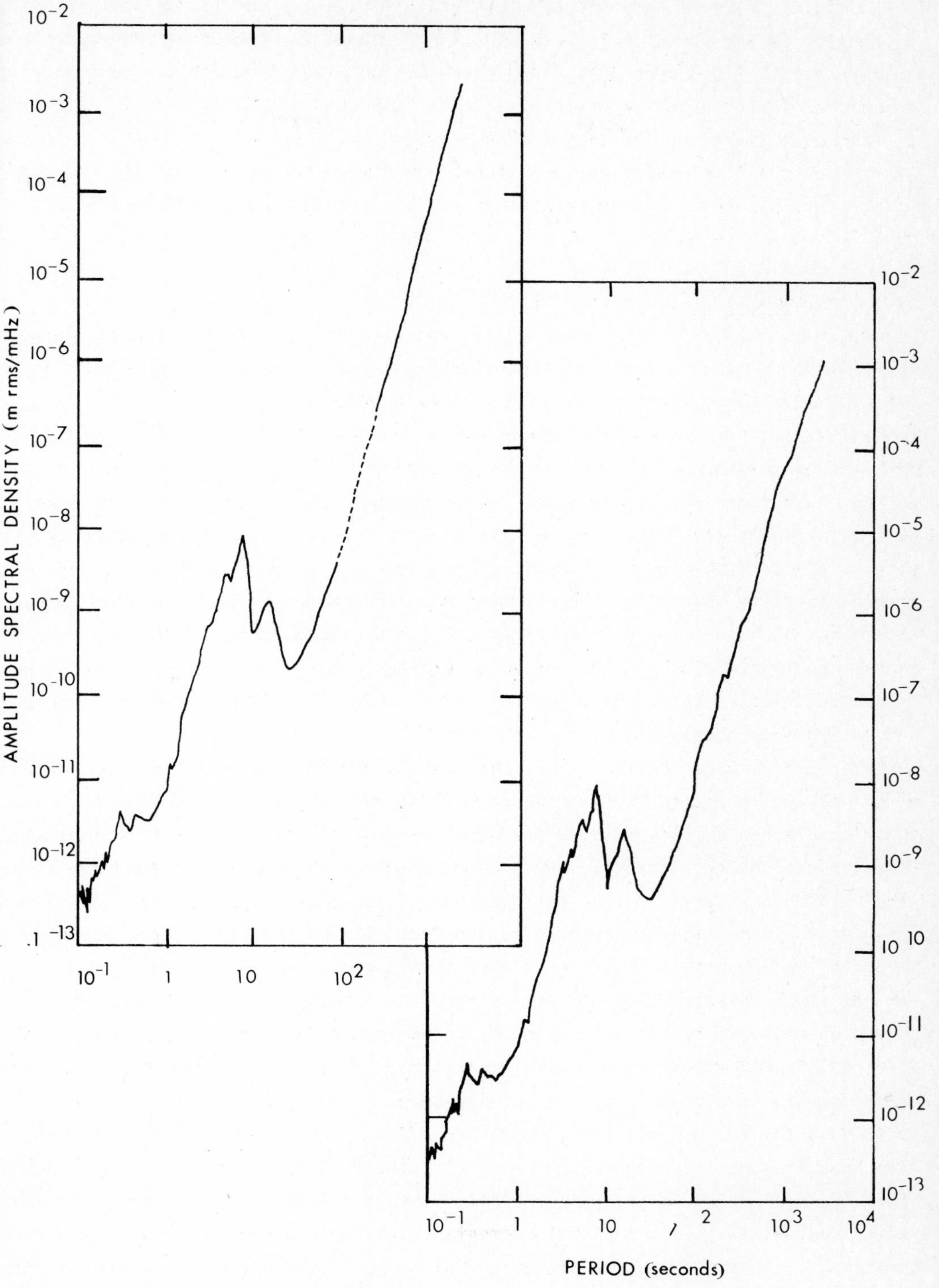
AMPLITUDE SPECTRAL DENSITY (m rms/mHz)
10⁻²
10⁻³
10⁻⁴
10⁻⁵
10⁻⁶
10⁻⁷
10⁻⁸
10⁻⁹
10⁻¹⁰
10⁻¹¹
10⁻¹²
10⁻¹³
10⁻¹
1
10
10²
10³
10⁴
PERIOD (seconds)

Turning to the observations, Figure 32-1, the peaks due to the swell-generated microseisms stand out clearly. In addition to these two features, there is a small feature at 0.3 - 0.5 sec. period and there is the steep rise at long periods. The former is at such a low level that not much has been published about it, although Frantti et al (1962) noted that it had not been satisfactorily explained. The latter interferes critically with the use of high-gain long-period instruments and considerable effort has been made to understand the source of energy. We shall say more abou this in the next section.

32.2. The Nature of Seismic Noise.

Because the swell-generated microseisms were the strongest noise component on th then available seismographic equipment, efforts were directed in the 1960's to the determination of the sources of both primary and double frequency microseisms. Haubrich et al (1963) used power spectral techniques to discover the source of both sets of microseisms. During a particularly high level of noise, it was found that not only was there the 2:1 frequency ratio between primary and double frequency, but also both frequencies were varying with time in such a manner as to suggest a distant source. Dispersive oceanic waves from this source then arrived at the Southern California coastline generating the microseisms. The mechanism of generation in coastal waters was such that the double frequency microseisms contained 100 times more energy than the primary frequency microseisms. Similar results were found by Haubrich and MacKenzie (1965). This second paper notes a coherence between atmospheric pressure and seismometer noise.

We have already mentioned that Savino et al (1972a) felt that the long-period noise was due to interactive atmospheric pressure changes with the ground. Neither o these papers had sufficient data to make a clear definition of the relationship. Sorrells and others (Sorrells, 1971; Sorrells et al, (1971a) investigated the effects o wind-related noise. Attention was then turned to long-period noise remaining after the wind was accounted for. Sorrells and Douze (1974) found evidence for infrasonic waves in the atmosphere as the source of this component.

Although these studies of swell-generated microseisms and of atmospherically generated noise were fairly conclusive, the development of large seismic arrays (Sec. 33.5) with their beam-forming capabilities more or less completed the studies of the origin of microseisms. There have been three large studies (Lacoss et al, 1969; Haubrich and McCamy, 1969; and Capon, 1973a) plus a number of smaller ones. Their results are in general agreement but details differ. This is to be expected since they all tried to separate the noise into discrete period ranges. The long-period range (>10 sec) contains fundamental Rayleigh and Love modes with the former dominant. In addition, a certain amount of non-propagating noise is present which may be due to the atmospheric loading of the earth (Capon, ibid). In the swell-generated microseismic range (6-8 sec) there is a mixture of higher and fundamental mode Rayleigh-waves with the former dominant. The short-period range (<2 sec) seems to be largely body-waves, some propagating and some non-propagating. In the in-between regions there is an overlapping of wave types.

It might be pointed out also that arrays have shown that the swell-generated microseisms originate in shallow coastal waters. The studies of Haubrich and MacKenzie (1965) found that the seismic energy originated along the Southern California coast; Haubrich and McCamy (1969) were able to use body-wave data (direction and $dT/d\Delta$) to locate most centers of energy generation as storms interacting with coastal waters; Capon (1973a) was able to use two arrays to triangulate upon the source region — again, a storm in near coastal waters. All this data seems to show that deep-water sources are minimal in their contribution to microseismic noise.

32.3. <u>The Elimination of Seismic Noise</u>.

Since early inertial seismographs were not responsive at periods greater than 100 seconds, the very long-period noise was no problem. Only the 6-8 second microseisms were worrisome. Instrument designers solved this problem by constructing two classes of instruments: short-period, centered about 1 second; long-period, centered about 20 seconds. Thus, by instrument design, they effectively filtered out the microseismic noise in all but the worst cases, such as along coastlines. Modern broad-band instruments record both ranges, but use analog or digital filters to accomplish the same result. For those who had older instrumentation that was bothered by the 6-8 second microseisms, Pomeroy and Sutton (1960) conceived of an ingenious scheme to get rid of them. By inserting a 6-8 second galvanometer in series with the long-period galvanometer (Figure 32-2), they were largely able to eliminate such noise from the recordings.

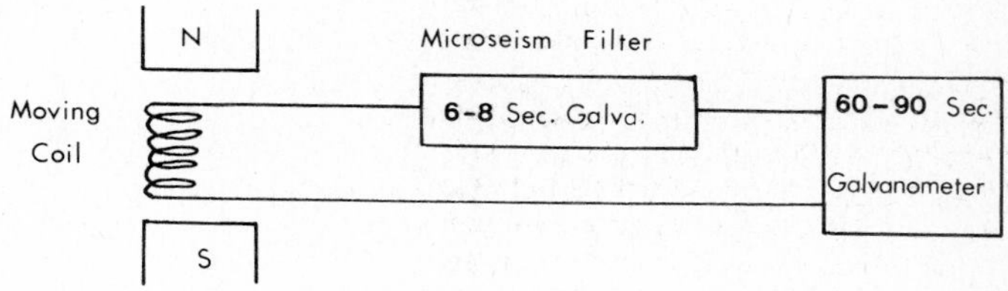

Figure 32-2. A practical microseism filter

For the most recent designs of high sensitivities at very long periods (>100 sec), the atmospheric component of noise had to be considered. A number of investigators have found that this noise can be largely eliminated by burying the seismic instruments a few hundred meters. This can be accomplished in mines or in boreholes drilled for the purpose. Ziolkowski (1973) and Douze and Sorrells (1975) have considered using the correlation of this earth noise with atmospheric pressure variations as a means of elimination. This method, however, requires extensive instrumentation and sophisticated data processing.

Lastly, the noise contributions of the instrumentation itself must be considered. Melton (1976) gives a review and Fix (1973) gives a large number of experimental results for high-gain instruments. It might be noted that instrument noise for inertial seismographs is well below ground noise, so there is not too much emphasis (at present) in trying to reduce the observed instrument noise to its theoretical minimum.

CHAPTER 33

SEISMOGRAPHS AND EARTHQUAKE SEISMOLOGY

33.1. General.

When one is studying earthquake phenomena, the receiving instrumentation is called a seismograph. There is such a variety of instrumentation that the part of the instrument which responds to the earthquake motion is called the seismometer, while a seismograph consists of the seismometer plus its associated recording device. This final chapter on seismometry has been included for two reasons: to familiarize the reader who is not instrumentally orientated with the capabilities and limitations of data gathering devices and to provide the necessary expressions so that instrumental response curves can be calculated.

The first factor we consider is dynamic range. This is the range of amplitudes that one may record without the signal being lost in instrumental noise (including finite width of graphic recording) or on the other hand driving the instrument off scale. One wants to be able to record seismic signals from a great distance (teleseisms) as well as larger signals from nearby earthquake events. In the case of the latter, the actual ground displacements and accelerations are tremendously important to the design of structures. It is easily seen that an extremely sensitive instrument would overload on large signals, while a less sensitive instrument would not be capable of recording the signals from distant events. Recordable ground motion ranges from 0.001 micron to several centimeters, i.e., a range of 10^8 to 1. Modern instruments can be built having a magnification (recorded displacement to actual ground displacement) of up to 10^7. The dynamic range, however, is usually limited by the recording instrument; a factor of 10^3 is attainable on a graphic recording while something in the order of 10^6 is obtainable by digital recording.

In a well equipped seismic observatory, there should be at least one instrument operating at maximum gain for local conditions in order to detect as many faint signals as possible. In addition, one needs instrumentation to record nearby events, and in the case of a strongly seismic region, one should have a strong-motion instrument capable of recording the largest expected ground motion.

The maximum magnification is limited by ground noise (microseisms) at present; this level being higher than the instrumental noise of modern apparatus. The use of arrays of seismometers helps to improve the signal to noise ratio by making use of the fact that the seismic signal is more coherent from recording site to recording site than is the microseismic background. In Chapter 32 we noted that the earth noise peaks at 6 sec period with a minor peak at 0.5 sec period. Consequently one would like to have maximum gain in regions excluding these bands. Seismographs with maximum amplification for periods less than 6 seconds are known as short-period instruments, while long-period devices record periods longer than this.

There are several additional factors involved in the choice of the operating band. First of all, most seismographs operate over a bandwidth of less than two decades due to their principles of construction, although the center of this band can be chosen over a large range. Secondly, the human interpreter has grown used to seeing events recorded in a relatively narrow frequency interval and cannot decipher a broad-band record. The advent of digital recording has made it possible to record broad-band (a range of four decades) and use a computer to filter the digital record in such a manner as to facilitate interpretation. These problems are merely technical and most any reasonable set of requirements can be implemented. The earth, however, puts two more stringent requirements upon the choice of frequency band.

In Chapter 27 we saw that between periods of several minutes and frequencies of several hundreds of Hertz, the attenuation of seismic waves is proportional to frequency, with the constant of proportionality varying by something like a factor of 10 over a frequency variation of 10^6 to one. This means that long-period waves will propagate further than short-period waves. In Chapter 29 we saw that the proportion of long-period relative to short-period energy grows with the size of the earthquake in a non-linear manner. The combination of these two factors leads to the observation of short-period waves for a small near earthquake and of long-period waves for a large distant earthquake both with the same trace amplitude. Small distant events are lost in station noise, while large nearby events quite often overload the sensitive instruments.

An additional constraint is imposed by the spatial resolution required by the investigator. If one is to use ray-path techniques successfully, the wavelengths of the seismic signal must be small compared to the distance between interfaces of interest. This is necessary to avoid multiple reflection interference effects. Ideally, then, one would want to use short-period waves to X-ray the interior of the earth. However, we have seen that attenuation makes this difficult. The story does not end there. Short-period seismic waves are severely scattered by inhomogeneities whose characteristic dimension approaches the wavelength of the seismic impulse. Hence a short-period seismic pulse arriving from deep in the interior of the earth will be scattered by the "geology" at the surface, making detail at depth difficult to ascertain. One should, however, be aware that such scattering may also take place at depth.

Detailed studies of modern seismometers and recording instrumentation can be found in Benioff (1955), and Willmore (1960). Press et al (1958) describe a set of matched long-period instruments commonly in use today.

33.2. The Pendulum Seismometer.

The most common type of seismograph uses a pendulum seismometer. This instrument works on the inertia principle — essentially having a large mass remaining fixed in space while the surface of the earth moves under it.

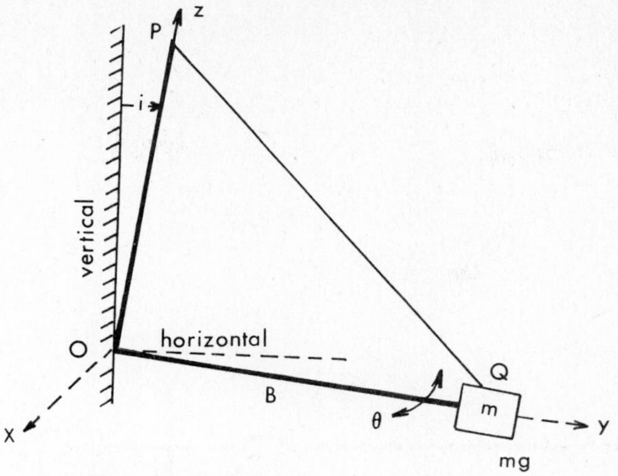

Figure 33-1. The horizontal seismometer.

Horizontal Seismometer. This instrument is simplest in theory and also simplest to construct. The forces acting at the center of gravity of the boom are

$$f_x = -m\ddot{u}_x,$$
$$f_y \simeq mgi. \tag{33-1}$$

The torque calculated about the almost vertical z-axis is

$$-L_z = mR^2\ddot{\theta} = -mB\ddot{u}_x - mgiB\theta; \tag{33-2}$$

where m is the total mass of the boom, R is the radius of gyration about $\overline{OP}$, θ is the angular displacement of the boom in the x-y plane, B is the distance from O to the center of mass of the boom (Q), and i is the inclination of the right-angled suspension POQ. Rewriting (33-2) we have

$$\ddot{\theta} + \frac{giB}{R^2}\theta = -\frac{B}{R^2}\ddot{u}_x,$$

or

$$\ddot{\theta} + \omega_s^2\theta = -\ddot{u}_x/\ell; \tag{33-3}$$

where $\ell = R^2/B$ is the equivalent length and $\omega_s^2 = gi/\ell$ is the square of the frequency of oscillation. If we add a damping force proportional to the boom velocity and oppositely directed, i.e.,

$$f_D = -DB\dot{\theta},$$

the torque will be in such a sense as to decrease $\ddot{\theta}$ and we can write

$$\ddot{\theta} + 2\lambda_s\omega_s\dot{\theta} + \omega_s^2\theta = -\ddot{u}_x/\ell, \tag{33-4}$$

where the dimensionless constant

$$\lambda_s = \frac{D}{2m(gi/\ell)^{\frac{1}{2}}}\,\frac{B}{\ell}, \tag{33-5}$$

has been introduced for convenience. The horizontal pendulum is not only sensitive to horizontal accelerations, but also to tilts of the ground. Rodgers (1968) gives an extensive study.

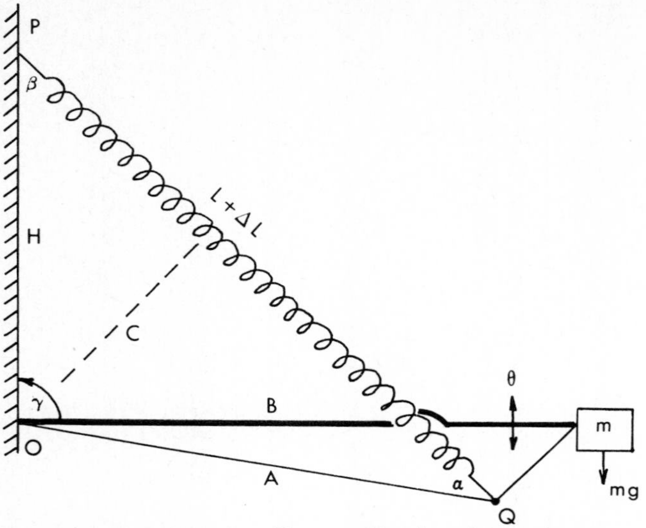

Figure 33-2. The vertical seismometer.

Vertical Seismometer. If we consider the torques about O perpendicular to the plane OPQ, we have

$$mR^2\ddot{\theta} = -mB\ddot{u}_z - mgB\text{Cos}\theta + kC\Delta L; \tag{33-6}$$

where m is the total mass of the boom, R is the radius of gyration about O, θ is the angular displacement of the boom, B is the distance from O to the center of mass of the boom, k is the spring constant, ΔL is the extension of a spring of original length L, and C is the perpendicular distance from the spring to the point O. At the position of equilibrium we have

$$kC_o\Delta L_o = mgB. \tag{33-7}$$

Hence we can write

$$kC\Delta L - mgB\text{Cos}\theta = -k\frac{d(C\Delta L)}{d\gamma}\theta + O(\theta^2); \tag{33-8}$$

where γ is the angle between the vertical support OP and the suspension arm OA. Substituting (33-8) into (33-6) and rewriting, we have

$$mR^2\ddot{\theta} + k\frac{d}{d\gamma}(C\Delta L)\theta = -mB\ddot{u}_z,$$

or

$$\ddot{\theta} + \omega_s^2\theta = -\ddot{u}_z/\ell. \tag{33-9}$$

Here $\ell = R^2/B$, and

$$\omega_s^2 = \frac{k\frac{d}{d\gamma}(C\Delta L)}{mR^2}. \tag{33-10}$$

We can investigate the properties of (33-10) further. From the law of sines, we have that

$$\frac{\text{Sin}\gamma}{L+\Delta L} = \frac{\text{Sin}\alpha}{H} = \frac{\text{Sin}\beta}{A}. \tag{33-11}$$

Using standard trigonometrical relationships we can also write that

$$(L + \Delta L)^2 = H^2 + A^2 - 2HACos\gamma, \tag{33-12}$$

and

$$C = HSin\beta = \frac{HASin\gamma}{L + \Delta L} = \frac{HASin\gamma}{(H^2 + A^2 - 2HACos\gamma)^{\frac{1}{2}}} . \tag{33-13}$$

We shall now compute the value of

$$K = \frac{1}{C\Delta L}\frac{d}{d\gamma}(C\Delta L) = \frac{Cos\gamma}{Sin\gamma} + \frac{L}{\Delta L}\ \frac{HASin\gamma}{H^2 + A^2 - 2HACos\gamma} . \tag{33-14}$$

Then we can write

$$\omega_s^2 = \frac{kC_o\Delta L_o K}{mR^2} = \frac{g}{\ell} K + O(\theta^2), \tag{33-15}$$

i.e., the square of the frequency is proportional to K. If we examine (33-14), we see that K will be small if either

$$\gamma \simeq \pi/2 \text{ and } L/\Delta L << 1;$$

or if

$$Cot\gamma \simeq -\frac{L}{\Delta L}\ \frac{HASin\gamma}{H^2 + A^2 - 2HACos\gamma} .$$

The former condition is the principle of the Lacoste (1934, 1935) vertical seismometer. Additional considerations have been given by Melton (1971). To determine the constants, we choose the frequency ω_s at which we wish to operate, find K from (33-15) and given L, the original length of the spring, we then find γ from (33-14).

Adding a damping term to (33-9) we can write

$$\ddot{\theta} + 2\lambda\omega_s\dot{\theta} + \omega_s^2\theta = -\ddot{u}_z/\ell. \tag{33-16}$$

<u>Sensitivity and Free Period</u>. Sensitivity will be defined as the ratio of boom displacement to the perturbing force in a static situation, i.e.,

$$\text{Sensitivity} = \ell\theta/f, \tag{33-17}$$

where <u>f</u> is the perturbing force. From (33-3) or (33-9) we see that in the static case

$$\omega_s^2\theta = \frac{f}{m\ell} .$$

Consequently we have

$$\text{Sensitivity} = \frac{1}{m\omega_s^2} = \frac{(\text{Period})^2}{m(2\pi)^2} . \tag{33-18}$$

One does not obtain this increased sensitivity. for free. For the horizontal seismometer one may lengthen the period by decreasing the inclination (i) of the boom from the horizontal. However, unless the frame is extremely stable with temperature, the ∠POQ can change giving large percentage changes in <u>i</u>, and possibly raising the boom $\overline{OQ}$ above the horizontal. Then the instrument collapses. Tilting of the ground underneath will also accomplish the same effect. For the vertical instrument, we see that small changes in γ will give large changes in sensitivity, so one must use extremely stable springs. However, the vertical instrument is not so sensitive to tilting. Thus the site need not be as carefully chosen as with a horizontal instrument.

Solutions of the Seismometer Equation. If we call the recorded displacement x, we find that

$$\ddot{x} + 2\lambda\omega_s\dot{x} + \omega_s^2 x = -V_s\ddot{u}, \tag{13-19}$$

where we have written for convenience

$$x = V_s \ell\theta, \tag{33-20}$$

and V_s is the statical magnification. If we assume the driving function to be of the form $u = \mathrm{Cos}\,\omega t$, the corresponding response x (as may be shown by direct substitution) is

$$x(t) = \frac{V_s\omega^2 \mathrm{Cos}(\omega t - \delta)}{\{(\omega_s^2 - \omega^2)^2 + 4\lambda^2\omega_s^2\omega^2\}^{\frac{1}{2}}} \; ; \; \delta = \mathrm{Tan}^{-1}\left(\frac{2\lambda\omega_s\omega}{\omega_s^2 - \omega^2}\right). \tag{33-21}$$

The following relationships hold (for small λ):

$$\begin{aligned} x(t) &\xrightarrow[\omega \to o]{} V_s \frac{\omega^2}{\omega_s^2}\mathrm{Cos}\,\omega t, \\ x(t) &\xrightarrow[\omega = \omega_s]{} V_s \frac{1}{2\lambda}\mathrm{Cos}(\omega t - \pi/2), \\ x(t) &\xrightarrow[\omega \to \infty]{} V_s\, \mathrm{Cos}(\omega t - \pi). \end{aligned} \tag{33-22}$$

The first of equations (33-22) shows the relationship between sensitivity and the square of the period. In this range the seismometer behaves as an accelerometer. It should also be noted that all seismometers of the same statical magnification give the same response at high frequencies, the resonant period and damping constant ($\lambda < 1$) having no effect.

From a study of the transient response of equation (33-19), one finds that a value of the parameter λ equal to one corresponds to critical damping. Any value of $\lambda < 1$ gives a damped oscillatory motion while $\lambda > 1$ gives a damping that is increased over that of critical. A value of $\lambda = 2^{-\frac{1}{2}}$ will give an optimally flat frequency response curve with no resonant peak.

33.3. Pendulum Seismometer with Moving Coil Sensing and Galvanometric Recording.

Events registered by means of a photographically recorded light spot reflected from a mirror mounted on a galvanometer have advantages of low cost and small loading of the seismometer in comparison with a mechanical magnification system. For this reason, it is probably the most common instrumentation in use today. We shall see that one obvious deficiency in such a system is its extremely low gain at low frequencies. The second is that the photographic records have to be processed before they can be read. At an increase in cost, this difficulty can be overcome by replacing the galvanometer system by a servo-operated, high input impedence, visible recorder. Besides the two analog methods of recording just considered, one may also record digitally. This has advantages of greater dynamic range, but many times the data have to be plotted for the observer to make full use of them. The analysis below largely follows Hagiwara (1958).

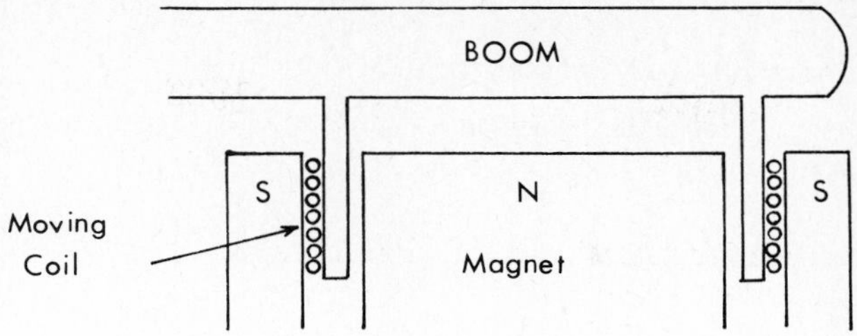

Figure 33-3. Moving Coil Sensor.

Let us now consider the response of such a pendulum-galvanometer system to the actual motion of the earth. The usual method of getting an electrical signal is to attach a coil to the seismometer boom as in Figure 33-3. If we calculate the EMF generated around such a simple one-loop coil, we find that

$$\text{EMF} = (2\pi r)\ B\ (\ell \frac{d\theta}{dt}) = G_s \frac{d\theta}{dt}, \qquad (33\text{-}23)$$

where we have introduced G_s as a convenient symbol for the product of factors $2\pi rB\ell$. In (33-23) $2\pi r$ is the circumference of the loop, B is the magnetic field (of radial symmetry and perpendicular to the plane of the loop), and $\ell \frac{d\theta}{dt}$ is the linear velocity of the loop element. Such a sensing device is known as a velocity transducer and its voltage or current output is proportional to the velocity of the boom.

If we also compute the torque about the boom hinge due to a current I_s in the loop, we find that the

$$\text{Torque} = (2\pi r)\ I \cdot B \cdot \ell = G_s I, \qquad (33\text{-}24)$$

and is in such a direction as to try to produce a motion that will reduce the current I. Thus we see that G_s is the seismometer generator constant and relates both the output voltage to the boom angular velocity as well as the torque about the hinge to the current flowing in the coil.

With a velocity transducer, we find that the seismometer equation becomes

$$\ddot{\theta} + \frac{D_s B_s^2}{M_s}\dot{\theta} + \omega_s^2\theta = -\frac{\ddot{u}}{\ell} - \frac{G_s I_s}{M_s}, \qquad (33\text{-}25)$$

where $M_s = mR^2$ is the moment of inertia about the hinge. If all the damping is electrical, we find that for critically damped motion

$$\frac{G_s I_s}{M_s} = \frac{G_s^2}{M_s R_{cr}} \frac{d\theta}{dt} = 2\omega_s \frac{d\theta}{dt}; \qquad (33\text{-}26)$$

i.e.,

$$G_s = (2\omega_s M_s R_{cr})^{\frac{1}{2}}. \qquad (33\text{-}27)$$

This is one way to determine G_s.

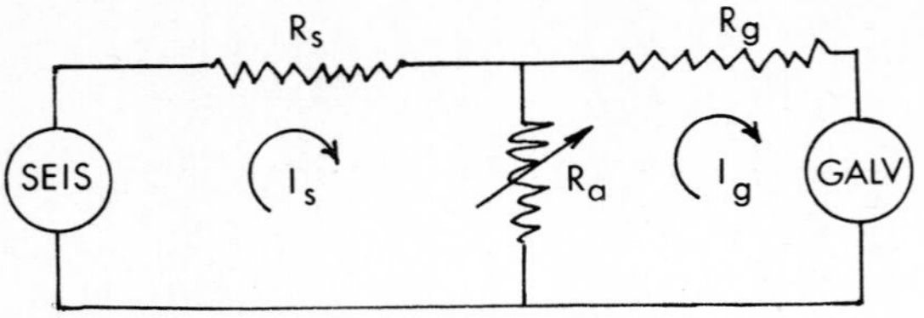

Figure 33-4. Electrical circuit for seismometer and recording galvanometer. The variable resistance R_a is a gain control.

A galvanometer works essentially in the same way, so for its governing equation we have

$$\ddot{\phi} + \frac{D_g B_g^2}{M_g}\dot{\phi} + \omega_g^2 \phi = -\frac{G_g I_g}{M_g} . \tag{33-28}$$

Usually an attenuating circuit is inserted for control of magnification as in Figure 33-4. Here we have

$$I_s R_s + (I_s - I_g) R_a = G_s \frac{d\theta}{dt} ,$$

and (33-29)

$$I_g R_g + (I_g - I_s) R_a = -G_g \frac{d\phi}{dt} .$$

Solving, we find that

$$I_s = \frac{G_s}{Z_s}\frac{d\theta}{dt} - \frac{a_s G_g}{Z_g}\frac{d\phi}{dt} , \tag{33-30}$$

$$I_g = \frac{a_g G_s}{Z_s}\frac{d\theta}{dt} - \frac{G_g}{Z_g}\frac{d\phi}{dt} ; \tag{33-31}$$

where

(33-32)

$$Z_s = \frac{R_s R_g + R_a R_g + R_a R_s}{R_g + R_a}$$ apparent input impedance of network to seismometer,

$$Z_g = \frac{R_s R_g + R_a R_g + R_a R_s}{R_s + R_a}$$ apparent input impedance of network to galvanometer,

$$a_g = \frac{I_g}{I_s}\bigg|_{\dot{\phi} = 0} = \frac{R_a}{R_g + R_a}$$ forward attenuation factor,

$$a_s = \frac{I_s}{I_g}\bigg|_{\dot{\theta} = 0} = \frac{R_a}{R_s + R_a}$$ backward attenuation factor.

Substituting (33-30) and (33-31) into (33-25) and (33-28) we have

$$\ddot{\theta} + \left(\frac{D_s B_s^2}{M_s} + \frac{G_s^2}{M_s Z_s} \right)\dot{\theta} + \omega_s^2 \theta = \frac{a_s G_s G_g}{M_s Z_g}\dot{\phi} - \frac{\ddot{u}}{\ell} ,$$

and

$$\ddot{\phi} + \left(\frac{D_g B_g^2}{M_g} + \frac{G_g^2}{M_g Z_g} \right)\dot{\phi} + \omega_g^2 \phi = \frac{a_g G_g G_s}{M_g Z_s}\dot{\theta}.$$

Introducing some new symbols, this becomes

$$\ddot{\theta} + 2\lambda_s \omega_s \dot{\theta} + \omega_s^2 \theta = 2\lambda_s \omega_s \sigma_s \dot{\phi} - \frac{\ddot{u}}{\ell}, \tag{33-33}$$

and

$$\ddot{\phi} + 2\lambda_g \omega_g \dot{\phi} + \omega_g^2 \phi = 2\lambda_g \omega_g \sigma_g \dot{\theta}; \tag{33-34}$$

where

$$\lambda_s = \frac{1}{2\omega_s}\left(\frac{D_s B_s^2}{M_s} + \frac{G_s^2}{M_s Z_s} \right),$$
$$2\lambda_s \omega_s \sigma_s = \frac{a_s G_s G_g}{M_s Z_g}, \tag{33-35}$$

etc. Eliminating θ by means similar to those used in obtaining (5-5) and (5-6), we find that

$$\frac{d^4\phi}{dt^4} + (2\lambda_s\omega_s + 2\lambda_g\omega_g)\frac{d^3\phi}{dt^3} + \left[\omega_s^2 + \omega_g^2 + 4\lambda_s\lambda_g\omega_s\omega_g(1 - \sigma_s\sigma_g)\right]\frac{d^2\phi}{dt^2}$$
$$+ (2\lambda_s\omega_s\omega_g^2 + 2\lambda_g\omega_g\omega_s^2)\frac{d\phi}{dt} + \omega_s^2\omega_g^2\phi = \frac{2\lambda_g\omega_g\sigma_g}{\ell}\frac{d^3u}{dt^3}. \tag{33-36}$$

It is apparent from this equation, that the seven constants $\lambda_s, \omega_s, \sigma_s, \lambda_g, \omega_g, \sigma_g, \ell$ reduce to five groups. Hence the same seismograph response can be set up by more than one set of parameters. This is more fully discussed by Willmore (1960).

If we set $u(t) = u_o e^{i\omega t}$, we see that

$$\phi(t) = u_o e^{i\omega t} \frac{-i\,\frac{2\lambda_g\omega_g\sigma_g}{\ell}\,\omega^3}{\left\{\begin{array}{l}\omega^4 - i(2\lambda_s\omega_s + 2\lambda_g\omega_g)\omega^3 - \left[\omega_s^2 + \omega_g^2 + 4\lambda_s\lambda_g\omega_s\omega_g(1 - \sigma_s\sigma_g)\right]\omega^2 \\ \qquad + i(2\lambda_s\omega_s\omega_g^2 + 2\lambda_g\omega_g\omega_s^2)\omega + \omega_s^2\omega_g^2\end{array}\right\}}. \tag{33-37}$$

Amplitude and phase curves for this response are given by Hagiwara (1958). We see that

$$\frac{\phi(t)}{u_o e^{i\omega t}} \begin{array}{l} \xrightarrow[\omega \to 0]{} -i\,\frac{2\lambda_g\omega_g\sigma_g}{\ell}\,\frac{\omega^3}{\omega_s^2\omega_g^2}, \\ \xrightarrow[\omega \to \infty]{} -i\,\frac{2\lambda_g\omega_g\sigma_g}{\ell\omega}, \end{array} \tag{33-38}$$

with a shift of 2π in between.

If D_s and D_g are small, one can solve the second of equations (33-35) for σ_s and substitute in the approximate value of λ_s from the first of equations (33-35) where the term containing D_s has been neglected. A similar operation for σ_g leads to

$$\sigma_s\sigma_g \simeq a_s a_g. \tag{33-39}$$

Consequently, the galvanometer reaction on the seismometer is measured by the product of the attenuation factors. For $\sigma_s\sigma_g \equiv 0$ we can write

$$\frac{\phi(t)}{u_o e^{i\omega t}} = \frac{-i2\lambda_g\omega_g\sigma_g\ \omega^2 \cdot \omega}{(\omega_s^2 - \omega^2 + 2i\lambda_s\omega_s\omega)(\omega_g^2 - \omega^2 + 2i\lambda_g\omega_g\omega)}\ ; \qquad (33\text{-}40)$$

ie., the response of the system is the product of the responses of the seismometer and galvanometer alone. Hagiwara (ibid) shows that this is approximately true as long as the product $\sigma_s\sigma_g$ remains less than one-third.

Other Types of Sensing Elements. The variable reluctance transducer was developed by Benioff (1932). This uses a variation in the gap of a magnet to change the magnetic flux within the core of a pickup coil. This device achieved great electrical gain but required a large mass to drive the moving element. As with the moving coil element discussed above, this sensor is also a velocity transducer.

Both these devices have the disadvantage of a 6 db per octave loss at low frequencies due to the factor of ω introduced by their velocity dependent output. This would suggest that a generated voltage proportional to the seismometer displacement would help increase the magnification of the seismograph in the low frequency range.

A number of such devices have been investigated, and one problem is inherent to them all. Since they are displacement sensitive, all slow drifts will be recorded and quite often they will either cause the system to become non-linear or else drive the recording device off scale. Such sensors as a differential transformer or a differential capacitor operating in a bridge, or a differential capacitor operating in a frequency modulation discriminator or a variable frequency oscillator have to be extremely stable electrically also. One way to remove this excessive need for circuit stability is to use electronic (or optical) sensing to keep the system at zero displacement using an electro-mechanical servo-loop and to record the correction signal applied by the servo-loop. Lastly, for very small displacements, optical interferometry techniques will work as long as large rapid displacements do not cause one to lose count of the fringes.

33.4. The Strain Seismometer.

An entirely different approach to sensing long-period seismic waves was taken by Benioff in the 1930's by measuring the strain in the earth directly. Over the years, improvements were made and a final version was described by Benioff (1959). Strain was measured by a fuzed-quartz tube with one end fixed to the earth as in Figure 33-5. At the other end the displacement Δx divided by the length L gave a measure of the strain. A quantitative analysis of its operation follows below.

Consider a plane-wave polarized in the direction of propagation and traveling parallel to the axis of the seismometer. We then have

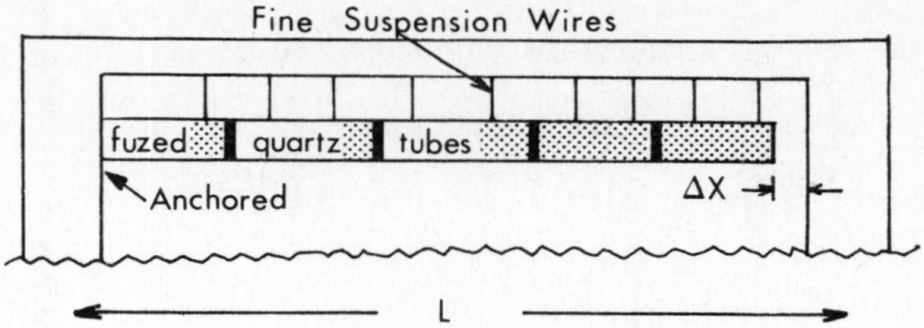

Figure 33-5. The fuzed-quartz strainmeter.

$$\begin{aligned}\Delta x &= L \frac{\partial}{\partial x} u_x \Big|_{x=0} \\ &= L \frac{\partial}{\partial x} \{u_o e^{i(kx - \omega t)}\} \Big|_{x=0} \\ &= Liku_o e^{-i\omega t} \\ &= Li \frac{\omega}{v} u_o e^{-i\omega t} .\end{aligned} \quad (33\text{-}41)$$

Consequently we see that instead of the ω^2 dependence of the inertial seismometer, we have an ω dependence. For permanent strains associated with tectonic deformations, there is no frequency dependence whatsoever, Δx being proportional to the strain itself.

If a plane P-wave is traveling at an angle θ with respect to the axis of the seismometer, we can write

$$u_x = u_o \mathrm{Cos}\theta e^{i(kx\mathrm{Cos}\theta - \omega t)} .$$

Then

$$\Delta x = L \frac{\partial u_x}{\partial x} \Big|_{x=0} = Li \frac{\omega}{v} \mathrm{Cos}^2\theta u_o e^{-i\omega t} , \quad (33\text{-}42)$$

i.e., we see that there is a $\mathrm{Cos}^2\theta$ dependence on angle of approach. As a result of this, we see that the sum of the motions on a pair of strain seismometers at right angles to each other will be

$$\Delta x + \Delta x_\perp = \frac{Li\omega}{v} \{\mathrm{Cos}^2\theta + \mathrm{Cos}^2(\theta + \pi/2)\} u_o e^{-i\omega t} = \frac{Li\omega}{v} u_o e^{-i\omega t} . \quad (33\text{-}43)$$

That is, motion parallel to the wave direction will be recorded at equal sensitivity regardless of the angle of approach.

On the other hand, if we have a plane S-wave, the motion will be at right angles to the propagation direction and in the plane of the earth's surface we have

$$u_x = u_o \mathrm{Sin}\theta\, e^{i(kx\mathrm{Cos}\theta - \omega t)} .$$

The displacement measured by the seismometer is then

$$\Delta x = \frac{Li\omega}{v} \operatorname{Sin}\theta \operatorname{Cos}\theta u_o e^{-i\omega t}, \qquad (33\text{-}44)$$

and in this case the sum of the components at right angles is

$$\Delta x + \Delta x_{\perp} = \frac{Li\omega}{v} \{\operatorname{Sin}\theta\operatorname{Cos}\theta + \operatorname{Sin}(\theta + \pi/2)\operatorname{Cos}(\theta + \pi/2)\} \; u_o e^{-i\omega t} = 0. \qquad (33\text{-}45)$$

This says that a horizontal strain seismometer pair at right angles is insensitive to waves polarized perpendicularly to the direction of motion and with the particle motion in the plane of the earth's surface. Therefore, this combination can discriminate between Rayleigh-waves and Love-waves, behaving, in this respect, like a vertical pendulum seismometer.

In the late 1960's, two other instruments were developed for measuring strain. There was an "Invar" wire strainmeter (King and Bilham, 1976) which used a wire under constant tension as a measuring device. The displacement of the tensioning element was used to determine strain. Such instruments are less expensive and more portable than the fuzed-quartz installation while giving comparable performance. The second device was the laser strainmeter (Levine and Hall, 1972). The laser instrument has considerably more sensitivity but is more costly than the fuzed-quartz installation. To achieve maximum performance, the laser beam must be sent down an evacuated tube. This restricts the portability of the laser system.

The limiting factor in accurately measuring strain seems to be conditions at the site. To avoid thermal fluctuations, deep mines are used, but long-term temperature changes (seasonal) and atmospheric pressure changes penetrate deeply, limiting the sensitivity of the installation to tectonic strain. A large number of papers relating to strain and its measurement in the earth can be found in the Philosophical Transactions of the Royal Society of London, Vol. A 274, pp. 181-433.

33.5. Special Installations, Seismic Networks, and Seismic Arrays.

Because of environmental constraints, only a few strainmeters are operating throughout the world. They might be considered special installations. Another type of instrumentation is required for the 78% of the world covered by water. Detailed studies of oceanic ridges, fracture zones and trenches require instruments nearby which mean that they must be placed underwater. Two recent articles describing ocean-bottom seismographs are Prothero (1974) and Cranford et al (1976). These devices are sunk to the ocean floor, leveled automatically, record seismic events on magnetic tape, and are raised at periodic intervals. For work closer to shore, telemetering is sometimes used. A review of their capabilities and a goodly number of references are given by Bolt (1977).

Another type of special installation can sometimes be used in noisy environments. Since much of the microseismic noise consists of Rayleigh-waves, its level can be reduced by placing the seismometer deep underground in a borehole. The cost of drilling a borehole increases rapidly with its diameter, so small instruments are at a premium. Melton and Kirkpatrick (1970) describe a rather unique borehole seismometer with three identical instruments orthogonally mounted but with their axes all at angles of about 55° with the vertical. Such an instrument is not as sensitive to tilting.

In the 1960's, in response to a number of needs, a world-wide network of standardized seismographs, WWSSN, was established. Approximately 120 stations were planned, but those actually operating in any given year have been near the 100 number. These stations were outfitted with three short-period and three long-period instruments and the recording was done photographically. The role that this network played in determining earth structure is immense and is in large part due to a far-sighted committment to make the records from this global network available to scientists all around the world at nominal cost. Every interested laboratory could have records from over 100 stations at less cost than operating a small local network. A short description of the system and a summary of early results based upon research using WWSSN records is given by Oliver and Murphy (1971).

Although the WWSSN provided and still is providing much valuable data, a need was recognized for improved sensitivity. This meant instruments of longer period, increased gain, and sites with less microseismic background. A High-Gain Long-Period electromagnetic seismograph system (HGLP) was developed by workers at Lamont-Doherty Observatory (Pomeroy et al, 1969) and ultimately eleven such systems were deployed. A summary of results was given by Savino et al (1972b). The improvement noted at these specially selected sites made further increases in sensitivity seem possible. By using newly designed instruments lowered in a borehole to a depth of 100 meters, sensitivity was increased significantly. Thirteen Seismic Research Observatories (SRO) using this instrumentation are planned. Details are given by Peterson et al (1976) and McCowan and Lacoss (1978). A different line of approach was taken by a group at UCSD (La Jolla). They have developed an ultralong-period instrument (Agnew et al, 1976) based upon an electrostatic sensing, feedback controlled zeroing, modification of the LaCoste-Romberg gravimeter (Block and Moore, 1966). Deployment of 15 of these instruments is planned — the International Deployment of Accelerometers (IDA) project. The records for these instruments will be used primarily for studies involving very long-period surface-waves and the periods of oscillation of the earth.

It might be noted that in addition to the specialized networks listed above, there are many local systems operated by both university groups and governmental agencies whose primary responsibility is the study of local seismicity. One such network operated by the U.S. Geological Survey is shown in Figure 23-7.

Seismic Arrays. The distribution of stations in Figure 23-7 is somewhat random and is largely dependent upon logistical considerations such as easy access, availability of power, communications, etc. However, a network of stations set out in an orderly fashion has nice mathematical properties which allow the discrimination of waves with differing velocities and azimuths of propagation. Such a net will be

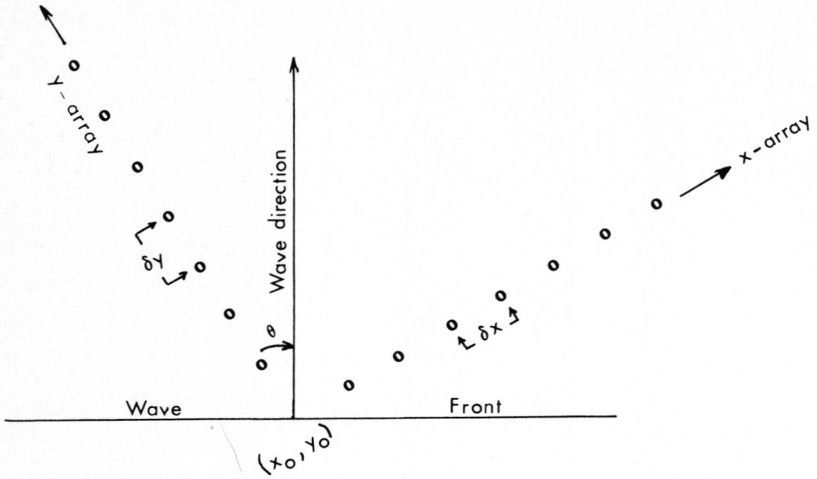

Figure 33-6. Map-view of a simple "L" shaped seismic array.

called a <u>seismic array</u>. Over a dozen of these arrays are in operation, charged mainly with the detection and identification of low-level seismic events. The largest of these arrays have dimensions comparable to the network of Figure 23-7.

To see how some of the basic array processing techniques work, consider the diagram in Figure 33-6. Here a straight line wavefront with surface velocity $d\Delta/dT$ is moving across the array. Let us further assume that the wave can be represented by the function

$$f(x,y,t) = \mathrm{Cos}\omega\{(x\mathrm{Sin}\theta + y\mathrm{Cos}\theta)dT/d\Delta - t\}. \quad (33\text{-}46)$$

Along the x-array, the signal at each seismograph would be given by

$$g_n(x,t) = A_n(\omega)\mathrm{Cos}\{k_x(x_o + \delta x_n) - \omega t\}, \quad (33\text{-}47)$$

where $A_n(\omega)$ gives the response of the seismometer at angular frequency ω. We have also written

$$k_x = \omega\mathrm{Sin}\theta dT/d\Delta, \quad (33\text{-}48)$$

as the apparent wave-number along the array. A thoughtful look at Figure 33-6 should convince the reader that it would be beneficial if we took a sum of all the seismometer responses with each response delayed by a time

$$\delta t_n = k_x(x_N - x_o - \delta x_n);$$

i.e., the time necessary to go from the nth seismometer to the last seismometer along the x-leg of the array. Here x_N and x_o represent the position of the last and first seismometer. Carrying out this operation, we would have

$$G(t) = \sum_{o}^{N} A_n(\omega)\mathrm{Cos}\{k_x(x_o + \delta x_n) - \omega t + k_x(x_N - x_o - \delta x_n)\}$$

$$= \mathrm{Cos}(k_x x_N - \omega t)\sum A_n(\omega),$$

which is just the signal received at the last station multiplied by the sum of all the seismometer responses.

Let us now look at the effect of a particular time delay along an array of evenly spaced seismometers with identical responses. In this case the time delay shall be prescribed as

$$\delta t_n = (N - n)\delta t, \tag{33-49}$$

i.e., greatest at the first seismometer and least at the Nth. The nth component of the sum will be

$$\begin{aligned} g_n(x,t) &= A(\omega)\mathrm{Cos}\{k_x(x_o + n\delta x) - \omega t + \omega(N - n)\delta t\} \\ &= A(\omega)\mathrm{Cos}\{(k_x x_N - \omega t) - (N - n)(k_x\delta x - \omega\delta t)\}. \end{aligned} \tag{33-50}$$

But then the sum of outputs will be a finite trigonometric series which can be summed giving:

$$G(t) = A(\omega)\mathrm{Cos}\{(k_x x_N - \omega t) - \frac{N}{2}(k_x\delta x - \omega\delta t)\}\ \frac{\mathrm{Sin}\ \frac{N+1}{2}\ (k_x\delta x - \omega\delta t)}{\mathrm{Sin}\ \frac{1}{2}\ (k_x\delta x - \omega\delta t)}\ . \tag{33-51}$$

The term $\mathrm{Cos}(k_x x_N - \omega t)$ represents the original signal recorded at the Nth station, the term $(N/2)(k_x\delta x - \omega\delta t)$ is a phase factor which varies with wavelength and period and comes as a result of the summation. The ratio of the two sine factors is a measure of the efficiency of the summation. If δt is chosen so that $k_x\delta x - \omega\delta t = 0$, then $G(t)$ is N+1 times the individual response. The ratio of the two sine factors plotted against the argument $(k_x\delta x - \omega\delta t)$ is shown in Figure 33-7. Because of the discrete nature of the array, there is an aliasing problem, i.e., maximum response is achieved whenever the argument is a multiple of 2π, not onlv for a zero value.

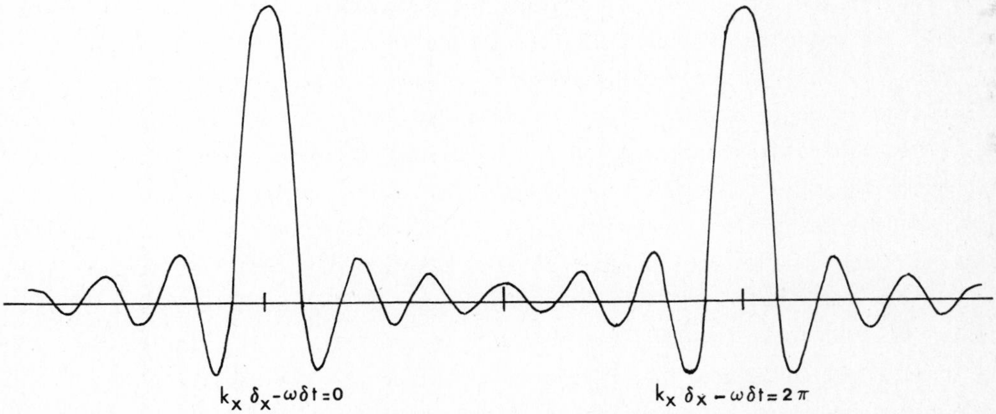

Figure 33-7. Linear array response as a function of the argument $(k_x\delta x - \omega\delta t)$. This figure is applicable to a thirteen element array. i.e., N = 12.

In terms of apparent surface velocity, maximum response in the x-leg occurs when

$$\delta t_x = \delta x \mathrm{Sin}\theta dT/d\Delta = \delta x/v_x, \tag{33-52}$$

where we have used (33-48). Thus we can use the choice of delay interval to pick out waves of apparent velocity v_x down the x-leg of the array. In a similar manner;

$$\delta t_y = \delta y \mathrm{Cos}\theta dT/d\Delta = \delta y/v_y. \tag{33-53}$$

If we now set $\delta x = \delta y = d$, we can combine (33-52) and (33-53) in such a way as to obtain maximum response for a wave with velocity

$$v = \frac{d}{\{(\delta t_x)^2 + (\delta t_y)^2\}^{\frac{1}{2}}}, \tag{33-54}$$

and appraching at an angle

$$\theta = \mathrm{Tan}^{-1}(\delta t_x/\delta t_y). \tag{33-55}$$

This process is known as <u>beam forming</u>. After forming beams at many angles and velocities, the results for a seismic surface wavetrain might appear as in Figure 33-8. Here the velocity changes (as a function of frequency) can be clearly seen. Such diagrams may be used to locate seismic sources insofar as $d\Delta/dT$ for body-waves is an accurate function of distance. The time interval $T_s - T_p$ can also be used.

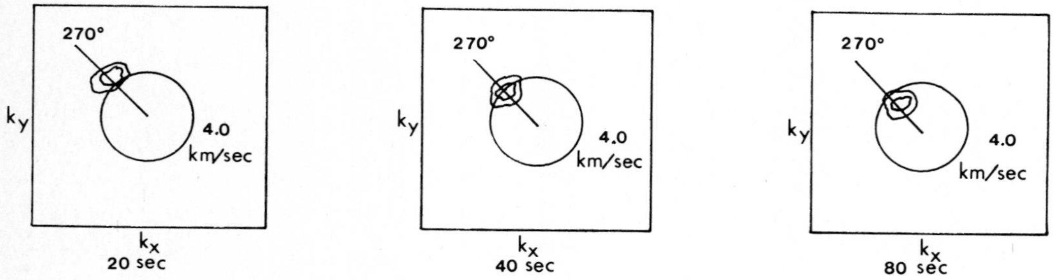

Figure 33-8. Wave-number diagrams as a function of period.

In forming the sum (33-51), there are two parameters left. If the A_n's are not identical, the response of Figure 33-7 will be altered. Specifically, by a proper choice of A_n's, one can place the zero-crossings at positions where there will be a null response to certain velocities. In this way some coherent noise can be eliminated. The spectral content of $A_n(\omega)$ is the second parameter, and it can be so chosen as to maximize the signal-to-noise ratio. These latter two operations are not simple, and the reader is referred to Capon (1973b) for a summary of array data processing techniques as applied to the Large Aperture Seismic Array (LASA) in Montana. Bungum and Husebye (1974) give a discussion of the characteristics of the Norwegian seismic array (NORSAR) and Weichert and Henger (1976) describe the Yellowknife Array (YKA) in Canada. Additional descriptions of the capabilities of arrays and the many uses which have been found for them are given in a volume edited by Beauchamp (1975).

PAPERS RELATING TO ANISOTROPIC MEDIA

See also general references given in Section 5.2.

General.

Backus, G. E., 1970. A geometrical picture of anisotropic elastic tensors. Rev. Geophys. Space Phys. 8, 633-671.

Buchwald, V. T., 1959. Elastic wave propagation in anisotropic media. Proc. Roy. Soc., London, A 253, 563-580.

Crampin, S., 1977. A review of the effects of anisotropic layering on the propagation of seismic waves. Geophys. J. 49, 9-27.

Duff, G. F. D., 1960. The Cauchy problem for elastic waves in an anisotropic medium. Phil. Trans. Roy. Soc., London, A 252, 249-273.

Lighthill, M. J., 1960. Anisotropic wave motions. Phil. Trans. Roy. Soc., London, A 252, 397-470.

Musgrave, M. J. P., 1959. The propagation of elastic waves in crystals and other anisotropic media. Rept. Prog. Phys. 22, 74-96.

Musgrave, M. J. P., 1961. The propagation of elastic waves in crystals and other anisotropic media. In: I. N. Sneddon and R. Hill (Editors), Progress in Solid Mechanics, Vol. 2. North-Holland, Amsterdam, pp. 64-85.

Synge, J. L., 1956. Flux of energy for elastic waves in anisotropic media. Proc. Roy. Irish Acad. A 58, 13-21.

Synge, J. L., 1957. Elastic waves in anisotropic media. J. Math. Phys. 35, 323-335.

Wheeler, L. T., 1970. Some results in the linear dynamical theory of anisotropic elastic solids. Q. Appl. Math 28, 91-101.

Lamb's Problem, Surface and Stoneley Waves.

Burridge, R., 1971. Lamb's problem for an anisotropic half-space. Q. J. Mech. Appl. Math 24, 81-98.

Abubakar, I. 1962a. Disturbance due to a line source in a semi-infinite transversely isotropic elastic medium. Geophys. J. 6, 337-359.

Abubakar, I., 1962b. Motion of the surface of a transversely isotropic half-space excited by a buried line source. Geophys. J. 7, 87-101.

Johnson, W. W., 1970. The propagation of Stoneley and Rayleigh waves in anisotropic elastic media. Bull., Seis. Soc. Amer. 60, 1105-1122. (This paper cites much previous work.)

Flat Layered Medium — Rays.

Daley, P. F., and F. Hron, 1977. Reflection and transmission coefficients for transversely isotropic media. Bull., Seis. Soc. Amer. 67, 661-675.

Keith, C. M. and S. Crampin, 1977a. Seismic body waves in anisotropic media: Reflection and refraction at a plane interface. Geophys. J. 49, 181-208.

Keith, C. M. and S. Crampin, 1977b. Seismic body waves in anisotropic media: Propagation through a layer. Geophys. J. 49, 209-223.

Keith, C. M. and S. Crampin, 1977c. Seismic body waves in anisotropic media: Synthetic seismograms. Geophys. J. 49, 225-243.

Musgrave, M. J. P., 1960. Reflexion and refraction of plane elastic waves at a plane boundary between aeolotropic media. Geophys. J. 3, 406-418.

Flat Layered Medium — Modes.

Anderson, D. L., 1961. Elastic wave propagation in layered anisotropic media. J. Geophys. Res. 66, 2953-2963.
Crampin, S., 1970. The dispersion of surface waves in multilayered anisotropic media. Geophys. J. 21, 387-402.
Crampin, S., 1975. Distinctive particle motion of surface waves as a diagnostic of anisotropic layering. Geophys. J. 40, 177-186.
Crampin, S., 1976. A comment on 'The early structural evolution and anisotropy of the oceanic upper mantle.' Geophys. J. 46, 193-197.
Crampin, S., and D. B. Taylor, 1971. The propagation of surface waves in anisotropic media. Geophys. J. 25, 71-87.
Smith, M. L., and F. A. Dahlen, 1973. The azimuthal dependence of Love and Rayleigh wave propagation in a slightly anisotropic medium. J. Geophys. Res. 78, 3321-3333.

Anisotropy due to prestress.

See also references to effects of prestress on source radiation patterns, page 365.

Dahlen, F. A., 1972. Elastic velocity anisotropy in the presence of an anisotropic initial stress. Bull., Seis. Soc. Amer. 62, 1183-1193.
Walton, K., 1973a. Seismic waves in pre-strained media. Geophys. J. 31, 374-394.
Walton, K., 1973b. Ray theory for a pre-strained medium. Geophys. J. 33, 223-235.
Walton, K., 1974. The seismological effects of elastic pre-straining within the earth. Geophys. J. 36, 651-677.

Anisotropy due to a magnetic field in a conductor.

Baños, A., 1956. Normal modes characterizing magnetoelastic plane waves. Phys. Rev. 104, 300-305.
Bazer, J., 1971. Geometrical magnetoelasticity. Geophys. J. 25, 207-237.
Bazer, J., and F. Karal, 1971. Simple wave motion in magnetoelasticity. Geophys. J. 25, 127-156.
Knopoff, L., 1955. The interaction between elastic wave motions and a magnetic field in electrical conductors. J. Geophys. Res. 60, 441-456.

Asymptotic Ray-Theory in anisotropic media.

Cerveny, V., 1972. Seismic rays and ray intensities in inhomogeneous anisotropic media. Geophys. J. 29, 1-13.
Cerveny, V., and I. Psencik, 1972. Rays and travel time curves in inhomogeneous anisotropic media. Z. Geophysik 38, 565-577.
Richards, P. G., 1971. An elasticity theorem for heterogeneous media, with an example of body wave dispersion in the earth. Geophys. J. 22, 453-472.

Measurements of elastic wave velocity in anisotropic material.

Baker, D. W., and N. L. Carter, 1972. Seismic velocity anisotropy calculated for ultramafic minerals and aggregates. In: H. C. Heard, I. Y. Borg, N. L. Carter, and C. B. Raleigh (Editors), Flow and fracture of rocks. Amer. Geophys. Union Monograph No. 16, Washington D.C. pp. 157-166.
Bennett, H. F., 1972. A simple seismic model for determining principal anisotropic directions. J. Geophys. Res. 77, 3078-3080.
Tillmann, S. E., and H. F. Bennett, 1973a. Ultrasonic shear wave birefringence as a test of homogeneous elastic anisotropy. J. Geophys. Res. 78, 7623-7629.
Tillmann, S. E., and H. F. Bennett, 1973b. A sonic method for petrographic analysis. J. Geophys. Res. 78, 8463-8469.

PAPERS RELATING TO INHOMOGENEOUS MEDIA

See also references given in Section 12.4 and Chapters 20-23. In addition, Ewing, Jardetzky, and Press (1957, Chapter 7) give many references to earlier literature.

General.

Alverson, R. C., F. C. Gair, and J. F. Hook, 1963. Uncoupled equations of motion in nonhomogeneous elastic media. Bull., Seis. Soc. Amer. 53, 1023-1030.

Ben-Menahem, A. and M. Weinstein, 1970. The P-SV decoupling condition and its bearing on the structure of the earth. Geophys. J. 21, 131-135.

Hook, J. F., 1961. Separation of the vector wave equation of elasticity for certain types of inhomogeneous isotropic media. J. Acoust. Soc. Amer. 33, 302-313.

Hook, J. F., 1962a. Generalization of a method of potentials for the vector wave equation of elasticity for inhomogeneous media. J. Acoust. Soc. Amer. 34, 354-355

Hook, J. F., 1962b. Contributions to a theory of separability of the vector wave equation of elasticity for inhomogeneous media. J. Acoust. Soc. Amer.34, 946-953.

Hook, J. F., 1962c. Green's function for axially symmetric elastic waves in an unbounded inhomogeneous medium having constant velocity gradients. J. Appl. Mech. 84, 293-298.

Hook, J. F. 1965. Determination of inhomogeneous media for which the vector wave equation of elasticity is separable. Bull., Seis. Soc. Amer. 55, 975-988.

Lock, M., 1963. Axially symmetric elastic waves in an unbounded inhomogeneous medium with exponentially varying properties. Bull., Seis. Soc. Amer. 53, 527-538.

Richards, P. G., 1971. Potentials for elastic displacement in spherically symmetric media. J. Acoust. Soc. Amer. 50, 188-197.

Richards, P. G., 1974. Weakly coupled potentials for high frequency elastic waves in continuously stratified media. Bull., Seis. Soc. Amer. 64, 1575-1588.

Singh, S. J., and A. Ben-Menahem, 1969a. Decoupling of the vector wave equation of elasticity for radially heterogeneous media. J. Acoust. Soc. Amer.46, 655-660.

Singh, S. J., and A. Ben-Menahem, 1969b. Asymptotoc theory of body waves in a radially heterogeneous earth. Bull., Seis. Soc. Amer. 59, 2039-2059.

Lamb's Problem.

Karlsson, T., and J. F. Hook, 1963. Lamb's problem for an inhomogeneous medium with constant velocities of propagation. Bull., Seis. Soc. Amer. 53, 1007-1022.

Layered Media — Modes.

Dutta, S., 1963. Rayleigh waves in a two-layer heterogeneous medium. Bull., Seis. Soc. Amer. 53, 517-526.

Hudson, J. A., 1961. Love waves in a heterogeneous medium. Geophys. J. 6, 131-147.

Knopoff, L., 1969. Phase and group slownesses in inhomogeneous media. J. Geophys. Res. 74, 1701.

Newlands, M., 1950. Rayleigh waves in a two-layer heterogeneous medium. Mon. Not. Roy. Astr. Soc., Geophys. Suppl. 6, 109-124.

Scott, R. A., 1970. Transient elastic waves in an inhomogeneous layer. Bull., Seis. Soc. Amer. 60, 383-392.

Singh, B. M., S. J. Singh, S. D. Chopra and M. L. Gognee, 1976. On Love waves in laterally and vertically heterogeneous layered media. Geophys. J. 45, 357-370.

Singh, V. P., 1974. Love-wave dispersion in a transversely isotropic and laterally inhomogeneous crustal layer. Bull., Seis. Soc. Amer. 64, 1967-1978.

Singh, V. P., 1977. SH waves in multilayered laterally heterogeneous media. Bull., Seis. Soc. Amer. 67, 331-343.

Vlaar, N. J., 1966a. The field from an SH point source in a continuously layered inhomogeneous medium: I. The field in a layer of finite depth. Bull., Seis. Soc. Amer. 56, 715-724.

Vlaar, N. J., 1966b. The field from an SH point source in a continuously layered inhomogeneous half-space: II. The field in a half-space. Bull., Seis. Soc. Amer. 56, 1305-1315.

APPENDIX C

PAPERS RELATING TO THE SCATTERING OF ELASTIC WAVES

Papers marked with a star (*) use either finite-element (Lysmer and Drake, 1972) or finite-difference (Boore, 1972) methods.

General.

See also Eringen and Suhubi (1975, Chapter 9) and Achenbach (1973, Chapter 9) for additional theory and references.

Gangi, A. F., and B. B. Mohanty, 1973. Babinet's principle for elastic waves. J. Acoust. Soc. Amer. 53, 525-534.

Knopoff, L., 1956. Diffraction of elastic waves. J. Acoust. Soc. Amer. 28, 217-229.

Scattering from Small Obstacles.

See also references in Section 22.4 for scattering and diffraction by a sphere.

Kennett, B. L. N., 1972a. Seismic waves in laterally inhomogeneous media. Geophys. J. 27, 301-325.

Kennett, B. L. N., 1972b. Seismic wave scattering by obstacles on interfaces. Geophys. J. 28, 249-266.

Kennett, B. L. N., 1973a. The effects of scattering on seismic wave pulses. Geophys. J. 32, 389-408.

Kennett, B. L. N., 1973b. Scattering approximations for thick and thin scatterers. Bull., Seis. Soc. Amer. 63, 1321-1325.

Pao, Y. H., and C. C. Mow, 1973. Diffraction of elastic waves and dynamic stress concentrations. Crane, Russak and Co., New York. (This book has a very large number of examples and references, particularly for scattering by circular, elliptic, and parabolic cylinders. However, only P-waves incident upon a sphere are treated, not SV-waves.)

Schwab, F., 1965. Scattering of shear waves by small transeismic obstacles. Geophysics 30, 24-31.

Teng, T-L., and P. G. Richards, 1969. Diffracted P, SV, and SH waves and their shadow boundary shifts. J. Geophys. Res. 74, 1537-1555.

Scattering from Cracks, Slits, and their Rigid Counterparts.

Cracks and slits are extensively treated by Eringen and Suhubi (1975, Sections 9.4, 9.5, 9.6, 9.17, and 9.18.) The references here are not included in their bibliography.

Achenbach, J. D., and A. K. Gautesen, 1977. Geometrical theory of diffraction for 3-D elastodynamics. J. Acoust. Soc. Amer. 61, 413-421.

Achenbach, J. D., A. K. Gautesen, and H. McMaken, 1978. Diffraction of point-source signals by a circular crack. Bull., Seis. Soc. Amer. 68, 889-905.

Datta, S. K., 1970. The diffraction of a plane compressional elastic wave by a rigid circular disc. Q. Appl. Math. 28, 1-14.

Fehler, M., and K. Aki., 1978. Numerical study of diffraction of plane elastic waves by a finite crack with application to location of a magma lens. Bull., Seis. Soc. Amer. 68, 573-598.

Fredericks, R. W., 1961. Diffraction of an elastic pulse in a loaded half-space. J. Acoust. Soc. Amer. 33, 17-22.

Fredericks, R. W. and L. Knopoff, 1960. The reflection of Rayleigh waves by a high impedance obstacle on a half-space. Geophysics 25, 1195-1202.

Gautesen, A. K., J. D. Achenbach, and H. McMaken, 1978. Surface waves in elastodynamic diffraction by cracks. J. Acoust. Soc. Amer. 63, 1824-1831.
Mal, A. K., 1971. Motion of a rigid disc in an elastic solid. Bull., Seis. Soc. Amer. 61, 1717-1729.
Mal, A. K., D. D. Ang, and L. Knopoff, 1968. Diffraction of elastic waves by a rigid circular disc. Proc. Cambridge Phil. Soc. 64, 237-247.

Scattering from Wedges.

Abo-Zena, A. M. and C-Y. King, 1973. SH pulse in an elastic wedge. Bull., Seis. Soc. Amer. 63, 1571-1582.
Alsop, L. E., A. S. Goodman, and S. Gregersen, 1974. Reflection and transmission of inhomogeneous waves with particular application to Rayleigh waves. Bull., Seis. Soc. Amer. 64, 1635-1652.
*Alterman, Z., and R. Nathaniel, 1975. Seismic waves in a wedge. Bull., Seis. Soc. Amer. 65, 1697-1719. (Several earlier papers of the senior author and her colleagues are referenced here.)
Forristall, G. Z., and J. D. Ingram, 1971. Elastodynamics of a wedge. Bull., Seis. Soc. Amer. 61, 275-287.
Gangi, A. F., 1967. Experimental determination of P wave / Rayleigh wave conversion coefficients at a stress-free wedge. J. Geophys. Res. 72, 5685-5692.
Hudson, J. A., 1963. SH waves in a wedge-shaped medium. Geophys. J. 7, 517-546.
Hudson, J. A., and L. Knopoff, 1964a. Transmission and reflection of surface waves at a corner, 1. Love waves. J. Geophys. Res. 69, 275-280.
Hudson, J. A., and L. Knopoff, 1964b. Transmission and reflection of surface waves at a corner, 2. Rayleigh waves (theoretical). J. Geophys. Res. 69, 281-290.
Kane, J., and J. Spence, 1963. Rayleigh wave transmission on elastic wedges. Geophysics 28, 715-723.
Kane, J., and J. Spence, 1965. The theory of surface wave diffraction by symmetric crustal discontinuities. Geophys. J. 9, 423-438.
*Kelley, K. R., R. W. Ward, S. Treitel, and R. M. Alford, 1976. Synthetic seismograms: a finite-difference approach. Geophysics 41, 2-27.
Kraut, E. A., 1968. Diffraction of elastic waves by a rigid 90° wedge. Bull., Seis. Soc. Amer. 58, 1083-1115.
Lapwood, E. R., 1961. The transmission of a Rayleigh pulse round a corner. Geophys. J. 4, 174-196.
Lewis, D., and J. W. Dally, 1970. Photoelastic analysis of Rayleigh wave propagation in wedges. J. Geophys. Res. 75, 3387-3398.
*Loewenthal, D., and Z. Alterman, 1972. Theoretical seismograms for the two welded quarter-planes. Bull., Seis. Soc. Amer. 62, 619-630.
McGarr, A., and L. E. Alsop, 1967. Transmission and reflection of Rayleigh waves at vertical boundaries. J. Geophys. Res. 72, 2169-2188.
Mal, A. K., and L. Knopoff, 1966. Transmission of Rayleigh waves at a corner. Bull., Seis. Soc. Amer. 56, 455-466.
*Munasinghe, M., and G. W. Farnell, 1973. Finite difference analysis of Rayleigh wave scattering at vertical discontinuities. J. Geophys. Res. 78, 2454-2466.
*Ottaviani, M., 1971. Elastic-wave propagation in two evenly-welded quarter-spaces. Bull., Seis. Soc. Amer. 61, 1119-1152.
Pilant, W. L., L. Knopoff, and F. Schwab, 1964. Transmission and reflection of surface waves at a corner, 3. Rayleigh waves (experimental). J. Geophys. Res. 69, 291-298.
Sato, R., 1963. Diffraction of SH waves at an obtuse-angled corner. J. Phys. Earth 11, 1-17.
Viswanathan, K., 1966. Wave propagation in welded quarter-spaces. Geophys. J. 11, 293-322.
Viswanathan, K., J. T. Kuo, and E. R. Lapwood, 1971. Reflection and transmission of Rayleigh waves in a wedge — I. Geophys. J. 24, 401-414.
Viswanathan, K., and A. Roy, 1973. Reflection and transmission of Rayleigh waves in a wedge — II. Geophys. J. 32, 459-478.

Scattering from Corrugated Surfaces and Interfaces.

*Alterman, Z. and J. Aboudi, 1971. Propagation of elastic waves caused by an impulsive source in a half space with a corrugated surface. Geophys. J. 24, 59-76.

Asano, S., 1966. Reflection and refraction of elastic waves at a corrugated interface. Bull., Seis. Soc. Amer. 56, 201-221. (Several earlier papers by this author are referenced here.)

Bhattacharyya, S., 1973. Effect of symmetric undulations on the surface on incident elastic waves. Bull., Seis. Soc. Amer. 63, 457-467.

Brekhovskikh, L. M., 1960. Propagation of surface Rayleigh waves along the uneven boundary of an elastic body. Soviet Physics, Acoustics, 5, 288-295.

Levy, A., and H. Deresiewicz, 1967. Reflection and transmission of elastic waves in a system of corrugated layers. Bull., Seis. Soc. Amer. 57, 393-419.

Onda, I., 1967. Rayleigh wave propagation along an undulatory surface — comparison with waves through a heterogeneous medium with a periodic structure. Tokyo Univ., Bull. Earthquake Res. Inst. 45, 589-600.

Scattering from Topographic Irregularities.

*Aboudi, J., 1971. The motion excited by an impulsive source in an elastic half-space with a surface obstacle. Bull., Seis. Soc. Amer. 61, 747-763.

Abubakar, I., 1962. Scattering of plane elastic waves at rough surfaces, I. Proc. Cambridge Phil. Soc. 58, 136-157.

Abubakar, I., 1963. Scattering of plane elastic waves at rough surfaces, II. Proc. Cambridge Phil. Soc. 59, 231-248.

Aki, K., and K. L. Larner, 1970. Surface motion of a layered medium having an irregular interface due to incident plane SH waves. J. Geophys. Res. 75, 933-954.

Boore, D. M., 1972. A note on the effect of simple topography on seismic SH waves. Bull., Seis. Soc. Amer. 62, 275-284.

Boore, D. M., K. L. Larner, and K. Aki, 1971. Comparison of two independent methods for the solution of wave-scattering problems: response of a sedimentary basin to vertically incident SH waves. J. Geophys. Res. 76, 558-569.

Bouchon, M., 1973. Effect of topography on surface motion. Bull., Seis. Soc. Amer. 63, 615-632.

Bouchon, M., and K. Aki, 1977. Near-field of a seismic source in a layered medium with irregular interfaces. Geophys. J., 50, 669-684.

Dally, J. W., and D. Lewis, 1968. A photoelastic analysis of propagation of Rayleigh waves past a step change in elevation. Bull., Seis. Soc. Amer. 58, 539-563.

Gilbert, F., and L. Knopoff, 1960. Seismic scattering from topographic irregularities. J. Geophys. Res. 65, 3437-3444.

Hudson, J. A., 1967. Scattered surface waves from a surface obstacle. Geophys. J. 13, 441-458.

McIvor, I. K., 1969. Two-dimensional scattering of a plane compressional wave by surface imperfections. Bull., Seis. Soc. Amer. 59, 1349-1364.

Mal, A. K., and L. Knopoff, 1965. Transmission of Rayleigh waves past a step change in elevation. Bull., Seis. Soc. Amer. 55, 319-334.

Martel, L., M. Munasinghe, and G. W. Farnell, 1977. Transmission and reflection of Rayleigh waves through a step. Bull., Seis. Soc. Amer. 67, 1277-1290.

Rogers, A. M., L. J. Katz, and T. J. Bennett, 1974. Topographic effects on ground motion for incident P waves: a model study. Bull., Seis. Soc. Amer. 64, 437-456.

Sabina, F. J., and J. R. Willis, 1975. Scattering of SH waves by a rough half-space of arbitrary slope. Geophys. J. 42, 685-703.

Singh, S. K., and F. J. Sabina, 1977. Ground-motion amplification by topographic depressions for incident P waves under acoustic approximation. Bull., Seis. Soc. Amer. 67, 345-352.

*Smith, W. D., 1975. The application of finite element analysis to body wave propagation problems. Geophys. J. 42, 747-768.

Tharpar, M. R., 1970. Rayleigh wave propagation and perturbed boundaries. Canad. J. Earth Sci. 7, 1449-1461.

Trifunac, M. D., 1971. Surface motion of a semi-cylindrical alluvial valley for incident plane SH waves. Bull., Seis. Soc. Amer. 61, 1755-1770.
Trifunac, M. D., 1973. Scattering of plane SH waves by a semi-cylindrical canyon. Intern. J. Earthquake Eng. and Stress Dynamics 1, 267-281.
Wong, H. L., and M. D. Trifunac, 1974a. Scattering of plane SH waves by a semi-elliptical canyon. Intern. J. Earthquake Eng. and Stress Dynamics 3, 152-169.
Wong, H. L., and M.D. Trifunac, 1974b. Surface motion of a semi-elliptical alluvial valley for incident plane SH waves. Bull., Seis. Soc. Amer. 64, 1389-1408.
Wong, H. L., and P. C. Jennings, 1975. Effects of canyon topography on strong ground motion. Bull., Seis. Soc. Amer. 65, 1239-1257.

Scattering Accompanying Propagation in Non-Uniform Waveguides.

Alsop, L. E., 1966. Transmission and reflection of Love waves at a vertical discontinuity. J. Geophys. Res. 71, 3969-3984.
*Boore, D. M., 1970. Love waves in nonuniform wave guides: finite difference calculations. J. Geophys. Res. 75, 1512-1527.
Drake, L. A., 1972a. Love and Rayleigh waves in nonhorizontally layered media. Bull., Seis. Soc. Amer. 62, 1241-1258.
*Drake, L. A., 1972b. Rayleigh waves at a continental boundary by the finite element method. Bull., Seis. Soc. Amer. 62, 1259-1268.
Gjevik, B., 1973. A variational method for Love waves in nonhorizontally layered structures. Bull., Seis. Soc. Amer. 63, 1013-1023.
Gregersen, S., and L. E. Alsop, 1974. Amplitudes of horizontally refracted Love waves Bull., Seis. Soc. Amer. 64, 535-553.
Gregersen, S., and L. E. Alsop, 1976. Mode conversion of Love waves at a continental margin. Bull., Seis. Soc. Amer. 66, 1855-1872.
Herrera, I., 1964. A perturbation method for elastic wave propagation, 1. Nonparallel boundaries. J. Geophys. Res. 69, 3845-3851.
Herrera, I., and A. K. Mal, 1965. A perturbation method for elastic wave propagation, 2. Small inhomogeneities. J. Geophys. Res. 70, 871-883.
Hudson, J.A., 1977. The passage of elastic waves through an anomalous region — IV. Transmission of Love waves through a laterally varying structure. Geophys. J. 49, 645-654.
Kazi, M. H., 1978a. The Love wave scattering matrix for a continental margin (theoretical). Geophys. J. 52, 25-44.
Kazi, M. H., 1978b. The Love wave scattering matrix for a continental margin (numerical). Geophys. J. 53, 227-243.
Kennett, B. L. N., 1973. The interaction of seismic waves with horizontal velocity contrasts. Geophys. J. 33, 431-450.
Kennett, B. L. N., 1974. The interaction of seismic waves with horizontal velocity contrasts — II. Diffraction effects for SH wave pulses. Geophys. J. 37, 9-22.
Kennett, B. L. N., 1975a. The interaction of seismic waves with horizontal velocity contrasts — III. The effects of horizontal transition zones. Geophys. J. 41, 29-36.
Kennett, B. L. N., 1975b. Theoretical seismogram calculations for laterally varying crustal structures. Geophys. J. 42, 579-589.
Knopoff, L., and J. A. Hudson, 1964. Transmission of Love waves past a continental margin. J. Geophys. Res. 69, 1649-1654.
Knopoff, L., and A. K. Mal, 1967. Phase velocity of surface waves in the transition zone of continental margins. J. Geophys. Res. 72, 1769-1776.
Kuo, J. T., and J. E. Nafe, 1962. Period equation for Rayleigh waves in a layer overlying a half space with a sinusoidal interface. Bull., Seis. Soc. Amer. 52, 807-822.
*Lysmer, J., 1970. Lumped mass method for Rayleigh waves. Bull., Seis. Soc. Amer. 60, 89-104.
*Lysmer, J., and L. A. Drake, 1971. The propagation of Love waves across nonhorizontally layered structures. Bull., Seis. Soc. Amer. 61, 1233-1251.
McGarr, A., 1965. Amplitude variations of Rayleigh waves — Propagation across a continental margin. Bull., Seis. Soc. Amer. 59, 1281-1305.

McGarr, A., 1969. Amplitude variations of Rayleigh waves — Horizontal refraction. Bull., Seis. Soc. Amer. 59, 1307-1334.

Mal, A. K., and I. Herrera, 1965. Scattering of Love waves by a constriction in the crust. J. Geophys. Res. 70, 5125-5133.

Obukhov, G. G., 1963a. The effect of periodic irregularities in relief on the dispersion curves of seismic surface waves. Izv., Akad. Nauk, SSSR, Ser. Geofiz. 546-551 (in Russian); Bull., Acad. Sci., USSR, Geophys. Ser., 340-342 (English translation).

Obukhov, G. G., 1963b. Propagation of Love surface waves in the earth's crust as derived from a model with periodic irregularities. Izv., Akad. Nauk, SSSR, Ser. Geofiz., 1649-1657 (in Russian); Bull., Acad. Sci., USSR, Geophys. Ser., 996-1000 (English translation).

Takahashi, T., 1964. Transmission of Love waves in a half-space with a surface layer whose thickness varies hyperbolically. Bull., Seis. Soc. Amer. 54, 611-625.

Wolf, B., 1970. Propagation of Love waves in layers with irregular boundaries. Pure Appl. Geophys. 78, 48-57.

Woodhouse, J. H., 1974. Surface waves in a laterally varying layered structure. Geophys. J. 37, 461-490.

Scattering from Random Inhomogeneities and Irregularities.

Aki, K., 1969. Analysis of the seismic coda of local earthquakes as scattered waves. J. Geophys. Res. 74, 615-631.

Aki, K., 1973. Scattering of P waves under the Montana LASA. J. Geophys. Res. 78, 1334-1346.

Berteussen, K. A., A. Christofferson, E. S. Husebye, and A. Dahle, 1975. Wave scattering theory in analysis of P-wave anomalies at NORSAR and LASA. Geophys. J. 42, 403-417.

Capon, J., 1974. Characterization of crust and upper mantle structure under LASA as a random medium. Bull., Seis. Soc. Amer. 64, 235-266.

Capon, J., and K. A. Berteussen, 1974. A random medium analysis of crust and upper mantle structure under NORSAR. Geophys. Res. Lett. 1, 327-328.

Greenfield, R. J., 1971. Short-period P-wave generation by Rayleigh-wave scattering at Novaya Zemlya. J. Geophys. Res. 76, 7988-8002.

Hudson, J., 1977. Scattered waves in the coda of P. J. Geophys. 43, 359-374.

Hudson, J. A. and L. Knopoff, 1966. Signal generated seismic noise. Geophys. J. 11, 19-24.

Hudson, J. A., and L. Knopoff, 1967. Statistical properties of Rayleigh waves due to scattering by topography. Bull., Seis. Soc. Amer. 57, 83-90.

Knopoff, L., and J. A. Hudson, 1964. Scattering of elastic waves by small inhomogeneities. J. Acoust. Soc. Amer. 36, 338-343.

Knopoff, L., and J. A. Hudson, 1967. Frequency dependence of amplitude of scattered elastic waves. J. Acoust. Soc. Amer. 42, 18-20.

ADDITIONAL REFERENCE SOURCES

Bibliography of Seismology.

Bull., Seis. Soc. Amer. 17, 150-182, 218-248.
Bull., Seis. Soc. Amer. 18, 16-23, 110-125, 214-235, 267-283.
Bull., Seis. Soc. Amer. 19, 206-227.
Publ. Dom. Obs. 10, 12, 13, 14, 22, 30.

Bibliography of Seismology, New Series. (Intern. Seis. Centre, Edinburgh)

This is a continuing series starting in 1965 with Vol. 1.

Symposia on Theory and Computers in Geophysics, Proceedings.

1.	Moscow-Leningrad, 1964	Rev. Geophys. 3, No. 1, 1965.
2.	Rehovoth, 1965	Geophys. J., Roy. Astr. Soc. 11, Nos. 1-2, 1966.
3.	Cambridge, Englend, 1966	Geophys. J., Roy. Astr. Soc. 13, Nos. 1-3, 1967.
4.	Trieste, 1967	Suppl. al Nuovo Cimento 6, No. 1, 1968.
5.	Tokyo-Kyoto, 1968	J. Phys. Earth 16, 1968.
6.	Copenhagen, 1969	Geophys. J., Roy. Astr. Soc. 21, Nos. 3-4, 1970.
7.	Södergarn, 1970	Geophys. J., Roy. Astr. Soc. 25, Nos. 1-3, 1971.
8.	Moscow, 1971	No proceedings published.
9.	Banff, 1972	Geophys. J., Roy. Astr. Soc. 35, Nos. 1-3, 1973.
10.	Cambridge, England, 1974	Geophys. J., Roy. Astr. Soc. 42, Nos. 2-3, 1975.
11.	Seeheim/Odenwald, 1976	J. Geophys. 43, 1977.
12.	Caracas, 1978	To be published.

U. S. National Reports to the IUGG.

This is a continuing series.

Adams, L. H. (Editor), 1960.
U. S. National Report, 1957-1960, Twelfth General Assembly, IUGG: Seismology and Physics of the Earth's Interior. Trans., Amer. Geophys. Union 41, 145-176.

Adams, L. H. (Editor), 1963.
U. S. National Report, 1960-1963, Thirteenth General Assembly, IUGG: Seismology and Physics of the Earth's Interior. Trans., Amer. Geophys. Union 44, 327-364.

Adams, L. H., and J. F. Schairer (Editors), 1967.
U. S. National Report, 1963-1967, Fourteenth General Assembly, IUGG: Seismology and Physics of the Earth's Interior. Trans., Amer. Geophys. Union 48, 389-426.

Bell, P. M. (Editor), 1971.
U. S. National Report, 1967-1971, Fifteenth General Assembly, IUGG: Seismology, EΦS, Trans., Amer. Geophys. Union 52, IUGG 157-194 (Special page numbering, following p. 440).

Bell, P. M. (Editor), 1975.
U. S. National Report, 1971-1975, Sixteenth General Assembly, IUGG: Seismology. Rev. Geophys. Space Phys. 13, 295-325.

BIBLIOGRAPHY

Numbers at the end of the entry are pages on which there is reference to the entry.

Geophys. J. refers to the Geophysical Journal of the Royal Astronomical Society.

Abramovici, F., 1968. Mode theory versus theoretical seismograms for a layered solid. Geophys. J. 16, 9-20. 181

Abramovici, F., 1970. Numerical seismograms for a layered elastic solid. Bull., Seis. Soc. Amer. 60, 1861-1876. 229

Abramovici, F., and Z. Alterman, 1965. Computations pertaining to the problem of propagation of a seismic pulse in a layered solid. In: B. Adler, S. Fernbach, and M. Rotenberg (Editors), Methods in Computational Physics, Vol.4, Applications in Hydrodynamics. Academic, New York, pp. 349-379. 230

Abramovitz, M., and I. A. Stegun, 1965. Handbook of mathematical functions (Eighth Printing). Dover, New York. 91, 332

Achenbach, J. D., 1973. Wave propagation in elastic solids. North-Holland, Amsterdam. 38, 52, 70, 80

Aggarwal, Y. P., L. R. Sykes, J. Armbruster, and M. L. Sbar, 1973. Premonitory changes in seismic velocities and prediction of earthquakes. Nature 241, 101-104. 410

Aggarwal, Y. P., L. R. Sykes, D. W. Simpson, and P. G. Richards, 1975. Spatial and temporal variations in t_s/t_p and in P wave residuals at Blue Mountain Lake, New York: Application to earthquake prediction. J. Geophys. Res. 80, 718-732. 410

Agnew, D., J. Berger, R. Buland, W. Farrell, and F. Gilbert, 1976. International deployment of accelerometers: a network for very long period seismology. E⊕S, Trans., Amer. Geophys. Union 57, 180-188. 435

Agrawal, R. C., 1978. The GH branch of PKP from deep-focus earthquakes. Geophys. J. 53, 459-465. 302

Aki, K., 1960a. Study of earthquake mechanism by a method of phase equalization applied to Rayleigh and Love waves. J. Geophys. Res. 65, 729-740. 256, 375

Aki, K., 1960b. Interpretation of source functions of circum-Pacific earthquakes obtained from long-period Rayleigh waves. J. Geophys. Res. 65, 2405-2417. 375

Aki, K., 1960c. Further study of the mechanism of circum-Pacific earthquakes from Rayleigh Waves. J. Geophys. Res. 65, 4165-4172. 375

Aki, K., 1961. Crustal structure in Japan from the phase velocity of Rayleigh Waves. Tokyo University, Bull. Earthquake Res. Inst. 39, 255-283. 250

Aki, K., 1967. Scaling law of seismic spectrums. J. Geophys. Res. 72, 1217-1231. 384, 385, 389

Aki, K., 1977. Three dimensional seismic velocity anomalies in the lithosphere. J. Geophys. 43, 235-242. 301

Aki, K., and B. Chouet, 1975. Origin of Coda waves: source, attenuation, and scattering effects. J. Geophys. Res. 80, 3322-3342. 384

Aki, K., and W. H. K. Lee, 1976. Determination of three-dimensional velocity anomalies under a seismic array using P arrival times from local earthquakes: 1. A homogeneous initial model. J. Geophys. Res. 81, 4381-4399. 301

Aki, K., and Y-B.Tsai, 1972. Mechanism of Love-wave excitation by explosive sources. J. Geophys. Res. 77, 1452-1475. 365

Aki, K., M. Bouchon, and P. Reasenberg, 1974. Seismic source function for an underground nuclear explosion. Bull., Seis. Soc. Amer. 64, 131-148. 394, 395, 396

Aki, K., A. Christoffersson, and E. S. Husebye, 1976. Three-dimensional seismic structure of the lithosphere under Montana LASA. Bull., Seis. Soc. Amer. 66, 501-524. 302

Aki, K., A. Christoffersson, and E. S. Husebye, 1977. Determination of the three-dimensional seismic structure of the lithosphere. J. Geophys. Res. 82, 277-296. 302

Alexander, S. S., and R. A. Phinney, 1966. A study of the core-mantle boundary using P waves diffracted by the earth's core. J. Geophys. Res. 71, 5943-5958. 302

Alsop, L. E., 1968. An orthonormality relation for elastic body waves. Bull., Seis. Soc. Amer. 58, 1949-1954. 183

Alsop, L. E., 1970. The leaky-mode period equation - a plane-wave approach. Bull., Seis. Soc. Amer. 60, 1989-1998. 194

Alsop, L. E., G. H. Sutton, and M. Ewing, 1961. Measurement of Q for very long period free oscillations. J. Geophys. Res. 66, 2911-2915. 343

Alsop, L. E., A. S. Goodman, and S. Gregersen, 1974. Reflection and transmission of inhomogeneous waves with particular application to Rayleigh waves. Bull., Seis. Soc. Amer. 64, 1635-1652. 102

Alterman, Z. S., and J. Aboudi, 1970. Source of finite extent, applied force and couple in an elastic half-space. Geophys. J. 21, 47-64. 356

Alterman, Z. S., and D. Loewenthal, 1972. Computer generated seismograms. In: B. A. Bolt (Editor), Methods in computational physics, Vol. 12. Seismology: Body waves and sources. Academic, New York, pp. 35-164. 230, 302, 321

Alterman, Z. S., H. Jarosch, and C. L. Pekeris, 1959. Oscillations of the earth. Proc. Roy. Soc. (London) A 252, 80-95. 305

Anderson, D. L., 1964. Universal Dispersion Tables I. Love waves across oceans and continents on a spherical earth. Bull., Seis. Soc. Amer. 54, 681-726. 258, 312

Anderson, D. L. and C. B. Archambeau, 1964. The anelasticity of the earth. J. Geophys. Res. 69, 2071-2084. 345

Anderson, D. L. and D. G. Harkrider, 1968. Universal Dispersion Tables II. Variational parameters for amplitudes, phase velocity and group velocity for first four Love modes for an oceanic and a continental earth model. Bull., Seis. Soc. Amer. 58, 1407-1499. 239, 258, 312

Anderson, D. L. and R. S. Hart, 1976. An earth model based on free oscillations and body waves. J. Geophys. Res. 81, 1461-1475. (Corrections, J. Geophys. Res. 81, 5348, 1976.) 322

Anderson, D. L. and R. S. Hart, 1978a. Attenuation models of the earth. Phys. Earth and Planet. Inter. 16, 289-306. 335, 342, 343, 345

Anderson, D. L. and R. S. Hart, 1978b. Q of the earth. J. Geophys. Res. 83, 5869-5882. 335, 340, 342, 343, 344, 345, 346

Anderson, D. L. and R. L. Kovach, 1964. Attenuation in the mantle and rigidity of the core from multiply reflected core phases. Proc. Nat. Acad. Science., U.S.A. 51, 168-172. 340, 341

Anderson, D. L. and R. L. Kovach, 1969. Universal Dispersion Tables III. Free oscillation variational parameters. Bull., Seis. Soc. Amer. 59, 1667-1693. 320

Anderson, D. L. and M. N. Toksöz, 1963. Surface waves on a spherical earth. J. Geophys. Res. 68, 3483-3500. 35, 243, 258, 312

Anderson, D. L., A. Ben-Menahem, and C. B. Archambeau, 1965. Attenuation of seismic energy in the upper mantle. J. Geophys. Res. 70, 1441-1448. 343, 344, 345, 348

Anderson, D. L., H. Kanamori, R. S. Hart, and H-P. Liu, 1977. The earth as a seismic absorption band. Science 196, 1104-1106. 348, 349

Anderssen, R. S., 1977. The effect of discontinuities in density and shear velocity on the asymptotic overtone structure of torsional eigenfrequencies of the earth. Geophys. J. 50, 303-309. 315

Archambeau, C. B., 1968. General theory of elastodynamic source fields. Rev. Geophys. 6, 241-288. 351, 360

Archambeau, C. B., 1972. The theory of stress wave radiation from explosions in prestressed media. Geophys. J. 29, 329-366. (Append. in 31, 361, 1973) 365

Archambeau, C. B., 1975. Developments in seismic source theory. Rev. Geophys. and Space Phys. 13, 304-306. 351

Archambeau, C. B., and C. Sammis, 1970. Seismic radiation from explosions in prestressed media and the measurement of tectonic stress in the earth. Rev. Geophys. and Space Phys. 8, 473-499. 365

Archambeau, C. B., E. A. Flinn, and D. G. Lambert, 1969. Fine structure of the upper mantle. J. Geophys. Res. 74, 5825-5865. 340

Arons, A. B., and D. R. Yennie, 1950. Phase distortion of acoustic pulses obliquely reflected from a medium of higher sound velocity. J. Acoust. Soc. Amer. 22, 231-237. 90

Asada, T., 1957. Observations of near-by microearthquakes with ultrasensitive seismometers. J. Phys. Earth 5, 83-113. 406

Asada, T., S. Suyehiro, and K. Akamatu, 1958. Observations of near-by microearthquakes with ultra-sensitive seismometers at Matsushiro, Japan. J. Phys. Earth 6, 23-33. 406

Auld, B. A., 1973. Acoustic Fields and Waves in Solids (2 Vols.). Wiley-Interscience, New York. 34, 38

Bache, T. C., and D. G. Harkrider, 1976. The body waves due to a general seismic source in a layered earth model: 1. Formulation of the theory. Bull., Seis. Soc. Amer. 66, 1805-1819. 364

Backus, G. E., 1962. Long wave elastic anisotropy produced by horizontal layering. J. Geophys. Res. 67, 4427-4440. 34

Backus, G. E., 1977a. Interpreting the seismic glut moments of total degree two or less. Geophys. J. 51, 1-25. 360

Backus, G. E., 1977b. Seismic sources with observable glut moments of spatial degree two. Geophys. J. 51, 27-45. 360

Backus, G. E., and F. Gilbert, 1961. The rotational splitting of the free oscillations of the earth. Proc. Nat. Acad. Sci., U.S.A. 47, 362-371 307

Backus, G. E., and J. Gilbert, 1967. Numerical applications of a formulation for geophysical inverse problems. Geophys. J. 13, 247-276. 260, 319

Backus, G. E., and F. Gilbert, 1968. The resolving power of gross earth data. Geophys. J. 16, 169-205. 260, 319

Backus, G. E., and F. Gilbert, 1970. Uniqueness in the inversion of inaccurate gross earth data. Phil. Trans. Roy. Soc. London, A 266, 123-192. 260, 319

Backus, G. E., and M. Mulcahy, 1976a. Moment tensors and other phenomenological descriptions of seismic sources — I. Continuous displacements. Geophys. J. 46, 341-361. 360

Backus, G. E., and M. Mulcahy, 1976b. Moment tensors and other phenomenological descriptions of seismic sources — II. Discontinuous displacements. Geophys. J. 47, 301-329. 360

Baker, R. G., 1970. Determining magnitude from LG. Bull., Seis. Soc. Amer. 60, 1907-1919. 390

Bakun, W. H., C. G. Bufe, and R. M. Stewart, 1976. Body-wave spectra of central California earthquakes. Bull., Seis. Soc. Amer. 66, 363-384. 385

Bamford, D., 1973. An example of the iterative approach to time-term analysis. Geophys. J. 31, 365-372. 225

Bamford, D., 1976. MOZAIC time-term analysis. Geophys. J. 44, 433-446. 225, 226

Bamford, D., and S. Crampin, 1977. Seismic anisotropy — the state of the art. Geophys. J. 49, 1-8. 35

Barazangi, M., and J. Dorman, 1969. World seismicity maps compiled from ESSA, Coast and Geodetic Survey, Epicenter data, 1961-1967. Bull., Seis. Soc. Amer. 59, 369-380. 399

Barr, K. G., 1971. The statistics of the time-term method. Bull., Seis. Soc. Amer. 61, 1853-1854. 225

Basham, P. W., 1969. Canadian magnitudes of earthquakes and nuclear explosions in southwestern North America. Geophys. J. 17, 1-13. 397

Båth, M., 1958. The energies of seismic body waves and surface waves. In: H. Benioff, M. Ewing, B. F. Howell, Jr., and F. Press (Editors), Contributions in Geophysics: In honor of Beno Gutenberg. Pergamon, New York, pp. 1-16 392

Båth, M., 1966. Earthquake energy and magnitude. Phys. Chem. Earth 7, 115-165. 386

Båth, M., 1974. Spectral Analysis in Geophysics. Elsevier Sci. Publ. Co., Amsterdam. 228, 247, 248, 316, 335

Beauchamp, K. G., (Ed.), 1974. Exploitation of Seismograph Networks. (NATO Adv. Study Inst., Sandefjord, Norway, 1974.) Noordhoff, Leiden. 438

Beloussov, V. G., B. S. Volvovsky, I. S. Volvovsky, and V. S. Ryaboi, 1962. Experimental data on recording of deep reflected waves. Izv. Akad. Nauk SSSR, Ser. Geofiz., 1034-1044. (In Russian) Bull., Acad. Sci. USSR, Geophys. Ser., 662-669. (English trans.) 216, 217

Benioff, H., 1932. A new vertical seismograph. Bull., Seis. Soc. Amer. 22, 155-169. 432

Benioff, H., 1954. Orogenesis and deep crustal structure — additional evidence from seismology. Bull. Geol. Soc. Amer. 65, 385-400. 401
Benioff, H., 1955. Earthquake seismographs and associated instruments. In: H. E. Landsberg (Editor), Advances in Geophysics, Vol. 2, Academic, New York, pp. 219-275. 424
Benioff, H., 1959. Fuzed-quartz extensometer for secular, tidal, and seismic strains. Bull. Geol. Soc. Amer.,70, 1019-1032. 432
Ben-Menahem, A., 1961. Radiation of seismic surface waves from a finite moving source. Bull., Seis. Soc. Amer. 51, 401-435. 360
Ben-Menahem, A., 1962. Radiation of seismic body waves from a finite moving source in the earth. J. Geophys. Res. 67, 345-350. 360
Ben-Menahem, A., 1964. Mode-ray duality. Bull., Seis. Soc. Amer. 54, 1315-1321. 313, 314
Ben-Menahem, A., 1964a. Spectral response of an elastic sphere to dipolar point-sources. Bull., Seis. Soc. Amer. 54, 1323-1340. 306
Ben-Menahem, A., 1965. Observed attenuation and Q values of seismic surface waves in the upper mantle. J. Geophys. Res. 70, 4641-4651. 343
Ben-Menahem, A., 1971. The force system of the Chilean earthquake of 1960 May 22. Geophys. J. 25, 407-417. 377
Ben-Menahem, A., and D. G. Harkrider, 1964. Radiation patterns of seismic surface waves from buried dipolar sources in a flat stratified earth. J. Geophys. Res. 69, 2605-2620. 364
Ben-Menahem, A., and S. J. Singh, 1968. Multipolar elastic fields in a layered half-space. Bull., Seis. Soc. Amer. 58, 1519-1572. 364
Ben-Menahem, A., and S. J. Singh, 1972. Computation of models of elastic dislocations in the earth. In: B. A. Bolt (Editor), Methods in computational physics, Academic, New York, pp. 299-375. 351, 360, 363
Ben-Menahem, A., and M. N. Toksöz, 1962a. Source-mechanism from spectra of long-period seismic surface waves, 1. The Mongolian earthquake of December 4, 1957. J. Geophys. Res. 67, 1943-1955. 375
Ben-Menahem, A., and M. N. Toksöz, 1962b. Source-mechanism from spectrums of long-period surface waves, 2. The Kamchatka earthquake of November 4, 1952. J. Geophys. Res. 68, 5207-5222. 375, 377
Ben-Menahem, A., and M. N. Toksöz, 1963. Source-mechanism from spectra of long-period seismic surface waves, 3. The Alaskan earthquake of July 10, 1958. Bull., Seis. Soc. Amer. 53, 905-920. 375
Ben-Menahem, A., and M. Vered, 1973. Extension and interpretation of the Cagniard-Pekeris method for Dislocation Sources. Bull., Seis. Soc. Amer. 63, 1611-1636. 363
Ben-Menahem, A., S. W. Smith and T-L. Teng, 1965. A procedure for source studies from spectrums of long-period body waves. Bull., Seis. Soc. Amer. 55, 203-235. 364
Ben-Menahem, A., H. Jarosch, and M. Rosenman, 1968. Large scale processing of seismic data in search of regional and global stress patterns. Bull., Seis. Soc. Amer. 58, 1899-1932. 368
Ben-Menahem, A., S. J. Singh, and F. Solomon, 1969. Static Deformation of a spherical earth model by internal dislocations. Bull., Seis. Soc. Amer. 59, 813-853. 365
Ben-Menahem, A., M. Rosenman, and D. G. Harkrider, 1970. Fast evaluation of source parameters from isolated surface wave signals. Bull., Seis. Soc. Amer. 60, 1337-1387. 364
Ben-Menahem, A., M. Israel, and U. Levité, 1971. Theory and computation of amplitudes of terrestrial line spectra. Geophys. J. 25, 307-406. 364, 377
Ben-Menahem, A., M. Rosenman and M. Israel, 1972. Source mechanism of the Alaskan earthquake of 1964 from amplitudes of free oscillations and surface waves. Phys. Earth and Planet. Inter. 5, 1-29. 377
Berger, J., and J. Levine, 1974. The spectrum of earth strain from 10^{-8} to 10^{2} hz. J. Geophys. Res. 79, 1210-1214. 418
Berry, M. J., 1971. Depth uncertainties from seismic first-arrival refraction studies. J. Geophys. Res. 76, 6464-6468. 221

Berry, M. J., and G. F. West, 1966. An interpretation of the first arrival data of the Lake Superior experiment by the time-term method. Bull., Seis. Soc. Amer. 56, 141-171. 225
Berry, M. J., and G. F. West, 1966a. Reflected and head wave amplitudes in a medium of several layers. In: J. S. Steinhart and T. J. Smith (Editors), The Earth Beneath the Continents. Amer. Geophys. Union, Monograph #10, Washington, D. C., pp. 464-481. 230
Berteussen, K. A., 1975. Array analysis of lateral inhomogeneities in the deep mantle. Earth Planet. Sci. Lett. 28, 212-216. 302
Berteussen, K. A., 1976. The origin of slowness and azimuth anomalies at large arrays. Bull., Seis. Soc. Amer. 66, 719-741. 302
Berzon, I. S., I. P. Passechnik, and A. M. Polikarpov, 1974. The determination of P-wave attenuation values in the earth's mantle. Geophys. J. 39, 603-611. 340
Bessonova, E. N., V. M. Fishman, and G. A. Sitnikova, 1970. Determination of the limits for velocity distribution from discrete travel time observations. Paper presented at the Seventh Upper Mantle Symposium on Geophysical Theory and Computers, Södergarn, Sweden. 277, 288
Bessonova, E. N., V. M. Fishman, V. Z. Ryaboyi, and G. A. Sitnikova, 1974. The tau method for the inversion of travel times - I. Deep seismic sounding data. Geophys. J. 36, 377-398. 277, 288
Bessonova, E. N., V. M. Fishman, M. G. Shnirman, G. A. Sitnikova, and L. R. Johnson, 1976. The tau method for the inversion of travel times - II. Earthquake data. Geophys. J. 46, 87-108. 276, 277, 288, 303
Bhattacharya, S. N., 1976. Extension of the Thompson-Haskell method to non-homogeneous spherical layers. Geophys. J. 47, 411-444. 313
Bhattacharya, S. N., 1978. Rayleigh waves from a point source in a spherical medium with homogeneous layers. Bull., Seis. Soc. Amer. 68, 231-238. 364
Biot, M. A., 1957. General theorems on the equivalence of group velocity and energy transport. Phys. Rev. 105, 1129-37. 176
Birch, F., 1966. Compressibility; Elastic constants. In: S. P. Clark (Editor), Handbook of Physical Constants, Geol. Soc. Amer. Mem. 97. pp. 97-173. 51
Biswas, N. N., 1972. Earth-flattening procedure for the propagation of Rayleigh waves. Pure and Appl. Geophys. 96, 61-74. 313
Biswas, N. N., and L. Knopoff, 1970. Exact earth-flattening calculation for Love waves. Bull., Seis. Soc. Amer. 60, 1123-1137. 311
Biswas, N. N., and L. Knopoff, 1974. The structure of the upper mantle under the United States from the dispersion of Rayleigh waves. Geophys. J. 36, 515-539. 256
Bland, D. R., 1960. The theory of linear viscoelasticity. Pergamon, New York. 326
Blandford, R. R., 1975. A source theory for complex earthquakes. Bull., Seis. Soc. Amer. 65, 1385-1405. 396
Blandford, R. R., 1977. Discrimination between earthquakes and underground explosions. Ann. Rev. Earth and Planet. Sci., 5, 111-122. 394
Block, B., and R. D. Moore, 1966. Measurements in the earth mode frequency range by an electrostatic sensing and feedback gravimeter. J. Geophys. Res. 71, 4361-4375. 435
Boatwright, J. and D. M. Boore, 1975. A simplification in the calculation of motion near a propagating dislocation. Bull., Seis. Soc. Amer. 65, 133-138. 385
Bogert, B. P., M. J. R. Healy, and J. W. Tukey, 1963. The quefrency alanysis of time series for echoes: Cepstrum, pseudo-autocovariance, cross-cepstrum and saphe cracking. In: M. Rosenblatt (Editor), Time series analysis. Wiley, New York, pp. 209-243. 298
Bohm, D., 1951. Quantum Theory. Prentice-Hall, New York. 176
Bolt, B. A., 1960. The revision of earthquake epicentres, focal depths and origin-times using a high-speed computer. Geophys. J. 3, 433-440. 297
Bolt, B. A., 1973. A proposal for the global calibration of group earthquake locations. Geophys. J. 33, 249-250. 297
Bolt, B. A., 1976. Nuclear explosions and earthquakes: the parted veil. W. H. Freeman, San Francisco. 368, 394
Bolt, B. A., 1977. Ocean bottom seismometry: a new dimension to seismology. Boll. Geofis., 19, 107-116. 434

Bolt, B. A., and D. R. Brillinger, 1975. Estimation of uncertainities in fundamental frequencies of decaying geophysical time series (Abstract). E⊕S, Trans. Amer. Geophys. Union 56, 403. 343

Bolt, B. A., and R. G. Currie, 1975. Maximum entropy estimates of earth torsional eigenperiods from 1960 Trieste data. Geophys. J. 40, 107-114. 316

Bolt, B. A., and J. Dorman, 1961. Phase and group velocities of Rayleigh waves in a spherical gravitating earth. J. Geophys. Res. 66, 2965-2981. 313

Bolt, B. A., and R. D. Miller, 1975. Catalogue of earthquakes in Northern California and adjoining areas, 1 January 1910 - 31 December 1972. Seismograph stations, University of California, Berkeley. 399

Bolt, B. A., and R. Uhrhammer, 1975. Resolution techniques for density and heterogeneity in the earth. Geophys. J. 42, 419-435. 322

Boore, D. M., 1969. Effect of higher mode contamination on measured Love wave phase velocities. J. Geophys. Res. 74, 6612-6616. 248

Boore, D. M., 1972. Finite difference methods for seismic wave propagation in heterogeneous materials. In: B. A. Bolt (Editor), Methods in computational physics, Vol. 11, Seismology: surface waves and earth oscillations. Academic, New York, pp. 1-37. 241

Boore, D. M., and W. S. Dunbar, 1977. Effect of the free surface on calculated stress drops. Bull., Seis. Soc. Amer. 67, 1661-1664. 385

Booth, D. C., P. D. Marshall, and J. B. Young, 1974. Long and short period P-wave amplitudes from earthquakes in the range 0°-114°. Geophys. J. 39, 523-537. 387

Borcherdt, R. D., and J. H. Healy, 1968. A method of estimating the uncertainty of seismic velocities measured by refraction techniques. Bull., Seis. Soc. Amer. 58, 1769-1790. 219

Brace, W. F., B. W. Paulding Jr., and C. H. Scholz, 1966. Dilatancy in the fracture of crystalline rocks. J. Geophys. Res. 71, 3939-3953. 410

Bradley, J. J., and A. N. Fort Jr., 1966. Internal friction in rocks. In: S. P. Clark Jr., (Editor), Handbook of Physical Constants. Geol. Soc. Amer. Mem. 97, pp. 175-193. 324, 335, 336

Braile, L. W., 1973. Inversion of crustal seismic refraction and reflection data. J. Geophys. Res. 78, 7738-7744. 231

Braile, L. W., and R. B. Smith, 1975. Guide to the interpretation of crustal re-refraction profiles. Geophys. J. 40, 145-176. 231

Broberg, K. B., 1960. The propagation of a brittle crack. Arkiv För Fysik 18, 159-192. 366

Brune, J. N., 1962. Attenuation of dispersed wave trains. Bull., Seis. Soc. Amer. 52, 109-112. 328

Brune, J. N., 1964. Travel times, body waves, and normal modes of the earth. Bull., Seis. Soc. Amer. 54, 2099-2128. 313, 314

Brune, J. N., 1966. P and S wave travel times and spheroidal normal modes of a homogeneous sphere. J. Geophys. Res. 71, 2959-2965. 315

Brune, J. N., 1969. Surface waves and crustal structure. In: P. J. Hart (Editor), The earth's crust and upper mantle. Amer. Geophys. Union, Monograph #13, Washington, D. C., pp. 230-242. 253

Brune, J. N., 1970. Tectonic stress and the spectra of seismic shear waves from earthquakes. J. Geophys. Res. 75, 4997-5009. (Correction, J. Geophys. Res. 76, 5002, 1971.) 381, 382, 389, 393

Brune, J. N., and C. R. Allen, 1967. A micro-earthquake survey of the San Andreas Fault system in Southern California. Bull., Seis. Soc. Amer. 57, 277-296. 406, 407

Brune, J. N., and J. Dorman, 1963. Seismic waves and earth structure in the Canadian Shield. Bull., Seis. Soc. Amer. 53, 167-209. 258, 259, 263

Brune, J. N., and G. R. Engen, 1969. Excitation of mantle Love waves and definition of mantle wave magnitude. Bull., Seis. Soc. Amer. 59, 923-933. 389, 393

Brune, J. N., and F. Gilbert, 1974. Torsional overtone dispersion from correlation of S waves to SS waves. Bull., Seis. Soc. Amer. 64, 313-320. 315

Brune, J. N., and J. Oliver, 1959. The seismic noise of the earth's surface. Bull., Seis. Soc. Amer. 49, 349-353. 418

Brune, J. N., J. Nafe, and J. Oliver, 1960. A simplified method for the analysis and synthesis of dispersed wave trains. J. Geophys. Res. 65, 287-304. 246

Brune, J. N., J. E. Nafe and L. E. Alsop, 1961. The polar phase shift of surface waves on a sphere. Bull., Seis. Soc. Amer. 51, 247-257. 309

Buchbinder, G. G. R. and G. Poupinet, 1973. Problems related to PcP and the core-mantle boundary illustrated by two nuclear events. Bull., Seis. Soc. Amer. 63, 2047-2070. 302

Buland, R., 1976. The mechanics of locating earthquakes. Bull., Seis. Soc. Amer. 66, 173-187. 297

Buland, R., and F. Gilbert, 1978. Improved resolution of complex eigenfrequencies in analytically continued seismic spectra. Geophysical J., 52, 457-470. 344

Bullen, K. E., 1963. An introduction to the theory of seismology, Third Edition. Cambridge University Press. 285, 294

Bullen, K. E., and R. A. W. Haddon, 1970. Evidence from seismology and related sources on the earth's present internal structure. Phys. Earth and Planet. Inter., 2, 342-349. 323

Bungum, H., and E. S. Husebye, 1974. Analysis of the operational capabilities for detection and location of seismic events at NORSAR. Bull., Seis. Soc. Amer. 64, 637-656. 438

Burg, J. P., 1972. The relationship between maximum entropy spectra and maximum likelihood spectra. Geophysics 37, 375-376. 316

Burkhard, N. R., and D. D. Jackson, 1976. Density and Surface wave inversion. Geophys. Res. Lett. 3, 637-638. 258

Burridge, R., 1973. Admissible speeds for plane-strain self-similar shear cracks with friction but lacking cohesion. Geophys. J. 35, 439-455. 366

Burridge, R., 1975. The effect of sonic rupture velocity on the ratio of S to P corner frequencies. Bull., Seis. Soc. Amer. 65, 667-675. 385

Burridge, R., and Z. Alterman, 1972. The elastic radiation from an expanding spherical cavity. Geophys. J. 30, 451-477. 365

Burridge, R., and G. S. Halliday, 1971. Dynamic shear cracks with friction as models for shallow focus earthquakes. Geophys. J. 25, 261-283. 366

Burridge, R., and C. Levy, 1974. Self-similar circular shear cracks lacking cohesion. Bull., Seis. Soc. Amer. 64, 1789-1808. 366

Burridge, R., and L. Knopoff, 1964. Body force equivalents for seismic dislocations. Bull., Seis. Soc. Amer. 54, 1875-1888. 80, 358

Burridge, R., and L. Knopoff, 1967. Model and theoretical seismicity. Bull., Seis. Soc. Amer. 57, 341-371. 367

Burridge, R., and J. Willis, 1969. The self-similar problem of the expanding elliptical crack in an anisotropic solid. Proc. Cambr. Phil. Soc. 66, 443-468. 366

Burridge, R., E. R. Lapwood, and L. Knopoff, 1964. First motions from seismic sources near a free surface. Bull., Seis. Soc. Amer. 54, 1889-1913. 357

Burton, P. W., 1977. Inversions of high frequency $Q_\gamma^{-1}(f)$. Geophys. J. 48, 29-51. 346

Cagniard, L., 1939. Réflexion et réfraction des ondes séismiques progressives. Gauthier-Villars, Paris. 103, 146

Cagniard, L., 1962. Reflection and refraction of progressive seismic waves, translated and revised by E. A. Flinn and C. H. Dix, McGraw-Hill, New York. 103

Calcagnile, G., and G. F. Panza, 1974. Vertical and SV components of S_a. Geophys. J. 38, 317-325. 257

Calcagnile, G., G. F. Panza, F. Schwab and E. G. Kausel, 1976. On the computation of theoretical seismograms for multimode surface waves. Geophys. J. 47, 73-81. 256

Canitez, N., and M. N. Toksöz, 1971. Focal mechanism and source depth of earthquakes from body- and surface-wave data. Bull., Seis. Soc. Amer. 61, 1369-1379. 298, 377

Capon, J., 1970. Analysis of Rayleigh-wave multipath propagation at LASA. Bull., Seis. Soc. Amer. 60, 1701-1731. 252

Capon, J., 1971. Comparison of Love- and Rayleigh-wave multipath propagation at LASA. Bull., Seis. Soc. Amer. 61, 1327-1344. 252

Capon, J., 1973a. Analysis of microseismic noise at LASA, NORSAR, and ALPA. Geophys. J. 35, 39-54. 420, 421

Capon, J., 1973b. Signal processing and frequency-wavenumber analysis for a large aperature seismic array. In: B. A. Bolt (Editor), Methods in Computational Physics, Vol. 13, Geophysics. Academic, New York, pp. 1-59. 438

Capon, J., and J. F. Evernden, 1971. Detection of interfering Rayleigh Waves at LASA. Bull., Seis. Soc. Amer. 61, 807-849. 252

Caputo, M., 1977. A mechanical model for the statistics of earthquakes, magnitude, and fault distribution. Bull., Seis. Soc. Amer. 67, 849-861. 402

Cerveny, V. and R. Ravindra, 1971. Theory of Seismic Head Waves. University of Toronto Press. 131, 146, 231, 280, 281, 338

Cerveny, V., J. Langer, and I. Psencik, 1974. Computation of geometric spreading of seismic body waves in laterally inhomogeneous media with curved interfaces. Geophys. J. 38, 9-19. 279

Chandra, U., 1972. Angles of Incidence of S-waves. Bull., Seis. Soc. Amer. 62, 903-915. 369

Chapman, C. H., 1973. The earth flattening transformation in body wave theory. Geophys. J. 35, 55-70. 303, 313

Chapman, C. H., 1974a. The turning point of elastodynamic waves. Geophys. J. 39, 613-621. 290

Chapman, C. H., 1974b. Generalized ray theory for an inhomogeneous medium. Geophys. J. 36, 673-704. 303

Chapman, C. H., 1976a. Exact and approximate generalized ray theory in vertically inhomogeneous media. Geophys. J. 46, 201-233. 303

Chapman, C. H., 1976b. A first-motion alternative to geometrical ray theory. Geophys. Res. Lett. 3, 153-156. 303

Chapman, C. H., and R. A. Phinney, 1970. Diffraction of P waves by the core and an inhomogeneous mantle. Geophys. J. 21, 185-205. 290

Chapman, C. H., and R. A. Phinney, 1972. Diffracted seismic signals and their numerical solution. In: B. A. Bolt (Editor), Methods in Computational physics, Vol. 12. Seismology: body waves and sources. Academic, New York, pp. 165-230. 289, 290

Chapman, S., and R. S. Lindzen, 1970. Atmospheric tides: thermal and gravitational. D. Reidel, Dordrecht. 242

Cherry Jr., J. T., 1962. The azimuthal and polar radiation patterns obtained from a horizontal stress applied at the surface of an elastic half space. Bull., Seis. Soc. Amer. 52, 27-36. 127

Chinnery, M. A., 1961. The deformation of the ground around surface faults. Bull., Seis. Soc. Amer. 51, 355-372. 378

Chinnery, M. A., 1963. The stress changes that accompany strike-slip faulting. Bull., Seis. Soc. Amer. 53, 921-932. 378

Chinnery, M. A., 1964. The strength of the earth's crust under horizontal shear stress. J. Geophys. Res. 69, 2085-2089. 378

Chinnery, M. A., 1965. The vertical displacements associated with transcurrent faulting. J. Geophys. Res. 70, 4627-4632. 378

Chinnery, M. A., 1975. The static deformation of an earth with a fluid core: a physical approach. Geophys. J. 42, 461-475. 365

Chouet, B., K. Aki, and M. Tsujiura, 1978. Regional variation of the scaling law of earthquake source spectra. Bull., Seis. Soc. Amer. 68, 49-79. 384

Choy, G. L., 1977. Theoretical seismograms of core phases calculated by frequency-dependent full wave theory, and their interpretation. Geophys. J. 51, 275-312. 303

Choy, G. L., and P. G. Richards, 1975. Pulse distortion and Hilbert transformation in multiply reflected and refracted body waves. Bull., Seis. Soc. Amer. 65, 55-70. 279

Cisternas, A., O. Betancourt, and A. Leiva, 1973. Body waves in a "real earth". Part I. Bull., Seis. Soc. Amer. 63, 145-156. 303

Claerbout, J. F., 1968. Synthesis of a layered medium from its acoustic transmission response. Geophysics 33, 264-269. 230

Claerbout, J. F., 1976. Fundamentals of geophysical data processing: with application to petroleum prospecting. McGraw-Hill, New York. 217

Clayton, R. W., B. McClary, and R. A. Wiggins, 1976. Comments on the paper: "Phase distortion and Hilbert transformation in multiply reflected and refracted body waves", by G. L Choy and P. G. Richards. Bull., Seis. Soc. Amer. 66, 325-352. 92

Cleary, J., 1967. Analysis of the amplitudes of short-period P waves recorded by Long Range Seismic Measurement stations in the distance range 30 to 102 degrees. J. Geophys. Res. 72, 4705-4712. 386

Cleary, J., 1974. The D" region. Phys. Earth and Planet. Inter. 9, 13-27. 298

Cleary, J., and A. L. Hales, 1966. An analysis of the travel times of P waves to North American stations, in the distance range 32° to 100°. Bull., Seis. Soc. Amer. 56, 467-489. 299, 300

Closs, H., 1969. Explosion seismic studies in Western Europe. In: P. J. Hart (Editor), The earth's crust and upper mantle. American Geophysical Union Monograph #13, Washington, D. C., pp. 178-188. 222

Cochran, M. D., A. F. Woeber, and J-Cl. DeBremaecker, 1970. Body waves as normal and leaking modes, 3. Pseudo modes and partial derivatives on the (+,-) sheet. Rev. Geophys. and Space Phys. 8, 321-357. 182

Cohen, T. J., 1970. Source-depth determinations using spectral, pseudo-autocorrelation and cepstral analysis. Geophys. J. 20, 223-231. 298

Corbishley, D. J., 1970. Multiple array measurement of P-wave travel-time derivative. Geophys. J. 19, 1-14. 301, 302

Costain, J. K., K. L. Cook, and S. T. Algermissen, 1963. Amplitude, energy, and phase angles of plane SV waves and their application to earth crustal studies. Bull., Seis. Soc. Amer. 53, 1039-1074. 131

Costain, J. K., K. L. Cook, and S. T. Algermissen, 1965. Corrigendum: Amplitude, energy, and phase angles of plane SV waves and their application to earth crustal studies. Bull., Seis. Soc. Amer. 55, 567-575. 131

Crampin, S., 1964a. Higher modes of seismic surface waves: preliminary observations. Geophys. J. 9, 37-57. 256

Crampin, S., 1964b. Higher modes of seismic surface waves: Phase velocities across Scandinavia. J. Geophys. Res. 69, 4801-4811. 256

Crampin, S., 1966a. Higher-mode seismic surface waves from atmospheric nuclear explosions over Novaya Zemlya. J. Geophys. Res. 71, 2951-2958. 256

Crampin, S., 1966b. Higher modes of seismic surface waves: propagation in Eurasia. Bull., Seis. Soc. Amer. 56, 1227-1239. 256

Cranford, M. D., S. H. Johnson, J. E. Bowers, R. A. McAlister, and B. T. Brown, 1976. A direct-recording ocean-botton seismograph. Bull., Seis. Soc. Amer. 66, 607-615. 434

Crosson, R. S., 1976. Crustal structure modeling of earthquake data 1. Simultaneous least squares estimation of hypocenter and velocity parameters. J. Geophys. Res. 81, 3036-3046. 297

Crough, S. T., and R. Van der Voo, 1973. A method for relocating seismic events using surface waves. Bull., Seis. Soc. Amer. 63, 1305-1313. 297

Cummings, D., and G. I. Shiller, 1971. Isopach map of the earth's crust. Earth Sci. Rev. 7, 97-125. 223

Dahlen, F. A., 1968. The normal modes of a rotating, elliptical earth. Geophys. J. 16, 329-367. 307, 308

Dahlen, F. A., 1969. The normal modes of a rotating, elliptical earth - II. Near-resonance multiplet couplet coupling. Geophys. J. 18, 397-436. 307

Dahlen, F. A., 1972a. The effect of an initial hypocentral stress upon the radiation patterns of P and S waves. Bull., Seis. Soc. Amer. 62, 1173-1182. 365

Dahlen, F. A., 1972b. Elastic dislocation theory for a self-gravitating elastic configuration with an initial static stress field. Geophys. J. 28, 357-383. 365

Dahlen, F. A., 1973a. A correction to the excitation of the Chandler Wobble by earthquakes. Geophys. J. 32, 203-217. 365

Dahlen, F. A., 1973b. Elastic dislocation theory for a self-gravitating elastic configuration with an initial static stress field, II. Energy release. Geophys. J. 31, 469-484. 365

Dahlen, F. A., 1974. On the ratio of P-wave to S-wave corner frequencies for shallow earthquake sources. Bull., Seis. Soc. Amer. 64, 1159-1180. 385

Dahlen, F. A., 1975. The correction of great circular surface wave phase velocity measurements for the rotation and ellipticity of the earth. J. Geophys. Res. 80, 4895-4903. 307
Dahlen, F. A., 1976. Reply. J. Geophys. Res. 81, 4951-4956. 308, 319
Dahlen, F. A., and M. L. Smith, 1975. The influence of rotation on the free oscillations of the earth. Phil. Trans. Roy. Soc., London, A 279, 583-624. 307
Dahlman, O., and H. Israelson, 1977. Monitoring underground nuclear explosions. Elsevier, Amsterdam. 368, 394
Dainty, A. M., and C. N. G. Dampney, 1972. A comparison of leaking modes and generalized ray theory. Geophys. J. 28, 147-161. 191
Dampney, C. N. G., 1971. The relationship between two- and three-dimensional elastic-wave propagation. Bull., Seis. Soc. Amer. 61, 1583-1588. 127
Dampney, C. N. G., 1972. A line source on a solid-solid interface — a study of the pseudo-Stoneley wave. Bull., Seis. Soc. Amer. 62, 1017-1027. 151, 152
Darbyshire, J., 1962. Microseisms. In: M. N. Hill (Editor), The Sea, Vol. 1. Interscience, New York, pp. 700-719. 412
Darbyshire, J., and E. O. Okeke, 1969. A study of primary and secondary microseisms recorded in Anglesey. Geophys. J. 17, 63-92. 412
Das, S., and K. Aki, 1977a. A numerical study of two-dimensional spontaneous propagation. Geophys. J. 50, 643-668. 367
Das, S., and K. Aki, 1977b. Fault plane with barriers: a versatile earthquake model. J. Geophys. Res. 82, 5658-5670. 367
Davies, L. M., and C. H. Chapman, 1975. Some numerical experiments on the inversion of discrete travel-time data. Bull., Seis. Soc. Amer. 65, 531-539. 278
DeBremaecker, J-Cl., 1956. Remark on Byerly's fault-plane method. Bull., Seis. Soc. Amer. 46, 215-216. 372, 374
DeBremaecker, J-Cl., 1967. Body waves as normal and leaking modes, 1. Introduction. Bull., Seis. Soc. Amer. 57, 191-198. 182
DeBremaecker, J-Cl., 1968. Body waves as normal and leaking modes, 2. The Canadian Shield. Nuovo Cimento, Suppl. 6, 98-104. 182
DeBremaecker, J-Cl., R. H. Godson, and J. S. Watkins, 1966. Attenuation measurements in the field. Geophysics 31, 562-569. 337
deHoop, A. T., 1960. A modification of Cagniard's method for solving seismic pulse problems. Appl. Sci. Res. B 8, 349-356. 56, 66, 103
deHoop, A. T., 1961. Theoretical determination of the surface motion of a uniform elastic half-space produced by a dilatational, impulsive, point-source. In: La Propagation des Ebranlements dans les Milieux Heterogenes, Colloques Internationaux du Centre National de la Recherche Scientifique, Marseille, 11-16 Septembre, pp. 21-32. 103
Der, Z. A., and T. W. McElfresh, 1977. The relationship between anelastic attenuation and regional amplitude anomalies of short-period P waves in North America. Bull., Seis. Soc. Amer. 67, 1303-1317. 340
Der, Z. A., R. Massé, and M. Landisman, 1970. Effects of observational errors on the resolution of surface waves at intermediate distances. J. Geophys. Res. 75, 3399-3409. 260, 261
Der, Z. A., R. Massé, and J. P. Gurski, 1975. Regional attenuation of short-period P and S waves in the United States. Geophys. J. 40, 85-106. 340
Derr, J. S., 1969a. Free oscillation observations through 1968. Bull., Seis. Soc. Amer. 59, 2079-2099. 316, 320
Derr, J. S., 1969b. Internal structure of the earth inferred from free oscillations. J. Geophys. Res. 74, 5202-5220. 323
Deschamps, A., 1977. Inversion of the attenuation data of free oscillations of the earth (fundamental and first higher modes). Geophys. J. 50, 699-722. 345
Dewey, J. W., 1972. Seismicity and tectonics of Western Venezuela. Bull., Seis. Soc. Amer. 62, 1711-1751. 297
Dieterich, J. H., 1972. Time-dependent friction as a possible mechanism for aftershocks. J. Geophys. Res. 77, 3771-3781. 367
Dieterich, J. H., 1974. Earthquake mechanisms and modeling. Ann. Rev. Earth Planet. Sci. 2, 275-301. 366
Dillinger, W. H., S. T. Harding, and A. J. Pope, 1972. Determining maximum likelihood body wave focal plane solutions. Geophys. J. 30, 315-329. 374

Dix, C. H., 1954. The method of Cagniard in seismic pulse problems. Geophysics 19, 722-738. 103

Dix, C. H., 1955. Seismic velocities from surface measurements. Geophysics 20, 68-86. 197

Dobrin, M. B., 1976. Introduction to geophysical prospecting, Third Edition. McGraw-Hill, New York. 198, 217

Dohr, G. P., and R. Meissner, 1975. Deep Crustal Reflections in Europe. Geophysics 40, 25-39. 217

Doornbos, D. J., 1976. Characteristics of lower mantle inhomogeneities from scattered waves. Geophys. J. 44, 447-470. 302

Doornbos, D. J., 1977. The excitation of normal modes of the earth by sources with volume changes. Geophys. J. 51, 465-474. 364

Dorman, J., 1962. Period equation for waves of Rayleigh type on a layered, liquid-solid half space. Bull., Seis. Soc. Amer. 52, 389-397. 214

Dorman, J., 1969. Seismic surface-wave data on the upper mantle. In: P. J. Hart (Editor), The earth's crust and upper mantle. Amer. Geophys. Union, Monograph #13, Washington, D. C., pp. 257-265. 253

Dorman, L. M., 1968. Anelasticity and the spectra of body waves. J. Geophys. Res. 73, 3877-3883. 340

Douglas, A., 1967. Joint epicenter determination. Nature 215, 47-48. 297

Douglas, A., J. A. Hudson, and V. K. Kembhavi, 1971. The relative excitation of seismic surface and body waves by joint sources. Geophys. J. 23, 451-460. 396

Douglas, A., J. A. Hudson, and C. Blamey, 1973. A quantitative evaluation of seismic signals at teleseismic distances - III. Computed P and Rayleigh wave seismograms. Geophys. J. 28, 385-410. 396

Douglas, B. M., and A. Ryall, 1972. Spectral characteristics and stress drop for micro-earthquakes near Fairview Peak, Nevada. J. Geophys. Res. 77, 351-359.407

Douglas, B. M., A. Ryall, and R. Williams, 1970. Spectral characteristics of Central Nevada microearthquakes. Bull., Seis. Soc. Amer. 60, 1547-1559. 407

Douze, E. J., and G. G. Sorrells, 1975. Prediction of pressure-generated earth motion using optimum filters. Bull., Seis. Soc. Amer. 65, 637-650. 421

Dowling, J. J., 1970. Uncertainties in velocities determined by Seismic refraction. J. Geophys. Res. 75, 6690-6992. 221

Dratler, J. W., E. Farrell, B. Block, and F. Gilbert, 1971. High Q overtone modes of the earth. Geophys. J. 23, 399-410. 344

Duda, S. J., 1965. Secular seismic energy release in the circum-Pacific belt. Tectonophysics 2, 409-452. 399

Dunkin, J. W., 1965. Computation of modal solutions in layered, elastic media at high frequencies. Bull., Seis. Soc. Amer. 55, 335-358. 241

Duschenes, J. D., and S. C. Solomon, 1977. Shear wave travel time residuals from oceanic earthquakes and the evolution of oceanic lithosphere. J. Geophys. Res. 82, 1985-2000. 300

Duwalo, G., and J. A. Jacobs, 1959. Effects of a liquid core on the propagation of seismic waves. Canad. J. Phys. 37, 109-128. 289

Dziewonski, A. M., 1970a. On regional differences in dispersion of mantle Rayleigh waves. Geophys. J. 22, 289-325. 318

Dziewonski, A. M., 1970b. Correlation properties of free period partial derivatives and their relation to the resolution of gross earth data. Bull., Seis. Soc. Amer. 60, 741-768. 320

Dziewonski, A. M., and F. Gilbert, 1972. Observations of normal modes from 84 recordings of the Alaskan earthquake of 1964, March 23. Geophys. J. 27, 293-446. 318

Dziewonski, A. M., and F. Gilbert, 1973. Observations of normal modes from 84 recordings of the Alaskan earthquake of 1964, March 28 - II. Further remarks based on new spheroidal overtone data. Geophys. J. 35, 401-437. 306, 315, 318

Dziewonski, A. M., and F. GILbert, 1976. The effect of small, aspherical pertubations on travel times and a re-examination of the corrections for ellipticity. Geophys. J. 44, 7-17. 295

Dziewonski, A. M., and R. A. W. Haddon, 1974. The radius of the core-mantle boundary inferred from travel time and free oscillation data; a critical review. Phys. Earth and Planet. Inter. 9, 28-35. 302

Dziewonski, A. M., and A. L. Hales, 1972. Numerical analysis of dispersed seismic waves. In: B. A. Bolt (Editor), Methods in Computational Physics; Vol II, Seismology: surface waves and earth oscillations. Academic, New York, pp. 39-85. 247
Dziewonski, A. M., and M. Landisman, 1970. Great circle Rayleigh and Love wave dispersion from 100 to 900 seconds. Geophys. J. 19, 37-91. 318
Dziewonski, A. M., and R. V. Sailor, 1976. Comments on 'the correction of great circular surface wave phase velocity measurements from the rotation and ellipticity of the earth' by F. A. Dahlen. J. Geophys. Res. 81, 4947-4950. 307
Dziewonski, A. M., J. Mills, and S. Block, 1972. Residual dispersion measurement - a new method of surface wave analysis. Bull., Seis. Soc. Amer. 62, 129-139. 247
Dziewonski, A. M., A. L. Hales, and E. R. Lapwood, 1975. Parametrically simple earth models consistent with geophysical data. Phys. Earth and Planet. Inter. 10, 12-48. 323
Eason, G., J. Fulton, and I. N. Sneddon, 1956. The generation of waves in an infinite elastic solid by variable body forces. Phil. Trans. Roy. Soc., London, A 248, 575-607. 63
Engdahl, E. R., and L. E. Johnson, 1974. Differential PcP travel times and the radius of the core. Geophys. J. 39, 435-456. 302
Engdahl, E. R., E. A. Flinn, and R. P. Massé, 1974. Differential PKiKP travel times and the radius of the inner core. Geophys. J. 39, 457-463. 302
Erdelyi, A. (Editor), 1954. Tables of Integral Transforms, Vol. I. McGraw-Hill, New York. 62
Erdelyi, A., 1956. Asymptotic Expansions. Dover, New York. 176
Eringen, A. C., and E. S. Suhubi, 1974. Elastodynamics: Vol I., Finite Motions. Academic, New York. 38
Eringen, A. C., and E. S. Suhubi, 1975. Elastodynamics: Vol II., Linear Theory. Academic, New York. 39, 42, 43, 48, 49, 52, 70, 75, 77, 80, 102, 103, 127, 304, 360, 394
Evernden, J. F., 1953. Direction of approach of Rayleigh waves and related problems, Part I. Bull., Seis. Soc. Amer. 43, 335-374. 250, 252
Evernden, J. F., 1954. Direction of approach of Rayleigh waves and related problems, Part II. Bull., Seis. Soc. Amer. 44, 159-184. 250
Evernden, J. F., 1969a. Identification of earthquakes and explosions by use of teleseismic data. J. Geophys. Res. 74, 3828-3856. 297
Evernden, J. F., 1969b. Precision of epicenters obtained by small numbers of world-wide stations. Bull., Seis. Soc. Amer. 59, 1365-1398. 298
Evernden, J. F., 1970. Study of regional seisimcity and associated problems. Bull., Seis. Soc. Amer. 60, 393-446. 401
Evernden, J. F., 1971a. Location capability of various seismic networks. Bull., Seis. Soc. Amer. 61, 241-273. 298
Evernden, J. F., 1975. Further studies on seismic discrimination. Bull., Seis. Soc. Amer. 65, 359-391. (Corrections: Bull., Seis. Soc. Amer. 66, 349-352, 1976.) 389, 390
Evernden, J. F., 1976a. Study of seismological evasion: Part I. General discussion of various evasion schemes. Bull., Seis. Soc. Amer. 66, 245-280. 397
Evernden, J. F., 1976b. Study of seismological evasion: Part II. Evaluation of evasion possibilities using normal microseismic noise. Bull., Seis. Soc. Amer. 66, 281-324. 397
Evernden, J. F., 1976c. Study of seismological evasion: Part III. Evaluation of evasion possibilities using codas of large earthquakes. Bull., Seis. Soc. Amer. 66, 549-592. 397
Evernden, J. F., 1977a. Adequacy of routinely available data for identifying earthquakes of $m_b \geq 4.5$. Bull., Seis. Soc. Amer. 67, 1099-1151. 394, 396
Evernden, J. F., 1977b. Spectral characteristics of the P codas of Eurasian earthquakes and explosions. Bull., Seis. Soc. Amer. 67, 1153-1171. 384, 395
Evernden, J. F., and D. M. Clark, 1970. Study of Teleseismic P. II - Amplitude data. Phys. Earth and Planet. Inter. 4, 24-31. 387
Evernden, J. F., W. J. Best, P. W. Pomeroy, T. W. McEvilly, J. M. Savino, and L. R. Sykes, 1971. Discrimination between small-magnitude earthquakes and explosions. J. Geophys. Res. 76, 8042-8055. 397

Ewing, M., and F. Press, 1959. Determination of crustal structure from phase velocity of Rayleigh waves. Part III: the United States. Bull., Geol. Soc. Amer. 70, 229-244. 243
Ewing, M., and J. L. Worzel, 1948. Long-range sound transmission. In: Memoir 27, Geol. Soc. Amer., New York. 242
Ewing, M., W. S. Jardetzky, and F. Press, 1957. Elastic Waves in Layered Media. McGraw-Hill, New York. 38, 72, 180, 181
Fernandez, L. M., 1967. Master curves for the response of layered systems to compressional seismic waves. Bull., Seis. Soc. Amer. 57, 515-543. 227
Fix, J. E., 1972. Ambient earth motion in the period range from 0.1 to 2560 seconds. Bull., Seis. Soc. Amer. 62, 1753-1760. 418
Fix, J. E., 1973. Theoretical and observed noise in a high-sensitivity long-period seismograph. Bull., Seis. Soc. Amer. 1979-1998. 422
Flinn, E. A., 1965. Confidence regions and error determinations for seismic event location. Rev. Geophys. 3, 157-185. (Corrections: Rev. Geophys. 7, 664, 1969.) 297, 298
Forsyth, D. W., 1975. The early structural evolution and anisotropy of the oceanic mantle. Geophys. J. 43, 103-162. 34, 256
Forsyth, D. W., 1975a. A new method for the analysis of multi-mode surface wave dispersion: application to Love-wave propagation in the East Pacific. Bull., Seis. Soc. Amer. 65, 323-342. 248, 256, 319, 321
Frantti, G. E., D. E. Willis, and J. T. Wilson, 1962. The spectrum of seismic noise. Bull., Seis. Soc. Amer. 52, 113-121. 418, 420
Frazer, L. N., 1977. Use of the spherical layer matrix in inhomogeneous media. Geophys. J. 50, 743-749. 306
Frazier, C. W., and R. G. North, 1978. Evidence for ω-cube scaling from amplitudes and periods of the Rat Island sequence (1965). Bull., Seis. Soc. Amer. 68, 265-282. 384
Freedman, H. W., 1967a. A statistical discussion of P residuals from explosions, Part II. Bull., Seis. Soc. Amer. 57, 545-561. 297
Freedman, H. W., 1967b. Estimating the accuracy of source "parameters". Bull., Seis. Soc. Amer. 57, 373-379. 298
Freedman, H. W., 1967c. Estimating earthquake magnitude. Bull., Seis. Soc. Amer. 57, 747-760. 390
Frez, J., and F. Schwab, 1976. Structural dependence of the apparent initial phase of Rayleigh Waves. Geophys. J. 44, 311-331. 250
Fuchs, K., and G. Müller, 1971. Computation of synthetic seismograms with the reflectivity method and comparison with observations. Geophys. J. 23, 417-433. 230
Futterman, W. I., 1962. Dispersive body waves. J. Geophys. Res. 67, 5279-5291. 328, 347
Gakenheimer, D. C., and J. Miklowitz, 1969. Transient excitation of an elastic half space by a point load traveling on the surface. J. Appl. Mech. 36, 505-515. 360
Gangi, A. F., 1970. A derivation of the seismic representation theorem using seismic reciprocity. J. Geophys. Res. 75, 2088-2095. 80
Gardner, J. K., and L. Knopoff, 1974. Is the sequence of earthquakes in Southern California, with aftershocks removed, Poissonian? Bull., Seis. Soc. Amer. 64, 1363-1367. 405, 407
Garvin, W. W., 1956. Exact transient solution of the buried line source problem. Proc. Roy. Soc. London, A 234, 528-541. 103
Gaulon, R., N. Jobert, G. Poupinet, and G. Roult, 1970. Application de la méthod de Haskell au calcul de la dispersion des ondes de Rayleigh sur un modèle sphérique. Ann. Geophys. 26, 1-8. 313
Geller, R. J., 1976a. Body force equivalents for stress-drop seismic sources. Bull., Seis. Soc. Amer. 66, 1801-1804. 360
Geller, R. J., 1976b. Scaling relations for earthquake source parameters and magnitudes. Bull., Seis. Soc. Amer. 66, 1501-1523. 389, 402
Geller, R. J., and S. Stein, 1977. Split free oscillation amplitudes for the 1960 Chilean and 1964 Alaskan Earthquakes. Bull., Seis. Soc. Amer. 67, 651-660. 316

Gerver, M., and V. Markushevich, 1966. Determination of a seismic wave velocity from the travel-time curve. Geophys. J. 11, 165-173. 269, 277, 283

Gerver, M., and V. Markushevich, 1967. On the characteristic properties of travel-time curves. Geophys. J. 13, 241-246. 277

Gibowicz, S. J., 1973. Stress deop and aftershocks. Bull., Seis. Soc. Amer. 63, 1433-1446. 408

Gilbert, F., 1956. Seismic wave propagation in a two-layer half-space: PhD thesis, Mass. Inst. of Technology, Cambridge, Massachusetts. 103, 143

Gilbert, F., 1964. Propagation of transient leaking modes in a stratified elastic waveguide. Rev. Geophys. 2, 123-153. 181

Gilbert, F., 1971a. Ranking and winnowing gross earth data for inversion and resolution. Geophys. J. 23, 125-128. 260, 319

Gilbert, F., 1971b. The diagonal sum rule and averaged eigenfrequencies. Geophys. J. 23, 119-123. 307

Gilbert, F., 1971c. Excitation of the normal modes of the earth by earthquake sources. Geophys. J. 21, 223-226. 377

Gilbert, F., 1972a. Inverse problems for the earth's normal modes. In: E. C. Robertson, J. F. Hays, and L. Knopoff (Editors), The nature of the solid earth. McGraw-Hill, New York, pp. 125-146. 260, 319

Gilbert, F., 1973a. Derivation of source parameters from low-frequency spectra. Phil. Trans. Roy. Soc. London, A 274, 369-371. 377

Gilbert, F., 1973b. The relative efficiency of earthquakes and explosions in exciting surface waves and body waves. Geophys. J. 33, 487-488. 396

Gilbert, F., 1975. Some asymptotic properties of the normal modes of the earth. Geophys. J. 43, 1007-1011. 315

Gilbert, F., 1976a. The representation of seismic displacements in terms of traveling waves. Geophys. J. 44, 275-280. 310

Gilbert, F., 1976b. Differential kernels for group velocity. Geophys. J. 44, 649-660. 320

Gilbert, F., and G. Backus, 1965. The rotational splitting of the free oscillations of the earth, 2. Rev. Geophys. 3, 1-9. 307

Gilbert, F., and G. Backus, 1966. Propagator matrices in elastic wave and vibration problems. Geophysics 31, 326-332. 205, 303

Gilbert, F., and R. Buland, 1976. An enhanced deconvolution procedure for retrieving the seismic moment tensor from a sparse network. Geophys. J. 47, 251-255. 378

Gilbert, F., and A. M. Dziewonski, 1975. An application of normal mode theory to the retrieval of structural parameters and source mechanisms from seismic spectra. Phil. Trans. Roy. Soc. London, A 278, 187-269. 316, 318, 320, 321, 322, 368, 377

Gilbert, F., and D. V. Helmberger, 1972. Generalized ray theory for a layered sphere. Geophys. J. 27, 57-80. 303

Gilbert, F., and S. J. Laster, 1962. Excitation and propagation of pulses on an interface. Bull., Seis. Soc. Amer. 52, 299-319. 123, 150, 152

Gilbert, F., and S. J. Laster, 1962a. Experimental investigation of PL modes in a single layer. Bull., Seis. Soc. Amer. 52, 59-66. 181

Gilbert, F., and G. J. F. MacDonald, 1960. Free oscillations of the earth: I. Toroidal oscillations. J. Geophys. Res. 65, 675-693. 305

Gilbert, F., S. J. Laster, M. M. Backus, and R. Schell, 1962. Observations of pulses on an interface. Bull., Seis. Soc. Amer. 52, 847-868. 152

Gilbert, F., A. M. Dziewonski, and J. Brune, 1973. An informative solution to a seismological inverse problem. Proc. Nat. Acad. Sci. (U.S.) 70, 1410-1413. 320

Ginzbarg, A. S., and E. Strick, 1958. Stoneley-wave velocities for a solid-solid interface. Bull., Seis. Soc. Amer. 48, 51-63. 147

Gogna, M. L., 1973. Travel times of S, PcP and ScS from Pacific earthquakes. Geophys. J. 33, 103-126. 301

Gossard, E. E., and W. H. Hooke, 1975. Waves in the Atmosphere. Elsevier, Amsterdam. 242

Goupillaud, P. L., 1961. An approach to inverse filtering of near-surface layer effects from seismic records. Geophysics 26, 754-760. 230

Grant, F. S., and G. F. West, 1965. Interpretation theory in Applied geophysics. McGraw-Hill, New York. 198, 217

Green, A. E., R. S. Rivlin, and R. T. Shield, 1952. General theory of small deformations superimposed on finite elastic deformation. Proc. Roy. Soc. London, A 211, 128-154. 35

Green, R., 1962. The hidden layer problem. Geophys. Prosp. 10, 166-170. 200

Gross, B., 1953. Mathematical structure of the theories of viscoelasticity. Hermann et Cie, Editeurs. Paris. 331

Gross, B., and E. P. Braga, 1961. Singularities of linear system functions. Elsevier, Amsterdam. 331

Gupta, R. N., 1966a. Reflection of elastic waves from a linear transition layer. Bull., Seis. Soc. Amer. 56, 511-526. 139

Gupta, R. N., 1966b. Reflection of plane elastic waves from transition layers with arbitrary variation of velocity and density. Bull., Seis. Soc. Amer. 56, 633-642. 139

Gutenberg, B., 1944. Energy ratio of reflected and refracted seismic waves. Bull., Seis. Soc. Amer. 34, 85-102. 131

Gutenberg, B., 1945. Amplitudes of surface waves and magnitudes of shallow earthquakes. Bull., Seis. Soc. Amer. 35, 3-12. 389

Gutenberg, B., 1958. Microseisms. In: H. E. Landsberg and J. Van Mieghem (Editors), Adv. Geophys. 5, 54-92. Academic, New York. 412

Gutenberg, B., 1959. Physics of the earth's interior. Academic, New York. 291, 292

Gutenberg, B., and C. F. Richter, 1954. Seismicity of the earth and associated phenomena. Princeton University Press, New Jersey. 399

Gutenberg, B., and C. F. Richter, 1956. Magnitude and energy of earthquakes. Ann. Geofis. 9, 1-15. 386, 389, 392

Haddon, R. A. W., and K. E. Bullen, 1969. An earth model incorporating free earth oscillation data. Phys. Earth and Planet. Inter. 2, 37-49. 323

Hagiwara, T., 1958. A note on the theory of the electromagnetic seismograph. Tokyo Univ., Bull. Earthquake Res. Inst. 36, 139-164. 428, 431

Hagiwara, T., and T. Iwata, 1968. Summary of the seismographic observation of Matsushiro swarm earthquakes. Tokyo Univ., Bull. Earthquake Res. Inst. 46, 485-515. 408

Hales, A. L., 1972. The travel times of P seismic waves and their relevance to the upper mantle velocity distribution. Tectonophysics 13, 447-482. 298

Hales, A. L., 1974. Eigenperiods of earth models and the determination of travel time base lines. J. Geophys. Res. 79, 422-423. 350

Hales, A. L., and E. Herrin, 1972. Travel Times of Seismic Waves. In: E. C. Robertson, J. F. Hays, and L. Knopoff (Editors), The nature of the solid earth. McGraw-Hill, New York, pp. 172-215. 298, 300, 301

Hales, A. L., and J. L. Roberts, 1970. The travel times of S and SKS. Bull., Seis. Soc. Amer. 60, 461-489. 299

Hall, D. H., and Z. Hajnal, 1973. Deep seismic crustal studies in Manitoba. Bull., Seis. Soc. Amer. 63, 885-910. 217

Hanks, T. C., and W. Thatcher, 1972. A graphical representation of seismic source parameters. J. Geophys. Res. 77, 4393-4405. 382, 384, 385

Hanks, T. C., and M. Wyss, 1972. The use of body-wave spectra in the determination of seismic-source parameters. Bull., Seis. Soc. Amer. 62, 561-589. 382, 385

Hannon, W. J., 1964. An application of the Haskell-Thomson matrix method to the synthesis of the surface motion due to dilatational waves. Bull., Seis. Soc. Amer. 54, 2067-2079. 227

Harkrider, D. G., 1964. Surface waves in multilayered elastic media I. Rayleigh and Love waves from buried sources in a multilayered elastic half-space. Bull., Seis. Soc. Amer. 54, 627-679. 202, 204, 232, 234, 239, 241

Harkrider, D. G., 1968. The perturbation of Love wave spectra. Bull., Seis. Soc. Amer. 58, 861-880. 258

Harkrider, D. G., 1970. Surface waves in multilayered elastic media. Part II. Higher mode spectra and spectral ratios from point sources in plane layered earth models. Bull., Seis. Soc. Amer. 60, 1937-1987. 241, 245

Harkrider, D. G., and D. L. Anderson, 1966. Surface wave energy from point sources in plane layered earth models. J. Geophys. Res. 71, 2967-2980. 234, 239

Hart, P. J. (Editor), 1969. The earth's crust and upper mantle. Amer. Geophys. Union, Monograph #13, Washington, D. C. 219

Hart, R. S., 1975. Shear velocity in the lower mantle from explosion data. J. Geophys. Res. 80, 4889-4894. 299

Hashizume, M., 1976. Surface-wave study of the Canadian Shield. Phys. Earth. Planet. Inter. 11, 333-351. 256

Haskell, N. A., 1953. The dispersion of surface waves on multilayered media. Bull., Seis. Soc. Amer. 43, 17-34. 205, 233

Haskell, N. A., 1960. Crustal reflections of plane SH waves. J. Geophys. Res. 65, 4147-4150. 209

Haskell, N. A., 1962. Crustal reflections of plane P and SV waves. J. Geophys. Res. 67, 4751-4767. 207, 208, 227

Haskell, N. A., 1963. Radiation patterns of Rayleigh waves from a fault of arbitrary dip and direction of motion in a homogeneous medium. Bull., Seis. Soc. Amer. 53, 619-642. 356, 357

Haskell, N. A., 1964a. Radiation pattern of surface waves from point sources in a multilayered medium. Bull., Seis. Soc. Amer. 54, 377-393. 364

Haskell, N. A., 1964b. Total energy and energy spectral density of elastic wave radiation from propagating faults. Bull., Seis. Soc. Amer. 54, 1811-1841. 384, 385, 390, 393

Haskell, N. A., 1966. Total energy and energy spectral density of elastic wave radiation from propagating faults. Part II: A statistical source model. Bull., Seis. Soc. Amer. 56, 125-140. 384, 385

Haskell, N. A., 1967. Analytic approximation for the elastic radiation from a contained underground explosion. J. Geophys. Res. 72, 2583-2587. 394

Haskell, N. A., 1969. Elastic displacements in the near-field of a propagating fault. Bull., Seis. Soc. Amer. 59, 865-908. (Corrections, Bull., Seis. Soc. Amer. 67, 1215, 1977.) 385

Haskell, N. A. and K. C. Thomson, 1972. Elastodynamic near-field of a finite propagating tensile fault. Bull., Seis. Soc. Amer. 62, 675-697. (Corrections, Bull., Seis. Amer. 67, 1215, 1977.) 385

Hasselmann, K., 1963. A statistical analysis of the generation of microseisms. Rev. Geophys. 1, 177-210. 412, 414

Haubrich, R. A. and K. McCamy, 1969. Microseisms: coastal and Pelagic sources. Rev. Geophys. 7, 539-571. 420, 421

Haubrich, R. A., and G. S. MacKenzie, 1965. Earth noise, 5 to 500 millicycles per second; 2: Reaction of the earth to oceans and atmosphere. J. Geophys. Res. 70, 1429-1440. 420, 421

Haubrich, R. A., W. H. Munk, and F. E. Snodgrass, 1963. Comparative spectra of microseisms and swell. Bull., Seis. Soc. Amer. 53, 27-37. 420

Healy, J. H., and D. H. Warren, 1969. Explosion seismic studies in North America. In: P. J. Hart (Editor), The Earth's crust and upper mantle. Amer. Geophys. Union Monograph #13, Washington D. C., pp. 208-220. 221, 222

Heaton, T. H., 1975. Tidal triggering of earthquakes. Geophys. J. 43, 305-326. 404

Helmberger, D. V., 1968. The crust-mantle transition in the Bering Sea. Bull., Seis. Soc. Amer. 58, 179-214. 229

Helmberger, D. V., 1972. Long-period body-wave propagation from 4° to 13°. Bull., Seis. Soc. Amer. 62, 325-341. 229

Helmberger, D. V., 1973a. On the structure of the low velocity zone. Geophys. J. 34, 251-263. 299, 350

Helmberger, D. V., 1973b. Numerical seismograms of long-period body waves from seventeen to forty degrees. Bull., Seis. Soc. Amer. 63, 633-646. 303

Helmberger, D. V., 1974. Generalized ray theory for shear dislocations. Bull., Seis. Soc. Amer. 64, 45-64. 364

Helmberger, D. V., and G. R. Engen, 1974. Upper mantle shear structure. J. Geophys. Res. 79, 4017-4028. 298, 340, 350

Herrera, I., 1964. On a method to obtain a Green's function for a multilayered half-space. Bull., Seis. Soc. Amer. 54, 1087-1096. 183

Herrin, E., 1968. 1968 Seismological tables for P phases. Bull., Seis. Soc. Amer. 58, 1193-1241. 292, 369

Herrin, E., and J. Taggart, 1962. Regional variations in P_n velocity and their effect on the location of epicenters. Bull., Seis. Soc. Amer. 52, 1037-1046. 224

Herrin, E., and J. Taggart, 1968. Regional variations in P travel times. Bull., Seis. Soc. Amer. 58, 1325-1337. 223, 297, 299

Herrmann, R. B., 1975. The use of duration as a measure of seismic moment and magnitude. Bull., Seis. Soc. Amer. 65, 899-913. 389

Herrmann, R. B., 1976. Focal depth determination from the signal character of long-period P waves. Bull., Seis. Soc. Amer. 66, 1221-1232. 298

Herrmann, R. B., and B. J. Mitchell, 1975. Statistical analysis and interpretation of surface-wave anelastic attenuation data for the stable interior of North America. Bull., Seis. Soc. Amer. 65, 1115-1128. 344

Hileman, J. A., C. R. Allen, and J. M. Nordquist, 1973. Seismicity of the Southern California region, 1 January 1932 to 31 December 1972. Calif. Inst. of Tech., Pasadena. 399

Hill, D. P., 1971. Velocity gradients and anelasticity from crustal body wave amplitudes. J. Geophys. Res. 76, 3309-3325. 338

Hill, D. P., 1972. An earth-flattening transformation for waves from a point source. Bull., Seis. Soc. Amer. 62, 1195-1210. 303

Hill, D. P., 1973. Critically refracted waves in a spherically symmetric radially heterogeneous earth model. Geophys. J. 34, 149-177. 338

Hill, D. P., 1974. Phase shift and pulse distortion in body waves due to internal caustics. Bull., Seis. Soc. Amer. 64, 1733-1742. 279

Hill, D. P., and D. L. Anderson, 1977. A note on the earth-stretching approximation for Love waves. Bull., Seis. Soc. Amer. 67, 551-552. 312

Hirasawa, T., and W. Stauder, 1965. On the seismic body waves from a finite moving source. Bull., Seis. Soc. Amer. 55, 237-262. 360

Hong, T-L., and D. V. Helmberger, 1977. Generalized ray theory for dipping structure. Bull., Seis. Soc. Amer. 67, 995-1008. 230

Howell Jr., B. F., 1972. Precision of different varieties of magnitude. Bull., Seis. Soc. Amer. 62, 789-792. 389

Hron, F., 1972. Numerical methods of ray generation in multilayered media. In: B. A. Bolt (Editor), Methods in computational physics, Vol 12. Seismology: Body waves and sources. Academic, New York, pp. 1-34. 230

Hron, F., and E. R. Kanasewich, 1971. Synthetic seismograms for deep seismic sounding studies using asymptotic ray theory. Bull., Seis. Soc. Amer. 61, 1169-1200. 280

Hron, F., E. R. Kanasewich, and T. Alpaslan, 1974. Partial ray expansion required to suitably approximate the exact wave solution. Geophys. J. 36, 607-625. 230

Hron, M., and C. H. Chapman, 1974a. The "shadow" from a velocity reversal. Bull., Seis. Soc. Amer. 64, 25-32. 278

Hron, M., and C. H. Chapman, 1974b. Seismograms from Epstein transition zones. Geophys. J. 37, 305-322. 278

Hron, M., and M. Razavy, 1977. Solution of the inverse problem for wave propagation using J-fraction expansion. Geophys. J. 51, 545-554. 230

Hudson, J. A., 1962. The total internal reflection of SH waves. Geophys. J. 6, 509-531. 90

Husebye, E. S., A. Christofferson, K. Aki, and C. Powell, 1976. Preliminary results on the 3-dimensional seismic structure under the USGS Central California Seismic Array. Geophys. J. 46, 319-340. 301

Isacks, B., and P. Molnar, 1971. Distribution of stresses in the descending lithosphere from a global survey of focal-mechanism solutions of mantle earthquakes. Rev. Geophys. and Space Phys. 9, 103-174. (Corr., 10, 847, 1972) 374, 403

Isacks, B., and J. Oliver, 1964. Seismic waves with frequencies from 1 to 100 cycles per second recorded in a deep mine in Northern New Jersey. Bull., Seis. Soc. Amer. 54, 1941-1979. 406

Isacks, B., J. Oliver, and L. R. Sykes, 1968. Seismology and the new global tectonics. J. Geophys. Res. 73, 5855-5900. 351, 403, 404

Isacks, B., L. R. Sykes, and J. Oliver, 1969. Focal mechanisms of deep and shallow earthquakes in the Tonga-Kermadec region and the tectonics of island arcs. Bull., Geol. Soc. Amer. 80, 1443-1470. 403

Ishii, H., and R. M. Ellis, 1970a. Multiple reflection of plane SH waves by a dipping layer. Bull., Seis. Soc. Amer. 60, 15-28. 230

Ishii, H., and R. M. Ellis, 1970b. Multiple reflection of plane P and SV waves by a dipping layer. Geophys. J. 20, 11-30. 230
Israel, M., and A. Ben-Menahem, 1975. Changes in the earth's inertial tensor due to earthquakes by MacCullagh's formula. Geophys. J. 40, 305-307. 365
Israel, M., A. Ben-Menahem, and S. J. Singh, 1973. Residual deformation of real earth models with application to the Chandler Wobble. Geophys. J. 32, 219-247. 365
Iyer, H. M., and J. H. Healy, 1972. Teleseismic residuals at the LASA-USGS extended array and their interpretation in terms of crust and upper-mantle structure. J. Geophys. Res. 77, 1503-1527. 300
Jackson, D. D., 1972. Interpretation of inaccurate, insufficient, and inconsistent data. Geophys. J. 28, 97-109. 262
Jackson, D. D., 1973. Marginal solutions to quasi-linear inverse problems in geophysics: the Edgehog Method. Geophys. J. 35, 121-136. 262
Jackson, D. D., 1976. Most squares inversion. J. Geophys. Res. 81, 1027-1030. 262
Jackson, D. D., and D. L. Anderson, 1970. Physical mechanisms of seismic wave attenuation. Rev. Geophys. and Space Phys. 8, 1-63. 333, 335
Jacob, K. H., 1970. Three-dimensional seismic ray tracing in a laterally heterogeneous spherical earth. J. Geophys. Res. 75, 6675-6689. 279
Jahnke, E., and F. Emde, 1938. Tables of functions with formulae and curves. Dover (reprint 1945), New York. 309
James, D. E., 1971. Anomalous Love wave phase velocities. J. Geophys. Res. 76, 2077-2083. 248
Jeffreys, H., 1939. The times of PcP and ScS. Mon. Not. Roy. Soc., Geophys. Suppl. 4, 537-547. 302
Jeffreys, H., 1961. Small corrections in the theory of surface waves. Geophys. J. 6, 115-117. 258
Jeffreys, H., 1970. The earth: Its origin, history and physical constitution, 5th edition. Cambridge Univ. Press. 276, 287, 292
Jeffreys, H., 1977. P and S beyond 95°. Geophys. J. 51, 387-392. 301
Jeffreys, H., and K. E. Bullen, 1940. Seismological tables. Brit. Assoc. Adv. Sci., London. 292
Jeffreys, H., and B. S. Jeffreys, 1956. Methods of Mathematical Physics. Cambridge, Univ. Press. 1
Jobert, G., 1974. On earth flattening transformations in body wave theory. Geophys. J. 39, 189-193. 303
Jobert, G., 1975. Propagator and Green matrices for body force and dislocation. Geophys. J. 43, 755-762. 364
Jobert, G., 1976. Matrix methods for generally stratified media. Geophys. J. 47, 351-362. 241
Jobert, G., 1977. 'Gravitating earth flattening' approximations of propagators and Green matrices for body force and dislocation sources. Geophys. J. 48, 123-130. 364
Jobert, G., N. Jobert, and B. L. N. Kennett, 1975. Comments on the paper "On variational principles and matrix methods in elastodynamics" by B. L. N. Kennett. Geophys. J. 43, 721-725. 241
Johnson, L. E., and F. Gilbert, 1972a. A new datum for use in the body wave travel time inverse problem. Geophys. J. 30, 373-380. 278, 288
Johnson, L. E., and F. Gilbert, 1972b. Inversion and inference for teleseismic ray data. In: B. A. Bolt (Editor), Methods in computational physics, Vol 12, Seismology: body waves and sources. Academic, New York, pp. 231-266. 278, 288, 303
Johnson, L. R., 1974. Green's function for Lamb's problem. Geophys. J. 37, 99-131. 127
Johnson, T. L., and C. H. Scholz, 1976. Dynamic properties of stick-slip friction of rock. J. Geophys. Res. 81, 881-888. 367
Jovanovich, D. B., 1975. An inversion method for estimating the source parameters of seismic and aseismic events from static strain data. Geophys. J. 43, 347-365. 378
Julian, B. R., and D. Gubbins, 1977. Three-dimensional seismic ray tracing. J. Geophys. 43, 95-113. 279
Julian, B. R., D. Davies, and R. M. Sheppard, 1972. PKJKP. Nature 235, 317-318. 302

Kagan, Y., and L. Knopoff, 1976. Statistical search for non-random features of the seismicity of strong earthquakes. Phys. Earth and Planet. Inter. 12, 291-318. 406, 408

Kagan, Y., and L. Knopoff, 1977. Earthquake risk prediction as a stochastic process. Phys. Earth and Planet. Inter. 14, 97-108. 406, 408

Kagan, Y., and L. Knopoff, 1978. Statistical study of the occurrence of shallow earthquakes. Geophys. J. 55, 67-86. 406, 408

Kaila, K. L., and N. M. Rao, 1975. Seismotectonic maps of the European area. Bull., Seis. Soc. Amer. 65, 1721-1732. 402

Kaila, K. L. and D. Sarkar, 1975. P-wave amplitude variation with epicentral distance and the magnitude relations. Bull., Seis. Soc. Amer. 65, 915-926. 387

Kanamori, H., 1967a. Spectrum of P and PcP in relation to the mantle-core boundary and attenuation in the mantle. J. Geophys. Res. 72, 559-571. 342

Kanamori, H., 1967b. Spectrum of short-period core phases in relation to the attenuation in the mantle. J. Geophys. Res. 72, 2181-2186. 342

Kanamori, H., 1970. Velocity and Q of mantle waves. Phys. Earth and Planet. Inter. 2, 259-275. 318

Kanamori, H., 1970a. Synthesis of long-period surface waves and its application to earthquake source studies — Kurile Islands earthquake of October 13, 1963. J. Geophys. Res. 75, 5011-5027. 375, 376

Kanamori, H., 1970b. The Alaska Earthquake of 1964: Radiation of long-period surface waves and source mechanism. J. Geophys. Res. 75, 5029-5040. 375

Kanamori, H., 1972. Determination of effective tectonic stress associated with earthquake faulting — Tottori earthquake of 1943. Phys. Earth and Planet. Inter. 5, 426-434. 393

Kanamori, H., 1973. Mode of strain release associated with major earthquakes in Japan. Ann. Rev. Earth. Planet Sci. 1, 213-239. 366

Kanamori, H., 1977. The energy release in great earthquakes. J. Geophys. Res. 82, 2981-2987. 393, 402

Kanamori, H., and D. L. Anderson, 1975. Theoretical basis of some empirical relations in seismology. Bull., Seis. Soc. Amer. 65, 1073-1095. 378, 380, 381, 390, 391, 392, 402

Kanamori, H., and D. L. Anderson, 1977. Importance of physical dispersion in surface wave and free oscillation problems: Review. Rev. Geophys. and Space Phys. 15, 105-112. 333, 347

Kanasewich, E. R., 1973. Time sequence analysis in geophysics. U. Alberta Press, Edmonton. 217

Kanasewich, E. R., T. Alpaslan and F. Hron, 1973. The importance of S-wave precursors in shear-wave studies. Bull., Seis. Soc. Amer. 63, 2167-2176. 230

Kane, J., 1966. Teleseismic response of a uniform dipping crust. Bull., Seis. Soc. Amer. 56, 841-859. 230

Karal, F. C., Jr., and J. B. Keller, 1959. Elastic wave propagation in homogeneous and inhomogeneous media. J. Acoust. Soc. Amer. 31, 694-705. 280

Karnik, V., 1969. Seismicity of the European Area, Vol. 1. D. Reidel, Dordrecht, Holland. 399

Karnik, V., 1971. Seismicity of the European Area, Vol. 2. D. Reidel, Dordrecht, Holland. 399

Kaufman, H., 1953. Velocity functions in seismic prospecting. Geophysics 18, 289-297. 275

Kausel, E. G., A. R. Leeds, and L. Knopoff, 1974. Variations of Rayleigh wave phase velocities across the Pacific Ocean. Science 186, 139-141. 253

Kausel, E. G., F. Schwab, and E. Mantovani, 1977. Oceanic S_a. Geophys. J. 50, 407-440. 257

Keilis-Borok, V. I., 1959. An estimation of the displacement in an earthquake source and of source dimensions. Ann. Geofis. 12, 205-214. 379, 380

Keilis-Borok, V. I., and T. B. Yanovskaja, 1967. Inverse problem of seismology (structural review). Geophys. J. 13, 223-234. 303

Kelleher, J. A., 1970. Space-time seismicity of the Alaska-Aleutian seismic zone. J. Geophys. Res. 75, 5745-5756. 406

Kelleher, J. A., 1972. Rupture zones of large South American earthquakes and some predictions. J. Geophys. Res. 77, 2087-2103. 406

Kennett, B. L. N., 1972. The connection between elastodynamic representation theorems and propagator matrices. Bull., Seis. Soc. Amer. 62, 973-983. 205

Kennett, B. L. N., 1974a. On variational principles and matrix methods in elastodynamics. Geophys. J. 37, 391-405. 241

Kennett, B. L. N., 1974b. Reflections, rays, and reverberations. Bull., Seis. Soc. Amer. 64, 1685-1696. 230

Kennett, B. L. N., 1975. The effects of attenuation on seismograms. Bull., Seis. Soc. Amer. 65, 1643-1651. 350

Kennett, B. L. N., 1976. A comparison of travel-time inversions. Geophys. J. 44, 517-536. 278

Kennett, B. L. N., and G. Nolet, 1978. Resolution analysis for discrete systems. Geophys. J. 53, 413-425. 262

Kennett, B. L. N., and J. A. Orcutt, 1976. A comparison of travel time inversions for marine refraction profiles. J. Geophys. Res. 81, 4061-4070. 278

Khattri, K. N., 1969. Focal mechanism of the Brazil deep focus earthquake of November 3, 1965, from the amplitude spectra of isolated P waves. Bull., Seis. Soc. Amer. 59, 691-704. 374

Khattri, K., 1973. Earthquake focal mechanism studies — a review. Earth Sci. Rev. 9, 19-63. 368

Kind, R., and G. Müller, 1977. The structure of the outer core from SKS amplitudes and travel times. Bull., Seis. Soc. Amer. 67, 1541-1554. 302

King, G., and R. Bilham, 1976. A geophysical wire strainmeter. Bull., Seis. Soc. Amer. 66, 2039-2047. 434

Klein, F. W., 1976. Earthquake swarms and the semidiurnal solid earth tide. Geophys. J. 45, 245-295. 404

Knopoff, L., 1958. Surface motions of a thick plate. J. Appl. Physics 29, 661-670. 109

Knopoff, L., 1958a. Love waves from a line SH source. J. Geophys. Res. 63, 619-630. 190

Knopoff, L., 1958b. Energy release in earthquakes. J. Geophys. Res. 1, 44-52. 379

Knopoff, L., 1961a. Analytical calculation of the fault-plane problem. Publ. Dom. Obs., Ottawa, 24, 309-315. 374

Knopoff, L., 1961b. Statistical accuracy of the fault-plane problem. Publ. Dom. Obs., Ottawa, 24, 316-319. 374

Knopoff, L., 1964a. A matrix method for elastic wave problems. Bull., Seis. Soc. Amer. 54, 431-438. 241

Knopoff, L., 1964b. "Q". Rev. Geophys. 2, 625-660. 329, 335, 336

Knopoff, L., 1964c. Earthtides as a triggering mechanism for earthquakes. Bull., Seis. Soc. Amer. 54, 1865-1870. 404

Knopoff, L., 1964d. The statistics of earthquakes in Southern California. Bull., Seis. Soc. Amer. 54, 1871-1873. 405

Knopoff, L., 1971. A stochastic model for the occurrence of main-sequence earthquakes. Rev. Geophys. and Space Phys. 9, 175-188. 406

Knopoff, L., 1972a. Observation and inversion of surface-wave dispersion. Tectonophysics 13, 497-519. 253, 254, 263

Knopoff, L., 1972b. Model for aftershock occurrence. In: H. C. Heard, I. Y. Borg, N. C. Carter, and C. B. Raleigh (Editors), Flow and Fracture of Rocks. Amer. Geophys. Union Monograph #16, Washington D. C., pp. 259-263. 367

Knopoff, L., and A. F. Gangi, 1959. Seismic Reciprocity. Geophysics 24, 681-691. 77

Knopoff, L., and J. K. Gardner, 1972. Higher seismic activity during local night on the raw worldwide earthquake catalog. Geophys. J. 28, 311-313. 405

Knopoff, L., and F. Gilbert, 1959. First Motion Methods in Theoretical Seismology. J. Acoust. Soc. Amer. 31, 1161-1168. 127

Knopoff, L., and F. Gilbert, 1959a. Radiation from a strike-slip fault. Bull., Seis. Soc. Amer. 49, 163-172. 358

Knopoff, L., and F. Gilbert, 1960. First motions from seismic sources. Bull., Seis. Soc. Amer. 50, 117-134. 358

Knopoff, L., and F. Gilbert, 1961. Diffraction of elastic waves by the core of the earth. Bull., Seis. Soc. Amer. 51, 35-49. 289

Knopoff, L., and G. J. F. MacDonald, 1958. Attenuation of small amplitude stress waves in solids. Rev. Modern Phys. 30, 1178-1192. 328

Knopoff, L., and J. O. Mouton, 1975. Can one determine seismic focal parameters from the far-field radiation? Geophys. J. 42, 591-606. 367, 385

Knopoff, L., and M. J. Randall, 1970. The compensated linear-vector dipole: a possible mechanism for deep earthquakes. J. Geophys. Res. 75, 4957-4963. 356, 368

Knopoff, L., and F. Schwab, 1968. Apparent initial phase of a source of Rayleigh waves. J. Geophys. Res. 73, 755-760. 250

Knopoff, L., and T. L. Teng, 1965. Analytical calculation of the seismic travel-time problem. Rev. Geophysics 3, 11-24. 221

Knopoff, L., R. W. Fredricks, A. F. Gangi, and L. D. Porter, 1957. Surface amplitudes of reflected body waves. Geophysics 22, 842-847. 96

Knopoff, L., F. Gilbert and W. L. Pilant, 1960. Wave propagation in a medium with a single layer. J. Geophys. Res. 65, 265-278. 181

Knopoff, L., K. Aki, C. Archambeau, A. Ben-Menahem, and J. A. Hudson, 1964. Attenuation of dispersed waves. J. Geophys. Res. 69, 1655-1657. 327, 328

Knopoff, L., S. Mueller, and W. L. Pilant, 1966. Structure of the crust and upper mantle in the Alps from the phase velocity of Rayleigh waves. Bull., Seis. Soc. Amer. 56, 1009-1044. 251

Knopoff, L., M. J. Berry, and F. A. Schwab, 1967. Tripartite phase velocity observations in laterally heterogeneous regions. J. Geophys. Res. 72, 2595-2601. 252

Knopoff, L., C. L. Drake, and P. J. Hart (Editors), 1968. The crust and upper mantle of the Pacific Area. American Geophysical Union Monograph #12, Washington, D. C. 219

Knopoff, L., F. Schwab and E. Kausel, 1973. Interpretation of Lg. Geophys. J. 33, 389-404. 257

Knopoff, L., J. O. Mouton, and R. Burridge, 1973a. The dynamics of a one-dimensional fault in the presence of friction. Geophysics J. 35, 169-184. 367

Knowles, J. K., 1966. A note on elastic surface waves. J. Geophys. Res. 71, 5480-5481. 102

Kosloff, D., 1975. A perturbation scheme for obtaining partial derivatives of Love-wave group velocity dispersion. Bull., Seis. Soc. Amer. 65, 1753-1760. 258

Kosminskaya, I. P., 1971. Deep seismic sounding of the earth's crust and upper mantle (Tr. from the Russian edition of 1968). Consultants Bureau, New York. 217

Kostrov, B. V., 1964a. The axisymmetric problem of propagation of a tension crack. J. Appl. Math. Mech. 28, 793-803. 366

Kostrov, B. V., 1964b. Self-similar problems of propagation of shear cracks. J. Appl. Math. Mech. 28, 1077-1087. 366, 367

Kostrov, B. V., 1966. Unsteady propagation of longitudinal shear cracks. J. Appl. Math. Mech. 30, 1241-1248. 366

Kostrov, B. V., 1974. Crack propagation at variable velocity. J. Appl. Math. Mech. 38, 511-519. 366

Kovach, R. L., 1965. Seismic surface waves: Some observations and recent developments. Phys. Chem. Earth 6, 251-314. 243, 256

Kovach, R. L., 1978. Seismic surface waves and crustal and upper mantle structure. Rev. Geophys. and Space Phys. 16, 1-13. 243

Kovach, R. L., and D. L. Anderson, 1964. Higher mode surface waves and their bearing on the structure of the earth's mantle. Bull., Seis. Soc. Amer. 54, 161-182. 173, 257

Kovach, R. L. and D. L. Anderson, 1964a. Attenuation of shear waves in the upper and lower mantle. Bull., Seis. Soc. Amer. 54, 1855-1864. 342

Kraut, E. A., 1963. Advances in the theory of anisotropic elastic wave propagation. Rev. Geophys. 1, 401-448. 35

Kunetz, G., and I. D'Erceville, 1962. Sur certaines propriétés d'une onde acoustique plan de compression dans un milieu stratifié. Ann. Geophys., 18, 351-359. 230

Kurita, T., 1973a. A procedure for elucidating fine structure of the crust and upper mantle from seismological data. Bull., Seis. Soc. Amer. 63, 189-209. 227, 228

Kurita, T., 1973b. Regional variations in the structure of the crust in the central United States from P-wave spectra. Bull., Seis. Soc. Amer. 63, 1663-1687. 227
Kurita, T., 1974. Upper-mantle structure in the central United States from P- and S-wave spectra. Phys. Earth and Planet. Inter. 8, 177-201. 227
Kurita, T., 1976. Crustal and upper-mantle structure in the central United States of America from body-wave spectra, surface-wave dispersion, travel-time residuals and synthetic seismograms. Phys. Earth and Planet. Inter. 12, 65-86. 228
Lacoss, R. T., E. J. Kelley, and M. N. Toksöz, 1969. Estimation of seismic noise structure using arrays. Geophysics 34, 21-38. 420
LaCoste, L. J. B., 1934. A new type long period vertical seismograph. Physics (now J. Appl. Phys.) 5, 178-180. 427
LaCoste, L. J. B., 1935. A simplification in the conditions for the zero-length spring seismograph. Bull., Seis. Soc. Amer. 25, 176-179. 427
Lamb, H., 1904. On the propagation of tremors over the surface of an elastic solid. Phil. Trans. Roy. Soc., London, A 203, 1-42. 103
Lanczos, C., 1961. Linear differential operators. Van Nostrand, London. 262
Landisman, M., T. Usami, Y. Sato, and R. Massé, 1970. Contributions of theoretical seismograms to the study of modes, rays, and the earth. Rev. Geophys. and Space Phys. 8, 533-589. 302, 321
Landisman, M., S. Mueller, and B. J. Mitchell, 1971. Review of evidence for velocity inversions in the continental crust. In: J. G. Heacock (Editor), the structure and physical properties of the earth's crust. Amer. Geophys. Union Monograph #14, Washington D. C., pp. 11-34. 223
Langston, C. A., 1977. The effect of planar dipping structure on source and receiver responses for constant ray parameter. Bull., Seis. Soc. Amer. 67, 1029-1050. 230
Langston, C. A., 1978. Moments, corner frequencies, and the free surface. J. Geophys. Res. 83, 3422-3426. 385
Langston, C. A. and D. V. Helmberger, 1975. A procedure for modelling shallow dislocation sources. Geophys. J. 42, 117-130. 364
Lapwood, E. R., 1975. The effect of discontinuities in density on torsional eigenfrequencies of the earth. Geophys. J. 40, 453-464. 315
Lapwood, E. R., and J. A. Hudson, 1975. The passage of elastic waves through an anomalous region — III. Transmission of obliquely incident body waves. Geophys. J. 40, 255-268. 139
Lapwood, E. R., J. A. Hudson, and V. K. Kembhavi, 1973. The passage of elastic waves through an anomalous region — I. Transmission of body waves through a soft layer. Geophys. J. 31, 457-467. 139
Lapwood, E. R., J. A. Hudson, and V. K. Kembhavi, 1975. The passage of elastic waves through an anomalous region — II. Transmission through a layer between two different media. Geophys. J. 40, 241-254. 139
Laster, S. J., J. G. Foreman, and A. F. Linville, 1965. Theoretical investigation of modal seismograms for a layer over a half-space. Geophysics 30, 571-596. 174, 181, 182, 183
Lee, K., 1975. The dispersion of Rayleigh waves in Eurasia. Ph.D. Dissertation, U. Pittsburgh. 255
Lee, T-C. and T-L. Teng, 1973. Polar radiation patterns of P and SV waves in a multilayered medium. Bull., Seis. Soc. Amer. 63, 529-547. 363
Lee, W. B., and S. C. Solomon, 1975. Inversion schemes for surface wave attenuation and Q in the crust and the mantle. Geophys. J. 43, 47-71. 346
Lee, W. B., and S. C. Solomon, 1978. Simultaneous inversion of surface wave phase velocity and attenuation: Love waves in western North America. J. Geophys. Res. 83, 3389-3400. 344, 346, 347
Leeds, A. R., 1973. Rayleigh wave dispersion in the Pacific Basin. Ph.D. Dissertation, U. California at Los Angeles. 253, 254, 320
Leeds, A. R., L. Knopoff, and E. G. Kausel, 1974. Variations of upper mantle structure under the Pacific Ocean. Science 186, 141-143. 253
Lehmann, I., 1970. The reading of earthquake diagrams. Geophys. J. 20, 391-396. 294
LePichon, X., J. Francheteau, and J. Bonnin, 1973. Plate Tectonics. Elsevier Sci. Publ. Co., Amsterdam. 399

Levine, J., and J. L. Hall, 1972. Design and operation of a methane absorption stabilized laser strainmeter. J. Geophys. Res. 77, 2595-2609. 434

Liao, A. H., F. Schwab, and E. Mantovani, 1978. Computation of complete theoretical seismograms for torsional waves. Bull., Seis. Soc. Amer. 68, 317-324. 257

Lilwall, R. C., and A. Douglas, 1970. Estimation of P-wave travel times using the joint epicentre method. Geophys. J. 19, 165-181. 297, 299

Lilwall, R. C., and R. Underwood, 1970. Seismic network bias maps. Geophys. J. 20, 335-339. 298

Lindh, A., G. Fuis, and C. Mantis, 1978a. Seismic amplitude measurements suggest foreshocks have different focal mechanisms than aftershocks. Science 201, 56-59. 408

Lindh, A. G., D. A. Lockner, and W. H. K. Lee, 1978b. Velocity anomalies: an alternative explanation. Bull., Seis. Soc. Amer. 68, 721-734. 410

Liu, H-P., D. L. Anderson and H. Kanamori, 1976. Velocity dispersion due to anelasticity; implications for seismology and mantle composition. Geophys. J. 47, 41-58. 333, 347, 348

Lomnitz, C., 1956. Creep measurements in igneous rocks. J. Geology 64, 473-479. 328

Lomnitz, C., 1957. Linear dissipation in solids. J. Appl. Phys. 28, 201-205. 328, 347

Lomnitz, C., 1966a. Magnitude stability in earthquake sequences. Bull., Seis. Soc. Amer. 56, 247-249. 407

Lomnitz, C., 1971. Travel-time errors in the laterally inhomogeneous earth. Bull., Seis. Soc. Amer. 61, 1639-1654. 295

Lomnitz, C., 1974. Global tectonics and earthquake risk. Elsevier Sci. Publ. Co., Amsterdam. 405, 406, 407, 409

Lomnitz, C., 1977a. A fast epicenter location program. Bull., Seis. Soc. Amer. 67, 425-431. 297

Lomnitz, C., 1977b. A procedure for eliminating the indeterminacy in focal depth determinations. Bull., Seis. Soc. Amer. 67, 533-535. 297

Longuet-Higgins, M. S., 1950. A theory of the origin of microseisms. Phil. Trans. Roy. Soc., London, A 243, 1-35. 412

Love, A. E. H., 1911. Some problems of geodynamics. Dover (reprint 1967), New York. 304

Luh, P. C., 1973. Free oscillations of the laterally inhomogeneous earth: quasi-degenerate multiplet coupling. Geophys. J. 32, 187-202. 308

Luh, P. C., 1974. Normal modes of a rotating, self-gravitating inhomogeneous earth. Geophys. J. 38, 187-224. 308

Luh, P. C., and A. M. Dziewonski, 1975. Theoretical seismograms for the Colombian earthquake of 1970 July 31. Geophys. J. 43, 679-695. 321

Luh, P. C., and A. M. Dziewonski, 1976. Theoretical normal-mode spectra of a rotating elliptical earth. Geophys. J. 45, 617-645. 308

Lysmer, J., 1970. Lumped mass method for Rayleigh waves. Bull., Seis. Soc. Amer. 60, 89-104. 241

Lysmer, J., and L. A. Drake, 1972. A finite element method for seismology. In: B. A. Bolt (Editor), Methods in computational physics, Vol. 11, Seismology: surface waves and earth oscillations. Academic, New York, 181-216. 241

McConnell, Jr., R. K., R. N. Gupta, and J. T. Wilson, 1966. Compilation of deep crustal seismic refraction profiles. Rev. Geophys. 4, 41-100. 223

McCowan, D. W., 1976. Moment tensor representation of seismic surface wave studies. Geophys. J. 44, 595-599. 360

McCowan, D. W., and R. T. Lacoss, 1978. Transfer function for the seismic research observatory seismograph system. Bull., Seis. Soc. Amer. 68, 501-512. 435

McCracken, L. G., 1957. Ray theory vs. normal mode theory in wave propagation problems. IRE Trans. on Antennas and Propagation, AP-5, 137-140. 190

McDonal, F. J., F. A. Angona, R. L. Mills, R. L. Sengbush, R. G. Van Nostrand, and J. E. White, 1958. Attenuation of shear and compressional waves in Pierre shale. Geophysics 23, 421-438. 331

Macdonald, J. R., 1959. Rayleigh-wave dissipation functions in low-loss media. Geophys. J. 2, 132-135. 334

McEvilly, T. V. and L. R. Johnson, 1974. Stability of P and S velocities from Central California quarry blasts. Bull., Seis. Soc. Amer. 64, 343-353. 410

McGarr, A., and L. E. Alsop, 1967. Transmission and reflection of Rayleigh waves at vertical boundaries. J. Geophys. Res. 72, 2169-2180. 183

McMechan, G. A., 1974. P-wave train synthetic seismograms calculated by quantized ray theory. Geophys. J. 37, 407-421. 303

McMechan, G. A., 1976a. Mantle structure and short period body wave spectra at intermediate epicentral distances. Geophys. J. 44, 269-274. 227

McMechan, G. A. 1977. Low-velocity zone parameters as functions of focal depth. Geophys. J. 51, 217-228. 278

McMechan, G. A., and S. K. Dey-Sarker, 1976. Quantized ray theory for non-zero focal depths. Geophys. J. 46, 235-246. 303

McMechan, G. A. and R. A. Wiggins, 1972. Depth limits in body wave inversions. Geophys. J. 28, 459-473. 278

McMechan, G. A. and W. G. Workman, 1974. P-wave amplitudes and magnitudes between 10° and 35°. Bull., Seis. Soc. Amer. 64, 1887-1899. 387

Madariaga, R. I., 1972. Toroidal free oscillations of the laterally heterogeneous earth. Geophys. J. 27, 81-100. 308

Madariaga, R., 1976. Dynamics of an expanding circular fault. Bull., Seis. Soc. Amer. 66, 639-666. 367, 385

Madariaga, R., 1978. The dynamic field of Haskell's rectangular dislocation fault model. Bull., Seis. Soc. Amer. 68, 869-887. 385

Madariaga, R., and K. Aki, 1972. Spectral splitting of toroidal free oscillations due to lateral heterogeneity of the earth's structure. J. Geophys. Res. 77, 4421-4431. 318

Mansinha, L., and D. E. Smylie, 1967. Effect of earthquakes on the Chandler Wobble and the secular polar shift. J. Geophys. Res. 72, 4731-4743. 364

Mantovani, E., F. Schwab, H. Liao, and L. Knopoff, 1977. Teleseismic Sn: a guided wave in the mantle. Geophys. J. 51, 709-726. 257

Mantovani, E., F. Schwab, H. Liao, and L. Knopoff, 1977a. Generation of complete theoretical seismograms for SH — II. Geophys. J. 48, 531-536. 321

Margenau and Murphy, 1943. The mathematics of physics and chemistry. Van Nostrand, New York. 1

Maruyama, T., 1963. On the force equivalents of dynamical elastic dislocations with reference to the earthquake mechanism. Tokyo Univ., Bull. Earthquake Res. Inst. 41, 467-486. 358

Maruyama, T., 1964. Statistical elastic dislocations in an infinite and semi-infinite medium. Tokyo Univ., Bull. Earthquake Res. Inst. 42, 289-368. 378

Mason, Warren P., 1964-1976. Physical Acoustics (Vol. 1-12) Academic Press, New York. This is a continuing series. 34, 38

Massé, R. P., 1973. Compressional velocity distribution beneath Central and Eastern North America. Bull., Seis. Soc. Amer. 63, 911-935. 219, 220

Massé, R. P., 1974. Compressional velocity distribution beneath Central and Eastern North America in the depth range 450 to 800 Km. Geophys. J. 36, 705-716. 219

Massé, R. P., 1975. Baltic Shield crustal velocity distribution. Bull., Seis. Soc. Amer. 65, 885-897. 219

Massé, R. P. and S. S. Alexander, 1974. Compressional velocity distribution beneath Scandinavia and Western Russia. Geophys. J. 39, 587-602. 219

Massé, R. P., M. Landisman and J. B. Jenkins, 1972. An investigation of the upper mantle compressional velocity distribution beneath the Basin and Range Province. Geophys. J. 30, 19-36. 219

Massé, R. P., D. G. Lambert and D. G. Harkrider, 1973. Precision of the determination of focal depth from the spectral ratio of Love/Rayleigh surface waves. Bull., Seis. Soc. Amer. 63, 59-100. 298

Massé, R. P., E. A. Flinn, and R. M. Seggelke, 1974. PKIIKP and the average velocity of the inner core. Geophys. Res. Lett. 1, 39-42. 302

Mellman, G. R. and D. V. Helmberger, 1974. High-frequency attenuation by a thin high-velocity layer. Bull., Seis. Soc. Amer. 64, 1383-1388. 200

Melton, B. S., 1971. The LaCoste suspension — principles and practice. Geophys. J. 22, 521-543. 427

Melton, B. S., 1976. The sensitivity and dynamic range of inertial seismographs. Rev. Geophys. and Space Phys. 14, 93-116. 422

Melton, B. S., and B. M. Kirkpatrick, 1970. The symmetrical triaxial seismometer — its design for application to long-period seismometry. Bull., Seis. Soc. Amer. 60, 717-739. 434

Mencher, A. G., 1953. Epicentral displacement caused by elastic waves in an infinite slab. J. Appl. Phys. 24, 1240-1246. 103

Mendiguren, J. A., 1973. Identification of free oscillation spectral peaks for 1970 July 31 Colombian deep shock using the excitation criterion. Geophys. J. 33, 281-321. 316, 318

Mendiguren, J. A., 1977a. Inversion of surface wave data in source mechanism studies. J. Geophys. Res. 82, 889-894. 377

Mendiguren, J. A., 1977b. A numerical method to determine physical source parameters from free oscillation data. Geophys. J. 51, 617-623. 378

Mereu, R. F., 1966. An iterative method for solving the time-term equations. In: J. S. Steinhart and T. J. Smith (Editors), The earth beneath the continents: a volume of geophysical studies in honor of Merle A. Tuve. Amer. Geophys. Union Monograph #10, Washington, D.C., 495-497. 225

Merzer, A. M., 1971. Head waves from different transition layers. Geophys. J. 24, 77-95. 145

Merzer, A. M., 1974. Head waves, normal waves and the shapes of transition layers. Geophys. J. 37, 1-7. 145

Meyer, S. L., 1975. Data analysis for scientists and engineers. Wiley, New York. 405

Michaels, P., 1977. A method for the separation of interfering surface waves and the computation of their dispersion. Bull., Seis. Soc. Amer. 67, 383-392. 252

Miller, G. F., and H. Pursey, 1954. The field and radiation impedance of mechanical radiators on the free surface of a semi-infinite isotropic solid. Proc. Roy. Soc. London, A 223, 521-541. 127

Mills, J. M. and A. L. Hales, 1977. Great-circle Rayleigh wave attenuation and group velocity, Part 1: Observations for periods between 150 and 600 seconds for seven great-circle paths. Phys. Earth and Planet. Inter. 14, 109-119. 343

Minster, J. B., 1976. Transformation of multipolar source fields under a change of reference frame. Geophys. J. 47, 397-409. 360

Mitchell, B. J., 1976. Anelasticity of the crustal upper mantle beneath the Pacific Ocean from the inversion of observed surface wave attenuation. Geophys. J. 46, 521-533. 346

Mitchell, B. J. and D. V. Helmberger, 1973. Shear velocities at the base of the mantle from observations of S and ScS. J. Geophys. Res. 78, 6009-6020. 301

Mitchell, B. J., L. W. B. Leite, Y. K. Yu, and R. B. Herrmann, 1976. Attenuation of Love and Rayleigh Waves across the Pacific at periods between 15 and 110 seconds. Bull., Seis. Soc. Amer. 66, 1189-1202. 344

Mogi, K., 1963. Some discussion on aftershocks, foreshocks, and earthquake swarms — the fracture of a semi-infinite body caused by an inner stress origin and its relation to the earthquake phenomena. Univ. Tokyo, Bull. Earthquake Res. Inst. 41, 615-658. 408

Mogi, K., 1974. Active periods in the world's chief seismic belts. Tectonophysics 22, 265-282. 406

Molnar, P., B. E. Tucker, and J. N. Brune, 1973. Corner frequencies of P and S waves and models of earthquake sources. Bull., Seis. Soc. Amer. 63, 2091-2104. 385

Moon, W., and R. A. Wiggins, 1977. Variational type finite element solution of normal modes of simple earth models. Geophys. J. 51, 327-348. 308

Mooney, H. M., 1974. Some numerical solutions for Lamb's Problems. Bull., Seis. Soc. Amer. 64, 473-491. 126

Morris, G. B., 1972. Delay-time-function method and its application to the Lake Superior refraction data. J. Geophys. Res. 77, 297-314. 225

Morris, G. B., R. W. Raitt, and G. G. Shor, Jr., 1969. Velocity anisotropy and delay-time maps of the mantle near Hawaii. J. Geophys. Res. 74, 4300-4316. 226

Morse, P. M., and H. Feshbach, 1953. Methods of Theoretical Physics. McGraw-Hill, New York. 1, 39, 54

Mueller, R. A. and J. R. Murphy, 1971. Seismic characteristics of underground nuclear detonations. Part 1: Seismic spectrum scaling. Bull., Seis. Soc. Amer. 61, 1675-1692. 394

Mueller, S. (Editor), 1973. The structure of the earth's crust based on seismic data. Proceedings of the I.U.M.C. Symposium on Crustal Structure, Moscow, July 30-31, 1971. Tectonophysics, 20, 1-391. 219

Mueller, S., and M. Landisman, 1971. An example of the unified method of interpretation for crustal seismic data. Geophys. J. 23, 365-371. 231

Müller, G., 1970. Exact ray theory and its application to the reflection of elastic waves from vertically inhomogeneous media. Geophys. J. 21, 261-283. 230

Müller, G., 1971. Approximate treatment of elastic body waves in media with spherical symmetry. Geophys. J. 23, 435-449. 303

Müller, G., 1973a. Amplitude studies of core phases. J. Geophys. Res. 78, 3469-3490. 302

Müller, G., and R. Kind, 1976. Observed and computed seismogram sections for the whole earth. Geophys. J. 44, 699-716. 294, 303

Müller, G., A. H. Mula, and S. Gregerson, 1977. Amplitudes of long-period PcP and the core-mantle boundary. Phys. Earth and Planet Inter. 14, 30-40. 302

Murphy, A. J. and J. M. Savino, 1975. A comprehensive study of long-period (20-200 seconds) earth noise at the high-gain worldwide seismograph stations. Bull., Seis. Soc. Amer. 65, 1827-1862. 418

Musgrave, M. J. P., 1970. Crystal Acoustics. Holden-Day, San Francisco. 34

Nakanishi, I., 1978. Regional differences in the phase velocity and quality factor Q of mantle Rayleigh waves. Science 200, 1379-1381. 344

Nakanishi, K. K., F. Schwab, and E. G. Kausel, 1976. Interpretation of S_a on a continental structure. Geophys. J. 47, 211-223. 257

Nakanishi, K. K., L. Knopoff and L. B. Slichter, 1976a. Observation of Rayleigh wave dispersion at very long periods. J. Geophys. Res. 81, 4417-4421. 318

Nakanishi, K., F. Schwab, and L. Knopoff, 1977. Generation of complete theoretical seismograms for SH - I. Geophys. J. 48, 525-530. 321

Ness, N. F., J. C. Harrison, and L. B. Slichter, 1961. Observations of the free oscillations of the earth. J. Geophys. Res. 66, 621-629. 343, 344

Nolet, G., 1975. Higher Rayleigh modes in Western Europe. Geophys. Res. Letters 2, 60-62. 256

Nolet, G., 1977. The upper mantle under Western Europe inferred from the dispersion of Rayleigh modes. J. Geophysics 43, 265-285. 320

Nolet, G., and B. Kennett, 1978. Normal-mode representation of multiple-ray reflections in a spherical earth. Geophys. J. 53, 219-226. 313

North, R. G., and A. M. Dziewonski, 1976. A note on Rayleigh-wave flattening corrections. Bull., Seis. Soc. Amer. 66, 1873-1879. 313

Nowroozi, A. A., 1968. Measurement of Q values from the free oscillations of the earth. J. Geophys. Res. 73, 1407-1415. 343

Nur, A., 1971. Effects of stress on velocity anisotropy in rocks with cracks. J. Geophys. Res. 78, 2022-2035. 35

Nur, A. 1972. Dilatancy, pore fluids, and premonitory variation of t_s/t_p travel times. Bull., Seis. Soc. Amer. 62, 1217-1222. 410

Nuttli, O. W., 1973. Seismic wave attenuation and magnitude relations for Eastern North America. J. Geophys. Res. 78, 876-885. 344

Nyman, D., 1975. Stacking and stripping for normal mode eigenfrequencies. Geophys. J. 41, 271-303. 316

Nyman, D. C., H. K. Gupta, and M. Landisman, 1977. The relationship between group velocity and phase velocity for finite, discrete observations. Bull., Seis. Soc. Amer. 67, 1249-1258. 249

O'Brien, P. N. S., 1968. Lake Superior crustal structure — a reinterpretation of the 1963 seismic experiment. J. Geophys. Res. 73, 2669-2689. 225

O'Connell, R. J. and B. Budiansky, 1978. Measures of dissipation in viscoelastic media. Geophys. Res. Letters 5, 5-8. 326

Odegard, M.E., and G.J. Fryer, 1977. Intensities and travel tines of seismic rays in a sphere with linear velocity function. Bull., Seis. Soc. Amer. 67, 33-42. 287

Officer, C. B., 1958. Introduction to the theory of sound transmission with applications to the ocean. McGraw-Hill, New York. 242, 271

Officer, C. B., 1974. Introduction to theoretical geophysics. Springer, New York. 1, 364

Ohnaka, M., 1976. A physical basis for earthquakes based on the elastic rebound model. Bull., Seis. Soc. Amer. 66, 433-451. 378

Okal, E. A., and D. L. Anderson, 1975. A study of lateral inhomogeneities in the upper mantle by multiple ScS travel-time residuals. Geophys. Res. Letters 2, 313-316. 300

Okeke, E. O., 1972. A theoretical model of primary frequency microseisms. Geophys. J. 27, 289-299. 412

Oliver, J., 1962. A summary of observed seismic surface wave dispersion. Bull., Seis. Soc. Amer. 52, 81-86. 253

Oliver, J., 1964. Propagation of PL waves across the United States. Bull., Seis. Soc. Amer. 54, 151-160. 256

Oliver, J., and M. Major, 1960. Leaking modes and the PL phase. Bull., Seis. Soc. Amer. 50, 165-180. 256

Oliver, J., and L. Murphy, 1971. WWNSS: Seismology's global network of observing stations. Science 174, 254-261. 435

Oliver, J., A. Ryall, J. N. Brune, and D. B. Slemmons, 1966. Microearthquake activity recorded by portable seismographs of high sensitivity. Bull., Seis. Soc. Amer. 56, 899-924. 406

Olson, P., 1977. Internal waves in the earth's core. Geophys. J. 51, 183-215. 307

Page, R., 1968. Aftershocks and microaftershocks of the great Alaska earthquake of 1964. Bull., Seis. Soc. Amer. 58, 1131-1168. 408

Panza, G. F., 1974. Focal-mechanism determination from multimode Rayleigh wave response. Phys. Earth and Planet. Inter. 8, 345-351. 377

Panza, G. F., and G. Calcagnile, 1975. Lg, Li, and Rg from Rayleigh modes. Geophys. J. 40, 475-487. 257

Panza, G. F., F. Schwab, and L. Knopoff, 1973. Multimode surface waves for selected focal mechanisms-I. Dip-slip sources on a vertical fault plane. Geophys. J. 34, 265-278. 364

Panza, G. F., F. Schwab, and L. Knopoff, 1975a. Multimode surface waves for selected focal mechanisms-II. Dip-slip sources. Geophys. J. 42, 931-943. 364

Panza, G. F., F. Schwab, and L. Knopoff, 1975b. Multimode surface waves for selected focal mechanisms-III. Strike-slip sources. Geophysics J. 42, 945-955. 364

Papadakas, E. P., 1976. Ultrasonic velocity and attennuation: Measurement methods with scientific and industrial applications. In: Mason, Warren P. (Editor), Physical Acoustics, Vol. 12, Academic, New York. 51

Parker, R. L., 1970. The inverse problem of electrical conductivity in the mantle. Geophys. J. 22, 121-138. 260

Parker, R. L., 1972. Inverse theory with grossly inadequate data. Geophys. J. 29, 123-138. 262

Parker, R. L., 1977. Understanding inverse theory. Ann. Rev. Earth Planet Sci. 5, 35-64. 262

Payton, R. G., 1964. An application of the dynamic Betti-Rayleigh reciprocal theorem to moving point loads in elastic media. Quart. Appl. Math. 21, 299-313. 360

Pekeris, C. L., 1941. A pathological case in the numerical solution of integral equations. Proc. Nat. Acad. Sci., U.S., 26, 433-437. 103

Pekeris, C. L., 1948. Theory of propagation of explosive sound in shallow water, In: Memoir No. 27, Geol. Soc. Amer., New York. 176, 183, 242

Pekeris, C. L., 1950. Ray theory vs. normal mode theory in wave propagation problems. Amer. Math. Soc., Proc. Symp. Appl. Math. 2, 71-75. 186, 189, 190

Pekeris, C. L., 1955a. The seismic surface pulse. Proc. Nat. Acad. Sci., U.S., 41, 469-480. 70, 103

Pekeris, C. L., 1955b. The seismic buried pulse. Proc. Nat. Acad. Sci., U. S., 41 629-639. 103

Pekeris, C. L., and H. Lifson, 1957. Motion of the surface of a uniform half-space produced by a buried pulse. J. Acoust. Soc. Amer. 29, 1233-1238. 126

Pekeris, C. L., and I. M. Longman, 1958. Ray-theory solution of the propagation of explosive sound in a layered liquid. J. Acoust. Soc. Amer. 30, 323-328. 191
Pekeris, C. L., Z. Alterman, F. Abramovici and H. Jarosch, 1965. Propagation of a compressional pulse in a layered solid. Rev. Geophys. 3, 25-47. 191, 229
Pelzer, H., 1957. Models of materials with loss per cycle nearly independent of frequency. J. Polymer Sci. 25, 51-60. 331
Peterson, J., H. M. Butler, L. G. Holcomb, and C. R. Hutt, 1976. The seismic research observatory. Bull., Seis. Soc. Amer. 66, 2049-2068. 435
Peterson, R. A., W. R. Fillippone, and F. B. Coker, 1955. The synthesis of seismograms from well log data. Geophysics 20, 516-538. 229
Phinney, R. A., 1964. Structure of the earth's crust from spectral behavior of long-period body waves. J. Geophys. Res. 69, 2997-3017. 227
Phinney, R. A., 1970. Reflection of acoustic waves from a continuously varying interfacial region. Rev. Geophys. Space Phys. 8, 517-532. 278
Phinney, R. A., and S. S. Alexander, 1966. P wave diffraction theory and the structure of the core-mantle boundary. J.Geophys. Res. 71, 5959-5975. 289, 306
Phinney, R. A. and S. S. Alexander, 1969. The effect of a velocity gradient at the base of the mantle on diffracted P waves in the shadow. J. Geophys. Res. 74, 4967-4971. 302
Phinney, R. A., and R. Burridge, 1973. Representation of the elastic-gravitational excitation of a spherical earth model by generalized spherical harmonics. Geophys. J. 34, 451-487. 360
Phinney, R. A., and L. M. Cathles, 1969. Diffraction of P by the core: a study of long-period amplitudes near the edge of the shadow. J. Geophys. Res. 74, 1556-1574. 302
Pho, H-T. and L. Behe, 1972. Extended distances and angles of incidence of P waves. Bull., Seis. Soc. Amer. 62, 885-902. 369
Pilant, W. L., 1972. Complex roots of the Stoneley-wave equation. Bull., Seis. Soc. Amer. 62, 285-299. 148, 150
Pilant, W. L., and L. Knopoff, 1964. Observation of multiple seismic events. Bull., Seis. Soc. Amer. 54, 19-39. 247, 252
Pilant, W. L., and L. Knopoff, 1970. Inversion of phase and group slowness dispersion. J. Geophys. Res. 75, 2135-2136. 249
Pinney, E., 1954. Surface motion due to a point source in a semi-infinite elastic medium. Bull., Seis. Soc. Amer. 44, 571-596. 103
Pomeroy, P. W., and G. H. Sutton, 1960. The use of galvanometers as band-rejection filters in electromagnetic seismographs. Bull., Seis. Soc. Amer. 50, 135-151. 421
Pomeroy, P. W., G. Hade, J. Savino, and R. Chander, 1969. Preliminary results from high-gain wide-band long-period electromagnetic seismographic systems. J. Geophys. Res. 74, 3295-3298. 435
Pope, A. J., 1972. Fiducial regions for body wave focal plane solutions. Geophys. J. 30, 331-342. 374
Poupinet, G., and C. Wright, 1972. The generation and properties of shear-coupled PL waves. Bull., Seis. Soc. Amer. 62, 1699-1710. 256
Press, F., 1956. Determination of crustal structure from phase velocity of Rayleigh waves. Part I: Southern California. Bull., Geol. Soc. Amer. 67, 1647-1658. 250
Press, F., 1956a. Rigidity of the earth's core. Science 124, 1204. 340
Press, F., 1965. Displacements, strains, and tilts at teleseismic distances. J. Geophys. Res. 70, 2395-2412. 364
Press, F., 1966. Seismic velocities. In: S. P. Clark, Jr. (Editor), Handbook of Physical Constants, Geol. Soc. Amer. Memoir No. 97, New York. 51
Press, F., 1970a. Earth models consistent with geophysical data. Phys. Earth and Planet. Inter. 3, 3-22. 323
Press, F., 1970b. Regionalized earth models. J. Geophys. Res. 75, 6575-6581. 323
Press, F., and D. Harkrider, 1962. Propagation of acoustic-gravity waves in the atmosphere. J. Geophys. Res. 67, 3889-3902. 242
Press, F., and J. Healy, 1957. Absorption of Rayleigh waves in low-loss media. J. Appl. Phys. 28, 1323-1325. 333

Press, F., M. Ewing, and F. Lehner, 1958. A long-period seismograph system. Trans., Amer. Geophys. Union 39, 106-108. 424

Press, F., A. Ben-Menahem, and M. N. Toksöz, 1961. Experimental determination of earthquake fault length and rupture velocity. J. Geophys. Res. 66, 3471-3485. 375

Prothero Jr., W. A., 1974. An ocean-bottom seismometer capsule. Bull., Seis. Soc. Amer. 64, 1251-1262. 434

Qamar, Anthony, 1973. Revised velocities in the earth's core. Bull., Seis. Soc. Amer. 63, 1073-1105. 298, 302

Radovich, B. J., and J-Cl. DeBremaecker, 1974. Body waves as normal and leaking modes — leaking modes of Love waves. Bull., Seis. Soc. Amer. 64, 301-306. 180

Raitt, R. W., G. G. Shor, Jr., T. J. G. Francis, and G. B. Morris, 1969. Anisotropy of the Pacific upper mantle. J. Geophys. Res. 74, 3095-3109. 34, 225

Randall, M. J., 1971a. A revised travel-time table for S. Geophys. J. 22, 229-234. 292

Randall, M. J., 1971b. Travel time tables for S waves. Seismological Observatory Bulletin S-177, Wellington. 292, 369

Randall, M. J., 1971c. Shear invariant and seismic moment for deep-focus earthquakes. J. Geophys. Res. 76, 4991-4992. 379

Randall, M. J., 1972a. Multipolar analysis of the mechanisms of deep-focus earthquakes. In: B. A. Bolt (Editor), Methods in computational physics, Vol 12, Academic, New York, pp. 267-298. 351, 360, 368, 374

Randall, M. J., 1973. The spectral theory of seismic sources. Bull., Seis. Soc. Amer. 63, 1133-1144. 378

Randall, M. J., 1976. Attenuative dispersion and frequency shifts of the earth's free oscillations. Phys. Earth and Planet. Inter. 12, P1-P4. 347

Randall, M. J., and L. Knopoff, 1970. The mechanism at the focus of deep earthquakes. J. Geophys. Res. 75, 4965-4976. 374

Real, C. R., and T-L. Teng, 1973. Local Richter magnitude and total signal duration in Southern California. Bull., Seis. Soc. Amer. 63, 1809-1827. 389

Reid, H. F., 1910. The mechanics of the earthquake: Vol. 2 of the California Earthquake of April 18, 1906. Carnegie Inst., Washington, Publ. No. 87. 357

Reiter, L., 1970. An investigation into the time term method in refraction seismology. Bull., Seis. Soc. Amer. 60, 1-13. 225

Rice, J. R., and M. A. Chinnery, 1972. On the calculation of changes in the earth's inertia tensor due to faulting. Geophys. J. 29, 79-90. 365

Richards, P. G., 1971. Elastic wave solutions in stratified media. Geophysics 36, 798-809. 205

Richards, P. G., 1973. Calculation of body waves, for caustics and tunnelling in core phases. Geophys. J. 35, 243-264. 290

Richards, P. G., 1973a. The dynamic field of a growing plane elliptical shear crack. Intern. J. Solids Structures 9, 843-861. 366

Richards, P. G., 1974. Weakly coupled potentials for high-frequency elastic waves in continuously stratified media. Bull., Seis. Soc. Amer. 64, 1575-1588. 36

Richards, P. G., 1976. On the adequacy of plane-wave reflection/transmission coefficients in the analysis of seismic body waves. Bull., Seis. Soc. Amer. 66, 701-717. 140, 290

Richards, P. G., 1976a. Dynamic motions near an earthquake fault: a three-dimensional solution. Bull., Seis. Soc. Amer. 66, 1-32. 367

Richter, C. F., 1958. Elementary Seismology. W. H. Freeman, San Francisco. 366, 385, 386

Rikitake, T., 1976. Earthquake prediction. Elsevier Sci. Publ. Co., Amsterdam. 366, 408, 409, 411

Ringdal, F., 1976. Maximum-likelihood estimation of seismic magnitude. Bull., Seis. Soc. Amer. 66, 789-802. 390

Robertson, H., 1965. Physical and topographic factors as related to short-period wind noise. Bull., Seis. Soc. Amer. 55, 863-877. 418

Robinson, E. A., 1967. Multichannel time series analysis with digital computer programs. Holden-Day, San Francisco. 217

Robinson, E. A. and S. Treitel, 1978. The fine structure of the normal incidence synthetic seismogram. Geophys. J. 53, 289-309. 230

Robinson, R., and H. M. Iyer, 1976. Temporal and spatial variations of travel-time residuals in Central California for Novaya Zemlya events. Bull., Seis. Soc. Amer. 66, 1733-1747. 300

Robinson, R., and R. L. Kovach, 1972. Shear wave velocities in the earth's mantle. Phys. Earth. and Planet Int. 5, 30-44. 301

Rodgers, P. W., 1968. The response of the horizontal pendulum seismometer to Rayleigh and Love waves, tilt, and the free oscillation of the earth. Bull., Seis. Soc. Amer. 58, 1384-1406. 425

Rodi, W. L., P. Glover, T. M. C. Li, and S. S. Alexander, 1975. A fast, accurate method for computing group-velocity partial derivatives for Rayleigh and Love Modes. Bull., Seis. Soc. Amer. 65, 1105-1114. 258

Roever, W. L. and T. F. Vining, 1959. Propagation of elastic wave motion from an impulsive source along a fluid/solid interface: 1. Experimental pressure response. Phil. Trans. Roy. Soc., London, A 251, 455-465. 151

Rogers Jr., A. M. and C. Kisslinger, 1972. The effect of a dipping layer on P-wave transmission. Bull., Seis. Soc. Amer. 62, 301-324. 227, 230

Rosenbaum, J. H., 1961. Refraction arrivals along a thin elastic plate surrounded by a fluid medium. J. Geophys. Res. 66, 3899-3906. 200

Rosenbaum, J. H., 1965. Refraction arrivals through thin high-velocity layers. Geophys. 30, 204-212. 200

Rothé, J. P., 1969. The seismicity of the earth 1953-1965. UNESCO, Paris. 399

Roy, A., 1974. Surface displacements in an elastic half space due to a buried moving point source. Geophys. J. 40, 289-304. 360

Rundle, J. B., and D. D. Jackson, 1977. Numerical simulation of earthquake sequences. Bull., Seis. Soc. Amer. 67, 1363-1377. 406

Ryall, A., and S. D. Malone, 1971. Earthquake distribution and mechanism of faulting in the Rainbow Mountain-Dixie Valley-Fairview Peak area, Central Nevada. J. Geophys. Res. 76, 7241-7248. 406, 407

Ryall, A., J. D. Van Wormer, and A. E. Jones, 1968. Triggering of microearthquakes by earth tides, and other features of the Truckee, California, earthquake sequence of September, 1966. Bull., Seis. Soc. Amer. 58, 215-248. 404

Sabatier, P. C., 1977. On geophysical inverse problems and constraints. J. Geophys. 43, 115-137. 262

Sailor, R. V., and A. M. Dziewonski, 1978. Measurements and interpretation of normal mode attenuation. Geophys. J. 53, 559-581. 344, 345

Saito, M., 1967. Excitation of free oscillations and surface waves by a point source in a vertically heterogeneous earth. J. Geophys. Res. 72, 3689-3699. 364

Sato, R., and A. F. Espinosa, 1967. Dissipation in the earth's mantle and rigidity and viscosity in the earth's core determined from waves multiply reflected from the mantle-core boundary. Bull., Seis. Soc. Amer. 57, 829-856. 342

Sato, Y., 1955. Analysis of dispersed surface waves by means of Fourier transform, 1. Tokyo Univ., Bull. Earthquake Res. Inst. 33, 33-47. 247

Sato, Y., 1956. Analysis of dispersed surface waves by means of the Fourier transform, 2. Synthesis of movement near the origin. Tokyo Univ., Bull. Earthquake Res. Inst. 34, 9-18. 375

Sato, Y., and T. Usami, 1962. Basic study on the oscillations of a homogeneous elastic sphere I. Frequency of the free oscillations. Geophys. Mag. 31, 15-24. 304, 310

Sato, Y., T. Usami and M. Ewing, 1962. Basic study on the oscillation of a homogeneous elastic sphere IV. Propagation of disturbances on the sphere. Geophys. Mag. 31, 237-242. 304

Sauter, F., 1950. Der elastische Halbraum bei einer mechanischen Beeinflussung seiner Oberfläche. Z. angew. Math. u. Mech. 30, 203-215. 103

Savage, J. C., 1965. The effect of rupture velocity upon seismic first motions. Bull., Seis. Soc. Amer. 55, 263-275. 362

Savage, J. C., 1972. Relation of corner frequency to fault dimensions. J. Geophys. Res. 77, 3788-3795. 384

Savage, J. C., 1974. Relation between P- and S-wave corner frequencies in the seismic spectrum. Bull., Seis. Soc. Amer. 64, 1621-1627. 385

Savage, J. C., and L. M. Hastie, 1969. A dislocation model for the Fairview Peak, Nevada, earthquake. Bull., Seis. Soc. Amer. 59, 1937-1948. 378

Savage, J. C., and M. E. O'Neill, 1975. The relation between the Lomnitz and Futterman theories of internal friction. J. Geophys. Res. 80, 249-251. 347

Savino, J., K. McCamy, and G. Hade, 1972a. Structures in earth noise beyond twenty seconds — a window for earthquakes. Bull., Seis. Soc. Amer. 62, 141-176. 418, 420

Savino, J. M., A. J. Murphy, J. M. W. Rynn, R. Tatham, L. R. Sykes, G. L. Choy and K. McCamy, 1972b. Results from the high-gain long-period seismograph experiment. Geophys. J. 31, 179-203. 435

Scheidegger, A. E., and P. L. Willmore, 1957. The use of a least squares method for the interpretation of data from seismic surveys. Geophysics 22, 9-22. 225

Scholte, J. C., 1947. The range of existence of Rayleigh and Stoneley waves. Mon. Not. Roy. Astr. Soc., Geophys. Suppl. 5, 120-126. 147

Scholz, C. H., L. R. Sykes, and Y. P. Aggarwal, 1973. Earthquake prediction: a physical basis. Science 181, 803-809. 410

Schreiber, E., O. L. Anderson and N. Soga, 1973. Elastic constants and their measurement. McGraw-Hill, New York. 51

Schwab, F. and E. Kausel, 1976. Quadripartite surface wave method: development. Geophys. J. 45, 231-244. 252

Schwab, F., and E. Kausel, 1976a. Long-period surface wave seismology: Love wave phase velocity and polar phase shift. Geophys. J. 45, 407-435. 310

Schwab, F., and L. Knopoff, 1972. Fast surface wave and free mode computations. In: B. A. Bolt (Editor), Methods in computational physics, Vol. 11, Seismology: surface waves and earth oscillations. Academic, New York, pp. 87-180. 241, 256, 313

Schwab, F., E. Kausel, and L. Knopoff, 1974. Geophys. J. 36, 737-742. 257

Semenov, A. M., 1969. Variations in the travel-time of transverse and longitudinal waves before violent earthquakes. Izv. Akad. Nauk., Fiz. Zemlii, No. 4, 72-77 (in Russian) Bull., Acad. Sci. USSR, Phys. Sol. Earth. No. 4, 245-248 (English Transl.). 410

Sengupta, M. K., and B. R. Julian, 1976. P-wave travel times from deep earthquakes. Bull., Seis. Soc. Amer. 66, 1555-1579. 299, 301

Sengupta, M. K. and M. N. Toksöz, 1977. The amplitudes of P-waves and magnitudes — corrections for deep focus earthquakes. J. Geophys. Res. 82, 2971-2980. 387

Sherwood, J. W. C., 1958. Elastic wave propagation in a semi-infinite medium. Proc. Phys. Soc., London, B 71, 207-219 (plus plates following p.292) 103, 116

Sherwood, J. W. C., and A. W. Trorey, 1965. Minimum-phase and related properties of the response of a horizontally stratified absorptive earth to plane acoustic waves. Geophysics 30, 191-197. 230

Shimshoni, M., Y. Sylman, and A. Ben-Menahem, 1973. Theoretical amplitudes of body waves from a dislocation source in the earth, II. Core phases. Phys. Earth and Planet. Inter. 7, 59-91. 364

Shlien, S., 1972. Earthquake — tide correlation. Geophys. J. 28, 27-34. 404

Shlien, S., and M. N. Toksöz, 1970. A clustering model for earthquake occurrences. Bull., Seis. Soc. Amer. 60, 1765-1787. 405

Shlien, S., and M. N. Toksöz, 1975. A branching Poisson-Markov model of earthquake occurrances. Geophys. J. 42, 49-59. 406

Shudde, R. H., and D. R. Barr, 1977. An analysis of earthquake frequency data. Bull., Seis. Soc. Amer. 67, 1379-1386. 404

Simmons, G., and A. Nur, 1968. Granites: relation of properties in situ to laboratory measurements. Science 168, 789-791. 337

Simon, R., 1969. Earthquake interpretations. Colo. School of Mines, Golden. 294

Singh, S. J., 1970. Static deformation of a multilayered half-space by internal sources. J. Geophys. Res. 75, 3257-3263. 364

Singh, S. J., 1973. Generation of SH-type motion by torsion-free sources. Bull., Seis. Soc. Amer. 63, 1189-1200. 365

Singh, S. J., A. Ben-Menahem, and M. Shimshoni, 1972. Theoretical amplitudes of body waves from a dislocation source in the earth, I. Core reflections. Phys. Earth and Planet. Inter. 5, 231-263. 364

Sipkin, S.A., and T.H. Jordan, 1975. Lateral heterogeneity of the upper mantle determined from the travel times of ScS. J. Geophys. Res. 80, 1474-1484. 300

Sipkin, S. A., and T. H. Jordan, 1976. Lateral heterogeneity of the upper mantle determined from the travel times of multiple ScS. J. Geophys. Res. 81, 6307-6320. 300
Slichter, L. B., 1932. The theory of the interpretation of seismic travel-time curves in horizontal structures. Physics (now J. Appl. Phys.) 3, 273-295. 277
Slichter, L. B., 1961. The fundamental free mode of the earth's inner core. Proc. Nat. Acad. Sci., USA, 47, 186-190. 307
Slichter, L. B., 1967. Spherical oscillations of the earth. Geophys. J. 14, 171-177. 343, 344
Smirnov, V. I., and S. L. Sobolev, 1932. Sur une méthode nouvelle dans le problème plan des vibrations élastiques. Akad. Nauk. SSSR, Seis. Inst. Trudy 20, 1-37. 103
Smith, E. G. C., 1976. Scaling the equations of condition to improve conditioning. Bull., Seis. Soc. Amer. 66, 2075-2076. 297
Smith, M. L., 1974. The scalar equations of infinitesimal elastic-gravitational motion from a rotating, slightly elliptical earth. Geophys. J. 37, 491-526. 307
Smith, M. L., 1976. Translational inner core oscillations of a rotating, slightly elliptical earth. J. Geophys. Res. 81, 3055-3065. 307
Smith, M. L., and J. F. Franklin, 1969. Geophysical application of generalized inverse theory. J. Geophys. Res. 74, 2783-2785. 262
Smith, S. W., 1961. An investigation of the earth's free oscillations. Ph.D. Thesis, Calif. Inst. of Technology, Pasadena. 343, 344
Smith, S. W., 1966. Free oscillations excited by the Alaskan earthquake. J. Geophys. Res. 71, 1183-1193. 317
Smith, S. W., 1972. The anelasticity of the mantle. Tectonophysics 13, 601-622. 335, 343, 344
Smith, T. J., J. S. Steinhart, and L. T. Aldrich, 1966. Lake Superior crustal structure. J. Geophys. Res. 71, 1141-1172. 225
Smylie, D. E., and L. Mansinha, 1971. The elasticity theory of dislocations in real earth models and changes in the rotation of the earth. Geophys. J. 23, 329-354. 365
Sobolev, S. L., 1932. Application de la théorie des ondes à la solution du problème de H. Lamb. Akad. Nauk. SSSR, Seis. Inst. Trudy 41, 1-41. 103
Sokolnikoff, I. S., 1956. Mathematical Theory of Elasticity. McGraw-Hill, New York. 20, 53
Solanki, J. J., 1974. Surface wave investigations of the crustal structure of the Western United States. Ph.D. Dissertation, U. Pittsburgh. 249
Solomon, S. C., 1973. Shear wave attenuation and melting beneath the Mid-Atlantic Ridge. J. Geophys. Res. 78, 6044-6059. 340
Solomon, S. C., and M. N. Toksöz, 1970. Lateral variation of attenuation of P and S waves beneath the United States. Bull., Seis. Soc. Amer. 60, 819-838. 340
Sorrells, G. G., 1971. A preliminary investigation into the relationship between long-period noise and local fluctuations in the atmospheric pressure field. Geophys. J. 26, 71-82. 420
Sorrells, G. G., and E. J. Douze, 1974. A preliminary report on infrasonic waves as a source of long-period seismic noise. J. Geophys. Res. 79, 4908-4917. 420
Sorrells, G. G., J. B. Crowley, and K. F. Veith, 1971. Methods for computing ray paths in complex geological structures. Bull., Seis. Soc. Amer. 61, 27-53. 279
Sorrells, G. G., J. A. MacDonald, Z. A. Der, and E. Herrin, 1971a. Earth motion caused by local atmospheric pressure changes. Geophys. J. 26, 83-98. 420
Spencer, T. W., 1960. The method of generalized reflection and transmission coefficients. Geophysics 25, 625-641. 143
Spencer, T. W., 1965a. Refraction along a layer. Geophysics 30, 369-388. 200
Stalmach, D. and J-Cl. De Bremaecker, 1973. Body waves as normal and leaking modes: dispersion and excitation on the (+-) sheet. Bull., Seis. Soc. Amer. 63, 995-1011. 182
Stamou, P., and M. Båth, 1974. The caustic and other properties of SKP. Phys. Earth Planet. Int. 8, 317-331. 302
Stauder, W., 1960. S waves and focal mechanisms: the state of the question. Bull., Seis. Soc. Amer. 50, 333-346. 358

Stauder, W., 1962. The focal mechanism of earthquakes. In: H. E. Landsberg and J. van Miegham (Editors). Advances in Geophysics, Vol. 9. Academic, New York, pp. 1-76. 368, 370, 371, 373, 374

Stauder, W., and A. Ryall, 1967. Spatial distribution and source mechanism of microearthquakes in Central Nevada. Bull., Seis. Soc. Amer. 57, 1317-1345. 406, 407

Stein, S., and R. J. Geller, 1977. Amplitudes of the split normal modes of a rotating, elliptical earth excited by a double couple. J. Phys. Earth 25, 117-142. 308

Stein, S., and R. G. Geller, 1978. Time-domain observation and synthesis of split spheroidal and torsional free oscillations of the 1960 Chilean earthquake: Preliminary results. Bull., Seis. Soc. Amer. 68, 325-332. 316

Stein, S., and R. J. Geller, 1978a. Attenuation measurements of split normal modes for the 1960 Chilean and 1964 Alaskan earthquakes. Bull., Seis. Soc. Amer. (In Press.) 343

Steinhart, J. S., and R. P. Meyer, 1961. Minimum statistical uncertainty of the seismic refraction profile. Geophys. 26, 574-587. 219

Steinhart, J. S., and T. J. Smith (Editors), 1966. The earth beneath the continents: A volume of geophysical studies in honor of Merle A. Tuve. American Geophysical Union Monograph No. 10, Washington, D. C. 219

Steketee, J. A., 1958. Some geophysical applications of the elasticity theory of dislocations. Canad. J. Phys. 36, 1168-1198. 358

Stevens, A. E., 1964. Earthquake mechanism determination by S-wave data. Bull., Seis. Soc. Amer. 54, 457-474. 374

Stevens, A. E., 1967. S-wave earthquake mechanism equations. Bull., Seis. Soc. Amer. 57, 99-112. 374

Stoneley, R., 1924. Elastic waves at the surface of separation of two solids. Proc. Roy. Soc., London A 106, 416-428. 147, 151

Stratton, J. A., 1941. Electromagnetic Theory. McGraw-Hill, New York. 53, 56

Street, R. L., 1976. Scaling Northeastern United States / Southeastern Canadian earthquakes by their Lg waves. Bull., Seis. Soc. Amer. 66, 1525-1537. 390

Street, R. L., R. B. Herrmann, and O. W. Nuttli, 1975. Spectral characteristics of the Lg wave generated by Central United States earthquakes. Geophys. J. 41, 51-63. 390

Strick, E., 1959a. Propagation of elastic wave motion from an impulsive source along a fluid/solid interface: II. Theoretical pressure response. Phil. Trans. Roy. Soc., London, A 251, 465-488. 151

Strick, E., 1959b. Propagation of elastic wave motion from an impulsive source along a fluid/solid interface: III: The pseudo-Rayleigh wave. Phil. Trans. Roy. Soc., London, A 251, 488-523. 151

Strick, E., 1967. The determination of Q, dynamic viscosity, and transient creep curves from wave propagation measurements. Geophys. J. 13, 197-218. 328, 331

Strick, E., 1976. Wave propagation in consolidated sedimentary rock. Paper given at the 46th annual meeting of the Society of Exploration Geophysicists, October 24-28, Houston, Texas. 332

Strick, E., 1978. A physically realizable model for the absorption of seismic waves and its application to the direct detection problem. Geophys. Explor. (In Press.) 331, 347

Strobach, K., 1965. Origin and properties of microseisms from the standpoint of oscillator theory. Bull., Seis. Soc. Amer. 55, 365-390. 414, 416

Su, S. S., and J. Dorman, 1965. The use of leaking modes in seismogram interpretation and in studies of crust-mantle structure. Bull., Seis. Soc. Amer. 55, 989-1021. 181, 256

Sykes, L. R., 1967. Mechanism of earthquakes and nature of faulting on the mid-oceanic ridges. J. Geophys. Res. 72, 2131-2154. 374

Sylman, Y., M. Shimshoni, and A. Ben-Menahem, 1974. Theoretical amplitudes of body waves from a dislocation source in the earth, III. P, S, and surface reflections. Phys. Earth and Planet. Inter. 8, 130-147. 364

Takeuchi, H., and M. Saito, 1972. Seismic surface waves. In: B. A. Bolt (Editor), Methods in computational physics, Vol. 11, Seismology: Surface waves and earth oscillations. Academic, New York, pp. 217-295. 241, 308

Takeuchi, H., and K. Sudo, 1968. Partial derivatives of free oscillation period with respect to physical parameter changes within the earth. J. Geophys. Res. 73, 3801-3806. 320

Takeuchi, H., M. Saito, and N. Kobayashi, 1962. Study of shear velocity distribution in the upper mantle by mantle Rayleigh and Love Waves. J. Geophys. Res. 67, 2831-2839. 258

Tanyi, G. E., 1967a. The generalized problem of Lamb: II. Generalized Stoneley waves in a fluid-solid spherical interface. Geophys. J. 12, 149-160. 152

Tanyi, G. E., 1967b. The generalized problem of Lamb: III. On a space-time representation for generalized Rayleigh and Stoneley waves. Geophys. J. 12, 161-163. 152

Tarr, A. C., 1969. Rayleigh-wave dispersion in the North Atlantic Ocean, Caribbean Sea, and Gulf of Mexico. J. Geophys. Res. 74, 1591-1607. 256

Tarr, A. C., 1970. New maps of polar seismicity. Bull., Seis. Soc. Amer. 60, 1745-1747. 399

Telford, W. M., L. P. Geldart, R. E. Sheriff, and D. A. Keys, 1976. Applied Geophysics. Cambridge Univ. Press. 198, 217

Teng, T-L., 1968. Attenuation of body waves and the Q structure of the mantle. J. Geophys. Res. 73, 2195-2208. (Corr., J. Geophys. Res. 74, 6720-6721, 1969.) 338, 339

Teng, T-L., 1970. Inversion of the spherical layer-matrix. Bull., Seis. Soc. Amer. 60, 317-320. 306

Teng, T-L., and A. Ben Menahem, 1965. Mechanism of deep earthquakes from spectrums of isolated body-wave signals, I. The Banda-Sea earthquake of March 21, 1964. J. Geophys. Res. 70, 5157-5170. 374

Thatcher, W., and J. N. Brune, 1969. Higher mode interferences and observed anomalous apparent Love wave phase velocities. J. Geophys. Res. 74, 6603-6611. 248

Thatcher, W., and J. N. Brune, 1973. Surface waves and crustal structure in the Gulf of California region. Bull., Seis. Soc. Amer. 63, 1689-1698. 248

Thatcher, W., and T. C. Hanks, 1973. Source parameters of Southern California earthquakes. J. Geophys. Res. 78, 8547-8576. 384, 385

Thau, S. A., and Y-H. Pao, 1970. On the derivation of point source responses from line source solutions. Intern. J. Eng. Sci. 8, 207-218. 127

Thomas, J. E., A. D. Pierce, E. A. Flinn and L. B. Craine, 1971. Bibliography on infrasonic waves. Geophys. J. 26, 399-426. 242

Thomas, J. E., A. D. Pierce, E. A. Flinn and L. B. Craine, 1972. Supplement to "Bibliography on infrasonic waves". Geophys. J. 30, 1-9. 242

Thomas, J. H., P. D. Marshall, and A. Douglass, 1978. Rayleigh-wave amplitudes from earthquakes in the range 0°-150°. Geophys. J. 53, 191-200. 389

Thomson, K.C. and N.A. Haskell, 1972. Elastodynamic near field of a finite propagating transverse shear fault. J. Geophys. Res. 77, 2574-2582. (Corr. 77, 7171, 1972) 385

Thomson, W. T., 1950. Transmission of elastic waves through a stratified solid medium. J. Appl. Phys. 21, 89-93. 204

Thrower, E. N., 1965. The computation of the dispersion of elastic waves in layered media. J. Sound. Vib. 2, 210-226. 241

Titchmarsh, E. C., 1937. Introduction to the theory of Fourier integrals. Oxford University Press. 90, 189

Toksöz, M. N. and D. L. Anderson, 1966. Phase velocities of long-period surface waves and structure of the upper mantle: 1. Great-circle Love and Rayleigh wave data. J. Geophys. Res. 71, 1649-1658. 318, 319, 321

Toksöz, M. N., J. Arkani-Hamed, and C. A. Knight, 1969. Geophysical data and long-wave heterogeneities of the earth's mantle. J. Geophys. Res. 74, 3751-3770. 223

Toksöz, M. N., K. C. Thomson, and T. J. Ahrens, 1971. Generation of seismic waves by explosions in prestressed media. Bull., Seis. Soc. Amer. 61, 1589-1623. 365

Tolstoy, I., 1968. Phase changes and pulse deformation in acoustics. J. Acoust. Soc. Amer. 44, 675-683. 279

Tolstoy, I., and C. S. Clay, 1966. Ocean acoustics: theory and experiment in underwater sound. McGraw-Hill, New York. 242

Tolstoy, I., and E. Usdin, 1953. Dispersive properties of stratified elastic and liquid media: a ray theory. Geophysics 18, 844-870. 172, 191, 192, 194

Tolstoy, I., and E. Usdin, 1957. Wave propagation in elastic plates: low and high mode dispersion. J. Acoust. Soc. Amer. 29, 37-42. 183

Trorey, A. W., 1962. Theoretical seismograms with frequency and depth dependent absorption. Geophysics 27, 766-785. 229

Tsai, Y-B., and K. Aki, 1970. Precise focal depth determination from aplitude spectra of surface waves. J. Geophys. Res. 75, 5729-5743. 298

Udias, A., 1964. A least squares method for earthquake mechanism determination using S-wave data. Bull., Seis. Soc. Amer. 54, 2037-2047. 374

Udias, A. and J. Rice, 1975. Statistical analysis of microearthquake activity near San Andreas Geophysical Observatory, Hollister, California. Bull., Seis. Soc. Amer. 65, 809-827. 407, 408

Underwood, R., and R. C. Lilwall, 1969. The systematic error in seismic location. Geophys. J. 17, 521-526. 298

U. S. National Academy of Sciences, 1975. Earthquake prediction and public policy. Washington, D. C. 408

Usami, T., and Y. Sato, 1964. Propagation of spheroidal disturbances on a homogeneous elastic sphere. Tokyo Univ., Bull. Earthquake. Res. Inst. 42, 273-284. 304

Utsu, T., and A. Seki, 1954. A relation between the area of aftershock region and the energy of main shock. J. Seis. Soc. Japan 7, 233-240. 391

Veith, K. F., and G. E. Clawson, 1972. Magnitude from short-period P-wave data. Bull., Seis. Soc. Amer. 62, 435-452. 386, 387

Vered, M., and A. Ben-Menahem, 1976. Generalized multipolar ray theory for surface and shallow sources. Geophys. J. 45, 195-198. 360

Vere-Jones, D., 1975. Stochastic models for earthquake sequences. Geophys. J. 42, 811-826. 405

von Seggern, D., 1972. Relative location of seismic events using surface waves. Geophys. J. 26, 499-513. 297

von Seggern, D., 1973. Joint magnitude determination and the analysis of variance for explosion magnitude estimates. Bull., Seis. Soc. Amer. 63, 827-845. 390

von Seggern, D., 1977. Amplitude-distance relation for 20-second Rayleigh Waves. Bull., Seis. Soc. Amer. 67, 405-411. 388

von Seggern, D., and R. Blandford, 1972. Source time functions and spectra for underground nuclear explosions. Geophys. J. 31, 83-97. 394, 395, 396

Walker, R. A., J. Z. Menard, and B. P. Bogert, 1964. Real-time high resolution spectroscopy of seismic background noise. Bull., Seis. Soc. Amer. 54, 501-509. 418

Warren, D. H., J. H. Healy, J. C. Hoffman, R. Kempe, S. Rauula, and D. J. Stuart, 1968. Project Early Rise: Traveltimes and amplitudes. U. S. Geol. Survey Open File Report, Menlo Park, Calif. 220

Wason, H. R., and S. J. Singh, 1971a. Transformation of the earthquake displacement for a spherical earth. Bull., Seis. Soc. Amer. 61, 289-295. 360

Wason, H. R. and S. J. Singh, 1971b. Transformation of the earthquake displacement field for spherical earth — static case. Bull., Seis. Soc. Amer. 61, 861-874. 365

Wason, H. R., and S. J. Singh, 1972. Static deformation of a multilayered sphere by internal sources. Geophys. J. 27, 1-14. 365

Watson, G. N., 1952. Treatise on the theory of Bessel functions (2nd Edition). Cambridge Univ. Press. 64

Watson, T. H., 1970. High and low frequency behavior of leaking modes in the layered half-space. Ph.D. Dissertation, U. Pittsburgh. 171, 172, 173, 180, 181

Watson, T. H., 1970a. A note on fast computation of Rayleigh wave dispersion in the multilayered elastic half-space. Bull., Seis. Soc. Amer. 60, 161-166. 241

Watson, T. H., 1972. A real frequency, complex wave-number analysis of leaking modes. Bull., Seis. Soc. Amer. 62, 369-384. 176, 181, 182

Weichert, D. H., and M. Henger, 1976. The Canadian Seismic Array Monitor Processing system (CANSAM). Bull., Seis. Soc. Amer. 66, 1381-1403. 38

Weidner, D. J., 1974. Rayleigh Wave Phase velocities in the Atlantic Ocean. Geophys. J. 36, 105-139. 256

Weidner, D. J., 1975. The effect of oceanic sediments on surface-wave propagation. Bull., Seis. Soc. Amer. 65, 1531-1552. 256

Weidener, D. J., and K. Aki, 1973. Focal depth and mechanism of mid-ocean ridge earthquakes. J. Geophys. Res. 78, 1818-1831. 377

Wheeler, L. T., and E. Sternberg, 1968. Some theorems in classical elastodynamics. Arch. Rational Mech. Anal. 31, 51-90. 77

Whitcomb, J. H., and D. L. Anderson, 1970. Reflection of P'P' Seismic waves from discontinuities in the mantle. J. Geophys. Res. 75, 5713-5728. 301

White, J. E., 1965. Seismic Waves. McGraw-Hill, New York. 38, 48, 70, 96, 334, 335, 337, 352, 356

Whitney, J., and R. Merrill, 1974. The use of Fisher statistics in fault plane solutions. Bull., Seis. Soc. Amer. 64, 279-283. 374

Whittaker, E. T., and G. N. Watson, 1935. Modern Analysis. Cambridge Univ. Press. 190

Wickens, A. J., 1971. Variations in litospheric thickness in Canada. Canad. J. Earth Sci. 8, 1154-1162. 256

Wickens, A. J., and J. H. Hodgson, 1967. Computer re-evaluation of earthquake mechanism solutions. Publ. Dom. Obs., Ottawa, 33. 374

Wiggins, R. A., 1968. Terrestrial variational tables for the periods and attenuation of the free oscillations. Phys. Earth and Planet. Inter. 1, 201-266. 320

Wiggins, R. A., 1969. Monte Carlo inversion of body-wave observations. J. Geophys. Res. 74, 3171-3181. 303

Wiggins, R. A., 1972. The general linear inverse problem: implication of surface waves and free oscillations for earth structure. Rev. Geophys. Space Phys. 10, 251-285. 262

Wiggins, R. A., 1976. A fast, new computational algorithm for free oscillations and surface waves. Geophys. J. 47, 135-150. 241, 305

Wiggins, R. A., 1976a. Body wave amplitude calculations — II. Geophys. J. 46, 1-10. 303

Wiggins, R. A., and D. V. Helmberger, 1974. Synthetic seismogram computation by expansion in generalized rays. Geophys. J. 37, 73-90. 230

Wiggins, R. A. and J. A. Madrid, 1974. Body wave amplitude calculations. Geophys. J. 37, 423-433. 303

Wiggins, R. A., G. A. McMechan and M. N. Toksöz, 1973. Range of earth structure nonuniqueness implied by body wave observations. Rev. Geophys. Space Phys. 11, 87-113. 303

Willmore, P. L., 1960. The detection of earth movements. In: S. K. Runcorn (Editor), Methods and techniques in geophysics. Interscience, New York, pp. 230-276. 424, 431

Willmore, P. L., and A. M. Bancroft, 1960. The time-term approach to refraction seismology. Geophys. J. 3, 419-432. 225

Wilson, J. T., 1965. A new class of faults and their bearing on continental drift. Nature 207, 343-347. 374

Woodhouse, J. H., 1976. On Rayleigh's principle. Geophys. J. 46, 11-22. 308

Woodhouse, J. H., and F. A. Dahlen, 1978. The effect of a general aspherical perturbation on the free oscillations of the earth. Geophys. J. 53, 335-354. 308

Woollard, G. P., 1959. Crustal Structure from Gravity and Seismic measurements. J. Geophys. Res. 64, 1521-1544. 258

Worzel, J. L., and M. Ewing, 1948. Explosion sounds in shallow water. In: Memoir No. 27, Geol. Soc. Amer., New York. 242

Wright, C., and J. R. Cleary, 1972. P wave travel-time gradient measurements for the Warramunga seismic array and lower mantle structure. Phys. Earth and Planet. Int. 5, 213-230. 301

Wu, F. T., 1972. Mantle Rayleigh wave dispersion and tectonic provinces. J. Geophys. Res. 77, 6445-6453. 318

Wuenschel, P. C., 1960. Seismogram synthesis including multiples and transmission coefficients. Geophysics 25, 106-129. 229

Wuenschel, P. C., 1965. Dispersive body waves, an experimental study. Geophysics 30, 539-551. 330, 331
Wyss, M., 1973. Towards a physical understanding of the earthquake frequency distribution. Geophys. J. 31, 341-359. 402
Wyss, M., and L. J. Shamey, 1975. Source dimensions of two deep earthquakes estimated from aftershocks and spectra. Bull., Seis. Soc. Amer. 65, 403-409. 385
Yacoub, N. K., and B. J. Mitchell, 1977. Attenuation of Rayleigh wave amplitudes across Eurasia. Bull., Seis. Soc. Amer. 67, 751-769. 344
Yamaguchi, R., and Y. Sato, 1955. Stoneley wave — its velocity, orbit, and distribution of amplitude. Tokyo Univ., Bull. Earthquake Res. Inst., 33, 549-560. 147
Yeatts, F. R., 1973. A multipole representation of earthquake source mechanisms. Bull., Seis. Soc. Amer. 63, 211-225. 364
Yeh, K. C. and C. H. Liu, 1974. Acoustic-Gravity Waves in the upper atmosphere. Rev. Geophysics Space Phys. 12, 193-216. 242
Young, G. B., and L. W. Braile, 1976. A computer program for the application of Zoeppritz' amplitude equation and Knott's energy equations. Bull., Seis. Soc. Amer. 66, 1881-1885. 131
Ziolkowski, A., 1973. Prediction and suppression of long-period nonpropagating seismic noise. Bull., Seis. Soc. Amer. 63, 937-958. 421

INDEX